Advances in Airborne Lidar Systems and Data Processing

Edited and Selected by
Jie Shan
Juha Hyyppä

MDPI • Basel • Beijing • Wuhan • Barcelona • Belgrade

MDPI

Edited and Selected by
Jie Shan
Purdue University
USA

Juha Hyyppä
Finnish Geospatial Research Institute
Finland

Editorial Office
MDPI
St. Alban-Anlage 66
Basel, Switzerland

This edition is a reprint of the Special Issue published online in the open access journal *Remote Sensing* (ISSN 2072-4292) from 2016–2017 (available at: http://www.mdpi.com/journal/remotesensing/ special_issues/rs_ALS), complemented by selected articles published in *Remote Sensing*.

For citation purposes, cite each article independently as indicated on the article page online and as indicated below:

Lastname, F.M.; Lastname, F.M. Article title. *Journal Name* **Year**, *Article number*, page range.

First Edition 2018

ISBN 978-3-03842-673-8 (Pbk)
ISBN 978-3-03842-674-5 (PDF)

Table of Contents

Part 1 Single Photon and Geiger-Mode Lidar

Part 2 Multispectral Lidar

Part 3 Waveform Lidar

Part 4 Registration of Point Clouds

Part 5 Trees and Terrain

Part 6 Building Extraction

About the Special Issue Editors

Jie Shan is currently a Professor with the Lyles School of Civil Engineering, Purdue University. He received his Ph.D in photogrammetry and remote sensing from Wuhan University, China. He has been a faculty at universities in China and Sweden, and Research Fellow in Germany. His areas of interests include sensor geometry and positioning, object extraction and reconstruction from images and point clouds, urban remote sensing, automated digital mapping, and pattern recognition of spatial, temporal and semantic data. He has authored/co-authored over 200 scientific publications, and is a recipient of multiple best paper awards, including the Talbert Abrams Grand Award and the Environmental Systems Research Institute Award. He is a Senior Member of IEEE and an elected ASPRS Fellow.

Juha Hyyppä, Professor and Director of Centre of Excellence in Laser Scanning Research, at Finnish GeoSpatial Research Institute, Finland. His references in science are represented by about 180 ISI Web of Science papers (5500 citations and H-index of 38) and more than 12,000 citations with an H-index of 58 in Google Scholar. He has educated 10 professors, and more than 20 doctors. He has coordinated more than 10 international science projects. He has focus on laser scanning systems, their performance and new applications, especially related to mobile laser scanning and point cloud processing.

Preface to "Advances in Airborne Lidar Systems and Data Processing"

Presented to audience as "Airborne Laser Systems and Data Processing", this book collects the papers in the *Remote Sensing* special issue " Airborne Laser Scanning" (November 2016) and several other papers selected from recent, previous *Remote Sensing* issues.

We were approached by *Remote Sensing* late 2015 to edit a special issue on Lidar remote sensing. After some literature review and careful thinking, we selected the theme for the special issue as "Airborne Laser Scanning". Having this focus in mind, we made a call for papers and spread it well over Internet and social media. In the meantime, a number of leading experts in their fields were reached out for contribution. The effort has led to the special issue (November 2016) with 15 papers. Beyond our expectation, the special issue was able to cover a variety of advanced, emerging subjects which reflect not only leading edge technologies, but their experimental results for large scale, real-world applications. Many of the contributions are either one of the first, comprehensive and systematic, or new development for the state-of-the-art technology, all of which made us believe this collection would be classical and durable. This motivates us to expand it to a printed book by selecting an additional number of relevant and representative papers from recent, previous *Remote Sensing* issues. It comprehends a wider spectrum of laser remote sensing techniques, while we also expect it is self-contained.

The book consists of 23 papers in six subject areas: (1) single photon and Geiger-mode lidar; (2) multispectral lidar; (3) waveform lidar; (4) registration of point clouds; (5) trees and terrain; and (6) building extraction. It is our expectation that it will be a valuable resource for scientists, engineers, developers, instructors, and graduate students interested in lidar systems and data processing. We hope it will be a gateway to the future for advanced laser remote sensing and data processing.

Finally, we would like to thank the excellent *Remote Sensing* editorial team, without their persistent effort and professional patience we would not be able to have the special issue and this expanded, printed book. Our gratitude also goes to the authors who have generously agreed their papers be included in this book.

Jie Shan and Juha Hyyppä
Special Issue Editors

Part 1
Single Photon and Geiger-Mode Lidar

remote sensing

MDPI

Article

Scanning, Multibeam, Single Photon Lidars for Rapid, Large Scale, High Resolution, Topographic and Bathymetric Mapping

John J. Degnan

Sigma Space Corporation, 4600 Forbes Blvd., Lanham, MD 20706, USA; john.degnan@sigmaspace.com;
Tel.: +1-301-552-6004

Academic Editors: Jie Shan, Juha Hyyppä, Guoqing Zhou and Prasad S. Thenkabail
Received: 29 July 2016; Accepted: 10 November 2016; Published: 18 November 2016

Abstract: Several scanning, single photon sensitive, 3D imaging lidars are herein described that operate at aircraft above ground levels (AGLs) between 1 and 11 km, and speeds in excess of 200 knots. With 100 beamlets and laser fire rates up to 60 kHz, we, at the Sigma Space Corporation (Lanham, MD, USA), have interrogated up to 6 million ground pixels per second, all of which can record multiple returns from volumetric scatterers such as tree canopies. High range resolution has been achieved through the use of subnanosecond laser pulsewidths, detectors and timing receivers. The systems are presently being deployed on a variety of aircraft to demonstrate their utility in multiple applications including large scale surveying, bathymetry, forestry, etc. Efficient noise filters, suitable for near realtime imaging, have been shown to effectively eliminate the solar background during daytime operations. Geolocation elevation errors measured to date are at the subdecimeter level. Key differences between our Single Photon Lidars, and competing Geiger Mode lidars are also discussed.

Keywords: airborne multibeam lidar; single photon lidar; 3D imaging; photon-counting; surveying; forestry; bathymetry; cryosphere

1. Introduction

Conventional mapping lidars fall into two broad categories—discrete return lidars and digitized waveform lidars. As can be seen from Figure 1, discrete return lidars provide one or more event times (ranges) where the received intensity exceeds a common threshold but there is no other vertical spatial information in between. Digitized waveform lidars, on the other hand, provide intensity information over the entire vertical structure but each point in the profile represents the sum of the returns over the transverse extent of the laser beam at a given range. Digitized waveform lidars typically require hundreds of detected photons and are most useful when mapping areas where multiple vertical returns are expected from complex semi-porous structures such as tree canopies, very rough terrain, or even manmade structures. The earliest version of our Single Photon Lidar (SPL), the NASA "Microaltimeter" described in Section 2.1, used a single beam with only 2 microjoules of energy per pulse and a multistop timing receiver to record tree canopies and the underlying terrain from altitudes above ground level (AGLs) as high as 7 km [1]. In effect, it was a degraded version of a digitized waveform lidar and, if one were to repeat the low energy measurements many times and create histograms versus range, one would expect to generate a profile comparable to that of the waveform digitizer. On the other hand, individual photon returns originate at a specific scattering point within the canopy and, unlike waveform lidars, are isolated from nearby returns which occur at the same range but originate from other points within the transverse extent of the laser beam. Later SPL generations, to be described in this article, take advantage of the receiver's single photon sensitivity by splitting a single laser beam into 100 beamlets, arranged in a 10×10 array. Each beamlet is then imaged onto a pixel in a matching 10×10 array detector which, in turn, is input to a timing channel able to record multiple stop events per pixel with few picosecond accuracy. This alone increases the surface measurement rate by two orders of magnitude relative to the laser fire rate. When the source laser is operating at tens of kHz, surface measurement rates of several megapixels per second are achieved. Furthermore, our lidars are designed to provide a mean of 3 photoelectrons per pixel for green vegetation (10% surface reflectance at 532 nm) when operated at their design AGL. Thus, a tree canopy will result in approximately 300 photoelectrons being detected per pulse, a number not dissimilar to some Digitized Waveform lidars, but with the added benefit that the transverse coordinates of the scattering points are identified as well as the range, thereby providing more detailed 3D vs. 1D maps of the canopy. It must also be mentioned that a competing single photon sensitive technique based on Geiger Mode Avalanche Photodiode Arrays has also recently been introduced to the commercial market, by Harris Corporation and their different characteristics will be discussed in Section 5.

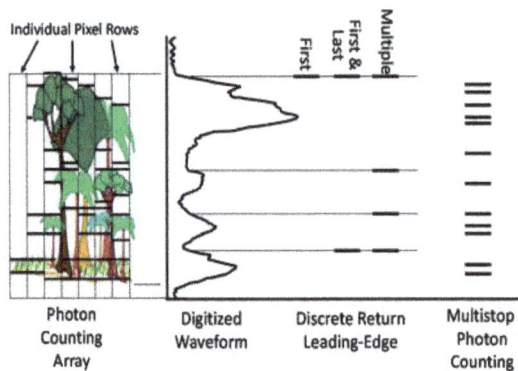

Figure 1. A comparison of Single Photon Lidars with conventional Discrete Return and Digitized Waveform lidars in interacting with a tree canopy (Courtesy of D. Harding, NASA GSFC).

Single photon sensitive 3D imaging lidars have multiple advantages relative to conventional multiphoton lidars. They are the most efficient 3D imagers possible since each range measurement requires only one detected photon as opposed to hundreds or even thousands in conventional laser pulse time of flight (TOF) or waveform altimeters. Their high efficiency enables orders of magnitude more imaging capability (e.g., higher spatial resolution, larger swaths and greater areal coverage). In our Single Photon Lidars (SPLs), single photon sensitivity is combined with a 1.6 nanosecond receiver recovery time (often referred to as "deadtime"), and is therefore capable of recording returns from objects differing by only 24 cm in range. This enables our lidars to operate effectively in daylight and to penetrate semi-porous obscurations such as vegetation, ground fog, thin clouds, etc. Furthermore, unlike most lidars which operate at the fundamental Nd:YAG wavelength of 1064 nm in the Near InfraRed (NIR), the SPL 532 nm operating wavelength is highly transmissive in water, thereby permitting shallow water bathymetry and 3D underwater imaging. In order to enhance the range resolution of SPLs, FWHM laser pulsewidths on the order of 100 to 700 picoseconds are used whereas conventional lidars typically employ few nanosecond pulsewidths and rely on large photon counts from the surface to improve the precision of the range measurement.

On the other hand, sensitivity to single photon surface returns also make SPLs sensitive to background noise originating from: (1) dark counts from the sensitive detectors; (2) solar backscatter from the surface being examined and the intervening atmosphere within the pixel field of view (FOV); and (3) laser backscatter from the atmosphere within the selected range gate. Sources (1) and (3) occur during both day and night mapping operations but are relatively inconsequential compared to the solar scatter encountered in daylight. Conventional lidars have multiphoton thresholds and therefore do not record single photon solar or dark count events. The solar count rate per pixel is proportional to the pixel FOV and the receive telescope aperture [2]. Thus, shrinking the pixel FOV not only reduces the solar count rate in an SPL, but it also improves the horizontal spatial resolution of the lidar. Furthermore, the single photon sensitivity of the receiver allows a substantial reduction in receive aperture, thereby further reducing the number of noise events [2]. Finally, these sources have been effectively mitigated through the use of highly effective noise filtering algorithms such as the Differential Cell Count method [2]. For further insights into the characteristics and relative merits of the various lidar types, the reader is referred to the book chapter by Harding [3].

In Section 2 of this paper, we present an overview of our multibeam scanning airborne SPLs to date and the manner in which they have been adapted to operate at higher AGLs and cruise speeds for faster areal coverage. In Section 3, we briefly discuss progress in developing fast and autonomous data editing software for extracting surface data from the solar background during daylight operations and the potential for near real time 3D image generation for cockpit display and/or transmission to a ground station. Section 4 provides examples of different data types in order to demonstrate their relevance to applications such as large scale surveying, Cryospheric studies, forestry, and shallow water bathymetry. Section 5 discusses the relative advantages and disadvantages of SPL vs Geiger Mode technology, which was developed over two decades by the US military but has recently been introduced into the civilian market by Harris Corporation. Finally Section 6 provides some concluding remarks about ongoing research and field activities to provide improved data products, including the possibility of globally contiguous mapping of planets and moons from orbital altitudes between 100 and 500 km.

2. SPL Instrument Overview and Heritage

2.1. NASA "Microaltimeter"

NASA's Microlaser Altimeter or "Microaltimeter" provided the first airborne demonstration of a scanning Single Photon Lidar (SPL) in early 2001 [1]. Although several natural properties (e.g., atmospheric transmission, natural surface reflectivity, solar background) favor use of the fundamental Nd:YAG wavelength at 1064 nm, 532 nm was chosen as the operating wavelength for technology reasons (e.g., higher efficiency COTS array detectors with nanosecond recovery times, high transmission narrowband filters, etc.) [2]. A side benefit of the choice was the instrument's demonstrated ability to see the bottom of the Atlantic Ocean off the coast of Virginia to a depth of about 3 m from an altitude of 4 km. The lidar also successfully penetrated tree canopies to see the underlying surface. The 532 nm operating wavelength has been maintained through the successive generations of lidar described here.

With less than 2 microjoules per pulse at a laser repetition rate of 3.8 kHz (~7.6 mW average power), the single beam "Microaltimeter" produced high resolution 2D profiles or low resolution 3D images over narrow swaths (~60 m) while operating mid-day at altitudes up to 6.7 km. Although the passively Q-switched, microchip Nd:YAG laser was incredibly small (~2.3 mm in length) and pumped by a single diode laser, the overall lidar was quite large and flew in the cabin of NASA's P-3 aircraft. Nevertheless, this 1st generation system demonstrated the feasibility of: (1) making accurate surface measurements with single photon returns under conditions of full solar illumination; and (2) developing high resolution spaceborne laser altimeters and imaging lidars operating from orbital altitudes of several hundred km [2].

2.2. Second Generation SPL ("Leafcutter")

From 2004 to 2007, Sigma developed its first multibeam Single Photon Lidar (SPL), dubbed "Leafcutter" [4]. Leafcutter, shown in Figure 2, was designed to fit into the nose cone of an Aerostar Mini-UAV and provide contiguous decimeter resolution images on a single overflight from AGLs between 1 and 2.5 km, depending on surface reflectance. The overall system, including GPS receiver and Inertial Measurement Unit (IMU), consisted of two units (optical bench and electronics box), weighed 33 kg, occupied a volume of less than 0.07 m³, and drew ~170 W of aircraft 28 VDC prime power. In parallel to this activity, Sigma Space also provided hardware and technical support to two other single photon systems, i.e., the University of Florida's Coastal Area Tactical-Mapping System or CATS [5,6] and NASA Goddard Space Flight Center's Slope Imaging Multi-polarization Photon-counting Lidar or SIMPL [7].

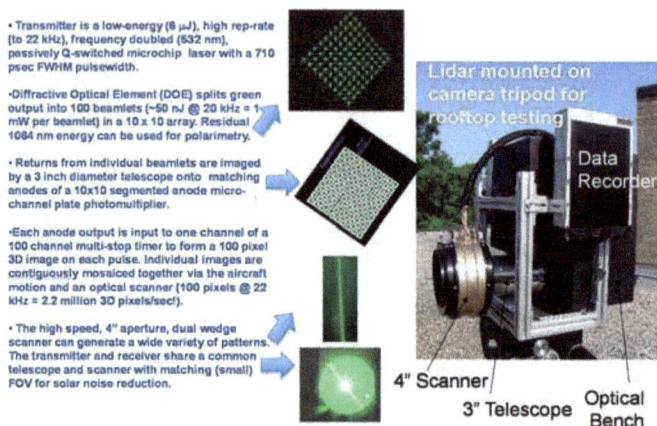

• Transmitter is a low-energy (6 μJ), high rep-rate (to 22 kHz), frequency doubled (532 nm), passively Q-switched microchip laser with a 710 psec FWHM pulsewidth.

•Diffractive Optical Element (DOE) splits green output into 100 beamlets (~50 nJ @ 20 kHz = 1 mW per beamlet) in a 10 x 10 array. Residual 1064 nm energy can be used for polarimetry.

• Returns from individual beamlets are imaged by a 3 inch diameter telescope onto matching anodes of a 10x10 segmented anode micro-channel plate photomultiplier.

•Each anode output is input to one channel of a 100 channel multi-stop timer to form a 100 pixel 3D image on each pulse. Individual images are contiguously mosaiced together via the aircraft motion and an optical scanner (100 pixels @ 22 kHz = 2.2 million 3D pixels/sec!).

• The high speed, 4" aperture, dual wedge scanner can generate a wide variety of patterns. The transmitter and receiver share a common telescope and scanner with matching (small) FOV for solar noise reduction.

Lidar mounted on camera tripod for rooftop testing

Data Recorder

4" Scanner
3" Telescope   Optical Bench

Figure 2. Leafcutter was the first Sigma Single Photon Lidar (SPL) to split the laser beam into 100 beamlets. In early mapping missions, the dual wedge scanner was used to generate either linear raster scans at 45° to the flight line or a conical scan with cone half angles up to 13.5 degrees. At the design AGL of 1 km, pixels on the ground were separated by 15 cm. Contiguous alongtrack and crosstrack mapping on a single pass was achieved by ensuring: (1) that the distance traveled by the aircraft during one scan cycle did not exceed the 1.5 m dimension of the single pulse array; and (2) that ground array patterns from subsequent pulses overlapped along the full circumference of the conical scan and the length of the linear scans.

A 10 × 10 square array of 100 beamlets was generated by passing the 140 mW COTS laser transmitter beam through an 80% efficient Diffractive Optical Element (DOE). Each beamlet contained approximately 1 mW of laser power in a 22 kHz stream of 700 ps FWHM, 50 nJ pulses. At the design AGL of 1 km, the interbeam spacing between beamlets was 15 cm, and the ground images of the beamlets were optically matched to a COTS 10 × 10 segmented anode, MicroChannel Plate PhotoMultiplier Tube (MCP/PMT). The individual anode outputs were then input to an inhouse multichannel timing receiver with an RMS timing/range precision of 23 ps/3.4 mm. Most importantly, the detector/receiver subsystem can record the arrival times of multiple, closely-spaced photons per channel with an event recovery time of only 1.6 ns. This made Leafcutter impervious to shut-down by random solar events and also permitted multiple returns per channel from semi-porous volumetric scatterers such as tree canopies. The solar noise per pixel was kept to a minimum through the use of a 0.3 nm FWHM spectral filter, a small receive telescope 7.5 cm in diameter, and a Field-of-View (FOV) limited by the nominal 15 cm × 15 cm ground pixel dimension, which over a nominal 1 km range amounts to a solid angle of only 2.2 × 10⁻⁸ steradians per pixel.

The use of the 10 × 10 beamlet array increased the surface measurement rate by two orders of magnitude to 2.2 million multistop pixels per second. The array also allowed high resolution contiguous maps of the underlying surface to be generated on a single overflight at relatively high air speeds with modest scan speeds on the order of 20 Hz or less, which were easily achieved with the relatively small receive aperture. A further advantage is that, for each of the spatially separated pixels, there is only one pulse in the air per measurement until the surface slant range exceeds 6.8 km. This is in contrast to some commercial linear mode lidar designs which attempt to achieve higher measurement rates using a single beam at very high repetition rates (~200 kHz). At these frequencies, complications associated with multiple pulses in flight begin at surface slant ranges an order of magnitude smaller (~700 m).

Leafcutter employs a dual wedge optical scanner, which is common to both the transmitter and receiver. By adjusting the rotation direction and/or the rotational phase differences between the two wedges, one can generate a wide variety of scan patterns including: (1) linear scans at arbitrary orientations to the flight line (see Figure 2); (2) conical scans of varying radius; (3) spiral scans, etc. Maximum angular offset from nadir when the two wedges are coaligned is 14 degrees, corresponding to a maximum swath of about 0.5 km at a 1 km AGL (Altitude above Ground Level). During rooftop testing, a "3D camera mode", i.e., a rotating line scan, shown in Figure 2, was used to generate a contiguous high resolution 3D image within a circular perimeter.

NASA funded several test flights to assess SPL capabilities in the areas of biomass (forest cover), cryospheric, and bathymetric measurements. A collage of sample results from Leafcutter is presented in Figure 3. A second similar unit, labeled "Icemapper", was later delivered to the University of Texas at Austin to participate in Antarctica ice-mapping missions. As can be seen in Figure 3, the 532 nm operating wavelength allowed bathymetry to a depth of 15 m in glacier melt ponds and to about 4 m depth in the more turbid waters of the Chesapeake Bay near Annapolis, MD, USA. Note also that the surface of the melt pond is well defined even at a beam incidence angle of 14 degrees, indicating that Lambertian scattering from water molecules at and just below the water surface, rather than specular reflections, are creating the surface signal. Furthermore, what appears to be excess noise at the pond bottom is in reality variations in the bottom elevation when the entire pond is viewed from the side.

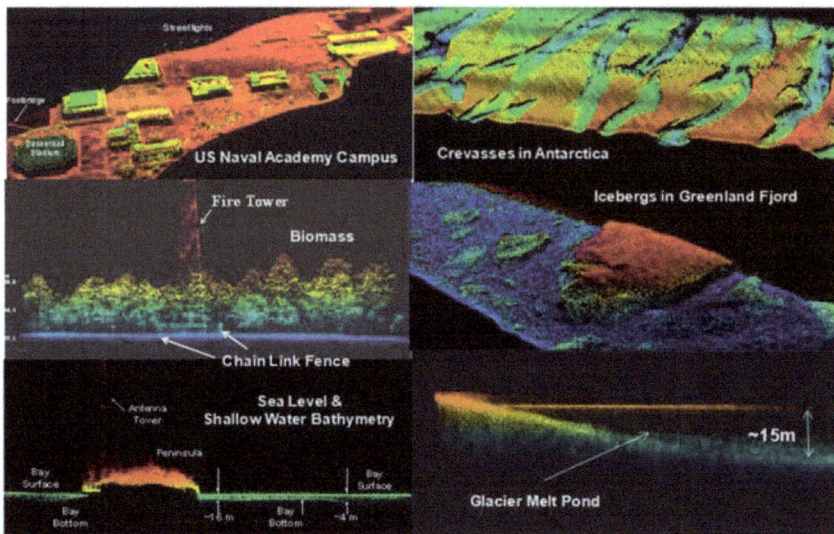

Figure 3. A collage of daytime images created on a single overflight by the Leafcutter SPL. The images in the left half were over low reflectance (10% to 15%) surfaces at above ground levels (AGLs) of 1 km or less while those in the right half were high reflectance cryospheric measurements in Greenland and Antarctica from AGLs up to 2.5 km. The images are color-coded according to the lidar-derived surface elevation (blue = low, red = high). Note the bathymetry results in the bottom two images.

2.3. NASA Mini-ATM

Subsequent to the highly successful cryospheric results obtained by Leafcutter, NASA funded development of an even smaller 100 beamlet system, imaged onto 25 pixels (4 beamlets per pixel), to potentially replace the highly successful, but much larger and heavier, P-3 based Airborne Topographic Mapper (ATM), which had mapped the Greenland ice sheets for many years. "Mini-ATM" reused most Leafcutter components and subsystems but was light-weighted and reconfigured to fit into the payload bay of a Viking 300 Micro-UAV (see Figure 4). The current version of the multiphoton ATM lidar has a nominal spacing between measurements of 2.5 m ($0.16/m^2$ point density) which generally met the needs of Cryospheric scientists tracking changes in ice sheet thickness in support of NASA Global Climate Change programs. Thus, to maximize swath and thereby minimize the time required to map large ice sheets, Mini-ATM features a 90° full conical holographic scanner. For the nominal Viking 300 velocity of 104 km/h and altitude ceiling of 3 km, the system is designed to autonomously map up to 600 km^2/h with a mean measurement point density in excess of $1.5/m^2$—a density about 10 times higher than that achieved by the current man-assisted ATM. Including a dedicated IMU, Mini-ATM has a cubic configuration (see Figure 4) with a volume of 0.03 m^3, weighs 12.7 kg, and consumes ~168 W of 28 VDC prime power. Mini-ATM completed its first successful test flight in a manned aircraft over California's Mojave Desert in October 2010.

Figure 4. NASA Mini-ATM (Airborne Topographic Mapper) and its designated host aircraft, the Viking 300 micro-UAV.

2.4. High Resolution Quantum Lidar System (HRQLS 1 and 2)

Development of the moderate altitude High Resolution Quantum Lidar Systems, (HRQLS-1) and its upgraded successor, HRQLS-2, were self-funded by Sigma and are shown in Figure 5. Both systems follow the same design philosophy as "Leafcutter", i.e., 100 beamlets in a 10 × 10 array, but the spacing between pixels at the ground is increased to 50 cm at their nominal AGLs as described in Table 1. The primary technical goal of HRQLS-1 was to map larger areas more quickly via a combination of higher air speeds and wider swaths while still permitting the experimenter to tailor the measurement point density to fit his or her individual needs. The wider swath is achieved by: (1) flying at a higher altitude; (2) increasing the laser power to about 1.7 W to compensate for the larger $1/R^2$ signal loss (where R is the slant range to the target); and (3) increasing the maximum half-cone angle of the scanner to 20 degrees.

Figure 5. Moderate altitude HRQLS-1 and HRQLS-2 lidars and the King Air B200 host aircraft.

Table 1. Summary table of design and performance properties for the current suite of Sigma scanning SPL lidars.

	Low Altitude SPLs		Medium Altitude SPLs		High Altitude SPLs
Instrument Name	USAF "Leafcutter"	NASA Mini-ATM	HRQLS-1	HRQLS-2	HAL
Prototype Completion Dates	2007	2010	2013	2016	2012
Units/Customers	2/USAF & Univ. of Texas	1/NASA	1/Sigma	6/Sigma	3/DoD
Primary Application	Military Prototype & Antarctic Cryosphere	Greenland Cryosphere	Civilian Surveying and mapping, Biomass Measurement, Bathymetry, Military Surveillance		Military Surveillance
Design Platform	Aerostar Mini-UAV	Viking 300 UAV	King Air	King Air	Various
# beams/pixels, N_p	100/100	100/25	100/100	100/100	100/100
Wavelength	532 nm		532 nm		532 nm
Laser Repetition Rate, f_{ps}	22 kHz		25 kHz	60 kHz	32 kHz
Laser Pulse Width (FWHM)	0.7 ns		0.7 ns	0.5 ns	0.1 ns
Laser Output Power	0.14 W		1.7 W	5 W	15 W
Maximum Measurements/s	2,200,000	550,000	2,500,000	6,000,000	3,200,000
Multiple Return Capability	Yes		Yes		Yes
Pixel Recovery Time	1.6 ns		1.6 ns		1.6 ns
RMS Range Precision	5 cm		5.7 cm	4.8 cm	3.6 cm
Telescope Diameter	7.5 cm		7.5 cm	14 cm	14 cm
# Scanner Wedges	2	DOE	2	1 Wedge or DOE	1 Wedge
Scan Width (FOV)	Variable 0° to 28°	Fixed 90° cone	Variable 0° to 40°	20°, 30°, 40° or 60°	Fixed 18°
Nominal A/C Velocity, v_g	161 km/h	104 km/h	370 km/h	370 km/h	370 km/h
Design AGL	1 km	2.5 km	2.3 km	3.4 km	7.6 km
Nominal AGL Range, h	1.0 to 2.5 km	0.55 to 3 km	2 to 3 km	3 to 5.5 km	6 to 11 km
Swath, S	0.0015 to 1.247 km	1.1 to 6 km	0.005 to 2.184 km	1.058 to 6.351 km	1.901 to 3.484 km
Areal Coverage, $S\eta$	0.242 to 201 km²/h	114 to 624 km²/h	2 to 808 km²/h	391 to 2350 km²/h	703 to 1289 km²/h
Mean Measurement Attempts per m² per pass, D_m	39 to 32,795/m²	3 to 17/m²	11 to 4865/m²	9 to 55/m²	9 to 16/m²
# of Modules	2	1	1 (rack-mounted)	1 (rack-mounted)	1 (rack mounted) 1 (pod mounted)
Instrument Volume/Dimensions	0.071 m³	0.027 m³ Quasi-cube (0.3 m)	0.26 m³ 48 × 64 × 84 cm³	0.139 m³ 82.5 × 48.25 × 35 cm³	0.52 m³ 49 × 64 × 163 cm³
Weight	33 kg	13 kg	57 kg	68 kg (sensor head) 22 kg (e-rack)	113 kg (est.)
Prime Power (28VDC)	266 W	~168 W	555 W	700 W	<900 W (est)
Status	2 Delivered	1 Delivered	1 Operational	2 Operational, 4 in fab	2 Delivered, 1 in fab

In order to accommodate a large range of measurement point densities, HRQLS-1 also features an external dual wedge scanner at the output of the 7.5 cm diameter telescope, which allows a range of full cone angles between 0 and 40 degrees, resulting in swath widths as small as 5 m and as large as 1.66 km at the nominal 2.3 km AGL. This feature allows measurement point density (or spatial resolution) to be traded off against swath and areal coverage. However, because of the longer pulse times-of-flight (TOFs) and high scan speeds, the images of the beamlet array become displaced relative to their assigned pixel centers unless one implements an optical TOF correction [7]. Thus, in HRQLS, annular corrector wedges are attached to each of the main scanner wedges in order to bring the transmitter and receiver FOVs into alignment at the nominal AGL. Maintaining alignment between the transmitter and receiver FOVs at different AGLs is accomplished by adjusting the angular speed of the scanner—faster for AGLs lower than nominal and slower for AGLs higher than nominal.

The upgraded HRQLS-2 was subsequently developed to allow high point density operation at AGLs above 3.1 km where FAA regulations permit more flexibility on flight lines. Instead of a dual wedge scanner, however, HRQLS-2 uses a variety of interchangeable single wedge or holographic scanners with full cone angles ranging from 20 to 60 degrees.

2.5. High Altitude Lidar (HAL)

Two versions of Sigma's High Altitude Lidar (HAL) currently exist to operate from either an internal cabin or an external pod environment. HAL was designed to produce contiguous, few decimeter resolution, topographic maps on a single pass from AGLs between 6.4 and 11 km. At these high AGLs, the importance of using scanner corrector wedges to compensate for finite speed of light effects is even more crucial since the overlap between transmit beamlet arrays and detector FOVs can, under some operational scenarios, be reduced to zero with the result that no surface signals are detected.

Depending on the operating AGL, there are either 2 or 3 pulses simultaneously in flight, and this can be taken into account during data processing by simply pairing the proper start pulse with the observed stop pulses. HAL can provide contiguous maps at aircraft speeds in excess of 407 km/h. The single wedge scanner has a $9°$ half-cone angle. Thus, at a maximum AGL of 11 km, the swath is 3.48 km and the maximum rate of areal coverage is 1415 km^2/h. The HAL images are comparable in quality and resolution to the HRQLS images in Section 4 of this paper [8].

2.6. NASA's Multiple Altimeter Beam Experimental Lidar (MABEL)

Sigma provided all of the electronics modules, including the multichannel timing receiver, as well as key mechanical, thermal, integration, testing and flight operations support to NASA's Multiple Altimeter Beam Experimental Lidar (MABEL) instrument, which was developed as a precursor and testbed for the Advanced Topographic Laser Altimeter System (ATLAS) SPL, scheduled to be launched in 2017 into a 500 km orbit on NASA's second generation Ice, Cloud, and land Elevation Satellite-2 (ICESat-2) mission [9]. MABEL is a nonscanning, 24 beam (8 @ 1064 nm, 16 @ 532 nm) pushbroom lidar hosted on NASA's ER-2 Research aircraft (see Figure 6). It has successfully demonstrated single photon surface profiling at AGLs of 20 km in both California and Greenland [10].

2.7. Summary Table of Sigma Scanning Lidar Properties

Table 1 provides a summary of the physical (Size, Weight, and Power or SWaP) and performance properties of the various scanning SPLs described in previous subsections. Dual wedge scanner systems such as "Leafcutter" and HRQLS-1 can vary the cone angle from $0°$ to some maximum cone angle, i.e., $28°$ for Leafcutter and $40°$ for HRQLS-1. HRQLS-2 is equipped to alternate between 4 distinct cone angles while HAL currently only has one ($18°$ full cone angle). All of the systems can operate effectively over a range of AGLs about the "Design AGL", which is defined as the AGL where, from Poisson statistics, the expected pixel Photon Detection Efficiency (PDE) = $1 - \exp(-n_p)$ = 95% (mean detected photoelectrons per pixel n_p equals 3) at the largest scan cone angle over a 10% reflectance Lambertian surface (e.g., green vegetation at 532 nm operating wavelength). The per-pixel PDE is over 99% for surface reflectances greater than 15% at 532 nm (e.g., soil and dry vegetation).

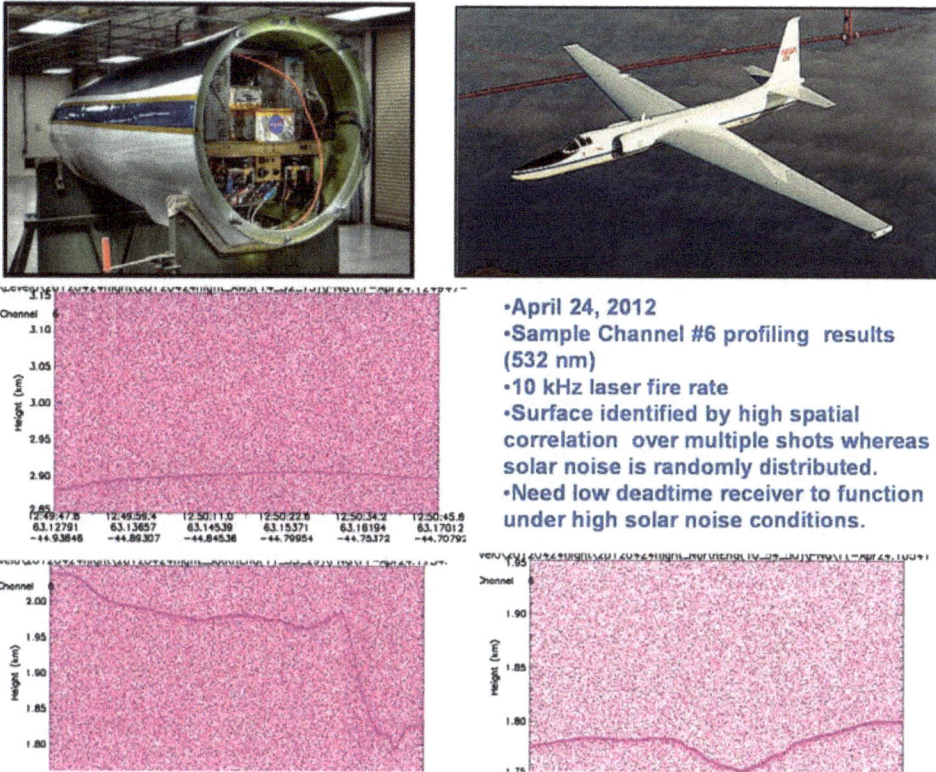

- •April 24, 2012
- •Sample Channel #6 profiling results (532 nm)
- •10 kHz laser fire rate
- •Surface identified by high spatial correlation over multiple shots whereas solar noise is randomly distributed.
- •Need low deadtime receiver to function under high solar noise conditions.

Figure 6. The NASA MABEL pushbroom lidar, jointly developed by NASA Goddard Space Flight Center and Sigma Space Corporation, has successfully generated 2D surface profiles in Greenland from an AGL of 20 km. The surface returns are highly spatially correlated and stand out against the dense "salt-and-pepper" solar noise background resulting from the high reflectance (typically 80% to 96%) of snow and ice at 532 nm.

By deviating from the design AGL, one can generate a greater density of measurements over a smaller swath (lower AGL) or a lower density of measurements over a wider swath (higher AGL) for faster areal coverage. When operating at the design AGL, the nominal pixel spacing at the ground is 50 cm for both HRQLS models and HAL The minimum swath in the table corresponds to the minimum cone angle from the minimum AGL while the maximum swath is obtained by using the maximum cone angle at the maximum AGL. At all AGLs at or below the design AGL, the percentage of returns from a 10% reflectance Lambertian surface is greater than 95%. At AGLs higher than the design AGL, the percentage of surface returns will decrease only slowly due to the fact that our systems are designed to operate in the highly nonlinear portion of the Poisson probability curve.

The mean number of range measurement attempts per square meter made by the lidar can be easily estimated by dividing the total number of measurement attempts by the total surface area scanned during the same time interval, i.e.,

$$D_m = \frac{N_p f_{qs}}{S v_g} = \frac{N_p f_{qs}}{2hv_g \tan\alpha} \tag{1}$$

where N_p = 100 is the number of beamlets/pixels per pulse, f_{qs} is the pulse repetition rate of the laser, S is the swath width, v_g is the ground velocity of the aircraft, h is the operating AGL, and α is the scanner cone half angle. As one can easily see from Table 1, all SPL lidar models listed can meet USGS Quality Level 1 data densities (8 pts/m^2) over some portion of their aircraft AGL and scan angle ranges.

The RMS instrument range errors listed in the table are computed based on a convolution of the RMS errors introduced by the laser, the detector, and the range receiver and do not include additional RMS contributions due to non-zero incidence angles of the beamlets on the surface [2]. Since all of the systems use virtually identical detectors and receivers, the small differences in RMS between systems are due to differing laser pulse widths.

3. Data Editing

Unlike conventional multi-photon lidars that nullify solar noise by operating at high detection thresholds, SPLs require a substantial amount of noise editing during daytime operations. Early in our development program, data editing approaches were implemented only after the complete point cloud (signal plus noise counts) was generated by inhouse software and viewed via a commercial program such as QT-Modeler®. Early editing approaches often involved substantial human intervention to generate acceptably "clean" images. However, we have developed and successfully tested highly automated data editing software which acts on either the returns from a single pulse or alternatively a sequence of consecutive pulses. This is made possible by the large number (~100) of simultaneous and spatially correlated surface measurements within a single pulse. Furthermore, such an approach lends itself well to real time editing, leading to substantial savings in onboard data storage capacity, data download times, point cloud processing times, and near real time 3D image generation for cockpit display and/or inflight transmission to a ground terminal.

The current denoising filter acts in two stages as illustrated in Figure 7. The raw/unfiltered lidar data taken by HRQLS-1 over a residential community in Oakland, Maryland, contains a great deal of solar noise which fills the nominal 4.6 microsecond (690 m) range gate. The 1st stage filter breaks the range gate into 23 30-meter bins, searches through the entire range gate, and, based on the Differential Cell Count (DCC) Algorithm [2], determines which bins are likely to contain surface returns. The bin sizes are large to allow for tall tree canopies, buildings, etc. It then keeps the data in those bins plus the two adjacent bins to yield a much smaller range interval (90 m) likely to contain all of the surface returns and provides an estimate of the mean solar noise per range interval for use by later filtering stages. Thus, for a typical 4.6 microsecond range gate, the 1st filter stage eliminates all but 90 m/690 m = 13% of the original solar noise. The first stage also allows for the presence of multiple surfaces such as street level returns and rooftop returns within a single pulse or short sequence of pulses. In the second stage, the surviving range intervals are divided into smaller range bins (~5 m) which are then retained or discarded based on the number of observed counts per bin relative to the estimated noise counts derived from the first stage. The second stage count threshold per bin is chosen such that it typically eliminates well over 90% of the noise counts retained following the first stage of filtering, leaving less than 1% of the original noise counts. Both stages are based on the DCC algorithm [2] which is designed to maximize the probability of detecting the actual surface while simultaneously minimizing the probability of detecting a false surface. Algorithms for a third stage filter have been developed to eliminate the very small amount of residual solar or other noise lying in close proximity to actual surfaces.

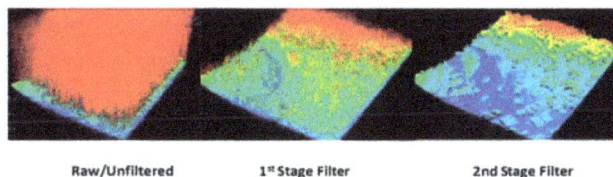

| Raw/Unfiltered | 1st Stage Filter | 2nd Stage Filter |

Figure 7. The automated filtering of HRQLS-1 lidar data taken on a single overflight of a residential community in Oakland, MD. The raw/unfiltered point cloud data is taken with a range gate of 4.6 microseconds corresponding to a total range interval of 690 m. The color scheme is deep blue to red in order of increasing elevation, and it should be mentioned that the solar noise is equally dense below the surface but does not show up as well in the raw unfiltered image because of the poor contrast against the black background. The first stage filter isolates a 90 m interval that contains the surface data as well as roughly 13% of the total noise, and the second stage filter uses narrower range bins (~5 m) to eliminate the vast majority of the remaining noise.

4. Sample HRQLS-1 Data

4.1. Garrett County, MD

Test flights of HRQLS have been funded by several interested customers to assess its capabilities for general surveying, tree height and biomass estimation, and bathymetry. For example, the University of Maryland, under a NASA grant, recently funded the airborne survey of Garrett County in Northwestern Maryland. The county—which is mountainous, heavily wooded, and has a total area of about 1700 km^2—was surveyed in approximately 12 h of flight time, which included a 50% overlap of flight lines, four roundtrips from the host airport, and turns. Because of the low (10%) reflectance and density of the dominant green vegetation, HRQLS was operated from a nominal altitude of 2.29 km with a half-cone scanner angle of 17° (1.36 km swath) rather than the maximum value of 20° (1.62 km swath). At an aircraft velocity of 278 km/h, the resulting areal coverage was 378 km^2/h. The full lidar data set for the county, color-coded from blue to red with increasing surface elevation, is shown superimposed on a Google Earth map of Garrett County in Figure 8. All flights were conducted during daylight hours.

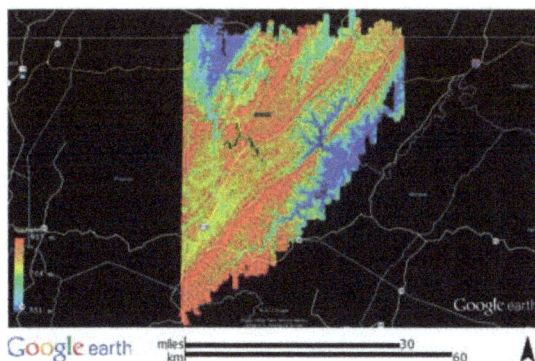

Figure 8. This color-coded elevation map of Garrett County, occupying approximately 1700 km^2 in the state of Maryland, was generated by HRQLS-1 from an AGL of 2.3 km. Total flight duration was approximately 12 h at an air speed of 278 km/h which included a 50% overlap between flight line, ferries, and turn maneuvers. The scanner was operating with a cone half angle of 17° resulting in a swath of 1.36 km and a mapping rate of 378 km^2/h. Highest and lowest elevations are: red = 857 m, blue = 551 m.

13

One can get a sense of the surface spatial resolution by looking more closely at subsets of data within the map. Figure 9 provides different lidar views of a Garrett County coal mine. Details of the coal mining operation such as buildings, conveyor belts and coal piles can be clearly seen.

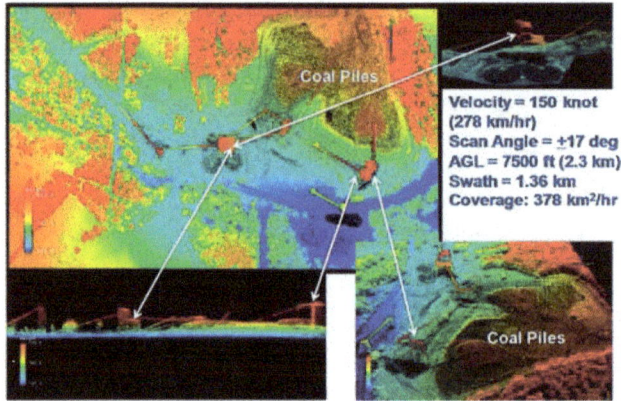

Figure 9. A Garrett County coal mine in which buildings, conveyor belts, and even black coal piles are clearly visible. Elevation Scales: Top Left red = 803.4 m, blue = 759.8 m; Bottom Left and Bottom Right red = 795.2 m, blue = 767.3 m.

Figure 10 demonstrates the ability of the HRQLS-1 lidar to see through heavy forest canopy to the underlying surface and to distinguish between different canopy growth patterns [11].

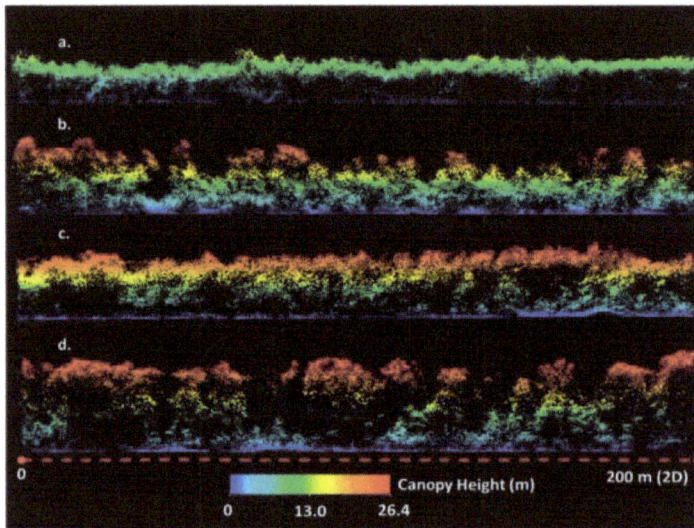

Figure 10. HRQLS-1 SPL point cloud profiles showing different growth patterns within a 1 square kilometer of forested area in Garrett County, MD. (**a**) Short even aged stand with little understory vegetation; (**b**) Uneven aged stand composed of tall trees and dense midstory vegetation; (**c**) Even aged stand with some mid and understory growth; (**d**) Tall open stand with distinct understory vegetation (Courtesy of the University of Maryland [11]).

4.2. Monterey/Pt. Lobos, California

Another set of test flights was conducted in the vicinity of the Naval Post Graduate School in Monterey, California. Figure 11 provides a side-by-side view of the HRQLS-1 lidar data with a digital photograph of the same area. When the lidar data is fused with the digital imagery, one can generate color 3D images, as in Figure 12, or even fly-through movies of the area.

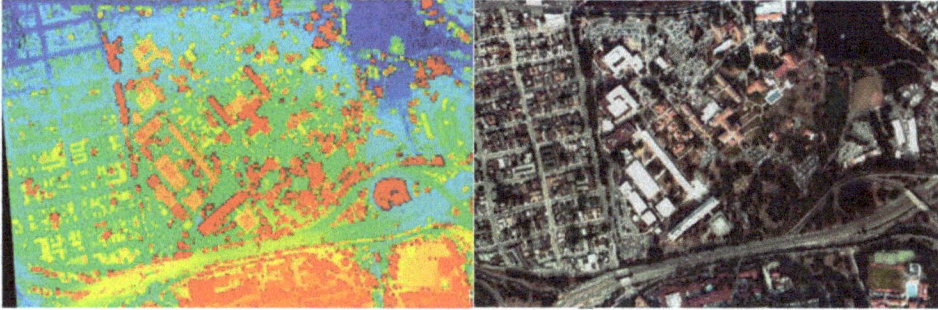

Figure 11. HRQLS-1 lidar image and digital color photograph of the area surrounding the Naval Post Graduate School in Monterey, California.

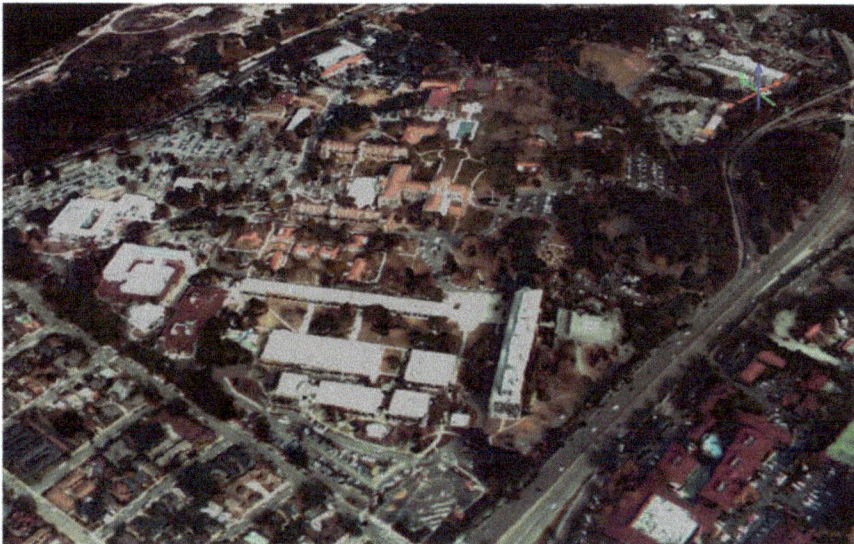

Figure 12. "Fused" HRQLS-1 lidar-photographic 3D image of the Naval Post Graduate School in Monterrey, California.

The Monterey flights also included topo-bathymetric experiments over the Pacific Ocean near Pt. Lobos. HRQLS-1, still operating at 2.3 km above the ocean surface to preserve the high speed contiguous mapping capability, was able to see the ocean bottom to an optical depth of roughly 18 m, as illustrated in Figure 13 This corresponds to an actual physical depth of about 13.5 m when one accounts for the refractive index of sea water. The low level of laser backscatter from the water and the large depth of penetration suggests very low turbidity. Water refraction effects have not been accounted for in the bottom image.

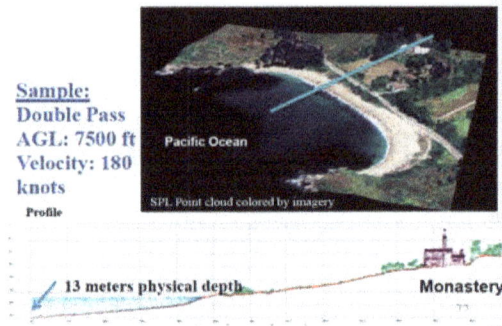

Figure 13. Top: Colored HRQLS-1 lidar topo-bathymetric 3D pointcloud of a hilltop monastery and the beach at Pt. Lobos near Monterey, CA; **Bottom**: The bottom image shows the 2D lidar profile along the blue line in the top figure and extending from the monastery to the beach and into the Pacific Ocean to an optical depth of 17.3 m or a physical depth of 13 m. Vertical grid size = 10 m, Horizontal grid size = 50 m.

4.3. High Density Images

Two or more passes over the target area can produce extremely detailed images. In Figure 14, we show an image of a cruise ship docked at Ft. Lauderdale, FL which was obtained in only two HRQLS-1 passes and a multipass view of an electrical power line grid in North Carolina having a mean measurement density greater than 40 points per square meter.

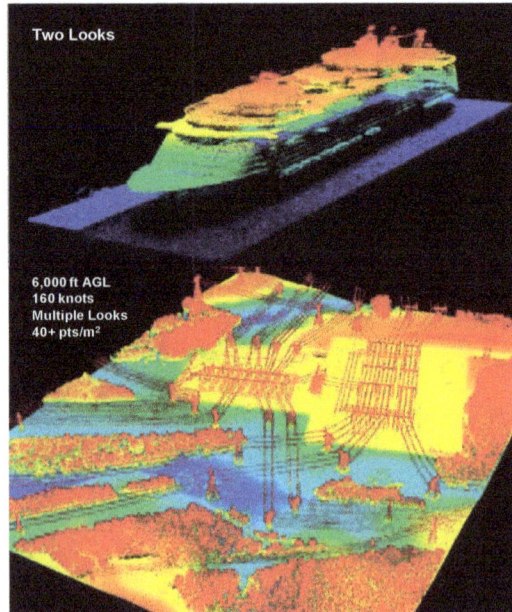

Figure 14. Top: Two passes of HRQLS-1 over a cruise ship docked at Ft. Lauderdale, Florida; **Bottom**: Multiple HRQLS-1 passes over a power line grid in North Carolina yielding over 40 points per square meter from an AGL of 1.83 km and an aircraft velocity of 296 km/h.

5. Single Photon Lidar (SPL) vs. Geiger Mode (GM) Lidar

It should be mentioned that there is much interest within the lidar user community with regard to the characteristics and relative merits of the SPL systems described here vs. competing Geiger Mode Avalanche PhotoDiode (GMAPD) systems, which also utilize single photon detection. The earliest airborne GM lidar, Jigsaw, was developed with DARPA funding at the Massachusetts Institute of Technology/Lincoln Laboratories (MIT/LL) and was designed to image targets of military interest under tree canopies from low altitudes [12]. Later generations included the medium altitude Airborne Lidar Research Testbed (ALIRT), and the High Altitude Lidar Operations Experiment (HALOE) [13,14]. More recently, DARPA transferred GMAPD technology from MIT/LL to commercial entities, and Harris Corporation has introduced the first commercial GM system, The IntelliEarth™ Geospatial Solutions Geigermode Lidar Sensor [15].

5.1. Key Differences between SPL and GM Lidars

While both system types are capable of generating highly detailed 3D images, the following three bullets describe important differences between these two emerging single photon sensitive lidar technologies:

1. Laser Wavelength: Current GM systems utilize the fundamental Nd:YAG wavelength at 1064 nm in the Near InfraRed (NIR) whereas Sigma SPLs use the frequency doubled green wavelength of Nd:YAG at 532 nm. The 1064 nm wavelength is sometimes touted as having several natural advantages including: (1) a factor of 3 lower solar background; (2) generally higher reflectances from natural surfaces such as soil/dry vegetation (25% vs. 15%) and green vegetation (65% vs. 10%); (3) slightly better atmospheric transmission; and (4) no frequency conversion losses in laser power which are typically on the order of 40% to 50% [2,15]. The 532 nm wavelength benefits from: (1) the availability of relatively mature and inexpensive, high efficiency array detectors and narrowband spectral filters; (2) detector dark count contributions to background noise are typically much lower in the visible spectrum; and (3) good transmission in water columns which allows solid land topography and bathymetry to be performed by a single instrument at a single wavelength as in Figure 13.

2. Detector Array Size: Sigma SPLs use relatively inexpensive and compact COTS segmented anode microchannel plate photomultipliers which are currently available in 10×10 formats or 100 pixels per laser pulse. The Harris GM systems, on the other hand, currently utilize relatively expensive InP/InGaAsP SPAD 128×32 arrays/cameras containing 4096 pixels with on-chip readout rates in excess of 100 kHz [16]. In SPL systems, each pixel/anode essentially has a zero recovery time since each 1.6 mm $\times$ 1.6 mm pixel contains tens of thousands of microchannels, and therefore a single photon entering the photocathode activates a very small percentage of the available microchannels in the immediate vicinity of the photon strike. Thus, photons entering at slightly different spatial locations within the pixel experience the same amplification unless the microchannels become saturated, which generally has not been the case in field operations to date. In effect, a single SPL detector pixel behaves much like highly pixelated Geiger Mode array with the exception that all of the microchannel outputs are tied to a common anode and input to a common multistop timing channel capable of recording all of the photon events within the range gate and the pixel FOV. This limits the ground horizontal resolution to the FOV of the pixel which was 15 cm for Leafcutter and 50 cm square for the moderate to high altitude lidars. The current Sigma SPL receiver design typically accepts ten surface and/or noise events per pixel per pulse, but this is not a hard limitation. In effect, each SPL pixel acts as if it was a large array of individual GM SPADs covering the same FOV but tied to a common anode so that the timing of all photon events occurring within a given beamlet and pulse can be measured by a single, fast recovery, timing channel.

3. Receiver Recovery Times: As just discussed, the SPL pixel recovery time of 1.6 nanoseconds (sometimes referred to as "deadtime" or "blanking loss" [15]) is limited not by the detector but by the timing receiver, whereas current GMAPD recovery times are typically in the range of 50 to 1600 nanoseconds depending on whether the Single Photon Avalanche Diodes (SPADs) making up the array are actively or passively quenched. This implies that SPLs can detect, within the same pixel, objects which are separated by only 0.24 m in range. In contrast, detected surfaces must be separated by 7.5 m or 240 m to be seen by an actively or passively quenched GMAPD respectively. Furthermore, each GMAPD in the array, as currently implemented in the Harris system, has only one measurement opportunity per imaging cycle although Harris claims that future asynchronous readout integrated circuits (ROICs) will enable multiple Time of Flight (TOF) measurements per APD per cycle [15]. While this would greatly enhance GM lidar performance, the detection rates within the small FOV of a given APD will still be limited by the longer quenching times.

We will now examine the impact of GMAPD recovery times for two very different daytime mapping scenarios, i.e., one in which the path between the aircraft and the solid target surface is unobstructed and one in which the target is obscured by a semi-porous obscurant (such as a tree canopy or ground fog). Night operations are not an issue for GMAPD lidars.

5.2. Mapping Unobstructed Solid Surfaces in the Presence of Solar Noise

A theoretical model for the solar background rate, Λ, is given in [2]

$$\Lambda = N_\lambda^0 \left(\Delta\lambda \right) \Omega_r \frac{\eta_c \eta_r A_r}{\pi h \nu} \left[\rho T_0^{1+\sec\theta_z} \cos\psi + \frac{1 - T_0^{1+\sec\theta_z}}{4\left(1 + \sec\theta_z\right)} \right] = \frac{0.05\delta^2}{cm^2 \mu sec} \tag{2}$$

where the first and second terms respectively correspond to the background rates due to scattered solar radiation from the surface under study and the intervening atmosphere respectively. In obtaining the numerical value, we have ignored the atmospheric contribution and used numerical values pertinent to the Harris GM lidar [17]. The quantity $N_\lambda^0 = 0.67$ W/m^2/nm is the extraterrestrial solar irradiance impinging on the Earth's atmosphere at 1064 nm, $\Delta\lambda = 3$ nm is the width of the best spectral bandpass filter at 1064 nm based on a short web search, $\Omega = (\delta/R)^2$ is the solid angle viewed by a single GMAPD where $R = 7.62$ km is the range to the target and δ is the ground resolution, $\eta_c = 0.3$ is the Photon Detection Efficiency (PDE) of the detector, $\eta_r = 0.75$ is an estimated optical throughput efficiency of the receiver optics, $A_r = 0.057$ m^2 is the area of the Harris receive telescope, $\rho = 0.65$ is the surface reflectance of green vegetation at the laser wavelength, $h\nu = 1.87 \times 10^{-19}$ J is the laser photon energy, $T_0 \sim 0.9$ is the one-way atmospheric transmission at nadir from the aircraft, $\theta_z = 0$ deg is the worst case solar zenith angle, and $\psi = 0$ is the worst case subtended angle between the Sun and the surface normal.

In either system type (SPL or GM), the solar background counts during daylight operations can be substantially reduced by installing narrowband spectral filters and minimizing the range gate, the collecting area of the telescope and/or the pixel FOV. This is especially important for the single stop GM system, however, since a noise count occurring within the range gate prior to the surface return will result in the loss of that surface measurement for one full array mapping cycle.

For the SPL, a single timing event is generated by a solid surface, irregardless of the number of photons received, since the subnanosecond laser pulsewidth is short compared to the pixel recovery time of 1.6 ns. As a result, the amplitude of the SPL anode output will vary if the "simultaneous" surface returns are spread over multiple microchannels and summed within the pixel/anode. Single photon noise counts within the range window are recorded as separate random events displaced in range and time from the surface returns and later eliminated via noise editing algorithms described previously. Unlike individual GM/APDs, the detection of a solar photon in an SPL pixel does not prevent the pixel from detecting the surface.

For unobscured hard targets, the principal concern raised by the single stop limitation of current GM sensors is the impact of the solar background on the surface probability of detection which, in daylight operations, dominates other noise sources such as detector dark counts. There are only three possible outcomes for a given GM pixel per imaging cycle: (1) a surface photon is detected; (2) no photon is detected; or (3) a background count is detected. If the range gate is approximately centered on the surface and Λ is the solar count rate observed by a single APD, the probability of detecting the surface is given by

$$P_s(n, \Lambda\tau_g) = \exp\left(-\frac{\Lambda\tau_g}{2}\right) P_s = \exp\left(-\frac{\Lambda\tau_g}{2}\right) [1 - \exp(-n)] \tag{3}$$

where the first term is the probability that the APD is not triggered by solar noise in the first half of the gate, and the second term is the Poisson probability that the surface return, consisting of n detected photoelectrons, is detected by a receiver with a single photon threshold. Similarly, the probability that zero counts are detected is given by

$$P_z(n, \Lambda\tau_g) = \exp\left(-\frac{\Lambda\tau_g}{2}\right) [1 - P_s(n, \Lambda\tau_g)] \exp\left(-\frac{\Lambda\tau_g}{2}\right) = \exp(-\Lambda\tau_g)\exp(-n) \tag{4}$$

and, since the three probabilities must sum to 1, the probability that a solar background count is detected is given by

$$P_b(n, \Lambda\tau_g) = 1 - P_s(n, \Lambda\tau_g) - P_z(n, \Lambda\tau_g) = \left[1 - \exp\left(-\frac{\Lambda\tau_g}{2}\right)\right]\left[1 + \exp\left(-\frac{\Lambda\tau_g}{2}\right)\exp(-n)\right] \tag{5}$$

The ratio of signal counts to solar counts is obtained by dividing (3) by (5) to yield

$$SNR = \frac{P_s(n, \Lambda\tau_g)}{P_b(n, \Lambda\tau_g)} = \frac{1 - \exp(-n)}{\left[\exp\left(\frac{\Lambda\tau_g}{2}\right) - 1\right]\left[\left[1 + \exp\left(-\frac{\Lambda\tau_g}{2}\right)\exp(-n)\right]\right]} \tag{6}$$

Figure 15a shows plots of Equation (3) for the fraction of GMAPDs recording surface returns over a range of surface signal strengths, $n = 0.1$ to 3, and the mean number of noise photons occurring in the first half of the range gate, $x = \Lambda\tau_g/2 = 0$ to 1. Figure 15b, plotting Equation (6), shows the ratio of signal to noise counts versus the same parameters. It is worthwhile to note that achieving even the lowest value of $n = 0.1$ in all 4096 pixels (one detected surface photon per APD in 10 pulses) would require a total signal strength of 410 photoelectrons (pe) detected across the array. This is comparable to what many Digitizer Waveform lidars require and what the HAL lidar achieves from similar AGLs for ground reflectances of 15%, or higher, i.e., 4 pe per pixel over 100 pixels, but the per pixel probability of detection is close to 100% as compared to 10% or less as in Figure 15a. Nevertheless, the GM lidar has the potential of recording surface returns from 4 times as many pixels provided the solar noise counts can be adequately suppressed.

The plots in Figure 15a would suggest that we strive for a value of

$$x \equiv \frac{\Lambda\tau_g}{2} = \frac{0.05\delta^2\tau_g}{2\,\mathrm{cm}^2\,\mathrm{\mu sec}} < 0.2 \tag{7}$$

in order to avoid severely diminishing the probability of detecting a surface return. A typical range gate in our high altitude flights is $\tau_g = 5\,\mu s$, which, from (7), would suggest a pixel dimension at the ground of $\delta < 1.3$ cm or a maximum array area FOV at the ground of AGM = 4096 $(1.3\,\mathrm{cm})^2 = 0.7\,\mathrm{m}^2$. The latter area is 36 times smaller than the HAL area for a single pulse and, while a 1.3 cm horizontal spatial resolution would be outstanding, the effect on the rate of areal coverage would not.

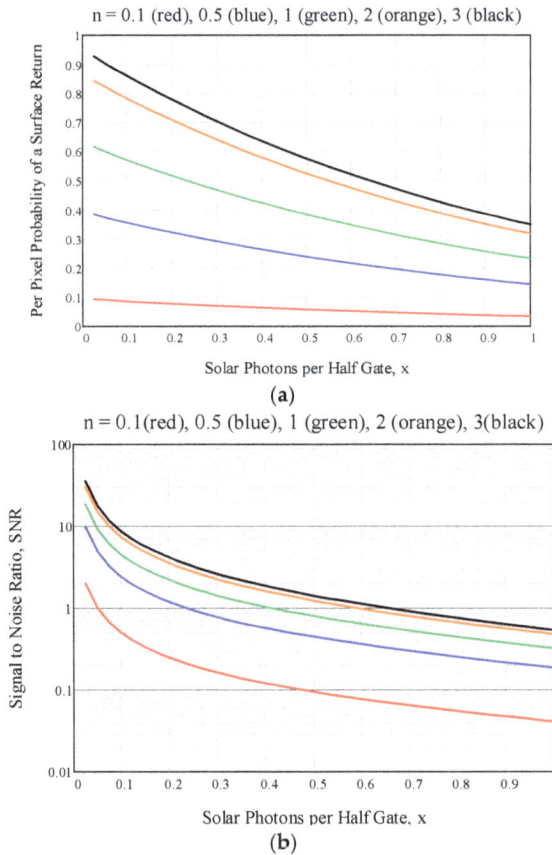

Figure 15. (**a**) Fraction of pixels recording surface returns as a function of surface signal strength, n, and mean number of noise photons detected within a half range gate; (**b**) ratio of signal to noise counts as a function of the same two parameters.

5.3. Viewing the Surface through Semi-Porous Obscurants Such as Tree Canopies

It has long been known from early experiments conducted with the Jigsaw GM system at MIT/LL that, in order to detect military vehicles under dense tree canopies, one had to reduce the pulse energy per pixel in order to reduce the probability that a photon from the canopy would disable the pixel and prevent observation of the surface under the canopy. Therefore, since the tree elements (leaves, branches, etc.) are opaque to laser light, a recognizable surface image could only be generated by making many low energy measurements from a wide variety of aspect angles in order to take advantage of any existing canopy "holes" between the aircraft and the surface. This can be mathematically represented by multiplying the probability of detecting the surface by the probability that the measurement is not disabled by a photon reflected off a tree element. From Poisson statistics, the probability of detecting a target beneath a canopy (or fog bank) with one-way transmission T_c is given by

$$PD(n_s, \gamma) = \exp\left[-\frac{\gamma n_s}{2}\left(1 - T_c^2\right)\right]\left[1 - \exp\left(-T_c^2 n_s\right)\right] \qquad (8)$$

where n_s is the expected number of detected photoelectrons from the unobscured target,

$$\gamma = \frac{\rho_c}{\rho_t} \tag{9}$$

and ρ_c and ρ_t are the reflectances of the canopy and target respectively. The second term in the equation gives the probability of detecting the surface signal with a single photon sensitive receiver while the first term gives the probability of disabling the receiver due to the detection of a canopy return. As mentioned previously, the Sigma SPL detector/receiver has a very short recovery time on the order of 1.6 ns and therefore only the second term in the equation is relevant. Figure 16 shows the surface detection probability for a tree canopy with a one way transmission $T_c = 0.4$ as a function of the unobscured signal strength ($T_c = 1$). The black curve in the plot shows the probability of detecting the unobscured target vs. the mean signal strength expressed in detected photoelectrons. The red curve gives the same probability for a low deadtime single photon sensitive receiver in the presence of a tree canopy with a one-way transmission of $T_c = 0.4$ ($T_c^2 = 0.16$). The remaining three curves show the Geiger Mode probabilities for different values of γ. Note that, for each value of γ, there is an optimum unobscured signal strength for detecting the underlying surface in qualitative agreement with the early Jigsaw experiments. In all cases, the peak Geiger Mode detection probabilities fall substantially below the SPL values, especially when the canopy has a higher reflectance than the final target ($\gamma > 1$). The 6.5 times stronger reflectance of vegetation at 1064 nm vs. 532 nm (65% vs. 10%) mentioned in Bullet 1 of Subsection 5.1 increases the value of γ substantially and further reduces the probability of seeing the under-canopy surface. In addition, the higher tree reflectance creates a 6.5/3 = 2.2 times stronger solar background during daylight operations which was not included in the plots of Figures 16 and 17.

The theoretical performance of SPL and GM lidars over a wide range of one-way tree canopy transmissions ($T_c = 0.1$ to 1) is provided by Figure 17. The reduction in canopy transmission could be due to more dense foliage or a longer slant range through a canopy with higher one-way transmission when viewed from nadir. This is an important consideration since the key to under canopy observations is finding "all the available" holes". The curves in the top left of the figure demonstrate how the GM probability of detecting the surface falls as the one-way canopy transmission decreases due to the fact that the lidar cannot "power" its way through the canopy by increasing the unobscured signal strength because of the one return per APD limitation. The curves in the bottom right of the figure show the relative surface detection rate of the SPL and GM systems for the same range of tree canopy transmissions where the SPL can, in fact, "power" its way through the canopy as in Figure 16.

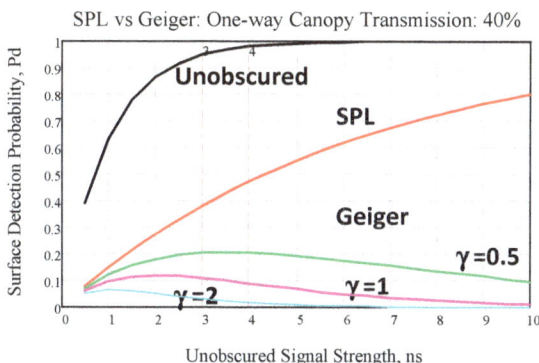

Figure 16. Surface detection probabilities for SPL and Geiger Mode (GM) lidars as a function of the unobscured signal strength for a tree canopy having a one way transmission of 40%. Unlike the GM lidar which has an unobscured signal strength that optimizes the surface detection probability, the SPL lidar can "power" through the canopy by increasing the laser pulse energy.

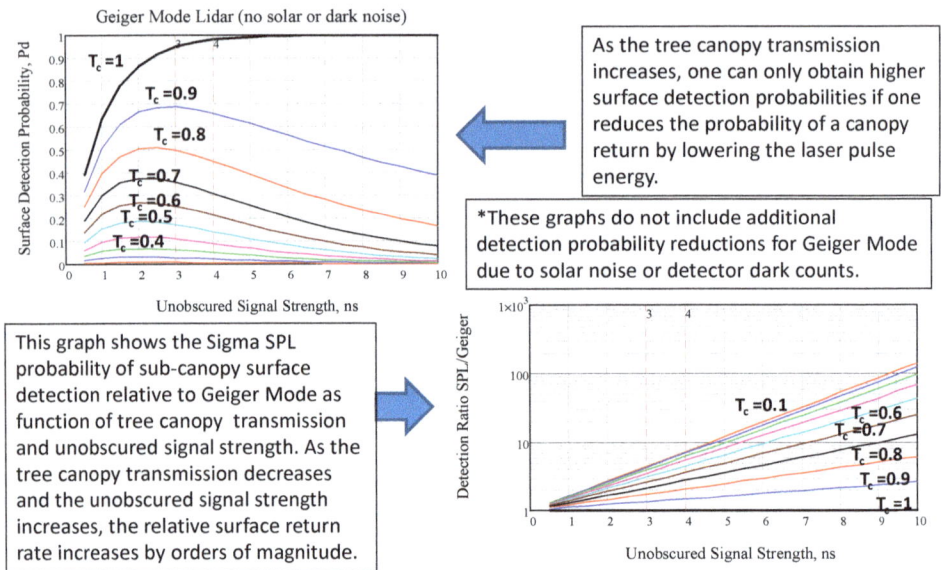

Figure 17. The relative performance of SPL and GM lidars over a wide range of one-way tree canopy transmissions (T_c = 0.1 to 1) A value γ = 1 is assumed. The top left graph demonstrates that, as the tree canopy transmission decreases, the optimum unobscured signal for maximum penetration decreases, further reducing the detectability of the under canopy surface by the GM lidar. The bottom right graph describes the increasing advantages of the SPL technique in detecting the under canopy surface as the one way canopy transmission decreases.

5.4. Brief Summary of the Theoretical Results

While there are clearly no issues with night operations of GM lidars as indicated by a large number of highly detailed 3D images posted on the Harris web site [17], the present analysis suggests that the expected solar noise over high reflectance surfaces, such as green vegetation (ρ = 0.65) could greatly reduce the PDE of individual GMAPD pixels having reasonably sized FOVs. This in turn would greatly reduce the rate at which large areas can be mapped in daylight. It must be mentioned, however, that Harris Corporation strongly claims an acceptable daylight capability on their web site [17] and a limited amount of daytime data was included in a recent USGS study [18]. Furthermore, the current analysis indicates that GM lidars would appear to be far inferior to SPLs when probing dense tree canopies.

6. Summary

Imaging SPLs operating at the 532 nm wavelength can provide seamless topographic and bathymetric maps from a single instrument. Single photon sensitivity allows a moderate power laser beam to be split, by a passive holographic element, into a 10 × 10 array of individual beamlets, whose images in the receiver plane fall onto a matching array of single photon sensitive, high bandwidth pixels. In addition to increasing the surface measurement rate to several megapixels per second for subnanosecond pulse lasers operating in the tens of kHz range, the arrays allow contiguous, decimeter resolution, alongtrack and crosstrack mapping of the surface on a single overflight with modest telescope and scanner apertures (7.5 cm to 15 cm) and scanner speeds on the order of 20 Hz (1200 RPM). Higher transverse spatial resolutions can be achieved by reducing the swath width or by making multiple overlapping passes over the site. The fast recovery times (1.6 ns) of the pixels and their individual timing channels provide a multistop capability that allows the SPLs to operate effectively under conditions of strong solar illumination and to penetrate semiporous obscurants such as tree canopies, ground fog, etc. The SPLSs are designed to have per pixel probabilities of detection on the order of 95% for a 10% reflectance surface (mean of 3 pe per pixel from green vegetation) and greater than 99% for reflectances greater than 15% (mean of 4 pe per pixel from soil or dry vegetation). Thus, the mean 300 photoelectron returns from a tree canopy over the full array is comparable to that of some conventional Digitized Waveform lidars but with the added advantage that the location of the scattering source within the canopy is identified in all three dimensions as opposed to being lumped together into a single dimension, range, thereby resulting in a more realistic image of the canopy.

As the SPLs have progressed to higher operational altitudes in order to provide wider swaths and faster areal coverage, we have had to address new technical challenges such as the availability of COTS laser/telescope/detector combinations to meet the higher link demands. Table 1 in Section 2.7 provides a summary of the key subsystem parameters for the various SPLs flown to date. Solar background has been minimized through the use of a 0.3 nm FWHM spectral filter and a small pixel Field of View, typically defined by a 50 cm × 50 cm square on the ground. The vast majority of the detected noise is eliminated via noise editing algorithms as described in Section 3. Also, at the higher AGLs, a corrector wedge is added to the common transmit/receive optical scanner to reimage the transmit beamlet array onto the detector array at the nominal AGL and scan speed. For alternative AGLs, the scan speed must be adjusted from its nominal value to achieve maximum overlap of the transmit and receive FOVs. In addition, angular biases, as well as atmospheric refraction and pulse group velocity effects, play a bigger role in achieving the necessary geolocation accuracy due to the longer slant range distances. As a result, we have developed algorithms and software to find and eliminate biases based on multiple looks at distinct features in the overall point cloud such as the corners of buildings.

Lidar users in the mapping community are most concerned about geolocation errors and spatial resolution. Geolocation is assessed by comparing lidar elevation products to surveyed ground control points. The HAL system was flown over a 400 square kilometer area at an AGL of approximately 7.6 km. A total of 22 ground points were surveyed and compared to elevations derived from the point cloud. After removing bias, an elevation RMS of 9 cm, meeting the highest USGS QL-1 requirement of 10 cm, was obtained [8]. In an earlier flight experiment over Monterey CA, HRQLS-1 point cloud results were compared to 21 points measured to 3 cm vertical accuracy by the Naval Postgraduate School and resulted in a similar 9.3 cm RMS Standard deviation. In addition, both HAL and HRQLS-1 easily meet the USGS QL-1 requirements on measurement point densities (>8 pts/m^2).

In 2015, Sigma's HRQLS-1 SPL and Harris GM systems participated in a series of USGS-sponsored field trials in the state of Connecticut in which the point clouds were analyzed by two independent and highly experienced lidar analysis groups (Woolpert and Dewberry) and presented at the International Laser Mapping Forum (ILMF2016) in Denver, Colorado. For recent field evaluations of the HRQLS-1 SPL and/or the Harris GM system over a wide variety of terrain types and opinions on their future operational role with respect to conventional linear mode lidars, the reader is referred to the following papers [19–21]. The current SPL and GM lidars are generally viewed as being highly competitive with conventional lidars when it comes to large scale mapping missions over unobscured terrain. On the other hand, as discussed in Section 5, the fast pixel recovery times would appear to give the SPL approach a significant advantage over GM systems for daytime mapping missions requiring wide range gates and/or the penetration of semi-porous obscurants such as tree canopies, ground fog, etc. As mentioned previously, use of the green wavelength also permits topo-bathymetric measurements to be carried out by a single, compact SPL instrument. Our newest moderate altitude SPL, HRQLS-2, and presumably the latest version of the Harris Corporation GM lidar, are expected to participate in a second set of USGS-sponsored experiments to be carried out over large areas in South Dakota in late 2016.

Commercial users of conventional lidars also ask whether or not SPLs can generate intensity information. In principle, aggregated single photon returns collected over a sufficient surface area could be used to ascertain reflectance but our SPL systems are designed to collect as many surface measurements as possible per square meter by multiphoton surface returns. This provides little discrimination between different surface reflectances since all surfaces over 10% reflectance provide 95 to 100 pixel returns per pulse. However, one Sigma colleague has developed an as yet unpublished but highly successful procedure applicable to daytime operations [22] while a second is experimenting with a second hardware approach applicable to both day and night missions [23].

In preparation for 3D imaging that can be viewed by the aircraft crew or transmitted to a ground station in near real time, we are currently implementing inflight algorithms and onboard processors that edit out solar and/or electronic noise and correct for atmospheric effects. Furthermore, analyses conducted for NASA have shown that the scanning SPL technique can even be extended to orbital altitudes for the globally contiguous mapping of extraterrestrial planets and moons [2,4,16] using space-qualified transmitters and timing receivers being developed for NASA's ATLAS SPL lidar on the ICESat-2 mission scheduled to be launched in 2017 [9]. For example, the three moons of Jupiter of most interest to NASA can each be globally mapped with 5 m horizontal resolution and decimeter vertical resolution in as little as two months for the larger moons, Ganymede and Callisto, and one month for Europa [24].

Acknowledgments: The author wishes to acknowledge both current and former Sigma Space employees who, over many years of development, have made important contributions to the funding, design, flight operations, and data processing efforts associated with the various lidars presented here. Program Management: J. Marcos Sirota, Katie Fitzsimmons; Electronics: Roman Machan, Ed Leventhal, Cesar Ventura, Gabriel Jodor, Jose Tillard; Optical: Yunhui Zheng, James Lyons, Robert Upton; Mechanical/Flight Operations: Spencer Disque, Steven Mitchell, David Lawrence, Nicholas Bellis; Thermal: Rodney Falkner; Data Processing: Christopher Field, Terence Barrett, Ivana Williams, David Yancich, Biruh Tesfaye, Sean Howell, Borislav Karaivanov, Chris Innannen, Ruben Nieves, Elias Waggoner. The Garrett County HRQLS data acquisition was funded by NASA Grant NNX12AN07G to the University of Maryland (Ralph Dubayah, PI) as part of the NASA Carbon Monitoring System Program.

Conflicts of Interest: The author declares no conflict of interest.

References

1. Degnan, J.; McGarry, J.; Zagwodzki, T.; Dabney, P.; Geiger, J.; Chabot, R.; Steggerda, C.; Marzouk, J.; Chu, A. Design and performance of an airborne multikilohertz, photon-counting microlaser altimeter. *Int. Arch. Photogramm. Remote Sens.* **2001**, *XXXIV-3/W4*, 9–16.

2. Degnan, J. Photon-Counting Multikilohertz Microlaser Altimeters for Airborne and Spaceborne Topographic Measurements. *J. Geodyn.* **2002**, *4*, 503–549. [CrossRef]

3. Harding, D. Pulsed Laser Altimeter Ranging Techniques and Implications for Terrain Mapping. In *Topographic Laser Ranging and Scanning: Principles and Processing*; Shan, J., Toth, C.K., Eds.; CRC Press: Boca Raton, FL, USA, 2009.

4. Degnan, J.; Wells, D.; Machan, R.; Leventhal, E. Second Generation 3D Imaging Lidars based on Photon-Counting. In Proceedings of SPIE Optics East, Boston, MA, USA, 12 September 2007.

5. Carter, W.; Shrestha, R.; Slatton, K. Photon counting airborne laser swath mapping (PC-ALSM). In *Gravity, Geoid, and Space Missions*; Jekeli, C., Bastos, L., Fernandez, J., Eds.; Springer: Berlin/Heidelberg, Germany, 2004; pp. 214–217.

6. Cossio, T.; Slatton, C.; Carter, W.; Shrestha, K.; Harding, D. Predicting topographic and bathymetric measurement performance for low-SNR airborne lidar. *IEEE Trans. Geosci. Remote Sens.* **2009**, *47*, 2298–2315. [CrossRef]

7. Degnan, J.J. A Conceptual Design for a Spaceborne 3D Imaging Lidar. *E&I Electrotechnik und Informationstechnik (Austria)* **2002**, *4*, 99–106.

8. Gluckman, J. Design of the processing chain for a high-altitude, airborne, single photon lidar mapping instrument. *Proc. SPIE* **2016**. [CrossRef]

9. Abdalati, W.; Zwally, H.; Bindschadler, R.; Csatho, B.; Farrell, S.; Fricker, H.; Harding, D.; Kwok, R.; Lefsky, M.; Markus, T.; et al. The ICESat-2 laser altimetry mission. *Proc. IEEE* **2010**, *98*, 735–751. [CrossRef]

10. McGill, M.; Markus, T.; Scott, V.S.; Neumann, T. The Multiple Altimeter Beam Experimental Lidar (MABEL): An Airborne Simulator for the ICESat-2 Mission. *J. Atmos. Ocean. Technol.* **2013**, *30*, 345–352. [CrossRef]

11. Swatantran, A.; Tang, H.; Barrett, T.; DeCola, P.; Dubayah, R. Rapid, High-Resolution Forest Structure and Terrain Mapping over Large Areas using Single Photon Lidar. *Sci. Rep.* **2016**. [CrossRef] [PubMed]
12. Vaidyanathan, M.; Blask, S.; Higgins, T.; Clifton, W.; Davidsohn, D.; Carson, R.; Reynolds, V.; Pfannenstiel, J.; Cannata, R.; Marino, R.; et al. JIGSAW Phase III: A miniaturized airborne 3-D imaging laser radar with photon-counting sensitivity for foliage penetration. *Proc. SPIE* **2007**. [CrossRef]
13. Knowlton, R. *Airborne Ladar Imaging Testbed*; Tech Notes; MIT Lincoln Laboratory: Lexington, MA, USA, 2011; Available online: http://www.princetonlightwave.com/products/geiger-mode-cameras/ (accessed on 17 November 2016).
14. Gray, G. High Altitude Lidar Operations Experiment (HALOE)—Part 1, System Design and Operation. In Proceedings of the Military Sensing Symposium, Active Electro-Optic Systems, Paper AH03, San Diego, CA, USA, 12–14 September 2011.
15. Clifton, W.; Steele, B.; Nelson, G.; Truscott, A.; Itzler, M.; Entwhistle, M. Medium altitude airborne Geiger-mode mapping Lidar system. *Proc. SPIE* **2015**, *9465*, 946506. [CrossRef]
16. Falcon IR Camera. Available online: http://www.princetonlightwave.com/products/geiger-mode-cameras/ (accessed on 17 November 2016).
17. Harris Corporation. Available online: http://www.apsg.info/resources/Documents/Presentations/APSG34/IntelliEarth_Presentation_APSG34_Oct_2015.pdf (accessed on 14 November 2016).
18. Higgins, S. Single Photon and Geiger Mode vs. Linear Mode Lidar. *SPAR 3D*. 2016. Available online: http://www.spar3d.com/news/lidar/single-photon-and-geiger-mode-vs-linear-mode-lidar/ (accessed on 14 November 2016).
19. Li, Q.; Degnan, J.; Barrett, T.; Shan, J. First Evaluation on Single Photon-Sensitive Lidar Data. *Photogramm. Eng. Remote Sens.* **2016**, *82*, 455–463. [CrossRef]
20. Abdullah, Q. A Star is Born: The State of New Lidar Technologies. *Photogramm. Eng. Remote Sens.* **2016**, *82*, 307–312. [CrossRef]
21. Higgins, S. Single Photon Lidar Proven for Forest Mapping. *SPAR 3D*. 2016. Available online: http://www.spar3d.com/news/lidar/single-photon-lidar-proven-forest-mapping/ (accessed on 14 November 2016).
22. Field, C. Sigma Space Corporation, Lanham, MD, USA. Private communication, 2014.
23. Machan, R. Sigma Space Corporation, Lanham, MD, USA. Private communication, 2016.
24. Degnan, J. Rapid, Globally Contiguous, High Resolution 3D Topographic Mapping of Planetary Moons Using a Scanning, Photon-Counting Lidar. *International Workshop on Instrumentation for Planetary Missions, GSFC*. 2012. Available online: http://www.lpi.usra.edu/meetings/ipm2012/pdf/1086.pdf (accessed on 14 November 2016).

remote sensing

MDPI

Article

Voxel-Based Spatial Filtering Method for Canopy Height Retrieval from Airborne Single-Photon Lidar

Hao Tang [1],*, Anu Swatantran [1], Terence Barrett [2], Phil DeCola [2] and Ralph Dubayah [1]

[1] Department of Geographical Sciences, University of Maryland, College Park, MD 20742, USA;
aswatan@umd.edu (A.S.); dubayah@umd.edu (R.D.)
[2] Sigma Space Corporation, 4600 Forbes Blvd, Lanham, MD 20706, USA;
terence.barrett@sigmaspace.com (T.B.); phil.decola@sigmaspace.com (P.D.)
* Correspondence: htang@umd.edu

Academic Editors: Jie Shan, Juha Hyyppä, Nicolas Baghdadi and Prasad S. Thenkabail
Received: 27 June 2016; Accepted: 12 September 2016; Published: 19 September 2016

Abstract: Airborne single-photon lidar (SPL) is a new technology that holds considerable potential for forest structure and carbon monitoring at large spatial scales because it acquires 3D measurements of vegetation faster and more efficiently than conventional lidar instruments. However, SPL instruments use green wavelength (532 nm) lasers, which are sensitive to background solar noise, and therefore SPL point clouds require more elaborate noise filtering than other lidar instruments to determine canopy heights, particularly in daytime acquisitions. Histogram-based aggregation is a commonly used approach for removing noise from photon counting lidar data, but it reduces the resolution of the dataset. Here we present an alternate voxel-based spatial filtering method that filters noise points efficiently while largely preserving the spatial integrity of SPL data. We develop and test our algorithms on an experimental SPL dataset acquired over Garrett County in Maryland, USA. We then compare canopy attributes retrieved using our new algorithm with those obtained from the conventional histogram binning approach. Our results show that canopy heights derived using the new algorithm have a strong agreement with field-measured heights ($r^2 = 0.69$, bias = 0.42 m, RMSE = 4.85 m) and discrete return lidar heights ($r^2 = 0.94$, bias = 1.07 m, RMSE = 2.42 m). Results are consistently better than height accuracies from the histogram method (field data: $r^2 = 0.59$, bias = 0.00 m, RMSE = 6.25 m; DRL: $r^2 = 0.78$, bias = −0.06 m and RMSE = 4.88 m). Furthermore, we find that the spatial-filtering method retains fine-scale canopy structure detail and has lower errors over steep slopes. We therefore believe that automated spatial filtering algorithms such as the one presented here can support large-scale, canopy structure mapping from airborne SPL data.

Keywords: single photon lidar; canopy height; noise removal

1. Introduction

Single photon lidar (SPL) is a new technology for rapid three-dimensional mapping of terrain and forest structure over large areas at high resolution [1]. SPL requires only one detected photon at each ranging measurement, instead of hundreds in the case of conventional sensors [2,3]. It therefore allows enhanced 3D mapping with greater coverage, spatial resolution, and photon density, and reduced acquisition time. Despite these advantages, SPL data includes more solar background noise than conventional near-infrared lidar instruments because of its use of green wavelength lidar and high photon detecting sensitivity [4–6]. This complicates the retrieval of terrain and canopy structural information from SPL data. Algorithms for efficiently handling noise in photon counting data are only beginning to be developed [7–9]. Existing methods largely rely on detecting maximum canopy height through histogram-based filtering algorithms [4,8,10,11]. This is achieved by aggregating point clouds into pseudo-waveforms at coarse spatial and vertical resolutions. Traditional waveform processing algorithms are then applied to the waveforms to derive canopy structure metrics including canopy height and cover. Although these methods have proven effective in filtering solar noise [4,8,10,11], the 3D structural detail in the dataset is reduced. It means that these methods forfeit a key advantage of SPL data over other systems—the ability to provide fine scale topographic and structural details. Applying histogram based filtering methods to SPL data would therefore drastically reduce its usefulness for applications such as high-quality topographic mapping [12], individual tree-level classification [13] and high resolution carbon monitoring [14,15]. Furthermore, histogram based filtering methods could potentially have errors over complex terrain because of the mixing between ground/canopy signals and noise points. To overcome these hurdles, we develop a spatial filtering algorithm that filters noise points using a variable 3D pixel (voxel) binning approach while retaining the spatial and vertical fidelity of the dataset.

We first describe data collection and the development of the spatial filtering algorithm. Next, we test it with airborne SPL data collected over Garrett County in Maryland [16] using an experimental High-Resolution Quantum Lidar System (HRQLS) instrument. We then compare canopy height metrics derived from SPL data, using both our algorithm and the conventional histogram filtering method, with the reference dataset collected from field campaigns and discrete return lidar (DRL). Lastly, we discuss the strengths and weaknesses of the two approaches and their applicability to large-scale forest mapping.

2. Materials and Methods

2.1. HRQLS

The High-Resolution Quantum Lidar System (HRQLS) is a moderate altitude SPL instrument with 10×10 laser beamlets and an operating wavelength of 532 nm. It is equipped with a dual wedge scanner allowing a full conical observing angle from $0°$ to $40°$. HRQLS was flown over Garrett county of Maryland, USA in early September of 2013 to support a NASA Carbon Monitoring System (CMS) project over the state of Maryland [15,16]. The flight survey required less than 12 h to cover the entire county (~1700 km^2) with 50% overlap of flight lines. At a nominal above ground level (AGL) of 2.3 km, HRQLS produced a swath of 1.62 km and a target spot of 5×5 m array on the ground, resulting in a ground-pixel dimension of 0.5 m and a mean point density of 12 per m^2 per conical scan (including both signal and noise, see Figure 1). More details about the HRQLS data can be found in [2]

2.2. Reference Data

Field measurements were collected during the summer of 2013 with a main survey focus on aboveground biomass estimates [15]. Field plots (fixed and variable radius) were selected using a stratified random sampling based on distributions of land cover type and canopy height. We measured diameter at breast height (DBH) of each individual tree to estimate plot level biomass. Field heights were recorded using a vertex hypsometer with sonic transponder attached to dbh of the largest tree within 71-forested plots. Non-forested plots were excluded because no field height measurement was recorded.

Discrete return lidar (DRL) data were collected by USGS over Garrett County in 2005 as part of their floodplain mapping efforts. These data were collected in leaf-off season, and the point clouds included first returns only with an average density of 1 per m^2. Both maximum canopy height (p100) and mean plot-level slope were extracted over the 71 field plots. Studies have suggested that those leaf-off lidar data sets can largely provide reliable canopy height measurements with accuracy similar to leaf-on data [17,18]. The comparison between leaf-off DRL and leaf-on field data produced $r^2 = 0.63$, bias = -0.26 m and RMSE = 5.37 m despite ~10 year difference [1]. These data sets might not be an optimal reference for assessing SPL-derived canopy heights at highest possible precession, but they could still serve as a benchmark for comparison when used together.

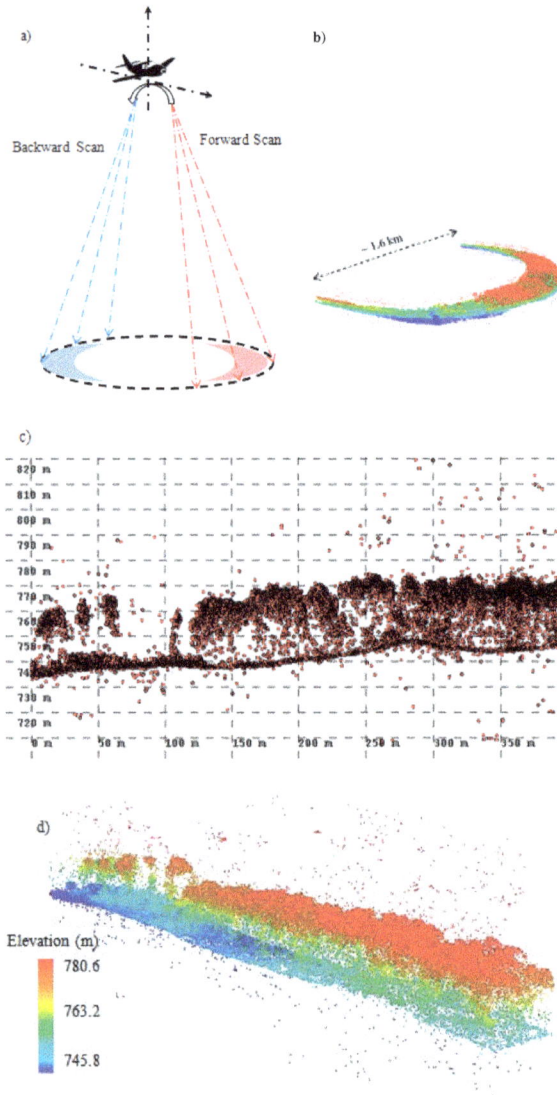

Figure 1. An overview of single photon lidar (SPL) data acquired from the High-Resolution Quantum Lidar System (HRQLS): (**a**) The conical scanning mechanism of HRQLS with both forward and backward scans; (**b**) An example of point clouds acquired in an individual scan; (**c**) An example of cross sectional profile composited from multiple scans; (**d**) 3D point clouds including photons from canopy, terrain and solar noise.

2.3. HRQLS Data Processing

2.3.1. Preprocessing

HRQLS data was collected and pre-processed using proprietary software by Sigma Space Corporation (Lanham, MD, USA) [2]. A level 1 filter was applied to filter noise from the atmospheric column extending up to flight AGL. The filter searched the entire ranging measurements using a 30 m histogram of photon density and identified the peak interval as ground surface range. The level 1 filtered SPL dataset was then produced using points extracted within one interval above and below the peak to include signals from low terrain surfaces and tall canopies. It was essentially a 90 m vertical range subset of raw HRQLS, and reduced the mean point density to about 22 per m^2. The Level 1 filter removed more than 90% of the solar noise from the entire vertical column [2] but considerable noise points remained in the 90 m vertical range subset and required further filtering.

2.3.2. The Voxel-Based Spatial Filtering Method

A point density based spatial filter was developed to filter noise from the Level 1 SPL dataset (Figure 2). This filter identified noise points by searching isolated points in 3D space based on following assumptions: First, SPL points were treated as either signal or noise, and the majority of noise points were randomly distributed in 3D space. Second, noise level over a given area of interest was dependent on the incident laser energy indicated by the total amount of point clouds. For example, an overlap of two flight lines would double the amount of both signal and noise. Third, the density of signal point was significantly higher within the canopy-terrain layer than in the atmosphere. Last, all points in a voxel would be classified as signals if their total number exceeding a certain threshold. This voxel-based approach was similar to methods previously applied in processing airborne or terrestrial lidar data [19–21]. A detailed implementation of the algorithm was described as follows:

1. Calculating the maximum-noise-level threshold as the mean volume point density (points per m^3) of the Level 1 SPL dataset at each 30 m × 30 m horizontal grid;
2. Splitting the area of interest into 3D cells at a given size;
3. Counting the number of points in each cell and its surrounding cells (a total of 27 cells);
4. Labeling the points as noise if the number was less than the pre-calculated noise threshold multiplied by the volume of 27 cells.

Figure 2. A flowchart of deriving canopy height from SPL data using two independent methods: the histogram based method and the spatial filtering method.

We determined the optimal filter by experimenting with different voxel sizes. In each experimental run, canopy heights were retrieved after noise filtering with different 3D voxel sizes and then compared with reference data while keeping other conditions the same. A $3 \times 3 \times 0.2$ m voxel resulted in the lowest bias value of mean canopy height and was determined to be the optimal filter in this study. The optimal filter was applied to the Garrett county SPL dataset to produce a relatively 'noise-free' dataset for further analyses [1]. We then classified the point clouds into ground and canopy and calculated canopy height percentiles over the field plots. Ground points were identified based on a progressive Triangular Irregular Network (TIN) densification method by iteratively selecting points within lowest elevation values [22]. Canopy height percentile here was defined as the height below which a specified percentage of total point clouds were located, and it was calculated from the unclassified remaining points above the ground reference. For this study, canopy height percentiles were calculated at every 5 percentage increment starting from ground to canopy top (e.g., p05, p10, ..., p100), and at every 1 percentage increment starting from p96 to p100. The entire procedure from de-noising to canopy metric calculations was automated by customizing scripts from (lasnoise, lasground and lascanopy) the LAStools software [23].

2.3.3. The Histogram Based Method

Next, we applied the histogram method to calculate canopy height from the level 1 dataset (Figure 2). The histogram method was a typical approach used to process photon counting lidar in many recent studies [4,8,10,11]. In this approach, photon-counting data were converted into pseudo-waveforms by aggregating lidar returns in user-defined elevation bins and footprint sizes. Canopy height metrics were then calculated from the pseudo-waveform using conventional waveform processing methods (e.g., identifying canopy top and ground elevation and height percentiles within each waveform). Here we did not apply any correction method to adjust a possible slope-induced error (e.g., usually requires DRL-derived slope map), because we aimed to compare their performances using information readily available from SPL only (the spatial-filtering vs. the histogram method).

In this study, we created SPL histograms over 71 field plots using a 0.15 m vertical resolution and a footprint radius of 15 m. The applied resolution and scale was identical to both field survey and previous studies [4,8,10,11]. We first normalized each histogram between 0 and 1 to compensate for spatial difference in photon density. We then subtracted the mean value from the normalized histogram and smoothed it using a Hann function with a window size of 8 (Hann function is a typical smoothing window method applied in signal processing) [24]. Bins with negative values in the smoothed histogram were identified as noise and were flattened to zero for further analysis. The subtraction and smoothing procedure separated signal ranges (ground and canopy) from noise ranges using the assumption that photons at ground and canopy range had a significantly higher density.

We then detected elevations of ground peak and canopy top using waveform lidar processing methods. Mean ground elevation was identified as the lowest mode of local maximum values in the smoothed histogram. Canopy top elevation was detected using a cut-off threshold value, calculated as the greater number between 0.01 and the maximum value within the highest 10 histogram bins. Finally, plot level canopy height was calculated as the difference between mean ground and canopy top elevations. In addition, we calculated canopy height percentiles by aggregating the histogram bins between ground and canopy top.

2.4. Canopy Height Comparison

We compared canopy heights calculated from the two methods using field measurements and discrete return lidar (DRL) as references. This was because of the lack of a ground reference for differentiating and validating noise/signal points. For both two methods above, we chose the p99 metric, rather than p100, to represent the true canopy height derived from SPL data. This was because p99 was a better indicator of canopy top height, particularly when anomalous laser returns may be present (e.g., birds or thin clouds for conventional lidar, and unfiltered noise points in this case) [25–28].

Comparisons were made using the following statistical variables: coefficient of determination (r^2), bias (mean difference between SPL height and reference data) and root mean square error (RMSE). The height differences between SPL and reference data were analyzed as a function of the topographic slope using an ordinary least square (OLS) linear regression method. We also assessed potential impacts of using different voxels sizes on the performance of the spatial filtering method.

3. Results

Canopy height percentiles (p96 to p100) derived from the spatial-filtering method showed strong agreement with field and DRL data as demonstrated by the r^2, bias, and RMSE values. Most of the height percentiles, except for the p100, achieved $r^2 > 0.7$ in comparison with field measured canopy heights and $r^2 > 0.9$ when compared with DRL derived heights (Table 1). The r^2 between the 100th height percentile (i.e., p100) and reference heights was much lower than those between the other canopy height percentiles (p96~99) and reference data. Additionally, the p100 metric presented positive bias with values greater than ~3 m against the field height. Such high bias values were not observed in the other percentiles (p96~99), all of which presented consistently better agreements with reference data than p100. For example, the 98th percentile height showed the highest agreement with field data $r^2 = 0.70$, bias = -0.12 m and RMSE = 4.77 m, while p97 had the lowest bias of 0.07 when compared with DRL heights.

Table 1. Comparisons between reference heights (from field data and discrete return lidar) and canopy height percentiles derived from the spatial-filtering method and the histogram method.

Spatial-Filtering		SPL—Field Height			SPL—DRL Height	
Method	r^2	Bias	RMSE	r^2	Bias	RMSE
p100	0.54	2.88	6.99	0.83	3.77	5.46
p99	0.69	0.42	4.85	0.94	1.07	2.42
p98	0.70	−0.12	4.77	0.94	0.49	2.28
p97	0.70	−0.52	4.78	0.93	0.07	2.35
p96	0.70	−0.88	4.82	0.92	−0.31	2.52
Histogram		SPL—Field Height			SPL—DRL Height	
Method	r^2	Bias	RMSE	r^2	Bias	RMSE
p100	0.59	0.77	6.41	0.78	0.68	5.05
p99	0.59	0.00	6.25	0.78	−0.06	4.88
p98	0.60	−0.52	6.24	0.78	−0.55	4.89
p97	0.60	−0.93	6.22	0.78	−0.93	4.92
p96	0.60	−1.31	6.25	0.78	−1.29	4.99

Correspondingly, canopy height percentiles derived from the histogram method also achieved good agreements with the reference data. All height percentiles (including p100) showed highly similar results in their comparisons with reference data ($r^2 = 0.60$ and RMSE ≈ 6.3 m for field heights, and $r^2 = 0.78$ and RMSE ≈ 4.9 m for DRL respectively). The r^2 values were 10%–15% lower than those from the filtering method, and the RMSE values were almost doubled in comparisons with DRL (Table 1). Unlike the filtering method, p100 values derived using the histogram approach had a smaller bias (~0.7 m) than the reference data and other height percentiles (p96~p99).

Canopy heights (the p99) derived from the spatial-filtering method (H_{filter}) achieved consistently better agreements with reference data than those derived from the histogram method (H_{hist}) (Figures 3 and 4, and Table 1). We found no significant impact on height measurement error among different forest types (all $p > 0.1$ using Welch's t-test). However, canopy height differences (ΔH) between H_{hist} and the reference data were positively correlated with the topographic slope based on the OLS linear regression analysis (red-dotted line in Figures 3b and 4b). The relationship was $\Delta H = 0.43 \times$ Slope $- 4.05$ with $r^2 = 0.28$ ($p < 0.01$) when compared to field data, and $\Delta H = 0.21 \times$ Slope $- 1.88$ with $r^2 = 0.11$ ($p < 0.01$) for DRL comparisons. In contrast, H_{filter} showed a mild correlation with slope in field height comparisons: $\Delta H = 0.26 \times$ Slope $- 2.07$ with $r^2 = 0.18$ ($p < 0.01$); and there was no significant effect of slope spotted on the height difference between DRL and the H_{filter} ($p = 0.77$). To better illustrate the effects of slope, we analyzed the point cloud in one of the field plots as an example (Figure 5). This plot was located on a steep slope of approximately 30°, canopy height measured in the field was 30.3 m and that from DRL data was 33.87 m. We found no discernible topographical effect on characterizing canopy structure using the spatial-filtering method with a H_{filter} value of 35.46 m. Instead, the slope had a pronounced effect on the H_{hist} derived from the histogram method, leading to a highly overestimated value of 42.3 m.

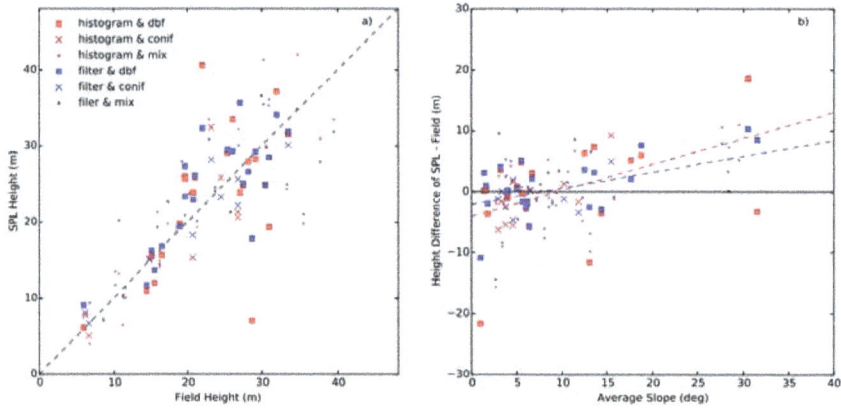

Figure 3. (**a**) Comparisons between field heights and canopy heights (p99) derived from SPL data (using both histogram method and the spatial-filtering method); and (**b**) height differences between field data and SPL as a function of averaged slope value at each plot. For the histogram method $\Delta H = 0.43 \times$ Slope $- 4.05$ with $r^2 = 0.28$ ($p < 0.01$), and for the spatial-filtering method $\Delta H = 0.26 \times$ Slope $- 2.07$ with $r^2 = 0.18$ ($p < 0.01$). Symbols of different color and shape stand for different processing methods (red: histogram method, and blue: spatial-filtering method) over different types of forests (dbf: deciduous broadleaf forests, conif: coniferous forests, and mix: mixed forests).

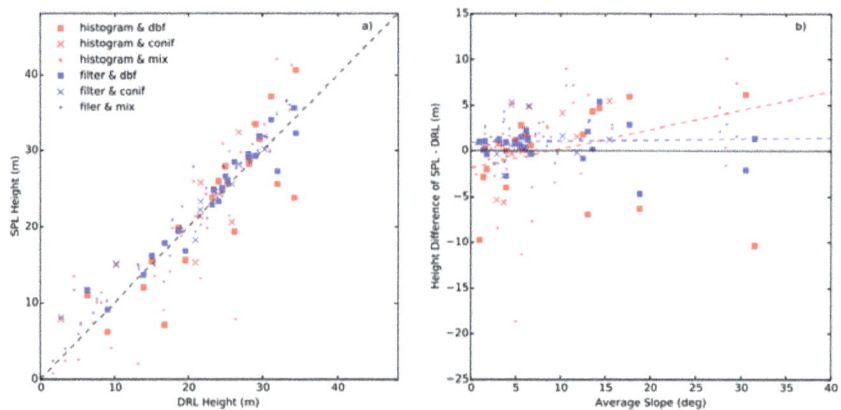

Figure 4. (**a**) Comparisons between canopy heights derived from discrete return lidar (DRL) and SPL data (using both histogram method and the spatial-filtering method); and (**b**) height differences between DRL and SPL as a function of averaged slope value at each plot. For the histogram method $\Delta H = 0.21 \times$ Slope $- 1.88$ with $r^2 = 0.11$ ($p < 0.01$). There was no significant relationship between ΔH and slope for the spatial-filtering method ($p = 0.77$). Same legend as Figure 3.

Even on low-slope plots, the spatial-filtering method outperformed the histogram method, as seen from our analysis after excluding the plots of slope greater than $10°$. Accuracies of the H_{filter} did not change drastically, with $r^2 = 0.74$, bias = -0.65 m and RMSE = 4.48 m against field heights and $r^2 = 0.95$, bias = 0.79 m and RMSE = 1.95 m against DRL heights. The agreement between H_{hist} and DRL heights had a very slight improvement with $r^2 = 0.79$, bias = -1.22 and RMSE = 4.43 m, whereas the comparison between H_{hist} and field heights was greatly improved with $r^2 = 0.68$, bias = -1.79 m and RMSE = 5.37 m.

$H_{drl} = 33.87$ m, $H_{field} = 30.3$ m $H_{hist} = 42.3$ m $H_{filter} = 35.46$ m

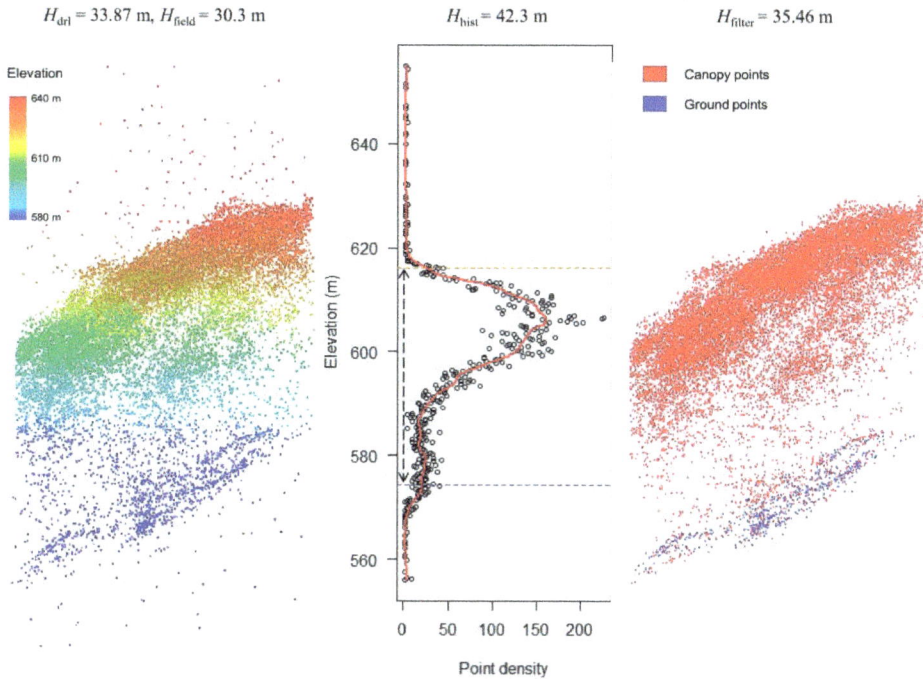

Figure 5. A comparison example of plot-level canopy height products derived from SPL data using both histogram method and the spatial-filtering method. The plot is on a slope of about $30°$ with a DRL measured canopy height of 33.87 m and a field measured height of 30.3 m. The left part shows the raw level 1 HRQLS data over the plot with noises distributed both above and below the canopy-terrain layer. The center shows the pseudo-waveform generated from the histogram method, with identified canopy top (616.15 m), ground peak (574.35 m) and canopy height (42.3 m) in dash lines. The right part shows the noise-removal and point-classification results of HRQLS data using the spatial-filtering method. The ground points are in blue, and canopy points are in red with an estimated canopy height (p99) of 35.46 m.

The use of different voxel sizes in the spatial-filtering method had no pronounced impact on height retrieval performance at the plot level (Table 2). Comparison results with field measurements were similar among all voxel sizes except for the ultra-fine ones (e.g., a $1 \times 1 \times 0.1$ m voxel). However, it may affect retrieval results at the individual tree level (Figure 6). Note that the vertical size of a voxel is not equivalent to vertical resolution, because voxels were only applied to remove noise photons (not for height extraction), and the remaining signal photons still have a continuous spatial distribution with a varying vertical distance (can be narrower or wider than the vertical voxel size [20]).

Table 2. Performance of the spatial-filtering methods across different cell sizes when compared with field heights. The columns stand for different vertical resolutions (unit: m) and the rows are different horizontal resolutions (unit: m). Each cell describes the lowest RMSE achieved at a certain voxel size (optimal radius, r^2, bias, RMSE).

	z = 0.1	z = 0.2	z = 0.3	z = 0.4
xy = 1	(11, 0.67, −1.17, 5.09)	(9, 0.74, −1.02, 4.44)	(7, 0.75, −0.67, 4.44)	(9, 0.73, 0.01, 4.50)
xy = 2	(11, 0.71, −0.71, 4.65)	(8, 0.77, −0.75, 4.17)	(8, 0.76, −0.62, 4.22)	(8, 0.75, −0.48, 4.37)
xy = 3	(8, 0.77, −0.83, 4.22)	(8, 0.76, −0.64, 4.22)	(9, 0.75, −0.32, 4.28)	(9, 0.75, −0.33, 4.35)
xy = 4	(8, 0.76, −0.60, 4.23)	(9, 0.75, −0.30, 4.35)	(9, 0.75, −0.29, 4.30)	(9, 0.75, −0.26, 4.30)
xy = 5	(9, 0.74, −0.26, 4.41)	(9, 0.74, −0.30, 4.44)	(9, 0.75, −0.29, 4.38)	(9, 0.74, −0.25, 4.39)

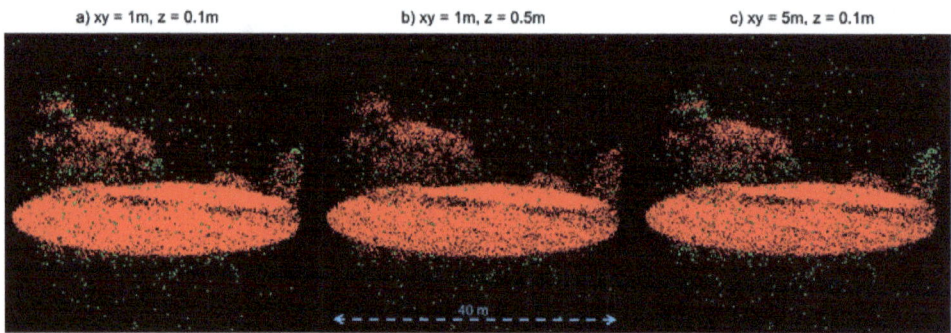

a) xy = 1m, z = 0.1m   b) xy = 1m, z = 0.5m   c) xy = 5m, z = 0.1m

40 m

Figure 6. An illustrative example of the impact of different voxel sizes on noise removal at the individual tree level. The voxel sizes are expressed as combinations of different horizontal resolutions (xy, unit: m) and vertical resolutions (z, unit: m) in (**a–c**). All the three voxels of different sizes can identify the majority of noise photons (green points) both above canopy and below ground. However, an extra-fine resolution voxel may fail to capture the top of individual trees (**a**), and an extra-coarse horizontal resolution voxel may miss the entire small tree in open space (**c**).

4. Discussion

There is considerable interest in developing effective algorithms to obtain accurate canopy height measurements from daytime SPL data. This is important yet challenging for automated mapping over large areas because of the presence of solar noise [6]. The spatial filtering method developed in this study overcomes some of the drawbacks of conventional histogram binning [4,8,10,11], offering an alternative for processing large scale airborne SPL datasets.

Comparisons between SPL-derived canopy heights and reference data suggested that the two methods were able to derive canopy height data with reasonable accuracy. Both methods achieved good agreements with most r^2 values greater than 0.6 and RMSE values less than 6 m. No significant bias was found either (<1 m). Our comparisons further indicated that the new method developed in this study could be more effective than histogram based methods, particularly for high point density SPL data, for several reasons:

First, the spatial-filtering method resulted in consistently better canopy height accuracies (higher r^2, lower RMSE and low bias) than the histogram method as noted in comparisons with field and DRL data. The primary reason for this was that the spatial filtering method identified and filtered noise on an individual point basis, while the histogram-based method did not identify noise, per se, but rather eliminated it by identifying the highest mode at the canopy top and first mode at the bottom of the waveform. Because of this, it was expected to have errors over steep terrain where the ground was difficult to identify in the waveform. Second, canopy heights derived from the spatial-filtering method were only slightly impacted by the slope, a major contributor to errors in canopy height measurements using waveform lidar. Steep topography was known to affect ground detection and led to overestimated canopy heights [29–31]. It can also lead to an underestimation when having a higher noise threshold for a given waveform [32]. Potential correction methods (e.g., slope estimates from high-resolution DEM or waveform leading and trailing extent metrics) may help improve the waveform-based processing techniques [30]; however, their performance would largely depend on the availability of high-quality topography maps that were mostly derived from DRL. Such maps or robust correction approaches are not available nationally or globally. We noted that comparisons between H_{filter} and field data were slightly affected by slope but believe it was more likely because of slope-induced errors in field height measurements (i.e., vertex method could be less accurate with large viewing angle) rather than in the spatial-filtering method. This was further confirmed in height comparisons between H_{filter} and DRL. Comparisons with DRL proved that SPL heights derived using the spatial filtering method were clearly not affected by slope (Figures 4 and 5). This could be because there was no aggregation in the spatial-filtering method and therefore fine-scale terrain variability did not affect it as strongly as in the case of the histogram approach. Third, the spatial filtering method did not transform the point cloud and retained its original 3D structure for downstream point cloud analyses, similar to conventional DRL data. This was critical in fine-scale analysis, such as crown delineation and individual tree identification [13,33], which may not be possible after histogram based filtering. This is because the transforming process of photon point clouds can lead to substantial loss of detailed spatial information (e.g., small scale topography variation), which cannot be retrieved back regardless of any further data processing method applied (e.g., the histogram method here). It also allowed fast batch process and direct visualization with no further requirement for format transformation (all files were stored in ASPRS las 1.2 format).

Interestingly, the histogram method was better at identifying canopy top, especially over low-slope plots, while the spatial-filtering method had errors (as noted from the lower accuracies of p100 heights). This was probably because canopies were more clustered and resulted in a strong peak in the pseudo-waveform at plot level. On the other hand, clustering at the canopy–atmosphere boundary made it difficult to separate signal from noise in the spatial-filtering method and even a few wrongly identified noise points could lead to large errors in the 100th percentile height. This was because the spatial filtering method was designed to remove noise at voxel level which had a higher spatial variation of photon density than the plot level. When the signal photon density was extremely low and had a similar level to the noise (e.g., from individual small tree), either a coarse horizontal resolution voxel or an extra-fine resolution one might fail to identify part of the crown (e.g., Figure 6). Overcoming this problem would require more research and improvements in the algorithm, but an effective and widely used solution was to use lower (p99 or p98) heights to represent canopy top and moderate voxel size as we did in this study [25–28].

Ultimately, the choice of processing method depends on the point density, quality, and extent of SPL data, particularly when SPL technology is still in the experimental stage. The histogram method is useful for a rapid analysis with moderate accuracy but will result in loss of spatial resolution and integrity. It can be mostly aimed at process datasets with low photon density (e.g., ICESat-2), which requires aggregation of points over transects to determine the profile of the canopy surface and the underlying topography. For applications involving airborne SPL, the spatial-filtering method is clearly a better choice. It can largely retain the 3D details with a much higher measurement accuracy, and is thus extremely valuable in fine-scale analysis. As an example, this method has been successfully implemented over the entire county of Garrett in Maryland, demonstrating its usefulness for large-scale terrain and canopy structure mapping [1]. However, this method may not be fully applicable to a dataset with extremely low point density because it violates the basic assumptions in the spatial-filtering method (see Section 2.3.2). Future studies would explore improvements in the density-dependent algorithm, particularly at the canopy-atmosphere boundary and over highly reflective urban targets. It would allow the process of ever-increasing SPL datasets from different platforms, and can extract canopy structure information efficiently over broad geographical areas.

When fully developed, the coupling of low acquisition cost and high processing efficiency together can revolutionize the use of airborne SPL in routinely forest monitoring. It will then serve as a relatively low cost and high efficient alternative to current airborne DRL systems, especially for large-area forest mapping [1]. For example, the area coverage rates of current HRQLS (~224 km^2/h) [2] are more than three times higher than a DRL system, typically varying from about 50 to 80 km^2/h in a bathymetry survey [34]. The density of lidar point clouds per scan is also much higher in a SPL data set (e.g., ~ 100 for HRQLS vs. < 5 for DRL). However, a direct cost-effective comparison between airborne SPL and DRL surveys is complicated because of both high variety in their platforms and sensor-related characteristics and the desired accuracy and density of lidar point clouds as well [35]. Considering the SPL system is still under fast development aiming to improve the quality and capacity of large-area mapping, we envision that there will be more large-scale high point density SPL data available in the near future [36,37]. These data sets, in together with existing high-quality DRL data [38,39], can allow large-scale analysis of ecosystem dynamics and carbon flux at high spatial resolution.

5. Conclusions

In this study, we have shown how high background solar noise in SPL can be efficiently filtered using voxel-based spatial filtering. The algorithm presented here can help derive standardized lidar products (e.g., canopy height model) under both high accuracy and great efficiency. This method could be applied to other photon-counting lidar acquisitions with high photon density as well. Further development of this can support fully automated processing of high point density SPL data over broad geographical areas, and facilitate routine monitoring of forest structure dynamics.

Acknowledgments: This study was funded by a NASA Carbon Monitoring System Grant (NNX12AN07G—Dubayah), a NASA Terrestrial Ecology Grant (NNX12AK07G—Dubayah) and a NASA Earth System Fellowship (NNX12AN43H—Dubayah/Tang). We also want to thank Martin Isenburg for his valuable feedback and technical support with the LAStools software package.

Author Contributions: H.T., A.S. and R.D. designed the experiments; P.D. and T.B. provided pre-processed and geo-referenced SPL data; H.T. and A.S. processed and analyzed the data; H.T., A.S. and R.D. wrote the paper with contributions from all other co-authors.

Conflicts of Interest: The authors declare no conflict of interest.

References

1. Swatantran, A.; Tang, H.; Barrett, T.; DeCola, P.; Dubayah, R. Rapid, high-resolution forest structure and terrain mapping over large areas using single photon lidar. *Sci. Rep.* **2016**, *6*, 28277. [CrossRef] [PubMed]
2. Degnan, J.J.; Field, C.T. Moderate to high altitude, single photon sensitive, 3D imaging lidars. *SPIE Proc.* **2014**, *9114*. [CrossRef]
3. Degnan, J. Photon-counting multikilohertz microlaser altimeters for airborne and spaceborne topographic measurements. *J. Geodyn.* **2002**, *34*, 503–549. [CrossRef]
4. Gwenzi, D.; Lefsky, M.A. Prospects of photon counting lidar for savanna ecosystem structural studies. *Int. Arch. Photogramm. Remote Sens. Spat. Inf. Sci.* **2014**, *40*, 141–147. [CrossRef]
5. Rosette, J.; Field, C.; Nelson, R.; DeCola, P.; Cook, B. A new photon-counting lidar system for vegetation analysis. In Proceedings of the 11th International Conference on LiDAR Applications for Assessing Forest Ecosystems (SilviLaser 2011), Hobart, Australia, 16–19 October 2011.
6. Glenn, N.F.; Neuenschwander, A.; Vierling, L.A.; Spaete, L.; Li, A.; Shinneman, D.J.; Pilliod, D.S.; Arkle, R.S.; McIlroy, S.K. Landsat 8 and ICESat-2: Performance and potential synergies for quantifying dryland ecosystem vegetation cover and biomass. *Remote Sens. Environ.* **2016**. [CrossRef]
7. Awadallah, M.S.; Abbott, A.L.; Thomas, V.A.; Wynne, R.H.; Nelson, R.F. Estimating forest canopy height and biophysical parameters using photon-counting laser altimetry. In Proceedings of the 13th International Conference on LiDAR Applications for Assessing Forest Ecosystems (SilviLaser 2013), Beijing, China, 9–11 October 2013.

8.   Herzfeld, U.C.; McDonald, B.W.; Wallin, B.F.; Neumann, T.A.; Markus, T.; Brenner, A.; Field, C. Algorithm for detection of ground and canopy cover in micropulse photon-counting lidar altimeter data in preparation for the ICESat-2 mission. *IEEE Trans. Geosci. Remote Sens.* **2014**, *52*, 2109–2125. [CrossRef]

9.   Magruder, L.A.; Wharton, M.E.; Stout, K.D.; Neuenschwander, A.L. Noise filtering techniques for photon-counting ladar data. *SPIE Proc.* **2012**, *8379*. [CrossRef]

10.   Moussavi, M.S.; Abdalati, W.; Scambos, T.; Neuenschwander, A. Applicability of an automatic surface detection approach to micro-pulse photon-counting lidar altimetry data: Implications for canopy height retrieval from future ICESat-2 data. *Int. J. Remote Sens.* **2014**, *35*, 5263–5279. [CrossRef]

11.   Montesano, P.M.; Rosette, J.; Sun, G.; North, P.; Nelson, R.F.; Dubayah, R.O.; Ranson, K.J.; Kharuk, V. The uncertainty of biomass estimates from modeled ICESat-2 returns across a boreal forest gradient. *Remote Sens. Environ.* **2015**, *158*, 95–109. [CrossRef]

12.   Lukas, V.; Eldridge, D.F.; Jason, A.L.; Saghy, D.L.; Steigerwald, P.R.; Stoker, J.M.; Sugarbaker, L.J.; Thunen, D.R. *Status Report for the 3D Elevation Program, 2013–2014*; U.S. Geological Survey: Reston, VA, USA, 2015; p. 17.

13.   O'Neil-Dunne, J.; MacFaden, S.; Royar, A. A versatile, production-oriented approach to high-resolution tree-canopy mapping in urban and suburban landscapes using geobia and data fusion. *Remote Sens.* **2014**, *6*, 12837–12865. [CrossRef]

14.   Goetz, S.; Dubayah, R. Advances in remote sensing technology and implications for measuring and monitoring forest carbon stocks and change. *Carbon Manag.* **2011**, *2*, 231–244. [CrossRef]

15.   Huang, W.; Swatantran, A.; Johnson, K.; Duncanson, L.; Tang, H.; O'Neil Dunne, J.; Hurtt, G.; Dubayah, R. Local discrepancies in continental scale biomass maps: A case study over forested and non-forested landscapes in Maryland, USA. *Carbon Balance Manag.* **2015**, *10*, 19. [CrossRef] [PubMed]

16.   Dubayah, R.; Swatantran, A.; Huang, W.; Duncanson, L.; Johnson, K.; Tang, H.; Dunne, J.O.; Hurtt, G. *CMS: LiDAR-Derived Aboveground Biomass, Canopy Height and Cover for Maryland, 2011*; ORNL Distributed Active Archive Center: Oak Ridge, TN, USA, 2016.

17.   Andersen, H.E.; McGaughey, R.J.; Reutebuch, S.E. Forest measurement and monitoring using high-resolution airborne lidar. In *USDA Forest Service—General Technical Report PNW*; USDA Forest Service: Washington, DC, USA, 2005; pp. 109–120.

18.   Wasser, L.; Day, R.; Chasmer, L.; Taylor, A. Influence of vegetation structure on lidar-derived canopy height and fractional cover in forested riparian buffers during leaf-off and leaf-on conditions. *PLoS ONE* **2013**, *8*, e54776. [CrossRef] [PubMed]

19.   Popescu, S.C.; Zhao, K. A voxel-based lidar method for estimating crown base height for deciduous and pine trees. *Remote Sens. Environ.* **2008**, *112*, 767–781. [CrossRef]

20.   Hancock, S.; Essery, R.; Reid, T.; Carle, J.; Baxter, R.; Rutter, N.; Huntley, B. Characterising forest gap fraction with terrestrial lidar and photography: An examination of relative limitations. *Agric. For. Meteorol.* **2014**, *189*, 105–114. [CrossRef]

21.   Cote, J.F.; Widlowski, J.L.; Fournier, R.A.; Verstraete, M.M. The structural and radiative consistency of three-dimensional tree reconstructions from terrestrial lidar. *Remote Sens. Environ.* **2009**, *113*, 1067–1081. [CrossRef]

22.   Axelsson, P. Dem generation from laser scanner data using adaptive tin models. *Int. Arch. Photogramm. Remote Sens.* **2000**, *33*, 111–118.

23.   Isenburg, M. *LAStools—Efficient Lidar Processing Software (Version 140929)*; rapidlasso GmbH: Gilching, Germany, 2014.

24.   Harris, F.J. Use of windows for harmonic-analysis with discrete Fourier-transform. *Proc. IEEE* **1978**, *66*, 51–83. [CrossRef]

25.   D'Oliveira, M.V.N.; Reutebuch, S.E.; McGaughey, R.J.; Andersen, H.E. Estimating forest biomass and identifying low-intensity logging areas using airborne scanning lidar in Antimary State Forest, Acre State, western Brazilian Amazon. *Remote Sens. Environ.* **2012**, *124*, 479–491. [CrossRef]

26.   Riano, D.; Chuvieco, E.; Condes, S.; Gonzalez-Matesanz, J.; Ustin, S.L. Generation of crown bulk density for *Pinus sylvestris* L. from lidar. *Remote Sens. Environ.* **2004**, *92*, 345–352. [CrossRef]

27.   Frazer, G.W.; Wulder, M.A.; Niemann, K.O. Simulation and quantification of the fine-scale spatial pattern and heterogeneity of forest canopy structure: A lacunarity-based method designed for analysis of continuous canopy heights. *For. Ecol. Manag.* **2005**, *214*, 65–90. [CrossRef]

28. Jaskierniak, D.; Lane, P.N.J.; Robinson, A.; Lucieer, A. Extracting lidar indices to characterise multilayered forest structure using mixture distribution functions. *Remote Sens. Environ.* **2011**, *115*, 573–585. [CrossRef]
29. Duncanson, L.I.; Niemann, K.O.; Wulder, M.A. Estimating forest canopy height and terrain relief from glas waveform metrics. *Remote Sens. Environ.* **2010**, *114*, 138–154. [CrossRef]
30. Lee, S.; Ni-Meister, W.; Yang, W.Z.; Chen, Q. Physically based vertical vegetation structure retrieval from ICESat data: Validation using LVIS in white mountain national forest, New Hampshire, USA. *Remote Sens. Environ.* **2011**, *115*, 2776–2785. [CrossRef]
31. Miller, M.E.; Lefsky, M.; Pang, Y. Optimization of geoscience laser altimeter system waveform metrics to support vegetation measurements. *Remote Sens. Environ.* **2011**, *115*, 298–305. [CrossRef]
32. Hancock, S.; Disney, M.; Muller, J.P.; Lewis, P.; Foster, M. A threshold insensitive method for locating the forest canopy top with waveform lidar. *Remote Sens. Environ.* **2011**, *115*, 3286–3297. [CrossRef]
33. Duncanson, L.I.; Dubayah, R.O.; Cook, B.D.; Rosette, J.; Parker, G. The importance of spatial detail: Assessing the utility of individual crown information and scaling approaches for lidar-based biomass density estimation. *Remote Sens. Environ.* **2015**, *168*, 102–112. [CrossRef]
34. Guenther, G.C. Airborne lidar bathymetry. In *Digital Elevation Model Technologies and Applications: The DEM Users Manual*, 2nd ed.; American Society for Photogrammertry and Remote Sensing: Bethesda, MD, USA, 2007; pp. 253–320.
35. Wulder, M.A.; White, J.C.; Nelson, R.F.; Næsset, E.; Ørka, H.O.; Coops, N.C.; Hilker, T.; Bater, C.W.; Gobakken, T. Lidar sampling for large-area forest characterization: A review. *Remote Sens. Environ.* **2012**, *121*, 196–209. [CrossRef]
36. Abdullah, Q.A. A star is born: The state of new lidar technologies. *Photogramm. Eng. Remote Sens.* **2016**, *82*, 307–312. [CrossRef]
37. Li, Q.; Degnan, J.; Barrett, T.; Shan, J. First evaluation on single photon-sensitive lidar data. *Photogramm. Eng. Remote Sens.* **2016**, *82*, 455–463. [CrossRef]
38. Asner, G.P.; Knapp, D.E.; Martin, R.E.; Tupayachi, R.; Anderson, C.B.; Mascaro, J.; Sinca, F.; Chadwick, K.D.; Higgins, M.; Farfan, W.; et al. Targeted carbon conservation at national scales with high-resolution monitoring. *Proc. Natl. Acad. Sci. USA* **2014**, *111*, E5016–E5022. [CrossRef] [PubMed]
39. Cook, B.D.; Corp, L.A.; Nelson, R.F.; Middleton, E.M.; Morton, D.C.; McCorkel, J.T.; Masek, J.G.; Ranson, K.J.; Ly, V.; Montesano, P.M. NASA Goddard's LiDAR, Hyperspectral and Thermal (G-LiHT) airborne imager. *Remote Sens.* **2013**, *5*, 4045–4066. [CrossRef]

![remote sensing logo] *remote sensing*

MDPI

Article

Evaluation of Single Photon and Geiger Mode Lidar for the 3D Elevation Program

Jason M. Stoker [1,*], Qassim A. Abdullah [2], Amar Nayegandhi [3] and Jayna Winehouse [4]

1   U.S. Geological Survey, Reston, VA 20192, USA
2   Woolpert, Arlington, VA 22206, USA; Qassim.Abdullah@woolpert.com
3   Dewberry Consultants LLC, Fairfax, VA 22031, USA; anayegandhi@dewberry.com
4   U.S. Geological Survey, Lakewood, CO 80225, USA; jwinehouse@usgs.gov
*   Correspondence: jstoker@usgs.gov; Tel.: +1-970-226-9227

Academic Editors: Jie Shan, Juha Hyyppä, Lars T. Waser and Prasad S. Thenkabail
Received: 28 June 2016; Accepted: 8 September 2016; Published: 19 September 2016

Abstract: Data acquired by Harris Corporation's (Melbourne, FL, USA) Geiger-mode IntelliEarth™ sensor and Sigma Space Corporation's (Lanham-Seabrook, MD, USA) Single Photon HRQLS sensor were evaluated and compared to accepted 3D Elevation Program (3DEP) data and survey ground control to assess the suitability of these new technologies for the 3DEP. While not able to collect data currently to meet USGS lidar base specification, this is partially due to the fact that the specification was written for linear-mode systems specifically. With little effort on part of the manufacturers of the new lidar systems and the USGS Lidar specifications team, data from these systems could soon serve the 3DEP program and its users. Many of the shortcomings noted in this study have been reported to have been corrected or improved upon in the next generation sensors.

Keywords: lidar; Geiger-mode; single photon

1. Introduction

While not new in terms of technology, the commercialization of two new types of lidar instruments, Geiger-mode Lidar (GML) and Single Photon Lidar (SPL) offer the promise of utilizing high-altitude lidar collections for large area mapping across the United States for the 3D Elevation Program (3DEP). The 3D Elevation Program is accelerating the rate of three-dimensional (3D) elevation data collection in response to a call for action to address a wide range of urgent needs nationwide [1]. With constantly changing lidar technology, the 3DEP needs to keep up with emerging trends and instruments that could fulfill the goals of the program at a reduced cost to the taxpayer. Currently, the 3DEP relies on the mature discrete multiple-return lidar systems for data collection; for the purposes of this paper, we will refer to these as "linear-mode lidar" (LML) systems.

Both GML and SPL utilize focal plane array detectors, where the returned pulse is recorded using an array of receivers instead of single receiver as is the case in LML systems. The transmitted laser pulses for both GML and SPL are low energy. These low energy pulses are detected by receivers that are sensitive to individual photons; thereby enabling the added advantage of higher flying altitudes.

Over the past 15 years, the Massachusetts Institute of Technology, Lincoln Laboratory (MIT/LL), Defense Advanced Research Projects Agency (DARPA) and private industry have been developing airborne lidar systems based on arrays of Geiger-mode Avalanche Photodiode (GmAPD) detectors capable of detecting a single photon [2,3]. The extreme sensitivity of GmAPD detectors allows operation of lidar sensors from very high altitudes and acquisition efficiency rates in excess of 1000 km^2/h. Up until now the primary emphasis of this technology has been limited to defense applications, despite the significant benefits of applying this technology to non-military uses such as mapping, monitoring critical infrastructure and disaster relief. The first commercial Geiger-mode lidar system, the IntelliEarth™ system by Harris Corporation, began advertising its capability in early 2015.

Photon-counting lidar systems were in development shortly after some of the first commercially available linear-mode systems began operating [4]. The first successful photon-counting airborne laser altimeter was demonstrated in 2001 under NASA's Instrument Incubator Program (IIP) [5]. This instrument was flown from altitudes up to 6700 m above ground level (AGL), operated at a wavelength of 532 nm, and imaged terrestrial and shallow water bathymetry to depths of a few meters over the Atlantic Ocean and Assawoman Bay off the Virginia coast. This SPL system modeled the solar noise background and developed simple algorithms, based on post-detection Poisson filtering (PDPF) to optimally extract the weak altimeter signal from a high noise background during daytime operations [6]. The theoretical results are reinforced by data from an airborne microlaser altimeter, developed under NASA's Instrument Incubator Program. This instrument was designed primarily to produce, over a mission life of three years, a globally contiguous map of the Martian surface, with 5 m horizontal resolution and decimeter vertical accuracy, from an altitude of 300 km. This type of SPL instrument is planned to replace the Geoscience Laser Altimeter System (GLAS) on the ICESAT-II mission, slated for launch in 2017 [7–9]. Currently, the only commercially available SPL system is the High-Resolution Quantum Lidar System (HRQLS) by Sigma Space Corporation.

While there are numerous potential benefits of these newly commercialized systems to collect data for 3DEP, such as increased flying height and higher point density, there are documented issues of concern for these types of systems as well. Williams Jr. [10] demonstrated through physics-based Monte Carlo simulations of avalanche photodiode (APD) lidar receivers that under typical operating scenarios, GmAPD with array-based receivers may often be ineffective in detecting partially occluded targets, such as bare earth under vegetation. Due to their ability to detect only one photon per laser pulse, the target detection efficiency of GmAPD receivers was shown to respond nonlinearly to the specific conditions including range, laser power, detector efficiency, and target occlusion, which caused the GmAPD target detection capabilities to vary unpredictably over standard mission conditions. In the detection of partially occluded targets, Williams Jr. [10] found that GmAPD lidar receivers performed optimally within only a narrow operating window of range, detector efficiency, and laser power; outside this window performance degraded sharply. He concluded that the inability of the GmAPD to detect target signal present at the receiver's aperture may lead to a loss of operational capability, may have undesired implications for the equivalent optical aperture, laser power, and/or system complexity, and may incur other costs that can affect operational efficacy.

Past history of using GmAPD detectors suggest an issue specifically in their ability to penetrate foliage [11]. Most APDs operating in Geiger-mode report only one range measurement per transmitted laser pulse. If a GmAPD makes a foliage range measurement, it cannot make a range measurement to a target concealed by the foliage. When too much laser energy is received, the vast majority of range measurements are from the foliage and only a small percentage are from the target.

Given the ability of these detectors to measure a single photon, minimizing solar background is an important consideration during the design of the system. There is a general concern by traditional LML users that these data are too noisy for commercial use, especially during daylight operations. The choice of system wavelength also has a significant impact on the solar background level and overall system efficiency. Other effective methods of reducing solar background is minimizing the system aperture, installation of a narrow bandpass filter in the receive path, reducing the detector instantaneous field of view, minimizing the range gate duration, or by simply operating at night. By designing and operating the system such that solar background is minimized significantly increases data quality while increasing the operating range of the sensor and reducing or eliminating the need for noise filtering.

Lower energy also translates into fewer photons reaching the receiver, and as a result these systems do not have full waveform digitization capability. Current LML technology relies on a flux of photons (500 to 1000 photons) to record the returned signal. This energy is much larger than the one needed for the GML and SPL technologies. Such large fluxes of energy make it possible to digitize the full waveform and to produce an intensity image with reasonable radiometric resolution to represent a black and white image at the lidar wavelength. This is not the case with the GML and SPL systems. To overcome such problem while capitalizing on the high density of its point cloud, the GML system in particular produce what is called a "reflectivity image" or "relative reflectivity image", by computing the ratio of the numbers of incident photons to the returned photons.

Currently, there are only two companies with instruments that are commercially available that claim they can collect data that meets 3DEP's requirements in terms of data quality and utility. These are the SPL High-Resolution Quantum Lidar System (HRQLS) system by Sigma Space Corporation, and the Harris IntelliEarth™ GML lidar. While both instruments are unique as described above, both claim that they can provide data that meets the requirements for the 3D Elevation Program. This paper will evaluate these particular sensors as they relate to the 3D Elevation Program, and are not necessarily reflective of the potential of both GML and SPL in general.

1.1. Sigma Space Single Photon Technology

Sigma Space utilized its single photon sensor, HRQLS, to collect the data for the study [12]. HRQLS has a 10 × 10 array of 100 beamlets, generated by a passive diffractive optical element (DOE) in front of a 25 kHz laser. The returned 100 beamlets are imaged into 10 × 10 micro-channel plate photomultiplier detectors with low jitter and very fast recovery time. The fast 1.6 nanosecond/24 centimeter recovery time/range of the combined detector and timing receiver allow the systems to view multiple photon events per pixel per pulse, making them capable of daylight operation. They also can penetrate semi-porous volumetric scatterers, such as tree canopies, turbid water bodies, thin clouds or fog. In addition, the 10 × 10 beamlet array, combined with a proper choice of aircraft velocity and altitude, allows the generation of contiguous few decimeter resolution maps on a single overflight while operating at altitudes up to 9000 m AGL and at speeds up to 250 knots yielding an aerial coverage of 1340 km²/h. Due to the finite speed of light and the high ground speed of the conical scanner, initially co-aligned transmit and received field of views (FOVs) become increasingly displaced at the higher altitudes. As a result, the scanner optical design and scan speed must compensate for the displacement over a wide range of potential operating altitudes. As an added benefit, the surface returns and receiver range gates are automatically paired with the correct laser start pulse even when multiple laser pulses are simultaneously in flight. Table 1 lists the technical specifications for the HRQLS sensor used by Sigma Space Corporation to collect the data.

Table 1. Technical specifications for the HRQLS sensor.

Parameter	Specification
Number of beams	100
Wavelength	532 nm
Laser Repetition Rate	25 kHz
Laser Pulsewidth	700 ps
Laser Output Power	1.5 W
Pixels/s	2.5 Million
Eye Safety	Eye safe by FAA standards
Multiple Return Capability	Yes
Pixel Recovery Time	1.6 ns
RMS Range precision	±5 cm
Scan Patterns	Linear, conical
Scan Width	0 to 40 degrees (selectable)
Operational Altitude Range	6500–10,000 ft
Swath vs. AGL (at maximum scan angle)	1.3 to 2 km
Area Coverage versus AGL (at maximum scan angle and 200 knots)	400 to 640 km²/h single pass
Mean Point Density	12 to 8 per square meter, single pass with 15% reflectivity
Size	19 W × 25 D × 33 H inches
Weight	50 lbs
Prime Power	555 W

1.2. Harris IntelliEarth™ Technology

The Harris IntelliEarth™ Geospatial Solutions Geiger-mode lidar sensor is the first commercial airborne lidar system that takes advantage of the single photon capabilities of the Geiger-mode avalanche photodiode [3]. Harris' system uses an array of 32 × 128 detectors. This concept of pulse splitting results in a much higher point density as compared to the current linear lidar, where the pulse may have few returns in vegetated areas as a best case scenario.

The primary components of the system include:

- 128 × 32 InP/InGaAsP Geiger-mode camera capable of readout rates in excess of 100 kHz.
- Compact Nd:YAG diode pumped solid state laser.
- 270 mm Holographic Optical Element (HOE) scanner with 15° scan half angle capable of rotational speeds in excess of 2000 rotations per minute. (US Patent US 2015/0029571 A1 29 January 2015).
- Real time transmit line of sight adjustment which compensates for scanner motion during pulse round trip time.
- High efficiency narrow bandpass filter that reduces solar background noise.
- Transmit beam shaping optics that optimizes illumination pattern on the ground.
- Nadir looking Ritchey—Chrétien telescope for collection of returned light.
- Inertial navigation system including an inertial measurement unit.
- High speed flash detector for precision laser pulse timing.
- Data acquisition electronics.
- Sensor controller.

The Harris Geiger-mode lidar sensor uses a conical Palmer scan pattern produced by a direct drive, hub driven HOE scanner. HOE scanners have been used in the past [13] but this particular implementation has several advantages over previous designs. When flown with 50 percent overlap, the scan pattern provides four looks from four different directions. With a scan half angle of 15°, the sides as well as roofs of structures are sampled, which significantly increases the potential for interpretability of the point-cloud data.

There have been very few studies comparing the performance of these new sensors to LML or ground survey measurements [12,14], and this study is the first of its kind to evaluate the performance of LML to GML and SPL systems as they relate to collecting data that is adequate for the 3D Elevation Program. We evaluated both commercial instruments: the SPL HRQLS system by Sigma Space Corporation, and the Harris IntelliEarth™ GML. Comparing data collected by these instruments over an area in Northern Connecticut with both existing and new LML and good survey ground measurements helped determine the current capability of these instruments to meet 3DEP's requirements.

2. Data Sets

We designed a 500 mi^2 project area for this evaluation study that overlapped the northern third of Connecticut Sandy QL2 Lidar collected for the 3DEP, which was flown in April/May 2014 by Dewberry. This coverage included all desired land cover and terrain variability, portions of Hartford including the main airport, and rivers and lakes to test hydroflattening in select areas. Three sub-areas were selected within the main project area—Urban, Mixed Use, and Forested (Figure 1). For all three sub-areas, two independent teams (Woolpert and Dewberry) processed the data acquired by Sigma Space and Harris using their own internal proprietary methods to create final deliverables that were intended to meet the United States Geological Survey (USGS) v1.2 specifications [15], including classified LAS v1.4 point cloud data, 1-m hydroflattened DEMs, intensity images (where applicable) and associated reports and metadata.

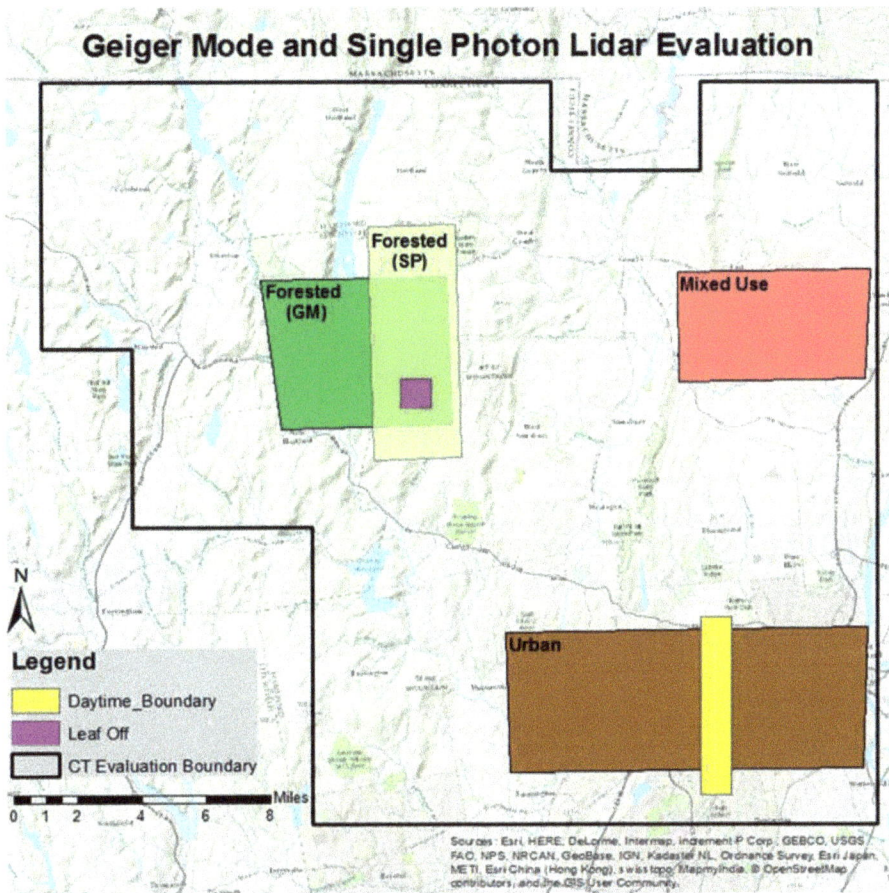

Figure 1. Areas of interest used for processing and assessing bare earth.

Data processing and analysis was done independently by Dewberry and Woolpert. Woolpert acquired leaf-on data in a small (~77 km^2) area in the Forested sub-area for this study around the same time as the HRQLS and IntelliEarth™ data acquisition. Woolpert acquired lidar data according to the USGS QL2 specifications over an area densely covered by vegetation. The purpose of this data acquisition was to have QL2 data during leaf-on condition close in time to the IntelliEarth™ and HRQLS data acquisition. Woolpert used a Leica ALS70 500 kHz Multiple Pulses in Air lidar sensor system.

Sigma Space and Harris Corporation acquired data in the project area during leaf-on conditions in August/September 2015. Calibrated, unclassified LAS data and supporting acquisition reports were delivered for this evaluation study. Harris acquired IntelliEarth™ data in the project area during nighttime conditions at 7950 m above ground; as requested, they provided data acquired in daytime conditions along two overlapping swaths in the Urban sub-area albeit acquired at a lower altitude (2293 m above ground) (Figure 2). Sigma Space acquired HRQLS data in nighttime and daytime conditions at 2293 m above ground, and delivered one set of LAS v1.2 data for the entire project area. In the Forested sub-area, there were some areas acquired in the early morning where fog was present. As a result, these areas were not considered for the evaluation. Table 2 lists all the data sets used in this evaluation study; the abbreviated names for each dataset are used throughout this paper.

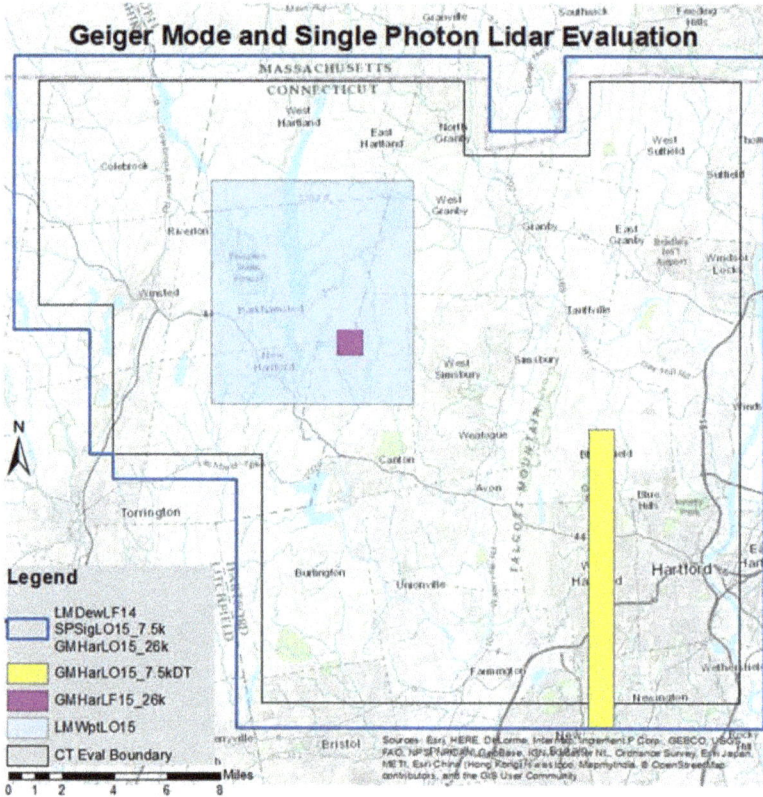

Figure 2. Location of leaf-on linear mode collection (LMWptLO15), IntelliEarth™ daytime collection (GMHarLO15_7.5kDT), and IntelliEarth™ sensor 2 leaf-off collection (GMHarLF15_26k).

Table 2. List of datasets and their attributes that were used in this study.

Abbreviation	Data Type	Acquired By	Type of Collect	Date/Year Collected	Collection Altitude (AGL)
LMDewLF14	Linear Mode	Dewberry	Leaf-Off	April/May 2014	917 m
LMWptLO15	Linear Mode	Woolpert	Leaf-On	September 2015	2140 m
GMHarLO15_26k	Geiger Mode	Harris	Leaf-On	September 2015	7950 m
GMHarLO15_7.5kDT	Geiger Mode	Harris	Leaf-On, Day Time	September 2015	2293 m
SPSigLO15_7.5k	Single Photon	Sigma Space	Leaf-On	August 2015	2293 m
GMHarLF15_26k	Geiger Mode	Harris	Leaf-Off	December 2015	7950 m

Harris Corporation acquired IntelliEarth™ data on 2 September and 16 September 2015 during leaf-on conditions. The primary evaluation data were collected using the Harris IntelliEarth™ sensor 1 from an altitude of 7950 m above ground. This dataset is referred to as GMHarLO15_26k (Table 2). The flights were conducted at night and with an overlap of approximately 55%. It should be noted, however, that while the IntelliEarth™ sensor collects data in swaths, it processes all swaths together for a final solution, and as a result Harris states it is not possible to identify the swaths within the final point cloud. In addition to the high altitude collection, Harris also collected a low altitude dataset during the day for a small overlapping area. These data were collected at 2293 m above ground in the Urban sub-area (Figure 2) and referred to as GMHarLO15_7.5kDT (Table 2). These data were tested for improved foliage penetration and differences between day and night collections. The unclassified data were received on 1 November 2015 for this evaluation study. The following data/reports were delivered by Harris for this study:

- Control Points and Report for Control
- Intensity Imagery
- Acquisition and Processing Report
- SBET files
- Raw Tiled LAS v1.4

The IntelliEarth™ data were processed to a LAS point-cloud and calibrated using a photogrammetric bundle adjustment method.

In addition to the leaf-on data acquired in September 2015, Harris also acquired IntelliEarth™ data in leaf-off conditions at 7950 m on 6 December 2015 within a small sample forested area using their new IntelliEarth™ Geiger-mode sensor 2. The first sensor (serial number 1), had a slight optical blur which defocuses the returning light into roughly a 3 × 3 pixel area. This blur made it more difficult for the system to match returning photons from the multiple looks in to a final XYZ point solution. It also reduced the spatial resolving power for data reflecting off of natural objects, buildings, and other man-made structures. Sensor 2 used a new Holographic Optical Element (HOE) that focused the light correctly onto 1 pixel and thereby significantly improving canopy penetration. Although this leaf-off dataset, referred to as GMHarLF15_26k (Table 2) was not part of the original scope, we evaluated the sample data for canopy penetration.

Data Classification and Editing

The data were processed by using Global Mapper (v16.2.7), GeoCue (v14.1.21.2), TerraScan (v15.031), TerraModeler (v15.007), and Microstation (v8i v08.02.04) software utilizing independent proprietary methods by both Dewberry and Woolpert. The acquired 3D laser point clouds, in LAS binary format, were imported and tiled according to the project tile grid. Once tiled, the points were classified using proprietary routines. These routines classify any obvious low outliers in the dataset to class 7 and high outliers in the dataset to class 18. After points that could adversely affect the ability to derive a ground model were removed from class 1, the ground layer was extracted from this remaining point cloud. The ground extraction process built an iterative surface model. For this evaluation study, it was important to maintain consistency between the ground models in order to ensure the comparisons were not skewed by incorrect classification. As a result, the classified ground points from the Dewberry 2014 lidar collection were used as a starting point in the macro and classified points that were within 20 cm to an initial ground class. This step minimized the amount of manual editing required. It was noted that for the IntelliEarth™ dataset, the density of points actually penetrating through the foliage was minimal and a traditional macro would have difficulty discerning ground and non-ground points with such low density. By using the 2014 data, we were able to retain valid ground points within the forested area with reduced manual effort.

Each tile was then imported into proprietary software and surface models were created to examine the ground classification. Analysts visually reviewed the ground surface models and corrected errors in the ground classification such as vegetation, buildings, bridges, and noise caused by random photon events that were present after the automated classification. 3D visualization techniques were employed to view the point-cloud data at multiple angles and in profile/transect mode to ensure that non-ground points were removed from the ground classification. After the ground classification corrections were completed, the dataset was processed through a water classification routine that utilized breaklines compiled to automatically classify hydro features. The water classification routines selected ground points within the breakline polygons and automatically classified them as class 9 (water). During these water classification routines, points that were within 1 × nominal point spacing (NPS) or less of the hydrographic features were moved to class 10 (ignored ground) due to breakline proximity. The lidar tiles were classified to the following classification schema:

- Class 1 = Unclassified, used for all other features that do not fit into the Classes 2, 7, 9, or 10 including vegetation, buildings, etc.
- Class 2 = Bare − Earth Ground
- Class 7 = Low Noise
- Class 9 = Water, points located within collected breaklines
- Class 10 = Ignored Ground due to breakline proximity

After manual classification, the LAS tiles were peer reviewed and then underwent a final QA/QC. After the final QA/QC and corrections, the LAS files were then converted from LAS v1.2 to LAS v1.4. All headers, appropriate point data records, and variable length records, including spatial reference information, were updated and verified using proprietary Dewberry and Woolpert tools.

All three sets of lidar data (IntelliEarth™, HRQLS and linear mode) derived products were processed using Woolpert's and Dewberry's individual workflows and production procedures and processes.

Once all hydro-flattened data was imported, surveyed ground control data was imported and calculated for positional accuracy assessments. As a quality control measure, Woolpert and Dewberry have developed routines to generate accuracy statistical reports by comparisons against the points and the DEMs using surveyed checkpoints of higher accuracy.

3. Methods

In total 83 ground checkpoints were acquired: Thirty-four were acquired in 2014 during the leaf-off survey and 49 additional checkpoints were acquired in 2015 to support this study (Figure 3).

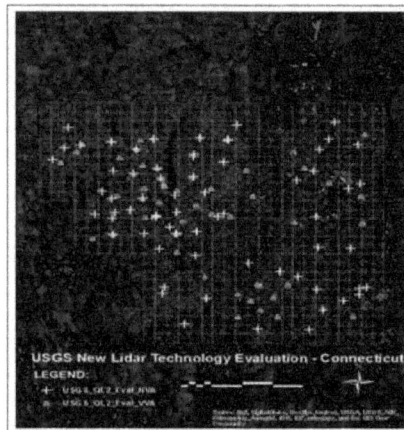

Figure 3. Location of checkpoints used for Vertical Accuracy Assessments. Plus signs are NVA checkpoints. Triangles are VVA checkpoints.

Data from both the HRQLS and IntelliEarth™ systems were checked to determine if they complied with requirements in the USGS Lidar Base Specification v1.2 [15]. Assessments included testing attributes contained the required values, determining if the point density met USGS base specification QL2 levels (at least two points per meter squared), if relative accuracies were sufficient, and if absolute accuracies were within specification. Other qualitative tests included an assessment of day versus night data quality, quality of reflectance images (for IntelliEarth™ only), and the amount of noise points inherent in the data. The Non-vegetative Vertical Accuracy (NVA) and Vegetated Vertical Accuracy (VVA) of the data were evaluated according to the guidelines and the recommendations of the "ASPRS Positional Accuracy Standards for Digital Geospatial Data" using all the available checkpoints [16].

As the primary derivative product for the 3DEP is a bare earth DEM, an assessment of the differences between an accepted QL2 bare earth DEM versus derived bare earth DEMs were also performed independently by both Dewberry and Woolpert, and then results of those determinations were checked by USGS to make sure that both assessments and processing of bare earth DEMs were truly independent.

4. Results and Discussion

4.1. Attributes Required to Meet USGS Lidar Base Specification 1.2

Data from both the HRQLS and IntelliEarth™ systems were checked to determine if they complied with requirements in the USGS Lidar Base Specification v1.2 [15]. While compliant in many categories, there were several requirements that neither dataset met (Table 3).

Table 3. Compliance to USGS Lidar Base Spec v1.2.

Requirement	IntelliEarth	HRQLS	Comments
LAS Version 1.4	LAS v1.4	LAS v1.2	Both data sets LAS v1.4 compatible
Point Data Format	Compliant	Compliant	
Coordinate Reference System	Compliant	Compliant	
Global Encoder bit	Compliant	Compliant	
Time Stamp	Compliant	Not Compliant	IntelliEarth—unique but not based on acquired swaths. HRQLS—none provided.
System ID	Compliant	Compliant	
Multiple Returns	Not Compliant	Not Compliant	Both systems do not produce multiple returns.
Point Source ID	Not Compliant	Compliant	IntelliEarth—No flight swaths.
Intensity	Reflectance	Not Compliant	IntelliEarth—similar to linear-mode HRQLS—no intensity data
Overlap and withheld	Not Compliant	Compliant	IntelliEarth—No flight swaths.
Scan Angle	Not Compliant	Not Compliant	Spec not compatible with these sensors.
XYZ Coordinates	Compliant	Compliant	

4.2. Data Density

Both IntelliEarth™ and HRQLS datasets had Aggregate Nominal Pulse Spacings (ANPS) that greatly exceeded USGS QL1 requirements. The IntelliEarth™ data acquired at 7540 m above ground produced 25 points per square meter (ppsm), and the HRQLS data acquired at 2293 m above ground produced 23 ppsm. Although both these systems produced data analogous to a "single return only" linear mode sensor, the data density was much higher at these high altitudes than typical linear mode sensors currently being used for the 3DEP program. Current state-of-the-art linear mode sensors will produce data at 2–4 ppsm at a flying altitude of 2100 m above ground or lower.

4.3. Data Smoothness/Relative Accuracy

Both sensors used different processing and calibration procedures that are based on detecting individual photons using an array-based detector to determine the range to a target. LML sensors determine the range to the target by detecting the return signal from the entire transmitted pulse. As a result, the typically used swath-based relative accuracy method was not used for this evaluation study. Instead, the entire dataset was treated as a single swath, and the relative accuracy QL1/QL2 requirement of 6 cm was verified in a sample area. The IntelliEarth™ data easily met the 6 cm requirements in flat and gently sloping terrain and there was very little variability within the test areas (except over bright reflective targets summarized below). The HRQLS data displayed greater variability over flat terrain with significant low noise points. Since there were no timestamps in the LAS data, there was no way to differentiate each swath after the data were merged to a tile grid. Furthermore, while the single swath accuracy requirement was more stringent than the accuracy assessment between swaths, this method was chosen to ensure that the data could easily be compared with the IntelliEarth™ evaluation. The smooth surface repeatability test require the data to meet 6 cm relative accuracy for QL1 or QL2 data. This test was performed by computing the maximum difference within a 0.5 m cell. Because the validation is only relevant on smooth surfaces with little to no slope, a 0.5 m cell size was sufficient to determine the repeatability on those surfaces. This included open terrain, roads, and flat roof tops. Based on the overall evaluation, the single photon data tended to contain slightly more noise within the swath than would be typically seen in a linear mode system. On average the differences were between 5 and 7 cm within areas where the slope was less than 10 degrees. The HRQLS data required more manual editing to remove outliers and some of this variability in the data. There was significant variability over highly reflective flat targets with the HRQLS data as summarized below. The variability in the HRQLS data exceeded the relative accuracy of 6 cm in these locations.

The IntelliEarth™ and HRQLS data were examined to determine effects of range walk over highly reflective surfaces such as roadway markings and runway paint stripes (Figure 4). Range walk occurs when highly reflective targets appear higher than the surrounding less reflective targets. In traditional LML sensors, range is calculated by measuring the time of flight of laser pulses reflected from a target. The timing of the return is typically calculated by first converting the returning photon pulse into a proportional current and then precisely measuring some position along the rise of the current pulse, for example by simple power threshold, finding the Full Wave at Half Maximum of the strongest reflected pulses, or through various cross correlation techniques. When techniques such as level thresholding are used to determine range, reflections from extremely bright surfaces will cause current pulses rise above that threshold much sooner than for typical surfaces in the scene, leading to significant timing difference. As a result, those very bright surfaces appear to rise or "walk" above the surrounding surface area. Range walk is a very common source of range errors in linear mode sensors, and sensor manufacturers and production groups have various methods for adjusting their data.

The HRQLS and IntelliEarth™ data along Runway 33 at Bradley International Airport in Windsor Locks, CT in the "Mixed Use" sub-area were analyzed for range errors (Figure 4). The SPSigLO15_7.5k data showed a 50–60 cm offset across the paint stripes and lots of low noise points at the two ends and in the middle of the runway. There were also high noise points in the HRQLS data along the transect. The GMHarLO15_26k data showed some variation (~15 cm) due to range walk, but was able to define the shape of the runway along the transect. The LML dataset had a lower point density but clearly showed the shape of the runway without any range walk effects.

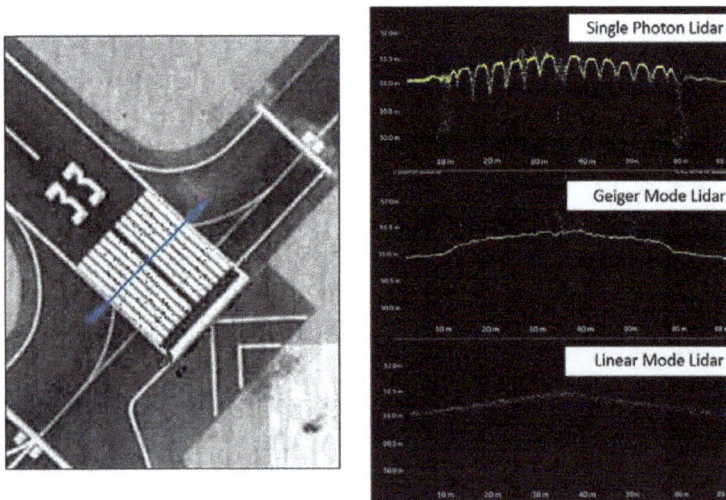

Figure 4. Example of range walk.

4.4. Absolute Accuracy Assessment

Two types of accuracy assessments were performed. First, we performed an assessment against point clouds. Secondly, we performed an assessment against derived bare earth DEMs (Table 4). For the point cloud data assessment, Vegetated Vertical Accuracy (VVA) were assessed only in areas that were processed to final classified point-cloud and where a valid lidar point was within 20 m of the checkpoint (Test 1). The accuracy assessment made against DEMs included interpolated areas as well (Test 2).

Table 4. Absolute accuracy tests using points and DEMs (* leaf on data).

Dataset/Sensor	Test #1—Points		Test #2—DEM	
	NVA	VVA	NVA	VVA
HRQLS (2293 m AGL) *	17.2 cm	17.4 cm	14.1 cm	40.6 cm
IntelliEarth (7950 m AGL) *	17.0 cm	25.6 cm	15.2 cm	92.0 cm
Existing, accepted 3DEP QL2 data (917 m AGL)	12.3 cm	19.8 cm	14.6 cm	25.0 cm

Both sensors met the absolute accuracy requirements for Non-vegetated Vertical Accuracy (NVA) and VVA for QL1/QL2 data in both tests. For IntelliEarth™ data, the NVA based on 32 checkpoints was 0.17 m and VVA based on 15 checkpoints was 0.256 m using the point clouds only. For HRQLS data, the NVA based on 31 checkpoints was 0.172 m and VVA based on 17 checkpoints was 0.174 m. The minimum requirements for QL1/QL2 accuracy are 0.196 m for NVA and 0.296 m for VVA. Using Test 2 shows that both HRQLS and IntelliEarth™ data are meeting the NVA of 19.6 cm at the 95 percent confidence level as required. However, both HRQLS and IntelliEarth™ datasets failed to meet the VVA requirements of 29.4 cm. The poor foliage penetration especially in the IntelliEarth™ data, and therefore the poor capability in delineating the ground under dense trees during leaf-on conditions compromised the quality of filtered DEM in vegetated areas.

4.5. Canopy Penetration

To understand canopy penetration, we analyzed the data that were collected by the IntelliEarth™ Sensor #1 at 7950 m above ground in leaf-on conditions (GMHar15LO15_26k) and the data acquired by the HRQLS sensor at 2293 m above ground in leaf-on conditions (SPSig15LO15_7.5k). These two datasets were compared with the linear-mode data acquired by Dewberry in 2014 leaf-off conditions (LMDewLF14) as well as the linear-mode data acquired by Woolpert in 2015 in leaf-on conditions (LMWptLO15) around the same time as the IntelliEarth™ and HRQLS collects (Figure 5).

Canopy penetration in the IntelliEarth™ data acquired using sensor 1 was very poor. The system was able to mostly capture only the top of the tree canopy in a vegetated terrain, with little or no returns from the canopy structure/understory or the ground. Although the acquisition was during leaf-on conditions, which can greatly reduce canopy penetration compared to a leaf-off collect, the canopy penetration was much poorer than the HRQLS and linear-mode collect during the same leaf-on season over the same area. Canopy penetration in the IntelliEarth™ data acquired using sensor 2 based on the sample data acquired in December 2015 and provided later was very good. The dataset produced by sensor 2 had an ANPS of over 107 ppsm that generated a bare earth (ground) density of 14 ppsm. The sensor 2 was able to generate returns from the canopy structure as well as the ground. It was unknown if canopy penetration improvement was due to the new sensor or the fact that data were collected during leaf-off conditions. It was hypothesized that it was both. Canopy penetration in the data acquired using the HRQLS sensor was qualitatively adequate and resembled the type of penetration expected from a LML dataset in leaf-on conditions; however, the HRQLS sensor did not pass the VVA assessment of bare earth DEMs, suggesting an influence from noisier points and larger than desired voids under vegetation.

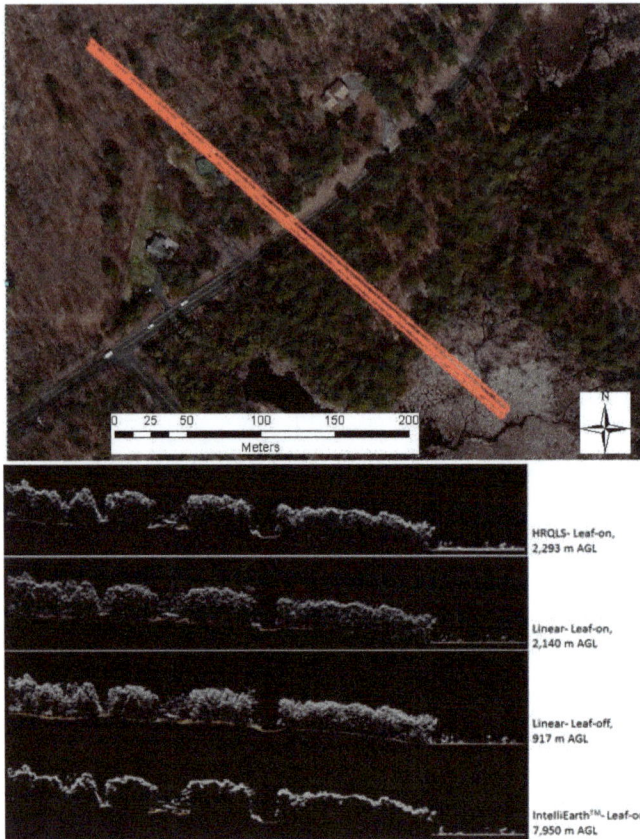

Figure 5. Cross section used for comparisons, overlaid on imagery (**image top**). Sample profiles of HRQLS, leaf on (**top profile**); linear-mode lidar, leaf-on (**middle-top profile**); linear-mode lidar, leaf-off; (**middle-bottom profile**); and IntelliEarth™ lidar, leaf on (**bottom profile**).

4.6. Derivative Products

4.6.1. Evaluation of Intensity Imagery

Intensity values were not collected with the HRQLS system. The lack of intensity was not a significant issue for the processing of data to ground/non-ground, however it did present challenges in collection of breaklines for hydro features as discussed in the section below.

The IntelliEarth™ system stored intensity data as a 16-bit unsigned value and was created from multiple aggregated measurements. These measurements were scaled over the 16-bit range via the number of returns observed for a given reflectance. In the IntelliEarth™ system intensity was referred to as "Reflectance" and exhibited similar grayscale characteristics as linear mode systems. We reviewed the reflectance data acquired by IntelliEarth™ to determine its usability and applicability. This evaluation focused on using the images for the extraction of feature data such as hydrographic features, roads, and buildings. In each case, the reflectance image produced by the IntelliEarth™ sensor offered the same type of capabilities as an intensity image and could be used as such. We did not identify any areas where the reflectance imagery would be less useful than a standard LML intensity image. Figure 6 shows an example of the reflectance imagery produced from the GMHar15LO_26k data.

Similar to linear mode intensity images, the IntelliEarth™ images have variability in the intensity values from the same or similar targets within each data set. The linear mode intensity images typically have variability across different swaths. For IntelliEarth™, the reflectance images tended to have a gridded appearance, which is based on the aggregation of points during processing of the data.

Figure 6. Example of an intensity image from: a linear-mode system (**top**); and the IntelliEarth™ system (**bottom**).

4.6.2. Evaluation of Breakline Development

The lack of intensity imagery from the HRQLS system primarily impacted the collection of breaklines required to develop products that meet USGS/NGP Base Specifications v1.2. Typically, the breaklines are derived using a combination of intensity imagery and the elevation data. The intensity imagery allows the users to reliably determine if features are water bodies or not. Without the use of the intensity imagery the analyst acquiring the breaklines will have to rely more heavily on the elevation values and ancillary imagery sources that were not collected coincidentally with the lidar. The result is that areas of marsh may lean more heavily toward being collected as water or temporal changes between ancillary imagery and the lidar collection could result in incorrect classification of water. For this exercise, a few sample features were collected using only the data available from the HRQLS collect. Since the HRQLS system uses the green wavelength (532 nm) as its laser source, the data may include bathymetry in water bodies. The presence of bathymetric data can cause greater error in collecting breaklines for hydro features. Some of the differences may also be temporal in nature and represent the variation in the water levels between collections.

4.6.3. Evaluation of Day/Night Collection

We performed an evaluation of the differences between the night and day collections provided by IntelliEarth™ (GMHarLO15_26k and GMHarLO15_7.5kDT). Harris collected an overlapping area within the Urban sub-area for this evaluation. The daytime dataset (GMHarLO15_7.5kDT) were acquired at 2293 m above ground compared to the 7950 m above ground of the nighttime collection. Based on a review of the overlapping area acquired in the night and day, we did not identify any major differences in canopy penetration or excessive noise due to solar radiance. The point density was higher on the daytime collection, but that difference is likely due to the difference in flying height and processing and is not a result of additional solar noise in the data. There was very little difference in the canopy structure or penetration of the data to the ground. There also did not appear to be any additional noise points in the daytime dataset.

4.7. Comparing Independent Evaluations

Both Woolpert and Dewberry independently bare earth processed the IntelliEarth™ and HRQLS data, and created bare earth DEMs from these points using internal proprietary methods. These DEMs were then differenced from the accepted QL2 Sandy 1m DEMs to determine differences between accepted DEMs and these test DEMs. We then tested the correlations between four difference grids to determine if they were truly created independently. Correlations between HRQLS DEM differences and IntelliEarth™ DEM differences were r = 0.74 and 0.55, respectively (Figure 7). These low correlations suggest that the derived bare earth DEMs developed by each group were created independently.

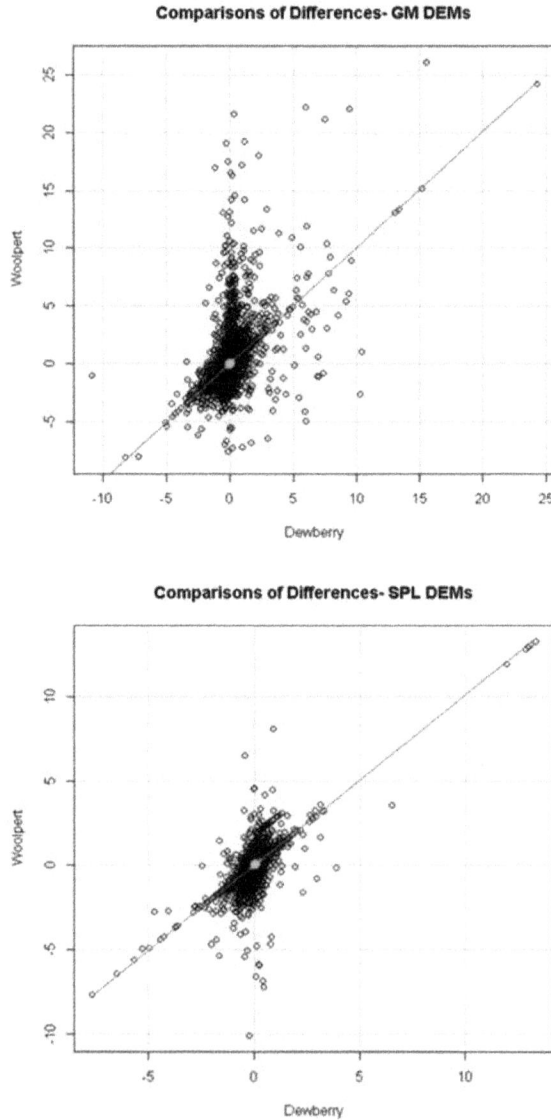

Figure 7. Correlations between Dewberry's and Woolpert's IntelliEarth™ DEM differences (**top**); and HRQLS DEM differences (**bottom**) from accepted 3DEP lidar. r = 0.74 and 0.55, respectively.

5. Conclusions

Elevation data are essential to a broad range of applications, including flood risk management, natural resource management, forest resources management, agriculture and precision farming, wildlife and habitat management, national security, recreation, and many others. The 3DEP is a national partnership program managed by the USGS to acquire high-resolution elevation data for the United States. The initiative is backed by a comprehensive assessment of requirements conducted by Dewberry in 2011 and is in the early stages of implementation. As part of the 3DEP program, we evaluated two new technologies that have the potential for rapid data collection rates with improved data densities and accuracy. These high-altitude airborne lidar sensors can acquire high-density data at much higher collection rates, thereby providing the possibility of QL1 data of 8 ppsm or higher at considerably lower costs.

Data acquired by Harris Corporation's IntelliEarth™ sensor and Sigma Space Corporation's HRQLS sensor were evaluated to assess the suitability of these new technologies for the 3D Elevation Program. While not able to collect data currently to meet USGS lidar base specification, some of this has to do with the fact that the specification was written for linear-mode systems specifically, and the next major version of the USGS lidar base specification will become more flexible to allow these instruments to be included. With little effort on part of the manufacturers of the new lidar systems and the USGS lidar specifications team, data from these systems could soon collect data that meets 3DEP requirements and serve the 3DEP program and its users. Many of the shortcomings noted in this study have been reported to have been corrected or improved upon by both companies in their next generation sensors.

Acknowledgments: We would like to thank both Harris Corporation and Sigma Space Corporation for flying and providing the point cloud data for this evaluation. We would also like to thank staff at Dewberry and Woolpert for flying linear mode lidar, as well as processing and helping assess these data.

Author Contributions: Q. Abdullah and A. Nayegandhi were responsible for processing and providing independent assessments. J. Stoker was responsible for validating results and checking for independence. J. Winehouse was responsible for directing the contract for both Woolpert and Dewberry.

Conflicts of Interest: The authors declare no conflict of interest. Any use of trade, firm, or product names is for descriptive purposes only and does not imply endorsement by the U.S. Government.

References

1. Sugarbaker, L.J.; Constance, E.W.; Heidemann, H.K.; Jason, A.L.; Lukas, V.; Saghy, D.L.; Stoker, J.M. The 3D Elevation Program initiative—A call for action. *U.S. Geol. Surv. Circ.* **2014**. [CrossRef]
2. Aull, B.F.; Loomis, A.H.; Young, D.J.; Stern, A.; Felton, B.J.; Daniels, P.J.; Landers, D.J.; Retherford, L.; Rathman, D.D.; Heinrichs, R.M.; et al. Three-dimensional imaging with arrays of geiger-mode avalanche photodiodes. *Proc. SPIE* **2004**, *5353*, 105–116.
3. Clifton, W.E.; Steele, B.; Nelson, G.; Truscott, A.; Itzler, M.; Entwistle, M. Medium altitude airborne geiger-mode mapping LIDAR system. *Proc. SPIE* **2015**, *9465*. [CrossRef]
4. Priedhorsky, W.C.; Smith, R.C.; Ho, C. Laser ranging and mapping with a photon-counting detector. *Appl. Opt.* **1996**, *35*, 441–452. [CrossRef] [PubMed]
5. Degnan, J.; Wells, D.; Machan, R.; Leventhal, E. Second generation airborne 3D imaging lidars based on photon counting. *Proc. SPIE* **2007**, *6771*. [CrossRef]
6. Degnan, J.J. Photon-counting multikilohertz microlaser altimeters for airborne and spaceborne topographic measurements. *J. Geodyn.* **2002**, *34*, 503–549. [CrossRef]
7. Moussavi, M.S.; Abdalati, W.; Scambos, T.; Neuenschwander, A. Applicability of an automatic surface detection approach to micro-pulse photon-counting lidar altimetry data: Implications for canopy height retrieval from future ICESat-2 data. *Int. J. Remote Sens.* **2014**, *35*, 5263–5279. [CrossRef]
8. Awadallah, M.; Abbott, A.L.; Thomas, V.; Wynne, R.H.; Nelson, R. Estimating Forest Canopy Height and Biophysical Parameters using Photon-counting Laser Altimetry. In Proceedings of the 13th International

Conference on LiDAR Applcation for Assessing Forest Ecosystems (SilviLaser 2013), Beijing, China, 9–11 October 2013; pp. 137–144.

9. Abdalati, W.; Zwally, H.J.; Bindschadler, R.; Csatho, B.; Farrell, S.L.; Fricker, H.A.; Harding, D.; Kwok, R.; Lefsky, M.; Markus, T.; et al. The ICESat-2 Laser Altimetry Mission. *Proc. IEEE* **2010**, *98*, 735–751. [CrossRef]

10. Williams, G.M., Jr. Limitations of geiger-mode arrays for flash ladar applications. *Proc. SPIE* **2010**, *7684*. [CrossRef]

11. Johnson, S.E. Foliage penetration optimization for geiger-mode avalanche photodiode lidar. *Proc. SPIE* **2013**, *8731*. [CrossRef]

12. Li, Q.; Dengan, J.; Barrett, T.; Shan, J. First Evaluation on Single Photon-Sensitive Lidar Data. *Photogramm. Eng. Remote Sens.* **2016**, *82*, 495–503. [CrossRef]

13. Schwemmer, G.K.; Wilkerson, T.D.; Guerra, D.V. Compact scanning lidar systems using holographic optics. *Proc. SPIE* **1998**. [CrossRef]

14. Kim, A.M.; Runyon, S.C.; Olsen, R.C. Comparison of full-waveform, single-photon sensitive, and discrete analog LIDAR data. *Proc. SPIE* **2015**, *9465*. [CrossRef]

15. Heidemann, H.K. Lidar base specification (ver. 1.2, November 2014). *U.S. Geological Survey Techniques and Methods*; Book 11, Chapter B4; USGS: Reston, VA, USA, 2014.

16. ASPRS Positional Accuracy Standards for Digital Geospatial Data. Available online: https://www.asprs.org/pad-division/asprs-positional-accuracy-standards-for-digital-geospatial-data.htm (accessed on 28 June 2016).

Part 2
Multispectral Lidar

remote sensing

MDPI

Article

NASA Goddard's LiDAR, Hyperspectral and Thermal (G-LiHT) Airborne Imager

Bruce D. Cook [1,*], Lawrence A. Corp [2], Ross F. Nelson [1], Elizabeth M. Middleton [1], Douglas C. Morton [1], Joel T. McCorkel [1], Jeffrey G. Masek [1], Kenneth J. Ranson [1], Vuong Ly [1] and Paul M. Montesano [2]

[1] NASA Goddard Space Flight Center, Greenbelt, MD 20771, USA;
E-Mails: ross.f.nelson@nasa.gov (R.F.N.); elizabeth.m.middleton@nasa.gov (E.M.M); douglas.morton@nasa.gov (D.C.M.); joel.mccorkel@nasa.gov (J.T.M.); jeffrey.g.masek@nasa.gov (J.G.M.); kenneth.j.ranson@nasa.gov (K.J.R.); vuong.t.ly@nasa.gov (V.L.)

[2] Sigma Space Corporation, Lanham, MD 20706, USA;
E-Mails: lawrence.a.corp@nasa.gov (L.A.C.); paul.m.montesano@nasa.gov (P.M.M.)

* Author to whom correspondence should be addressed; E-Mail: bruce.cook@nasa.gov; Tel.: +1-301-614-6689; Fax: +1-301-614-6695

Received: 20 June 2013; in revised form: 6 August 2013 / Accepted: 8 August 2013 / Published: 13 August 2013

Abstract: The combination of LiDAR and optical remotely sensed data provides unique information about ecosystem structure and function. Here, we describe the development, validation and application of a new airborne system that integrates commercial off the shelf LiDAR hyperspectral and thermal components in a compact, lightweight and portable system. Goddard's LiDAR, Hyperspectral and Thermal (G-LiHT) airborne imager is a unique system that permits simultaneous measurements of vegetation structure, foliar spectra and surface temperatures at very high spatial resolution (~1 m) on a wide range of airborne platforms. The complementary nature of LiDAR, optical and thermal data provide an analytical framework for the development of new algorithms to map plant species composition, plant functional types, biodiversity, biomass and carbon stocks, and plant growth. In addition, G-LiHT data enhance our ability to validate data from existing satellite missions and support NASA Earth Science research. G-LiHT's data processing and distribution system is designed to give scientists open access to both low- and high-level data products (http://gliht.gsfc.nasa.gov), which will stimulate the community development of synergistic data fusion algorithms. G-LiHT has been used to collect more than 6,500 km² of data for NASA-sponsored studies across a broad range of ecoregions in the USA and Mexico. In this paper, we document G-LiHT design considerations, physical specifications, instrument performance and calibration and acquisition parameters. In addition, we describe the data processing system and higher-level data products that are freely distributed under NASA's Data and Information policy.

Keywords: remote sensing; airborne scanning LiDAR; imaging spectroscopy; surface temperature; sensor fusion; data fusion; ecosystem structure; forest disturbance; forest health; primary production

1. Introduction

LiDAR, hyperspectral and thermal remote sensing are core areas of current and planned NASA remote sensing capability (e.g., Landsat Enhanced Thematic Mapper Plus, ETM+, and Operational Land Imager, OLI; Earth Observing-1, Hyperion; Earth Observing System's Advanced Spaceborne Thermal Emission and Reflection Radiometer, ASTER, Multi-angle Imaging SpectroRadiometer, MISR, and Moderate Resolution Imaging Spectroradiometer, MODIS; Suomi National Polar-orbiting

Partnership Visible Infrared Imaging Radiometer Suite, NPP VIIRS; Ice, Cloud, and Land Elevation Satellite-2 Advanced Topographic Laser Altimeter System, ICESat-2; and Hyperspectral Infrared Imager, HyspIRI), and fusion of complementary data from different sensors offers the potential for improved global remote sensing of terrestrial ecosystems. Formation flying missions, such as NASA's A-Train satellite constellation, offer near-simultaneous coverage from a number of Earth-observing systems; however, the spatial resolution of the current generation of NASA sensors limits some terrestrial ecology applications. Furthermore, efforts to merge 3D structure information from LiDAR with imaging spectrometer and thermal data is currently hindered by the absence of a LiDAR designed for vegetation in space [1] and an EO-1 satellite that is approaching the end of its useful life.

Airborne platforms are more flexible than satellite missions for developing and testing data fusion at fine spatial (1 to 10 m) and spectral resolutions, which have prompted a new generation of LiDAR and imaging spectrometer instrument packages [2–4]. Airborne platforms offer specific advantages for the study of terrestrial ecosystem, including targeted acquisitions of seasonal and diurnal processes (e.g., wetland inundation, plant phenology, drought and fire impacts) and coordinated field and remote sensing data collection needed to scale process-level understanding to the scale of airborne and satellite remote sensing observations [5]. Furthermore, the addition of coincident thermal data to LiDAR and imaging spectrometer data broadens the range of potential terrestrial ecology applications to include research on evapotranspiration, hydrology, forest health and urban applications [6,7].

Here, we describe the development of a unique multi-sensor instrument, Goddard's LiDAR, Hyperspectral and Thermal (G-LiHT) airborne imager, which advances previous concepts for data fusion by integrating LiDAR, hyperspectral and thermal sensors in a lightweight and portable system for worldwide research applications. G-LiHT's single-solution GPS-INS (Global Positioning System and Inertial Navigation System) avoids multi-dimensional data effects that are introduced when data is collected at different time and observational scales [2]. Imaging spectroscopy provides quantitative information on vegetation cover, species composition and biophysical and chemical properties that can be derived from measurements of reflected sunlight in the visible through shortwave infrared wavelengths [8–12]. Light detection and ranging provides quantitative, 3D information on terrain and vegetation cover, height and distribution of canopy elements, which can be used to characterize biodiversity and habitat [13] and estimate light interception in plant photosynthesis and production models [14]. Land surface temperature provides data needed to estimate evapotranspiration and other surface energy fluxes [6], and can be an indicator of soil or vegetation moisture status [15]. By using commercial off-the-shelf instrumentation and general aviation aircraft, G-LiHT also reduces development and operational costs. Low-cost deployment of the G-LiHT system opens a wide range of applications for targeted, airborne remote sensing, including diurnal and seasonal processes in terrestrial ecosystems at ~1 m spatial resolution (Figure 1).

The goal of this multi-sensor and data fusion effort is to characterize ecosystem form and function using remote sensing data, with a particular emphasis on the data products needed to develop a new generation of high resolution ecosystem and radiative transfer models. Analysis of G-LiHT data will be used to:

- provide new insight into photosynthetic functionality and vegetation productivity, including new, spatially-explicit remote sensing indicators of key dynamic biological processes;
- characterize fine-scale spatial and temporal heterogeneity in ecosystem structure and function under diverse environmental and climate conditions; and
- create new methods for data fusion to monitor ecosystem health and the effects of climate and human-induced changes on these ecosystems.

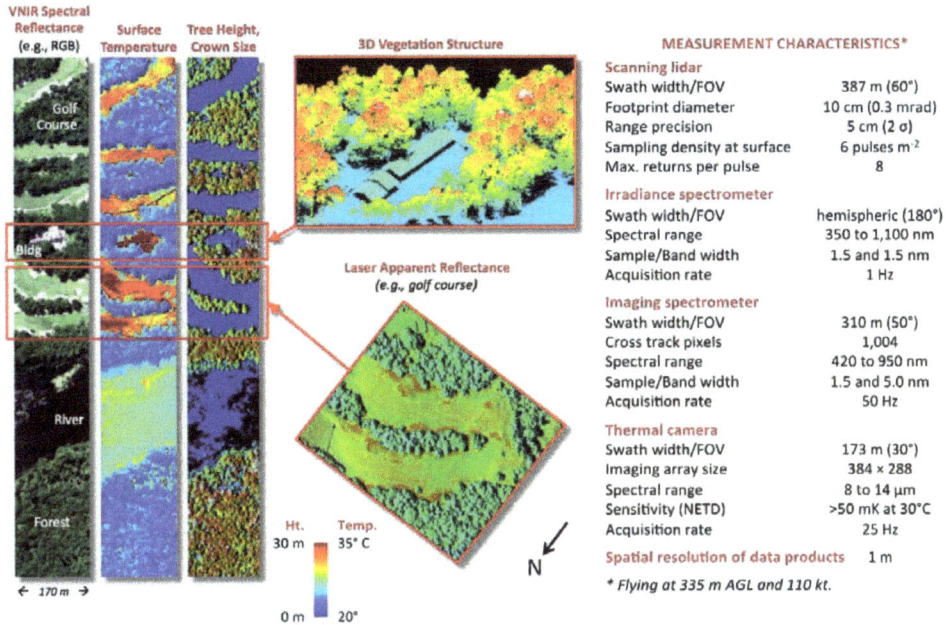

Figure 1. First coincident acquisition of passive optical, thermal and LiDAR data with G-LiHT (14 July 2011; 37.1839°N 76.5291°W) and key measurement characteristics of the instruments. Spectral and structural differences between a forest, river, golf course and buildings demonstrate the synergistic potential of data fusion for airborne remote sensing of ecosystem composition, structure, function and health. FOV, field of view; NETD, Noise Equivalent Temperature Difference.

2. G-LiHT Design and Instrumentation

2.1. Scientific Objectives and Design Considerations

G-LiHT was designed as a relatively inexpensive, robust and portable research tool for evaluating the potential benefits of data fusion for studies of terrestrial ecosystems. Table 1 lists the objectives and related measurement requirements for G-LiHT. One of the major obstacles to the development of science-based data fusion algorithms is the availability of accurately co-registered data of similar grain size for different information types. This is often the case when instruments are flown on different platforms and acquired on different dates. We believe that "instrument fusion" is a prerequisite to "data fusion" and conceived G-LiHT as a multi-sensor airborne imaging system that would simultaneously map the composition, structure and function of terrestrial ecosystems.

In addition, G-LiHT was designed to simplify worldwide deployment and minimize collection and data processing costs. As a result, G-LiHT features eye-safe lasers, a portable, low-power payload (37 kg; 30 × 30 × 60 cm; 210 W), a single-solution GPS-INS, compatibility with common, civilian-use aircraft (e.g., Cessna, Piper, Twin Otter; 12/28 VDC compatibility) using a standard camera port or custom wing-mounted pod and commercial off-the-shelf (COTS) instruments that are easy to replace and not regulated by ITAR (International Traffic in Arms Regulation). Both G-LiHT science and deployment objectives are traced to instrument and design requirements in Table 1.

Table 1. Traceability of Goddard's LiDAR, Hyperspectral and Thermal (G-LiHT) science and deployment objectives. VNIR, Visible and Near Infrared; PALS, Portable Airborne Laser System; VDC, Volts Direct Current; GPS-INS, Georeferenced Positioning System and Inertial Navigation System.

Objective	Requirement
Direct computation of at-sensor reflectance and record of solar illumination conditions	• downwelling irradiance spectrometer (VNIR) • clear sky solar irradiance model
Mapping species composition and variations in biophysical variables (e.g., photosynthetic pigments, nutrient content)	• VNIR imaging spectrometer
Mapping forest health and photosynthetic responses to environmental conditions	• spectral resolution ≤5 nm
Tree-Scale measurements with minimal atmospheric interference	• low-and-slow data acquisition (~335 m AGL, 110 kts)
Indicator of evapotranspiration and stress	• surface temperature observations
Mapping terrain, canopy height, and structural attributes (i.e., spatial distribution of canopy elements)	• scanning airborne lidar
Continuous canopy height profile	• profiling lidar
Continuity with PALS [16]	• Reigl range finder with 905 nm laser
High technology readiness and reliability	• Commercial Off-The-Shelf (COTS) instrumentation
Portable (ship or hand-carry)	• mass <50 kg; volume <0.2 m3
Suitable for international campaigns	• non-ITAR components
Ease of installation and flight certification	• ability to mount over camera port or in wing-mounted pod • low power (<250 W; 12 and 28 VDC capability) • FAA compliant design and materials • eye safe lasers
Accurate co-registration	• single solution GPS-INS and data acquisition computer • GPS time server and image time stamps • boresight alignment
Ability to collect large data volumes at high data acquisition rates	• removable solid state hard disks with eSATA interface • dedicated video capture card • gigabit Ethernet communication • on-board processing of lidar waveforms
Radiometrically calibrated data	• laboratory and vicarious calibration
Ability to operate under range of cloud conditions	• low altitude (<500 m AGL) data acquisition
Low acquisition and processing costs	• COTS instrumentation and acquisition software • compatibility with general aviation aircraft • automated data processing system • internet data distribution system

G-LiHT instruments were physically arranged to fit in a rigid, compact package (Figure 2), with the scanning LiDAR, imaging spectrometer, hyperspectral and thermal camera aligned along an optical axis parallel to the flight path. Wire rope isolators (WR4-200-10, Endine Inc., Orchard Park, NY, USA) are used to mount the instrument package to the aircraft and reduce the impact of high-frequency aircraft vibrations. The system is weather resistant and can be mounted either internally to the aircraft over an appropriately sized view port or externally attached to aircraft using a custom fabricated pod. A custom pod was designed and fabricated by NASA for any Cessna 206, using mounting points that are standard on this platform (Figure 3).

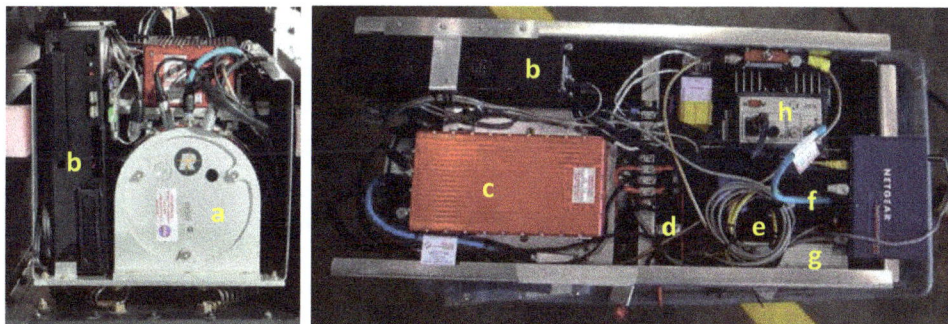

Figure 2. End and top views of G-LiHT instrument package, showing the (**a**) scanning LiDAR; (**b**) data acquisition computer; (**c**) GPS-INS; (**d**) irradiance spectrometer; (**e**) imaging spectrometer; (**f**) thermal infrared camera; (**g**) GPS time server; and (**h**) profiling LiDAR.

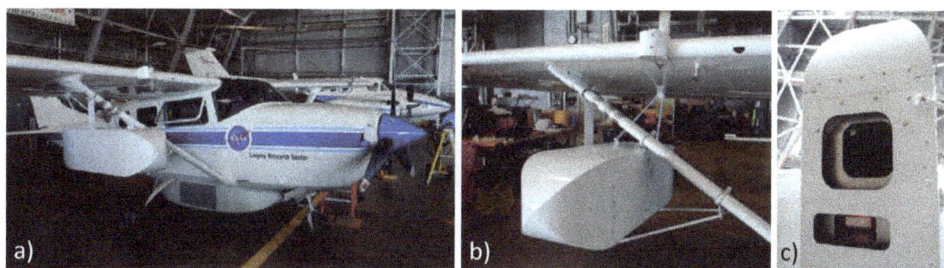

Figure 3. (**a**) G-LiHT installed on NASA's Cessna 206; (**b**) wing-mounted pod showing mounting points common to all Cessna 206s; (**c**) view ports on bottom of custom pod.

Specifications of the individual G-LiHT instruments are provided in the sections below.

2.2. GPS and Inertial Navigation System (INS)

A single-solution GPS-INS (RT-4041, Oxford Technical Solutions, Oxfordshire, UK) is used to obtain high precision position and attitude measurements for all G-LiHT sensors. The unit is directly attached to the airborne laser scanning (ALS), which shares the same mounting plate as the optical breadboard for the imaging spectrometer and thermal camera (Figure 2). The GPS-INS incorporates a six-axis inertial navigation system (three gyroscopes and three servo-grade accelerometers) and an L1/L2 GPS (Global Positioning System) and GLONASS (Global Navigation Satellite System) receiver with an OmniStar decoder to deliver 10 cm positioning, 0.1° heading and 0.03° roll and pitch accuracies. Measurements are acquired at 250 Hz, which is required to geolocate data from the optical imagers (25 to 50 Hz) and ALS (300 kHz laser). The real-time internal processing includes strap down algorithms, a WGS-84 and EGM96 (Earth Gravitational Model 1996) Earth model, Kalman filtering and in-flight alignment algorithms. The Kalman filter and in-flight algorithms monitor the performance of the system and update the measurements to correct for inertial sensor errors and maintain high positional accuracy.

2.3. Airborne Scanning LiDAR

The VQ-480 (Riegl USA, Orlando, FL, USA) airborne laser scanning (ALS) instrument was selected for use with G-LiHT, because it was affordable, compact, provided evenly distributed pulses on the ground and offered near-turnkey operation. The VQ-480 uses a high-performance laser rangefinder and a rotating polygon mirror with three facets to deflect a 1,550 nm Class 1 laser beam onto the ground. A user-selectable pulse repetition rate up to 300 kHz provides an effective measurement rate of up to 150 kHz along a 60° swath perpendicular to the flight direction (Figure 3).

A laser beam divergence of 0.3 mrad produces a 10 cm diameter footprint at the nominal operating altitude of 335 m. The small footprint laser beam allows detection of small gaps in the canopy and the ability to characterize fine scale disturbances, which are difficult to deconvolve from large footprint LiDAR waveforms (Figure 4, [17]). The mirror speed is set to a maximum of 100 rotations s^{-1} during G-LiHT acquisitions, whose points are spaced 0.23 m apart within a line perpendicular to the flight direction and 0.57 m between lines with a nominal aircraft speed of 110 knots. For each laser shot, the return waveform is digitized, and online processing algorithms are used to provide ranging data for multiple targets. Up to eight discrete ranging returns may be identified and recorded for a given pulse. Each return target is time tagged and accurately synchronized with the GPS-INS using a 1 Hz Transistor-Transistor Logic (TTL) pulse. Riegl's software, RiACQUIRE, provides a graphical user interface for scanner control and near real-time monitoring of scanning LiDAR and GPS-INS data.

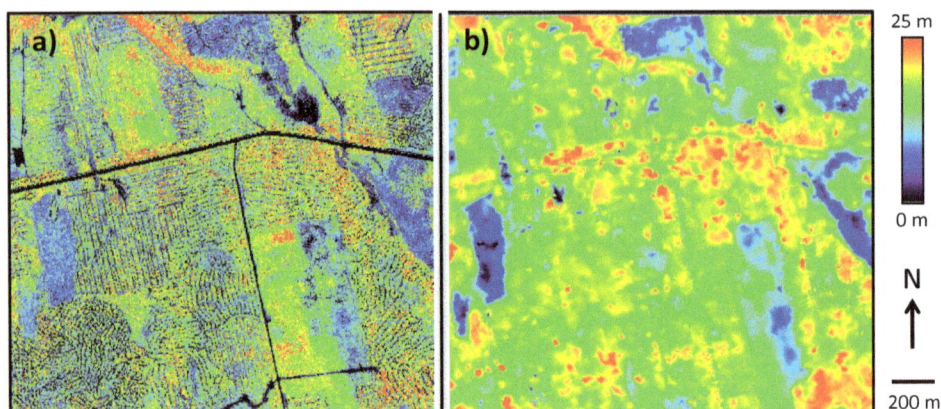

Figure 4. Canopy height model from (**a**) small footprint (10 cm) G-LiHT LiDAR during June 2012; and (**b**) large footprint (25 m) Land, Vegetation and Ice Sensor (LVIS) LiDAR during August 2009 [17], for a commercial forest near Howland, ME, USA (45.2220°N 68.7423°W). Discrete returns from small footprint LiDAR are able to detect small gaps and characterize fine-scale disturbance (*i.e.*, strip harvesting), which are challenging to deconvolve from large footprint LiDAR waveforms.

2.4. Profiling LiDAR

G-LiHT's profiling LiDAR is an LD321-A40 (Riegl USA, Orlando, FL, USA), multi-purpose laser distance meter that is similar to the LD90-3800-VHS used in the Portable Airborne Laser System (PALS) [16]. The LD321-A40 provides a continuous profile of canopy height measurements, which, from a regional sampling perspective, provides data similar to other space-based profilers that have been flown (Ice, Cloud, and Land Elevation Satellite Geoscience Laser Altimeter System, ICESat GLAS), proposed for flight (Deformation, Ecosystem Structure and Dynamics of Ice, DESDynI) or which may be flown in the near future (ICESat-2 ATLAS). In addition, profiling data is important for studying horizontal landscape patch structure [13] and ensuring continuity between ALS and PALS datasets that have been collected worldwide. Real-time digital echo signal processing with the LD321-A40 enables precise distance measurement for complex multi-target situations, resolving up to five target distances per pulse. Distance measurements parallel to the flight line are used to continuously measure vegetation height and structure along a sampling transect. A Class 1M laser diode emits a 905 nm beam with a divergence of 1.5 mrad to produce a 50 cm diameter footprint at the nominal operating altitude of 335 m.

2.5. Irradiance Spectrometer

Downwelling irradiance is measured with an Ocean Optics (Dunedin, FL, USA) USB4000-VIS-NIR spectrometer. Light energy is transmitted to the spectrometer through an upward looking opaline glass cosine diffuser with a 180° field of view (FOV). The cosine diffuser is mounted on top of the aircraft, where there is an unobstructed view of the sky. A custom mount was designed for the leading edge of the Cessna 206 (Figure 5), which integrates both the GPS antenna and cosine diffuser. A 3 m long, 100 μm diameter optical fiber delivers the light energy through a 25 μm entrance slit and a multi-bandpass order-sorting filter, where it is dispersed with a fixed grating across a 3,648-element Toshiba linear Charge Coupled Device (CCD) array. The spectrometer covers the spectral range from 350 to 1,100 nm with a native optical resolution to ~1.5 nm (full width half maximum, FWHM). Power and communications are transferred through a USB 2.0 connection to the computer. The downwelling radiometer is operated continuously during flight with a 33 ms integration time, a 30 scan average and 3× spectral binning, which is equivalent to a 1 Hz data acquisition rate. The unit is cross-calibrated with the imaging spectrometer to enable atmospheric characterization of downwelling irradiance, which is used to compute the at-sensor reflectance product.

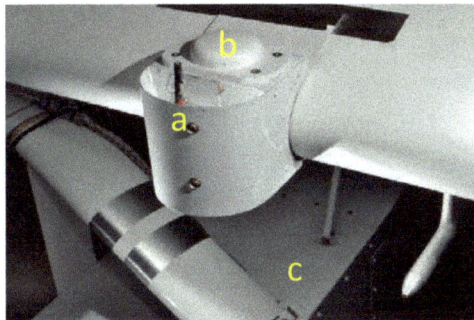

Figure 5. Irradiance cosine diffuser (**a**) and GPS antenna (**b**) attached to the leading edge of a Cessna 206 wing with a custom-mounting device. The wing-mounted pod containing the G-LiHT instrument package is seen below (**c**).

2.6. Imaging Spectrometer

The Hyperspec imaging spectrometer (Headwall Photonics, Fitchburg, MA, USA) enables high spectral and spatial resolution imaging by using f/2.0 telecentric optics and a high efficiency aberration-corrected convex holographic diffraction grating, providing an optical dispersion of 100 nm per mm over a 7.4 mm spatial by 6.0 mm spectral focal plane. The imaging spectrometer is based on the Offner form and is designed to operate in the 400–1,000 nm spectral region with a 50° full field of view. The Hyperspec imaging spectrometer accepts a C-mount objective lens (Cinegon f/1.4 8 mm, Schneider Optics, Hauppauge, NY, USA) with high optical performance using ultralow dispersion glass and broadband anti-reflection coating designed for the visible to near-IR spectrum. Coupled to the spectrometer is the RA1000 m/D high speed rugged megapixel focal plane array that allows 50 progressive frames per second to be acquired through an EPIX (Buffalo Grove, IL, USA) PIXCI ECB1 PCI Express CameraLink interface. The camera uses a 1,004 × 1,004 pixel 2/3-inch format interline CCD with 7.4 μm square pixels and 12-bit radiometric resolution. Other camera features include a digital fine gain for adjustable camera sensitivity over a 60 dB dynamic range, electronic shuttering and low smear characteristics. The camera is controlled with the serial communication channel of the CameraLink interface. Each image frame is coded with a computer timestamp synchronized with a Time Tools (Dudley, UK) LC2750 GPS Timing Receiver for post-processing geolocation using data from the GPS-INS detailed in Section 2.2 (Figure 6).

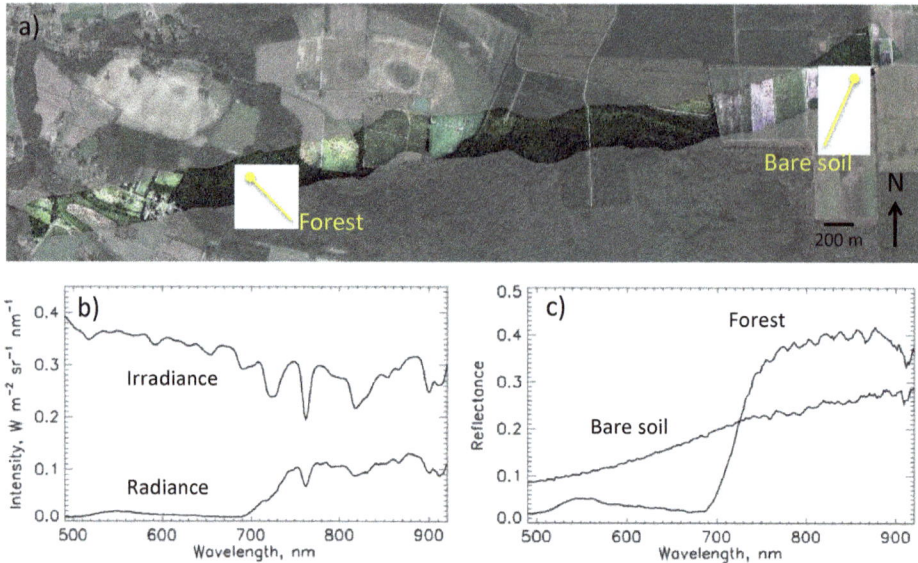

Figure 6. (a) True color quick look data product (Keyhole Markup Language (KML) format) viewed in Google Earth, illustrating image georegistration in a turbulent atmosphere near Plymouth, NC, USA (28 July 2011; 35.8437°N 76.6994°W); **(b)** Coincident downwelling solar irradiance and upwelling radiance spectra over a forested area in the swath; **(c)** reflectance spectra for bare soil and forest targets in **(a)**.

2.7. Thermal Imaging

G-LiHT's thermal remote sensing capability originates from the Gobi-384 thermal imaging camera (Xenics, Leuven, Belgium). The instrument measures long wave infrared (LWIR) radiation over a broad spectral range (8 to 14 µm) using an uncooled amorphous silica (ASi) microbolometer detector with 384 cross track pixels and a 30° FOV. This compact camera (72 × 60 × 50 mm) uses a Gigabit Ethernet TCP/IP (Transmission Control Protocol and the Internet Protocol) interface to deliver 16-bit radiometrically calibrated thermal imaging data at a 25 Hz frame rate to the computer running Xenics's Xeneth infrared camera software. Each image frame is coded with a computer timestamp synchronized to a GPS Timing Receiver (see Section 2.6).

3. Calibration

3.1. Boresight Alignment

Georeferenced LiDAR returns and optical observations are computed as a function of lever arm offsets, boresight angles and optical measurements. Bias in any of these parameters and GPS-INS errors will result in offset point clouds and distorted images. Boresight biases are computed using data collected over buildings with peaked roofs, where overlapping data are collected with different headings. Boresight corrections are computed between campaigns and consistently result in an accuracy of ~10 cm (1 σ).

3.2. Radiometric Calibration

The absolute spectral response (ASR) of G-LiHT's imaging spectrometer was calculated using measurements made with a portable version of the US National Institute of Standards and Technology's (NIST) Spectral Irradiance and Radiance Responsivity Calibrations Using Uniform Sources (SIRCUS) [18], which is currently housed at NASA Goddard Space Flight Center (GSFC). SIRCUS uses continuously tunable lasers coupled to an integrating sphere as a radiance source for the calibration of detectors operating in the solar reflective spectrum. SIRCUS and appropriately characterized transfer radiometers allow absolute radiometric calibration with uncertainties of 0.5% and traceability to national standards [19]. The availability of lasers determines the spectral coverage on SIRCUS, and the uncertainties achievable are determined by the quality of the transfer radiometers and measurement technique. The high power and wavelength stability of the laser-based sources enable large aperture instruments to be characterized and calibrated. All irradiance and radiance responsivity calibrations are traceable to NIST's Primary Optical Watt Radiometer (POWR) through regularly characterized transfer radiometers. The short calibration chain from NIST's Primary Optical Watt Radiometer (POWR) to the transfer radiometer and imaging spectrometer minimizes the uncertainty of the Spectral Radiometer Facility (SRF). Spectral radiance calibrations at power levels typical of solar irradiance values allowed for quantitative measurements of the spectral response functions and under flight-like conditions over the focal plane from 415 to 1,020 nm. In addition to providing spectral and radiometric calibration, the combination of full-field, full-aperture and near-monochromatic method of radiometric calibration used here allows the unique capability of assessing and correcting for the total system stray light of the imaging spectrometer, not possible with traditional broadband calibration sources.

3.3. Wavelength and Radiometric Stability

Both the downwelling irradiance and upwelling imaging spectrometer are monitored for radiometric and wavelength stability. A portable, 10 cm Teflon integrating sphere with a Hg and Ar pen lamp illumination sources produce nine pronounced emission lines (Figure 7) that are used periodically to verify the spectral channel to wavelength relationship over the detector spectral range with sub-nanometer precision. A large aperture, 1 m integrating sphere and an NIST traceable uniform source is used to monitor the radiometric stability and performance of the spectrometers between campaigns.

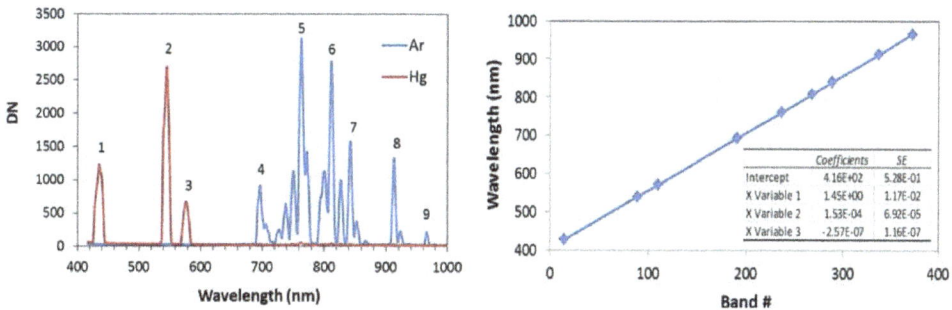

Figure 7. (**a**) Hg and Ar lamp emission lines as viewed through G-LiHT's imaging spectrometer; (**b**) relationship between band number and band center wavelength using Gaussian iterative curve fitting.

3.4. Thermal Radiometric Calibration

G-LiHT's thermal imaging camera has factory radiometric calibration for surface temperatures between −20 °C and 120 °C, with a Noise Equivalent Temperature Difference (NETD) >50 mK. This wide dynamic range allows for airborne operations of the thermal image over a diverse range of surface temperatures. Thermal calibration stability as a function of a microbolometer operating temperature range of 25 °C to 50 °C was verified against a GSFC Calibration Facilities blackbody for stable operation over the target temperature range of −5 °C to 85 °C (Figure 8).

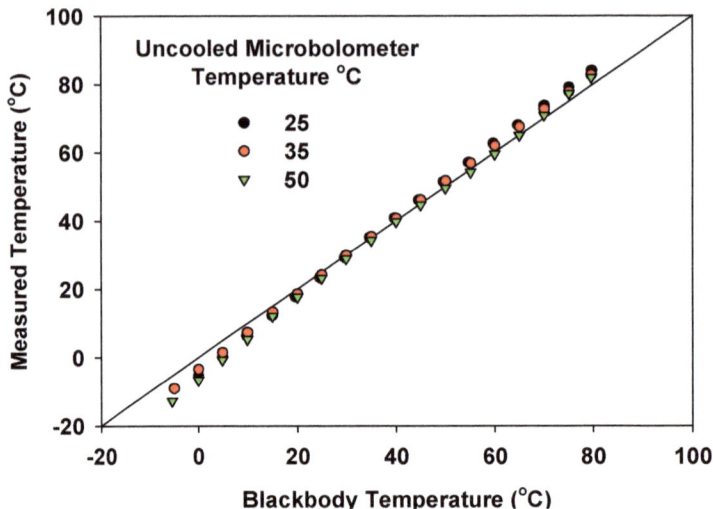

Figure 8. Radiometric response of the for the Xenics Gobi-384 long wave infrared (LWIR) thermal imaging camera as a function of instrument body temperature.

4. Flight Planning and Data Acquisition

Flight plans are made with X-TRACK Flight Management Software Suite (Track'Air Aerial Survey Systems, Oldenzaal, The Netherlands), using a nominal flight altitude of 335 m AGL, 60° FOV and 30% swath overlap. This configuration results in a LiDAR, spectrometer and thermal swath of 387, 310 and 173 m. The pixel size of the pushbroom imaging spectrometer is ~1 m and is limited by the aircraft speed and frame acquisition rate. The instantaneous per pixel resolution of the thermal camera is approximately 3 times coarser than the imaging spectrometer, but oversampling with the 2D detector allows us to create a product of equal resolution (Section 5.2.4).

Flight lines are uploaded onto a 696 GPS (Garmin International, Inc., Olathe, KS, USA) for pilot navigation and onto the instrument operator's computer, which is using Reigl's RiAcquire acquisition software. Each instrument is separately controlled and triggered to collect data using vendor-supplied software. Ejectable, solid state hard drives provide fast input/output and quick retrieval upon conclusion of the flight.

5. Data Products, Processing and Distribution

5.1. Data Products

NASA's Earth Science Data Systems program references data products based on their level of processing, ranging from Level 0 to Level 4 [20]. Level 0 (L0) data products include unprocessed instrument data; Level 1 (L1) products are time-referenced data that have been processed to at-sensor radiometric units; Level 2 (L2) products are geophysical variables derived from L1 products; and Level 3 (L3) products are geophysical variables mapped on a space-time grid scale. G-LiHT data products (L1 through L3) are listed in Table 2 for each of the instruments. Many of these products will help end users create additional higher-order products (e.g., solar and view angles are needed for computing bidirectional corrected reflectance; terrain and canopy heights are needed for orthorectification) and facilitate scientific data analysis (e.g., cloudiness and solar illumination conditions affect canopy photosynthesis) [21].

Table 2. G-LiHT data products for public distribution. FWHM, full width half maximum; PAR, Photosythetically Active Radiation; DTM, Digital Terrain Model; CHM, Canopy Height Model.

Instrument	L1	L2	L3
Oxford RT-4041 GPS-INS 250 Hz measurement rate	Trajectory data (coordinates, roll, pitch, yaw)	•Aircraft elevation •Aircraft altitude AGL •Geographic Look-Up Table (GLT)	•Aircraft elevation •Aircraft altitude AGL •View angle •View azimuth
Riegl VQ-480 Scanning Lidar 1550 nm laser discrete returns (≤8 pulse⁻¹) 150 kHz measurement rate	Return data (coordinates, scan angle, return number, apparent reflectance)	•Classified return data (ground, non-ground) •AGL heights	•LiDAR returns ("point clouds") •DTM •CHM •LiDAR metrics
Headwall Hyperspec Imaging Spectrometer 417 to 1,007 nm 402 bands, ≤5 nm FWHM 1,004 pixels per line 50 Hz measurement rate	At-sensor radiance spectra $(W \cdot m^{-2} \cdot sr^{-1} \cdot nm^{-1})$	•At-sensor reflectance computed with observed irradiance •Surface reflectance computed with atmospheric correction •Fluorescence [experimental]	•At-sensor reflectance computed w/observed irradiance •Surface reflectance computed w/atmospheric correction •Common vegetation indices •Fluorescence [experimental]
Ocean Optics USB 4000 Irradiance Spectrometer cosine diffuser 346 to 1,041 nm 1.5 nm FWHM 1 Hz measurement rate	Solar irradiance spectra $(W \cdot m^{-2} \cdot sr^{-1} \cdot nm^{-1})$	•Incoming PAR •Cloudiness index •Modeled solar zenith angle •Modeled solar azimuth angle	•Incoming PAR •Cloudiness Index •Modeled solar zenith angle •Modeled solar azimuth angle
Xenics Gobi 384 Thermal Camera 8 to 14 μm 25 Hz measurement rate	Temperature data (°C)	•Atmospherically corrected surface temperature	•Atmospherically corrected surface temperature

5.2. Data Processing System

An automated data processing system is key to releasing standardized products in a timely manner. Data processing begins with the GPS-INS and scanning LiDAR data, since some of the products are used in processing the imaging spectrometer and thermal data. Workflows for the scanning LiDAR and spectrometer data have been developed as illustrated in Figures 9 and 10. Manual pre-processing of L0 data with vendor supplied software is the first step in the workflows, followed by automated data processing algorithms that have been custom coded in the IDL-ENVI (Interactive Data Language and Environment for Visualizing Images) scientific programing language (Exelis Visual Information Solutions, Boulder, CO, USA). The following sections describe the specific data processing steps.

5.2.1. GPS and Inertial Data

GPS-INS data is stored on internal memory in a raw, unprocessed format. Lever arm and other offset coefficients are applied to the L0 data with Oxford's RT Post Process software, which converts the proprietary binary data to an ASCII format. During this step, the inertial data is processed forwards and backwards in time to minimize the effects of GPS data drift. During data acquisition, OmniStar HP differential correction service provides a real-time positioning error of <15 cm in extended periods of open sky. Increased precision can be achieved where base station GPS data is available.

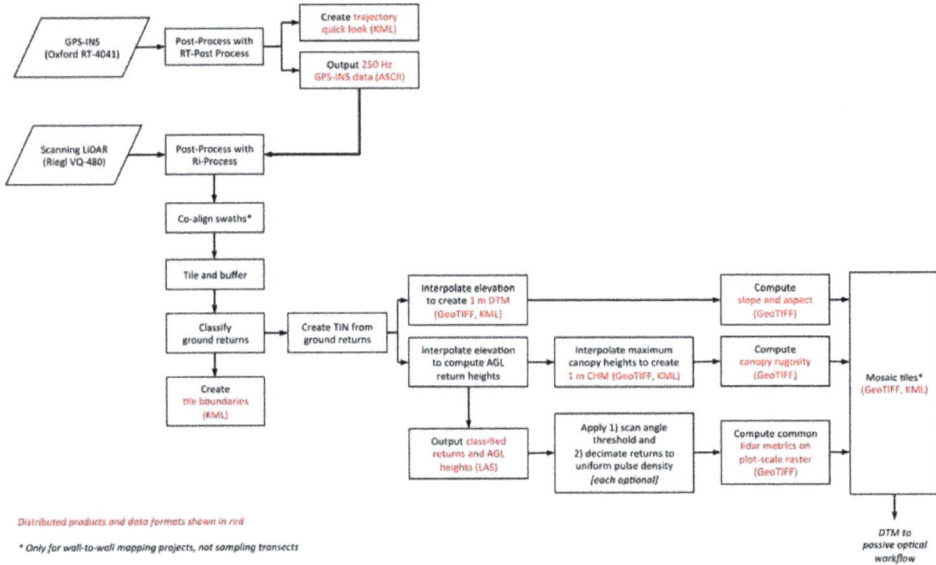

Figure 9. Data processing workflow for GPS-INS and airborne laser scanning (ALS) data. KLM, Keyhole Markup Language; ASCII, American Standard Code for Information Interchange.

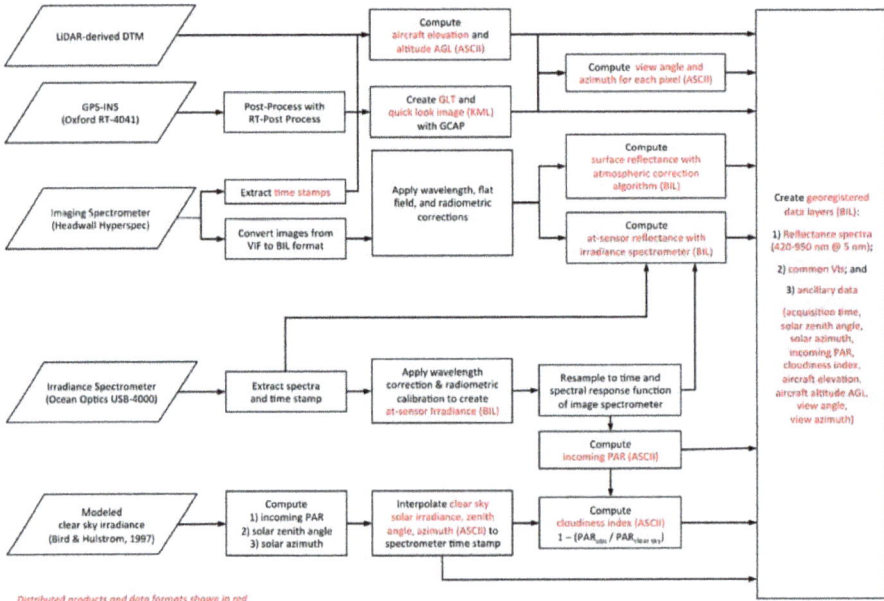

Figure 10. Data processing workflow for downwelling irradiance and upwelling image spectroscopy data. GLT, Geometric Look-up Table; BIL, Band Interleaved by Line.

5.2.2. Scanning LiDAR Data

Riegl's RiPROCESS software is used for managing, processing, analyzing and visualizing data acquired with the ALS system. RiPROCESS ingests raw laser scanner data and pre-processed GPS-INS data, applies calibration information, transforms the scan data into geographic coordinates and exports return data in LAS file format [22]. For mapping projects, individual swaths are co-aligned for small differences (typically <0.5 m) in elevation due to differences in position uncertainty, which is typically an effect of different satellite configurations and atmospheric conditions. A cross-track swath is typically collected at the conclusion of a mapping project to serve as a reference point for swath co-alignment.

LiDAR returns are resampled and processed as smaller data volumes (~1 km tiles for mapping and ~7 km segments for sampling transects) for efficient processing. Tiles are buffered to prevent edge artifacts during ground classification and creation of the Digital Terrain Model (DTM). Classification of ground returns is performed with a progressive morphological filter [23]. Delaunay triangulation is used to create a Triangulated Irregular Network (TIN) of ground hits, and the TIN is used to linearly interpolate DTM elevations on a 1 m raster grid. Additionally, the TIN is used to interpolate the base elevation of every non-ground return, and vegetation heights are computed by difference. A Canopy Height Model (CHM) is created by selecting the greatest return height in every 1 m grid cell, using these points to create a TIN and interpolating canopy heights on a 1 m raster grid. An example of the DTM and CHM models are shown in Figure 11. For mapped areas, tiles are mosaicked, and terrain slope and aspect, canopy rugosity (standard deviation of height) and common LiDAR metrics [24,25] are derived from the resulting data at the scale of a US Forest Service Forest Inventory and Analysis (FIA) field subplot (7.32 m radius).

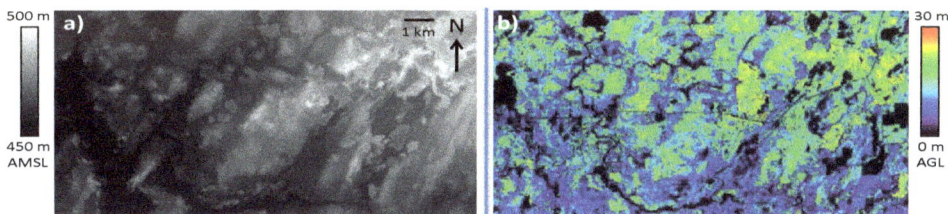

Figure 11. (**a**) Digital Terrain Model (DTM) and (**b**) Canopy Height Model (CHM) products for a 9 × 14 km study area in the Chequamegon-Nicollet National Forest, Park Falls District, WI, USA (5 June 2012; 45.9447°N 90.2519°W). The DTM shows lakes, rivers and landforms formed by glacial scouring and transported deposits during the Wisconsin glaciation, and variations in the CHM reflect a diversity of ecosystems and land use management practices.

5.2.3. Imaging Spectrometer Data

Pre-processing of the image spectrometer data requires the use of XCAP software (EPIX, Inc., Buffalo Grove, IL, USA) to convert L0 data in a proprietary VIF format to ASCII time stamps and an image cube in standard Band Interleaved by Line (BIL) format. Calibration coefficients (see below) are applied to both the downwelling irradiance data and upwelling imaging spectrometer data to compute at-sensor radiance (Figure 6). Prior to computing at-sensor reflectance, irradiance spectra are resampled spectrally to match the imaging spectrometer. Two reflectance products are computed as part of the workflow: at-sensor reflectance is computed as the fraction of irradiance detected by the imaging spectrometer, and surface reflectance is computed using the radiometric data and an atmospheric correction algorithm. Following the reflectance calculations, data are aggregated from the 1.5 nm native sampling interval using a Gaussian-based spectral resampling procedure to generate a distribution product with uniform 5 nm FWHM band spacing. This is a finer sampling and spectral resolution than is reported for NASA's Next Generation Airborne Visible/Infrared Imaging Spectrometer (AVIRISng) [26] and may be useful for measuring fine spectral features, such as solar induced fluorescence [27]. From the reflectance data, common vegetation indices for greenness (e.g., Normalized Difference Vegetation Index, (NDVI); Enhanced Vegetation Index, (EVI); red edge), light use efficiency (e.g., Photochemical Reflectance Index) and leaf pigments (e.g., carotenoid, anthocyanin) are computed.

Each pixel in the image cube is georeferenced using Geo-Correction for Airborne Platforms (GCAP) [28], a software package developed at NASA Goddard Space Flight Center that provides the user the capability to georeference a raster image using the image time stamps and GPS-INS data. Pixel coordinates are used to build a geographic lookup table in ENVI that is distributed with the radiance data cube and is used to create georeferenced layers for the reflectance data, vegetation indices and ancillary data products.

Ancillary data products include valuable information for each pixel of the image cube, including: acquisition time, solar zenith angle, solar azimuth, incoming PAR, cloudiness index [21], aircraft elevation, aircraft altitude AGL, view angle and view azimuth. These time-space variables are computed from the LiDAR-derived DTM and a clear sky irradiance model [29]. The products will help end users create additional higher-order products (e.g., solar and view angles are needed for computing bidirectional corrected reflectance; terrain and canopy heights are needed for orthorectification) and facilitate scientific data analysis (e.g., cloudiness and solar illumination conditions affect canopy photosynthesis).

Future development will focus on improved co-registration through automated control points and implementing orthorectification and atmospheric correction algorithms specifically designed for low altitude, wide FOV airborne imagers [30].

5.2.4. Thermal Data

By selecting a single line of cross-track pixels from the thermal image array, the thermal data can be treated as a line imager in the same manner as the imaging spectrometer. However, the number of cross track pixels in a single line and frame rate limits the product to a coarse spatial resolution (~3 m) using this method. Since the field of view of the full array is much greater than the spacing between images, we can use spatial oversampling techniques to create a smoothed, fine resolution (1 m) temperature product. We might also use additional information from the LiDAR and imaging spectrometer (e.g., land cover) to resolve non-linear edge effects, such as abrupt land cover boundaries [31]. Both of these methods for deriving fine-resolution surface temperature products are currently under development.

5.2.5. Profiling LiDAR Data

Level 0 profiling LiDAR data is captured in ASCII format through a 10/100 Mbit TCP/IP port, and can be processed in the same manner as PALS data [16].

5.3. G-LiHT Data Distribution

NASA's Earth Science Program promotes the full and open sharing of data with all users in accordance with NASA's Data and Information Policy [32], and this includes G-LiHT data products. Every effort is taken by the G-LiHT instrument team to ensure that accurate, well-calibrated data is released in a timely manner, and data is distributed in common, readily usable file formats. Classified LiDAR returns and feature heights are made available in ASPRS LAS file format (American Society for Photogrammetry and Remote Sensing, LASer file format), a non-proprietary, binary file industry standard [22]. This open data file format allows for the raw data and information specific to data collection to be recorded (e.g., scan angle, return number, classification, AGL height). Gridded LiDAR products are made available as GeoTIFF files that conform to established Tagged Image File Format (TIFF) interchange format for georeferenced raster imagery [33]. GeoTIFF files include geographic metadata formally, using compliant TIFF tags and structures. Multiband products from the imaging spectrometer will be distributed in standard Band Interleaved by Line (BIL) format. In addition to the data files, quick look images of key spatial variables (e.g., aircraft trajectory, DTM, CHM, true color image) created in Keyhole Markup Language (KML) [34] for visualization in Google Earth.

The data archive for G-LiHT products can be accessed through the G-LiHT webpage, http://gliht.gsfc.nasa.gov or by directly connecting to the anonymous ftp site, fusionftp.gsfc.nasa.gov/G-LiHT. The G-LiHT website contains additional instrument specifications, a description of various airborne campaigns, software tools and links to open source data analysis and visualization software. G-LiHT has been used to collect more than 6,500 km² of data for NASA-sponsored studies across a broad range of ecoregions in the USA during 2011–2012 (Figure 12, [35–37]), with plans to collect data in Mexico and interior Alaska in 2013–2014. Our automated data processing and distribution system is designed to give scientists open access to both low- and high-level data products (http://gliht.gsfc.nasa.gov), which will stimulate community development of synergistic data fusion algorithms.

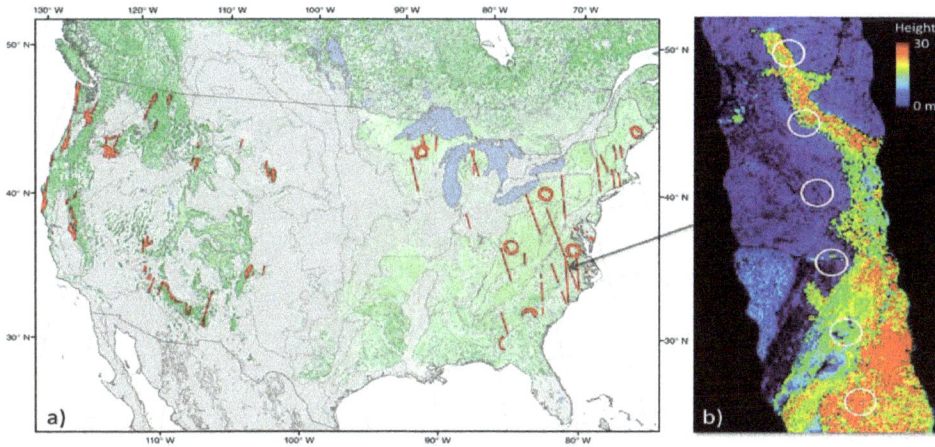

Figure 12. G-LiHT coverage (show as red lines) as of September 2012 (**a**). Long transects are coincident G-LiHT data along ICESat GLAS LiDAR tracks, enlarged in (**b**), which capture fine scale heterogeneity in large, space-base LiDAR footprints (65 m diameter). Other collections in the USA include Forest Inventory and Analysis (FIA) plots and intensive study sites. The green color in (a) represents percent forest cover [35,36]; dark gray lines indicate country borders, and light gray lines delineate terrestrial ecoregions [37].

6. Conclusions

Goddard's LiDAR, Hyperspectral and Thermal (G-LiHT) Airborne Imager is a unique system that permits simultaneous measurements of vegetation structure, foliar spectra and surface temperatures at very high spatial resolution (~1 m). The complementary nature of LiDAR, optical and thermal data provide an analytical framework for the development of new algorithms to map plant species composition, plant functional types, biodiversity, biomass and carbon stocks and plant growth. G-LiHT data will enhance our ability to validate data from existing satellite missions, design new missions and produce data products related to biodiversity and climate change.

The scientific rationale and motivation for G-LiHT is similar to other multi-sensor systems (e.g., Carnegie Airborne Observatory, CAO, Alpha and Beta Systems [3]; CAO Airborne Taxonomic Mapping System, AToMS [2]; National Ecological Observatory's Airborne Observation Platform, NEON AOP [4]), and differences largely exist due to specific mission objectives. G-LiHT was specifically designed to simplify worldwide deployment and minimize collection and data processing costs by using commercial off-the-shelf instruments and local general aviation aircraft. In contrast, NEON AOP is an operational observatory that will acquire annual acquisitions over selected NEON sites in the US and additional in-country flights requested by the scientific community. CAO systems have largely focused on sustainable forest management and habitat conservation in global tropical forests, but unlike G-LiHT and NEON, the data is not openly distributed. Other differences between these systems include choice of instrumentation. NEON and CAO systems use a custom-built, high fidelity visible to shortwave infrared spectrometer (380 to 2,510 nm) that covers a wider range of the electromagnetic spectrum and has a greater signal-to-noise ratio than the G-LiHT spectrometer (400 to 1,000 nm), but the instrument cost is one-hundred-fold greater, and a more capable aircraft is needed to accommodate the greater size, mass and power requirements. Additionally, G-LiHT is the only system that currently acquires downwelling irradiance and surface temperature measurements. Comparisons are currently under way at locations where both NEON and G-LiHT data have been collected to evaluate the performance and cross-calibrate the LiDAR and spectrometer data products.

Additional G-LiHT instrument specifications, campaign information and access to more than 6,500 km² of data can be obtained from the through the G-LiHT webpage, http://gliht.gsfc.nasa.gov.

Acknowledgments: This research was funded in part by NASA Goddard Space Flight Center's Internal Research and Development program and NASA's Terrestrial Ecology, Carbon Cycle and Carbon Monitoring System programs. The authors wish to thank NASA Langley Research Center, who provided engineering and aircraft support throughout this project, Riegl USA and Headwall Photonics, who provided technical assistance with instrument installation and testing, and Timothy Creech and Joshua Bronston for their participation in the development of GCAP.

Disclaimer of Endorsement: References in this manuscript to any specific commercial products, processes or services or the use of any trade, firm or corporation name are for the information and convenience of the reader and do not constitute endorsement, recommendation or favoring by the US government or National Aeronautics and Space Administration.

Conflicts of Interest: The authors declare no conflict of interest.

References and Notes

1. Goetz, S.; Dubayah, R. Advances in remote sensing technology and implications for measuring and monitoring forest carbon stocks and change. *Carbon Manag.* **2011**, *2*, 231–244.
2. Asner, G.P.; Knapp, D.E.; Boardman, J.; Green, R.O.; Kennedy-Bowdoin, T.; Eastwood, M.; Martin, R.E.; Anderson, C.; Field, C.B. Carnegie Airborne Observatory-2: Increasing science data dimensionality via high-fidelity multi-sensor fusion. *Remote Sens. Environ.* **2012**, *124*, 454–465.
3. Asner, G.P.; Knapp, D.E.; Kennedy-Bowdoin, Jones, M.O.; Martin, R.E.; Boardman, J.; Field, C.B. Carnegie Airborne Observatory: In-Flight fusion of hyperspectral imaging and waveform light detection and ranging (wLiDAR) for three dimensional studies of ecosystems. *J. Appl. Remote Sens.* **2007**, *1*, 013536.
4. Kampe, T.U.; Johnson, B.R.; Kuester, M.; Keller, M. NEON: The First continental-scale ecological observatory with airborne remote sensing of vegetation canopy biochemistry and structure. *J. Appl. Remote Sens.* **2010**, *4*, 043510.
5. Chambers, J.Q.; Asner, G.P.; Morton, D.C.; Anderson, L.O.; Saatchi, S.S.; Espírito-Santo, F.D.B.; Palace, M.; Souza, C. Regional ecosystem structure and function: Ecological insights from remote sensing of tropical forests. *Trends Ecol. Evol.* **2007**, doi:10.1016/j.tree.2007.05.001.
6. Anderson, M.C.; Norman, J.M.; Kustas, W.P.; Houborg, R.; Starks, P.J.; Agam, N. A thermal-based remote sensing technique for routine mapping of land-surface carbon, water and energy fluxes from field to regional scales. *Remote Sens. Environ.* **2008**, *112*, 4227–4241.
7. Kustas, W.; Anderson, M. Advances in thermal infrared remote sensing for land surface modeling. *Agric. For. Meteorol.* **2009**, *149*, 2071–2081.
8. Gamon, J.A.; Penuelas, J.; Field, C.B. A narrow-waveband spectral index that tracks diurnal changes in photosynthetic efficiency. *Remote Sens. Environ.* **1992**, *41*, 35–44.
9. Gitelson, A.A. Nondestructive Estimation of Foliar Pigment (Chlorophylls, Carotenoids, and Anthocyanins) Contents: Evaluating a Semi Analytical Three-Band Model. In *Hyperspectral Remote Sensing of Vegetation*; Thenkabail, P.S., Lyon, J.G., Huete, A., Eds.; Taylor and Francis: New York, NY, USA, 2011; pp. 141–166.
10. Kokaly, R.F.; Asner, G.P.; Ollinger, S.V.; Martin, M.E.; Wessman, C.A. Characterizing canopy biochemistry from imaging spectroscopy and its application to ecosystem studies. *Remote Sens. Environ.* **2009**, *113*, S78–S91.
11. Middleton, E.M.; Huemmrich, K.F.; Cheng, Y.-B.; Margolis, H.A. Spectral Bio-Indicators of Photosynthetic Efficiency and Vegetation Stress. In *Hyperspectral Remote Sensing of Vegetation*; Thenkabail, P.S., Lyon, J.G., Huete, A., Eds; Taylor and Francis: New York, NY, USA, 2011; pp. 265–288.
12. Ustin, S.L.; Roberts, D.A.; Gamon, J.A.; Asner, G.P.; Green, R.O. Using imaging spectroscopy to study ecosystem processes and properties. *Bioscience* **2004**, *54*, 523–534.

13. Bergen, K.M.; Goetz, S.J.; Dubayah, R.O.; Henebry, G.M.; Hunsaker, C.T.; Imhoff, M.L.; Nelson, R.F.; Parker, G.G.; Radeloff, V.C. Remote sensing of vegetation 3-D struture for biodiversity and habitat: Review and implications for lidar and radar spaceborne missions. *J. Geophys. Res.* **2009**, *114*, G00E06.

14. Cook, B.D.; Bolstad, P.V.; Næsset, E.; Anderson, R.S.; Garrigues, S.; Morisette, J.; Nickeson, J.; Davis, K.J. Using LiDAR and Quickbird data to model plant production and quantify uncertainties associated with wetland detection and land cover generalizations. *Remote Sens. Environ.* **2009**, *113*, 2366–2379.

15. Berni, J.A.; Zarco-Tejada, P.J.; Sepulcre-Canto, G.; Fereres, E.; Villalobos, F. Mapping canopy conductance and CWSI in olive orchards using high resolution thermal remote sensing imagery. *Remote Sens. Environ.* **2009**, *113*, 2380–2388.

16. Nelson, R.; Parker, G.; Hom, M. A Portable airborne laser system for forest inventory. *Photogramm. Eng. Remote Sensing* **2003**, *69*, 267–273.

17. Land, Vegetation, and Ice Sensor (LVIS). Available online: http://lvis.gsfc.nasa.gov (accessed on 24 April 2013).

18. NIST SIRCUS. Available online: http://www.nist.gov/pml/div685/grp06/sircus.cfm (accessed on 24 April 2013).

19. Brown, S.W.; Eppeldauer, G.P.; Lykke, K.R. Facility for spectral irradiance and radiance responsivity calibrations using uniform sources. *Appl. Opt.* **2006**, *45*, 8218–8237.

20. Processing Levels. NASA's Earth Observing System Data and Information System (EOSDIS). Available online: http://earthdata.nasa.gov/data/standards-and-references/processing-levels (accessed on 23 April 2013).

21. Cook, B.D.; Bolstad, P.V.; Martin, J.G.; Heinsch, F.A.; Davis, K.J.; Wang, W.; Desai, A.R.; Teclaw, R.M. Using light-use and production efficiency models to predict photosynthesis and net carbon exchange during forest canopy disturbance. *Ecosystems* **2008**, *11*, 26–44.

22. LASer (LAS) File Format Exchange Activities. Available online: http://asprs.org/Committee-General/LASer-LAS-File-Format-Exchange-Activities.html (accessed on 24 April 2013).

23. Zhang, K.; Chen, S.; Whitman, D.; Shyu, M.; Yan, J.; Zhang, C. A progressive morphological filter for removing nonground measurements from airborne LIDAR data. *IEEE. Trans. Geosci. Remot. Sens.* **2003**, *41*, 872–882.

24. Evans, J.; Hudak, A.; Faux, R.; Smith, A.M. Discrete return lidar in natural resources: Recommendations for project planning, data processing, and deliverables. *Remote Sens.* **2009**, *1*, 776–794.

25. Næsset, E. Predicting forest stand characteristics with airborne scanning laser using a practical two-stage procedure and field data. *Remote Sens. Environ.* **2002**, *80*, 88–99.

26. Hamlin, L.; Green, R.O.; Mouroulis, P.; Eastwood, M.; Wilson, D.; Dudik, M.; Paine, C. In *AERO'11*, Imaging Spectrometer Science Measurements for Terrestrial Ecology: AVIRIS and New Developments. In Proceedings of the 2011 IEEE Aerospace Conference, Big Sky, MT, USA, 5–12 March 2011; pp. 1–7.

27. Corp, L.A.; Middleton, E.M.; McMurtry, J.E.; Campbell, P.K.E.; Butcher, L.M. Fluorescence sensing techniques for vegetation assessment. *Appl. Opt.* **2006**, *45*, 1023–1033.

28. Goddard Space Flight Center—Innovative Partnerships Program Office. Available online: http://techtransfer.gsfc.nasa.gov (accessed on 25 April 2013).

29. Bird, R.E.; Hulstrom, R.L. *A Simplified Clear Sky Model for Direct and Diffuse Insolation on Horizontal Surfaces*; Solar Energy Research Institute Technical Report; SERI/TR-642-761; Solar Energy Research Institute: Golden, CO, USA, 1981.

30. Remote Sensing Applications by ReSe. Available online: http://www.rese.ch/index.html (accessed on 24 April 2013).

31. Limaye, A.; Crosson, W.L.; Laymon, C.A.; Njoku, E.G. Land cover-based deconvolution of PALS L-band microwave brightness temperatures. *Remote Sens. Environ.* **2004**, *92*, 497–506.

32. Data & Information Policy-NASA Science. Available online: http://science.nasa.gov/earth-science/earth-science-data/data-information-policy/ (accessed on 23 April 2013).

33. GeoTIFF. Available online: http://trac.osgeo.org/geotiff/ (accessed on 24 April 2013). Keyhole

34. Markup Language: Google Developers. Available online: https://developers.google.com/kml/ (accessed on 24 April 2013).

35. Multi-Resolution Land Characteritcs Consortium (MRLC). Available online: http://www.mrlc.gov/nlcd06_ref.php (accesssed on 21 May 2013).

36. Earth Observation for Sustainable Development of Forests. Available online: https://pfc.cfsnet.nfis.org/mapserver/eosd_portal/htdocs/eosd-cfsnet.phtml (accessed on 21 May 2013).
37. Olson, D.M.; Dinerstein, E.; Wikramanayake, E.D.; Burgess, N.D.; Powell, G.V.N.; Underwood, E.C.; D'Amico, J.A.; Itoua, I.; Strand, H.E.; Morrison, J.C.; *et al.* Terrestrial ecoregions of the world: A new map of life on Earth. *Bioscience* **2001**, *51*, 933–938.

remote sensing

MDPI

Article

Capability Assessment and Performance Metrics for the Titan Multispectral Mapping Lidar

Juan Carlos Fernandez-Diaz [1,2,*], William E. Carter [1,2], Craig Glennie [1,2], Ramesh L. Shrestha [1,2], Zhigang Pan [1,2], Nima Ekhtari [1,2], Abhinav Singhania [1,2], Darren Hauser [1,2] and Michael Sartori [1,2]

[1] National Center for Airborne Laser Mapping (NCALM), Houston, TX 77204, USA; carter4451@bellsouth.net (W.E.C.); clglennie@uh.edu (C.G.); rlshrestha@uh.edu (R.L.S.); panzhigang520@gmail.com (Z.P.); nima.ekhtari@gmail.com (N.E.); asinghan@central.uh.edu (A.S.); dlhauser@central.uh.edu (D.H.); msartori@central.uh.edu (M.S.)
[2] Department of Civil and Environmental Engineering, University of Houston, Houston, TX 77204, USA
* Correspondence: jfernan4@central.uh.edu; Tel.: +1-832-842-8884

Academic Editors: Jie Shan, Juha Hyyppä, Guoqing Zhou, Wolfgang Wagner and Prasad S. Thenkabail
Received: 17 August 2016; Accepted: 3 November 2016; Published: 10 November 2016

Abstract: In this paper we present a description of a new multispectral airborne mapping light detection and ranging (lidar) along with performance results obtained from two years of data collection and test campaigns. The Titan multiwave lidar is manufactured by Teledyne Optech Inc. (Toronto, ON, Canada) and emits laser pulses in the 1550, 1064 and 532 nm wavelengths simultaneously through a single oscillating mirror scanner at pulse repetition frequencies (PRF) that range from 50 to 300 kHz per wavelength (max combined PRF of 900 kHz). The Titan system can perform simultaneous mapping in terrestrial and very shallow water environments and its multispectral capability enables new applications, such as the production of false color active imagery derived from the lidar return intensities and the automated classification of target and land covers. Field tests and mapping projects performed over the past two years demonstrate capabilities to classify five land covers in urban environments with an accuracy of 90%, map bathymetry under more than 15 m of water, and map thick vegetation canopies at sub-meter vertical resolutions. In addition to its multispectral and performance characteristics, the Titan system is designed with several redundancies and diversity schemes that have proven to be beneficial for both operations and the improvement of data quality.

Keywords: airborne laser scanning; mapping lidar; multispectral lidar; lidar bathymetry; lidar accuracy; lidar range resolution; active imagery

1. Introduction

Over the past two decades airborne mapping light detection and ranging (lidar), also known as airborne laser scanning (ALS), has become one of the prime remote sensing technologies for sampling the Earth's surface and land cover in three dimensions (3D), especially in areas covered by vegetation canopies [1]. In addition to the range (spatial) information derived from time-of-flight (ToF) measurements, pulsed airborne lidar sensors deliver an arbitrarily scaled measure of the strength of the optical backscattered signal that is proportional to the radiance incident on the detector, typically referred to as intensity. Intensity is correlated with a target's reflectance at the given laser wavelength, making it useful in the interpretation of the lidar spatial information [2] or as a standalone data source for identifying the characteristics of the likely backscattering surface for each return [3]. The intensity also depends on other target characteristics such as roughness and the lidar cross-section and on sensor-target geometry parameters including range and incidence angle.

Currently the general practice is to digitally scale the "intensity" signal to normalize it to a given range. However, the quality of intensity information can be improved through geometric and radiometric corrections for other factors such as incidence angle, and atmospheric attenuation [4–6]. Taken a step further, radiometric calibration methods transform intensities to physical quantities such as target reflectance [7], allowing interpretation with respect to known material spectra.

In the past, intensity information has been used for various applications, including land cover classification [3,5]; enhancement of lidar ground return classification [8,9]; fusion with multispectral and hyperspectral data to enable a better characterization of terrain or land cover [10]; derivation of forest parameters and tree species mapping [11–13]; and production of greyscale stereo-pair imagery to generate breaklines traditionally used in photogrammetry through a technique named lidargrammetry [14]. However, the usefulness of lidar intensity information has been fundamentally limited because it provides a measure of backscatter at a single, narrow laser wavelength band. This is usually a near-infrared (NIR) wavelength (1064 or 1550 nm) for topographic lidar systems and a green-blue wavelength [15] for bathymetric lidar systems. Currently, the second harmonic of neodymium-doped yttrium aluminum garnet (Nd:YAG) lasers, 532 nm, is a common choice for bathymetric lidars [16]. This single wavelength spectral limitation has been previously recognized and several experiments have attempted to mitigate it by combining data obtained from individual sensors that operate at different laser wavelengths [17,18] or by observations from prototype multispectral lidar systems [19–21]. These multispectral lidar experiments are testament to the potential of this newly developing remote sensing technique, and as with any other technology, the potential is coupled with challenges. Some of these challenges are related to physical principles (atmospheric transparency, background solar radiation, etc.) and hardware limitations (available laser wavelengths, eye safety) [22], while other limitations are related to software and algorithms (e.g., radiometric calibration of the raw lidar intensities) [17].

This paper presents a general overview of the design and performance of the first operational multispectral airborne laser scanner that collects range, intensity and optionally full waveform return data at three different laser wavelengths (1550, 1064, 532 nm) through a single scanning mechanism. This airborne multispectral lidar scanner is the Teledyne Optech Titan MW (multi-wavelength) which was developed based on the specifications and technical requirements of the National Science Foundation (NSF) National Center for Airborne Laser Mapping (NCALM). The system was delivered to the University of Houston (UH) in October of 2014. Since then, the sensor has undergone significant testing, improvement and fine tuning in a wide range of environments [23].

This paper is intended to highlight the flexibility of the Titan to perform 3D and active multispectral mapping for different applications (bathymetry, urban mapping, ground cover classification, forestry, archeology, etc.) without delving too deeply into one specific application. Individual papers that provide more details for specific applications are currently in preparation. The basic research question addressed in this paper relates to the performance of the Titan system, which, given its operational flexibility, has to be assessed using a variety of metrics that vary according to the application. The performance metrics discussed within this work are: (a) the accuracy of experimental ground cover classification based on un-calibrated multispectral lidar return intensity and structural metrics in an urban environment (Houston, TX, USA); (b) maximum water penetration, and (c) accuracy of the measured water depths under ideal bathymetric mapping conditions (Destin, FL, USA, and San Salvador Island, Bahamas); (d) canopy penetration; (e) range resolution in tropical rain forests (Guatemala, Belize and Mexico); and (f) precision and accuracy of topographically derived elevations.

It is important to clarify that since the delivery of NCALM's original Titan system, the manufacturer has produced other Titan sensors; however, not all units have the same engineering specifications or system design. The design of the sensor is flexible, allowing the exact configuration and performance of the unit to be adapted to the specific needs of a customer. The discussion presented below is specific to the design and performance of the NCALM unit and may not be applicable or reproducible for other Titan units, even though they carry the same make and model designation.

This paper is structured as follows: Section 2 presents a high-level description of the Titan system design and operational characteristics; Section 3 presents the results and discussion of performance tests related to (a) ground cover classification based on multispectral intensity; (b) bathymetric capabilities; (c) canopy penetration and characterization; (d) redundancy and diversity design schemes; and (e) vertical positional precision and accuracy; finally, conclusions are presented in Section 4.

2. Titan Instrument Description

The first Titan MW lidar sensor (Serial number 14SEN/CON340) was developed to meet operational specifications and requirements established by NCALM. The specifications called for a multipurpose integrated multichannel/multispectral airborne mapping lidar unit, with an integrated high resolution digital camera that could seamlessly map terrain and shallow water bathymetry environments from flying heights between 300 and 2000 m above ground level (AGL). NCALM's operational experience with airborne lidar units operating at 1064 and 532 nm wavelengths, and the requirement to perform simultaneous terrestrial and bathymetry mapping, determined two of the three laser wavelengths. There were several laser wavelengths considered for the third channel, including 950 and 1550 nm; however, the 1550 nm option was selected because of the ease in complying with eye safety regulations, the proven reliability of the laser sources, and because it results in nearly equally spaced (500 nm) spectral sampling wavelengths when combined with the 1064 and 532 nm wavelengths (Figure 1).

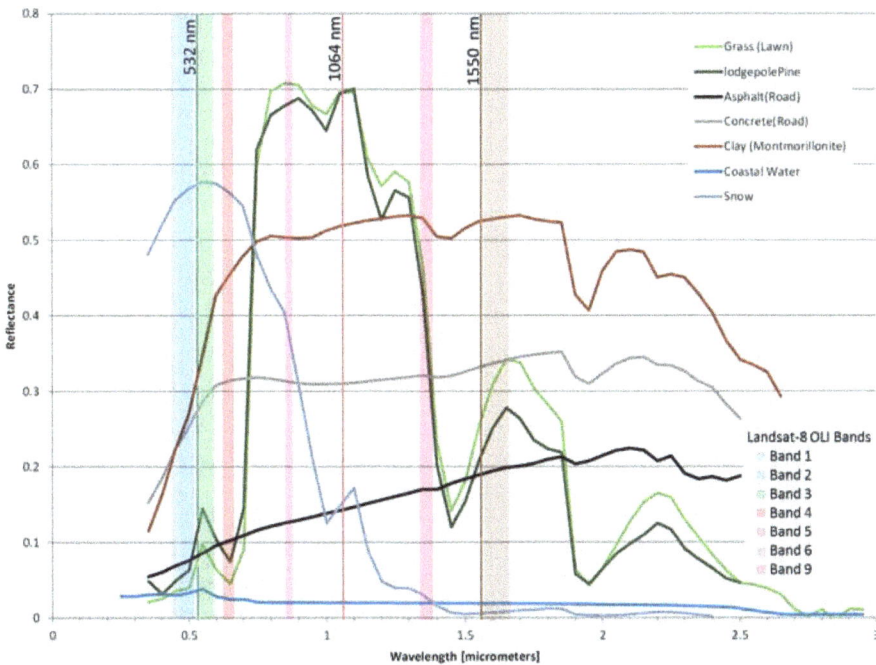

Figure 1. The Titan's operational wavelengths with reference to reflectance spectra of different land cover features and the Landsat 8 Operational Land Imager (OLI) passive imaging bands.

NCALM's Titan has two fiber laser sources. The first source "Laser A" has a primary output at 1064 nm. Part of the 1064 nm output is directly used as channel two for the lidar and another part of the output is passed through a frequency-doubling crystal to obtain 532 nm wavelength pulses for channel three of the lidar unit. The second laser source, "Laser B", has its output at the 1550 nm wavelength and is used as the source for lidar channel one. Both lasers are synchronized and can produce pulse rates between 50 and 300 kHz, programmable at 25 kHz intervals. The laser output of the Titan corresponds to Class IV as per United States Food and Drug Administration, 21 Code of Federal Regulations 1040.10 and 1040.11; International Electrotechnical Commission 60825-1. The characteristics of the individual laser sources as well as other characteristics of each lidar channel required for assessing system behavior and performance are presented in Table 1.

Table 1. Performance specifications of each of the Titan's channels.

	Channel 1	Channel 2	Channel 3
Laser Wavelength (nm)	1550	1064	532
Look angle (degrees)	3.5 forward	nadir	7.0 forward
Pulse Repetion Frequency (kHz)	50–300	50–300	50–300
Beam Divergence (mRad)	~0.36	~0.3	~1.0
Pulse Energy (µJ)	50–20	~15	~30
Pulse Width (ns)	~2.7	3–4	~3.7

Other equipment providers offer single-pass multispectral lidar systems by mounting, either in tandem or side by side, individual sensor heads operating at different wavelengths [22]. However, for the Titan system, we required the operation of the different wavelengths' lasers through a single scanning mechanism to provide near-identical swaths from the three wavelength channels. The channels are arranged such that the 1064 nm channel points at the nadir, and the 1550 and 532 nm channels are pointed 3.5° and 7° forward of the nadir, respectively. The primary reason for this configuration was to minimize returns from the water surface and maximize the probability of water penetration for the 532 nm pulses. A secondary reason was to maximize correlation with legacy lidar datasets collected with a 1064 channel pointing to the nadir. The Titan scanner has a ±30° field of scan and a maximum scanner product (half scan angle × scan frequency) of 800 degrees-Hertz. The beam divergence values are close to 0.3 milliradians for the 1064 and 1550 nm channels and one milliradian for the 532 nm channel (see Table 1). Boresight parameters for each channel are currently determined independently, and combined with the sensor model in the manufacturer's proprietary software to obtain a geometrically correct and consistent point cloud. Individual point cloud files are generated for each flight line and channel.

The laser return signal is analyzed in real time through an analogue constant fraction discriminator (CFD) [24,25] which detects and records discrete laser returns (range and intensities) at all pulse repetition frequencies. While the system can detect a large number of returns per pulse, it only records up to four returns for each outgoing laser pulse (first, second, third and last). In addition to the analogue CFD, the outgoing and return waveforms of all the channels can optionally be digitized at a 12 bit amplitude quantization resolution and at a rate of 1 gigasample/s. Currently, this digitization can only be done for each outgoing pulse and return waveforms at a maximum PRF of 100 kHz; for higher PRFs, full waveform digitization is only performed for a decimated subset of the emitted pulses.

The Titan is capable of ranging beyond the single pulse-in-the-air limit (range ambiguity), meaning that it is able to obtain accurate ranges at high PRFs when there are several laser pulses from each channel in the air simultaneously before a return from the first emitted pulse is obtained [26,27]. The Titan is capable of measuring in a fixed multi-pulse mode, which means that the sensor needs to be aware of how many pulses are planned to simultaneously be in the air in order to compute accurate ranges. This is done in the planning phase of the data collection. If, for some reason, the actual number of pulses-in-the-air (PIA) is different than the planned value, the system will produce erroneous range values. This fixed multi-pulse capability has certain combinations of ranges and PRFs where the sensor is not able to resolve the range ambiguities and which the manufacturer calls "blind zones".

The PRFs and range regions where the Titan can work without suffering range ambiguity for a given PRF are displayed in Figure 2 as the white regions between the colored bands and labeled according the number of pulses-in-the-air (PIA) at a given time.

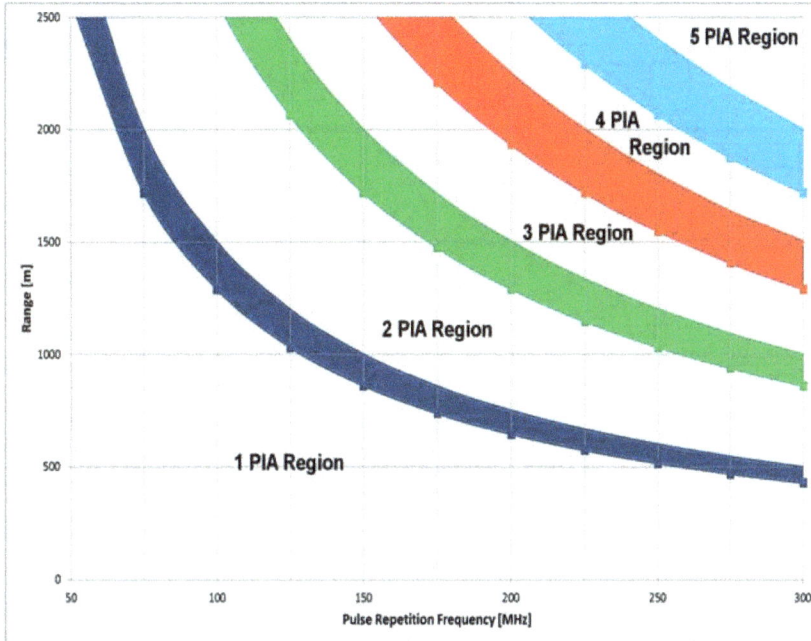

Figure 2. Pulse repetition frequency (PRF) versus range operation regions for NCALM's Titan sensor. The graph shows both regions of operation (**white**) as well as range ambiguity regions depicted as the solid colored bands.

Technically, the range ambiguity only occurs at a specific range or multiples of this range value. However, due to the wide scan angle and the forward-looking channels on the Titan (channel 1 and channel 3), the specific range at which the ambiguity occurs turns into a band of range values as it applies to the entire sensor. These blind zones, or, more accurately, ambiguity zones, are depicted as the solid color bands in Figure 2. Figure 2 also illustrates that as the PRF increases, the different PIA regions of operation get smaller. Another way of visualizing these regions of operation is to relate them to how much range variation the sensor can experience within a single flight line as the result of terrain variation or elevation of manmade structures. The higher the PRF, the less terrain relief can be tolerated by the sensor without entering into the range ambiguity regions depicted in the figure.

The laser shot densities obtainable with the Titan are mainly a function of instrument parameters (laser PRF, scan angle, scan frequency), and flying parameters (ground speed and flying height above terrain). However, eye safety regulations and range ambiguity also limit the maximum measurement density obtainable for a specific flying height. Figure 3 illustrates the laser shot density operational envelope of the Titan as a function of the flying height for a single channel and single pass (i.e., no swath overlap). The figure is for reference only and is based on the specific assumptions described below; density values outside the envelope can be obtained under specific circumstances but are not considered to be normal mapping operations.

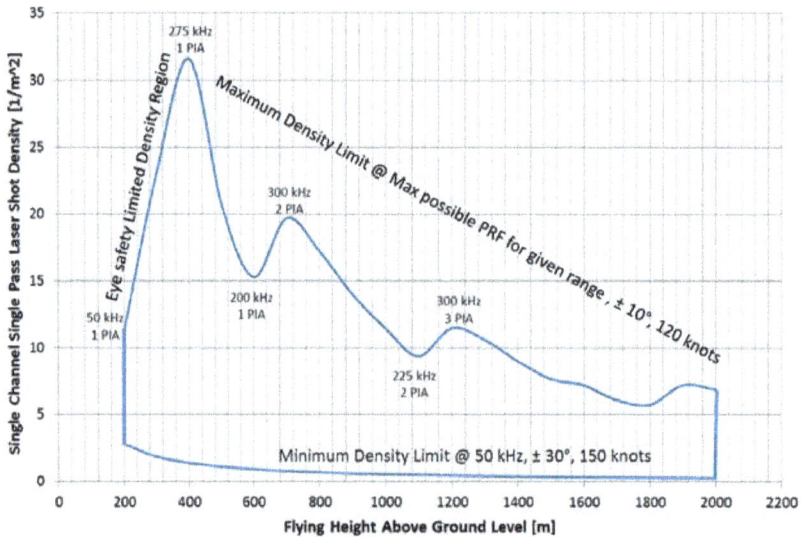

Figure 3. Laser shot density envelope for a single pass and single channel of the Titan sensor.

In Figure 3, the lower limit of the envelope is obtained by assuming the lowest PRF of 50 kHz, a ground speed of 150 knots and a scanner operating at 25 Hz at the maximum field of view ($\pm30°$). At lower flight heights the maximum density is limited by eye safety considerations. Each laser PRF has a nominal ocular hazardous distance (NOHD) that increases as the PRF increases. While it is technically possible to operate at higher PRFs at low altitudes, this does not comply with applicable eye safety regulations and thus it is necessary to use the maximum PRF that ensures exceeding the NOHD at ground level (see Figure 3). The upper limit of the envelope is obtained by assuming the highest possible PRF for a given flying height (taking into account the range ambiguity regions), a ground speed of 150 knots and a scanner operating at 70 Hz and a very narrow field of view ($\pm10°$). The peaks and valleys in the upper envelope limit are caused by the range ambiguity regions (Figure 2) given that for specific flying heights the Titan's maximum operation PRF of 300 kHz per channel falls within a range ambiguity region and thus the PRF has to be reduced to measure unambiguous ranges.

The digital camera integrated into the Titan sensor is a DIMAC Ultralight (also known as D-8900). The D-8900 is based on a charged coupled device (CCD) with 60 megapixels, each with a dimension of 6 μm × 6 μm. The pixels are arranged in an array of 8984 pixels oriented perpendicular to the flight direction and 6732 pixels along the flight direction, which translates to a CCD physical frame size of 5.39 cm × 4.04 cm. The image is formed on the focal plane through a compound lens with a nominal focal length of 70 mm. The combination of the lens and CCD array yields a total field-of-view (FOV) of 42.1 × 32.2 and a ground sample distance (GSD) of 0.0000825 × flying height. The position of the CCD is adjusted during flight by a piezo actuator to compensate for the motion of the aircraft during an exposure, reducing the pixel smear at very small GSDs. The digital camera can be triggered in a variety of modes (time interval, position) from the Titan control software, or independently through its own control software. In addition to its integrated camera, the Titan system can interface with other imaging sensors; for example, NCALM has integrated the Titan with an ITRES CASI-1500 hyperspectral camera.

Physically, the Titan consists of a sensor head and sensor control rack. The sensor rack in its base configuration consists of an electrical power unit and a control computer. Depending on what components are being operated in conjunction with the lidar sensor, the control rack can also incorporate up to three additional control computers for the waveform digitizers (one per channel) and the control computer for the digital camera. The sensor head is basically a cylinder with a box on the top; the cylinder houses the optical components (scanner, camera, lasers) and the system inertial measurement unit (IMU), and the box on top encloses the electronic components of the sensor. The cylinder has an approximate diameter of 45 cm and a height 55.5 cm. The sensor control rack is enclosed in an Edak transit case with a standard 19-in-wide rack and a footprint of 2600 cm^2. Figure 4 shows photos of the Titan sensor integrated on a DHC-6 Twin Otter airplane.

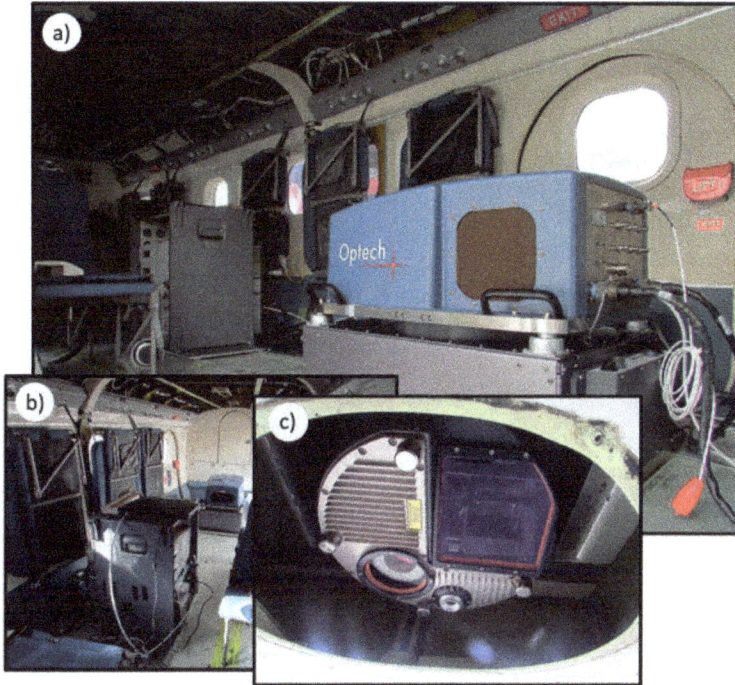

Figure 4. The Titan multispectral lidar sensor integrated into a DHC-6 Twin Otter aircraft: (**a**) Overview of installation layout from the port side of sensor head; (**b**) View from front looking aft, sensor control rack is in the foreground, sensor head in the background; (**c**) View of the sensor head through the mapping port of the aircraft. The laser output window is the rectangular window on the right, and the DIMAC camera lens is behind the circular window.

3. Field Testing of Capabilities

Since the delivery of the Titan sensor to NCALM in October 2014, the sensor has been used to collect data for more than 30 projects across the United States and in remote locations such as Antarctica and the Petén jungle in Guatemala. With over 140 h of laser-on operation, the Titan sensor has been tested in a wide variety of environments, for diverse applications and throughout its design operational envelope. Table 2 presents details of the mapping and test projects acquired with the Titan sensor. In addition to project data collection, NCALM has also performed tests to assess the performance of the sensor under different operational conditions. This paper presents results from some of these performance tests focused on ground cover classification based on multispectral lidar intensity and structural metrics in urban environments, bathymetry (maximum detectable depth and accuracy of bathymetric elevations), the ability to penetrate thick tropical forest canopies, the ability to finely resolve the vertical structure of vegetated environments (range resolution), and its vertical positional accuracy.

Table 2. Summary of projects and test collections to date with the Titan system.

Project/Test Location	Collection Year and Day of Year	Primary Application	Laser on Time (H)
Baytown, TX, USA	2014: 289, 300 2015: 47, 222 2016: 127–129	System test	
Houston, TX, USA	2014: 289	System test	0.4
Jordan, MT, USA	2014: 291	Geomorphology	0.7
Hebgen Lake, MT, USA	2014: 292	Tectonics	1.1
Big Creek River, ID, USA	2014: 293	River bathymetry	1.0
Greys River, WY, USA	2014: 295	River bathymetry	1.3
Bishop, CA, USA	2014: 296	Geomorphology	0.7
Wheeler Ridge, CA, USA	2014: 296	Geomorphology	0.8
Yucaipa, CA, USA	2014: 297	Forestry	0.9
Beaver, UT, USA	2014: 299	River morphology	0.7
McMurdo Dry Valleys, Antractica	2014: 338 to 2015: 19	Geomorphology	47.5
El Ceibal, Peten, Guatemala	2015: 77–82	Archaeology	5.5
Zacapu, Michoacan, Mexico	2015: 88	Archaeology	0.7
Angamuco, Michoacan, Mexico	2015: 88	Archaeology	0.8
Teotihuacan, Mexico	2015: 91–92	Archaeology	1.69
Laser Servicing			
Trinity River, TX, USA	2015: 219–222	River morphology	4.8
NASA JSC Clear Lake, TX, USA	2015: 223–227	Climate change resiliency	5.1
Barataria Bay, LA, USA	2015: 228–230	Marsh response to oil spill	6.8
Destin Inlet, FL, USA	2015: 231	Bathymetry test	1.5
Apalachicola, FL	2015: 232–233	Aquatic ecosystem	1.6
Redfish Bay, TX	2015: 235	Bathymetry test	0.8
Texas Gulf Coast, USA	2015: 235	Coastal morphology	0.5
Reynolds Creek, ID, USA	2015: 289, 294	Ecology	2.6
Santa Clara River, CA, USA	2015: 291–293	River morphology	3.4
Calhoun Creek, SC, USA	2016: 057	Ecology	2.2
Laser Servicing			
Campeche, Mexico	2016: 138–141	Archaeology	7.6
Lake Peak Fault, CA, USA	2016: 155	Tectonics	1
Inyo Domes, CA, USA	2016: 158	Geomorphology	0.5
Monterey, CA, USA	2016: 157, 159	Urban spectral classification	0.7
Bastrop, TX, USA	2016: 161	Orthophotos	0.3
NorthWestern Belize	2016: 184–186	Archaeology, Geomorphology	5.0
San Salvador, Bahamas	2016: 189–191	Island hydrology	7.0
Mayan Biosphere Reserve, Peten, Guatemala	2016: 197–207	Archeaology, Ecology	23.7

3.1. Multispectral Capabilities

The Titan senor was designed as a flexible multi-purpose and multi-application system; therefore, it may not outperform systems that were designed exclusively for bathymetry or for topographic mapping from high altitudes. However, one application for which it can excel is providing high resolution active multispectral data derived from lidar intensity. In theory and in practice, multispectral lidar intensity can be used for many applications, including return/target classification [28,29]; individual tree identification, parameterization and classification [30]; water/land interface identification; and archaeological feature detection [31], among many others. Several years will go by before scientists in each of the above-mentioned fields will have an opportunity to assess the utility of the Titan multispectral lidar datasets as compared to what has been theorized or experimented so far. For brevity, this paper will only briefly describe the use of Titan multispectral data to generate active lidar intensity images, describe the advantages and limitations of these intensity products and provide an example of ground cover classification using Titan data.

It does not take much imagination to realize that the simplest way to use multispectral lidar intensity is to enhance the visualization of lidar data, just as single-wavelength intensity values have been used in the past. As such, the most basic way of utilizing the multispectral intensity is for rendering false colored point clouds, where the intensity of each return is used to assign hue saturation and brightness values. In this application, the hue and saturation are assigned based on the laser wavelength and the lidar return intensity value determines or modulates the brightness value.

The advantage of this approach is that the renderings can retain the precise and complex 3D spatial nature of the lidar data despite the return locations being irregularly spaced.

Multispectral lidar intensity can also be used to generate two-dimensional (2D) active intensity images by combining the spatial and spectral information of each laser return. There are several approaches to produce intensity imagery from irregularly spaced lidar point cloud data. The first and simpler method consists of assigning a greyscale value corresponding to the lidar intensity to each return and then rendering the point cloud in two dimensions to create a top-view intensity image. A second approach, which is more elaborate and provides more control over the spatial integrity of the image, consists of interpolating the intensity values into a regular two-dimensional horizontal array (grid) using methods such as triangulation with linear interpolation, kriging, or inverse distance weighting, etc. This method is particularly useful when the horizontal distribution of the lidar returns is sparse as compared to the desired grid (raster) resolution.

When the return density is high enough to provide several laser returns per raster cell, it is possible to use a third approach. This approach consists of deriving a single intensity value for each raster cell from the intensity values of all the returns within the given raster cell. The resultant intensity raster can be produced by averaging intensity values using simple or weighted methods or determining minimum, maximum, standard deviation or any other statistical metric to characterize the returns' intensity within the raster [32]. Herein, we refer to this approach as "binning". One advantage of binning, especially when it uses some sort of averaging mechanism, is that it reduces the variability of intensity values due to factors that are difficult to account for, such as irregular incidence angles and varying surface roughness. However, the greatest advantage of this method is that it enables the generation of false color imagery from multi-wavelength lidar intensities even when the centers of the footprints at the different wavelengths may not be exactly collocated, which is the case for the Titan sensor.

Figure 5a–c show intensity images generated using the binning method for data collected with the Titan sensor over the campus of the University of Houston. When intensity information is available for three or more independent wavelengths, it is possible to generate false color RGB imagery. In the case of the Titan, with three different wavelengths, it is possible to combine the intensity images in six different arrangements. Figure 5d shows a false color RGB combination using the intensity from the 1550 nm wavelength as the red channel and the 1064 and 532 nm wavelengths' intensities for the green and blue image channels, respectively. By combining independent intensity images into a false color RGB image, it is possible to access the multidimensionality of the color space which can highlight features that perhaps are not easily identified in a single-wavelength grayscale intensity image.

In addition to using binning to generate intensity rasters, the same technique can be applied to the elevation values of the returns within a bin to generate what in this paper are referred to as "structural" images (Figure 5e–g), as they provide information on the vertical structure of the targets within a raster cell. Similar to the way the intensity images are produced, the structural images are produced by assigning a single value to the raster based on some statistical measure derived from the elevations of the returns within a cell. These structural rasters can be generated from either absolute geodetic elevations (ellipsoidal or orthometric) or relative elevations such as height above local ground. For land cover classification purposes, statistical dispersion metrics such as elevation spread and standard deviation can be computed from either absolute or relative elevation values. However, for averaged elevation metrics to be of any use for classification, they need to be computed from relative elevation above local ground values. These structural images have great potential for assisting land cover classification tasks because different kinds of targets (buildings, roads, trees, power lines, etc.) have very distinct elevation signatures. Other structural metrics can be derived not only from elevation values but from return statistics such as the number of returns detected from a given raster cell (return density) and the ratio of returns to laser shots (Figure 5g). These return metrics images can be used to segregate impervious (roads, ground, building roofs) and diffuse targets (vegetation and water). Structural metrics from different lidar wavelengths can also aid in land cover classification.

For example, a difference in mean elevation between structural images based on the 532 nm and either of the NIR wavelengths, for a given set of pixels, will indicate a potential water body, because the 532 nm wavelength may produce multiple returns from the water surface, the water column and the benthic layer, while the NIR wavelengths will detect returns only from the water surface or no returns at all.

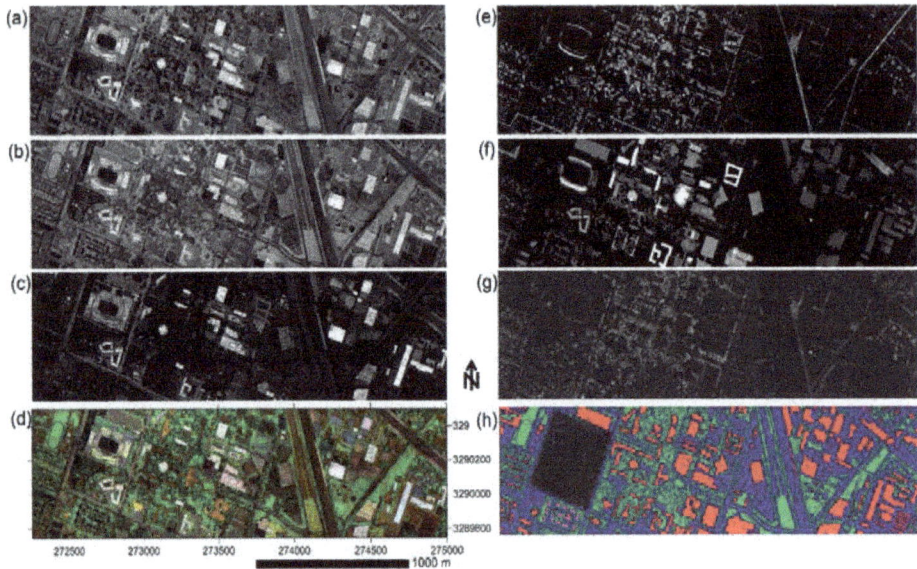

Figure 5. Intensity and structural images generated from the Titan multispectral data. (**a**) Intensity image generated from the 1550 nm channel; (**b**) intensity image for the 1064 nm channel; (**c**) intensity image for the 532 nm channel; (**d**) false color multispectral intensity image generated by using the 1550 nm intensity for the red channel and the 1064 and 532 nm intensities for the green and blue channels; (**e**) structural image based on the spread of the returns height; (**f**) structural image based on the height above ground; (**g**) structural image based on the number of returns per pulse; (**h**) ground cloud classification results map.

Active intensity and structural images have some advantages over traditional passive images; specifically they: (a) eliminate the dependency on solar illumination (imagery can be collected at night or below the cloud ceiling); (b) greatly reduce the effects of shadowing and occlusions caused by buildings and topography; (c) provide good knowledge of the illumination source (wavelength, amplitude, phase, polarization, etc.) which, in principle, should allow for easier calibration to remotely obtain target physical properties; and (d) provide images that are almost perfectly orthorectified, only limited by the positional accuracy of the lidar returns. However, there are also limitations or disadvantages of active intensity imagery, including: (a) the limited number of available laser wavelengths defined by the energy level transitions of the lasing materials and processes; (b) atmospheric attenuation of laser energy in the visible and near-infrared wavelengths by low altitude (below sensor flying level) atmospheric phenomena such as clouds, haze and fog; and (c) the target surface is only partially illuminated (spatial sampling) by the lidar beams rather than fully illuminated as in the case of passive imagery where almost the entire area (with the exception of areas with shadows) is illuminated by the sun. The active lidar intensity and structural images can be analyzed with the same techniques that are used to analyze images derived from passive optical sensors, for example to generate land cover classification maps (Figure 5h).

3.1.1. Land Cover Classification Based on Active Spectral and Structural Data

To illustrate the multispectral capabilities of the Titan sensor, this section presents and discusses results from an experiment conducted with the goal of assessing the value of lidar multispectral intensities for the purpose of ground cover classification. Data for this experiment was collected with the Titan sensor on 16 October 2014 over the campus of the University of Houston located in Houston, Texas. The collection was performed from 500 m AGL with the scanner running at full scan angle ($\pm30°$) and a PRF of 250 kHz per channel; the swaths had a lateral overlap of 50% (edge of swath over the centerline of the adjacent swath).

For this experiment, the raw intensities obtained by the sensor were only normalized by range and no further geometric or radiometric calibration or correction was performed. The intensity information from each of the Titan channels was binned into 2 m resolution images by averaging the intensity of all the returns within the bin. The raw images were then rescaled to a range of 0–255, assigning the 99th percentile of the intensity values a digital number of 255 (Figure 5a–c). In addition, five structural images with a 2 m resolution were generated by binning height values of the returns (Figure 5e–g). These structural images are based on the average height of returns above ground, the spread of the returns' height (max height–min height), the number of returns per pulse, the number of first returns and the number of total returns. All of these structural metrics were generated from Titan channel 1 data.

Five ground cover classes were targeted for classification: grass/lawn, road/parking, trees and short vegetation, commercial buildings and residential buildings. The training and validation samples used to run and assess the accuracy of the classification were selected from the training and validation datasets originally prepared for use in the 2013 Institute of Electrical and Electronics Engineers (IEEE) Geoscience and Remote Sensing Society (GRSS) data fusion contest [33]. The contest dataset was developed based on the analysis of high resolution imagery and ground verification collected in 2012, and contains samples from 15 ground cover classes. For the experiments presented here, the 15 ground classes were merged into the five classes previously described. Because of the temporal difference between the collection of the validation and the test data, some areas in the test data were masked due to the significant changes that occurred in the interim time period. Table 3 presents the number of pixels used for training and validation for each of the target classification classes.

Table 3. Detail of the number of training and validation samples used for each ground cover class.

Class	Training	Validation
Grass	169	1269
Tree	123	771
Residential	24	412
Commercial	172	1089
Road	520	1203

Eight different image stacks were generated, based on the combination of different spectral and structural images (refer to Table 4). The contents of these image stacks were varied and included stacks that contained: (a) only the five structural bands (what was available from the first-generation lidar systems without intensity measurements); (b) five structural bands plus one intensity band (what was available from the early generation of lidar systems); (c) five structural and the three intensity images (what is available with the Titan system); and (d) only the spectral information from the three Titan channels. The purpose of generating these sets of images that contained progressively more information was to assess how the quality of classification improves as more spectral information is made available, representing the technological progression of lidar systems.

Two supervised parametric classification methods were selected for the analysis: the Mahalanobis distance and the maximum likelihood classifiers. The Mahalanobis distance is the most rigorous of the minimum-distance-to-means supervised classifiers that do not consider training sample variance, while the maximum likelihood classifier not only considers the distance between training sample means but also considers the sample variance [34]. Using two different classification approaches allows an assessment of trends in the classification accuracy analysis. The results for the classification experiments are presented in Table 4. The results are consistent with initial expectations: having more independent spectral and structural data sources enables higher-accuracy classification results. With only structural information, the classification accuracy is only at 55%–65%; once spectral information from one intensity channel is added, the accuracy rises to 90% (for the maximum likelihood classifier and the image stack that contains the 1064 nm intensity). The results indicated that the intensity band that provides the most separability between the selected classes is 1064 nm. If only spectral information from the three different channels is utilized, the classification accuracy ranges between 74% and 78%. It is interesting to note that the best classification accuracy obtained (90.22%) was from an image stack that contained all structural images and only two of the intensity images (1064 and 532 nm). The best classification accuracy based on the image stack that contained all the spectral and structural images was 88.15%, which is just marginally below the overall best result.

The reason behind this reduction in the classification accuracy, when going from two to three intensity bands, has to do with the role of the intensity bands in training the classification algorithms for the residential and commercial building classes. These two classes are very similar in structural bands (most of the difference is from the height above ground band). With more intensity bands, the within-class variance of the commercial buildings increases drastically (given the wide variety of materials used in constructing or covering of the rooftops). This in turn increases the correlation between commercial and residential building classes and results in misclassification of residential buildings into the commercial buildings class and, therefore, a higher omission error for the residential buildings class and a higher commission error for the commercial buildings class.

Table 4. Accuracy assessment results from ground cover classification experiments.

Image Stack	Mahalanobis Distance		Maximum Likelihood	
	Overall Accuracy (%)	Kappa Coefficient	Overall Accuracy (%)	Kappa Coefficient
1550, 1064, 532 nm + 5 st	80.59	0.75	88.15	0.85
1550, 1064 nm + 5 st	76.94	0.7	87.27	0.83
1064, 532 nm + 5 st	82.33	0.77	90.22	0.87
1550 nm + 5 st	63.41	0.53	67.96	0.58
1064 nm + 5 st	77.16	0.71	89.89	0.87
532 nm + 5 st	60.15	0.5	80.31	0.74
5 st (only structural)	55.63	0.45	63.12	0.52
1550, 1064 and 532 nm	74.18	0.67	78.64	0.72

This classification experiment is relatively basic; however, it does showcase the usefulness and promising future of active intensity images and multispectral lidar even when the intensity values have not been calibrated or corrected to reflectance estimates. Performing more complex intensity corrections and ground cover classification experiments is outside of the scope of this paper. However, it is worth mentioning that other research groups are performing such experiments based on Titan data with very promising results [29,35], while others have been actively working on the field of radiometric calibration of single and multispectral lidar intensities [4,6,7,17,36].

3.1.2. Qualitative Multispectral Observations

Besides the analytical classification results described in Section 3.1.1, it is also possible to combine the false color active multispectral images with other structurally derived products to produce data visualizations that exploit the complementary nature of the three-dimensional spatial and three-wavelength spectral information of the Titan data. Figure 6c illustrates the combination of spectral and spatial information obtained with the Titan sensor for the Mesoamerican archaeological site of Teotihuacan in central Mexico. Figure 6c was generated by overlaying a false color image derived from the Titan's multispectral intensity (Figure 6a) over a 3D surface model derived from the lidar spatial data (Figure 6b). Figure 6a is the 2D active multispectral image where the information for the red, green and blue image channels was derived from the 1550, 532 and 1064 nm intensities, respectively.

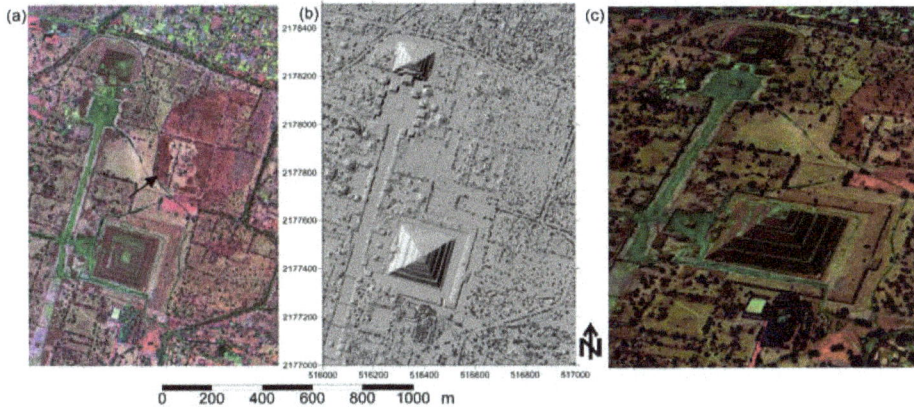

Figure 6. Spectral and spatial data products derived with the Titan sensor of the archaeological site of Teotihuacan in central Mexico. (**a**) False color multispectral lidar intensity image generated by using the 1550 nm intensity for the red channel and the 532 and 1064 nm intensities for the green and blue channels; (**b**) digital surface model (DSM) derived from the lidar spatial data; (**c**) perspective view generated by overlaying the false color multispectral intensity image over a 3D surface model based on the lidar DSM.

Renderings like the one presented in Figure 6c enable researchers to visualize the relationship between topography and spectral features in a single hybrid domain. In this case, because the archaeological site is only sparsely covered with vegetation, it is possible to visually discriminate among a multitude of surface differences. Areas with short shrubs and grasses appear in yellowish-brown tones, areas with loose gravel or bare compacted soil that are used for pedestrian traffic appear in greenish tones, and areas that are used for parking lots appear as bright red-pinkish areas. It is even possible to discriminate between vegetated areas that appear to be under different levels of water stress. One such area is the quasi-circular feature characterized by a reddish-brownish tone marked with a black arrow in Figure 6a. This type of combined height relief and spectral data and visualization are very promising for the field of archaeology, especially in deserts and sparsely vegetated areas, as illustrated here and by a previous experiment in a Roman cultural setting and European landscape [37]. As more multispectral lidar data is made available to the research community, its utility for identifying archaeological features will become more apparent.

Similar discrimination capabilities based solely on lidar intensity have been observed in other datasets. For instance, in data collected at the US McMurdo Antarctic Station it was possible to discriminate between snow that was mechanically compacted to make ice roads from snow that was not disturbed. Figure 7 illustrates this differentiation between compacted and loose snow. Figure 7c,d are active intensity images generated from channels 1 (1550 nm) and 2 (1064 nm), respectively, where the ice roads have been marked with the aid of yellow arrows. These roads are also evident in Figure 7b which is a perspective view of a 3D digital surface model that has been overlaid with a false color multispectral lidar image.

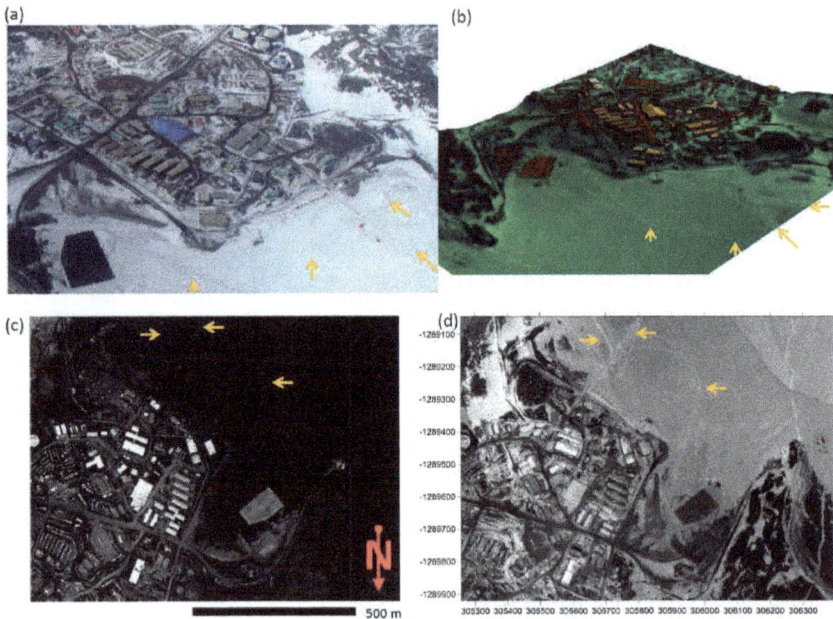

Figure 7. Potential spectral separability of loose and compacted snow, roads with compacted ice and snow are marked with yellow arrows in the figures. (**a**) Aerial oblique image of McMurdo Station at the time of lidar data collection; (**b**) Perspective view of a 3D surface model overlaid with a false color lidar intensity image; (**c**) Active intensity image generated from the 1550 nm channel; (**d**) Active intensity image generated from the 1064 nm channel.

Other spectral characteristic that will prove very useful in the future, specifically for automatic bathymetric processing, will be the ability to automatically delineate the water-land interface based solely on multispectral intensity. Several studies have already been conducted into this water-land interface delineation with single-wavelength lidar sensors [38,39], but the improvement based on multispectral intensity still needs to be investigated.

3.2. Bathymetric Capabilities

A principal design criterion for the Titan was the ability to perform seamless high resolution bathymetric and topographic mapping in areas where the terrestrial and aquatic domains overlap (marshes, rivers, lakes, coastal and estuarine). This criteria was meant to improve upon the capabilities obtained with the single-wavelength Aquarius lidar bathymetric sensor described in [16]. The Titan bathymetric channel is designed based on a performance specification of 1.5/Kd considering a flying height of 400 m above water and a benthic reflectivity of 20%. Kd is the diffuse attenuation coefficient [40] for water at the 532 nm wavelength and can be conceptually simplified as a measure of the "transparency" of the water. The 1.5/Kd performance specification establishes a theoretical limit for the water penetration capabilities of the Titan for a given benthic reflectivity and flying height. For example, if the Kd of the water body is 0.5 m^{-1}, then the maximum water penetration would be $1.5/0.5 = 3$ m. If the Kd is 0.1 m^{-1}, the maximum depths that could be mapped would be around 15 m, assuming the same bottom reflectivity and flying height.

Bathymetric projects have been performed in river, coastal and ocean environments (see Table 2). The results presented below are from experiments performed near Destin, Florida, USA, more specifically over the East Pass, which is the entrance to Choctawhatchee Bay from the Gulf of Mexico, and it is located to the west of Destin (Figure 8). The East Pass is protected from the shifting sands of the Gulf of Mexico by twin channel jetties, which were constructed by the U.S. Army Corps of Engineers (USACE). The East Pass channel is regularly dredged to maintain the mean channel depth. This is a good location for testing the bathymetric performance of the system under close to ideal conditions, meaning very clear water and a very reflective benthic layer. The East Pass is also a good test location because within a relatively small geographic area the water depth changes dramatically from very shallow to depths of more than 15 m. In addition, the constant flow of water moving in and out of the inlet produces complex sand features which are ideal for elevation accuracy assessment, and the subaquatic environment contains areas of bare sand and areas covered by sea grass which provide variation in the benthic reflectance.

The red line on Figure 8 represents the flight line track that was flown for the tests, with one end on the Gulf side of the inlet and the other on the bay side. Bathymetric testing in this area was conducted twice, the first time on 19 August 2015 and the second time on 11 May 2016. This double collection was due to the fact that, during the first data collection, it was discovered that the energy output of the 532 nm channel, which is the one responsible for the bathymetry measurement, was severely degraded. The laser source for the 1064 and 532 nm channels was eventually replaced and the test repeated. The results from the water penetration test come from the data collected in May 2016 and the data for the accuracy assessment of measured water depths comes from the August 2015 collection. It is important to note that the results presented in the sections below are derived from the discrete output of the bathymetric channel. Full waveform data were also collected during these bathymetric tests; however, processing and analysis of the waveform data is outside the scope of this paper.

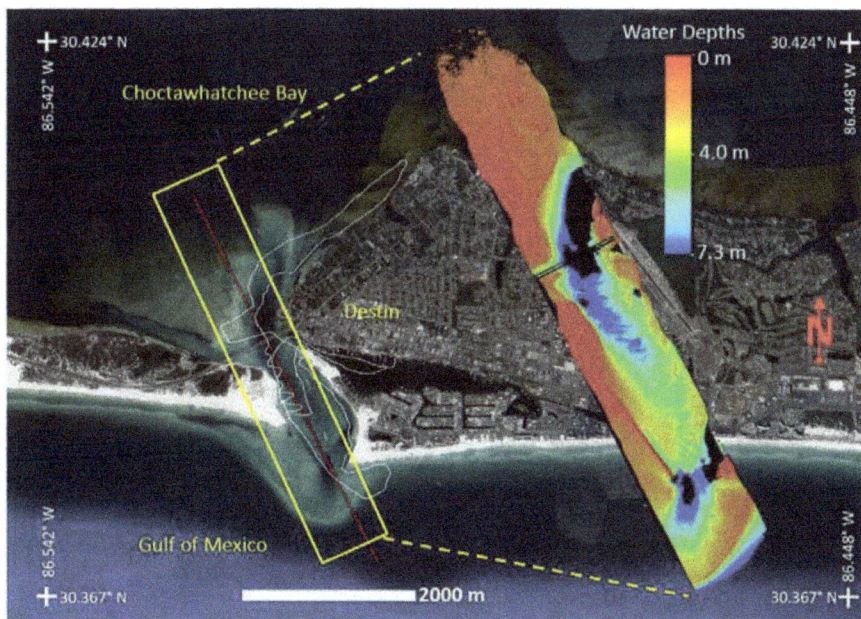

Figure 8. Illustration of the bathymetric test area: the East Pass near Destin, FL, USA. The solid red line represents the test line that was flown multiple times with different configurations. The white solid line represents the track of the validation samples obtained with an acoustic Doppler current profiler. The yellow rectangle represents the coverage of one the acquired test lines, and the bathymetric elevations derived from that test line dataset are presented as a color map that is offset to the east of the pass.

3.2.1. Maximum Water Penetration

For the water penetration tests, the flight line was acquired a total of six times from two different altitudes above the water level (300 and 500 m). These altitudes represent the upper and lower limits of the recommended bathymetric operational envelope for the Titan. Each time the line was flown, the system PRF was varied between 75 and 200 kHz. These configurations of altitudes and PRFs allow for an assessment of the bathymetric performance of the Titan across the bathymetric PRF-altitude envelope.

The results summarized in Table 5 indicate a fairly uniform performance across the PRF operational spectrum, which is a considerable improvement with respect to the Aquarius system, which displayed better penetration performance for lower PRFs [16] due to the higher energy per pulse at lower PRFs. As expected, there was a small decrease in the penetration performance as the flying height was increased from 300 to 500 m above the water. The maximum mapped depths obtained from the Titan are comparable to the ones obtained with the Aquarius sensor at its highest PRF of 70 kHz. Some deeper measurements were obtained by the Aquarius on the bay side of the pass at its lowest PRF of 33 KHz [16]. However, the Titan performance is superior to the Aquarius with respect to bathymetric return densities, doubling the Aquarius densities even when the Titan returns are spread over a wider swath. The penetration depths are lower on the bay side of the pass mainly for two reasons: (a) the Kd parameter value is higher on the bay side due to organic and other suspended solids; and (b) the presence of sea grass reduces the reflectivity of the benthic layer.

As previously discussed, the maximum detectable depth of the sensor is extremely dependent on both the water transparency and bottom reflectance. The 10.4 m obtained for the Destin test are by no means the deepest that the Titan sensor can map. On a mapping project on the island of San Salvador, Bahamas, depths greater than 15 m were detected from a flying height of 500 m and a PRF of 175 kHz for the bathymetric channel. Figure 9 presents a small sample of the bathymetric results obtained in the Bahamas. Figure 9a is a rendering of the first return point cloud for the Titan bathymetric channel, colored by flight line and intensity. This point cloud rendering illustrates segments of three overlapping flight lines. Figure 9b is an image map generated from the topographic and first-order-corrected bathymetric DEM. The map is color-coded by elevation from the water surface at the time of collection. Figure 9a illustrates that laser pulses were fired and produced water surface returns for the entire length of the flight line segment. However, bathymetric returns (Figure 9b) were only detected for a subsection of each of the swaths lengths. The bathymetric detection cutoff runs at an azimuth of about 35° east of north and it occurs at an acute angle with respect to the orientation of the flight swaths which have an azimuth of 10° east of north (Figure 9a). This bathymetric detection cutoff aligns with the San Salvador Island shelf, and the maximum detected depth along the cutoff boundary for this section of the project area was 16.8 m.

Table 5. Results from the water penetration tests under almost ideal bathymetric conditions near Destin, FL, USA.

Flying Height (m)	PRF (kHz)	Depth Cutoff (m)		Return Density (m)	
		Bay	Gulf	Bay	Gulf
300	75	5.9	10	2.8	3
300	150	5.7	10.1	5.8	6
300	200	6.0	10.4	7.5	8
500	75	5.8	9.1	2	2.1
500	150	5.8	9.6	4	4.1
500	175	5.7	9.0	4.6	5

Additional experiments that require complementary measurements, such as water turbidity, Kd values, Secchi depths and bathymetric reflectance and elevations, are necessary to rigorously determine if the Titan performs to the 1.5/Kd specification. However, the initial results from tests and mapping flights are very promising both in terms of depth penetration and bathymetric density.

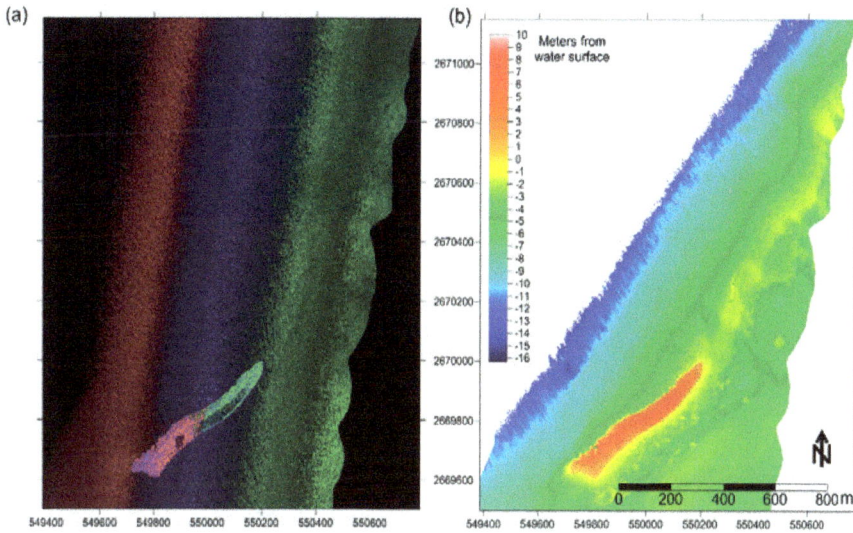

Figure 9. Small sample of a bathymetric survey surrounding the Green Cay, Bahamas: (**a**) Rendering of the point cloud of the first returns of the bathymetric channel colored by flight line and intensity; (**b**) Topographic and bathymetric color map showing water depths and island elevations.

3.2.2. Accuracy Assessment of Measured Water Depths and Bathymetric Elevations

The maximum lidar bathymetric mapping depth is just one of the important bathymetry performance metrics; another critical factor is the accuracy of the measured water depths and bathymetric elevations. The accuracy of the determined water depth is of importance, because the position of the bathymetric returns needs to be corrected for the refraction of the laser path in proportion to the distance the laser beam traveled through the water. To perform an independent water depth accuracy assessment, validation data were collected with a SonTek acoustic Doppler current profiler (ADCP) (San Diego, CA, USA) during the August 2015 bathymetric tests near Destin, FL, USA. The ADCP was mounted on a small catamaran that was towed with a pontoon boat. A geodetic-grade dual frequency marine antenna (Ashtech 700700; Sunnyvale, CA, USA) was mounted over the ADCP electronics box and connected to a Trimble NetR9 receiver to provide precise post-processed differential GPS positions using a GPS reference station no more than 10 km away (see Figure 10a). The ADCP sensor head has three transducers and takes depth and water flow measurements every 5 s. A total of 2489 measurements were collected along the boat track, the white trajectory in Figure 8. SonTek reports a depth resolution of 0.001 m and a best scenario accuracy of 1% over the range from 0.2 to 15 m.

The ADCP data were collected on 20 August 2015, while the lidar data were collected the previous day (19 August) around the same time of day, making the tidal difference negligible. The water height at the time of the airborne survey was determined based on the information from Titan channels 1 and 2. Water depths were determined by applying a 1.33 [41] first-order refraction correction (correction just for the vertical component) to the differences between the infrared channels and channel 3 (532 nm) elevations. A geographic subset contained within the main section of the East Pass was considered for the accuracy assessment. A total of 423 ADC samples and coincident lidar-derived depths were utilized. Figure 10b shows a dispersion plot of the validation and lidar depth samples; the agreement between these data sets is good with an R^2 of 0.92 and an root mean square error (RMSE) of 0.26 m. This RMSE value is consistent with other various bathymetric depth or elevation accuracy values reported for the Aquarius bathymetric Lidar system [16,42,43], and for the Experimental Advanced Airborne Research Lidar-B (EAARL-B) system [44].

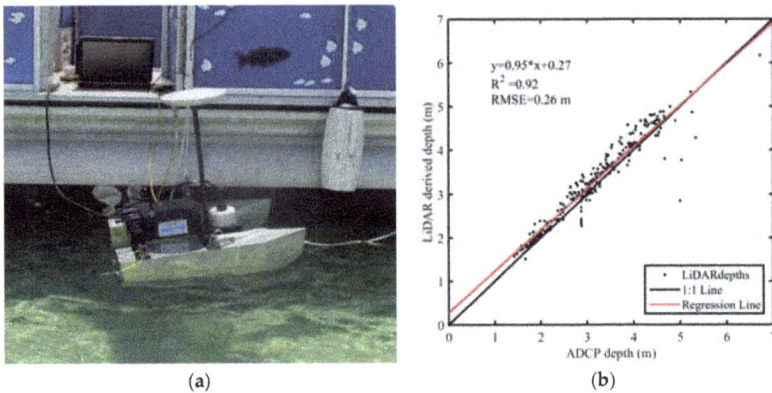

Figure 10. Bathymetric depth accuracy assessment equipment and results. (a) Photo of the SonTek acoustic Doppler current profiler (ADCP) and GPS antenna mounted on a small catamaran; (b) Dispersion plot showing results from the bathymetric depth accuracy assessment.

From Figure 10b it should be noted that there is a higher dispersion between the lidar-derived water depths and the reference ADCP data at larger depths. This is due to several factors which include the selection of the 1.33 refraction correction factor and only applying a refraction correction to the vertical component of the benthic return. The vertical component correction is a simplification that allows for a faster computation of the benthic elevations without introducing a significant vertical error for very shallow water depths. For the Destin project site, which has a gentle bathymetric slope that, on average, is less than 0.1 m/m, the maximum modeled vertical error induced by not considering the horizontal refraction correction is 0.08 m at a water depth of 10 m.

3.3. Canopy Penetration and Canopy Characterization Capabilities

ALS or airborne mapping lidar has become the de facto standard remote sensing technique to obtain ground surface elevations underneath forest and vegetation canopies and it has enabled multiple scientific applications in fields that range from archaeology to tectonics [1]. It has also been used extensively for forestry studies over both large scales [45] and for individual tree detection and biophysical parameter estimation [46,47]. Two important capabilities of ALS systems related to forested environments concern both the ability of the laser signal to penetrate the canopy to produce accurate and dense ground returns and the ability to finely and precisely define the vertical structure of the canopy (related to range resolution). The following sections will present and discuss experimental results aimed at characterizing the performance of the Titan sensor with respect to canopy penetration and laser range resolution in vegetated environments. The tests for these characteristics were performed in the tropical rain forests of Central America and Mexico. The complex and thick canopies of these forests represent the most challenging and, thus, the ideal location to perform these assessments. It is important to note that the data used for the following analyses comes from the analogue discrete detector of the Titan. In theory, if similar analyses were performed from data derived from the analysis of full waveform returns, the performance of the system could be better. However, the analyses using the waveform data are outside the scope of this paper.

3.3.1. Canopy Penetration

There have been several experiments aimed at understanding the influence of sensor configuration and forest physical parameters on the canopy penetration performance for lidar systems [48–52]. All of these experiments have produced interesting and promising results; however, these studies have been limited in several aspects: (a) they analyze canopy penetration by isolating one system or flight parameter at a time without considering the tradeoff between different parameters (altitude, PRF, divergence, etc.); (b) they have been performed for a single homogenous study area; (c) while they provide metrics on the detection of ground returns as a ratio of ground returns to total returns, these metrics do not account for how much of the target area was illuminated or how the energy per pulse varies as a function of the system PRF; (d) finally, they analyze data collected from single sensors that are from older technological generation(s) and are not representative of the current state-of-the-art.

NCALM has conducted similar experiments aimed at understanding how to optimize the configuration of lidar systems to maximize the detection of ground returns through canopies in different types of forests [53]. Besides conducting canopy penetration experiments in different kinds of forested environments, NCALM experiments are unique because emphasis has been placed on understanding the pulse energy characteristics of the laser source as a function of the pulse repetition frequency. Previously reported experiments [53] were performed with legacy lidar systems (Optech 3100, Gemini and Aquarius). These systems were powered by Q-switched solid-state laser sources. An operational characteristic of such laser sources is that the output laser energy of each laser pulse decreases when the PRF is increased [53]. Based on this characteristic, maximizing canopy penetration with these older systems is achieved by a tradeoff between illuminating as much of the target area as possible, which is directly proportional to the PRF while maintaining enough energy per pulse to ensure the round trip of the laser pulse through the canopy and back to the sensor (which is inversely proportional to the PRF). Of course, both of these factors are also affected by the flying height of the system. An advantage of the Titan fiber laser sources is that the energy per pulse does not degrade significantly as the PRF increases (Table 1).

The experiments conducted to date with the Titan, the results of which are presented in Table 6 below, reinforce the importance of the energy budget for canopy penetration. However, the energy characteristics of the Titan laser sources allow for good canopy penetration even at high PRFs. Similar to the bathymetric performance tests, these canopy penetration tests are conducted by flying the same flight line over a densely vegetated area several times with varying PRFs and/or flying heights. These experiments have been performed in the tropical forest near the archeological sites of Calakmul in Campeche, Mexico (test area 280,020 m^2), Lamanai in central Belize (test area 640,726 m^2) and El Ceibal in el Petén, Guatemala (test area 407,694 m^2). Once the point cloud data were produced for each of the test flight strips, data samples common to all of the test strips were cropped to produce identical areas that were then processed to obtain shot and return statistics including ground return statistics. The returns are processed to obtain ground returns using the Axelsson algorithm [54] implemented in the Terrasolid Terrascan software. For consistency, all the above described test areas have been processed using the same classification parameters (maximum building size 30 m, maximum terrain angle 88°, maximum iteration angle 12°, maximum iteration distance 3 m). The statistical results recorded for these experiments include: the number of laser shots fired, the number of first, second, third and last returns obtained, the number of returns per fired shot, the number of ground returns detected, the fraction of shots that produced secondary, tertiary, last and ground returns, as well as the associated densities. All of these metrics were analyzed, but for brevity only a subset of these are summarized in Table 6.

As previously stated, to ensure good canopy penetration and good ground sampling, it is important that there is enough energy per laser pulse to ensure two-way travel from the sensor to the ground and back. In a complex multi-story canopy this also translates into the ability of the sensor to detect multiple returns as the laser pulse propagates through the canopy. This is why Table 6 includes information related to the average number of returns produced per laser pulse. The higher the number of returns per pulse, the higher the probability that a fraction of those returns will be from the ground. The table also presents the number of returns that were classified as ground by the Axelsson algorithm as well as a fraction of returns for which ground returns were detected; it also summarizes the shot and ground return densities.

Table 6. Detailed results from the canopy penetration experiments conducted in tropical rain forests.

Configuration	Laser Shots	Shots/m^2	Returns/Shot	Ground Returns	Shots W Grnd	Grnd/m^2
			Calakmul			
125 kHz·W 500 m	1,732,770	6.19	1.54	162,860	9.4%	0.58
100 kHz·W 500 m	1,018,028	3.64	1.71	163,024	16.0%	0.58
70 kHz·W 500 m	725,078	2.59	1.98	195,382	26.9%	0.70
100 kHz 500 C1	591,352	2.11	2.58	395,748	66.9%	1.41
100 kHz 500 C2	596,128	2.13	3.04	448,277	75.2%	1.6
100 kHz 500 C3	587,335	2.10	2.18	350,696	59.7%	1.25
			Lamanai			
300 kHz 650 m C1	3,291,439	5.14	1.34	160,239	4.9%	0.25
300 kHz 650 m C2	3,314,884	5.17	1.85	234,873	7.1%	0.37
300 kHz 650 m C3	3,282,482	5.12	1.69	140,492	4.3%	0.22
175 kHz 550 m C1	2,709,159	4.23	1.57	198,713	7.3%	0.31
175 kHz 550 m C2	2,715,926	4.24	1.97	242,570	8.9%	0.37
175 kHz 550 m C3	2,708,299	4.23	1.79	147,443	5.4%	0.23
75 kHz 550 m C1	1,026,081	1.60	1.68	98,613	9.6%	0.15
75 kHz 550 m C2	1,035,340	1.62	2.00	107,097	10.3%	0.17
75 kHz 550 m C3	1,019,546	1.59	1.76	61,154	6.0%	0.10
			El Ceibal			
100 kHz 700 m C1	933,915	2.29	1.78	59,417	6.4%	0.15
100 kHz 700 m C2	933,550	2.29	1.75	40,572	4.3%	0.10
100 kHz 700 m C3	905,700	2.22	1.36	19,643	2.2%	0.05
150 kHz 700 m C1	1,382,953	3.39	1.71	76,700	5.5%	0.19
150 kHz 700 m C2	1,383,895	3.39	1.77	57,607	4.2%	0.14
150 kHz 700 m C3	1,333,620	3.27	1.33	26,178	2.0%	0.06
150 kHz 600 m C1	1,558,967	3.82	1.81	93,086	6.0%	0.22
150 kHz 600 m C2	1,557,677	3.82	1.91	74,269	4.8%	0.18
150 kHz 600 m C3	1,545,070	3.79	1.51	37,294	2.4%	0.09
150 kHz 400 m C1	2,978,301	7.31	2.11	188,587	6.3%	0.46
150 kHz 400 m C2	2,969,762	7.28	2.31	174,980	5.9%	0.42
150 kHz 400 m C3	2,958,431	7.26	2.17	125,777	4.3%	0.31

The first seven rows in Table 6 present results from tests performed in Calakmul and compare results obtained with the Optech Gemini (rows two to four) and Titan lidar sensors (rows five to seven). The Gemini data were collected on 23–24 May 2014 and the Titan data were collected on 18 May 2016. All data were collected from a flying height of 500 m above ground level. The results for the Gemini data are meant to illustrate how the canopy penetration performance for that sensor degraded as the system PRF was increased while all other parameters remained constant. The results indicate a reduction of 22% in the number of returns produced per laser shot when the PRF is increased from 70 kHz to 125 kHz. Perhaps more important was a reduction of 65% in the number of shots that produced ground returns which varied from 26.9% of shots at 70 kHz to only 9.4% of shots at 125 kHz. Another interesting comparison is that despite a shot density that was more than double for the 125 kHz (as compared to 70 kHz), the ground return density was 20% higher for the 70 kHz test line. This last comparison illustrates that for canopy penetration and ground return detection, it is not the quantity of fired laser shots that matters but, more importantly, the quality of the shots, which is determined by the energy contained in each laser pulse.

The Titan data for the Calakmul test area is not directly comparable to the Gemini data due to the temporal separation between the collections, which was compounded by a severe drought that hit the region after 2014. However, it is worth highlighting certain key points in the comparison of the Titan and Gemini data for Calakmul. First, despite the thinner canopy during the Titan collection as a result of the drought, the number of returns per laser pulse obtained in all of the channels of the Titan sensor is higher than the comparable results obtained with the Gemini at the same system PRF of 100 kHz, and for that matter, to all tested Gemini PRFs (70, 100 and 125 kHz). Second, the results from the Titan data are separated by channel. This is important because (a) each channel has different beam divergences and pulse energy values, which modify the energy density of their resultant footprint; and (b) each channel has different look angles, as previously described. Even if energy or power density were the same for all channels, the imaging geometry of the channel pointing at the nadir enables better canopy penetration performance as compared to the channels that look forward of the nadir. This factor can be observed in the test results summarized in Table 6 where, for a given test, the statistics for ground returns per laser shot and ground return density for channel 2 (C2, which points to the nadir) are higher than the results for the other channels. Note that this is true for the results presented for the Calkamul and the Lamanai test sites, but it is not the case for the El Ceibal test site due to technical malfunctions that will be expanded below.

The Lamani test was conducted to characterize the canopy penetration performance of the Titan as a function of the system PRF and to determine the optimal configuration (height and PRF) that maximized the detection of ground returns. The same test line was flown three different times at PRFs that are representative of the operational envelope of the Titan: 75 kHz, 175 kHz and 300 kHz. Because of the range ambiguity region, the 300 kHz test flight had to be collected at 650 m above ground which is considerably higher than the 550 m that was used for the test collections at 75 and 175 kHz. However, these combinations of PRF and height are close to the normal operational conditions that would be used in a survey project. The first conclusion from analyzing the Lamanai test results is that the number of returns produced per laser shot does not vary as much for the Titan as they did for the earlier-generation Gemini sensor. This is mostly true for channels 2 and 3, for which the energy per pulse characteristics do not vary much with the increasing PRF. The number of returns per pulse varied by only about 7.5% for channel 2 (1064 nm) and 4% for channel 3 (532 nm). The energy pulse characteristics of the laser source for channel 1 (1550 nm) do degrade by a small amount with increasing pulse repetition rates but not as widely as the solid-state source that powered the Gemini lidar. The variation in the number of returns produced per pulse for channel 1 only varied by 20% from 75 kHz all the way to 300 kHz (225 kHz), which is close to the 22% variation observed for the Gemini sensor for a 55 kHz variation of PRF.

A second lesson derived from the Lamanai results is that the fraction of laser shots that produce detectable ground returns is significantly affected by the lower energy per pulse and the need to fly at higher altitudes at the higher end of the PRF operational range (300 kHz) when compared to the fraction of laser shots obtained at the lower PRFs. For channel 1, for which the pulse energy characteristics are the most affected with increased PRF, the variation in the fraction of pulses with detectable ground returns was 49.3% between the test line flown at 650 m and 300 kHz and the line flown at 550 m and 75 kHz. For channels 2 and 3, the variation in the results for the same metric of the fraction of shots with ground returns and for the same configurations listed above was close of 31.5% and 28.6%, respectively. The same conclusion can be reached by analyzing the return and ground return density metrics. While the line flown at 300 kHz produces much higher shot densities, the highest ground return densities were obtained from the line flown at 175 kHz and at lower elevation. This demonstrates again that it is not the quantity of the shots that produce the higher number of ground returns but the quality of the shots, which is determined by the energy budget (energy per pulse and ranging distance).

The El Ceibal test was conducted to characterize the canopy penetration performance of the Titan sensor at varying flying heights above the ground. As mentioned earlier, when this test was performed, the optical energy output of the laser source that powers channels 2 and 3 was starting to degrade. The laser source was repaired by the laser manufacturer after the completion of the Guatemala and Mexico mapping campaign (see Table 1). In addition, the receiver optics of channel 2 were out of alignment (this issue was also corrected after the laser source was replaced), which also caused a degraded canopy penetration performance. The results from this test, while not representative in absolute terms of nominal sensor performance, do provide comparative performance metrics and important insights into canopy penetration. This test was conducted by flying the same line at different heights above the ground while maintaining the system PRF at 150 kHz. A pair of lines was also flown at the maximum test height of 700 m above the terrain but at two different PRFs (100 and 150 kHz).

By comparing the results of the test lines flown at 100 and 150 kHz from 700 m above ground level, a few of the previous conclusions are reinforced. Based on the returns per shot and shots that produced ground return metrics, it can be noted that while the performances for channels 2 and 3 do not vary significantly between PRF settings, the performance of channel 1 does vary slightly with better performance at the lower PRF. The lines flown at 700, 600 and 400 m confirm the expected trend: the metrics of returns per laser pulse and pulses with detected ground returns increase as the flying height decreases. This is due to the spreading of the laser energy during its two-way trip from the sensor to the ground and back, as a function of the range to the fourth power (R^4) as defined by the lidar equation [1]. These results from the El Ceibal test serve to reinforce the conclusion that canopy penetration for the production of ground returns is mostly dominated by energy considerations. There has to be enough energy within each pulse to withstand the two-way attenuation caused by its normal propagation through the atmosphere and by the scattering of the forest canopies. When trying to maximize ground detection (maximize ground return density), it is important to optimize the tradeoff between surface illumination, which is related to shot density (determined by system PRFs and flying height), and the laser energy budget, which is affected by the laser source pulse energy characteristics and by the flying height.

It is important to note that a research group in Canada has used Titan data and other multi-sensor lidar data to assess the impact of multispectral lidar data on forestry studies [35]. They have observed similar return ratio differences among the Titan channels as the ones presented in this section. However, they conclude that these differences are mainly due to the wavelength-dependent characteristics of penetration, absorption and reflection of the forest canopies. While the spectral dependence of the light and matter interactions are definitely a factor, the researchers neglect the hardware characteristics such as energy per pulse, beam divergence, power density and look angles that vary from channel to channel, and which also have a significant impact on the system's ability to map the forest canopies.

3.3.2. Range Resolution/Canopy Characterization

Another important operational performance characteristic of a lidar is the system range resolution, which is defined as the ability of a sensor to separate targets along the range direction within a single lidar footprint. This assumes that the targets are illuminated by the same laser pulse/footprint and that the first target(s) do not completely occlude the laser footprint. As a proxy, because of the narrow birds-eye-view scanning geometry of airborne lidar systems, the range resolution can be simplified to be approximately equivalent to the vertical resolution capability of the lidar. This characteristic is important for applications in forestry or ecology where researchers are interested in describing the canopy structure accurately to model habitats [55] or to assess biomass [56]. It is also important when trying to obtain reliable ground returns in areas covered by vegetation. It is well known that the range resolution of a sensor is mainly determined by the laser pulse width [57]; however, it also depends on other factors including the electronic characteristics of the detector sub-systems. While some laboratory experiments have been conducted to assess the range resolution of lidar systems [58], a field experimental approach was taken to assess the Titan's range resolution and ability to finely characterize forest canopies. This field approach consisted of computing the range separation between successive returns for the same laser pulse generated by tropical forest canopies. The same test data that were used to characterize canopy penetration obtained from the forest canopies near Calakmul, Lamanai and El Ceibal were used for this purpose. Data from all of the test flights were analyzed; however, for brevity, only some returns are presented and discussed in the following paragraphs.

First, the discrete return data were segregated and classified depending on the number of returns produced per pulse. Because of its four-stop recording capability, the Titan data can be segregated into four groups: pulses with single returns, pulses with only two returns, pulses with three returns and pulses with more than three returns. For this analysis, the groups of pulses with two returns and the pulses with more than three returns were selected as representing the extreme cases. The pulses that produced more than three returns were the ones that traveled through a significant cross-section of canopy, while the ones with only two returns and that have a relatively short inter-return separation (case analyzed here) likely did not interact much with the canopy structure. It is important to note that in this type of field experiment, the actual distribution of the separation between returns is determined by the canopy structure. However, the lowest percentile values for the separation between returns are determined by the system capabilities. In essence, this analysis is aimed at determining the minimum separations that were detected by the sensor.

Tables 7 and 8 summarize results from the analysis of the data collected near Calakmul and Lamani for different test configurations and for the different Titan channels. These tables present statistics for the number of shots that produced two, three and more than three returns. They also present the values for the minimum as well as the one and three percentile distribution values for the detected return separations for the two cases (shots with only two returns and shots with more than three returns). For the case of shots that produced more than three returns, the minimum and percentile values are presented for the separation between the first and second returns, the second and third returns, and the third and last returns. The last returns are not necessarily a fourth return; it could have been a fifth, sixth, seventh or even higher return. The Titan can detect multiple returns within its range gate, but only records the first three and the last return.

Table 7 presents results from the Calakmul test and also compares the range resolution of the Titan with respect to the older-generation Gemini sensor. The results summarized in the Table 7 show some significant trends. First, the values for the minimum separation between returns, while consistent with the theoretical minimum range resolution value (equivalent to half of the laser pulse width) can be outliers and have to be treated with caution. For this reason, first and third percentile measurements are reported. Second, irrespective of the minimum values, the separation between returns for the Gemini sensor increase as the PRF increases. These results are expected as the laser source for that system produces pulses with increasing width as the PRF is increased. Finally, from the last three columns that correspond to the results obtained from Titan data, it can be observed that while there is a small variation in the range resolution between the different channels, this variation is usually less than 10 cm in most cases and the range resolutions for channel 1 and channel 3 are almost the same.

Table 7. Comparative results from minimum separation between returns test obtained from the Gemini and Titan lidar systems near the Calakmul Mayan site in Campeche, Mexico. Range separation results are given in meters.

	2014 Gemini			2016 Titan @ 100 kHz		
	125 kHz	100 kHz	70 kHz	C1	C2	C3
Number of Shots	694,825	381,483	262,265	226,802	229,250	223,838
Shots with 2 Returns	214,904	129,270	92,123	67,336	47,747	82,178
Shots with 3 Returns	78,030	64,087	61,024	67,980	64,078	57,482
Shots with >3 Returns	8289	10,712	17,607	46,092	93,873	20,073
Only 2 Returns						
Min	1.535	1.577	1.502	0.678	0.666	0.662
1%	2.782	2.606	2.434	0.989	0.888	0.928
3%	3.217	3.015	2.828	1.208	1.044	1.103
1st–2nd						
Min	1.495	1.475	1.321	0.665	0.663	0.675
1%	2.122	2.031	1.894	0.832	0.789	0.83
3%	2.452	2.342	2.19	0.965	0.883	0.937
2nd–3rd						
Min	1.51	1.652	1.55	0.659	0.662	0.7
1%	2.43	2.247	2.084	0.863	0.774	0.848
3%	2.746	2.553	2.379	1.002	0.891	0.971
3rd–Last						
Min	1.683	1.73	1.582	0.691	0.653	0.71
1%	2.386	2.278	2.112	1.06	0.986	1.03
3%	2.73	2.555	2.36	1.241	1.197	1.256

Table 8 summarizes the results from the range resolution experiments conducted near Lamanai, Belize, which were aimed at assessing performance differences related to changing PRF values. Because results for all channels and all PRFs tested would not fit in the table below, only the results for channels 1 and 2 are presented, given that the range resolutions results for channel 1 and 3 have already been shown to be comparable (Table 7). The most important conclusion that can be drawn from the data presented in Table 8 is that the range resolution for the Titan system is not significantly affected by the selection of system PRF, which is different from the result for the Gemini sensor. Also, as noted from the data of the previous table, the variation between Titan channels is less than 10 cm in most cases.

Table 8. Comparative results from minimum separation between returns test obtained from the Titan sensor running at different PRFs near the Lamani Mayan site in Belize. Range separation results are given in meters.

	300 kHz C1	650 m C2	175 kHz C1	550 m C2	75 kHz C1	550 m C2
Number of Shots	3,292,063	3,315,493	2,709,669	2,716,424	1,026,408	1,035,563
Shots with 2 Returns	827,019	1,118,222	819,056	894,329	320,157	338,856
Shots with 3 Returns	133,990	544,008	272,968	501,007	128,830	196,336
Shots with >3 Returns	8973	208,025	62,132	250,047	38,421	102,401
Only 2 Returns						
Min	0.667	0.662	0.669	0.661	0.666	0.662
1%	0.907	0.804	0.896	0.81	0.898	0.808
3%	1.044	0.942	1.064	0.953	1.076	0.949
1st–2nd						
Min	0.705	0.658	0.668	0.663	0.673	0.666
1%	0.837	0.769	0.824	0.77	0.809	0.767
3%	0.936	0.881	0.952	0.885	0.953	0.883
2nd–3rd						
Min	0.67	0.664	0.643	0.648	0.661	0.659
1%	0.872	0.784	0.842	0.777	0.848	0.778
3%	0.968	0.896	0.976	0.894	0.981	0.894
3rd–Last						
Min	0.684	0.638	0.678	0.653	0.675	0.66
1%	0.879	0.803	0.89	0.804	0.879	0.804
3%	0.978	0.933	1.023	0.943	1.023	0.943

A final crucial observation that can be made based on the results presented in both Tables 7 and 8, particularly from the range resolution results of those pulses that produce more than three returns, is that as the system detects returns from deeper within the canopy, the range resolution is degraded. This would perhaps indicate that the ability to detect and discriminate closely spaced pulses is a function of the received signal strength.

3.4. Special Operational Capabilities

Besides the multispectral, bathymetric and canopy penetrating/mapping capabilities that have already been discussed, it is important to briefly highlight other operational advantages or capabilities that are enabled by the sensor's unique multispectral and multichannel design. The first advantage is significantly higher target surface illumination in a single instrument pass. Until recently, lidar systems have only sparsely sampled the mapped surface. A few returns per square meter has been considered sufficient to derive topographic maps for certain engineering or scientific applications [59]. However, certain applications, such as small target detection or archaeology, benefit from sampling or illuminating 100% or more of the surface of interest [59–61]. The combination of three different look angles (nadir, 3.5° and 7° forward of the nadir) for the different channels, the varied beam divergence and the larger scan product (product of scan angle and scan frequency) allow for almost full surface illumination in a single pass (Figure 11a). Figure 11b shows a detailed view of a graph that plots both the position and footprint of the laser returns for the Titan's three channels sampled from the central portion of a test flight swath from Figure 11a.

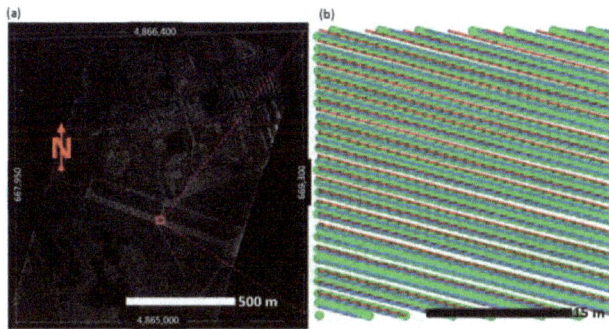

Figure 11. Illustration of Titan's footprints and surface illumination from a single pass of the sensor. (**a**) Intensity rendering of a test swath generated from Titan channel 1; (**b**) Graph that plots the position and footprints of returns from all of Titan's channels for the red square sample of the test swath. This graphs illustrated how much of the target surface is illuminated by the laser beams.

This test flight swath was collected by Teledyne Optech during the first test flight of the Titan system. The system was configured with a PRF of 200 kHz per channel (600 kHz combined), a scan angle of $\pm25°$ and a scan frequency of 32 Hz (scan product of 800 degrees-Hertz) and flown at an altitude of 1000 m above ground level. In Figure 11b, the footprints for each of the returns for the three Titan channels are plotted at a proper scale. The footprints of the returns from channel 1 (1550 nm) are depicted in blue, while the footprints for channels 2 and 3 are represented with red and green, respectively. The sample square that was cropped from the swath has an area of 1033 m^2, and the number of shots and the area illuminated by the laser footprints (not counting the overlap between footprints of the same channel) are summarized in Table 9. Due to the larger divergence of channel 3 (532 nm), the footprints from the laser shots illuminate 83.3% of the sample area, while a single infrared channel (channel 1) illuminates 25% of the sample area. When combining the footprints of the three channels, and not counting the area overlap, the fraction of the surface area that is illuminated by the main portion of the laser beams totals 88.2%. Of course, the surface illumination can be increased significantly by utilizing a 50% lateral overlap or by flying lines in an orthogonal fashion. This near-full-surface illumination is critical for other capabilities of the Titan sensor, for example canopy penetration, small target detection and active intensity image generation.

Table 9. Shot densities and fraction of surface illuminated in a single pass of the Titan sensor.

Channel	Number of Shots	Shot Density 1/m^2	Illuminated Surface m^2	% of Surface Illuminated
C1	2666	2.58	257.76	25.0%
C2	2650	2.56	186.37	18.0%
C3	2699	2.61	861.09	83.3%
All Channels	8015	7.75	910.73	88.2%

Some might characterize the multichannel and multi-laser source design of the Titan as being redundant, and while this is true, perhaps a better engineering descriptor for the Titan design would be "diversity". In telecommunications, a diversity scheme is a design feature that improves the reliability of a system by using two or more channels with different technical or physical characteristics [62,63]. There is equipment diversity when two or more radios are used to transmit the same information; there is frequency diversity where different radios will transmit the same information over two different frequency channels; and there is spatial diversity where multiple antennas are used to transmit the same signal. The Titan sensor has several diversity schemes built into its design. The way the different channels interrogate the target at different angles represents a diversity of look angles, the three different laser wavelengths represent spectral diversity and the two laser sources constitute equipment diversity.

The look angle diversity brings several operational advantages: as mentioned previously, it provides for a more complete and uniform surface illumination even in a single pass, it provides multiple points of view to pierce through vegetation covers, and it allows for a better mapping of vertical structures. Having equipment diversity or laser source redundancy ensures an operational sensor even if one of the laser sources fails. The sensor will continue to collect data, at two-thirds or one-third of the maximum desired measurement rate. Collecting a decimated data set is better than no data at all, and therefore, having this redundancy is valuable when mapping in remote areas where having manufacturer technical support might not be an option or for time-critical collections that cannot wait a couple of days for a repair. The spectral diversity provides some operational advantages besides providing the spectrally rich data sets described through this paper. The Titan will be able to detect relatively strong returns over a large variety of land covers and throughout its operational envelope for at least one of it channels, something that single-wavelength systems may have a hard time doing.

An example of how the diversity of equipment and the spectral diversity of the Titan system proved to be advantageous for the successful completion of a project comes from NCALM's experience mapping the McMurdo Dry Valleys in Antarctica [23]. The Dry Valleys are perhaps one of the most challenging places to map with airborne lidar systems. Besides its remoteness and harsh temperature conditions, the terrain relief can vary more than 1000 m in just a few kilometers which make it challenging to fly (see Figure 12a). The ground cover also alternates between frozen soil, snow and ice. A single-wavelength lidar system that operates in the 1550 nm wavelength would have a hard time obtaining strong returns from the snow and ice at safe flying heights (~2000 m AGL) because of the low reflectance of snow and ice at the 1550 nm wavelength (see Figure 1). Figure 12b shows a density map of lidar returns detected from the Titan system from one of these valleys; despite a uniform number of flight lines through the areas of interest, it is obvious that the return density varies significantly through the valleys. The lower-density areas are due to dropouts (laser shots with no return detected) that result from the combination of flying height and the low reflectance of ice and snow that cover the area. These lower-density areas would have been voids if not for the spectral diversity of the Titan system.

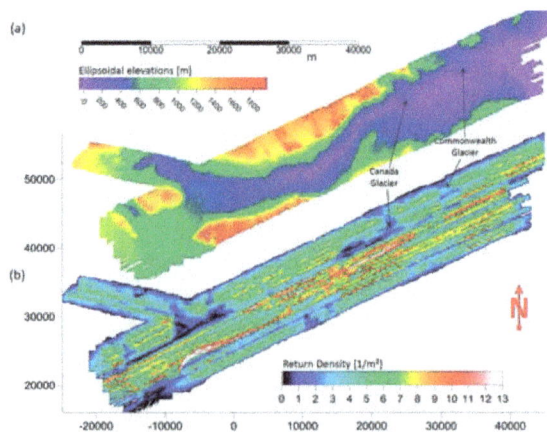

Figure 12. Image maps for the Taylor and Pearse valleys in Antarctica. (**a**) Image map showing the topographic relief of the valleys based on the lidar DEM; (**b**) Image map showing the laser return density obtained from the valleys.

3.5. Precision and Accuracy Assessments of Topographic Elevations

As a final performance analysis of the Titan system, this section examines the positional precision and accuracy of the lidar returns. It is important to note that while precision (also known as repeatability or internal accuracy) is an intrinsic performance characteristic of a lidar system, accuracy is extrinsic because it depends on several factors external to the system. Among these factors are the geometrical strength of the GNSS constellation at the time of collection, the distance from the aircraft to the reference base stations, the performance of the integrated navigation system (INS), the aircraft dynamics, and the INS data processing algorithms, among many others [64]. As such, it is important to note that the accuracy values presented below should be considered as representative values; the actual accuracy of a dataset will be dependent on the specific conditions particular to each project's collection and processing. It is also important to note that the positional accuracy of airborne lidar returns can be broken down into vertical and horizontal components. The accuracy results presented below relate only to the vertical component.

The airborne data for this analysis were collected in October 2014 (medium altitude) and August 2016 (low altitude) and consist of repeat passes of a flight line perpendicular to the runway at the Baytown airport (KHPY) located in Baytown, Texas. The test line was flown six times for three different PRFs and at flight heights in the middle and lower end of the Titan operational envelope and within the PIA regions of operation allowable for each PRF (see Figure 2). The precision (repeatability) of the datasets was computed by selecting planar surfaces within the dataset (the test surface had an area of 420 m^2), fitting a plane based on the XYZ coordinates of the returns that define the plane and then computing the distance between individual returns and the fitted plane. Dispersion statistics of the distances between returns and the fitted plane define the precision of the dataset.

To conduct an absolute vertical accuracy assessment of the airborne test data validation, elevations were collected by installing a geodetic grade dual frequency GPS antenna (Trimble Zephyr Model 2) and a Trimble NetR9 receiver on a vehicle and performing a kinematic survey along the runway and taxiways of the Baytown airport. A reference ground station was located at the airport during the time of the survey and its coordinates were determined through the online positioning user service (OPUS) of the US National Geodetic Survey (NGS). The distance between the rover and the reference station was less than 2 km at any time. The Novatel software GrafNav (Calgary, AB, Canada) was used to obtain dual-frequency carrier-based solutions for the rover GPS antenna trajectory that was then used as the validation data. An algorithm was used to perform a nearest neighbor search for airborne lidar returns that are located within a preset distance (30–50 cm) to the kinematic survey reference points. Once the pairs of lidar returns and reference measurements are identified, the vertical separations between these datasets were computed and RMSE values were reported. Table 10 presents vertical precision and accuracy results for these experiments for each of the different combinations of PRFs and flying heights, segregated by channel (C1, C2, C3) as well with all three channels combined (C123).

The results presented in Table 10 are within the expected limits, with precision values better than 2 cm and height accuracy values better than 10 cm. Some higher values were obtained from channel 2 with respect to the other channels. This was the result of the misalignment of the receiver optics for that channel. The results also indicate no significant variation in the precision or accuracy performance of the sensor for the different tested PRFs or flying heights. It can be seen that the precision and accuracy do degrade when data from all channels is considered for the assessment (column C123) as compared to analyzing a single channel at a time. These results are expected and are the product of the complexity of a multichannel system where the scanning geometries of each channel are so varied. However, these combined channel precision and accuracy metrics indicate that the channel by channel calibration parameters and the sensor geometric model are adequate and produce consistent and accurate data within very good limits exceeding the USGS lidar base specification quality level 1 (QL1) [65].

Table 10. Results from a precision and accuracy assessment of height values for different system PRFs at the middle of the sensor operational flying height envelope.

PRF (kHz)	Range (m)	Number of Samples			Height RMSE (m)			
		C1	C2	C3	C1	C2	C3	C123
			Precision					
100	900	658	-	667	0.02	-	0.018	0.030
200	900	1411	93	1542	0.018	0.028	0.018	0.020
300	800	2535	756	4766	0.019	0.026	0.017	0.048
100	500	1038	1061	1060	0.019	0.018	0.021	0.045
200	500	2208	2280	2255	0.020	0.018	0.021	0.022
300	350	7476	7218	7380	0.017	0.016	0.016	0.027
			Accuracy					
100	900	207	-	179	0.051	-	0.082	0.048
200	900	347	27	324	0.055	0.037	0.060	0.044
300	800	420	82	465	0.073	0.044	0.059	0.0649
100	500	219	236	242	0.022	0.037	0.022	0.042
200	500	434	487	500	0.018	0.035	0.020	0.030
300	350	917	895	914	0.019	0.033	0.021	0.026

One of the limitations of the experimental data presented in Table 10 is that it is based on single-swath data and not data from overlapping swaths that may cause degradation in data precision and accuracy. As also mentioned earlier, the accuracy of a lidar system is to subject to many external factors. Therefore, in order to provide some additional data points related to the Titan's accuracy, results from height accuracy assessment measurements performed for mapping projects executed with different system configurations are presented in Table 11. These accuracy assessment exercises follow the same procedures described above related to the collection and processing of the height validation data and the computations of vertical difference between the airborne and validation data. However, some differences include: (a) the assessment of data originating from different flight swaths; (b) processing of DGPS validation data with the Ashtech office suite (AOS) software instead of GrafNav; and (c) larger rover and reference station separations that may have extended to distances of up to 10 km. The results presented in Table 11 are similar to the results in Table 10, indicating good levels of height accuracy, which again exceed the quality level 1 (QL1) of the USGS lidar base specifications.

Table 11. Results from height accuracy assessments performed as part of mapping projects with different sensor PRFs and flying heights.

Location	Configuration	Number of Test Points	RMSE (m)
Teotihuacan, Mexico	250 × 3 kHz, 900 m	581	0.041
University of Houston, Texas	250 × 3 kHz, 500 m	1018	0.035
Orange Walk, Belize	175 × 3 kHz, 550 m	2238	0.044
NASA JSC, Texas	150 × 3 kHz, 750 m	2923	0.049
Calhoun Creek, South Carolina	100 × 3 kHz, 700 m	6314	0.033
Monterey, California	100 × 3 kHz, 700 m	1086	0.037

4. Conclusions

This paper presents an overview of the design and performance of the Teledyne Optech Titan multispectral lidar system based on almost two years of operational experience in varied conditions, landscapes and environments. The paper is intended to provide a wide overview of the capabilities and applications enabled by the unique multipurpose design of the Titan sensor in a way that is useful to researchers that work with data from this sensor and for lidar technology specialists. This paper describes and discusses experiments and results aimed at quantifying the performance of the Titan system as it applies to ground cover classification, bathymetry, forestry, ground return detection, and geometrical accuracy, among others. Because of brevity and scope considerations, this paper is not intended to explore in detail certain aspects of specific applications such as radiometric calibration of multispectral lidar intensity or three-dimensional refraction correction of bathymetric returns. Topics such as these merit research papers on their own and it is our hope that future papers on such topics will draw upon the general description presented here.

The Titan sensor has proven to be extremely flexible and reliable due to the equipment, spectral and view/scan geometry diversity scheme incorporated in it design. Its multispectral capabilities provide spectrally rich and consistent datasets that were not available before. The multispectral capabilities enable applications that include: land cover and target classification based on both spectral and 3D spatial information, and high resolution seamless topographic and very shallow water bathymetry. Ground cover classification accuracies of 90% have been achieved with simple methodologies and with intensity data that has not been corrected to represent reflectance values. It is expected that classifications with a larger number of classes and higher levels of accuracy can be obtained with more advanced methods and algorithms. Other researchers have written about the challenges of achieving good multispectral intensity radiometric calibration; while this was not explored in this work, it is definitely a challenge and future avenue to be investigated.

The bathymetric performance in terms of detected depths and accuracy of derived water depths and bathymetric elevations is within the performance expectations of a low-optical-power 1.5/Kd bathymetric lidar. The accuracy of benthic elevations can be improved with more rigorous three-dimensional refraction correction of the returns; this, however, is mainly limited by the ability to obtain a good representation of the water surface at the exact place where the 532 nm beam penetrates the water. The first-order approximation that corrects only for the vertical component of the returns is good for water depths less than 5 m, but it can introduce vertical bathymetric errors larger than a decimeter in deeper water (>10 m).

The combination of varied look angles, full surface illumination, almost-uniform pulse energy versus PRF characteristics of the laser sources, and short pulse widths enables characterizing and penetrating forest canopies at spatial resolutions and performance levels at least twice as effective as were possible with previous-generation airborne lidar mapping systems. The three-channel, three-wavelength and three-look angle design also provides redundancy and diversity which is beneficial not only from a technical/operational point of view but also reduces operational costs. One disadvantage of the design when compared to a single-channel system has to do with the complexity of achieving a good inter-channel calibration. While achieving this is not impossible, it has taken considerably more effort and attention to detail than what was required with older, single-channel systems. However, the precision and accuracy assessment experiments conducted for this study indicate that despite these challenges, good and accurate geometric calibrations have yielded point clouds that are better than USGS lidar base specifications quality level 1 (QL1). Visual and analytical inspection of shaded relief maps based on digital elevation models produced over the past two-year period have identified very little inter-channel or intra-channel elevation artifacts.

The capability to provide two independent streams of range/intensity data for each channel through both its analogue discrete return detector and the full waveform recording capabilities will enable a proper comparison of the advantages and disadvantages of both of these data processing and return detection approaches. Despite its advantages, the Titan system does have a few limitations. The 2 km operation range limit as well as the fixed multiple pulse mechanism limit the performance and applicability of the system in regions with extreme topographic variability. The analogue detection is also limited by only recording up to four returns, as the sensor has enough energy budget and range resolution to detect more than four returns in thick vegetation canopies. While this limitation is overcome by using the waveform digitizers, they unfortunately are currently limited to recording waveforms at PRF rates of values of only 100 kHz. It is clear that future iterations of the sensor will attempt to overcome these limitations.

Over the next few years the research community will develop new applications and new scientific approaches based on the unique capabilities and design characteristics of the Titan senor. However, there is no need to wait for the results of those new applications or approaches to be published to realize that the Titan sensor represents a significant leap forward in the evolution of airborne mapping lidar systems.

Acknowledgments: The authors would like to acknowledge Paul LaRocque, Brent Smith and Jon Henderson of Teledyne Optech for their constant support and communication during the design, integration, testing, operation and ongoing improvement of the Titan sensor. We would also like to acknowledge the principal investigators of the different mapping projects that have been performed with the Titan across the United States, Mexico, Central America, the Caribbean, and Antarctica; their projects have enabled a thorough testing of the system across multiple environments and applications. This work is supported by grant number EAR 1339015 provided by the United States National Science Foundation Division of Earth Sciences, Instrumentation and Facilities Program.

Author Contributions: All authors participated in the drafting and editing of the paper text and the review of the experimental results. Juan Carlos Fernandez-Diaz conceived and designed the experiments which were performed with the assistance of Zhigang Pan and Nima Ekhtari under the supervision of Craig Glennie, William E. Carter and Ramesh L. Shrestha. Collection and processing of the data used for the experiments was conducted by Michael Sartori, Abhinav Singhania, Darren Hauser and Juan Carlos Fernandez-Diaz.

Conflicts of Interest: The authors declare no conflicts of interest.

References

1. Glennie, C.L.; Carter, W.E.; Shrestha, R.L.; Dietrich, W.E. Geodetic imaging with airborne lidar: The earth's surface revealed. *Rep. Prog. Phys. Phys. Soc.* **2013**, *76*, 086801. [CrossRef]
2. Antonarakis, A.; Richards, K.S.; Brasington, J. Object-based land cover classification using airborne lidar. *Remote Sens. Environ.* **2008**, *112*, 2988–2998. [CrossRef]
3. Stevens, C.W.; Wolfe, S.A. High-resolution mapping of wet terrain within discontinuous permafrost using lidar intensity. *Permafr. Periglac. Process.* **2012**, *23*, 334–341. [CrossRef]
4. Höfle, B.; Pfeifer, N. Correction of laser scanning intensity data: Data and model-driven approaches. *ISPRS J. Photogramm. Remote Sens.* **2007**, *62*, 415–433. [CrossRef]
5. Yan, W.Y.; Shaker, A.; Habib, A.; Kersting, A.P. Improving classification accuracy of airborne lidar intensity data by geometric calibration and radiometric correction. *ISPRS J. Photogramm. Remote Sens.* **2012**, *67*, 35–44. [CrossRef]
6. Kashani, A.; Olsen, M.; Parrish, C.; Wilson, N. A review of lidar radiometric processing: From AD HOC intensity correction to rigorous radiometric calibration. *Sensors* **2015**, *15*, 28099. [CrossRef] [PubMed]
7. Kaasalainen, S.; Hyyppa, H.; Kukko, A.; Litkey, P.; Ahokas, E.; Hyyppa, J.; Lehner, H.; Jaakkola, A.; Suomalainen, J.; Akujarvi, A. Radiometric calibration of lidar intensity with commercially available reference targets. *IEEE Trans. Geosci. Remote Sens.* **2009**, *47*, 588–598. [CrossRef]
8. Goepfert, J.; Soergel, U.; Brzank, A. Integration of intensity information and echo distribution in the filtering process of lidar data in vegetated areas. In Proceedings of the SilviLaser, Edinburgh, UK, 17–19 September 2008.
9. Wang, C.; Glenn, N.F. Integrating lidar intensity and elevation data for terrain characterization in a forested area. *IEEE Geosci. Remote Sens. Lett.* **2009**, *6*, 463–466. [CrossRef]
10. Dalponte, M.; Bruzzone, L.; Gianelle, D. Fusion of hyperspectral and lidar remote sensing data for classification of complex forest areas. *IEEE Trans. Geosci. Remote Sens.* **2008**, *46*, 1416–1427. [CrossRef]
11. Hopkinson, C.; Chasmer, L. Using discrete laser pulse return intensity to model canopy transmittance. *Photogramm. J. Finl.* **2007**, *20*, 16–26.
12. Hopkinson, C.; Chasmer, L. Testing lidar models of fractional cover across multiple forest ecozones. *Remote Sens. Environ.* **2009**, *113*, 275–288. [CrossRef]
13. Donoghue, D.N.; Watt, P.J.; Cox, N.J.; Wilson, J. Remote sensing of species mixtures in conifer plantations using lidar height and intensity data. *Remote Sens. Environ.* **2007**, *110*, 509–522. [CrossRef]
14. Renslow, M.S. *Manual of Airborne Topographic Lidar*; American Society for Photogrammetry and Remote Sensing: Bethesda, MD, USA, 2012.

15. Feigels, V.; Kopilevich, Y.I. Lasers for lidar bathymetry and oceanographic research: Choice criteria. In Proceedings of the IEEE International Geoscience and Remote Sensing Symposium, IGARSS'94, Surface and Atmospheric Remote Sensing: Technologies, Data Analysis and Interpretation, Amherst, MA, USA, 8–12 August 1994; pp. 475–478.

16. Fernandez-Diaz, J.C.; Glennie, C.L.; Carter, W.E.; Shrestha, R.L.; Sartori, M.P.; Singhania, A.; Legleiter, C.J.; Overstreet, B.T. Early results of simultaneous terrain and shallow water bathymetry mapping using a single-wavelength airborne lidar sensor. *IEEE J. Sel. Top. Appl. Earth Obs. Remote Sens.* **2013**, *7*, 623–635. [CrossRef]

17. Briese, C.; Pfennigbauer, M.; Lehner, H.; Ullrich, A.; Wagner, W.; Pfeifer, N. Radiometric calibration of multi-wavelength airborne laser scanning data. *ISPRS Ann. Photogramm. Remote Sens. Spat. Inf. Sci.* **2012**, *1*, 335–340. [CrossRef]

18. Hartzell, P.; Glennie, C.; Biber, K.; Khan, S. Application of multispectral lidar to automated virtual outcrop geology. *ISPRS J. Photogramm. Remote Sens.* **2014**, *88*, 147–155. [CrossRef]

19. Chen, Y.; Räikkönen, E.; Kaasalainen, S.; Suomalainen, J.; Hakala, T.; Hyyppä, J.; Chen, R. Two-channel hyperspectral lidar with a supercontinuum laser source. *Sensors* **2010**, *10*, 7057–7066. [CrossRef] [PubMed]

20. Woodhouse, I.H.; Nichol, C.; Sinclair, P.; Jack, J.; Morsdorf, F.; Malthus, T.J.; Patenaude, G. A multispectral canopy lidar demonstrator project. *IEEE Geosci. Remote Sens. Lett.* **2011**, *8*, 839–843. [CrossRef]

21. Wei, G.; Shalei, S.; Bo, Z.; Shuo, S.; Faquan, L.; Xuewu, C. Multi-wavelength canopy lidar for remote sensing of vegetation: Design and system performance. *ISPRS J. Photogramm. Remote Sens.* **2012**, *69*, 1–9. [CrossRef]

22. Pfennigbauer, M.; Ullrich, A. Multi-wavelength airborne laser scanning. In Proceedings of the International Lidar Mapping Forum, ILMF, New Orleans, LA, USA, 7–9 February 2011.

23. Fernandez Diaz, J.C.; Carter, W.E.; Glenie, C.; Shrestha, R.L. Multicolor terrain mapping documents critical environments. *Eos Trans. Am. Geophys. Union* **2016**, *97*, 10–15. [CrossRef]

24. Spieler, H. Class Notes: Introduction to Radiation Detectors and Electronics. Available online: http://www-physics.lbl.gov/~spieler/physics_198_notes/ (accessed on 7 November 2016).

25. Lackowicz, J.R. *Principle of Fluorescence Spectroscopy*, 3rd ed.; Springer: New York, NY, USA, 2006; Volume 1, p. 954.

26. Optech_Inc. Overcoming the Timing Limit with Multipulse Technology Altm Gemini. 2007. Available online: http://www.geo-konzept.de/data/downloads/AltMaxPaperWEB.pdf (accessed on 7 November 2016).

27. Roth, R.; Thompson, J. Practical application of multiple pulse in air (mpia) lidar in large area surveys. *Int. Arch. Photogramm. Remote Sens. Spat. Inf. Sci.* **2008**, *37*, 183–188.

28. Wang, C.K.; Tseng, Y.H.; Chu, H.J. Airborne dual-wavelength lidar data for classifying land cover. *Remote Sens.* **2014**, *6*, 700–715. [CrossRef]

29. Morsy, S.; Shaker, A.; El-Rabbany, A.; LaRocque, P. Airborne multispectral lidar data for land-cover classification and land/water mapping using different spectral indexes. *ISPRS Ann. Photogramm. Remote Sens. Spat. Inf. Sci.* **2016**, 217–224. [CrossRef]

30. Wallace, A.M.; McCarthy, A.; Nichol, C.J.; Ren, X.; Morak, S.; Martinez-Ramirez, D.; Woodhouse, I.H.; Buller, G.S. Design and evaluation of multispectral lidar for the recovery of arboreal parameters. *IEEE Trans. Geosci. Remote Sens.* **2014**, *52*, 4942–4954. [CrossRef]

31. Doneus, M.; Briese, C. Airborne laser scanning in forested areas—Potential and limitations of an archaeological prospection technique. In *Remote Sensing for Archaeological Heritage Management*; Cowley, D.C., Ed.; Europae Archaeologica Consilium (EAC): Brussels, Belgium, 2011.

32. Hartzell, P.J.; Fernandez-Diaz, J.C.; Wang, X.; Glennie, C.L.; Carter, W.E.; Shrestha, R.L.; Singhania, A.; Sartori, M.P. Comparison of Synthetic Images Generated from Lidar Intensity and Passive Hyperspectral Imagery. In Proceedings of the 2014 IEEE International Geoscience and Remote Sensing Symposium (IGARSS), Quebec City, QC, Canada, 13–18 July 2014; pp. 1345–1348.

33. Debes, C.; Merentitis, A.; Heremans, R.; Hahn, J.; Frangiadakis, N.; van Kasteren, T.; Liao, W.; Bellens, R.; Pižurica, A.; Gautama, S. Hyperspectral and lidar data fusion: Outcome of the 2013 grss data fusion contest. *IEEE J. Sel. Top. Appl. Earth Obs. Remote Sens.* **2014**, *7*, 2405–2418. [CrossRef]

34. Richards, J.A.; Jia, X. *Remote Sensing Digital Image Analysis*; Springer: Berlin, Germany, 1999; Volume 3.

35. Hopkinson, C.; Chasmer, L.; Gynan, C.; Mahoney, C.; Sitar, M. Multisensor and multispectral lidar characterization and classification of a forest environment. *Can. J. Remote Sens.* **2016**, *42*, 501–520. [CrossRef]

36. Habib, A.F.; Kersting, A.P.; Shaker, A.; Yan, W.-Y. Geometric calibration and radiometric correction of lidar data and their impact on the quality of derived products. *Sensors* **2011**, *11*, 9069–9097. [CrossRef] [PubMed]
37. Briese, C.; Pfennigbauer, M.; Ullrich, A.; Doneus, M. Multi-wavelength airborne laser scanning for archaeological prospection. *Int. Arch. Photogramm. Remote Sens. Spat. Inf. Sci.* **2013**, *40*, 119–124. [CrossRef]
38. Starek, M.J.; Vemula, R.K.; Slatton, K.C.; Shrestha, R.L.; Carter, W.E. Shoreline based feature extraction and optimal feature selection for segmenting airborne lidar intensity images. In Proceedings of the 2007 IEEE International Conference on Image Processing, San Antonio, TX, USA, 16–19 September 2007; pp. IV-369–IV-372.
39. Crasto, N.; Hopkinson, C.; Forbes, D.; Lesack, L.; Marsh, P.; Spooner, I.; van der Sanden, J. A lidar-based decision-tree classification of open water surfaces in an arctic delta. *Remote Sens. Environ.* **2015**, *164*, 90–102. [CrossRef]
40. Baker, K.; Smith, R. Quasi-inherent characteristics of the diffuse attenuation coefficient for irradiance. In *Ocean Optics VI*; International Society for Optics and Photonics: Bellingham, WA, USA, 1980; pp. 60–63.
41. Austin, R.W.; Halikas, G. The Index of Refraction of Seawater. 1976. Available online: https://escholarship.org/uc/item/8px2019m#page-1 (accessed on 7 November 2016).
42. Pan, Z.; Glennie, C.; Hartzell, P.; Fernandez-Diaz, J.; Legleiter, C.; Overstreet, B. Performance assessment of high resolution airborne full waveform lidar for shallow river bathymetry. *Remote Sens.* **2015**, *7*, 5133–5159. [CrossRef]
43. Legleiter, C.; Overstreet, B.; Glennie, C.; Pan, Z.; Fernandez-Diaz, J.; Singhania, A. Evaluating the capabilities of the casi hyperspectral imaging system and aquarius bathymetric lidar for measuring channel morphology in two distinct river environments. *Earth Surf. Process. Landf.* **2015**. [CrossRef]
44. Wright, C.W. Eaarl-B missions, calibration and validation. In Proceedings of the 15th Annual JALBTCX Airborne Coastal Mapping and Charting Workshop, Mobile, AL, USA, 10–12 June 2014.
45. Wulder, M.A.; White, J.C.; Nelson, R.F.; Næsset, E.; Ørka, H.O.; Coops, N.C.; Hilker, T.; Bater, C.W.; Gobakken, T. Lidar sampling for large-area forest characterization: A review. *Remote Sens. Environ.* **2012**, *121*, 196–209. [CrossRef]
46. Jakubowski, M.K.; Li, W.; Guo, Q.; Kelly, M. Delineating individual trees from lidar data: A comparison of vector-and raster-based segmentation approaches. *Remote Sens.* **2013**, *5*, 4163–4186. [CrossRef]
47. Wallace, L.; Lucieer, A.; Watson, C.S. Evaluating tree detection and segmentation routines on very high resolution uav lidar data. *IEEE Trans. Geosci. Remote Sens.* **2014**, *52*, 7619–7628. [CrossRef]
48. Hopkinson, C. The influence of lidar acquisition settings on canopy penetration and laser pulse return characteristics. In Proceedings of the 2006 IEEE International Symposium on Geoscience and Remote Sensing, Denver, CO, USA, 31 July–4 August 2006; pp. 2420–2423.
49. Chasmer, L.; Hopkinson, C.; Treitz, P. Investigating laser pulse penetration through a conifer canopy by integrating airborne and terrestrial lidar. *Can. J. Remote Sens.* **2006**, *32*, 116–125. [CrossRef]
50. Hopkinson, C. The influence of flying altitude, beam divergence, and pulse repetition frequency on laser pulse return intensity and canopy frequency distribution. *Can. J. Remote Sens.* **2007**, *33*, 312–324. [CrossRef]
51. Massaro, R.; Zinnert, J.; Anderson, J.; Edwards, J.; Crawford, E.; Young, D. Lidar flecks: Modeling the influence of canopy type on tactical foliage penetration by airborne, active sensor platforms. In *SPIE Defense, Security, and Sensing*; International Society for Optics and Photonics: Bellingham, WA, USA, 2012; pp. 836008–836010.
52. Hsu, W.C.; Shih, P.T.Y.; Chang, H.C.; Liu, J.K. A study on factors affecting airborne lidar penetration. *Terr. Atmos. Ocean. Sci.* **2015**, *26*, 241–251. [CrossRef]
53. Fernandez-Diaz, J.C.; Lee, H.; Glennie, C.L.; Carter, W.E.; Shrestha, R.L.; Singhania, A.; Sartori, M.P.; Hauser, D.L. Optimizing ground return detection through forest canopies with small footprint airborne mapping lidar. In Proceedings of the 2014 IEEE International Geoscience and Remote Sensing Symposium (IGARSS), Quebec City, QC, Canada, 13–18 July 2014; pp. 1963–1966.
54. Axelsson, P. DEM generation from laser scanner data using adaptive tin models. *Int. Arch. Photogramm. Remote Sens.* **2000**, *33*, 111–118.
55. Vierling, K.T.; Vierling, L.A.; Gould, W.A.; Martinuzzi, S.; Clawges, R.M. Lidar: Shedding new light on habitat characterization and modeling. *Front. Ecol. Environ.* **2008**, *6*, 90–98. [CrossRef]
56. Zolkos, S.; Goetz, S.; Dubayah, R. A meta-analysis of terrestrial aboveground biomass estimation using lidar remote sensing. *Remote Sens. Environ.* **2013**, *128*, 289–298. [CrossRef]

57. Baltsavias, E.P. Airborne laser scanning: Basic relations and formulas. *ISPRS J. Photogramm. Remote Sens.* **1999**, *54*, 199–214. [CrossRef]
58. Parrish, C.E.; Jeong, I.; Nowak, R.D.; Smith, R.B. Empirical comparison of full-waveform lidar algorithms. *Photogramm. Eng. Remote Sens.* **2011**, *77*, 825–838. [CrossRef]
59. Slatton, K.C.; Carter, W.E.; Shrestha, R.L.; Dietrich, W. Airborne laser swath mapping: Achieving the resolution and accuracy required for geosurficial research. *Geophys. Res. Lett.* **2007**, *34*. [CrossRef]
60. Cossio, T.K.; Slatton, K.C.; Carter, W.E.; Shrestha, K.Y.; Harding, D. Predicting small target detection performance of low-snr airborne lidar. *IEEE J. Sel. Top. Appl. Earth Obs. Remote Sens.* **2010**, *3*, 672–688. [CrossRef]
61. Fernandez-Diaz, J.; Carter, W.; Shrestha, R.; Glennie, C. Now you see it ... now you don't: Understanding airborne mapping lidar collection and data product generation for archaeological research in mesoamerica. *Remote Sens.* **2014**, *6*, 9951–10001. [CrossRef]
62. Penttinen, J.T. *The Telecommunications Handbook: Engineering Guidelines for Fixed, Mobile and Satellite Systems*; John Wiley & Sons: Chichester, UK, 2015.
63. Laneman, J.N.; Martinian, E.; Wornell, G.W.; Apostolopoulos, J.G. Source-channel diversity for parallel channels. *IEEE Trans. Inf. Theory* **2005**, *51*, 3518–3539. [CrossRef]
64. Glennie, C. Rigorous 3D error analysis of kinematic scanning lidar systems. *J. Appl. Geod. Jag* **2007**, *1*, 147–157. [CrossRef]
65. Heidemann, H.K. Lidar Base Specification (Version 1.2, November 2014). Available online: https://pubs.usgs.gov/tm/11b4/pdf/tm11-B4.pdf (accessed on 7 November 2016).

remote sensing

MDPI

Article

Airborne Dual-Wavelength LiDAR Data for Classifying Land Cover

Cheng-Kai Wang, Yi-Hsing Tseng and Hone-Jay Chu *

Department of Geomatics, National Cheng Kung University, Tainan 701, Taiwan;
E-Mails: c.k.wang1983@gmail.com (C.-K.W.); tseng@mail.ncku.edu.tw (Y.-H.T.)
* Author to whom correspondence should be addressed; E-Mail: honejaychu@gmail.com;
  Tel.: +886-6-237-0876; Fax: +886-6-237-5764.

Received: 8 October 2013; in revised form: 16 December 2013 / Accepted: 31 December 2013 / Published: 8 January 2014

Abstract: This study demonstrated the potential of using dual-wavelength airborne light detection and ranging (LiDAR) data to classify land cover. Dual-wavelength LiDAR data were acquired from two airborne LiDAR systems that emitted pulses of light in near-infrared (NIR) and middle-infrared (MIR) lasers. The major features of the LiDAR data, such as surface height, echo width, and dual-wavelength amplitude, were used to represent the characteristics of land cover. Based on the major features of land cover, a support vector machine was used to classify six types of suburban land cover: road and gravel, bare soil, low vegetation, high vegetation, roofs, and water bodies. Results show that using dual-wavelength LiDAR-derived information (e.g., amplitudes at NIR and MIR wavelengths) could compensate for the limitations of using single-wavelength LiDAR information (*i.e.*, poor discrimination of low vegetation) when classifying land cover.

Keywords: dual-wavelength; LiDAR; land cover classification; support vector machine (SVM)

1. Introduction

Airborne light detection and ranging (LiDAR), which measures distance by illuminating a target with a laser, is used for the rapid collection of geolocated elevation data from the surface of the earth. The positions of the targets can be obtained based on a positioning and orientation system. Increasing numbers of researchers have used airborne LiDAR data in landscape mapping [1,2]. LiDAR data typically contain 3D spatial point clouds and the intensity of returns (echoes), and its penetration capabilities make it a better system for identifying vegetation compared with photogrammetry. LiDAR systems can automatically classify land cover from geometric properties [1,3]. Moreover, multispectral image and LiDAR data can provide a large amount of spectral and geometric information for land cover classification. The combination of LiDAR data with either multispectral [4–6] or hyperspectral [7] imagery has been demonstrated to improve land cover classification.

Recently, LiDAR technology has been developed into a full-waveform LiDAR system, which can record the complete waveform of a backscattered signal echo [8]. The full-waveform LiDAR collects a continuous signal for each pulse, whereas the discrete-return LiDAR only collects four to five discrete points. Previous studies [8–10] have indicated that waveform LiDAR data record more physical characteristics than discrete-return LiDAR data. These physical characteristics affect the shape of waveforms and potentially benefit the land cover classification. For example, the waveform of an echo is wider on the canopy or ploughed fields than that on the roads [8]. Each waveform is commonly represented by a mixed Gaussian model that is produced using a Gaussian decomposition process [11]. Each return echo is represented by a Gaussian function, and the Gaussian parameters can be used to characterize the physical features of the echoes. For example, the echo width (Gaussian standard deviation) obtained from full-waveform data after decomposition, which is unavailable to discrete-return LiDAR data, has proven useful for land cover classification [12–14]. The signal-processing step extracts various features from the waveforms, such as echo width [14,15], amplitude [15], intensity [15], rise/ fall time [9] and Fourier coefficients [10,16], which are used to classify land cover and identify tree species. Given these useful features, the application of waveform LiDAR data in land cover classification has been demonstrated.

Although most commercial airborne LiDAR systems emit laser radiation at a single wavelength, multi-spectral LiDAR (MSL) systems that emit laser radiation at various wavelengths have been recently developed. Given that the return laser intensities at various wavelengths are combined in the MSL data, these data can then be used to obtain several MSL indices, such as the normalized difference vegetation index (NDVI) [17] and tree structure segmentation [3], which cannot be obtained using single-wavelength LiDAR data [18,19]. Thus, multiple potential applications of MSL systems have been demonstrated. Chlorophyll content retrieval with hyperspectral LiDAR was reported by [20], and NDVI with multispectral LiDAR was studied by [21,22]. Morsdorf *et al.* [23] simulated an MSL waveform system to demonstrate its ability in capturing a vertical profile of leaf-level physiology. A dual-wavelength LiDAR can separate the canopy from ground returns [24]. The dual-wavelength LiDAR system, a current MSL system, has been used for specific applications, such as measuring coastal water depths by using green and near-infrared (NIR) bathymetric LiDARs [25], measuring NDVI by using red-NIR wavelength LiDARs [26] and measuring the moisture content of vegetation by using NIR and middle-infrared (MIR) wavelength LiDARs [27]. However, most dual-wavelength or MSL systems are commonly used for bench mounted test instruments or experimental terrestrial operations. MSL has not yet been used to measure the land from airborne platforms, as it is still at an experimental stage.

The classification of land cover in regional areas using remote sensing is essential. In this study, airborne dual-wavelength LiDAR data were obtained by combining two commercial airborne LiDAR systems that emit NIR and MIR laser pulses. The results demonstrated the potential of using dual-wavelength airborne LiDAR data to investigate land cover types. The dual-wavelength amplitude information and waveform features were used to classify land cover. A progressive classification test was conducted to demonstrate that using dual-wavelength LiDAR data resulted in more accurate land cover classification than using single-wavelength LiDAR data.

2. Methodology

2.1. Study Area and Remote Sensing Data

Figure 1a shows the study area, Namasha (Namaxia), which is located on a hillside in southern Taiwan. Namasha, which is a famous source of precious wood, is a suburban district in the northeastern part of Kaohsiung City, located upstream of the Kao-ping river watershed (Figure 1a). This area was severely damaged by Typhoon Morakot in 2009. The study area is 0.95 km², with an average elevation and slope of approximately 722 m and 18°, respectively. Table 1 shows the dual-wavelength data *configuration* in the two LiDAR systems. LiDAR data were acquired using the Optech ALTM Pegasus HD400 and the Riegl LMS-Q680i systems. The Optech system emits NIR laser pulses at a wavelength of 1,064 nm [28], whereas the Riegl system emits MIR laser pulses at a wavelength of 1,550 nm [29]. The proposed dual-wavelength LiDAR was obtained by integrating two LiDAR systems, because no airborne, dual-wavelength (e.g., NIR-MIR) LiDAR system was currently available. In the experimental period, most land cover did not change in study area. The radiometric correction for each LiDAR system has been determined [30]. Further correction of dual-wavelength LiDAR systems will be considered for advanced usage [31]. The accuracy of the collected LiDAR data can be verified by comparing with independently surveyed ground control points. Both systems yielded horizontal accuracy of less than 0.40 m and vertical accuracy of less than 0.10 m.

Table 1. Configuration of dual-wavelength data in the two light detection and ranging (LiDAR) systems. .

	Optech ALTM Pegasus HD400	Riegl LMS-Q680i
Laser wavelength (nm)	1,064	1,550
Pulse width (FWHM, full width at half maximum) (ns)	7	4
Beam divergence (mrad)	0.20	0.50
Field of view (degree)	40	60
Footprint size (m)	0.2 at 1 km	0.5 at 1 km
Pulse rate (kHZ)	150	220
Range accuracy (cm)	1	2
Date of survey	7 October 2011	8 January 2012
Flying height (m)	2,000	1,900
Point density (pts/m²)	1.81	2.07

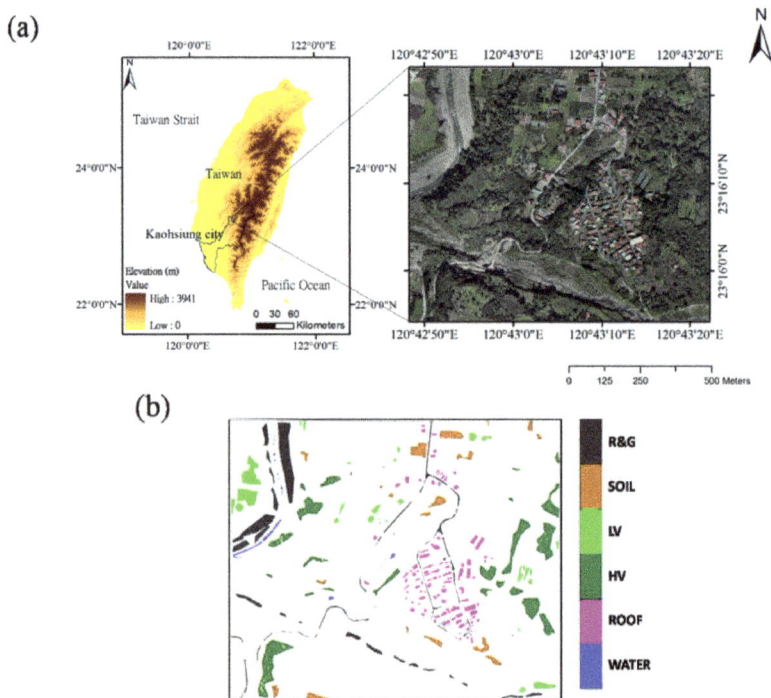

Figure 1. (a) Location of the study area; (b) location of the reference data for classification.

An IGI DigiCAM was used in the Riegl LMS-Q680i system to produce an orthoimage. To develop a reference dataset for validating the classification results, we identified six classes of land cover based on this orthoimage. The classes were selected based on the landscape of the test area: road and gravel (R&G), bare soil (SOIL), low vegetation (LV), high vegetation (HV), roofs (ROOF) and water bodies (WATER). R&G comprised the asphalt and gravel along the western side of the river and on the south side of the study area. LV comprised grass, low crops and other vegetation shorter than 2 m. HV comprised vegetation taller than 2 m, such as broadleaf evergreen forests. Water absorbs most of the incoming radiation [32]. This could result in the low intensity of LiDAR return points or few return points from water bodies. In this study, low-intensity points were returned from water bodies in the Optech system, whereas few return points from water bodies were observed in the Riegl system. Studies have applied the LiDAR data from water bodies to delineate the river boundaries [33].

Figure 1b shows the locations of the reference samples used for training and tests. Various classes of land cover within a small area are often mixed. For example, when LV is not dense, SOIL and LV may mix and become difficult to separate. Thus, two rules were used to assess the reference samples. First, the pixels must be clearly recognizable on the reference samples. Second, the reference samples must be pure, containing no more than one class of land cover. For example, an area containing a mixture of grass (LV) and trees (HV) would not be considered a reference sample.

2.2. Data Processing

Figure 2 shows the processes used in the classification model, namely, data processing, data integration, feature selection and classification. Both the Optech and Riegl LiDAR systems can provide waveform data, recording an intensity signal that represents the interactions between the emitted laser and the illuminated objects along the laser path. Multi-return echoes are recorded in the laser waveform information, and the waveform data can be decomposed into individual components to characterize the original waveform and echoes [34]. In the Gaussian decomposition method, which has been widely applied [11,13,14,35], a Gaussian function is used to represent a decomposed component; this method was used in this study to decompose a waveform into individual echo components. After decomposition, a Gaussian mixture representing a waveform with multiple distinct components was obtained. These components were described using three Gaussian parameters, namely, mean, amplitude and standard deviation. The Gaussian mean of each component was combined with the attitude information of the system when the laser was fired to map the 3D coordinates of each object. The echo amplitude and standard deviation were then attached to each 3D component as the attributes of the LiDAR points. The amplitude and standard deviation of the first LiDAR echo are termed "amplitude" and "echo width" hereafter.

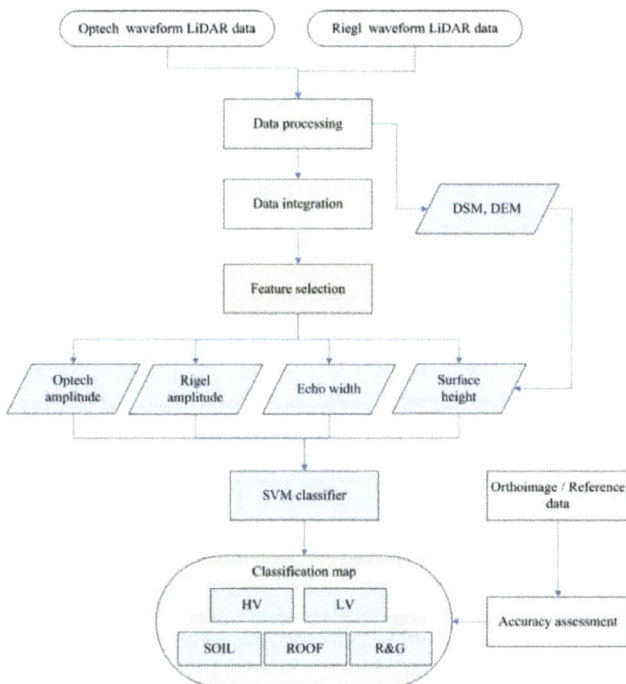

Figure 2. Flowchart of the approach. DSM, digital surface model; DEM, digital elevation model; SVM, support vector machine; HV, high vegetation; LV, low vegetation; SOIL, bare soil; ROOF, roofs; R&G, road and gravel.

Remote Sens. **2014**, *6*, 700–715

2.3. Data Integration and Feature Selection

Most land covers contain one major echo, except trees and building roofs. Only the first-return (echo) extracted from each full waveform was selected to analyze the land cover. To integrate the LiDAR data, the sample points from the two LiDAR systems were interpolated into gridded images at 1-m resolution and integrated for subsequent processing. The moving average in a circle with a 2-m radius was applied for the interpolation. Based on the LiDAR data characteristics, the following features were captured: (1) amplitude; (2) echo width; and (3) surface height from the Riegl and Optech systems. Surface height is the height of the land cover from the ground elevation and the digital surface model (DSM). The ground elevation was obtained from the digital elevation models (DEMs) that were, in turn, obtained by processing the point clouds by using TerraScan (TerraSolid software) and manual procedures. First, the TerraScan was applied to filter out non-ground points automatically. Manual inspection and editing were subsequently conducted to ensure the quality of the ground data points.

Major features were selected using the Bhattacharyya distance (separability) [36], which is widely used in feature selection and extraction studies. For feature selection, the Bhattacharyya distance, B_{ij}, has been used as a class-separability measurement between two land cover types based on the assumption of multivariate normality, and is expressed as follows:

$$B_{ij} = \frac{1}{8}(M_i - M_j)^T \left(\frac{C_i + C_j}{2}\right)^{-1}(M_i - M_j) + \frac{1}{2}\ln\left[\frac{|(C_i + C_j)/2|}{(|C_i| \cdot |C_j|)^{1/2}}\right] \tag{1}$$

where M_i and C_i are the mean vector and covariance matrices of class i, respectively. The lower values of Bhattacharyya distance represent less separable classes and higher classification errors. Based on the relation between the Bhattacharyya distance and classification error in the graph of [36], the criterion for the Bhattacharyya distance is 1 if the classification error is less than 10%.

2.4. Classification

The support vector machine (SVM), a supervised classification algorithm, is an effective classification method. SVM is capable of mixing data from diverse sources, responding robustly to dimensionality, and effectively functioning non-linearly in remote sensing applications [37]. The kernel of SVM used in this study was the Gaussian radial basis function. The SVM algorithm is implemented by using the functions from MATLAB (R2012a). Six classes (R&G, SOIL, LV, HV, ROOF and WATER) were chosen as the land cover categories. Amplitude, surface height and echo width from Riegl and Optech systems were used as the major features for classification. From the reference (sampling) data, 1% of samples in each class was selected as the training data in the SVM classifier. After the SVM classifier was trained, all reference data, except the training data, were treated as validation data. The various LiDAR feature sets were used for the progressive classification test. The confusion matrices for each feature set were calculated to assess the classification results.

3. Results and Discussion

3.1. Analysis of Features

Figure 3 shows the distribution of the amplitude, surface height and echo width of the image pixel elements from the six classes in the reference data. The amplitude values from the Riegl system allowed three groups, namely, WATER, {R&G, LV, HV} and {SOIL, ROOF}, to be distinguished. The amplitude feature from the Optech system improved the separation of WATER, R&G and the remaining classes. The merits of using both amplitudes for classifying land cover are reflected in the accuracy of the preliminary classification. The surface height from the Riegl and Optech systems provided information for separating {R&G, SOIL, LV, WATER} from {HV, ROOF}. The echo width information from the Riegl system indicated two groups, namely WATER, and {R&G, SOIL, LV, HV, ROOF}.

In summary (Figure 3), WATER can be readily classified using most of the features. R&G can be classified using amplitude information from the Optech system. SOIL can be separated from other classes by combining data on amplitude and surface height from the Riegl system. HV can be classified by combining amplitude and surface height information from the Riegl system, and ROOF can be classified by combining all features.

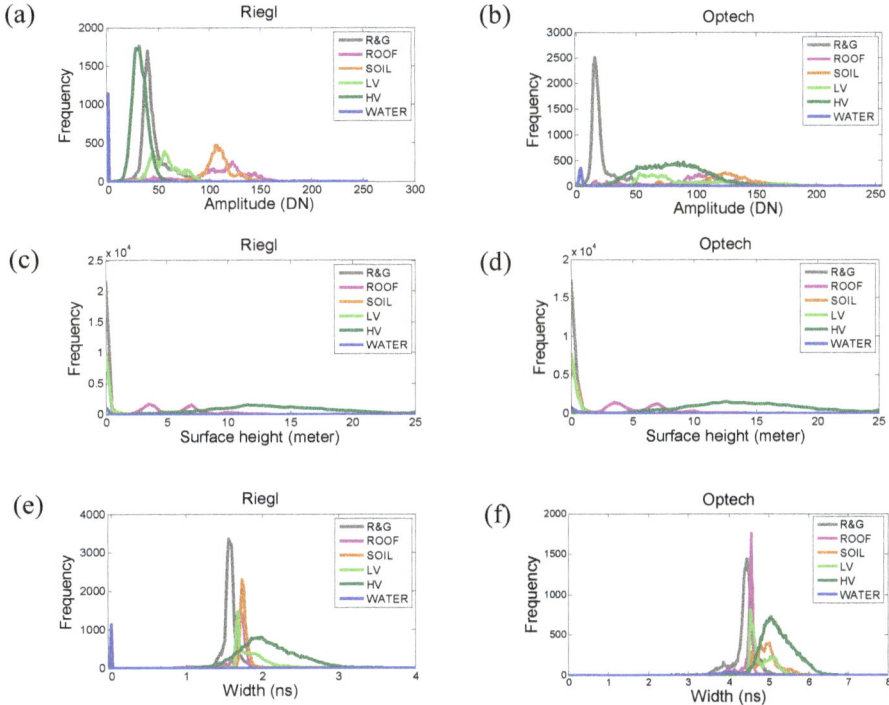

Figure 3. Frequency distribution of (**a**) the amplitude from the Riegl system, (**b**) the amplitude from the Optech system, (**c**) the surface height from the Riegl system, (**d**) the surface height from the Optech system, (**e**) the echo width from the Riegl system and (**f**) the echo width from the Optech system.

3.2. Feature Selection Using Bhattacharyya Distance

Table 2 lists the Bhattacharyya distances among the classes for different feature sets. The performances of the Riegl and Optech surface height and echo width were consistent. The Riegl surface height and echo width were eventually considered as the major features in the study based on the comparison of Bhattacharyya distance matrix determinants. The matrix determinants of the Riegl surface height and echo width were larger than those of the Optech ones. When the model considered the Riegl surface height information, the classes such as HV and ROOF could be separated from other classes. When the model considered the Riegl echo width information, the Bhattacharyya distances between HV and SOIL and between HV and R&G were 0.85 and 0.83, respectively. The Riegl and Optech systems provided complementary amplitude information for land cover discrimination. When the Optech amplitude information was used, the separability between LV and R&G was 1.68, and 0.21 between LV and SOIL. When the Riegl amplitude information was used, the separability between LV and R&G was 0.44, and 1.98 between LV and SOIL. The same situation in complementary amplitude information occurred between HV and R&G and between HV and SOIL. Compared with the separability values obtained using the Riegl amplitude information, those obtained using the Optech amplitude information were higher for HV and R&G but lower for HV and SOIL. However, when the model considered both sets of amplitude information, the separability between LV and R&G and between LV and SOIL increased. When the model considered both the

Riegl and Optech amplitude information, all land cover became separable, except between ROOF and SOIL and between ROOF and LV.

A feature is more critical if the separability among all land cover types is higher. Moreover, feature separability is highly related to classification accuracy. Amplitude is a dominant feature that varies based on the radiometric and geometric properties of the targets [38]. When classifying land cover, the measured amplitudes are high for bare soil and grass and low for water and roads. However, the amplitude varies for high vegetation and roofs of buildings depending on the materials and sensors. LiDAR-based features, such as laser intensity, amplitude, surface height, and topographic data, are primarily used to classify land cover [39]. The feature information of LiDAR data is critical to increase the discriminability of LV and HV classes because the information contains similar spectral signatures [40]. Numerous applications described in the introduction (e.g., chlorophyll or NDVI) are available from dual-wavelength LiDAR data. Future studies should examine the potential of dual-wavelength LiDAR data for extracting the details of vegetation species. When the commercial MSL becomes available for airborne platforms in the future, the MSL instruments will contain many more wavelengths to improve separability. Key information, such as the chlorophyll, NDVI and moisture content, about the vegetation can be derived from MSL data. The applications for vegetation species recognition and forest ecosystem estimation would be expected to benefit from the information.

Table 2. Bhattacharyya distance between land cover classes with different feature combinations.

Bhattacharyya Distance Using h *						
	R&G	**SOIL**	**LV**	**HV**	**ROOF**	**WATER**
R&G	0	0.28 (0.25)	0.00 (0.16)	3.21 (2.52)	2.27 (1.30)	19.52 (0.24)
SOIL		0	0.29 (0.02)	3.79 (3.10)	2.87 (1.94)	18.98 (0.11)
LV			0	3.20 (2.98)	2.26 (1.82)	19.54 (0.15)
HV				0	0.76 (0.92)	23.07 (3.41)
ROOF					0	22.16 (2.31)
WATER						0
Bhattacharyya Distance Using σ *						
	R&G	**SOIL**	**LV**	**HV**	**ROOF**	**WATER**
R&G	0	0.74 (0.70)	0.48 (0.31)	0.83 (0.56)	0.12 (0.05)	80.07 (0.28)
SOIL		0	0.30 (0.12)	0.85 (0.11)	0.30 (0.60)	301.67 (1.68)
LV			0	0.27 (0.20)	0.11 (0.23)	51.78 (1.06)
HV				0	0.47 (0.54)	28.45 (1.02)
ROOF					0	51.23 (0.53)
WATER						0
Bhattacharyya Distance Using A_{Optech} **						
	R&G	**SOIL**	**LV**	**HV**	**ROOF**	**WATER**
R&G	0	5.44	1.68	1.06	1.92	2.58
SOIL		0	0.21	0.63	0.14	7.66
LV			0	0.09	0.01	2.88
HV				0	0.14	2.12
ROOF					0	3.15
WATER						0
Bhattacharyya Distance Using A_{Riegl} **						
	R&G	**SOIL**	**LV**	**HV**	**ROOF**	**WATER**
R&G	0	5.34	0.44	0.29	1.42	28.47
SOIL		0	1.98	7.29	0.18	38.63
LV			0	1.06	0.68	26.63
HV				0	1.81	25.36
ROOF					0	24.70
WATER						0

Table 2. *Cont.*

Bhattacharyya Distance Using A_{Riegl}, A_{Optech}						
	R&G	SOIL	LV	HV	ROOF	WATER
R&G	0	8.94	1.69	1.55	2.17	29.98
SOIL		0	2.09	7.36	0.05	40.48
LV			0	1.17	0.48	26.10
HV				0	1.58	25.00
ROOF					0	24.95
WATER						0

Bhattacharyya Distance Using A_{Riegl}, A_{Optech}, h, σ						
	R&G	SOIL	LV	HV	ROOF	WATER
R&G	0	9.52	2.34	7.53	4.66	111.44
SOIL		0	2.91	13.79	4.03	376.15
LV			0	5.47	4.42	103.44
HV				0	5.21	56.45
ROOF					0	112.30
WATER						0

Bhattacharyya Distance Using A_{Riegl}, A_{Optech}, h						
	R&G	SOIL	LV	HV	ROOF	WATER
R&G	0	8.23	1.70	5.45	3.82	107.14
SOIL		0	2.55	12.07	3.58	60.54
LV			0	4.97	3.39	98.71
HV				0	3.22	46.58
ROOF					0	44.97
WATER						0

* The surface height (h) and echo width (σ) are from the Riegl system; *the number in parentheses represents those from the Optech system.* ** A_{Optech}, A_{Riegl}: the Optech and Riegl amplitude information.

3.3. Classification Accuracies

Table 3 shows the confusion matrices of the classification results using various feature sets. Based on the feature set, ϕ_1, which comprised the surface height and echo width, the overall accuracy of the classification reached 84.29%. However, the level of producer accuracy was extremely low for LV, and many LV pixels were misclassified into R&G and SOIL. Thus, the user accuracy was poor for R&G and SOIL. For the other classes (R&G, SOIL, HV, ROOF and WATER), the feature set, ϕ_1, provided sufficient information for classification. Based on the feature set, ϕ_2, including additional Optech LiDAR amplitude information, the overall accuracy reached 90.00%. By considering Riegl amplitude features, surface height and echo width in the feature set, ϕ_3, the overall accuracy reached 91.63%. LV was misclassified as R&G more frequently using Riegl amplitude information compared with using Optech amplitude information. However, SOIL was misclassified less using the Riegl amplitude than it was using the Optech amplitude (Table 3). User accuracy in separating SOIL and ROOF was higher using the Riegl amplitude than it was using the Optech amplitude, whereas user accuracy for R&G and LV was higher using the Optech amplitude information compared with using the Riegl amplitude information.

When the feature set, ϕ_4 (surface height, echo width and dual-wavelength amplitude), was used, the overall classification accuracy substantially increased compared with using a single system. When ϕ_4 was used, the producer accuracy for LV increased to 88.3% from 44.82% and 51.90% for single systems, and both the overall producer and user accuracies exceeded 90%, except the LV producer accuracy. The overall accuracy (97.4%) and Kappa (0.966) values were highest when features including the dual-wavelength amplitude were used. Without considering the echo width in ϕ_5 (surface height and dual-wavelength amplitude), the overall accuracy decreased to 96.8% and the

Kappa value decreased to 0.959. Thus, the echo width could be discarded because of its low effect on the classification. Figure 4 shows the land cover classification results based on various datasets. Most land covers were classified more accurately. These results indicate the effectiveness of using dual-wavelength airborne LiDAR data to classify land cover (Figure 4). Given that the reflectance of land cover objects varies based on wavelength, land cover objects (e.g., LV and HV, SOIL and LV) cannot be readily distinguished when amplitude information is used at a single wavelength. The features of dual-wavelength data are primarily responsible for the improvement in land cover classification demonstrated in this study.

Table 3. Confusion matrices between the reference and SVM classification using various feature sets. The feature sets are ϕ_1: {h, σ}, ϕ_2: {A_{Optech}, h, σ}, ϕ_3: {A_{Riegl}, h, σ}, ϕ_4:{A_{Riegl}, A_{Optech}, h, σ } and ϕ_5:{A_{Riegl}, A_{Optech}, h}, respectively. The user's, producer's and overall accuracies and the Kappa of the classifications are shown. A_{Riegl}, A_{Optech}, h, σ.

Feature Set	Reference Pixels	Classified Pixels						Producer's Accuracy (%)
		R&G	SOIL	LV	HV	ROOF	WATER	
ϕ_1	R&G	21,231	1,731	70	0	23	3	92.07
	SOIL	508	9,758	59	0	2	0	94.49
	LV	3,029	5,174	2,602	3	0	0	24.07
	HV	22	0	166	29,402	1,473	60	94.47
	ROOF	233	0	21	1,001	8,664	8	87.28
	WATER	0	0	0	0	0	1,254	100.00
	User's accuracy (%)	84.85	58.56	89.17	96.70	85.26	94.64	
	Overall accuracy (%)	84.29						
	Kappa	0.804						
ϕ_2	R&G	22,790	14	208	1	4	41	98.84
	SOIL	0	10,275	50	0	2	0	99.50
	LV	409	5,540	4,844	13	0	2	44.82
	HV	19	1	119	30,001	946	37	96.39
	ROOF	193	39	40	974	8,680	1	87.44
	WATER	0	0	0	0	0	1,254	100.00
	User's accuracy (%)	97.35	64.75	92.07	96.81	90.12	93.93	
	Overall accuracy (%)	90.00						
	Kappa	0.872						
ϕ_3	R&G	22,173	259	526	0	91	9	96.16
	SOIL	83	10,230	13	0	1	0	99.06
	LV	3,893	1,301	5,609	2	0	3	51.90
	HV	49	0	228	30,624	192	30	98.40
	ROOF	351	3	10	196	9,365	2	94.34
	WATER	0	0	0	0	0	1,254	100.00
	User's accuracy (%)	83.52	86.75	87.83	99.36	97.06	96.61	
	Overall accuracy (%)	91.63						
	Kappa	0.892						

Table 3. *Cont.*

Feature Set	Reference Pixels	Classified Pixels						Producer's Accuracy (%)
		R&G	SOIL	LV	HV	ROOF	WATER	
ϕ_4	R&G	22,924	12	93	0	8	21	99.42
	SOIL	0	10,283	43	0	1	0	99.57
	LV	309	943	9,543	6	0	7	88.30
	HV	30	0	249	30,767	54	23	98.86
	ROOF	360	19	55	14	9,477	2	95.47
	WATER	0	0	0	0	0	1,254	100.00
	User's accuracy (%)	97.04	91.35	95.59	99.93	99.34	95.94	
	Overall accuracy (%)	97.40						
	Kappa	0.966						
ϕ_5	R&G	22,942	24	73	1	1	17	99.50
	SOIL	0	10,289	38	0	0	0	99.63
	LV	307	869	9,631	0	0	1	89.11
	HV	8	0	251	30,702	131	31	98.65
	ROOF	658	79	41	204	8,944	1	90.10
	WATER	0	0	0	0	0	1,254	100.00
	User's accuracy (%)	95.93	91.37	95.98	99.34	98.55	96.17	
	Overall accuracy (%)	96.84						
	Kappa	0.959						

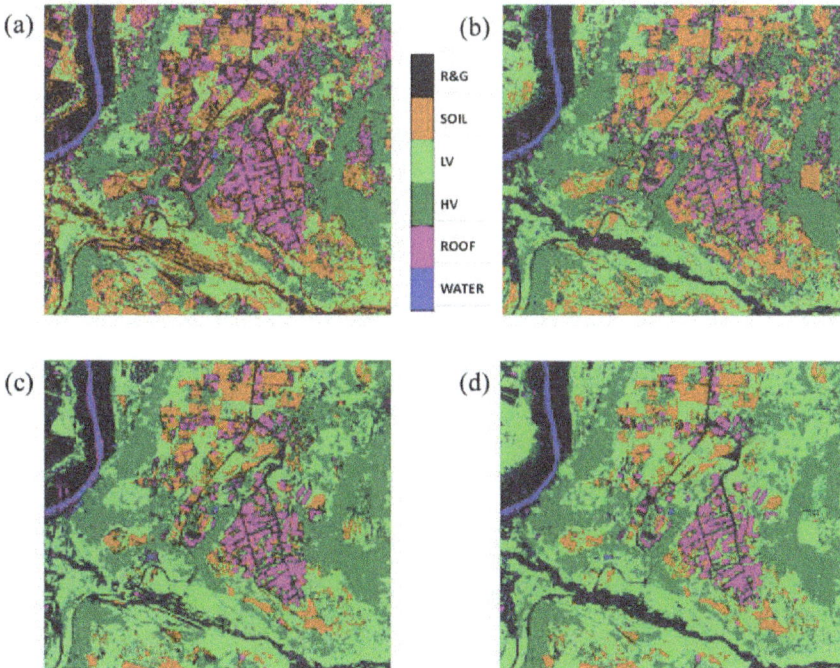

Figure 4. *Cont.*

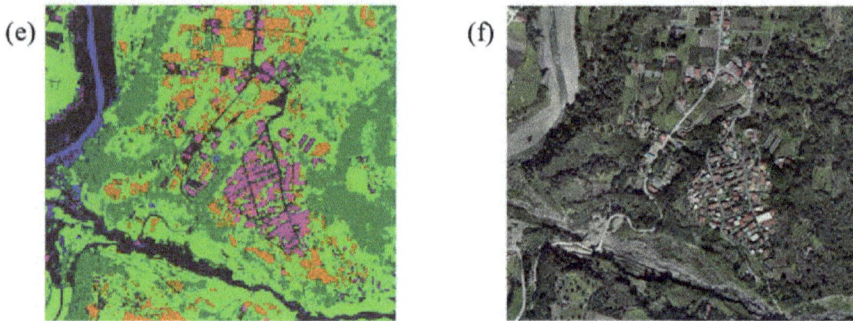

Figure 4. Results of the classifications using the five feature sets: (**a**) Riegl surface height, echo width (set ϕ_1); (**b**) Optech amplitude, Riegl surface height, echo width (set ϕ_2); (**c**) Riegl amplitude, Riegl surface height, echo width (set ϕ_3); (**d**) Riegl amplitude, Optech amplitude, Riegl surface height, echo width (set ϕ_4); (**e**) Riegl amplitude, Optech amplitude, Riegl surface height (set ϕ_5); and (**f**) the orthoimage.

The use of dual-wavelength LiDAR data offers effective geometry information to classify land cover. First, LiDAR data can provide 3D information. Thus, the DSM, DEM, and surface height can be directly obtained. Second, LiDAR data can record multiple returns in forest areas. The canopy reflectance information in spectral images is considerably influenced by the objects under the canopy. Dual-wavelength LiDAR amplitude and geometric information for the canopy, understory vegetation, soil, and other land cover types precisely represent the features of these covers. By contrast, based on the spectral image, the canopy signal cannot be readily separated from that of the understory vegetation and soil. Thus, the LiDAR data are potentially useful in classifying 3D tree species. Third, current LiDAR systems can record waveform data that allow physical features to be extracted, such as the echo width used in this study. These features cannot be obtained from discrete-return LiDAR. All these features, including dual-wavelength amplitude features, facilitate land cover classification, as clearly demonstrated by the current findings. Therefore, this study revealed the potential of dual-wavelength LiDAR applications, which can be developed when airborne LiDAR systems become available. From a practical perspective, the combination of LiDAR and multi-spectral images will be useful for land cover classification.

4. Conclusion

In this study, two airborne LiDAR systems were used to obtain dual-wavelength LiDAR data (i.e., amplitudes at NIR and MIR wavelengths) and classify land cover. The proposed processes involved waveform data processing, data integration, feature selection, and land cover classification. The findings show that using dual-wavelength airborne LiDAR systems could substantially improve land cover classification in large areas compared with using single-wavelength LiDAR. The dual-wavelength amplitude features facilitated the identification of vegetation, particularly LV, more accurately compared with using single-wavelength amplitude.

Based on the major features of LiDAR data, land cover was effectively classified in the absence of auxiliary remote sensing data, and the overall classification accuracy reached 97.4%. Additional applications can be designed for this method in the future until airborne dual-wavelength LiDAR systems are developed.

Acknowledgments: The authors wish to thank the editors and reviewers for their valuable comments and suggestions and Chi-Kuei Wang for the method discussion. This research also received funding from the Headquarters of University Advancement at the National Cheng Kung University, which is sponsored by the Ministry of Education, Taiwan, ROC.

Conflicts of Interest: The authors declare no conflict of interest.

References

1. Korpela, I.S. Mapping of understory lichens with airborne discrete-return LiDAR data. *Remote Sens. Environ.* **2008**, *112*, 3891–3897.
2. Miliaresis, G.; Kokkas, N. Segmentation and object-based classification for the extraction of the building class from LIDAR DEMs. *Comput. Geosci.* **2007**, *33*, 1076–1087.
3. Suomalainen, J.; Hakala, T.; Kaartinen, H.; Räikkönen, E.; Kaasalainen, S. Demonstration of a virtual active hyperspectral LiDAR in automated point cloud classification. *ISPRS J. Photogramm. Remote Sens.* **2011**, *66*, 637–641.
4. Bork, E.W.; Su, J.G. Integrating LIDAR data and multispectral imagery for enhanced classification of rangeland vegetation: A meta analysis. *Remote Sens. Environ.* **2007**, *111*, 11–24.
5. Hellesen, T.; Matikainen, L. An object-based approach for mapping shrub and tree cover on grassland habitats by use of LiDAR and CIR orthoimages. *Remote Sens.* **2013**, *5*, 558–583.
6. Hartfield, K.A.; Landau, K.I.; van Leeuwen, W.J.D. Fusion of high resolution aerial multispectral and LiDAR data: Land cover in the context of urban mosquito habitat. *Remote Sens.* **2011**, *3*, 2364–2383.
7. Dalponte, M.; Bruzzone, L.; Gianelle, D. Fusion of hyperspectral and LIDAR remote sensing data for classification of complex forest areas. *IEEE Trans. Geosci. Remote Sens.* **2008**, *46*, 1416–1427.
8. Mallet, C.; Bretar, F. Full-waveform topographic lidar: State-of-the-art. *ISPRS J. Photogramm. Remote Sens.* **2009**, *64*, 1–16.
9. Neuenschwander, A.L.; Magruder, L.A.; Tyler, M. Landcover classification of small-footprint, full-waveform lidar data. *J. Appl. Remote Sens.* **2009**, *3*, doi:10.1117/1.3229944.
10. Vaughn, N.R.; Moskal, L.M.; Turnblom, E.C. Fourier transformation of waveform Lidar for species recognition. *Remote Sens. Lett.* **2011**, *2*, 347–356.
11. Wagner, W.; Ullrich, A.; Ducic, V.; Melzer, T.; Studnicka, N. Gaussian decomposition and calibration of a novel small-footprint full-waveform digitising airborne laser scanner. *ISPRS J. Photogramm. Remote Sens.* **2006**, *60*, 100–112.
12. Alexander, C.; Tansey, K.; Kaduk, J.; Holland, D.; Tate, N.J. Backscatter coefficient as an attribute for the classification of full-waveform airborne laser scanning data in urban areas. *ISPRS J. Photogramm. Remote Sens.* **2010**, *65*, 423–432.
13. Wagner, W.; Hollaus, M.; Briese, C.; Ducic, V. 3D vegetation mapping using small-footprint full-waveform airborne laser scanners. *Int. J. Remote Sens.* **2008**, *29*, 1433–1452.
14. Hollaus, M.; Aubrecht, C.; Höfle, B.; Steinnocher, K.; Wagner, W. Roughness mapping on various vertical scales based on full-waveform airborne laser scanning data. *Remote Sens.* **2011**, *3*, 503–523.
15. Heinzel, J.; Koch, B. Exploring full-waveform LiDAR parameters for tree species classification. *Int. J. Appl. Earth Obs. Geoinf.* **2011**, *13*, 152–160.
16. Vaughn, N.R.; Moskal, L.M.; Turnblom, E.C. Tree species detection accuracies using discrete point lidar and airborne waveform lidar. *Remote Sens.* **2012**, *4*, 377–403.
17. Wei, G.; Shalei, S.; Bo, Z.; Shuo, S.; Faquan, L.; Xuewu, C. Multi-wavelength canopy LiDAR for remote sensing of vegetation: Design and system performance. *ISPRS J. Photogramm. Remote Sens.* **2012**, *69*, 1–9.
18. Rall, J.A.R.; Knox, R.G. Spectral ratio biospheric lidar. In Proceedings of the 2004 IEEE International, Geoscience and Remote Sensing Symposium, 2004, IGARSS'04, Anchorage, AK, USA, 20–24 September 2004; Volume 3, pp. 1951–1954.
19. Kaasalainen, S.; Lindroos, T.; Hyyppa, J. Toward hyperspectral lidar: Measurement of spectral backscatter intensity with a supercontinuum laser source. *IEEE Geosci. Remote Sens. Lett.* **2007**, *4*, 211–215.
20. Hakala, T.; Suomalainen, J.; Kaasalainen, S.; Chen, Y. Full waveform hyperspectral LiDAR for terrestrial laser scanning. *Opt. Express* **2012**, *20*, 7119–7127.
21. Woodhouse, I.H.; Nichol, C.; Sinclair, P.; Jack, J.; Morsdorf, F.; Malthus, T.J.; Patenaude, G. A multispectral canopy liDAR demonstrator project. *IEEE Geosci. Remote Sens. Lett.* **2011**, *8*, 839–843.
22. Wallace, A.; Nichol, C.; Woodhouse, I. Recovery of forest canopy parameters by Inversion of multispectral LiDAR data. *Remote Sens.* **2012**, *4*, 509–531.
23. Morsdorf, F.; Nichol, C.; Malthus, T.; Woodhouse, I.H. Assessing forest structural and physiological information content of multi-spectral LiDAR waveforms by radiative transfer modelling. *Remote Sens. Environ.* **2009**, *113*, 2152–2163.
24. Hancock, S.; Lewis, P.; Foster, M.; Disney, M.; Muller, J.-P. Measuring forests with dual wavelength lidar: A simulation study over topography. *Agric. For. Meteorol.* **2012**, *161*, 123–133.
25. Irish, J.L.; Lillycrop, W.J. Scanning laser mapping of the coastal zone: The SHOALS system. *ISPRS J. Photogramm. Remote Sens.* **1999**, *54*, 123–129.
26. Chen, Y.; Raikkonen, E.; Kaasalainen, S.; Suomalainen, J.; Hakala, T.; Hyyppa, J.; Chen, R. Two-channel hyperspectral LiDAR with a supercontinuum laser source. *Sensors* **2010**, *10*, 7057–7066.

27. Gaulton, R.; Danson, F.M.; Ramirez, F.A.; Gunawan, O. The potential of dual-wavelength laser scanning for estimating vegetation moisture content. *Remote Sens. Environ.* **2013**, *132*, 32–39.
28. Optech. Airborne Surveying. Available online: http://www.optech.ca/ (accessed on 20 March 2011).
29. Riegl Laser Measurement Systems. Products of Airborne Scanning. Available online: http://www.riegl.com/ (accessed on 20 March 2011).
30. Höfle, B.; Pfeifer, N. Correction of laser scanning intensity data: Data and model-driven approaches. *ISPRS J. Photogramm. Remote Sens.* **2007**, *62*, 415–433.
31. Briese, C.; Pfennigbauer, M.; Lehner, H.; Ullrich, A.; Wagner, W.; Pfeifer, N. Radiometric calibration of multi-wavelength airborne laser scanning data. In Proceedings of the ISPRS Annals of the Photogrammetry, Remote Sensing and Spatial Information Sciences, XXII ISPRS Congress, Melbourne, Australia, 25 August–1 September 2012.
32. Antonarakis, A.S.; Richards, K.S.; Brasington, J. Object-based land cover classification using airborne LiDAR. *Remote Sens. Environ.* **2008**, *112*, 2988–2998.
33. Cobby, D.M.; Mason, D.C.; Davenport, I.J. Image processing of airborne scanning laser altimetry data for improved river flood modelling. *ISPRS J. Photogramm. Remote Sens.* **2001**, *56*, 121–138.
34. Jinha, J.; Crawford, M.M. Extraction of features from LIDAR waveform data for characterizing forest structure. *IEEE Geosci. Remote Sens. Lett.* **2012**, *9*, 492–496.
35. Reitberger, J.; Schnörr, C.; Krzystek, P.; Stilla, U. 3D segmentation of single trees exploiting full waveform LIDAR data. *ISPRS J. Photogramm. Remote Sens.* **2009**, *64*, 561–574.
36. Choi, E.; Lee, C. Feature extraction based on the Bhattacharyya distance. *Pattern Recognit.* **2003**, *36*, 1703–1709.
37. Bretar, F.; Chauve, A.; Bailly, J.S.; Mallet, C.; Jacome, A. Terrain surfaces and 3-D landcover classification from small footprint full-waveform lidar data: Application to badlands. *Hydrol. Earth Syst. Sci.* **2009**, *13*, 1531–1544.
38. Mallet, C.; Bretar, F.; Roux, M.; Soergel, U.; Heipke, C. Relevance assessment of full-waveform lidar data for urban area classification. *ISPRS J. Photogramm. Remote Sens.* **2011**, *66*, S71–S84.
39. Ke, Y.; Quackenbush, L.J.; Im, J. Synergistic use of QuickBird multispectral imagery and LIDAR data for object-based forest species classification. *Remote Sens. Environ.* **2010**, *114*, 1141–1154.
40. Tooke, T.R.; Coops, N.C.; Goodwin, N.R.; Voogt, J.A. Extracting urban vegetation characteristics using spectral mixture analysis and decision tree classifications. *Remote Sens. Environ.* **2009**, *113*, 398–407.

remote sensing

MDPI

Article

Single-Sensor Solution to Tree Species Classification Using Multispectral Airborne Laser Scanning

Xiaowei Yu [1,*], Juha Hyyppä [1], Paula Litkey [1], Harri Kaartinen [1], Mikko Vastaranta [2] and Markus Holopainen [2]

[1] Finnish Geospatial Research Institute, National Land Survey, Geodeetinrinne 2, FI-02430 Masala, Finland; juha.hyyppa@nls.fi (J.H.); paula.litkey@nls.fi (P.L.); harri.kaartinen@nls.fi (H.K.)
[2] Department of Forest Sciences, University of Helsinki, FI-00014 Helsinki, Finland; mikko.vastaranta@helsinki.fi (M.V.); markus.holopainen@helsinki.fi (M.H.)
* Correspondence: xiaowei.yu@nls.fi; Tel.: +358-50-4126536

Academic Editors: Lars T. Waser, Randolph H. Wynne and Prasad S. Thenkabail
Received: 20 October 2016; Accepted: 20 January 2017; Published: 27 January 2017

Abstract: This paper investigated the potential of multispectral airborne laser scanning (ALS) data for individual tree detection and tree species classification. The aim was to develop a single-sensor solution for forest mapping that is capable of providing species-specific information, required for forest management and planning purposes. Experiments were conducted using 1903 ground measured trees from 22 sample plots and multispectral ALS data, acquired with an Optech Titan scanner over a boreal forest, mainly consisting of Scots pine (*Pinus Sylvestris*), Norway spruce (*Picea Abies*), and birch (*Betula* sp.), in southern Finland. ALS-features used as predictors for tree species were extracted from segmented tree objects and used in random forest classification. Different combinations of features, including point cloud features, and intensity features of single and multiple channels, were tested. Among the field-measured trees, 61.3% were correctly detected. The best overall accuracy (OA) of tree species classification achieved for correctly-detected trees was 85.9% (Kappa = 0.75), using a point cloud and single-channel intensity features combination, which was not significantly different from the ones that were obtained either using all features (OA = 85.6%, Kappa = 0.75), or single-channel intensity features alone (OA = 85.4%, Kappa = 0.75). Point cloud features alone achieved the lowest accuracy, with an OA of 76.0%. Field-measured trees were also divided into four categories. An examination of the classification accuracy for four categories of trees showed that isolated and dominant trees can be detected with a detection rate of 91.9%, and classified with a high overall accuracy of 90.5%. The corresponding detection rate and accuracy were 81.5% and 89.8% for a group of trees, 26.4% and 79.1% for trees next to a larger tree, and 7.2% and 53.9% for trees situated under a larger tree, respectively. The results suggest that Channel 2 (1064 nm) contains more information for separating pine, spruce, and birch, followed by channel 1 (1550 nm) and channel 3 (532 nm) with an overall accuracy of 81.9%, 78.3%, and 69.1%, respectively. Our results indicate that the use of multispectral ALS data has great potential to lead to a single-sensor solution for forest mapping.

Keywords: multispectral laser scanning; ALS; individual tree detection; tree species classification; random forest

---

1. Introduction

Knowledge of tree species plays an important role in forest management and planning. The optimum output, requested by forest companies from the forest mapping process, is the species-specific size distribution of the trees. The traditional method, based on field inventory work for tree species identification, is labor intensive, time consuming, and limited by spatial extent.

Therefore, remote sensing techniques were introduced, such as the interpretation of large-scale aerial color or infra-red images [1,2]. Although remotely-sensed data have been widely used for forest applications, traditional optical remote sensing techniques suffer from a lack of the ability to capture three-dimensional forest structures, particularly in unevenly-aged, mixed species forests with multiple canopy layers [3]. Recent developments in active remote sensing, particularly laser scanning techniques, have shown potential in forest mapping and other applications because of the capability to capture three-dimensional (3D) information of forests [4–11].

Airborne laser scanning (ALS) is a useful tool for retrieving biophysical variables and for updating forest inventory maps. The successful use of ALS data has been demonstrated for a variety of applications. For example, ALS has been used to estimate tree height [6,7], identify tree species [8–10], and estimate tree volume, biomass [11–13], and growth [14,15]. Tree species information at an individual tree level is particularly useful in growth and yield estimates, and has been primarily studied for forest applications, such as updating forest inventories. Tree species classification using ALS has not been intensively studied, when compared with studies on the successful use of ALS for other forest attribute mapping, because of the lack of spectral information. Brandtberg [9] classified three leaf-off individual deciduous tree species (oaks, red maple, and yellow poplar) in West Virginia, USA, using high density laser data, and reported 64% total accuracy. Holmgren and Persson [8] classified Norway spruce and Scots pine in Remningstorp, Sweden, using ALS-derived point and intensity features, and achieved an accuracy of 95%. Ørka et al. [16] classified three species (spruce, birch and aspen) at the Ostmarka natural forest in southern Norway. Suratno et al. [17] classified ponderosa pine, Douglas-fir, western larch, and lodgepole pine, in a western North American montane forest using low density ALS data, and achieved a classification accuracy of 95% at the dominant species level, and 68% for individual trees.

Intensity was also demonstrated to be useful information for tree species identification. Ørka et al. [18] reported an accuracy of 73% when classifying conifers and deciduous trees, solely based on intensity information. Korpela et al. [19] classified Scots pine, Norway spruce, and birch, by using intensity variables at Hyytiälä in southern Finland, and showed that intensity features can contribute to a classification accuracy of 88% among the three species. With full-waveform (FWF) lasers, the total received power corresponding to the backscattering cross-section can be calculated, which provides information on the objects, from the intensity waveform.

Previous studies have demonstrated that FWF data and the derived metrics can be used to improve the performance of tree species classification. For example, Yao et al. [20] demonstrated the usefulness of waveform features for the classification of deciduous and coniferous trees. Heinzel and Koch [21] analyzed a set of waveform features and identified the most predictive features for classifying up to six tree species. Cao et al. [22] demonstrated that full-waveform data and derived metrics have significant potential for tree species classification in the subtropical forests, and results demonstrated that all tree species were classified with relatively high accuracy (68.6% for six classes, 75.8% for four main species, and 86.2% for conifers and broadleaved trees).

Previous studies have also revealed that combining multispectral information with 3D ALS data can lead to improvement in the accuracy of tree extraction and tree species classification, as we can take advantage of both datasets. For example, Naidoo et al. [23] concluded that the use of ALS and hyperspectral data yielded the highest classification accuracy and prediction success for the eight savanna tree species, with an overall classification accuracy of 87.68%. Zhou et al. [24] demonstrated that the ALS intensity data can contribute to the classification of shaded areas in an urban environment where high resolution digital aerial imagery alone did not produce good results. The fusion of high resolution (satellite or aerial) remote sensing and ALS data can achieve mutual benefits for compensating the lack of 3D structure from imagery and multi-spectral information from ALS data. With respect to the success of these case studies, multi-sensor data fusion seems to be a feasible solution, especially for the mapping of land cover over large areas.

However, there are challenging factors that limit the effective operational use of the fused datasets [25,26]. For example, geometric and radiometric registration between two datasets is demanding, because of the fact that data are normally acquired at different times, using different sensors. It is also costly to make measurements with two sensors, particularly in the boreal forest zone where the measurements can seldom be carried out during a single flight, because ALS measurements can be taken two to four times longer than aerial/hyperspectral measurements during a day, since ALS does not depend on sun light illumination. Furthermore, in contrast to passive imagery, laser scanning always views the targets at the zero degree phase angle in a narrow off-nadir viewing geometry and the transmitted energy is also controllable, thus the interpretation of the laser intensity is less complex than in the case of passive airborne images [27]. The recently developed multispectral laser scanning technique is therefore becoming an attractive option for forest mapping, because it can provide not only a dense point cloud, but also spectral information which can simplify data processing and facilitate the interpretation of data. There are a couple of studies that have demonstrated the potential of multispectral ALS for classifying tree species [28,29]. In Lindberg et al [28], multispectral data were acquired with separate instruments and from different flights—an analogue to Titan multispectral data. The study described the characterization of tree species from ALS data, using three wavelengths: 1064 nm, 1550 nm, and 532 nm, and a point density over 20 point/m^2. However, classification accuracy was not reported. In St-Onge and Budei [29], values for the mean and standard deviation of the intensity in three channels of Titan multispectral ALS, were used in the classification of broadleave vs. needleleaf trees (level 1), and eight genera (level 2) in a suburb of the city of Toronto, Canada. Random forest classification produced a classification error of 4.59% in the case of the level 1 classification (broadleave vs. needleleaf trees), and of 24.29% in the case of the level 2 classification. The point density of the data used was 10.6 first returns/m^2 per channel. Currently, the cost of data acquisition of multispectral ALS is relatively higher than that of aerial images and ALS data, if they are acquired from the same flight. However, it is expected that this cost will decrease in the future, as the technology advances. Therefore, it is worth investigating the potential of multispectral laser scanning for forest inventories, particularly for tree species classification. The objectives of this study are to evaluate the feasibility of multispectral ALS data for tree species classification with intensive field measurements, and to investigate the information content of features derived from both point cloud and intensity. The study was conducted in a boreal forest using 1903 trees in 22 plots.

2. Study Area and Materials

2.1. Test Site

The 5 km × 5 km study area, located in Evo, southern Finland (61.19°N, 25.11°E), belongs to the southern Boreal Forest Zone. It contains approximately 2000 ha of managed boreal forest, having an average stand size of slightly less than 1 ha. The area comprises a broad mixture of forest stands, varying from natural to intensively managed forests. The elevation of the area varies from 125 m to 185 m above sea level. Scots pine (*Pinus sylvestris*) and Norway spruce (*Picea abies*) are the dominant tree species in the study area, and contribute 40% and 35% of the total volume, respectively, whereas the share of deciduous trees (mainly birches, *Betula* sp.) constitutes only 24% of the total volume.

2.2. Field Measurements

Field measurements were undertaken in the summer of 2014 and consisted of individual tree measurements for 91 plots in Evo. Sample plots, with dimensions of 32 m × 32 m, were selected, based on the prestratification of ALS data to distribute plots over various stand height and density classes. Sample plot locations were determined using the geographic coordinates of the plot center and its four corners. Plot center positions were measured using a total station (Trimble 5602), which was oriented to the local coordinate system using ground control points measured with VRS-GNSS (Trimble R8) in open areas, close to the plot. Terrestrial laser scanning was also used to assist tree mapping in the field.

After field measurements had been made, the tree map was further verified by comparing it with ALS data. If there was a discrepancy between the two data, the plot was manually corrected to match the ALS data, ensuring a positional accuracy of 0.5 m. Detailed information on the establishment of the sample plots can be found in Yu et al. [30].

From the sample plots, all trees with a diameter at beast height (DBH) exceeding 5 cm, were tallied with steel calipers from two directions perpendicular to each other, and a mean was taken as the value for the DBH. Tree height was measured using an electronic hypsometer. Height measurement accuracy is expected to be approximately 0.5 m. Tree species was also recorded. Among 91 sample plots, 22 plots were fully covered by the airborne laser scanning data and used in this study (Figure 1). The descriptive statistics of 22 sample plots, and the sample trees by species, are summarized in Tables 1 and 2, respectively.

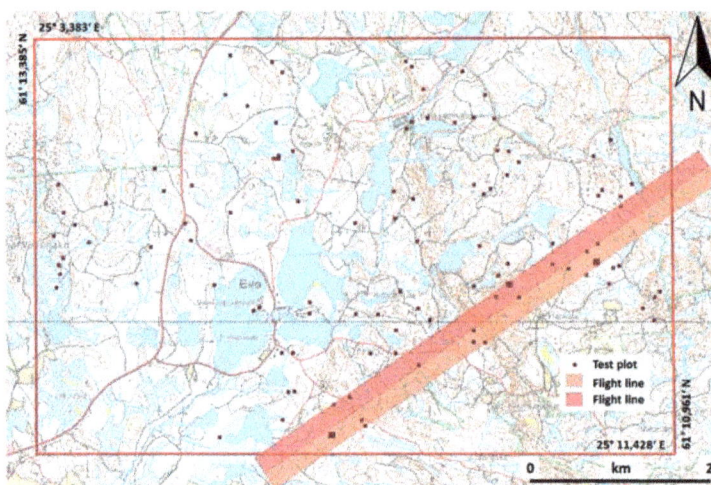

Figure 1. Study area, airborne laser scanning coverage, and sample plots.

Table 1. The descriptive statistics of Lorey's height (H_g), basal area weighted mean diameter (D_g), basal area (G), stem volume (VOL), aboveground biomass (AGB), and trees per hectare (TPH) in the 22 sample plots.

	Minimum	Maximum	Mean	Standard Deviation
H_g (m)	10.02	31.09	21.09	4.41
D_g (cm)	13.92	46.42	25.78	7.50
G (m^2/ha)	6.60	43.17	26.79	7.83
VOL (m^3/ha)	34.46	518.39	270.14	110.04
AGB (Mg/ha)	19.06	230.63	134.49	48.33
TPH (trees/ha)	342	3057	940	554

Table 2. The descriptive statistics of sample trees by tree species.

		Minimum	Maximum	Mean	Standard Deviation	Number of Trees
Pine	Tree height (m)	2.30	28.20	17.29	4.76	839
	DBH (cm)	5.00	39.80	19.37	6.92	
Spruce	Tree height (m)	2.20	35.30	14.32	8.94	630
	DBH (cm)	5.00	57.90	16.27	11.75	
Birch	Tree height (m)	2.00	30.20	16.89	4.80	434
	DBH (cm)	5.10	55.80	14.61	6.42	

2.3. Airborne Laser Scanning Data

Airborne laser scanning data were acquired on the 21st of August 2015, using an Optech Titan multispectral system, operating at a pulse rate of 300 kHz per channel. Optech Titan is the first commercial airborne laser scanner which operates with three channels. The spectral channels are two infrared ones, of 1550 nm (channel 1) and 1064 nm (channel 2), and a green channel of 532 nm (channel 3). The three channels are oriented in different directions, so that the 1064 nm channel is pointing nadir, the 1550 nm channel is positioned 3.5 degrees forward, and the 532 nm channel is positioned 7 degrees forward. As a result, laser pulses are not registered from exactly the same location in each channel. The data in this study were collected from an altitude of 400 m above sea level, resulting in an average pulse density of approximately 3×21 pulses per m^2, and the footprint sizes in diameter were 14 cm in channel 1 and 2, and 28 cm in channel 3 (beam divergence of 0.35 mrad in channel 1 and 2, and 0.7 mrad in channel 3). The system was configured to record up to five echoes per pulse, and intensities were also recorded for each return and channel.

3. Methods

3.1. Preprocessing of Multispectral ALS data

Recorded intensity is the amount of energy reflected back (i.e., backscattered) to the laser sensor, which is a function of several variables, such as target surface characteristics (reflectance, wetness and roughness), environmental effects (atmospheric transmittance, moisture), and acquisition parameters and instruments [31,32]. It is therefore necessary to calibrate intensity values for compensating the impact of these factors and achieving better classification accuracy. In this study, a simplified model was used for the return intensity calibration, in order to correct for range according to the Equation (1) with an exponential factor of 2.5 [27] for forest area, since the environmental factors can be considered stable, and the same acquisition parameters and instruments were maintained during the survey.

$$I_c = I * \left(\frac{R}{Rs} \right)^{2.5} \tag{1}$$

where I_c is the normalized intensity, I is the raw intensity, R is the sensor to target range, and Rs is the reference range or average flying height (in this study $Rs = 400$ m). The physical explanation for the exponential factor of 2.5 is that the laser beam is affected by the mixture of targets illuminated by the laser beam, such as leaves, dense needle groups (exponential factor close to 2), and the branches and needles (exponential factor close to 3). Correction was separately completed for each channel.

Strip matching between flight lines and between channels was performed by the data provider. Afterwards, the ALS point clouds were processed to separate ground returns from vegetation returns, using the progressive triangulated irregular network (TIN) densification method proposed by Axelsson [33]. Point cloud of channel 2 was used in this process, in order to reduce the amount of data provided that ground returns were dense enough to represent the variation of the terrain. The ALS data from the three channels were then normalised by removing ground elevation from the laser height measurements based on the digital terrain model created from classified ground points. The normalized point cloud was further processed for individual tree detection.

3.2. Individual Tree Detection

Individual trees were detected using a minimum curvature-based algorithm [34], which started with the creation of canopy height model (CHM). The method has two major steps: firstly, the tree tops were found by a local maximum filtering algorithm. Secondly, tree crowns were delineated using a watershed algorithm. CHM was created by taking the maximum value of normalized laser points within a grid cell of size 0.5 m. In the first step, CHM was smoothed by Gaussian filtering and stretched by minimum curvature, and then local maxima were detected from processed CHM.

These local maxima were considered as tree tops and used as seeds in the following step, where crown was delineated by a marker-controlled watershed algorithm, with a background mask of a 2 m height threshold, i.e., if the CHM value was less than 2 m, the pixel was classified as "background". During the segmentation processes, the tree crown shape and location of individual trees were determined, based on the segment outline and the location of maximum hit within the segment. In this study, points of first returns from all three channels were used to create CHM.

Detected trees were then linked with the trees measured in the field by an automatic matching algorithm based on the Hausdorff distance [14]. In the matching procedure, the distance in 3D space between the detected tree and the field-measured tree was used as a matching criterion. If a field-measured tree and a detected tree were the closest to each other, and the distance between them was less than a threshold, the tree was considered as correctly detected. Given the possible difference in tree location measurements from ALS data (at tree top), and in the field (at tree root) and tree height underestimation by laser scanning, a 5 m threshold distance was used to reject a match. The field-measured trees without any link to a tree segment were considered as non-detectable trees, resulting in an omission error, and a tree segment without a link to a reference tree resulted in a commission error.

3.3. Features Derivation from Multispectral ALS Data

In order to classify and characterize the object properly, we can use geometry (from point clouds) and spectral information. For each extracted tree segment, several features were derived from multispectral ALS data and used in tree species classification. They can be grouped into three categories: point cloud features, single-channel intensity (SCI) features and multi-channel intensity (MCI) features. For point cloud features, points falling within each individual tree segment were extracted from all returns, and the normalized heights of these points were used for deriving the tree features. The features were calculated based on points over a height threshold of 2 m above ground from all channels, including maximum height (Hmax), mean height (Hmean), standard deviation of height (Hstd), range of the height (Hrange) represented by the difference between the lowest and highest points, penetration rate as the ratio of points below 2 m to the total number of points, crown area (CA) and volume (CV) estimated by a 2D and 3D convex hull of the points, and crown diameter (CD). In addition, height percentiles (HP_{10} to HP_{90}) from 10% to 90%, with an increment of 10%, were calculated. Furthermore, density-related features were calculated by dividing the height into 10 equal intervals, and calculating the ratio of points within each interval to the total number of points (D_1 to D_{10}). As SCI features, we calculated the minimum, maximum, mean, standard deviation, skewness, kurtosis of the intensity, and percentiles of intensity at 5%, and from 10% to 90%, with a 10% increment for each channel. MCI features included the intensity ratio between each channel and the sum of all channels, and the subtraction of channel 2 and 3, divided by the sum of channel 2 and 3. In total, 145 tree features were generated and used in the analysis. More detailed definitions and explanations are given in Table 3.

Table 3. Tree features derived from normalized point data and spectral information. (superscript i = 1, 2, and 3 for channel 1, 2, and 3)

Feature	Definition
Point cloud features	
Hmax	Maximum of the normalized heights of all points
Hmean	Arithmetic mean of normalized height of all points above 2 m threshold
Hstd	Standard deviation of normalized height of all points above 2 m threshold
Hrange	Range of normalized height of all points above 2 m threshold
P	Penetration as a ratio between number of returns below 2 m and total returns
CA	Crown area as the area of convex hull in 2D
CV	Crown volume as the convex hull in 3D
CD	Crown diameter calculated from crown area considering crown as a circle.

Table 3. *Cont.*

Feature	Definition
HP_{10} to HP_{90}	10% to 90% percentiles of normalized height of all points above 2 m threshold with a 10% increment
D_1 to D_{10}	$D_i = N_i/N_{total}$, where i = 1 to 10, N_i is the number of points within ith layer when tree height was divided into 10 intervals starting from 2 m, N_{total} is the number of all points.
Single-channel Intensity features	
I^i_{min}	Minimum of intensity
I^i_{max}	Maximum of intensity
I^i_m	Mean of intensity
I^i_{std}	Standard deviation of intensity
I^i_{sk}	Skewness of intensity
I^i_{range}	Range of intensity
I^i_{kut}	Kurtosis of intensity
$I^i_{5,10\ to\ 90}$	Percentiles of intensity at 5% and from 10% to 90% with 10% increment.
Multi-channel intensity features	
$R^i_F = I^1_F/(I^1_F + I^2_F + I^3_F)$ $N_F = (I^2_F - I^3_F)/(I^2_F + I^3_F)$	Ratios of intensity features, F refers to different single-channel intensity features. Index of intensity features

3.4. Feature Selection and Tree Species Classification

Introduced by Breiman [35], random forests (RF) is a technique which consists of an ensemble of decision trees, using a majority vote for the final prediction. RF has shown successful performances in many applications, such as in the classification of urban scenes [36] and forest attribute prediction [34,37]. In this study, tree species were estimated based on prediction models by RF using tree features as predictors and tree species as a response for correctly detected trees. Although RF is able to deal with high dimensional data [38], the results of classification can be significantly improved if only the most important features are used [39]. Considering the number of observations and the correlation between the features in this study, it was necessary to reduce the feature dimension to avoid overfitting. The RF built-in measure of feature importance was used to search for a subset of predictors that optimally model responses, subject to constraints which minimize the correlation among the features. In this study, 15 of the most important features were selected for each experimental setting by measuring how influential the predictor was at predicting the response. The parameter settings for RF in each classification were as follows: 200 decision trees were built, with four predictors being randomly selected for the best splitting at the nodes, when decision trees were built.

3.5. Evaluation of Accuracy

The accuracy of tree species classification was evaluated by comparing the classified tree species with the reference tree species recorded in the field for correctly detected trees. The result of the comparison can be represented by an error matrix. Four widely-used measures, i.e., producer's accuracy, user's accuracy, overall accuracy (OA), and Kappa coefficient, were computed for evaluating the performance of the classification. To avoid overfitting of the classification model, independent validation was conducted by equally dividing available data into two sets: one for training the classification model, and the other for testing the performance.

We evaluated different combinations of extracted features for their predictive power as follows: (i) point cloud features as predictors, (ii) SCI features as predictors, (iii) MCI features as predictors, (iv) point cloud and SCI features as predictors, and (v) all features as predictors. The McNemar test was used to determine whether there are statistically significant differences between pairs of classifications, with the different predictor settings mentioned above (e.g., point cloud features vs. SCI features, SCI vs. MCI features vs. all features, and so on).

We also classified four categories of trees, to analyze how crown positioning affected classification accuracy. Thus, the field-measured trees were divided into four categories, based on the distance and height difference of neighbor trees as follows:

- Isolated or dominant trees that are well separated from other trees (distance to neighbor trees is greater than 3 m or tree height is greater than neighbor tree by 2 m) (referred to as C1).
- Group of trees: trees are growing closely to each other (distance less than 3 m) and have a similar height (height difference is less than 2 m) (referred to as C2).
- Tree next to a larger tree: the distance of a tree to a neighbor tree is greater than 1.5 m and the height is smaller than the height of neighbor tree by at least 2 m (referred to as C3).
- Tree under a larger tree: the distance of a tree to a neighbor tree is less than 1.5 m and the height is smaller than the height of neighbor tree by at least 2 m (referred to as C4).

The number of trees in each category was 580 in C1, 552 in C2, 590 in C3, and 181 in C4.

4. Results

4.1. Accuracy of Individual Tree Detection

The accuracy of individual tree detection was evaluated by comparing tree segments with field reference data. Overall, out of 1903 trees, 61.3% of trees were correctly detected. Most of the undetectable specimens were understory trees and trees that were near to a larger tree. At plot level, the detection rate varied between 50% and 98%. In the dense plots, the tree detection rate was lower than that in the sparse plots. The detection rate was also affected when the plot was located near the boundary of the data coverage, where the point distribution was not optimum, i.e., the points in one direction were denser than in a perpendicular direction. When considering the different categories of trees, the detection rate was 91.9% for C1, and 81.5%, 26.4%, and 7.2%, for C2, C3, and C4, respectively. A higher detection rate was expected for trees in C1, because the crown boundary was well defined. For trees in C2, there was a tendency to merge trees into one segment if they were close to each other, whereas in C3, trees were more likely to merge with neighbor trees, leading to a low detection rate. In C4, trees were often not detectable because individual tree detection was based on a CHM where taller trees overtopped the tree underneath. An example of individual tree detection is shown in Figure 2.

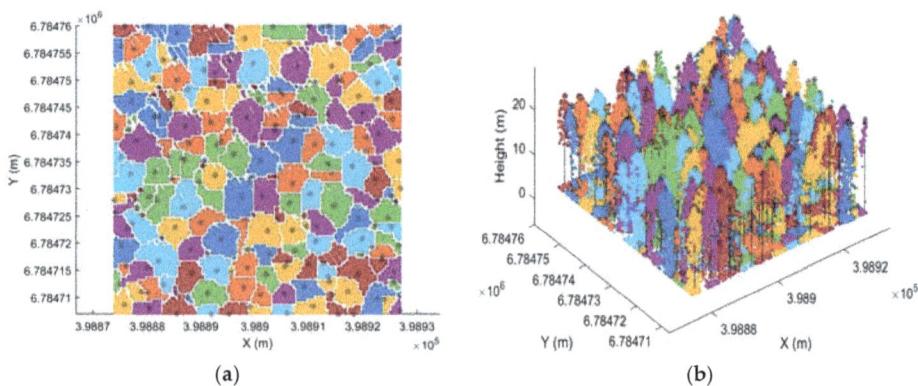

Figure 2. Result of individual tree detection for one plot. (**a**) top view, (**b**) 3D view.

4.2. Classification with Different Combinations of Features

The confusion matrix of classification and the result of accuracy evaluation are presented in Tables 4–8 for the species classification based on the different combination of features, i.e., point cloud features alone (Table 4), SCI features (Table 5), MCI features (Table 6), point cloud and SCI features (Table 7), and all features combined (Table 8). The highest level of accuracy (85.9%) was obtained with a combination of point cloud and SCI features. Point cloud features alone produced the lowest overall accuracy of 76.0%, while single-channel intensity features produced an overall accuracy of 85.4%. A McNemar test indicated no significant difference between classifications based on SCI and all features, at a 5% significant level ($p = 0.69$). Additionally, there was no difference between classifications based on SCI features, and the combination of point cloud and SCI features ($p = 0.58$). This suggested that point cloud features did not provide more information for classification. McNemar tests showed that the difference between classifications based on other pairs of features, were all significant at a 5% significant level (Table 9). Classification accuracy also varied between species. The best accuracies were obtained for pine trees with a 97.5% producer's accuracy using point cloud and SCI features, and for spruce trees with a 78.2% producer's accuracy using SCI features, followed by birch trees with a 71.8% producer's accuracy using all features. SCI features produced slightly better results than MCI features ($p = 0.03$). The results suggested that MCI features do not provide more information than SCI features.

Table 4. Confusion matrix and accuracy evaluation of classification with 15 selected point cloud features and test data.

		Predicted			Producer (%)
		Pine	Spruce	Birch	
Reference	Pine	294	10	22	90.18
	Spruce	24	84	11	70.59
	Birch	54	17	60	45.80
User (%)		79.03	75.68	64.52	OA = 76.04%, Kappa = 0.57

Table 5. Confusion matrix and accuracy evaluation of classification with 15 selected single-channel intensity features and test data.

		Predicted			Producer (%)
		Pine	Spruce	Birch	
Reference	Pine	306	8	12	93.87
	Spruce	15	93	11	78.15
	Birch	24	14	93	70.99
User (%)		88.70	80.87	80.17	OA = 85.42%, Kappa = 0.75

Table 6. Confusion matrix and accuracy evaluation of classification with 15 selected multi-channel intensity features and test data.

		Predicted			Producer (%)
		Pine	Spruce	Birch	
Reference	Pine	299	9	17	92.00
	Spruce	19	85	16	70.83
	Birch	23	22	86	65.65
User (%)		87.68	73.28	72.27	OA = 81.60%, Kappa = 0.68

Table 7. Confusion matrix and accuracy evaluation of classification with 15 selected point cloud and single-channel intensity feature combination and test data.

		Predicted			Producer (%)
		Pine	Spruce	Birch	
Reference	Pine	317	3	5	97.54
	Spruce	18	85	17	70.83
	Birch	29	9	93	70.99
User (%)		87.09	87.63	80.87	OA = 85.94%, Kappa = 0.75

Table 8. Confusion matrix and accuracy evaluation of classification with 15 selected features among all features and test data.

		Predicted			Producer (%)
		Pine	Spruce	Birch	
Reference	Pine	306	8	11	94.15
	Spruce	13	93	14	77.50
	Birch	28	9	94	71.76
User (%)		88.18	84.55	78.99	OA = 85.59%, Kappa = 0.75

Table 9. McNemar tests on pairs of the classifications using different combination of features. The number in the table is p value. The number with a superscript * indicated that the difference between classifications is significant at a 5% significant level.

Feature	Point Cloud	SCI	MCI	Point Cloud + SCI
SCI	$1.4 \times 10-7$ *			
MCI	$2.4 \times 10-4$ *	0.03 *		
Point cloud + SCI	$1.3 \times 10-11$ *	0.58	0.01 *	
All	$7.2 \times 10-8$ *	0.69	0.02 *	0.66

4.3. Classifications for Four Defined Categories of Trees

We also examined classification accuracy for four categories of trees based on the 15 best features among point cloud and SCI features, because the use of all features does not improve the accuracy. A 10-fold cross-validation strategy was applied in this case because the number of trees in C3 and C4 was low. The obtained accuracies varied widely. For isolated and dominant trees, an overall accuracy of 90.47% was achieved (Table 10). The corresponding accuracy was 89.80% for trees in C2 (Table 11), 79.09% for trees in C3 (Table 12), and 53.85% for trees in C4 (Table 13). As can be seen, very high accuracy was achieved for isolated and dominant trees, and for groups of trees. The accuracy was about 36 percentage points lower for suppressed trees. For dominant trees, both pine and spruce achieved a high accuracy of over 90%, because of their well identified conical shape. For birch, moderate accuracy was obtained. Overall, pines are classified with higher accuracy and less misclassifications, while birches tend to be misclassified as pine for all four categories of trees, resulting in a low user's accuracy for pine. Spruces are more likely to be mixed with both pine and birch.

Table 10. Confusion matrix of classification with point cloud and single-channel intensity features for isolated and dominant trees (C1).

		Predicted			Producer (%)
		Pine	Spruce	Birch	
	Pine	299	0	6	98.03
Reference	Spruce	10	128	4	90.14
	Birch	20	11	57	64.77
User (%)		90.88	92.09	85.07	OA = 90.47%, Kappa = 0.83

Table 11. Confusion matrix of classification with point cloud and intensity features for group of trees (C2).

		Predicted			Producer (%)
		Pine	Spruce	Birch	
	Pine	268	6	2	98.41
Reference	Spruce	4	34	10	67.57
	Birch	19	5	103	77.59
User (%)		92.10	75.56	89.57	OA = 89.80%, Kappa = 0.80

Table 12. Confusion matrix of classification with point cloud and intensity features for trees next to a larger tree (C3).

		Predicted			Producer (%)
		Pine	Spruce	Birch	
	Pine	63	1	2	95.45
Reference	Spruce	6	30	8	68.18
	Birch	10	5	28	65.12
User (%)		79.75	83.33	73.68	OA = 79.09%, Kappa = 0.67

Table 13. Confusion matrix of classification with point cloud and intensity features for trees under a larger tree (C4).

		Predicted			Producer (%)
		Pine	Spruce	Birch	
	Pine	4	0	0	100
Reference	Spruce	1	2	2	40.0
	Birch	2	1	1	25.0
User (%)		57.14	66.67	33.33	OA = 53.85%, Kappa = 0.32

4.4. Feature Importance

We also investigated which input features and channels are most relevant for tree species classification based on the measure provided by the RF algorithm for assessing feature importance. If a feature is influential in the prediction, then permuting its values should affect the model error. If a feature is not influential, then permuting its values should have little or no effect on the model error. Table 14 lists the top five features in the classifications based on different combinations of the features. In the classification based on point cloud features, the most important features were penetration and higher level percentiles. Two density-related features at higher and middle layers were also scored as important as higher percentiles. For the case of classification based on the SCI features, the wavelength of 1064nm (Channel 2) seems to contain more information for separating pine, spruce, and birch, followed by wavelengths of 1550nm (channel 1), and 532nm (channel 3). The classification based on the features of the three separate channels also confirmed analysis with an overall accuracy of 81.9%, 78.3%, and 69.1%, for channel 2, 1, and 3, respectively. The difference between the pairs of classification is significant at a 5% significant level based on McNemar tests. Minimum values and the 90% percentile of intensity are two of the most powerful predictors for all channels in such cases. In MCI-based classification, ratios at higher percentiles for the three channels were among the most important features. Overall, when all features were considered, the minimum intensity of channel 2 and 3, the ratio at 90% percentile for channel 2 and 3, and one point cloud features (P), are among the top five most importance features.

Table 14. The features have the most predictive power in different classification scenarios. A detailed explanation of the features can be found in Table 3. The number in parentheses is the score for the feature. The higher the score, the more important the feature.

Cases	Top 5 features
Point cloud features	P (3.8), D_9 (1.6), Hmax (1.5), D_5 (1.4), HP_{90} (1.3)
SCI features	I^2_{min} (1.9), I^2_{p90} (1.6), I^1_{sk} (1.4), I^1_{p90} (1.5), I^3_{p90} (1.5)
MCI features	R^3_{p90} (1.7), R^2_{p90} (1.4), R^2_{range} (1.4), R^1_{p80} (1.3), N_{p90} (1.3)
Point cloud and SCI features	I^2_{min} (2.0), Hmax (1.5), I^2_{p90} (1.5), I^3_{p90} (1.8), P (1.6)
All features	I^2_{min} (1.8), R^3_{p90} (1.7), P (1.5), I^3_{min} (1.4), R^2_{p90} (1.2)

5. Discussion

In this study, we explored the potential of multispectral ALS data in tree species classification of a boreal forest. Results showed that multispectral ALS data can be used to separate three main tree species, i.e., pine, spruce, and birch, with a high overall accuracy of 85.9% in the best case scenario, which was based on the combined use of point cloud and SCI features. Overall, the results indicated that the intensity of the three channels contains more information for tree species classification than point cloud data. When using the intensity of the three channels, both the producer's and user's accuracies for single tree species were improved, as well as the overall accuracy compared with the results obtained from point cloud data. However, different types of features are more influential on certain tree species. For example, intensity features are more powerful in separating birch from pine and spruce (produce's accuracy improved from 45.8% to 71%, and user's accuracy from 64.5% to 80.2%, when compared with those using point cloud features). With the inclusion of point cloud features, the classification accuracy of intensity features was improved by only 0.5 percentage point, while the corresponding value was 10 percentage points when adding intensity features to point cloud features.

The individual tree-detection rate was not very high in this study. Two factors influenced this. Firstly, individual tree detection was based on CHM, so most of the understory trees were not detectable and 3D information of the dense point cloud was not fully utilized. Secondly, distribution of the point cloud was not optimal, as it was denser in scanning direction than flight direction. The uneven distribution of points affected the results of individual tree detection, as the detection rate tended to decrease when the plot was located near the boundary of data coverage where uneven distribution was more severe. In order to improve individual tree detection, we recommend developing methods which can fully utilize the 3D information provided by point cloud. Multispectral information could also be useful for improving the accuracy of individual tree detection. When point cloud and spectral information are used in tree detection, a simultaneous classification is possible, such that the knowledge relevant to each can aid in the analysis of the other. Ultimately, this could lead to the improvement of accuracy of individual tree detection and classification, as well as computational advantages.

A large variance in feature values can be found, due to the irregular geometry of the canopy surface and varying degrees of penetration. There were more points penetrated, thus reaching the ground, in channels 1 and 3, than in channel 2. One potential factor that contributed to this was the forward viewing geometry for channels 1 and 3. There were also more returns in channels 2 and 3, than in channel 1. For the same point cloud features, the values in channel 3 were higher than those in channels 1 and 2, while channels 1 and 2 produced similar values. This trend was observed for all three species and could be one reason why the point cloud features did not significantly improve the classification, when used with SCI features. SCI features also overlapped between species. However, the degree of overlap varied among the features and channels. In general, higher percentiles of the intensity distribution and minimum intensity value were more separated than the lower percentiles. For example, the maximum intensity was smaller for pine than spruce in both channel 1 and channel 2, while similar values were observed for pine and spruce in channel 3. There were more overlapping values and variations at lower percentiles of intensity distribution, among tree species in all channels.

MCI features have been used to reduce the radiometric effects on multispectral images and improvements in classification have been reported. In this study, the use of similar ratios and indices did not improve the classification. The reason for this could be that the laser scanner is an active instrument, and recorded intensity mainly depends on the instrument design, measurement range, and reflectance of the targets. If the same instrument has been used for data acquisition and the range effect has been corrected, the major factor affecting recorded intensity is the targets illuminated by the laser. Therefore, the intensity itself is good enough to characterize the objects.

The results in this study are in agreement with previous results, in which tree species were classified using ALS combined with multi/hyper-spectral data, although the studies cannot be compared directly because of the differences in the data used, and the number and type of species identified. For example, Dalponte et al [40] reported a kappa accuracy of 0.89 when classifying three boreal tree species (pine, spruce and broad-leaves), using hyperspectral and ALS data with the manual detection of trees. The higher kappa coefficient obtained in their study could be a result of better delineation of individual trees by manual detection, and a higher spectral resolution. Jones et al [41] achieved an overall accuracy of 73% for classifying 11 species in coastal south-western Canada, using hyperspectral and ALS data. The lower accuracy could be explained by the higher number of species recognised in the study. This indicates that multispectral ALS data contains similar information to the fusion of multispectral images and ALS data. Compared with the previous study, which used a multispectral ALS of similar density for tree species classification, St-onge and Budei [29] reported a classification error of 4.59% in the case of the level 1 classification (broadleave vs. needleleaf trees), and of 24.29% in the case of level 2 classification (eight genera), using intensity features (mean and standard deviation of intensity in three channels). The different number of species could be the reason for the difference in accuracy.

The use of a single source of data apparently has advantages over the use of fused data, with respect to data processing. For example, geometric and radiometric calibrations between different data sources produce big challenges, and require much effort to compensate the changes in illumination conditions and vegetation [26]. Furthermore, previous studies have shown that background signal reduced classification accuracies when using multispectral/hyperspectral images [42–44]. In contrast, multispectral ALS data can easily separate the reflections of vegetation from the reflections of the ground, thus background influences on the results, like soil, could be minimised. Therefore, the accuracy of the classification could be improved with the use of multispectral ALS data. However, this issue needs to be explored further in order to investigate the extent to which the accuracy can be improved with multispectral ALS data.

The intensity values of different returns are affected by the vertical structures of trees. In theory, the intensity of only returns can be radiometrically calibrated with high reliability. The first of many returns is distorted by the signal penetrating to the second and other layers. However, there is still valuable information of all return intensities confirmed by this study. In the future, it should be studied whether it is possible to calibrate the intensities of multiple returns in a better way, by taking into account the attenuated part of the signal and the part that causes other returns.

The major drawback with applied Titan data was the inhomogeneous distribution of the point cloud. In the across track, the point spacing was significantly smaller than that in the along track. Either lower aircraft speed or higher scan frequency should be achieved to provide more homogenous point spacing. Another drawback is that the points from the three channels are not registered from the same location, which means that it is not multispectral data in the conventional sense. As a result, pixel/point wise classification cannot be performed; instead, object-based analysis has to be carried out, like in this study. The accuracy of the classification may also deteriorate, because the backscatter from different channels could come from different parts of the objects. The impact of such system design on classifications needs further investigation. Regardless of these drawbacks, multispectral ALS data are still a valuable data source for tree species classification, as shown in this study.

Currently, it is more expensive to acquire multispectral ALS data than aerial images and single-channel ALS. However, it is anticipated that the price will drop as technology develops, and the market is growing. Furthermore, ALS data can be acquired during both the day and night, which partly compensates for the cost of the data acquisition. Therefore, multispectral ALS data could be a cost-effective solution for species-specific mapping of forests in the future, and it has the potential to increase the automation of the whole processing chain.

6. Conclusions

In this study, we assessed the potential utility of single-sensor multispectral ALS data for tree species classification in mixed coniferous forests in a boreal zone. The results suggest that additional information, provided by multispectral laser scanning, may be a valuable source of information for tree species classification of pine, spruce, and birch, which are the main tree species found in boreal forest zones. The best overall classification accuracy achieved was 85.9% using point cloud and SCI features, which was not significantly different from the ones in which all features, or solely SCI features were used. Point cloud features alone achieved an accuracy of 76.0%. Channel 2 performed the best when separating pine, spruce, and birch, followed by channel 1 and channel 3, with overall accuracies of 81.9%, 78.3%, and 69.1%, respectively.

This preliminary study has demonstrated the potential of multispectral airborne laser scanning for possible future solutions for automatic single-sensor forest mapping. It is expected that multispectral airborne laser scanning can provide highly valuable data for forest mapping. However, there are many aspects of multispectral ALS that need to be investigated further, for example: how will multispectral ALS data perform in other forest zones where the number of species composition is higher? Is it possible to derive more useful features to improve the classification? From a practical point of view, future studies could explore the possibility to improve the accuracy of forest inventory mapping using species information obtained from this study.

Acknowledgments: The research leading to these results has received funding from the Academy of Finland projects "Interaction of LiDAR/Radar Beams with Forests Using Mini-UAV and Mobile Forest Tomography" (No. 259348), "Centre of Excellence in Laser Scanning Research (CoE-LaSR)", laserscanning.fi, (No. 272195) and "Competence-Based Growth Through Integrated Disruptive Technologies of 3D Digitalization, Robotics, Geospatial Information and Image Processing/Computing–Point Cloud Ecosystem", pointcloud.fi (No. 293389).

Author Contributions: X. Yu and J. Hyyppä designed the experiments. X. Yu carried out the research, analyzed the data and wrote the first draft of the paper. P. Litkey performed intensity calibration. M. Holopainen and M. Vastaranta were responsible for the field measurements. H. Kaartinen contributed material collection. All co-authors assisted in writing and improving the manuscript.

Conflicts of Interest: The authors declare no conflict of interest.

References

1. Gillis, M.; Leckie, D. Forest inventory update in Canada. *For. Chron.* **1996**, *72*, 138–156. [CrossRef]
2. Waser, L.T.; Ginzler, C.; Kuechler, M.; Baltsavias, E.; Hurni, L. Semi-automatic classification of tree species in different forest ecosystems by spectral and geometric variables derived from Airborne Digital Sensor (ADS40) and RC30 data. *Remote Sen. Environ.* **2011**, *115*, 76–85. [CrossRef]
3. Lovell, J.L.; Jupp, D.L.; Culvenor, D.S.; Coops, N.C. Using airborne and ground based ranging LiDAR to measure canopy structure in Australian forests. *Can. J. Remote Sens.* **2003**, *29*, 607–622. [CrossRef]
4. Coops, N.C.; Hilker, T.; Wulder, M.A.; St-Onge, B.; Newnham, G.; Siggins, A.; Trofymow, J.T. Estimating canopy structure of Douglas-fir forest stands from discrete-return LiDAR. *Trees* **2007**, *21*, 295–310. [CrossRef]

5. Wulder, M.A.; White, J.C.; Nelson, R.F.; Næsset, E.; Ørka, H.O.; Coops, N.C.; Hilker, T.; Bater, C.W.; Gobakken, T. LiDAR sampling for large-area forest characterization: A review. *Remote Sens. Environ.* **2012**, *121*, 196–209. [CrossRef]

6. Næsset, E.; Økland, T. Estimating tree height and tree crown properties using airborne scanning laser in a boreal nature reserve. *Remote Sens. Environ.* **2002**, *79*, 105–115. [CrossRef]

7. Clark, M.L.; Clark, D.B.; Roberts, D.A. Small-footprint LiDAR estimation of subcanopy elevation and tree height in a tropical rain forest landscape. *Remote Sens. Environ.* **2004**, *91*, 68–89. [CrossRef]

8. Holmgren, J.; Persson, Å. Identifying species of individual trees using airborne laser scanning. *Remote Sens. Environ.* **2004**, *90*, 415–423. [CrossRef]

9. Brandtberg, T. Classifying individual tree species under leaf-off and leaf-on conditions using airborne LiDAR. *ISPRS J. Photogramm. Remote Sens.* **2007**, *61*, 325–340. [CrossRef]

10. Lindberg, E.; Eysn, L.; Hollaus, M.; Holmgren, J.; Pfeifer, N. Delineation of tree crowns and tree species classification from full-waveform airborne laser scanning data using 3-D ellipsoidal clustering. *IEEE J. Sel. Top. Appl. Earth Obs. Remote Sens.* **2014**, *7*, 3174–3181. [CrossRef]

11. Hyyppä, J.; Kelle, O.; Lehikoinen, M.; Inkinen, M. A segmentation-based method to retrieve stem volume estimates from 3-D tree height models produced by laser scanners. *IEEE Trans. Geosci. Remote Sens.* **2001**, *39*, 969–975. [CrossRef]

12. Hollaus, M.; Wagner, W.; Maier, B.; Schadauer, K. Airborne laser scanning of forest stem volume in a mountainous environment. *Sensors* **2007**, *7*, 1559–1577. [CrossRef]

13. Ahmed, R.; Siqueira, P.; Hensley, S. A study of forest biomass estimates from LiDAR in the northern temperate forests of New England. *Remote Sens. Environ.* **2013**, *130*, 121–135. [CrossRef]

14. Yu, X.; Hyyppä, J.; Kukko, A.; Maltamo, M.; Kaartinen, H. Change detection techniques for canopy height growth measurements using airborne laser scanning data. *Photogram. Eng. Remote Sens.* **2006**, *72*, 1339–1348. [CrossRef]

15. Yu, X.; Hyyppä, J.; Kaartinen, H.; Maltamo, M.; Hyyppä, H. Obtaining plotwise mean height and volume growth in boreal forests using multi-temporal laser surveys and various change detection techniques. *Int. J. Remote Sens.* **2008**, *29*, 1367–1386. [CrossRef]

16. Ørka, H.O.; Næsset, E.; Bollandsås, O.M. Utilizing airborne laser intensity for tree species classification. *Int. Arch. Photogramm. Remote Sens. Spat. Inf. Sci.* **2007**, *36*, 300–304.

17. Suratno, A.; Seielstad, C.; Queen, L. Tree species identification in mixed coniferous forest using airborne laser scanning. *ISPRS J. Photogramm. Remote Sens.* **2009**, *64*, 683–693. [CrossRef]

18. Ørka, H.O.; Naesset, E.; Bollandsas, O.M. Classifying species of individual trees by intensity and structure features derived from airborne laser scanner data. *Remote Sens. Environ.* **2009**, *113*, 1163–1174. [CrossRef]

19. Korpela, I.; Ørka, H.O.; Maltamo, M.; Tokola, T. Tree species classification using airborne LiDAR-effects of stand and tree parameters, downsizing of training set, intensity normalization and sensor type. *Silva Fenn.* **2010**, *44*, 319–339. [CrossRef]

20. Yao, W.; Krzystek, P.; Heurich, M. Tree species classification and estimation of stem volume and DBH based on single. *Remote Sens. Environ.* **2012**, *123*, 368–380. [CrossRef]

21. Heinzel, J.; Koch, B. Exploring full-waveform LiDAR parameters for tree species classification. *Int. J. Appl. Earth Obs. Geoinf.* **2011**, *13*, 152–160. [CrossRef]

22. Cao, L.; Coops, N.C.; Innes, J.L.; Dai, J.; Ruan, H. Tree species classification in subtropical forests using small-footprintfull-waveform LiDAR data. *Int. J. Appl. Earth Obs. Geoinf.* **2016**, *49*, 39–51. [CrossRef]

23. Naidoo, L.; Cho, M.A.; Mathieu, R.; Asner, G. Classification of savanna tree species, in the Greater Kruger National Park region, by integrating hyperspectral and LiDAR data in a Random Forest data mining environment. *ISPRS J. Photogramm. Remote Sens.* **2012**, *69*, 167–179. [CrossRef]

24. Zhou, W.; Huang, G.; Troy, A.; Cadenasso, M. Object-based land cover classification of shaded areas in high spatial resolution imagery of urban areas: A comparison study. *Remote Sens. Environ.* **2009**, *113*, 1769–1777. [CrossRef]

25. Packalén, P.; Suvanto, A.; Maltamo, M. A two stage method to estimate species-specific growing stock. *Photogramm. Eng. Remote Sens.* **2009**, *75*, 1451–1460. [CrossRef]

26. Puttonen, E.; Suomalainen, J.; Hakala, T.; Räikkönen, E.; Kaartinen, H.; Kaasalainen, S.; Litkey, P. Tree species classification from fused active hyperspectral reflectance and LiDAR measurements. *For. Ecol. Manag.* **2010**, *260*, 1843–1852. [CrossRef]

27. Korpela, I.; Orka, H.; Hyyppa, J.; Heikkinen, V.; Tokola, T. Range and AGC normalization in airborne discrete-return LiDAR intensity data for forest canopies. *ISPRS J. Photogramm. Remote Sens.* **2010**, *65*, 369–379. [CrossRef]

28. Lindberg, E.; Briese, C.; Doneus, M.; Hollaus, M.; Schroiff, A.; Pfeifer, N. Multi-wavelength airborne laser scanning for characterization of tree species. In Proceedings of SilviLaser 2015, La Grande Motte, France, 28–30 September 2015; pp. 271–273.

29. St-Onge, B.; Budei, B.C. Individual tree species identification using the multispectral return intensities of the Optech Titan LiDAR system. In Proceedings of SilviLaser 2015, La Grande Motte, France, 28–30 September 2015; pp. 71–73.

30. Yu, X.; Hyyppä, J.; Karjalainen, M.; Nurminen, K.; Karila, K.; Vastaranta, M.; Kankare, V.; Kaartinen, H.; Holopainen, M.; Honkavaara, E.; et al. Comparison of laser and stereo optical, SAR and InSAR point clouds from air- and space-borne sources in the retrieval of forest inventory attributes. *Remote Sens.* **2015**, *7*, 15933–15954. [CrossRef]

31. Bright, B.C.; Hicke, J.A.; Hudak, A.T. Estimating aboveground carbon stocks of a forest affected by mountain pine beetle in Idaho using LiDAR and multispectral imagery. *Remote Sens. Environ.* **2012**, *124*, 270–281. [CrossRef]

32. Ahokas, E.; Hyyppä, J.; Yu, X.; Liang, X.; Matikainen, L.; Karila, K.; Litkey, P.; Kukko, A.; Jaakkola, A.; Kaartinen, H.; et al. Towards automatic single-sensor mapping by multispectral airborne laser scanning. In Proceedings of XXIII ISPRS Congress, Commission III, Prague, Czech Republic, 12–19 July 2016; pp. 155–162. [CrossRef]

33. Axelsson, P. DEM generation from laser scanner data using adaptive TIN models. *Int. Arch. Photogramm. Remote Sens. Spat. Inf. Sci.* **2000**, *33*, 110–117.

34. Yu, X.; Hyyppä, J.; Vastaranta, M.; Holopainen, M.; Viitala, R. Predicting individual tree attributes from airborne laser point clouds based on random forests technique. *ISPRS J. Photogramm. Remote Sens.* **2011**, *66*, 28–37. [CrossRef]

35. Breiman, L. Random forests. *Mach. Learn.* **2001**, *45*, 5–32. [CrossRef]

36. Guo, L.; Chehata, N.; Mallet, C.; Boukir, S. Relevance of airborne LiDAR and multispectral image data for urban scene classification using Random Forests. *ISPRS J. Photogramm. Remote Sens.* **2011**, *66*, 56–66. [CrossRef]

37. Hudak, A.T.; Crookston, N.L.; Evans, J.S.; Hall, D.E.; Falkowski, M.J. Nearest neighbour imputation of species-level, plot-scale forest structure attributes from LiDAR data. *Remote Sens. Environ.* **2008**, *112*, 2232–2245. [CrossRef]

38. Cutler, D.R.; Edwards, T.C., Jr.; Beard, K.H.; Cutler, A.; Hess, K.T.; Gibson, J.; Lawler, J.J. Random forests for classification in ecology. *Ecology* **2007**, *88*, 2783–2792. [CrossRef] [PubMed]

39. Millard, K.; Richardson, M. On the importance of training data sample selection in Random Forest image classification: A case study in peatland ecosystem mapping. *Remote Sens.* **2015**, *7*, 8489–8515. [CrossRef]

40. Dalponte, M.; Ørka, H.O.; Ene, L.T.; Gobakken, T.; Næsset, E. Tree crown delineation and tree species classification in boreal forests using hyperspectral and ALS data. *Remote Sens. Environ.* **2014**, *140*, 306–317. [CrossRef]

41. Jones, T.G.; Coops, N.C.; Sharma, T. Assessing the utility of airborne hyperspectral and LiDAR data for species distribution mapping in the coastal Pacific Northwest, Canada. *Remote Sens. Environ.* **2010**, *114*, 2841–2852. [CrossRef]

42. Shang, X.; Chisholm, L.A. Classification of Australian native forest species using hyperspectral remote sensing and machine-learning classification algorithms. *IEEE J. Sel. Top. Appl. Earth Obs. Remote Sens.* **2014**, *7*, 2481–2489. [CrossRef]

43. Adelabu, S.; Mutanga, O.; Adam, E.; Cho, M.A. Exploiting machine learning algorithms for tree species classification in a semiarid woodland using RapidEye image. *J. Appl. Remote Sen.* **2013**, *7*. [CrossRef]

44. Carleer, A.; Wolff, E. Exploitation of very high resolution satellite data for tree species identification. *Photogramm. Eng. Remote Sens.* **2004**, *70*, 135–140. [CrossRef]

Part 3
Waveform Lidar

![remote sensing logo] *remote sensing*

MDPI

Article

A Sparsity-Based Regularization Approach for Deconvolution of Full-Waveform Airborne Lidar Data

Mohsen Azadbakht [1,2,*], **Clive S. Fraser** [1,2] **and Kourosh Khoshelham** [2]

[1] Cooperative Research Centre for Spatial Information, Carlton VIC 3053, Australia; c.fraser@unimelb.edu.au
[2] Department of Infrastructure Engineering, University of Melbourne, Parkville VIC 3010, Australia;
k.khoshelham@unimelb.edu.au
* Correspondence: m.azadbakht@student.unimelb.edu.au; Tel.: +61-3-9035-8559

Academic Editors: Jie Shan, Juha Hyyppä, Guoqing Zhou and Prasad S. Thenkabail
Received: 30 May 2016; Accepted: 3 August 2016; Published: 8 August 2016

Abstract: Full-waveform lidar systems capture the complete backscattered signal from the interaction of the laser beam with targets located within the laser footprint. The resulting data have advantages over discrete return lidar, including higher accuracy of the range measurements and the possibility of retrieving additional returns from weak and overlapping pulses. In addition, radiometric characteristics of targets, e.g., target cross-section, can also be retrieved from the waveforms. However, waveform restoration and removal of the effect of the emitted system pulse from the returned waveform are critical for precise range measurement, 3D reconstruction and target cross-section extraction. In this paper, a sparsity-constrained regularization approach for deconvolution of the returned lidar waveform and restoration of the target cross-section is presented. Primal-dual interior point methods are exploited to solve the resulting nonlinear convex optimization problem. The optimal regularization parameter is determined based on the L-curve method, which provides high consistency in varied conditions. Quantitative evaluation and visual assessment of results show the superior performance of the proposed regularization approach in both removal of the effect of the system waveform and reconstruction of the target cross-section as compared to other prominent deconvolution approaches. This demonstrates the potential of the proposed approach for improving the accuracy of both range measurements and geophysical attribute retrieval. The feasibility and consistency of the presented approach in the processing of a variety of lidar data acquired under different system configurations is also highlighted.

Keywords: deconvolution; full-waveform; lidar; L-curve; sparse solution; target cross-section

1. Introduction

Laser remote sensing is being increasingly adopted as a useful tool for capturing 3D information in conjunction with other descriptive data (e.g., intensities) from targets, and it is now being utilized for a number of applications. These include classification [1,2], object extraction and 3D reconstruction [3], vegetation structure characterization [4,5] and digital elevation model (DEM) generation [6].

Discrete lidar systems estimate the distance between the sensor and the target through precise measurement of the transit time between the emitted and backscattered laser pulse using real-time echo detection. Such discrete systems generally do not specify the exploited pulse detection method [4,7,8] and only provide range information with a single backscatter intensity value, inherently neglecting the backscattering characteristics of the illuminated target. In contrast, highly accurate pulse detection and subsequent range measurement is potentially possible in full-waveform lidar, subject to the methods adopted for data post-processing. Full-waveform lidar systems record the entire backscattered signal, providing a range measurement, a measure of the energy reflected from the target and the distribution of the returned energy along the laser path. Physical characteristics of the underlying target can then be revealed, thus enhancing the capability of the lidar system to yield both geometric and physical information of targets for a broad range of applications.

Classical pulse detectors, in ideal conditions, can only provide the range and amplitude values of pulses, without allowing removal of the effect of the transmitted pulse or the delivery of echo parameters and radiometric information about targets [9,10]. Moreover, even though accurate range determination is an important aspect in processing full-waveform lidar data, restoring the target response is a key prerequisite step not only for range measurement but also for the extraction of geophysical attributes of the target. The target cross-section is an attribute that is independent of the instrument [11,12] and is more related to the effective target area (and its shape) inside the footprint, as well as to its reflectivity and the direction of the incoming light [13,14]. The scattering cross-section, which provides insight into geophysical characteristics and target roughness [11,15] is an issue of considerable significance in laser radar remote sensing [14,16] and it can be recovered in the case of full-waveform lidar [8].

In addition, more useful attributes can be extracted by analyzing the lidar signal geometry. Recovery of overlapping signals and weak echoes is also feasible and advantageous in full-waveform lidar to avoid both misinterpretation and incorrect estimation of the range. The additional pulses, which may not be detected by discrete return systems, along with their associated features, can provide useful inputs for further processing, e.g., for segmentation, classification and 3D modelling [1,17–19].

A lidar waveform is generally complex, particularly in vegetated areas. Overlapping waveforms are recorded as a consequence of a finite detection time, noise, impulse function of the receiver and changes in the transmitted pulse, leading to resolution degradation for closely located targets along the laser beam [20]. Critical to the success of full-waveform lidar systems is the ability to understand and interpret the shape of the recorded return laser pulse so as to separate fine object structures. A number of lidar waveform processing approaches have been developed [14,21–23] and these have generally concentrated on decomposition of the waveform into a series of parametric (e.g., Gaussian) pulses so as to extract the range, amplitude and pulse width. Since the lidar waveform is essentially the result of a convolution of the system transmitted laser pulse and the target cross-section, the retrieval of target properties requires a deconvolution of the returned waveform. In addition, account must be taken of the fact that incorporation of the system properties delivers a more meticulous analysis of the waveform [24]. In particular, deconvolution essentially removes the effect of the system signal, resulting in an unbiased estimate of the target cross-section.

This paper proposes a new signal deconvolution approach for efficient temporal target cross-section retrieval from full-waveform lidar. This is achieved through a regularization approach with sparsity constraints that can be efficiently solved when cast in the form of a convex optimization problem. An efficient way to determine the optimal regularization parameter value, a critical issue in regularization, through utilization of the L-curve method and a search for the bending point on such a curve, is introduced. In order to model and estimate the system waveform, in situations where it is unavailable, an appropriate approach is proposed through the use of blind deconvolution of waveforms recorded over nearly flat targets, without requiring information about the surface reflectivity.

The remainder of the paper is organized as follows: Section 2 provides a review of research on lidar waveform processing. Following this, in Section 3, the proposed lidar waveform deconvolution method is presented through reference to the fundamentals of laser propagation, regularization for deconvolution and estimation of an optimal regularization parameter value. In Section 4, the experimental tests with both synthetic and real data covering different test sites are presented, and performance of the proposed lidar waveform deconvolution approach is evaluated. Concluding remarks are offered in Section 5.

2. Review of Related Research

With full-waveform lidar systems increasingly being adopted, improved methods are being developed for the processing of lidar waveforms. Approaches adopted in the literature can be categorized as either decomposition- or deconvolution-based, where the former approaches fit parametric functions to the received signals before retrieval of the target cross-section.

2.1. Decomposition Methods

Gaussian waveform decomposition methods for the extraction of a parametric description of the pulse properties, including range, width and amplitude, have been developed for instance by Hofton et al. [21] for a laser vegetation imaging sensor (LVIS), by Persson et al. [22] for the TopEye Mark II lidar system and by Wagner et al. [14] for small footprint data. Generally, a set of Gaussian functions is considered to both fit to the received backscattered waveform and to characterize each pulse shape in this approach. Gaussian decomposition and other similar decomposition methods have been extensively used to interpret targets related to the backscattered waveform in urban and forested areas [1,2,17,25]. However, this method is considered challenging in the case of echoes with low signal strength (low SNR) and it is deficient in its calculation of the cross-section in complex waveforms. The method also requires initial determination of the number of targets [16,18,23,26], which is sometimes impractical or of high computational cost. In addition, a symmetric function might not always be sufficiently accurate to describe the target characteristics [17,27]. Furthermore, Gaussian decomposition has a vulnerability in that it can result in a null solution or even negative amplitudes in some cases [11,18]. The possibility of failure in the detection of peaks and in the fitting of Gaussian functions to complex waveforms, principally due to a high dependency on local maxima, has been reported [18]. These problems cause the range accuracy to deteriorate, which then impacts upon subsequent lidar processing.

More complicated parametric distribution functions for the extraction of pulse parameters have been investigated. Chauve et al. [27] extracted target attributes from pulses by utilizing a mixture of the generalized Gaussian and Lognormal functions, with an improved global fitting for the former, and a better pulse fitting result locally in asymmetric cases for the latter. This work was extended by Chauve et al. [17] to suppression of the ringing effect, and the results were utilized in DTM and Canopy Height Model (CHM) extraction, as well as for accurate determination of tree height. A low rate of height underestimation in the CHM generation was reported by the authors. A set of parametric pulse shapes (the Burr, Nakagami and generalized Gaussian models) have been considered by Mallet et al. [23], where it was pointed out that the efficacy of the approach when applied to object classification was inadequate. However, more accurate results could be achieved when these waveform features were utilized in combination with geometric attributes. Lin et al. [18] studied weak and overlapping pulses and reported higher obtainable accuracies in multi-target separation using a rigorous pulse fitting method, however at very high computational cost.

In general, decomposition methods all attempt to fit a function (or functions) to the received waveform and extract related attributes from the parametric function(s). However, parameter estimation based only on the received waveform and not on the perceived deviations in the transmitted pulse may not in general be valid [11]. In addition, direct extraction of the range from the maximal amplitude value is not recommended since the signal pulse may be altered as a result of superimposing different target responses [12,15]. It has therefore been suggested that recording both the received and the outgoing waveforms is indispensable [12]. Disentanglement of the received waveform from the instrument and estimation of a signal more related to the target response can be achieved by exploiting deconvolution approaches. Reconstruction of instrument-independent target features by deconvolution can result in precise characterization of the actual response of the targets [5,12].

2.2. Deconvolution Methods

The retrieval of target cross-section by signal deconvolution is an ill-posed problem and a unique solution is usually not obtainable without the imposition of appropriate solution constraints [16]. Unlike Gaussian decomposition, deconvolution does not require information regarding the number of signal peaks and there is no essential assumption regarding the pulse shape [7,26].

An encouraging reported approach for both retrieval of the target response and improved range estimation is Wiener filtering [5,12,19,20]. The Wiener filter minimizes the mean square error between the approximated cross-section and the true surface response. However, negative amplitudes and ringing effects usually appear in signals restored by using this method [19,20].

In another approach, reported in Roncat et al. [11], the differential cross-section is measured by using B-splines as a linear deconvolution approach, with the assumption that pulses are not necessarily symmetric. Negative amplitude values were reported in the resulting cross-section profile.

Wu et al. [28] defined four steps to form a pre-processing sequence for calibrating full-waveform lidar data based upon synthetic waveform generation from the DIRSIG simulation model. The defined steps are noise removal, signal deconvolution, geometric correction, and angular rectification. Among these, deconvolution has emerged as the most critical for waveforms from lidar with a nadir scan angle, while geometric calibration is the principal enhancement factor for off-nadir scanning. The Richardson-Lucy (R-L) method was adopted in the deconvolution investigation. While this method is efficient in most cases, underestimation of the amplitude values and the occasional missing of pulses, in addition to essential speculation on the number of iterations by the user, remain its main shortcomings. It is noticeable that increasing the number of iterations does not generally guarantee that the optimal results will be achieved because after some iterations the quality of the restored signal can deteriorate and spurious pulses will be generated as a result of noise amplification [29,30]. As also reported in Neilsen [31], increasing the number of iterations in the R-L algorithm has an adverse effect on the range resolution and can cause an increase in the range error.

Deconvolution of the received waveform can be accomplished by regularization techniques, as demonstrated in Wang et al. [16]. This approach can restrain noise propagation and improve computational efficiency. However, the method utilized to estimate the regularization parameter, namely the discrepancy principle, demands that the noise level in the waveform be known, which is not practical in the case of real data and hence the method is restricted in its application.

Negative amplitude values can be avoided by using iterative deconvolution methods, e.g., the Gold and R-L methods. Such approaches can also minimize smearing effects (signal distortion level) on the signal and they have been shown to be effective in retrieving pulses in lidar data [20,26,28], as well as in other fields of research [32]. Wu et al. [20] reported that not all peaks can be detected. This can cause errors in range estimation and lead to false target detection. Moreover, the solutions generated by these methods can be significantly affected by the number of iterations. Thus, any invalid assumptions can lead to inaccurate results.

Application of compressed sensing in lidar processing using synthetic data has also attracted limited attention [33,34]. In this case, inclusion of lidar data compression in the acquisition was endeavored in order to cope with large data volume. The transmitted waveform of the chaotic type presented in [34] was described to compress waveform data whilst allowing retrieval of the range information using the DIRSIG simulation model. Higher SNR were perceived in case of chaotic signals than that of using linear Frequency Modulated (LFM) signals. Reconstructing of randomly generated surfaces was investigated at different complexity levels through implementing compressed sensing in [33]. The edges connecting facets could not be retrieved in [33] when the proposed method is based on the total variation algorithm, while such an algorithm is known for successful retrieval of edges. In addition, the complexity of surfaces is required in [33] to appropriately retrieve the surface.

The present paper explores the theory of signal deconvolution for improved full-waveform lidar processing through a reliable sparsity-constrained regularization. In addition, the estimation of system waveforms based on blind deconvolution is presented and an approach to the determination of regularization parameters is proposed.

3. Methodology

The proposed approach explores signal deconvolution for target cross-section retrieval. Following a brief review of two major classes of deconvolution methods and a justification for selection of the most efficient group, the sparsity-constrained regularization for deconvolution is presented and the approaches to the sparse solution are detailed. A method based on the L-curve for the determination of the regularization parameter is then proposed and a new approach using blind deconvolution for estimation of the system waveform is developed.

3.1. Deconvolution for Cross-Section Retrieval

Deconvolution is the process of signal and image restoration in a system with linear and shift-invariant characteristics [35]. There are two major classes of deconvolution methods, namely direct and iterative. Of the latter group, the Gold [36] and R-L [37,38] are recognized as approaches that provide signals/images of positive values, provided that the input raw waveform values are also non-negative [32]. Direct deconvolution methods, on the other hand, involve a regularization parameter, which controls the level of restoration. The number of optimal iterations in the iterative methods is of the same level of significance as the regularization parameter in the direct deconvolution methods, and different values for each of these parameters will surely result in a different solution, and sometimes may deliver over-smooth or noisy results. Unfortunately, selection of the optimal number of iterations is still a challenge and changes in this number may result in varying levels of signal recovery (overestimation or underestimation). In this research, direct deconvolution methods are investigated for two main reasons. First, an optimal regularization parameter can be more appropriately determined, so controlling the solution will therefore not be problematic. Second, this class of deconvolution approaches makes it possible to incorporate additional constraints and conditions [39] and, in turn, is more flexible compared to the iterative approaches.

3.1.1. Target Cross-Section Retrieval

As mentioned, the process of target cross-section retrieval is an ill-posed problem, and it requires consideration of additional conditions. In this situation, substantial perturbations can occur in the restored signal as a consequence of either trivial changes in the received waveform or noise in the waveform. Therefore, regularization is a necessary way to achieve a stable solution to the deconvolution problem by providing an approximation after imposing additional constraints [40]. In order to accomplish the condition, it is necessary to determine an optimal regularization parameter by an appropriate method [41].

The target cross-section for an extended non-overlapping diffuse surface with reflectivity ρ, the solid angle of π steradians, and under incidence angle θ, will be [13]:

$$\sigma_{ex} = 4\rho d_A \cos(\theta) \tag{1}$$

where $d_A = \frac{\pi R^2 \beta_t^2}{4}$ is the effective target area illuminated by the laser footprint, for a transmitter with a typical disk-like aperture with β_t as its beamwidth, at the range R.

The received signal power $P_r(t)$ at the receiver is essentially a convolution of the point spread function (system waveform) $S(t)$ and the unknown differential target cross-section $\sigma(t)$, along with an additive noise term $n(t)$, and it can be expressed through the following equation [16]:

$$P_r(t) = S(t) * \sigma(t) + n(t) \tag{2}$$

In Equation (2), any possible loss of the transmitted pulse energy when intercepted by multi-return targets (e.g., vegetation) is ignored, and $\sigma(t)$ represents only the cross-section of the effective areas in multi-return targets. In order to recover the target response, the effect of convolution in Equation (2) should be compensated through deconvolution.

3.1.2. Sparsity-Constrained Regularization

Throughout the rest of this paper, waveforms are treated as discrete signals. The general form of the variational regularization, used to retrieve the temporal target cross-section σ in Equation (2), is expressed by Gunturk and Li [35] as:

$$\sigma = \underset{\sigma}{\text{argmin}}\{||P - S\sigma||^2 + \lambda\varphi(\sigma)\} \tag{3}$$

where, $\lambda \in (0, \infty)$ is the regularization parameter, $\varphi(.)$ is the regularization function, $||.||$ denotes the Euclidean norm, P is a vector representing the received waveform and S is the blur matrix comprised of the system waveform.

This objective function seeks a solution by minimizing the sum of squared dissimilarities between the estimated and the desired solution, while imposing a constraint on the solution to achieve desirable properties. The first term (residual) in Equation (3) measures how the solution fits the data, and the regularization term introduces prior information about the solution and imposes a penalty on large amplitude values of the retrieved signal [42]. The selection of λ should be made with caution, since it controls the level of signal restoration and different values result in different solutions, and sometimes may deliver over-smooth or noisy results.

The regularization function plays a critical role and different regularization functions, appropriate to different specific applications have been introduced in the literature [35]. Tikhonov regularization [43], which is the most common variational regularization method, makes use of the regularization function $\varphi(\sigma) = ||L\sigma||^2$, where L is a regularization operator. Through a consideration of prior information and adherence to the Gibbs distribution, the signal probability is determined by measuring the deviation from smoothness [44]. This is a convex quadratic optimization problem and can be solved with different approaches [42]. However, the Tikhonov regularization function, which utilizes the l_2-norm, exaggerates data smoothness. The shape of the penalty function affects the solution. When using the l_2-norm, as in Tikhonov regularization, large residuals are allocated a higher weight than small residuals [42], and also small elements are revealed in the obtained signal associated with small singular values [45]. As a result, Tikhonov regularization may cause oscillation around principal peaks, restoring signals with under-estimated amplitude values, and it may also generate negative amplitude values [32,46]. To avoid such problems, sparsity constraints have been introduced as an appropriate technique in regularization [35,47]. The term sparsity refers to the minimum number of nonzero elements within a desired solution that can be included in the objective function through use of the l_0-norm.

Sparse formulation of the regularization solution can be expressed as follows:

$$\sigma = \underset{\sigma}{\mathrm{argmin}}\{||P - S\sigma||^2 + \lambda ||L\sigma||_0\} \tag{4}$$

where $||.||_0$ represents the sparsity constraint on the solution.

Among the methods for sparse solution, greedy pursuit methods and convex optimization have proven to be effective and are broadly used whilst also being computationally tractable [48]. The effectiveness of convex relaxation techniques in achieving an appropriate solution for the sparsity-constrained regularization has been verified to a large extent, especially in cases involving high noise levels. More importantly, a local optimal point, where it exists in convex optimization, is also globally optimal [42]. In convex optimization methods the idea is to impose a relaxation and replace the l_0-norm in Equation (4) by its closest convex form, the l_1-norm, such that standard methods in convex optimization can be adopted [48]. The reason for the substitution of the l_0-norm with the l_1-norm is that the former results in impractical methods because of the high computational complexity [35]. The cost function is therefore altered to:

$$\sigma = \underset{\sigma}{\mathrm{argmin}}\{||P - S\sigma||^2 + \lambda ||L\sigma||_1\} \tag{5}$$

In contrast with Tikhonov regularization, substantial weight is assigned to small residuals in the case of the l_1-norm penalty functions in sparse regularization, hence resulting in even smaller residuals [42]. Additionally, the l_1-norm solution does not produce signal elements relevant to small singular values [45]. Large amplitude values can be recovered by imposing smaller penalties when adopting the l_1-norm solution, in lieu of that of the l_2-norm [44]. Also, the oscillations of Tikhonov regularization can be controlled to a large extent by this solution. The l_1-norm regularization method has been adopted in many applications, from signal/image deconvolution to feature selection, and to compressed sensing [35,49]. In the case of waveform deconvolution, small elements of the target response vector are encouraged to become zero, which is an aim of sparse solutions.

A sparse solution for Equation (5) always exists as a nonlinear function of the noisy observations, although it is not necessarily unique. However, solving this problem involves high computational complexity compared to its counterpart in the l_2-norm solution due to its non-differentiability [50].

By recasting waveform restoration into the form of the convex problem in Equation (5), advanced approaches in convex optimization can be explored. Iterative Shrinkage/Thresholding (IST) methods, based on either the proximal gradient or expectation-maximization [35,51], together with interior-point methods [42,50,51], are noteworthy examples. To exploit the latter group, the convex optimization should be converted to a quadratic form, second-order cone programming (SOCP) problem [50,51], where interior-point methods have proven to be effective in producing highly accurate results compared with other methods such as the barrier method [42].

3.1.3. Optimization with SOCP

Linear and convex quadratic problems can be expressed in the style of nonlinear convex SOCP problems, which can be efficiently solved by primal-dual methods [52,53]. The SOCP form of the main objective function in Equation (5), with σ, u and v as the variables, is written as [42]:

$$\min u + \lambda v \quad s.t. \ ||S\sigma - P|| \leqslant u, \ ||\sigma||_1 \leqslant v \tag{6}$$

Because of the convexity of the cost function and the second-order cone constraint, the SOCP problem in Equation (6) is a convex problem. In primal-dual interior-point methods, Newton's method is employed to calculate the search direction on the central path, from the modified Karush-Kuhn-Tucker (KKT) conditions. Assigning the residual components (the dual, centrality and primal residuals) to zero, the Newton step can be obtained as the primal-dual search direction by solving a set of equations [42].

3.1.4. Blur Matrix Structure

The blur matrix S in Equation (5) should be structured as a function of the system waveform elements. Undesirable oscillations will likely be revealed in cases where an inappropriate blur model is defined [54]. The blur matrix can be constructed in the following form [55], where h_i represents the ith element of the system waveform:

$$S = \begin{bmatrix} h_1 & 0 & 0 & \cdots & 0 \\ h_2 & h_1 & 0 & \cdots & 0 \\ \vdots & \vdots & \vdots & \ddots & \vdots \\ h_n & h_{n-1} & h_{n-2} & \cdots & 0 \\ 0 & h_n & h_{n-1} & \cdots & 0 \\ 0 & 0 & h_n & \ddots & \vdots \\ 0 & 0 & 0 & \ddots & h_1 \\ \vdots & \vdots & \vdots & \ddots & \vdots \\ 0 & 0 & 0 & h_n & h_{n-1} \\ 0 & 0 & 0 & 0 & h_n \end{bmatrix} \tag{7}$$

This blur matrix has a Toeplitz form and belongs to the class of zero boundary condition. The received waveform is a $(m + n - 1) \times 1$ vector; with the target cross-section vector of size $m \times 1$, the system waveform vector of size $n \times 1$ and the dimension of the blur matrix S is $(m + n - 1) \times m$. In Section 3.3, an approach for estimation of the system waveform in cases where it is unknown will be presented.

3.2. Determination of the Regularization Parameter

Optimal selection of the regularization parameter is important for an ideal adjustment between the regularization function and the solution residual, as the solution can be significantly affected by the selected value [35]. An optimal regularization parameter avoids under-/over-regularized results and in turn unreliable outputs. Therefore, a proper choice of the regularization parameter needs to be investigated thoroughly in order to achieve the optimal value.

Among the available approaches, some only deal with the residual term, without considering the regularization term. Of this type, the discrepancy principle [56] and generalized cross-validation (GCV) [57] could be mentioned. However, some other approaches, e.g., the L-curve [58], take care of both terms simultaneously [59].

The discrepancy principle method relies strictly upon a known data noise level and it estimates the regularization parameter such that the residual term is proportional to the predefined noise level [54,59]. The known noise level is considered as a highly impractical requirement when dealing with real data. Hansen et al. [54] highlighted an over-smoothed output yet with a known (or with a precise estimate of) noise level. Use of a large regularization parameter value makes the output signal over-smooth, and hence, the detailed description of the target response is not recovered [45,59]. These deficiencies make this approach inappropriate in the case of full-waveform lidar data processing.

Unlike the discrepancy principle, the GCV makes no assumption with regard to the noise level. The optimal value of the regularization parameter is obtained by minimizing the GCV function, where the effort is to achieve the minimal value of the residual term and eliminate the influence of small singular values on the output signal [54,59]. However, the occurrence of high correlation between noise elements impedes the achievement of an appropriate regularization parameter. Moreover, finding an absolute minimum value of the GCV function may not be achievable due to difficulty in locating the function's minimum value since it is near constant in the vicinity of the minimum [45]. This method is also known for resulting in an overestimated regularization parameter, even in cases of unnoticeable correlation between errors. When it is applied to data with high noise level, results can be unsatisfactory [59].

In the L-curve method, both the residual and regularization terms are considered concurrently in order to find the optimal regularization parameter. In search of the optimal value, these two terms are plotted against each other for a range of plausible regularization parameter values. The idea is to find a changing point on the curve, where both terms are minimized simultaneously. This point is actually the margin between over- and under-regularization of the results and it is a representative of the optimal value for the regularization parameter [35]. The L-curve method is adopted for estimation of the optimal regularization parameter in this research in the following way: Suppose $P = \{(Rg_1, Rs_1), \ldots, (Rg_n, Rs_n)\}$ is a set of n pairs from the regularization (Rg_i) and residual term (Rs_i) magnitudes that constitute the vertices of the L-curve for a range of regularization parameter values $(\lambda_1, \lambda_2, \ldots, \lambda_n)$. The optimal regularization value λ_{Opt} over the aforementioned set of points P_i can then be conveniently calculated by finding the maximum Euclidean distance. Figure 1 illustrates the location of the optimal regularization parameter on a sample L-curve.

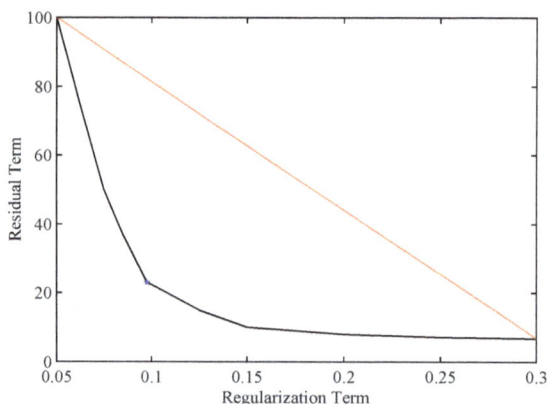

Figure 1. Optimal regularization parameter on the L-curve.

3.3. System Waveform Estimation

Waveform deconvolution requires the emitted signal to be known. This information is not always available in the recorded lidar data. In this section, a new method for estimation of the system waveform using blind deconvolution is proposed. Under the assumption of an ideally flat target (at nadir) and considering the measured transmitted signal $s_M(t)$ as a convolution of the transmitted pulse $s(t)$ and the impulse function of the transmitter $h_S(t)$ on one side; and assuming the measured received waveform $r_M(t)$ as a convolution of the transmitted pulse $s(t)$ and the receiver impulse function $h_R(t)$ on the other side; the measured received waveform can be expressed in the frequency domain as [24]:

$$R_M(f) = S_M(f) \frac{H_R(f)}{H_S(f)} = S_M(f) H(f) \tag{8}$$

The atmospheric propagation is treated as a constant value for similar targets located at similar ranges from the receiver. Therefore, the average of the system waveform can be obtained by deconvolution over a number of recorded samples. A prominent approach to eliciting a rough estimation of the unknown system waveform is to exploit blind deconvolution, which provides an estimate of both the point spread function (PSF) and the true signal [35,54]. In this research, only pulses captured at nadir over almost flat targets are taken into account for the calibration. In order to alleviate the effects of possible changes of the received waveforms from similar targets due to the noise, changes of the (unknown) system waveform or other unknown sources, the retrieved system waveforms are averaged. Estimation of the system waveform in this study is carried out using the iterative damped R-L method [60]. Once the system waveform is estimated, it will be applied to waveforms from other targets in order to achieve a rough estimate of the target cross-section via deconvolution. Possible changes in the system waveform, even between subsequent pulses, makes the retrieved target cross-section only a rough approximation.

3.4. Quantitative Assessment of the Retrieved Waveform

For quantitative comparison of the methods, the Fréchet distance, Spectral Angle Mapper (SAM), and Pearson correlation coefficient have been utilized to calculate the similarity of the results with the "true" cross-section signal.

The SAM metric measures the angle between two waveforms via treating them as vectors with identical lengths, and it can be expressed as follow:

$$\theta = cos^{-1} \left(\frac{\sum_{i=1}^{N} Rt_i Rf_i}{\sqrt{\sum_{i=1}^{N} Rt_i^2 \sum_{i=1}^{N} Rf_i^2}} \right) \tag{9}$$

Here, Rt and Rf represent the retrieved and reference waveforms, respectively, where i and N are the index referring to the waveform samples and the total number of samples in a waveform.

The Pearson correlation coefficient r considers the linear relationship between the retrieved waveform Rt and the reference signal Rf in terms of the magnitude and direction:

$$r = \frac{SS_{Rt,Rf}^2}{SS_{Rt,Rt} SS_{Rf,Rf}} \tag{10}$$

where, $SS_{Rt,Rf}$ refers to the covariance of the waveforms, and $SS_{Rt,Rt}$ and $SS_{Rf,Rf}$ represent the standard deviations of each signal.

The Fréchet distance between $P:[a,\,b] \rightarrow \mathbb{R}^2$ and $Q:[a',\,b'] \rightarrow \mathbb{R}^2$ as two planar curves, with $d(P(\alpha(t)), Q(\beta(t)))$ representing the Euclidean distance between these two curves at specified vertices, is defined as the minimum required length in connecting two separate paths, when navigating continuously from one endpoint to the other [61]:

$$\delta_F(P,Q) = \inf_{\alpha,\beta} \max_{t \in [0,1]} d(P(\alpha(t)), Q(\beta(t))) \tag{11}$$

where $a, b \in \mathbb{R}$ and $a \leqslant b$ (respectively a', b'), a (respectively β) is a function with continuous nondecreasing characteristics from $[0, 1]$ onto $[a, b]$ (respectively $[a', b']$), and $inf(.)$ is the "infimum" or the maximum lower bound of a set.

4. Experiments and Discussion

4.1. Experimental Data

A synthetic dataset and three different real datasets are investigated in order to evaluate the presented method.

4.1.1. Synthetic Data

Ten synthetic lidar waveforms were generated by convolution of a known system signal and the given differential target cross-sections. For simplicity, all samples were considered as mixtures of samples with Gaussian distribution, with various levels of complexity and also different arrangements of pulses along the signals. This simulated data enables direct comparison of the restored waveform with the known "truth" data, facilitating quantitative evaluation of the performance of the deconvolution methods. Different arrangements of pulses with different amplitude values were considered so as to model different complexities of targets. In order to investigate the robustness of the methods against noise, the simulated signals were contaminated by three different noise levels (Gaussian noise with $\sigma = 0.01, 0.02$, and 0.05), after imposing a convolution with the synthetic system waveform. The absolute value of the amplitude values was utilized in order to avoid negative values resulted from the Gaussian noise. Figure 2 illustrates an example of both the synthetic system signal and the received waveforms at three different noise levels for limited samples. The true signals have been generated by computing the probability density functions based on the normal distribution, with different values assigned to the standard deviation to control the pulse shape.

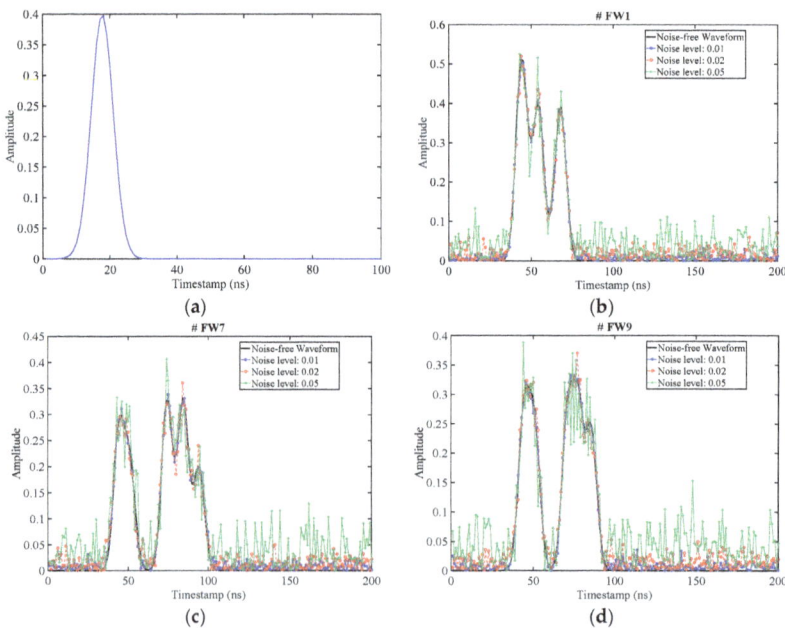

Figure 2. (a) The synthetic system waveform; and (b–d) selected sample waveforms at different noise levels.

4.1.2. Real LiDAR Datasets

Three different real datasets have been utilized to assess different aspects of the proposed method, including the approximation of the system waveform and the extraction of radiometric attributes for different targets. The QLD dataset was acquired over Rockhampton in Queensland, Australia, in December 2012. The NSW dataset was captured over the Namoi River in New South Wales, Australia in September 2013, and the third dataset was acquired over the Karawatha forest park in Queensland, Australia in October 2013. The laser footprint diameters at nadir were approximately 0.30 m, 0.50 m, and 0.15 m, respectively. A Riegl LMS-Q680i scanner, operating at 1550 nm, with the laser beam width of 0.5 mrad was used to acquire the datasets. Further details regarding the three lidar missions are provided in Table 1.

Table 1. Data specification.

Parameter	QLD Dataset	NSW Dataset	Karawatha Dataset
Flying altitude (AGL-m)	600	1000	300
Pulse rate (kHz)	150	240	240
Maximal scan angle ($^{\circ}$)	30	40	25

Although all datasets were recorded with the same instrument, the QLD data was provided without the system waveforms, while such information was delivered in the NSW and Karawatha data. With the latter two datasets, more satisfactory results are expected because of the recording of an equivalent system waveform for each received signal. Partial differences even between subsequent emitted pulses could then be observed.

As shown in Figure 3, the recorded system waveform in the NSW dataset was usually incomplete and there is no information at the signal tail. Interestingly, the maximum amplitude of the received waveform, shown in Figure 3, is higher than the value of the system waveform, which may seem contradictory. However, it is noticeable that amplitude values of the received waveform and the system waveform are not necessarily of identical units, which could be connected to the sampling procedure of the transmitted pulse [5]. The delayed backscattered energy due to multiple scattering within the tree crown is another possible explanation for this phenomenon.

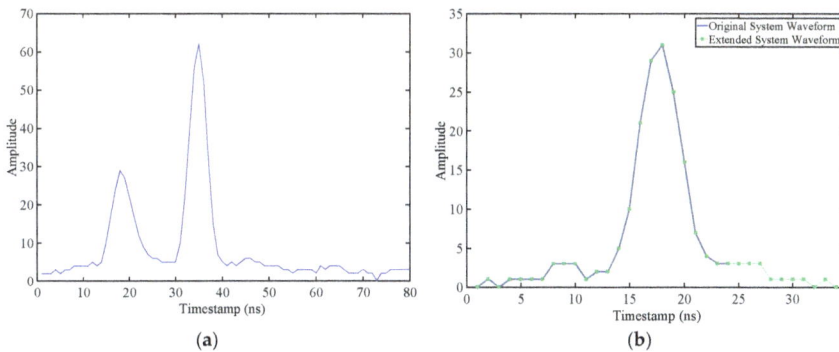

(a)   (b)

Figure 3. (**a**) A typical recorded waveform in the NSW dataset; and (**b**) its corresponding original system waveform.

The diversity of lidar datasets can present challenges in waveform deconvolution and thus the provision of the QLD, NSW and Karawatha datasets provided a good opportunity to evaluate the consistency of the proposed regularization method.

4.2. Preprocessing

The raw incoming waveform typically exhibits a certain level of noise due to sensor impacts and environment conditions. Moreover, deconvolution might amplify the noise due to its characteristics [31]. Therefore, a noise reduction method for full-waveform lidar based on the integration of the Savitzky-Golay (S-G) [62] and the Singular Value Decomposition (SVD) approaches was employed in the analysis [63]. This approach has advantages for signal pattern preservation, in addition to avoidance of waveform distortion.

4.3. Deconvolution Results for Synthetic Waveforms

The synthetic data were processed by the proposed regularization approach and the results were compared to those from the commonly used approaches in signal processing, namely the R-L, Wiener filter, Tikhonov regularization, and Gaussian decomposition methods. The regularization parameter was determined by the L-curve method. The performance of the utilized approaches was quantitatively evaluated through a comparison of the estimated target cross-section with the "truth" data.

Table 2 explains the parameter tuning for the methods, including the utilized algorithms and the associated references. The optimal iteration numbers of the R-L method were estimated by calculating the goodness of fit to the truth data by using the RMSE criterion, through consideration of a wide range of iteration numbers from 1 to 500. For the Wiener filter, the Noise to Signal Ratio (NSR) was calculated from the truth data at different noise levels by dividing the variance of the noise elements to the noise-free signal. Similar to the proposed deconvolution method, the regularization parameter was estimated by the L-curve method in the case of Tikhonov regularization. The exact number of expected pulses was assumed to be known in order to apply the Gaussian decomposition method through application of nonlinear optimization, e.g., the Levenberg-Marquardt algorithm [14].

Table 2. Parameter tuning of the methods.

Method	Algorithm	Parameter Tuning	Reference(s)
Gaussian decomposition	Levenberg-Marquardt/Thrust region (curve fitting in Matlab)	Visual inspection to specify the number of pulses	[14,64]
R-L deconvolution	deconvlucy function in Matlab	RMSE criterion to select optimal iteration numbers	-
Wiener filtering	deconvwnr function in Matlab	$\frac{Noise\ variance}{Signal\ Variance}$	-
Tikhonov regularization	CVX package	L-curve	[65,66]
Sparsity-based regularization	CVX package	L-curve	[65,66]

Qualitative comparisons of the proposed method against the other four methods are shown in Figure 4 for selected samples. The figure provides a detailed evaluation of their performance in cases where the quantitative measures do not necessarily show a better performance of the proposed method. As shown in Figure 4a, qualitative inspection reveals the weakness of the R-L method in the retrieval of the second pulse, where two separate pulses are instead generated. Gaussian decomposition overestimates the amplitude for all three pulses, as well as the area under the first pulse, while the proposed method marginally overestimates two pulses. The Wiener filter and Tikhonov regularization methods generate oscillatory pulses that underestimate amplitude values of all pulses. In Figure 4b, the fourth pulse was not recovered by the R-L method, while the second and the last pulse were also shifted, with an underestimation of the last pulse as well. The proposed method retrieved all the pulses at their correct positions and with a minimum of spurious pulses. Gaussian decomposition only retrieves the third pulse which is marginally shifted. The second and fourth pulses were not retrieved by the Wiener and Tikhonov methods, while oscillation and underestimation of amplitude values were apparent. The last pulse was shifted by both the R-L and proposed methods in Figure 4c. Also, the R-L method could not retrieve the first two pulses and the fourth pulse, and many additional pulses were generated, probably as a result of noise amplification. The first two pulses and the last three pulses were combined separately in two pulses by the Gaussian decomposition, where the amplitude values were significantly overestimated. The Wiener filter performed slightly better in the retrieval of pulses than the Tikhonov method, yet the amplitude values were underestimated. In Figure 4d, the sixth pulse was marginally shifted by the proposed method, while the amplitudes of three other pulses were overestimated. The R-L method in this case cannot place the fourth pulse, while the first two pulses were not correctly restored in terms of both their amplitude and pulse locations. In addition, the last two pulses were slightly shifted and some other false pulses were generated. Two pulses were retrieved as combinations of the first four pulses by the Gaussian decomposition, where the amplitude values were overestimated. In Figure 4e, the first, fourth and fifth pulses were not restored by the R-L method, with spurious pulses instead being generated. The Gaussian decomposition method retrieved the last two pulses as well as a combination of the first two pulses, with the amplitudes being overestimated. Tikhonov and Wiener filtering resulted in both oscillations and underestimation of amplitude values for all pulses. In Figure 4f, the second pulse was not restored in its position using the R-L method. Gaussian decomposition, on the other hand, does not retrieve the first pulse, while the last pulse is overestimated in terms of both amplitude and the area under the pulse. The Wiener filter produces additional pulses, in addition to oscillation around the main peaks, whereas less spurious pulses were generated using the Tikhonov method. Even though the proposed method occasionally overestimates some of the pulses, the area under the curves were almost similar to the true signal.

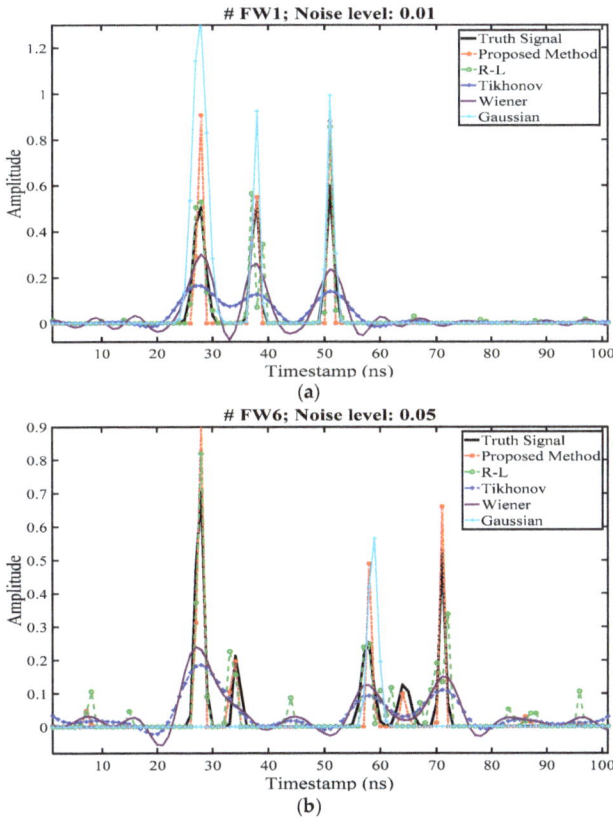

Figure 4. *Cont.*

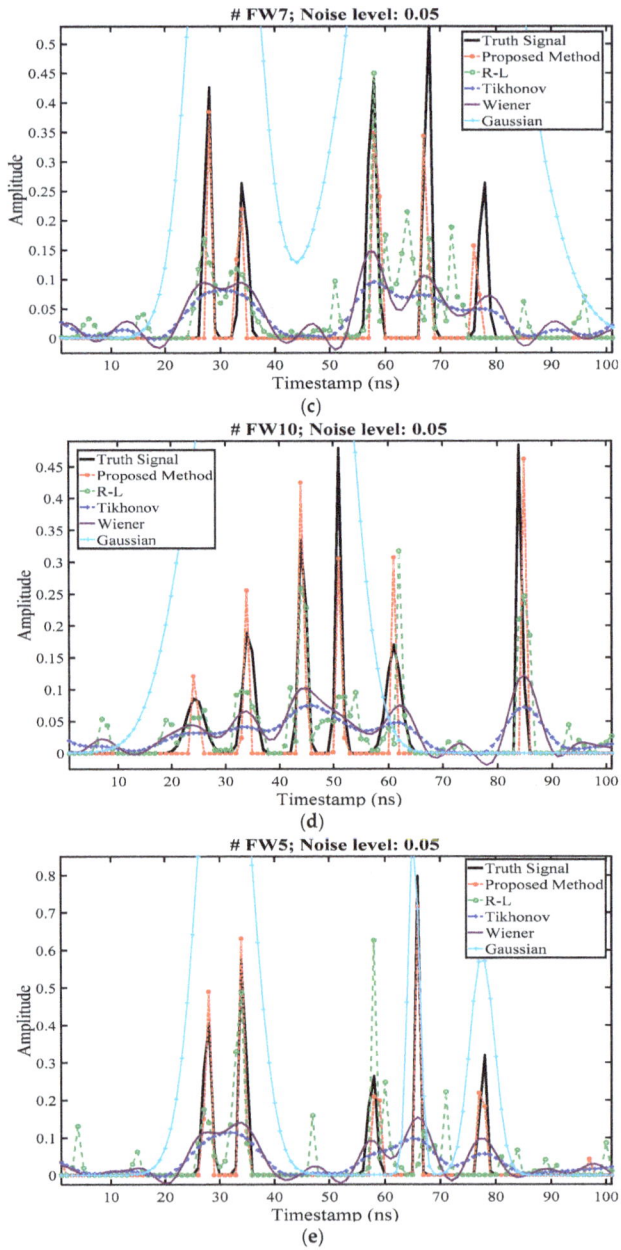

(c)

(d)

(e)

Figure 4. *Cont.*

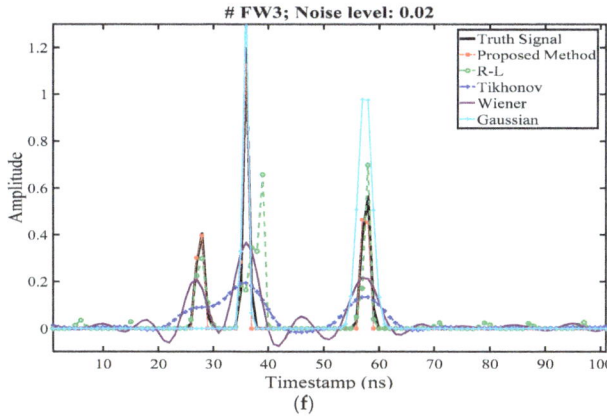

Figure 4. (a–f) Qualitative comparison of sample retrieved waveforms by the proposed method, and the R-L, Wiener filter, Tikhonov regularization and Gaussian decomposition methods.

It is noteworthy that the proposed method provides results of higher accuracy, retrieving all the peaks with minimal oscillations and minimum change in pulse position. By increasing the noise level to 0.05, better performance of the proposed method becomes more apparent, in comparison with the R-L method as the second best approach.

Figure 5 shows the comparison of the proposed method versus the R-L method, the Wiener filter, the Tikhonov regularization and Gaussian decomposition at three different noise levels based upon different quantitative metrics. Quantitatively, the accuracy of all methods is gradually diminished with increases in the noise level. According to the similarity metrics, shown in Figure 5, the proposed approach retrieves the synthetic target responses with higher accuracy in the majority of the cases. A slightly better performance is apparent for the R-L method at a σ value 0.02. However, the provided results of the R-L method are based on an ideal number of iterations related to the minimum RMSE against the true signals, which cannot be readily determined when dealing with real data.

As shown in Figure 5, Tikhonov regularization, Wiener filtering and Gaussian decomposition all provide signals with the lowest similarity to the truth waveforms. As seen in Figure 4, both Tikhonov regularization and the Wiener filter approaches result in ringing effects, with better separated pulses in the Wiener filter at the cost of revealing more spurious pulses and higher negative amplitude values. Both methods are deficient in regard to retrieval of the amplitude values and they both fail when dealing with complex waveforms and closely located pulses, with a higher level of pulse omission for the Tikhonov method. The inferior performance of Gaussian decomposition is mainly due to the inadequacy of the method in resolving closely located pulses. In addition, it generates pulses with negative amplitude values, as shown in Figure 6. Such pulses can be easily detected and removed from further processing, though at the cost of information deficiency. The poor performance of this method in the retrieval of the true signals is due to the smaller variance (pulse width) of the fitted pulse than the synthetic system waveform, which ultimately results in no solution for that pulse. The superior performance of the proposed method can be seen by looking at both the SAM distance and the correlation coefficient in Figure 5, while differences in terms of the Frechet distance are less noticeable.

Figure 7 shows the L-curves related to the l_1-norm and Tikhonov regularization methods. The L-curve provides an appropriate basis for determination of the optimal regularization parameter, which occurs at a balance point, thus demonstrating approximately reciprocal minimum values of both the residual and regularization terms. Although the optimal regularization parameter for solutions (at each noise levels) results in roughly similar residual norms (vertical axis), the solution magnitude (horizontal axis) is evidently larger in the case of the l_1-norm method at both noise levels. This implies that the solution amplitude can be retrieved with higher values closer to the true cross-section via the l_1-norm, and it confirms the underestimation of amplitude by the Tikhonov method. Overall, closer similarity to the truth data confirm the potential of the proposed method for retrieval of the cross-section signal.

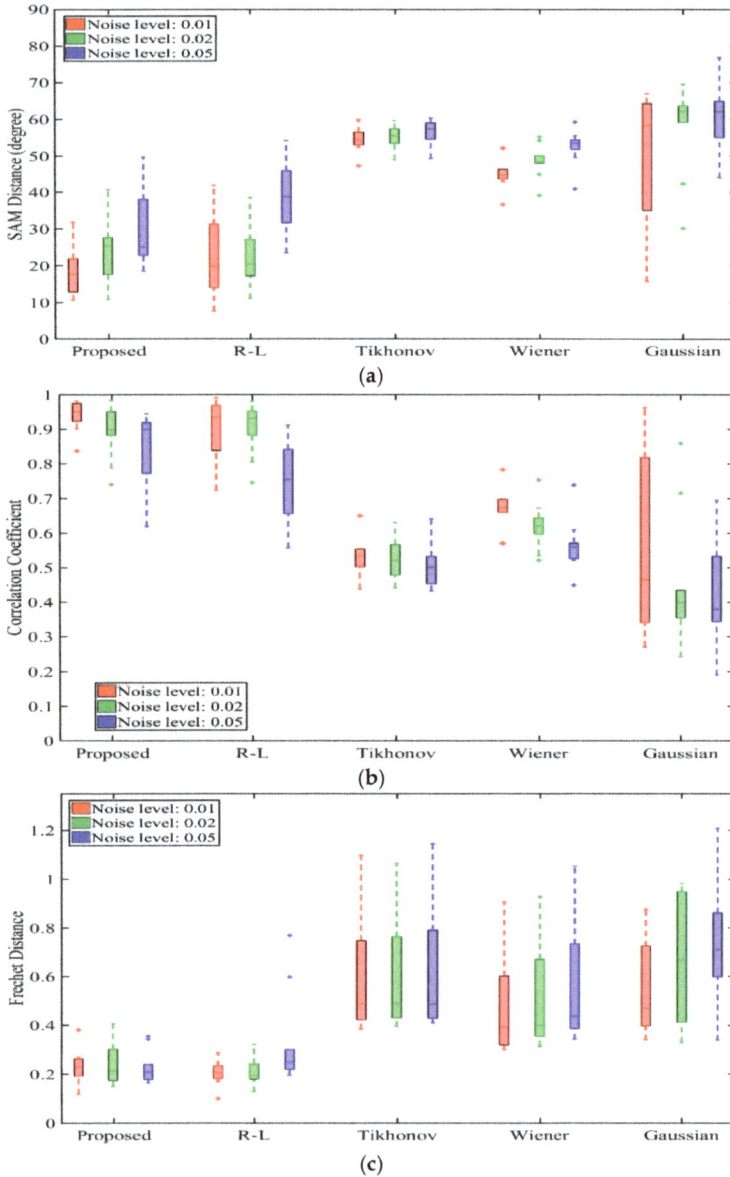

Figure 5. Comparison of the proposed method with others at different noise levels in terms of the three different metrics: (**a**) SAM distance; (**b**) Correlation Coefficient and (**c**) Fréchet distance.

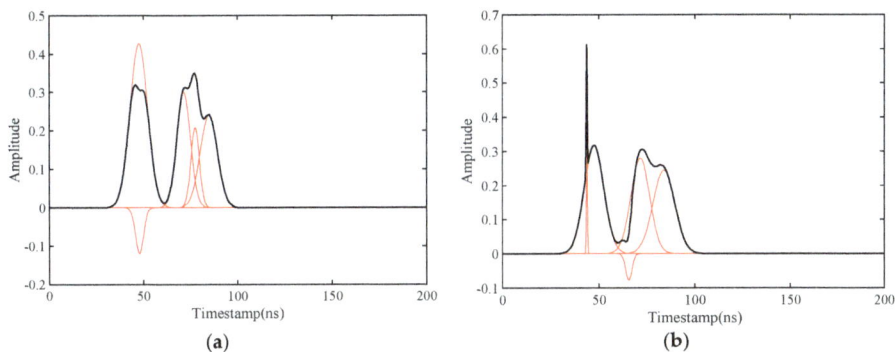

Figure 6. (a,b) Negative amplitudes resulting from Gaussian decomposition. Red curves represent individual scatterers, while the sum of all Gaussian functions are shown in black.

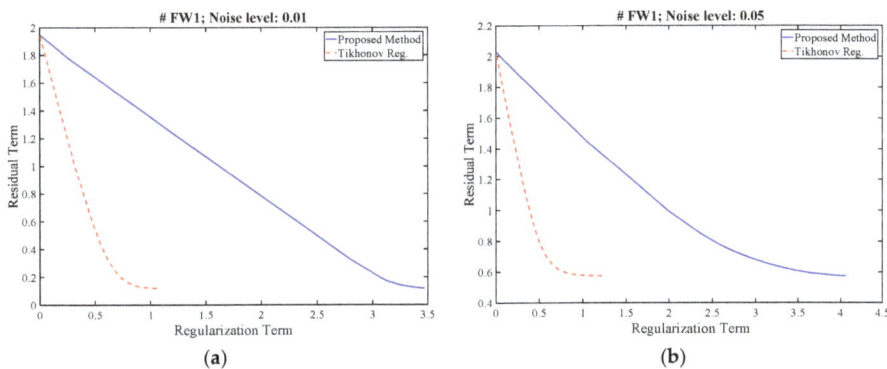

Figure 7. The L-curve plots related to the proposed and Tikhonov regularization approaches for the same waveform at two different noise levels: (**a**) σ = 0.01 and (**b**) σ = 0.05.

4.4. QLD Dataset

The system waveform was not available in the QLD dataset. Due to its necessity for target cross-section retrieval, an estimation of the system waveform was provided by averaging the waveforms calculated from more than 500 waveform samples through applying blind deconvolution. The selected asphalt targets from which the system waveform was estimated have very gentle slopes of 2.4 degrees on average, at nadir. These small slope values cannot be considered as detrimental to the determination of the system waveform, since the pulse shapes are not distorted significantly by these gentle slopes. The averaged system waveform estimated based on blind deconvolution is shown in Figure 8a, where it was applied with the Gaussian decomposition, R-L method, the proposed method and Tikhonov deconvolution. Based on the USGS spectral library [67] for different types of asphalt, the reflectivity value of 0.2 proposed by Wagner et al. [14] was adopted since none of the values in the spectral library were more than 0.2 for the desired wavelength. Figure 8b shows the restored sample waveforms, where the first two returns are from a canopy and the last pulse belongs to a grass surface underneath.

The waveforms restored by the considered methods based on the estimated system waveforms from blind deconvolution were then examined, where prominent peaks are shown in all methods and marginal fluctuations were revealed in the restored signal in some cases.

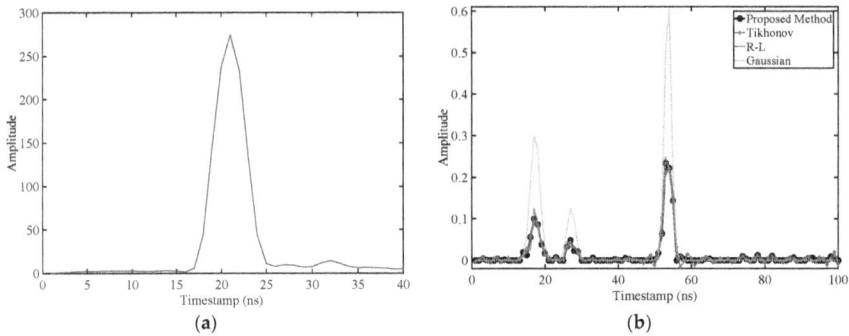

Figure 8. (**a**) Estimated system waveforms based on blind deconvolution; and (**b**) restored target responses using different methods.

From the restored signal, the target cross-section can be calculated. The overall target cross-section of the third pulse in Figure 8b, obtained simply by calculating the area under the pulse after applying the range correction [68], is 0.054 for the l_1-norm solution and 0.057 for the Tikhonov regularization method. These values are close to the equivalent value (0.071) obtained for a grass target with an approximate reflectivity of 0.25 [67] using Equation (1). The target cross-section estimates from Gaussian decomposition and the R-L methods were 0.082 and 0.046, respectively. This overestimation by Gaussian decomposition method is explained by the fact that such a function cannot describe all pulses appropriately. It is noticeable that the second pulse was not recovered when the number of expected pulses was set to three in the Gaussian decomposition. All three visible pulses could be restored by setting this value to four instead, while the restoration resulted in a null solution for the fourth pulse (located immediately after the third pulse).

Marginal underestimation of the total cross-section value by all methods, including the proposed method but excepting the case using Gaussian decomposition, can be attributed to either the footprint area being partially occluded by other targets (e.g., leaves) within the laser path which diminishes the transmitted pulse energy before it reaches the target, or to the approximated system waveform averaged over several samples. In addition, asphalt may not be ideally representative of a Lambertian target due to its variable optical characteristics [1]. Nevertheless, the experiment demonstrated the feasibility of blind deconvolution for estimation of the system waveform. This characteristic was confirmed in the NSW dataset where system waveforms were recorded.

4.5. NSW Dataset

The NSW dataset allowed further evaluation of the potential of the proposed approach in situations where the system waveform is recorded. Moreover, the availability of the system waveform provides an avenue to evaluate the proposed blind deconvolution approach. The effect of the L-curve method on the determination of regularization parameters can then be investigated and the performance of the proposed l_1-norm regularization approach can be compared to that of the other considered methods.

4.5.1. Evaluation of Blind Deconvolution for System Waveform Estimation

Over 4000 waveforms captured at nadir over non-asphalt flat targets were selected for the estimation of the system waveform, with the retrieved system waveform being plotted against the average of the recorded system waveforms in Figure 9a. The result shows that the shapes of both system waveforms are similar. The amplitude of the estimated waveform is slightly larger. This difference is explained by the fact that they are averaged over many samples, in which the arithmetic mean can be affected by extreme values. The SAM measure between the estimated and the averaged real waveforms is as small as 6.8 degrees. This result strongly supports the use of blind deconvolution for system waveform estimation when such information is not available. The retrieved temporal cross-section of a returned lidar waveform obtained using these two system waveforms in the proposed l_1-norm solution in the NSW dataset is shown in Figure 9b where similar amplitude values are seen in both pulses, through the first pulse is marginally underestimated when the estimated system waveform is adopted.

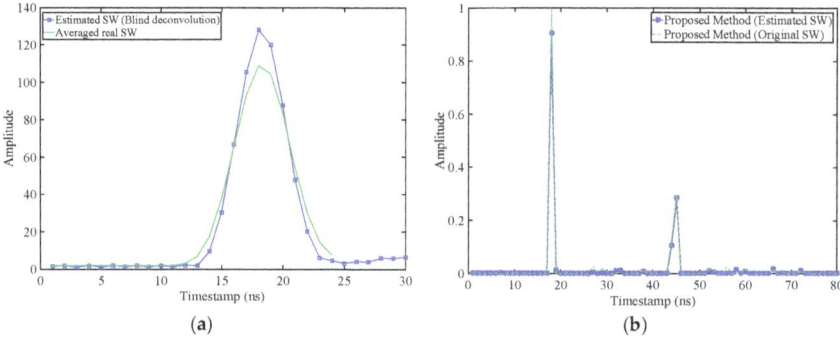

Figure 9. (**a**) The reconstructed system waveform by blind deconvolution versus the average of the recorded system waveforms; and (**b**) comparison of the restored signals based upon the original incomplete system waveform and that retrieved from blind deconvolution.

4.5.2. Target Cross-Section Recovery

Figure 10a illustrates the L-curve plots, which depict the associated magnitudes of the residual and the regularization terms for a range of expected regularization parameters of the waveform in Figure 3 for the l_1-norm and Tikhonov regularization approaches. As seen, the maximum magnitude of the regularization term (horizontal axis) in the proposed l_1-norm approach is larger than the magnitude of this term for Tikhonov regularization, implicitly confirming the potential of the former approach in the retrieval of a signal with higher amplitude values.

Larger regularization parameters for Tikhonov regularization ($\lambda = 32.37$) represent the placing of more emphasis on the regularization term and forcing the residual term to smaller values. Nevertheless, as shown in Figure 10b, the deconvolved signal was affected by noise and underestimation of the main amplitude values, together with fluctuations in between major amplitude values, which will occur as a consequence of the intrinsic characteristics of the l_2-norm, as well as smaller regularization terms, in comparison with the proposed l_1-norm solution. On the other hand, while a smaller regularization parameter for the l_1-norm method ($\lambda = 15.26$) indicates less emphasis on the regularization term, it allows the regularization magnitude to reach higher amplitude values. The deconvolved signal with the l_1-norm reveals the least artifacts or oscillations, especially around the main peaks.

Even though the nonlinear optimization results in the desired parameters for both pulses, Gaussian decomposition fails to retrieve the second pulse, mainly due to the smaller pulse width (variance) of the fitted function than that of the system waveform. This example clearly shows the deficiency of such a method in the retrieval of the target cross-section, even for pulses that are clearly separated. The R-L method results in a signal similar to that of the proposed method, when marginal over-/under-estimation of the pulses is evident.

Figure 11 presents a complex lidar waveform in the NSW dataset. The lidar signal penetrated through the tree canopy before reaching the ground, producing multiple peaks with varying amplitudes. The effect of multiple scattering reduces the strength of the pulses and causes delay in the returns, resulting in a long tail in the received waveform. The result of the implemented denoising approach in Azadbakht et al. [63], on the received waveform, is illustrated in Figure 11a.

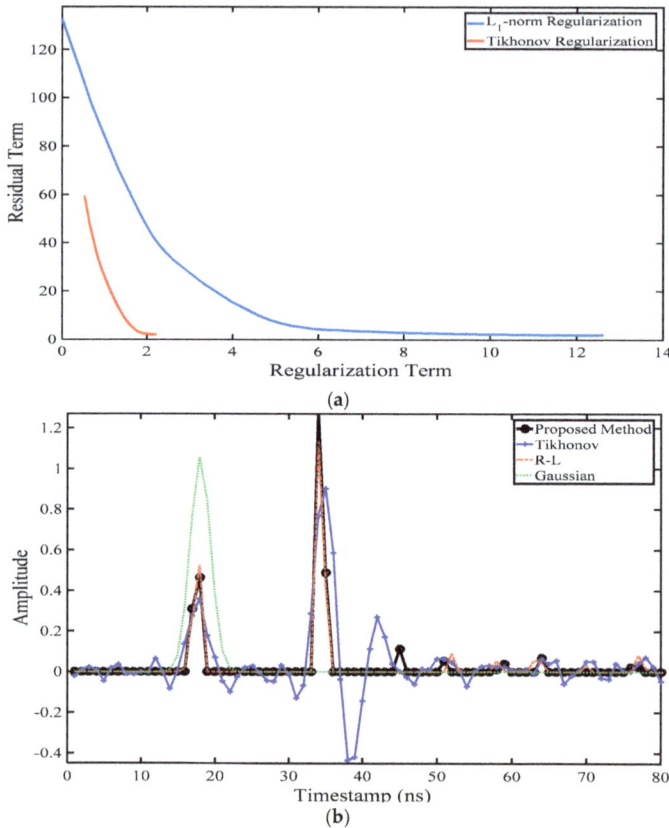

Figure 10. (a) The L-curve results corresponding to the l_1-norm regularization and Tikhonov regularization methods; and (b) restored signals from different methods.

The restored signals using the proposed l_1-norm method in comparison with those from the Tikhonov, Wiener and R-L approaches using the recorded system waveform are presented in Figure 11b. The l_1-norm regularization reveals all the peaks with no oscillation or negative amplitude. In the l_1-norm solution, small amplitude values between the main pulses, and after the ground return (last pulse), can be related to tree leaves and branches. The weak signals may be attributed to multiple scattering of tree foliage or to noise. Spikey-like pulses emerge in the case of the R-L approach, as the closest approach to the l_1-norm method in terms of cross-section retrieval, probably as a result of noise amplification. This confirms the vulnerability of the R-L method illustrated earlier using the synthetic dataset (Section 4.3). In comparison to all other methods, the R-L method produces a significantly overestimated amplitude for the first pulse. The largest oscillation and negative amplitudes in the restored waveforms are associated with the Tikhonov method, due to the inherent properties of the l_2-norm solution. Next is the Wiener filter with slightly smaller estimated amplitude values in addition to oscillation. In Figure 11, none of the pulses were retrieved using Gaussian decomposition, principally due to the system waveform being wider than the fitted Gaussian pulses in the received waveform. This method could also not fit a pulse to the second return when three pulses were considered as an initial value for the nonlinear optimization problem.

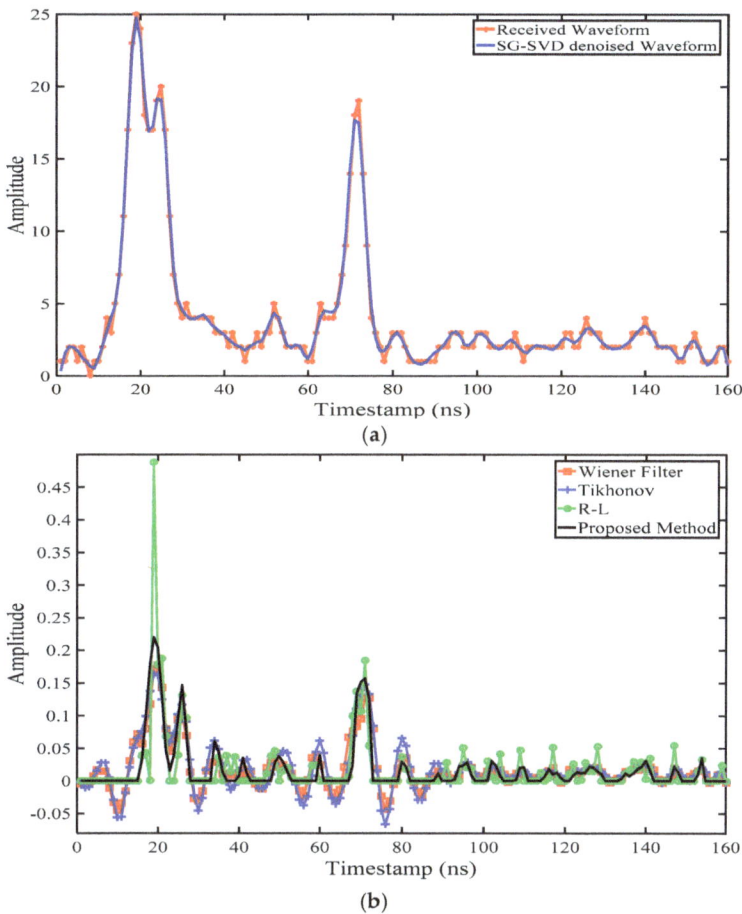

Figure 11. (**a**) Raw received waveform and its noise-reduced version; and (**b**) the recovered differential cross-sections for different methods.

4.6. Karawatha Dataset

The target cross-section, the reflectance and the backscatter coefficient [15] of three types of extended non-overlapping targets were calculated, after retrieval of the temporal cross-sections and application of further required corrections [68]. Samples were collected over a range of scan angles from nadir to 25 degrees by inspecting points in a high resolution aerial photograph, taken simultaneously. Except for water bodies that were only available at limited scan angles, a total of around 200 samples were collected for each target.

Diffuse target reflectance behaviour was assumed in the calculation of the radiometric features, although this assumption is not valid for some targets, e.g., water bodies. However, relatively small reflectance values are due to the low strength of the received signal at off-nadir, for instance in the case of water bodies with specular reflectance characteristics. The provided results from the boxplots in Figure 12 confirm the reliability and relevance of the proposed approach in lidar waveform restoration. These targets can be readily distinguished only by using the radiometric information, while their height is almost similar and, therefore, geometric attributes are not effective.

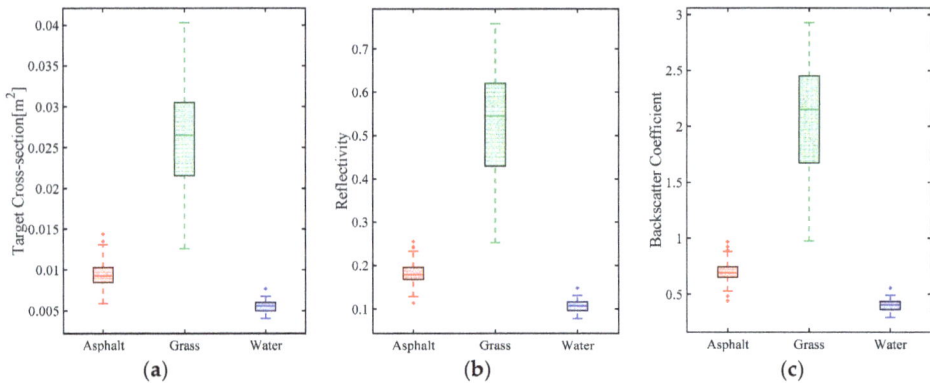

Figure 12. (a) Target cross-section; **(b)** the reflectance; and **(c)** the backscatter coefficient for the selected targets aggregated over different scan angles.

5. Conclusions

This paper has presented a new deconvolution approach for efficient target cross-section extraction from full-waveform lidar data. The deconvolution, which removes the lidar system signal from the returned waveform and reconstructs the target cross-section independently of the instrument, is achieved through a versatile regularization approach with sparsity constraints, which copes with different kinds of waveforms. The regularization parameter, a critical issue in deconvolution, is determined with the L-curve method in which the optimal parameter is found at the turning point of the curve. Blind deconvolution is proposed to estimate the system signal, which is necessary on lidar instruments that do not record the system waveform.

The proposed approach has been evaluated using both synthetic and real lidar datasets acquired with and without system waveforms. While the synthetic data allowed for evaluation with "true" target cross-section, the real data provided an avenue to assess the consistency of the proposed approach under different instrument and environmental conditions. The results with synthetic data showed that the l_1-norm regularization, with the regularization parameter derived by the L-curve method, is superior to Tikhonov regularization, the Wiener filter, Gaussian decomposition and the R-L methods, which currently attract wide usage in lidar signal processing. The restored waveforms are strongly similar to the truth data despite the interference of different noise levels (Section 4.3).

The effectiveness of the proposed deconvolution method has been further demonstrated in experimental testing with real data, with and without system signals. It has been shown that the l_1-norm approach always results in a deconvolved signal almost free of substantial oscillations and negative amplitude values, while such phenomena are usually present with other approaches. Taken together, the findings of this study support adoption of sparse regularization as a means to retrieve the target response from full-waveform lidar.

The L-curve method has been shown to be reliable in the determination of the regularization parameter, as demonstrated in the experiments with both synthetic and real data.

Another finding to emerge from this research is that blind deconvolution has been demonstrated to be useful in the estimation of the system waveform when it is not available. Comparisons between the restored lidar waveforms using the system waveform based on blind deconvolution has been reported. Although restoration of the exact target response is not anticipated, due to the approximate estimation of the system waveform by averaging over several selected samples, the results demonstrate the consistency of the reported method in retrieving all pulses. Moreover, investigation of the blind deconvolution in the NSW dataset showed that the retrieved system waveform is in close agreement with the average system waveform of more than 3000 waveform samples. These visual and qualitative assessments indicate the significance of blind deconvolution in the retrieval of the system waveform, and they confirm that the deconvolved signal is associated with the target cross-section.

It is concluded that the proposed regularization approach can efficiently process full-waveform lidar data to restore the target cross-section. The use of this approach will provide additional insight into the waveform data. The restored cross-section affords the potential identification of either desired reflecting surfaces or even specific targets in a more consistent manner than is achieved with other approaches, thus improving the accuracy of range measurement and target attribute retrieval.

Future research will concentrate on investigation of the extracted geometric and radiometric features using the proposed method, along with additional potential attributes from the restored signal. Moreover, since direct regularization methods support inclusion of new constraints to better control the solution, investigation of additional term(s) in the objective function may result in even higher accuracy.

Acknowledgments: This work was supported by Ph.D. research funding provided by the University of Melbourne, Australia. The authors thank the Aerial Topographic Laser Survey Systems Company (http://www.atlass.com.au/) and Geoscience Australia who provided laser scanning data. The authors are also thankful to three anonymous reviewers for through review with valuable comments.

Author Contributions: Mohsen Azadbakht designed and performed the experiments, and prepared the manuscript. All authors contributed to the analysis, interpretation of the results and manuscript revisions.

Conflicts of Interest: The authors declare no conflict of interest.

References

1. Mallet, C.; Bretar, F.; Roux, M.; Soergel, U.; Heipke, C. Relevance assessment of full-waveform lidar data for urban area classification. *ISPRS J. Photogramm. Remote Sens.* **2011**, *66*, S71–S84. [CrossRef]
2. Bretar, F.; Chauve, A.; Mallet, C.; Jacome, A. Terrain surfaces and 3-D landcover classification from small footprint full-waveform lidar data: Application to badlands. *Hydrol. Earth Syst. Sci.* **2009**, *13*, 1531–1545. [CrossRef]
3. Jutzi, B.; Stilla, U. Extraction of features from objects in urban areas using spacetime analysis of recorded laser pulses. *Int. Arch. Photogramm. Remote Sens. Spat. Inf. Sci.* **2004**, *35*, 1–6.
4. Hovi, A.; Korpela, I. Real and simulated waveform-recording LiDAR data in juvenile boreal forest vegetation. *Remote Sens. Environ.* **2014**, *140*, 665–678. [CrossRef]
5. Adams, T.; Beets, P.; Parrish, C. Extracting more data from lidar in forested areas by analyzing waveform shape. *Remote Sens.* **2012**, *4*, 682–702. [CrossRef]
6. Hollaus, M.; Aubrecht, C.; Höfle, B.; Steinnocher, K.; Wagner, W. Roughness mapping on various vertical scales based on full-waveform airborne laser scanning data. *Remote Sens.* **2011**, *3*, 503–523. [CrossRef]
7. Mallet, C.; Bretar, F. Full-waveform topographic lidar: State-of-the-art. *ISPRS J. Photogramm. Remote Sens.* **2009**, *64*, 1–16. [CrossRef]
8. Höfle, B.; Pfeifer, N. Correction of laser scanning intensity data: Data and model-driven approaches. *ISPRS J. Photogramm. Remote Sens.* **2007**, *62*, 415–433. [CrossRef]
9. Jutzi, B.; Stilla, U. Laser pulse analysis for reconstruction and classification of urban objects. In Proceedings of the International Society Photogrammetry and Remote Sensing (ISPRS) Archives, Munich, Germany, 17–19 September 2003.
10. Wagner, W.; Ullrich, A.; Melzer, T.; Briese, C.; Kraus, K. From single-pulse to full-waveform airborne laser scanners: Potential and practical challenges. *Int. Arch. Photogram. Remote Sens.* **2004**, *35*, 201–206.
11. Roncat, A.; Bergauer, G.; Pfeifer, N. B-spline deconvolution for differential target cross-section determination in full-waveform laser scanning data. *ISPRS J. Photogramm. Remote Sens.* **2011**, *66*, 418–428. [CrossRef]
12. Jutzi, B.; Stilla, U. Range determination with waveform recording laser systems using a Wiener Filter. *ISPRS J. Photogramm. Remote Sens.* **2006**, *61*, 95–107. [CrossRef]
13. Jelalian, A.V. *Laser Radar Systems*; Artech House: Boston, MA, USA; London, UK, 1992.
14. Wagner, W.; Ullrich, A.; Ducic, V.; Melzer, T.; Studnicka, N. Gaussian decomposition and calibration of a novel small-footprint full-waveform digitising airborne laser scanner. *ISPRS J. Photogramm. Remote Sens.* **2006**, *60*, 100–112. [CrossRef]

15. Wagner, W. Radiometric calibration of small-footprint full-waveform airborne laser scanner measurements: Basic physical concepts. *ISPRS J. Photogramm. Remote Sens.* **2010**, *65*, 505–513. [CrossRef]

16. Wang, Y.; Zhang, J.; Roncat, A.; Künzer, C.; Wagner, W. Regularizing method for the determination of the backscatter cross section in lidar data signals. *Opt. Soc. Am.* **2009**, *26*, 1071–1079. [CrossRef]

17. Chauve, A.; Vega, C.; Durrieu, S.; Bretar, F.; Allouis, T.; Deseilligny, M.P.; Puech, W. Advanced full-waveform lidar data echo detection: Assessing quality of derived terrain and tree height models in an alpine coniferous forest. *Int. J. Remote Sens.* **2009**, *30*, 5211–5228. [CrossRef]

18. Lin, Y.-C.; Mills, J.P.; Smith-Voysey, S. Rigorous pulse detection from full-waveform airborne laser scanning data. *Int. J. Remote Sens.* **2010**, *31*, 1303–1324. [CrossRef]

19. Parrish, C.E.; Jeong, I.; Nowak, R.D.; Smith, R.B. Empirical Comparison of Full-Waveform Lidar Algorithms: Range Extraction and Discrimination Performance. *Photogram. Eng. Remote Sens.* **2011**, *77*, 825–838. [CrossRef]

20. Wu, J.; Aardt, J.A.N.V.; Asner, G.P. A comparison of signal deconvolution algorithms based on small-footprint lidar waveform simulation. *IEEE Trans. Geosci. Remote Sens.* **2011**, *49*, 2402–2414. [CrossRef]

21. Hofton, M.A.; Minster, J.B.; Blair, J.B. Decomposition of Laser Altimeter Waveforms. *IEEE Trans. Geosci. Remote Sens.* **2000**, *38*, 1989–1996. [CrossRef]

22. Persson, A.; Söderman, U.; Töpel, J.; Ahlberg, S. Visualization and analysis of full-waveform airborne laser scanner data. In Proceedings of the ISPRS Workshop Laser Scanning, Enschede, The Netherlands, 12–14 September 2005; pp. 103–108.

23. Mallet, C.; Lafarge, F.; Roux, M.; Soergel, U.; Bretar, F.; Heipke, C. A marked point process for modeling lidar waveforms. *IEEE Trans. Image Process.* **2010**, *19*, 3204–3221. [CrossRef] [PubMed]

24. Jutzi, B.; Stilla, U. Characteristics of the measurement unit of a full-waveform laser system. *Int. Arch. Photogramm. Remote Sens. Spat. Inf. Sci.* **2006**, *36*, 17–22.

25. Alexander, C.; Tansey, K.; Kaduk, J.; Holland, D.; Tate, N.J. Backscatter coefficient as an attribute for the classification of full-waveform airborne laser scanning data in urban areas. *ISPRS J. Photogramm. Remote Sens.* **2010**, *65*, 423–432. [CrossRef]

26. Zhu, R.; Pang, Y.; Zhang, Z.; Xu, G. Application of the deconvolution method in the processing of full-waveform LiDAR data. In Proceedings of the 3rd International Congress on Image and Signal Processing (CISP2010), Yantai, China, 16–18 October 2010; pp. 2975–2979.

27. Chauve, A.; Mallet, C.; Bretar, F.; Durrieu, S.; Deseilligny, M.P.; Puech, W. Processing full-waveform lidar data: Modelling raw signals. In Proceedings of the ISPRS Workshop on Laser Scanning 2007 and SilviLaser 2007, Espoo, Finland, 12–14 September 2007; pp. 102–107.

28. Wu, J.; van Aardt, J.A.N.; McGlinchy, J.; Asner, G.P. A robust signal preprocessing chain for small-footprint waveform lidar. *IEEE Trans. Geosci. Remote Sens.* **2012**, *50*, 3242–3255. [CrossRef]

29. Reeves, S. Generalized cross-validation as a stopping rule for the Richardson-Lucy algorithm. *Int. J. Imaging Syst. Technol.* **1995**, *6*, 387–391. [CrossRef]

30. Khan, M.K.; Morigi, S.; Reichel, L.; Sgallari, F. Iterative methods of richardson-lucy-type for image debluring. *Numer. Math. Theory Methods Appl.* **2013**, *6*, 262–275.

31. Neilsen, K.D. Signal Processing on Digitized LADAR Waveforms for Enhanced Resolution on Surface Edges. Master's Thesis, Utah State University, Logan, UT, USA, 2011.

32. Morháč, M.; Matoušek, V. High-resolution boosted deconvolution of spectroscopic data. *J. Comput. Appl. Math.* **2011**, *235*, 1629–1640. [CrossRef]

33. Castorena, J.; Creusere, C.D.; Voelz, D. Modeling LiDAR scene sparsity using compressicve sensing. In Proceedings of the 2010 IEEE International Geoscience and Remote Sensing Symposium (IGARSS), Honolulu, HI, USA, 25–30 July 2010.

34. Verdin, B.; von Borries, R. Lidar compressive sensing using chaotic waveform. *Proc. SPIE Radar Sens. Technol.* **2014**. [CrossRef]

35. Gunturk, B.K. Fundamentals of image restoration. In *Image Restoration: Fundamentals and Advances*; Gunturk, B.K., Li, X., Eds.; CRC Press: Boca Raton, FL, USA, 2012; pp. 25–61.

36. Gold, R. *An Iterative Unfolding Method for Response Matrices*; Argonne National Laboratory: Argonne, IL, USA, 1964.

37. Richardson, W.H. Bayesian-based iterative method of image restoration. *J. Opt. Soc. Am.* **1972**, *62*, 55–59. [CrossRef]

38. Lucy, L.B. An iterative method for the rectification of the observed distributions. *Astron. J.* **1974**, *79*, 745–754. [CrossRef]

39. Vogel, C.R. *Computational Methods for Inverse Problems*; Frontiers in Applied Mathematics Series; Society of Industrial and Applied Mathematics (SIAM): Philadelphia, PA, USA, 2002; p. 183.

40. Stefan, W. Total Variation Regularization for Linear Ill-Posed Inverse Problems Extensions and Applications. Ph.D. Thesis, Arizona State University, Phoenix, AZ, USA, 2008.

41. Hansen, P.C. The discrete picard condition for discrete ill-posed problems. *BIT Numer. Math.* **1990**, *30*, 658–672. [CrossRef]

42. Boyd, S.; Vandenberghe, L. *Convex Optimization*; Cambridge University Press: New York, NY, USA, 2004; p. 727.

43. Tikhonov, A.N.; Arsenin, V.I.A. *Solutions of Ill-Posed Problems*; Winston & Sons: Washington, DC, USA, 1977; p. 258.

44. Elad, M.; Figueiredo, A.T.; Ma, Y. On the role of sparse and redundant representations in image processing. *IEEE Proc.* **2010**, *98*, 972–982. [CrossRef]

45. Hansen, P.C.; O'Leary, D.P. The use of the L-curve in the regularization of discrete ill-posed problems. *SIAM J. Sci. Comput.* **1993**, *14*, 1487–1503. [CrossRef]

46. Pan, T.C. Signal and Image Deconvolution: Algorithms and Applications. Ph.D. Thesis, Hong Kong Baptist University, Hong Kong, China, 2010.

47. Figueiredo, M.A.T.; Bioucas-Dias, J.M. Algorithms for imaging inverse problems under sparsity regularization. In Proceedings of the 3rd International Workshop on Cognitive Information Processing (CIP), Baiona, Spain, 28–30 May 2012; pp. 1–6.

48. Tropp, J.A.; Wright, S.J. Computational methods for sparse solution of linear inverse problems. *IEEE Proc.* **2010**, *98*, 948–958. [CrossRef]

49. Schmidt, M.; Fung, G.; Rosales, R. Fast optimization methods for L_1 regularization: A comparative study and two new approaches. In *Machine Learning: ECML 2007*; Springer: Berlin, Germany; Heidelberg, Germany, 2007; pp. 286–297.

50. Kim, S.-J.; Koh, K.; Lustig, M.; Boyd, S.; Gorinevsky, D. An interior-point method for large-scale l_1-regularized least squares. *IEEE J. Sel. Top. Signal Proc.* **2007**, *1*, 606–617.

51. Palomar, D.P.; Eldar, Y.C. *Convex Optimization in Signal Processing and Communications*; Cambridge University Press: New York, NY, USA, 2010; p. 498.

52. Alizadeh, F.; Goldfarb, D. Second-order cone programming. *Math. Program.* **2003**, *95*, 3–51. [CrossRef]

53. Sousa, M.; Vandenberghe, L.; Boyd, S.; Lebret, H. Applications of second-order cone programming. *Linear Algebra Appl.* **1998**, *284*, 193–228.

54. Hansen, P.C.; Nagy, J.G.; O'Leary, D.P. *Deblurring Images: Matrices, Spectra, and Filtering*; Society of Industrial and Applied Mathematics (SIAM): Philadelphia, PA, USA, 2006; p. 130.

55. Orfanidis, S.J. *Introduction to Signal Processing*; Prentice Hall: Upper Saddle River, NJ, USA, 2010; p. 783.

56. Morozov, V.A. On the solution of functional equations by the method of regularization. *Sov. Math. Dokl.* **1966**, *7*, 414–417.

57. Golub, G.; Heath, M.; Wahba, G. Generalized cross-validation as a method for choosing a good ridge parameter. *Technometrics* **1979**, *21*, 215–223. [CrossRef]

58. Hansen, P.C. Analysis of discrete ill-posed problems by means of the L-curve. *SIAM Rev.* **1992**, *34*, 561–580. [CrossRef]

59. Correia, T.; Gibson, A.; Schweiger, M.; Hebden, J. Selection of regularization parameter for optical topography. *J. Biomed. Opt.* **2009**, *14*. [CrossRef] [PubMed]

60. White, R.L. Image restoration using the damped Richardson-Lucy method. In Proceedings of the SPIE 2198, Instrumentation in Astronomy VIII, Kailua-Kona, HI, USA, 13 March 1994; pp. 1342–1348.

61. Rote, G. Computing the Fréchet distance between piecewise smooth curves. *Comput. Geom.* **2007**, *37*, 162–174. [CrossRef]

62. Savitzky, A.; Golay, M.J.E. Smoothing and differentiation of data by simplified least squares procedures. *Anal. Chem.* **1964**, *36*, 1627–1639. [CrossRef]

63. Azadbakht, M.; Fraser, C.S.; Zhang, C.; Leach, J. A signal denoising method for full-waveform LiDAR data. In Proceedings of the ISPRS Annals of Photogrammetry, Remote Sensing and Spatial Information Sciences, Antalya, Turkey, 11–13 November 2013; pp. 31–36.

64. Roncat, A. Backscatter Signal Analysis of Small-Footprint Full-Waveform Lidar Data. Ph.D. Thesis, Vienna University of Technology, Vienna, Austria, 2014.

65. Grant, M.; Boyd, S. Graph implementations for nonsmooth convex programs. In *Recent Advances in Learning and Control*; Blondel, V.D., Boyd, S., Kimura, H., Eds.; Springer: London, UK, 2008; pp. 95–110.

66. Grant, M.; Boyd, S. CVX: Matlab Software for Disciplined Convex Programming. Available online: http://cvxr.com/cvx (accessed on 30 May 2016).

67. Clark, R.N.; Swayze, G.A.; Wise, R.A.; Livo, K.E.; Hoefen, T.M.; Kokaly, R.F.; Sutley, S.J. USGS Digital Spectral Library Splib06a. Available online: http://speclab.cr.usgs.gov/spectral.lib06/ds231/ (accessed on 20 September 2007).

68. Azadbakht, M.; Fraser, C.S.; Zhang, C. Separability of targets in urban areas using features from full-waveform LiDAR data. In Proceedings of the IEEE International Geoscience and Remote Sensing Symposium (IGARSS), Milan, Italy, 26–31 July 2015; pp. 5367–5370.

remote sensing

MDPI

Article

Performance Assessment of High Resolution Airborne Full Waveform LiDAR for Shallow River Bathymetry

Zhigang Pan [1,2,*], **Craig Glennie** [1,2,†], **Preston Hartzell** [1,2,†], **Juan Carlos Fernandez-Diaz** [1,2,†], **Carl Legleiter** [3,†] and **Brandon Overstreet** [3,†]

[1] Department of Civil and Environmental Engineering, University of Houston, Houston, TX 77204, USA; E-Mails: clglennie@uh.edu (C.G.); pjhartzell@uh.edu (P.H.); jfernan4@central.uh.edu (J.C.F.-D.)

[2] National Center for Airborne Laser Mapping, University of Houston, 5000 Gulf Freeway Building 4 Room 216, Houston, TX 77204, USA

[3] Department of Geography, University of Wyoming, 1000 E University Ave, Laramie, WY 82071, USA; E-Mails: Carl.Legleiter@uwyo.edu (C.L.); boverstr@uwyo.edu (B.O.)

* Author to whom correspondence should be addressed; E-Mail: pzhigang@uh.edu; Tel.: +1-832-842-8881; Fax: +1-713-743-0186.

† These authors contributed equally to this work.

Academic Editors: Wolfgang Wagner and Prasad S. Thenkabail

Received: 17 February 2015 / Accepted: 20 April 2015 / Published: 24 April 2015

Abstract: We evaluate the performance of full waveform LiDAR decomposition algorithms with a high-resolution single band airborne LiDAR bathymetry system in shallow rivers. A continuous wavelet transformation (CWT) is proposed and applied in two fluvial environments, and the results are compared to existing echo retrieval methods. LiDAR water depths are also compared to independent field measurements. In both clear and turbid water, the CWT algorithm outperforms the other methods if only green LiDAR observations are available. However, both the definition of the water surface, and the turbidity of the water significantly influence the performance of the LiDAR bathymetry observations. The results suggest that there is no single best full waveform processing algorithm for all bathymetric situations. Overall, the optimal processing strategies resulted in a determination of water depths with a 6 cm mean at 14 cm standard deviation for clear water, and a 16 cm mean and 27 cm standard deviation in more turbid water.

Keywords: LiDAR; full waveform; bathymetry; wavelet transformation

1. Introduction

Airborne Light Detection and Ranging (LiDAR) is an active remote sensing technique used to acquire 3D representations of objects with very high resolution [1,2]. LiDAR systems emit short laser pulses to illuminate the Earth's surface and then capture the reflected light with photodiode detectors. By measuring the laser flight time propagating through the medium, a distance to target (range) can be determined assuming a known constant speed of light in the medium [3,4]. With superior performance for acquiring 3D measurements and easy deployment, LiDAR data has been used for many scientific applications such as biomass estimation, archaeological application, power line detection, and earth science applications including temporal change detection [5–8].

A conventional discrete-LiDAR system records only a few (≤ 5) discrete-returns for each outgoing laser pulse. A hardware ranging method called constant fractional discrimination (CFD) [9] is implemented in most current LiDAR systems to discriminate vertically cluttered illumination targets along the laser path [10]. With critical requirements for vertical resolution and the advancement of processing and storage capacity over last decade, Full Waveform LiDAR (FWL) has emerged as a viable alternative to discrete-LiDAR. FWL records the entire digitized backscatter laser pulse received by detector with very high sampling rate (1–2 GHz) [11].

FWL was introduced in commercial topographic LiDAR systems in 2004 and a number of LiDAR systems now have the capability to store the entire digitized return waveform [12–14]. FWL enables better vertical resolution because discrete-return LiDAR resolution is largely influenced by laser pulse width or Full Width at Half Maximum (FWHM) [10,15]. However, sophisticated digital signal processing methods are required to extract points and other target information (e.g., radiometry) from current FWL systems. To date, many full waveform processing algorithms have been proposed and are widely used in the research community. An overview of waveform processing techniques can be found in [11]. Gaussian decomposition [16–18] and deconvolution [15,19] are two processing strategies that have been applied in a number of previous studies. Hartzell et al. [20] proposed an empirical system response model which is estimated from single return waveforms over the dynamic range of the instrument. The empirical response is then used as a template for actual return waveforms, resulting in improved ranging.

Recently, the analysis of FWL processing has focused on evaluating the different methods in parallel to find superior algorithms for specific applications. For example, Wu et al. [21] compared three deconvolution methods: Richardson-Lucy, Wiener filter, and nonnegative least squares to determine the best performance using simulated full waveforms from radiative transfer modeling; the Richardson-Lucy method was found to have superior performance for deconvolution of the simulated full waveforms. Parrish et al. [10] presented an empirical technique to compare three different methods for full waveform processing: Gaussian decomposition, Expectation-Maximization (EM) deconvolution and a hybrid method (deconvolve-decompose). Using precisely located screen-targets in a laboratory, they arrived at the conclusion that no single best full waveform method can be found for all applications.

Despite the recent focus on applications of FWL for topographic studies, it was first evaluated for the processing of LiDAR bathymetry [22]. Recently, however, full waveform bathymetric LiDAR has not received much attention in the literature, especially compared to topographic FWL. This is likely due to the lack of available bathymetric LiDAR datasets for the scientific community and the more complicated modeling required for LiDAR bathymetry to compensate for factors such as water surface reflection and refraction, water volume scattering and turbidity that can complicate the propagation models and attenuate return strength resulting in a lower signal to noise ratio (SNR). Water volume scattering can be difficult to rigorously model, especially for shallow water environments where water surface backscatter, water volume scattering, and benthic layer backscattering are mixed into a single complex waveform that makes discrimination of individual responses from a single return difficult [23]. The complex waveform signals in a bathymetric environment demand a noise-resistant and adaptive signal processing methodology. In order to reduce the complexity of bathymetric LiDAR, multiple wavelengths (usually a NIR LiDAR system for water surface detection, and a green LiDAR system for water penetration) systems are normally used to facilitate benthic layer retrieval [12]. For example, Allouis et al. [24] compared two new processing methods for depth extraction by using near-infrared (NIR), green and Raman LiDAR signals. By combining NIR and green waveforms, significantly more points are extracted by full waveform processing and better accuracy is achieved. Even though multi-wavelength LiDAR systems are common for bathymetry, single band systems have emerged recently as well [14,25,26]. Wang et al. [27] has compared several full waveform processing algorithms for single band shallow water bathymetry using both simulated and actual full waveform data, and concluded that Richardson-Lucy deconvolution performed the best of the tested waveform processing techniques. However, the performance with the actual full waveform data was not verified with comparison to external high accuracy truth data. There have also been several studies which have examined the performance of single band full waveform bathymetry using simulated LiDAR datasets. Abady et al. [23] proposed a mixture of Gaussian and quadrilateral functions for bathymetric LiDAR waveform decomposition using nonlinear recursive least squares. Both satellite and airborne configurations were simulated and examined and showed significant improvement for bathymetry retrieval, however, the simulation has not to date been validated with observations from real-world studies, especially for very shallow water bathymetry in turbid conditions. The performance of full waveform LiDAR in shallow water has received little attention in the literature beyond the study by McKean et al. [25]. Limited water depths and significant turbidity impose challenges for bathymetric LiDAR, especially for longer pulse width laser systems where water surface, water column and benthic layer returns mix together. A bathymetric full waveform processing strategy to account for the longer pulse width and the excessive noise present in the bathymetric waveform would enable more accurate bathymetry determination.

In this paper, we first propose a novel full waveform processing algorithm using a continuous wavelet transformation (CWT) to decompose single band bathymetric waveforms. The seed peak locations acquired from CWT are then used as input to both an empirical system response (ESR) algorithm and a Gaussian decomposition method. As a benchmark for comparison, a common Gaussian decomposition algorithm is also used with candidate seed locations acquired from second derivative peaks, similar to that presented in [17,18]. The waveform processing methods are applied to two distinct fluvial environments with varying degrees of water turbidity. Water depths extracted from both a discrete point cloud and full waveform processed point clouds are then compared to water depths measured in the field with an Acoustic Doppler Current Profiler (ADCP). Finally, we analyze the accuracy of water surfaces extracted from discrete point clouds and full waveform processed point clouds using both green wavelength and near-infrared detected water surfaces compared to GNSS RTK field measurements. The rest of the paper is organized as follows: Section 2 presents the waveform processing mathematical models and the procedure used for water depth generation from the LiDAR observations, Section 3 provides a description of the airborne datasets and ground truthing used to evaluate the processing methodology, and Section 4 presents the experimental results. The paper closes with a discussion of the study conclusions and areas for future research.

2. Method and Mathematical Model

FWL return profiles are normally a fixed length discrete time signal containing backscatter information for a large region of interest. For return profiles where the echoes are clustered in a short range window, a significant portion of the full waveform does not carry useful information (i.e., the profile represents the noise threshold); an effective method to pre-process the full waveforms that removes this extraneous information from the original waveform will reduce the total amount of processing time required. A noise level can be defined as the minimum amplitude and can be estimated from the full waveform data itself; for example as the median absolute deviation for each waveform [28]. For our study, amplitudes within 10% of the return gate are considered as the noise level (Figure 1). The return gate is an instrument specific configuration parameter used to reduce the effect of sun glint and noise returns. Herein, all the bins below the noise level were removed, and only the remaining signal was examined. The removal of data below the noise threshold significantly speeds up the calculations due to the decreased data volume to be analyzed. It should be noted that bathymetric LiDAR waveforms can have quite complicated return energy profiles. To demonstrate this, representative samples of bathymetric waveform are given in Figure 2.

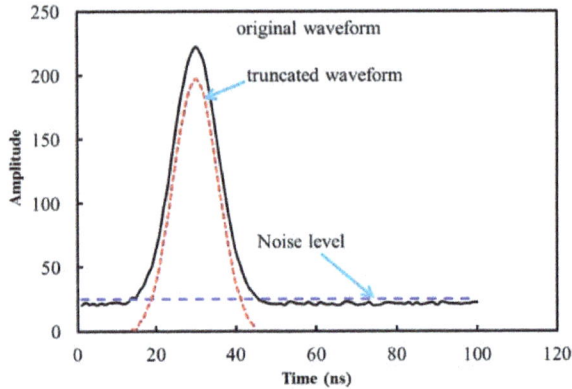

Figure 1. Pre-processing of the return waveform by removing data below the noise threshold of the original waveform. The noise level is defined as 10% above the return gate which is given by the manufacturer specifications. The truncated waveform is saved for posterior processing.

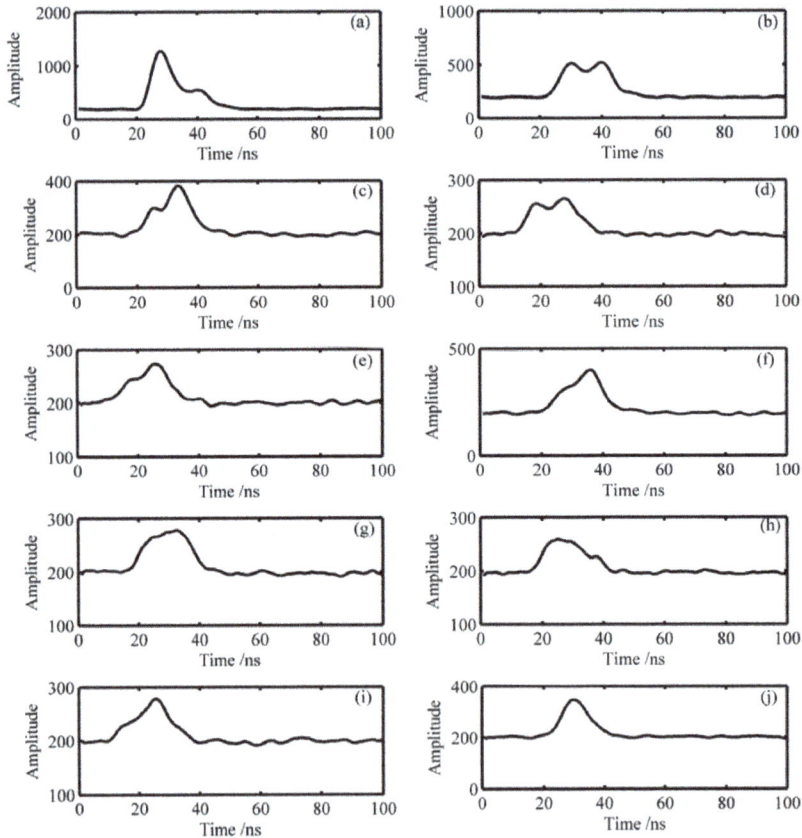

Figure 2. Typical bathymetric return waveforms. (**a**–**d**) are from clear water with multiple visible peaks of varying peak amplitude; (**e**–**f**) contain more subtle evidence of multiple peaks; (**g**–**i**) contain multiple peaks that overlap and are visibly not discernible; (**j**) contains a single peak. Multiple peaks are critical for benthic layer retrieval as the first return is normally the water surface and the latter return more probably from the benthic layer.

2.1. Continuous Wavelet Transformation

The wavelet transformation can be used to project a continuous time signal into multiple subspaces consisting of wavelets [29]. By examining this projection, objectives such as denoising, compression, filtering and other applications can be achieved. A continuous wavelet transformation (CWT) projects the signal into a continuous time and scale subspace (instead of discrete subspaces) whereby the signal can be reconstructed from the resulting continuous components [29,30]. CWT is a very effective method to detect the peak locations in an overlapped waveform [31]. Extending the use of CWT to FWL thus is natural since the return waveforms can be highly mixed due to potentially closely spaced backscatters along the laser path.

CWT can construct a time frequency representation of a signal that offers very good time and frequency localization, making it suitable to localize the peak locations as initial approximations for subsequent peak estimation algorithms. The mother wavelet template should be continuously differentiable and compactly support scaling and capture of a high vanishing moment. Considering that most FWL systems have Gaussian-like signals, the Lorentzian of Gaussian mother wavelet has been used in this study [31], and is given in Equation (1).

$$\omega_{a,b}(t) = \left[1 - \left(\frac{t-b}{a}\right)^2\right] \exp\left[-\left(\frac{t-b}{\sqrt{2}a}\right)^2\right] \tag{1}$$

Here, $\omega_{a,b}$ is the mother wavelet used in CWT, a dilates the mother wavelet and b translates the mother wavelet, t is time. Special caution is needed for determination of a and b. A smaller a can assist in discriminating highly overlapped peaks, but a slight undulation of the waveform may result in a false peak; larger values of a can resist disturbing undulations (i.e., noise) but could miss weak returns and result in single returns for multiple echoes; however, the smallest a cannot be less than the digitizing interval of signal. The ridge defined in [31] is a good implementation for the detection of peaks (and a determination of a) but requires a significant amount of computation, so instead we directly chose a single value for a to detect potential peaks. In our study, a is set to 1.0 ns because the interval of full waveform samples is 1.0 ns and b was set to 0.1 ns for both data types. A value of 0.1 ns for b is equivalent to 1.5 cm in air. Parameters a and b can be adjusted to fit different applications and different FWL systems. The wavelet decomposition process is a good noise-resistant subspace representation of a signal, and therefore a simple local maximum filter can be used to find the peak locations after a wavelet transformation. In our study, a window with a size of 15 ns was used to detect the local maxima for the peak locations as the Full Width at Half Maximum (FWHM) is 8.3 ns for the Optech Aquarius LiDAR system used in this study [14].

2.2. Gaussian Decomposition Method

Gaussian decomposition is a popular approach for FWL processing as it can simultaneously provide estimation of peak locations and widths. Gaussian decomposition is implemented using Expectation-Maximization (EM) in this study. EM is an iterative method, normally used in signal and image processing, to estimate the maximum probability for a set of parameters of a statistical model. As the name indicates, there should be an expectation (E) step and a maximization (M) step, and EM iterates between the E step and the M step until a convergence criterion is satisfied [28,32].

A LiDAR waveform return can be represented as the sum of multiple Gaussian distributions [17], and mathematically this can be expressed as:

$$f(t) \sim \sum_{i=1}^{n} N(\mu_i, \sigma_i) \tag{2}$$

Here, $f(t)$ is the full waveform that is the sum of the Gaussian components with multiple components (n) and t is time, $N(\mu_i, \sigma_i)$ represents a Gaussian component with an individual mean (μ_i) and a standard deviation (σ_i). The number of peaks and the initial peak locations are needed as initial values for the EM algorithm described by the following equations:

$$Q_{ij} = \frac{p_j f_i(i)}{\sum_{j=1}^{k} p_j f_i(i)} \tag{3}$$

$$p_j = \frac{\sum_{i=1}^{S} N_i Q_{ij}}{\sum_{i=1}^{S} N_i} \tag{4}$$

$$\mu_i = \frac{\sum_{i=1}^{S} N_i Q_{ij}\, i}{p_j \sum_{i=1}^{S} N_i} \tag{5}$$

$$\sigma_i = \sqrt{\frac{\sum_{i=1}^{S} N_i Q_{ij}\, (i - \mu_i)^2}{p_j \sum_{i=1}^{S} N_i}} \tag{6}$$

Here, p_j is the relative weight of the component distribution $f_i(x)$;
Q_{ij} is the probability that sample i belongs to component j;
N_i is the amplitude for sample i;
S is the number of samples in the waveform;
μ_i is the mean peak location; and
σ_i is the standard deviation for that component, proportional to the pulse width or FWHM.

As EM is a local maximum searching method, peaks with spurious μ_i or σ_i are removed to ensure a reasonable result. Also, extremely weak returns, for example, peaks with p_i less than 0.05 are removed to guarantee algorithm convergence. From Equations (2)–(5), it is evident that EM is actually a Gaussian decomposition because its underlying model is a Gaussian mixture model. For the purpose of assessing performance of Gaussian decomposition with different seeding peak locations, both CWT detected peak locations and peaks acquired from second derivative analysis [18] are applied to initialize EM estimation.

2.3. Empirical System Response Waveform Decomposition

An alternative to the Gaussian model for waveform decomposition is an empirical system response (ESR) model that represents the convolution of the emitted pulse shape and the sensor response. Decomposition with an ESR model has the potential to reduce decomposition residuals and improve ranging precision compared to Gaussian decomposition [20]. The method described in [20] requires an ESR model spanning the dynamic range of a terrestrial LiDAR sensor to accommodate nonlinear response characteristics. However, for a FWL sensor with a predominantly linear response, which includes the airborne systems used in this study, a simplified ESR waveform decomposition method can be derived.

In lieu of an ESR model spanning the sensor dynamic range, a single empirical response model can be approximated by averaging waveforms from a single, diffuse, extended target illuminated at normal incidence. Using standard nonlinear least squares, the model is iteratively shifted (μ parameter), scaled in amplitude ($\wr$ parameter), and scaled in width (w parameter) until the parameter corrections are negligible, i.e., the model is fit to the observed waveform in an optimal sense. An un-weighted Gauss-Newton least squares expression can be written in matrix form as [33]:

$$JX = K + V \tag{7}$$

$$X = (J^T J)^{-1} J^T K \tag{8}$$

where J is the $m \times 3$ matrix (m = number of waveform data points) of partial derivatives of the ESR model with respect to the unknown μ, $\wr$, and w parameters evaluated at each waveform data point; K is the column vector of differences between the observed waveform amplitudes and the amplitudes computed from the ESR model; V is the column vector of residuals; and X is the column vector of ESR model parameter corrections. The partial derivatives required to populate the J matrix are numerically computed from the ESR model using the current parameter values at each iteration in the adjustment. Figure 3 illustrates the numeric partial derivatives. As with Gaussian decomposition, the least squares algorithm can be extended to accommodate a superposition of multiple ESR models when overlapping return echoes are detected in the observed waveform.

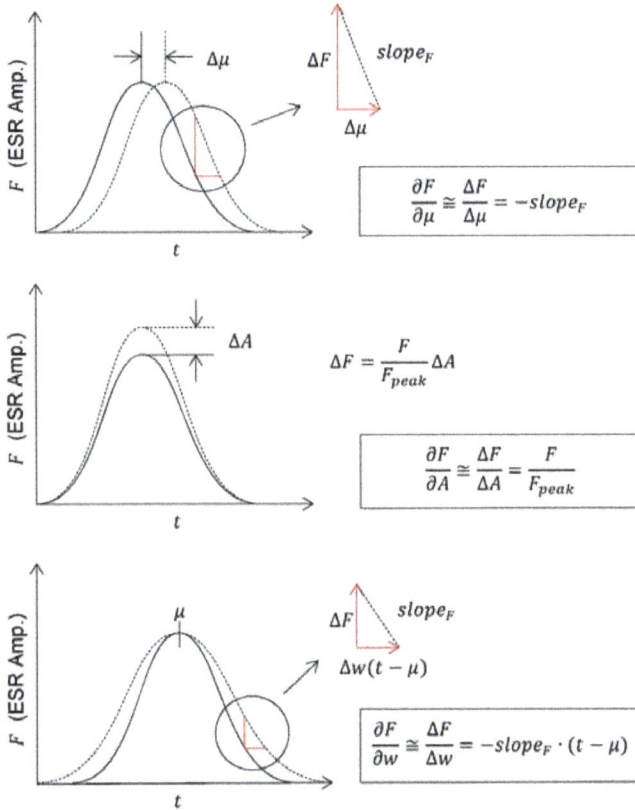

Figure 3. Graphical representation of numeric partial derivatives necessary for the empirical system response least squares waveform decomposition algorithm.

2.4. Water Depth Generation

As the field measurements used in the paper are water depths records collected with an Acoustic Doppler Current Profiler (ADCP), we need to infer water depths from the 3D LiDAR points as a basis of comparison. We also need to segment the raw point clouds from each of the target extraction techniques to separate water column and bottom returns and properly identify the benthic layer. The basic strategy for benthic classification is to first classify the last of multiple returns as initial candidate benthic returns, and then use a region growing method with the initial benthic points and regionally lowest elevation points to refine the total benthic surface points using the TerraScan software package. The classification algorithm is similar to that used to determine ground returns in topographic LiDAR surveys and is based on the methodology presented in [34]. It should be noted that each of the green LiDAR returns from below the surface of the water has been corrected for both refraction of the pulse at the air/water interface, and for the change in the speed of light within water [22]. To define the water boundary, a fluvial river-line is acquired from aerial orthoimages by visually identifying and digitizing the land/water border.

To convert the benthic layer points extracted from the point cloud to water depths, a water surface is required to subtract the benthic layer elevation from the water surface elevation for each benthic point. To highlight the differences in depth determination between single and multiple bands bathymetric LiDAR sensors we examine two realizations of the water surface for each river: The first water surface is extracted from alternative sources (NIR water surface for the Snake River, RTK water surface for the Blue/Colorado River) and the second water surface is extracted from each of the green LiDAR point clouds alone. For green LiDAR point clouds, the water surface can be defined as the remaining LiDAR returns within the boundary of the water body after benthic classification. The NIR water surface was acquired by extracting all NIR returns within the water boundary, as NIR LIDAR can theoretically only be retro-reflected from the water surface [35].

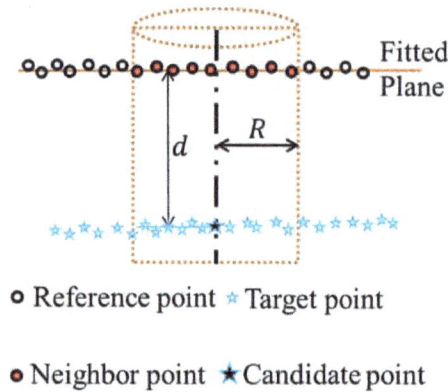

Figure 4. Definition of point to plane distance. For each target point, the neighbor points are those within the cylinder with a radius of R. A fitted plane is constructed by least squares estimation, and the distance of the candidate point to the fitted plane is defined as the point to plane distance.

Point clouds created by airborne LiDAR are generally irregularly distributed, and therefore conventional image processing techniques which assume raster input are not suitable for posterior analysis. As an alternative, we utilized a point to plane distance to compute the distances between an individual LiDAR returns and its neighbor points [36]. Figure 4 shows the schematic steps to compute the point to plane distance. For each specific candidate point, neighbor points are selected within the cylinder with specific searching radius R, and thus a fitted plane is constructed by least squares estimation. The distance from the candidate point to the fitted plane is defined as the point to plane distance d. The point to plane distance is used in this study to calculate the water depth given a cloud of water surface (reference points) and benthic points (target points).

3. Description of Datasets

3.1. Airborne Bathymetric LiDAR Datasets

To assess the performance of single band full waveform bathymetric LiDAR and the processing algorithms described in this paper, two datasets representing different river conditions are investigated: the Snake River in Wyoming's Grand Teton National Park and the confluence of the Blue and Colorado Rivers in north-central Colorado. Both the Snake and Blue/Colorado Rivers originate from the Rocky Mountains and their flow conditions are dominated by the annual snowmelt hydrograph; all remote sensing data collection occurred during low flow conditions in late summer. The Snake River is predominantly clear water after snowmelt runoff recedes. Here, the portion of river bed mainly consists of gravel and cobble (see Figure 5a) and is coated with varying degrees of periphyton and bright green filamentous algae. The varying water depths present in the study area of the Snake River are well suited for a performance assessment of bathymetric LiDAR. The Colorado River originates in Rocky Mountain Park and the Blue River enters the Colorado River from the south near the town of Kremmling, CO. The Blue/Colorado River has lower gradients than the Snake River and bed materials consist mainly of sand and fine sediment. This site has variable water conditions because the Colorado River is also joined by a smaller tributary called Muddy Creek, which as the name implies, was turbid due to rainfall and surface erosion a few days before the LiDAR data collection. The Blue River also contains dense aquatic vegetation extending into the water column from the bed (see Figure 5b). These varying water conditions present an opportunity to assess how water clarity influences bathymetric LiDAR performance.

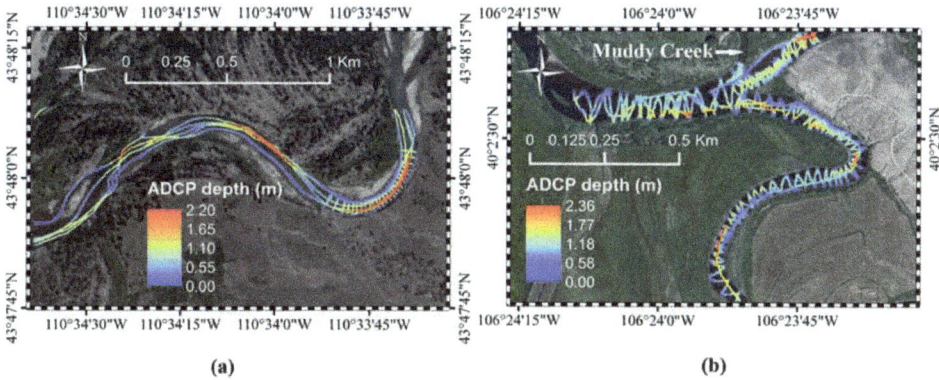

(a)   **(b)**

Figure 5. Study areas, (**a**) Overview of the Snake River with ADCP profiles locations highlighted and colored by depth. (**b**) Overview of the Blue/Colorado River study area, the circled area is turbid plume from Muddy Creek; Colorado River flows from east to west in the image and the Blue River enters the channel from the south. The ADCP profiles locations are highlighted and colored by depth.

The airborne datasets were collected by the National Center for Airborne Laser Mapping (NCALM) with Optech Aquarius and Gemini systems. The Aquarius sensor is a single band LiDAR based on a Q-switched frequency doubled Nd:YAG laser with a resultant wavelength of 532 nm, pulse repetition frequencies (PRFs) of 33, 50, and 70 kHz, a pulse energy of 30 µJ (at 70 kHz), and a beam divergence of 1 mrad. The scanner is a conventional side-to-side oscillating mirror (saw-tooth pattern) with an adjustable field of view up to ± 25° and a maximum mirror frequency of 70 Hz. The return signal is both analyzed in real time by a constant fraction discriminator (CFD) and stored using a waveform recorder with 12 bit amplitude quantization and a sampling speed of 1 GHz for post-mission processing. The Gemini system is similar to the Aquarius system with a Nd:YAG laser at 1064 nm, smaller and adjustable divergence angle and PRF up to 167 kHz. Table 1 shows the principal data acquisition parameters for both project sites.

Table 1. Airborne LiDAR acquisition parameters.

	Snake River	Snake River	Blue/Colorado River
Laser wavelength (nm)	532	1064	532
Pulse width (FWHM in ns)	8.3	12	8.3
Digitization frequency (GHz)	1	N/A	1
Resolution of full waveform (bits)	12	N/A	12
Field of View (FOV)	40°	46°	40°
Beam divergence (mrad)	1	0.25/0.8	1
Pulse rate (KHz)	33	100	33
Date of survey	August 2012	August 2012	September 2012
Flight height (AGL, m)	510	580	580
Point density (pts/m^2)	4.2	6.3	4.0

It should be noted in Table 1 that there is no NIR data listed for the Blue/Colorado River. NIR was collected for this study, but unfortunately was acquired at a high flight elevation (2600 m AGL); laser pulses on water surfaces were mostly absorbed. Effectively no water surface returns were found and therefore the NIR data for the Blue/Colorado River was not used. It should also be noted that flights with the Gemini system and Aquarius system cannot be performed at the same time. The

Aquarius data was collected three days after Gemini data collection for Snake River, but negligible water surface elevation change was found for the Snake River and verified using USGS river gauge station data. Both ADCP field data and Aquarius data were collected on the same day for Blue River, and thus the water surface elevation change is negligible.

3.2. Acoustic Doppler Current Profiler Data

To assess the ability of full waveform bathymetric LiDAR for measuring river morphology, ground reference datasets were collected with a Sontek RiverSurveyor S5 Acoustic Doppler Current Profiler (ADCP) deployed from a kayak. SonTek reports a depth resolution of 0.001 m and an accuracy of 1% over the range of 0.2–15 m. ADCP data is our primary ground reference data, as the accuracy should be better than 3 cm for these two projects because most water was shallower than 3 m. Real-Time Kinematic (RTK) GPS observations were collected concurrently with the ADCP observations to provide measurements of water surface elevation. The vertical datum difference between the LiDAR and RTK GPS was corrected by using common observed ground control points in a parking lot and then applying the offset to correct the LiDAR observations. The ADCP depth observation locations for both projects are shown in Figure 5. The distribution of ADCP water depths through the Snake River (Figure 6a) and Blue/Colorado River (Figure 6b) show that most water depths for the Snake River are less than 2 m while most water depths for the Blue/Colorado River are less than 1.5 m.

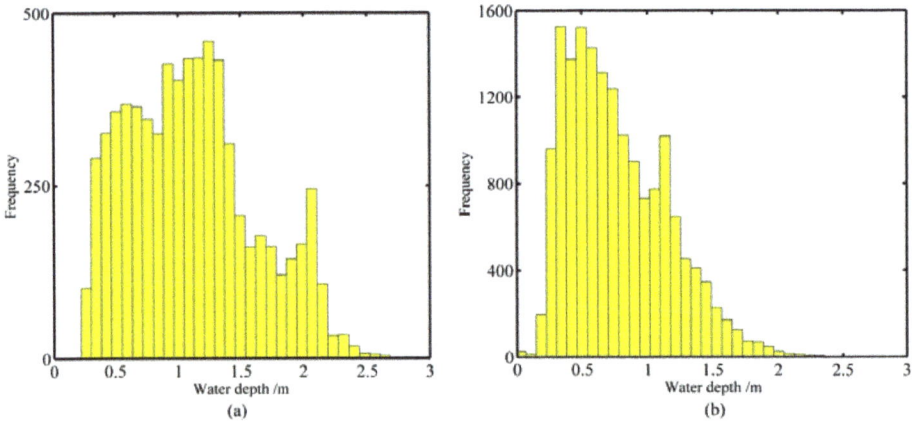

Figure 6. Field measured Acoustic Doppler Current Profiler (ADCP) water depth distribution for (**a**) Snake River (6968 measurements) and (**b**) Blue/Colorado River (16,681 measurements). Most water depths for the Snake River are less than 2 m while most water depths for the Blue/Colorado River are less than 1.5 m.

3.3. Water Turbidity Data

In this study, a WET Labs EcoTriplet was deployed from a kayak on the Blue/Colorado River to measure the portion of the total back-scattering associated with particulates (i.e., suspended sediment and organic material) in the water column. Turbidity, a common metric of water clarity, is derived from the measured backscatter. Figure 7 shows the spatial distribution of turbidity measurements across areas with distinct levels of water clarity at the Blue/Colorado River confluence site. The northern part of the river is distinctly more turbid than the southern portion of the river. Note that turbidity measurements and ADCP measurements were collected on separate deployments of the kayak. Figure 8 shows the bimodal distribution of the turbidity measurements.

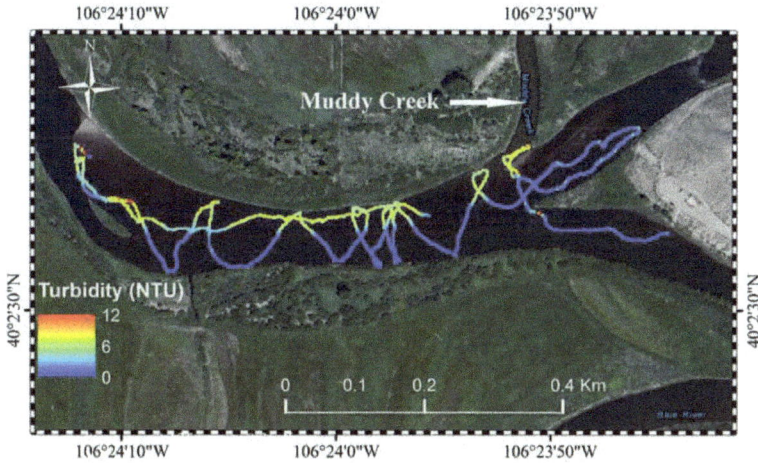

Figure 7. Turbidity measurements for the Blue/Colorado River (4663 measurements) with Muddy Creek entering from the north.

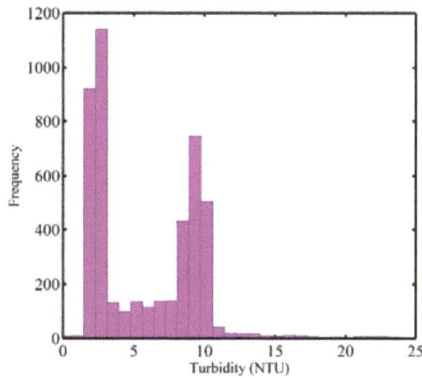

Figure 8. Distribution of turbidity data for the Blue/Colorado River confluence. An obvious bimodal distribution is displayed; more turbid water is present in the northern part of the river due to Muddy creek (Figure 7), and the clearer water is present in the southern portion.

4. Experimental Results

4.1. Experiment I: Snake River Bathymetry Study

4.1.1. Distribution of Number of Full Waveform Returns

Four different full waveform processing algorithms have been applied in this study. The full waveform data for the Snake River was first preprocessed to reduce computing load by thresholding (Section 2). To analyze the effect of the initial peak location estimates on nonlinear least square Gaussian decomposition, peak locations that were detected with a second derivative and peak locations that were detected with a CWT were both used as initial approximations for Gaussian decomposition. The resulting point clouds are referred to as s_G (Gaussian decomposition initiated with second derivatives) and c_G (Gaussian decomposition initiated with CWT), respectively. The peak locations detected by CWT are also used as initial seed values for the ESR pulse fitting. A point cloud was also generated by using just the peak locations derived from CWT without further

Gaussian or ESR refinement. The four point clouds from these full waveform fitting are then used for further analysis.

The CWT and s_G algorithms generated 24.32% and 43.35% more points, respectively, compared to the discrete points provided by the manufacturer software, for the Snake River. The distribution of the number of returns for discrete points, CWT and s_G are shown in Figure 9. This suggests that s_G performs better than CWT for peak detection in the fluvial environment of the Snake River. More importantly, both CWT and s_G methods are markedly better at resolving multiple returns; almost all discrete points are composed of single return points. More return points have a direct benefit for bathymetric mapping as better coverage and higher density data is the result. In addition, multiple returns are also critical for shallow water bathymetric mapping as the surface returns and benthic returns are more likely both represented by multiple reflections. It should be noted that the ESR and c_G methods are not given in Figure 9 because they were both seeded using the CWT peak locations and therefore theoretically have the same statistics as the CWT results.

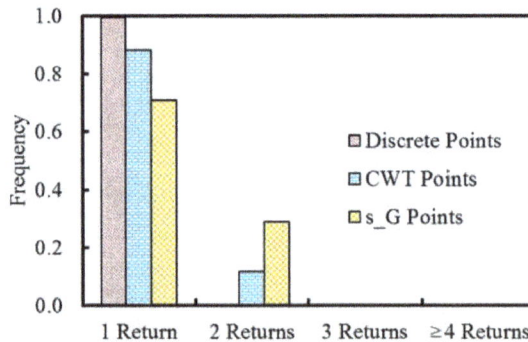

Figure 9. Distribution of the number of full waveform returns using different peak detection methods for the Snake River. CWT and s_G methods are better able to detect multiple returns while almost all discrete points are single return.

4.1.2. Water Depth Analysis

To avoid local anomalies (e.g., floating wood, submerged objects, facets of waves, *etc.*), for each benthic point, the point to plane distance (see Section 2.4) is calculated as water depth with a search radius of 10 m for both the NIR and green water surfaces. To evaluate full waveform bathymetric LiDAR performance, the retrieved water depths have been compared to field measured ADCP depths. Figure 10 shows all the possible combination of water depths compared to ADCP water depths and Table 2 shows the statistical comparison results for each water depth estimate.

Table 2. Comparison of LiDAR retrieved water depths to field measured ADCP water depths for the Snake River. Results in meters.

Point Type:	Discrete		s_G		c_G		CWT		ESR	
Water Surface:	NIR	Green	NIR	Green	NIR	Green	NIR	Green	NIR	Green
Mean(Z_f–Z_r) (m)	−0.02	0.13	−0.02	0.18	0.13	0.32	−0.11	0.06	−0.13	0.17
Std.(Z_f–Z_r) (m)	0.17	0.20	0.16	0.18	0.14	0.17	0.15	0.14	0.13	0.14
Slope	1.08	1.16	0.93	0.79	0.91	0.75	1.12	1.08	1.08	0.95
Intercept (m)	−0.06	−0.29	0.09	0.04	−0.04	−0.06	0.00	−0.14	0.05	−0.12
R^2	0.87	0.87	0.87	0.81	0.90	0.83	0.91	0.92	0.92	0.88

*Z_f is field measurement, Z_r is LiDAR derived water depth.

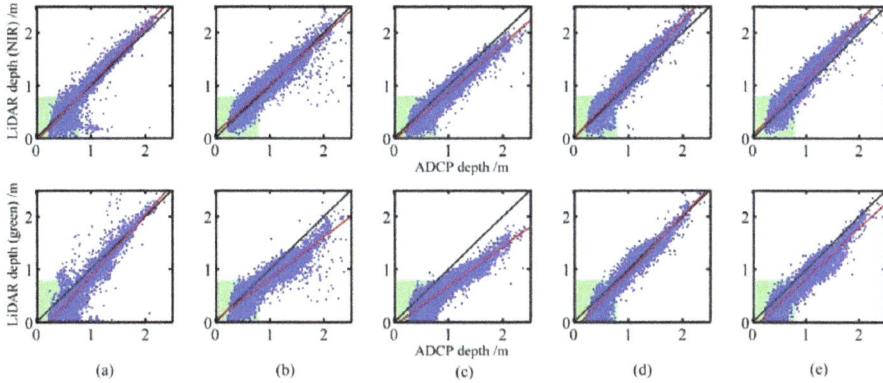

Figure 10. Comparison of LiDAR depth to ADCP depth for the Snake River (top row: LiDAR depths generated with NIR estimated water surface, bottom row: LiDAR depths with green return water surface). Black lines in the figures are the 1:1 line while the red lines are regression lines, and the green shaded areas highlight extremely shallow water (<0.8 m), (**a**) water depths from discrete points; (**b**) water depths from s_G points; (**c**) water depths from c_G points; (**d**) water depths from CWT points; and (**e**) water depths from ESR points.

With the NIR water surface, ESR performs the best with the lowest standard deviation (Std.) of 13 cm and the highest R^2 of 0.92; water depths retrieved from discrete points have slightly higher Std. of 17 cm and lower R^2 of 0.87. With a green water surface, CWT performs the best with a Std. of 14 cm and the highest R^2 of 0.92 while s_G water depths and c_G water depths have the worst performance with 18 cm and 17 cm for Std., 0.81 and 0.83 for R^2, respectively. The mean bias of water depth using a NIR water surface is lower than the mean bias with a green water surface except for CWT derived points; this is likely caused by water volume scattering and the overlap of benthic and surface returns for shallow water. In addition, the R^2 values for water depths retrieved with a NIR water surface are higher than those for water depths retrieved with a green water surface with the exception of the CWT points (0.87 vs. 0.87 for discrete points, 0.87 vs. 0.81 for s_G points, 0.90 vs. 0.83 for c_G points, 0.92 vs. 0.88 for ESR). These differences indicate that NIR returns give a more accurate water surface than green returns. The CWT methodology is the lone outlier, and shows the opposite performance as water depths retrieved with a green water surface are better than water depths retrieved with a NIR water surface (−11 cm vs. 6 cm for mean depth error, 15 cm vs. 14 cm for Std., 0.91 vs. 0.92 for R^2, respectively). This suggests that the CWT is more effective than the other methods for green LiDAR waveform processing as it provides a better estimate of the water surface.

The water depths retrieved from c_G points are slightly better than water depths retrieved from s_G points (with NIR water surface: 14 cm vs. 16 cm for Std., 0.90 vs. 0.87 for R^2, respectively; with green water surface: 17 cm vs. 18 cm for Std., 0.83 vs. 0.81 for R^2, respectively). This suggests that the initial peak location estimates have an effect on the final least square estimates, and that CWT provides marginally better seed locations.

The green shaded areas (depths < 0.8 m) in Figure 10 indicate that all shallow water depths retrieved from LiDAR observations have been underestimated. Theoretically, LiDAR can underestimate water depth because of overlap between the surface return and benthic return for extremely shallow water. Also, any suspended particulate matter in the water body, or a rough benthic layer can stretch the incident laser pulse. For very shallow water (green shaded area), the final laser return will be a superposition waveform of water surface backscatter, water volume backscatter and benthic layer backscatter.

Because Table 2 shows significant differences between water depths with either an NIR or green water surface definition, a further inspection of these water surface definitions is warranted. The NIR water surface shows the best overall internal consistency, with a Std. of 11.76 cm for planar fits of points within a 2 m search radius. Therefore the NIR water surface is used as a common basis for comparison for all the green water surfaces by calculating the point to plane distance with a 2 m search radius from the green LiDAR points to the NIR surface plane. As Table 3 shows, different green water surfaces have significantly different mean vertical errors with ESR having the largest at 45 cm and c_G the smallest at 17 cm. The discrete water surface has only a 10 cm of Std., indicating that the discrete point cloud estimates the water surface well (at least for the Snake River conditions). However, the overall performance (i.e., determining water depths) of discrete returns is not as good as CWT which has Std. of 24 cm for water surface; this implies that a CFD is unable to properly estimate benthic returns in the presence of water column backscatter. The c_G method performs better than s_G for water surface detection with 17 cm vs. 34 cm for mean vertical error, and 28 cm and 31 cm for Std, respectively. Again, this is further evidence that an accurate initial peak estimate is necessary for nonlinear Gaussian decomposition.

Table 3. Statistical mean vertical error and Std. for different green water surfaces. NIR water surface has an 11.76 cm Std.

Water Surface	Discrete	s_G	c_G	CWT	ESR
Mean (m)	0.18	0.34	0.17	0.29	0.45
Std. (m)	0.10	0.31	0.28	0.24	0.33

4.2. Experiment II: Blue/Colorado River Study

4.2.1. Distribution of Number of Full Waveform Returns

To further assess full waveform bathymetric LiDAR performance, we performed another study on the Blue/Colorado River, which has significantly more turbid water than the Snake River. Similar to the Snake River analysis, all four full waveform processing algorithms were applied to extract individual point clouds. Only 4.62% more points were detected with a CWT over discrete returns. The s_G method actually gave 2.07% fewer points than the discrete. The distribution of returns for this fluvial environment is shown in Figure 11. Note that, CWT extracted significantly more multiple returns while almost all discrete returns are single return. Again, more multiple returns in general mean better separation between water surface and benthic layer. The same region growing classification methodology described for the Snake River was also applied to the Blue/Colorado River point clouds.

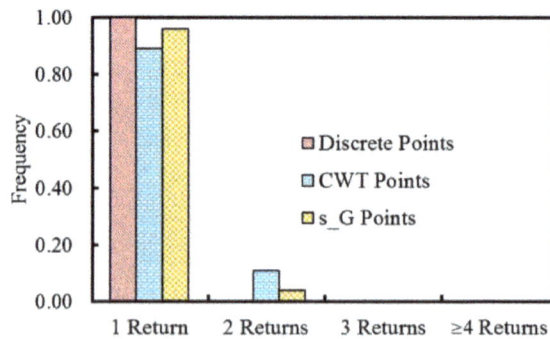

Figure 11. Distribution of the number of full waveform returns for different peak detection methods on the Blue/Colorado River. CWT and s_G methods are better able to detect multiple returns while almost all discrete points are single return.

4.2.2. Water Depth Analysis

After extracting benthic returns from the full waveform and discrete point clouds, a water surface was required to retrieve water depths for comparison with the ADCP measurements. In contrast to the Snake River, no effective NIR water surface was acquired during the airborne LiDAR data collection because of high flight altitude (2.6 km above ground) of the NIR collection (see Figure 3b in [14]). Therefore, instead of using a NIR water surface we have used a field measured RTK water surface. The RTK water surface locations were recorded during the ADCP water depth collection as shown in Figure 5b. In addition, the water surface returns from the discrete bathymetric points proved to have extremely low density, and therefore no water surface was estimated from the discrete returns. Therefore, for the Blue/Colorado River only four sets of water depths were compared with the green water surface. For each benthic point, the point to plane distance is calculated with a search radius of 10 m for both RTK water surface and green water surface. The comparison between the LiDAR and ADCP depths are given in Figure 12 and Table 4.

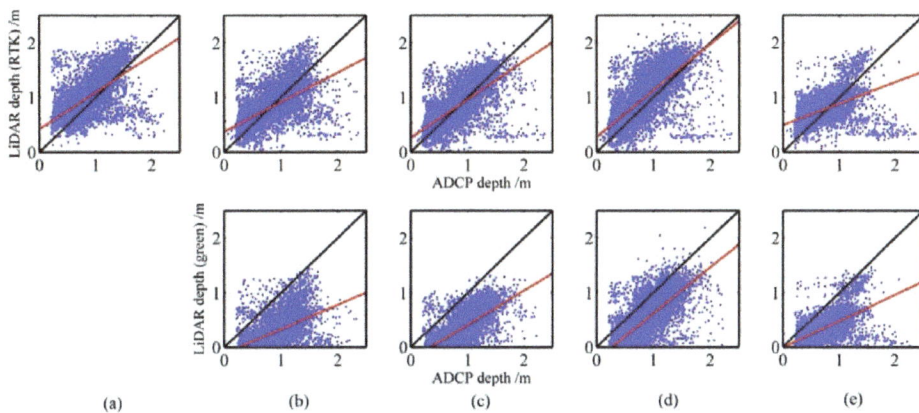

Figure 12. Comparison of LiDAR depth to ADCP depth for the Blue/Colorado River ((**top**) LiDAR depths generated with RTK estimated water surface, (**bottom**) LiDAR depths with green return water surface). Black lines in the figures are the 1:1 lines while the red lines are regression lines, (**a**) water depths from discrete points; (**b**) water depths from s_G points; (**c**) water depths from c_G points; (**d**) water depths from CWT points; and (**e**) water depths from ESR points.

Table 4. Comparison of LiDAR retrieved water depths to field measured ADCP water depths for the Blue/Colorado River. Results in meters.

Point Type:	Discrete		s_G		c_G		CWT		ESR	
Water Surface:	RTK	Green	RTK	Green	RTK	Green	RTK	Green	RTK	Green
Mean(Z_f–Z_r)(m)	–0.17	N/A	–0.03	0.60	–0.04	0.55	–0.16	0.35	–0.10	0.35
Std.(Z_f–Z_r)(m)	0.29	N/A	0.27	0.27	0.25	0.24	0.27	0.24	0.28	0.22
Slope	0.67	N/A	0.54	0.45	0.70	0.63	0.85	0.84	0.39	0.48
Intercept (m)	0.42	N/A	0.37	–0.12	0.26	–0.21	0.28	–0.22	0.49	–0.01
R^2	0.44	N/A	0.41	0.29	0.53	0.43	0.57	0.58	0.25	0.39

*Z_f is field measurement, Z_r is LiDAR derived water depth.

All waveform algorithms performance have degraded in the turbid water of Blue/Colorado River. The mean biases for s_G and c_G water depths with green water surface are significantly higher than that of the Snake River with values of 60 cm and 55 cm, respectively. The Std. for all water depths retrieved with a green water surface is slightly lower than the Std. of water depths with RTK water surface, but with significantly higher mean biases. The highest R^2 of 0.58 was achieved by CWT water depths with a green water surface while CWT still gave the highest R^2 of 0.57 with the RTK surface. The more consistent results from the purely peak finding CWT algorithm suggests that the water turbidity substantially distorts the return waveform shape, which causes significant problems for algorithms such as Gaussian decomposition or ESR that make assumptions about the shape of the return energy profile. ESR performed relatively poorly in the Blue/Colorado River with only a R^2

of 0.25 for water depths with an RTK water surface and R² of 0.39 for water depths with green water surface.

The overall Std. for the c_G method is slightly better than the s_G method (with RTK water surface: 25 cm vs. 27 cm, with green water surface: 24 cm vs. 27 cm) and has a higher R² value (with RTK water surface: 0.53 vs. 0.41, with green water surface: 0.43 vs. 0.29). This difference reinforces that accurate initial peak estimates are critical for nonlinear least square Gaussian decomposition.

The differences in depth estimation between an RTK water surface and a green laser water surface necessitates a further assessment of the water surfaces used to infer water depths. Given the water turbidity, we would expect the RTK water surface to have better performance. Therefore, we compare each green LiDAR water surface using the RTK surface as a common reference. For each green water surface point, the RTK points within 10 m are used to form a water surface plane and each green water surface point to plane distance to the RTK surface is defined as the planar uncertainty. Table 5 shows that all green water surfaces from the Blue/Colorado River have high mean error (s_G: 82 cm, c_G: 79 cm, CWT: 72 cm, ESR: 63 cm). The Std. (s_G: 16 cm, c_G: 13 cm, CWT: 17 cm, ESR: 18 cm) of all water surfaces are marginally better than those for the Snake River because the RTK water surface is less noisy than the NIR water surface used for comparison on the Snake River (NIR has 11.76 cm Std., RTK has 4.11 cm Std.). The significant mean vertical biases highlights the overall poorer performance of bathymetric LiDAR for the Blue/Colorado River. By comparing the results from Table 4, water depths calculated by using an RTK water surface have a smaller mean bias than green water surfaces. This suggests that the increasing amount of water volume scattering caused by the turbid water has skewed the mixture of water surface and volume scattering toward the bottom causing a larger mean error for green water surfaces. The relatively poor performance of green water surface extraction is troubling because it suggests that an independent accurate water surface, i.e., NIR water surface, is a necessity for turbid water depth determination.

Table 5. Statistical mean vertical error and Std. for different green water surfaces. RTK water surface has a 4.11 cm Std.

	s_G	c_G	CWT	ESR
Mean (m)	0.82	0.79	0.72	0.63
Std. (m)	0.16	0.13	0.17	0.18

4.2.3. Water Surface Detection Performance Analysis

In order to better study the impacts of water turbidity, we collected a few representative waveforms with CWT detected peaks and actual water surface locations calculated from RTK surveyed points. Figure 13 displays these individual bathymetric waveforms under different water conditions, varying from shallower to deeper water and also varying from lower to higher turbidity. A single peak can be detected for shallow water with lower turbidity (Figure 13a,b), shallow water with higher turbidity (Figure 13f–h) and deeper water with higher turbidity (Figure 13i,j). CWT detected peaks are closer to the actual water surface for more turbid water (Figure 13f–h) and they move away from the actual water surface for lower turbidity water (Figure 13a,b). The different behavior of full waveform detection in less turbid and more turbid water suggests that a significant amount of water volume scattering for more turbid water skewed the bathymetric returns toward the actual water surface.

However, a further analysis of Figure 13 shows the actual water surface (as measured by RTK) is located at the very beginning of the waveform. Therefore, it would appear that a simple leading edge detection method would be able to accurately estimate the actual water surface. We have set a leading edge detector with an amplitude threshold of 210 to define the water surface. Figure 14 shows the leading edge detected water surface as well as the CWT detected water surface. A significant vertical bias is present for the CWT detected water surface in profile A and profile B. This visual vertical bias confirms the significant increase of water surface error in Table 5. The leading edge detected water surface matches the RTK water surface very well, confirming that leading edge detection is effective for estimating the water surface in the Blue/Colorado River. In order to generalize the leading edge detection, the same method was also applied to the Snake River to independently assess performance. Figure 15 shows two profiles of the Snake River with leading edge water surface detection. A significant vertical bias is present in the Snake River leading edge water surface; the CWT detected water surface agrees much better with the NIR detected water surface. This result confirms that the biases in the waveform water surfaces for the Blue/Colorado River are caused by the increased water turbidity, and not by the waveform processing methodology. The different performance of leading edge water surface detection and the CWT water surface indicates that there may be no single solution that can be applied to all rivers to accurately estimate the water surface for single band LiDAR bathymetry.

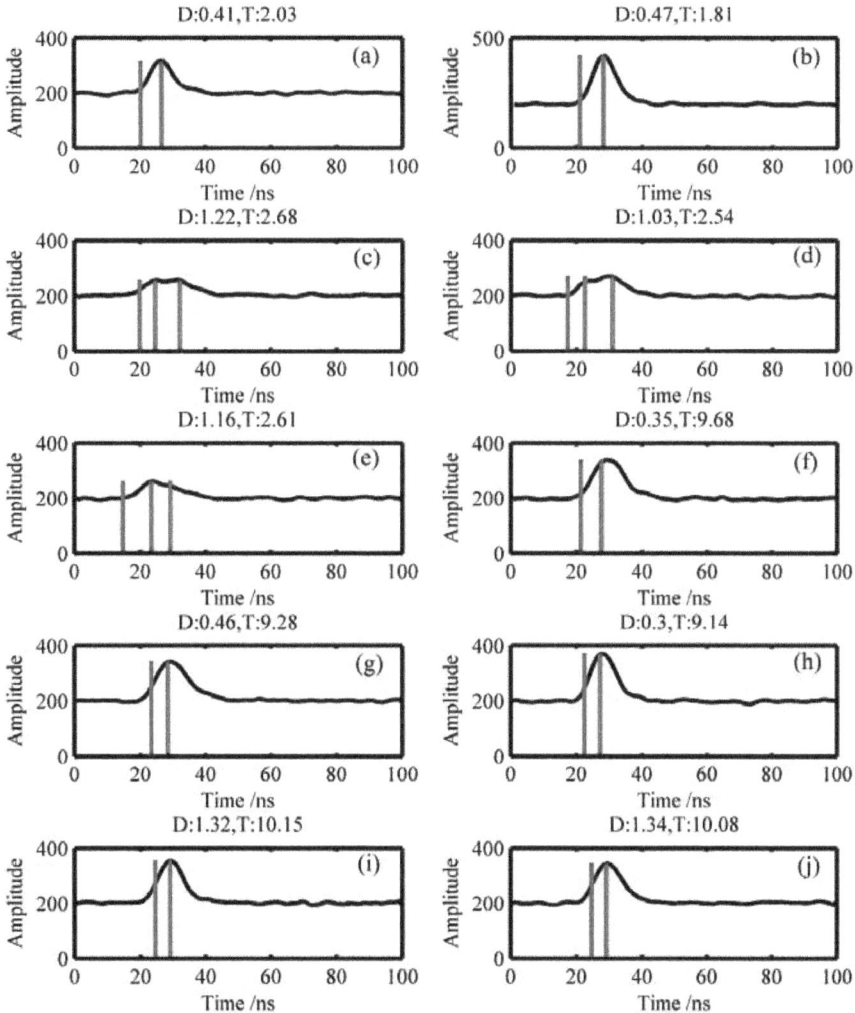

Figure 13. Individual waveform with CWT detected peaks (red lines) and actual water surface location (from RTK—black line) under different water conditions with depth (D, unit: m) and turbidity (T, unit: NTU). (**a,b**) are shallow water with lower turbidity, (**c–e**) deeper water with lower turbidity, (**f–h**) shallow water with higher turbidity, and (**i–j**) deeper water with higher turbidity.

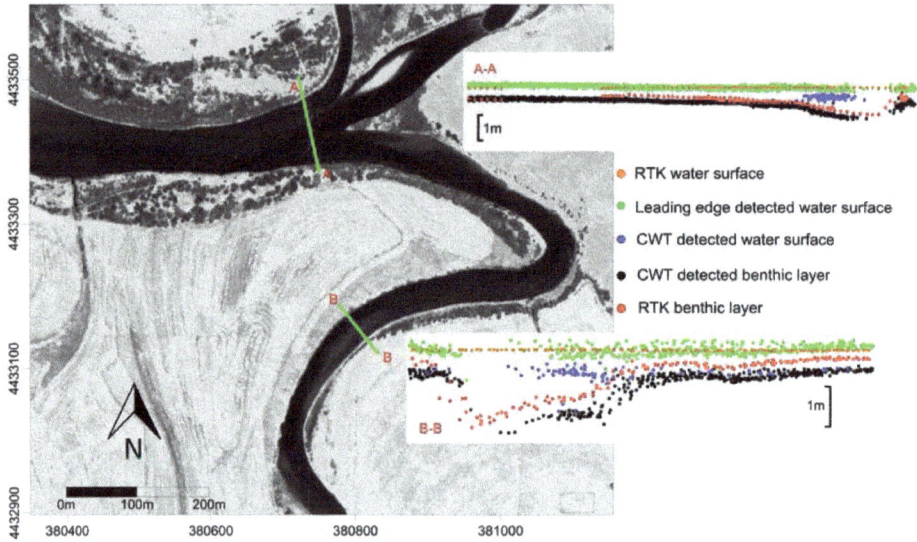

Figure 14. Profiles for leading edge detected water surface on the Blue/Colorado River. The CWT detected waveform shows a clear vertical bias from the RTK water surface while the leading edge detected water surface is much closer to the RTK water surface. Coordinates are in UTM 13N (NAD83).

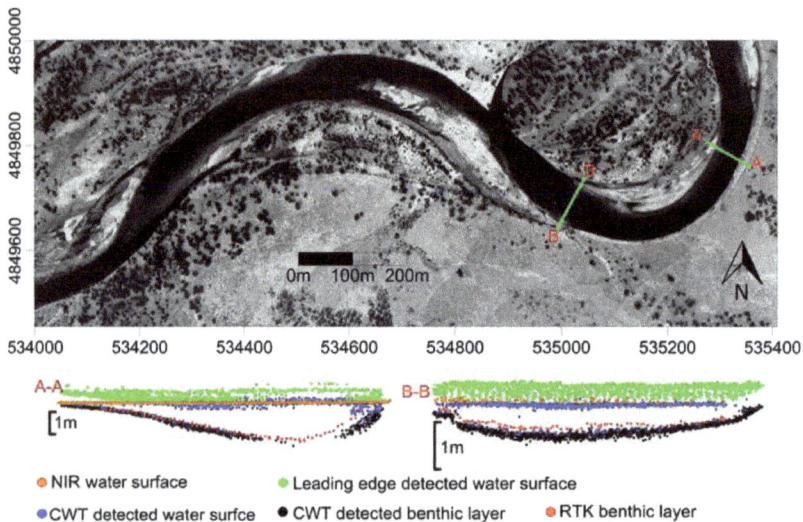

Figure 15. Profiles for leading edge detected water surface on the Rusty Bend of Snake River. The water surface detected with CWT matches the NIR water surface well. Coordinates are in UTM 12N (NAD83).

Table 6 lists the statistical results for leading edge water surface detection for both the Blue/Colorado and Snake River. For each leading edge detected point, the point to plane distances (RTK points are reference points for Blue/Colorado River, NIR points are reference points for Snake River) were used to form a plane by least square estimation and the point to water surface plane distance is defined as error. The leading edge detection is poorer than the waveform derived surfaces for the Snake River as the water volume scattering with low turbidity is not significant. However, if the water becomes more turbid, then leading edge detection performs better than the peak detection or waveform fitting methods.

Table 6. Statistical mean vertical error and Std. for leading edge detected water surfaces. RTK water surface and NIR water surface are used as reference for the Blue/Colorado and Snake Rivers, respectively.

	Blue/Colorado River	Snake River
Mean (m)	−0.01	−0.60
Std. (m)	0.19	0.27

4.3. Best Performance for Single Band Bathymetric LiDAR

If we specify only single band (green) LiDAR observations, then Table 3 shows that the most consistent water surface estimate for the Snake River is given by the discrete returns with an 18 cm mean bias and a 10 cm Std., and Table 6 indicates that leading edge detection yields the best representation of the water surface for the Blue/Colorado River with 1 cm of mean bias and 19 cm Std. Therefore, we can assess the best performance for single band green LiDAR in each study by combining the best estimate of water surface with the best discriminator of benthic layer returns. For the Snake River, we combined a discrete water surface with CWT benthic layer returns to infer water depth. For the Blue/Colorado River we used leading edge detection for the water surface and combined it with CWT benthic layer returns. The optimal single band water depth maps are shown in Figure 16. Both optimized estimates of depth were compared to field measured ADCP water depths and the results are shown in Table 7.

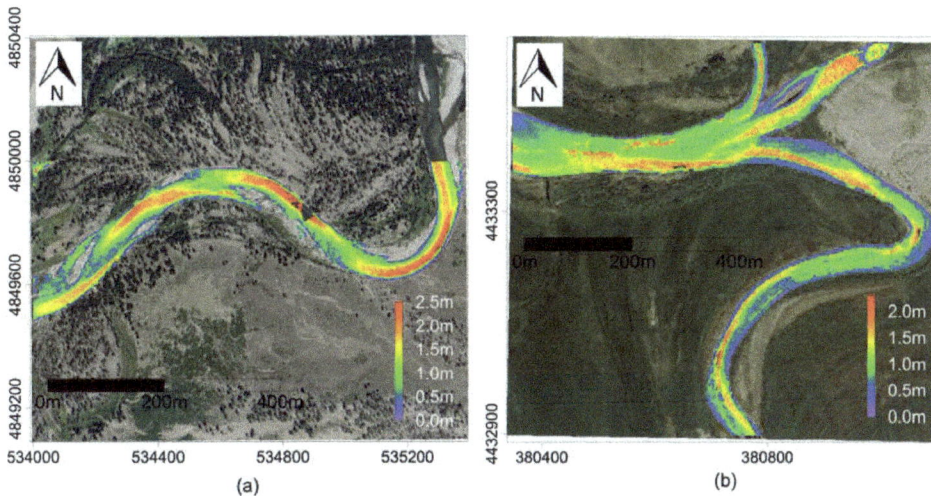

Figure 16. Optimal single band water depth map for: (**a**) the Snake River, coordinates are in UTM 12N (NAD83), and (**b**) the Blue/Colorado River, coordinates are in UTM 13N (NAD83)).

Table 7. Best Performance for single band bathymetric LiDAR for both the Snake River and The Blue/Colorado River.

	Snake River	Blue/Colorado River
Mean(Z_f–Z_r) (m)	0.06	−0.16
Std.(Z_f–Z_r) (m)	0.14	0.27
Slope	1.11	0.85
Intercept (m)	−0.06	0.27
R^2	0.93	0.58

The Snake River water depth inferred from a combination of discrete water surface and CWT benthic layer results in a 6 cm mean bias and 14 cm for Std. with an R^2 of 0.93. The results are comparable to the CWT water depth using a NIR water surface estimate in Table 2. The water depth inferred from a combination of leading edge detected water surface and CWT detected benthic returns for Blue/Colorado River also shows similar performance to the CWT water depth with RTK water surface given in Table 4 (16 cm for mean bias and 27 cm for Std. with R^2 of 0.58). This reinforces the fact that the leading edge detected water surface is close to the RTK water surface and the relatively significant errors present in the Blue/Colorado River results are heavily dependent on the accuracy of the benthic layer estimation.

5. Discussion

Full waveform LiDAR processing is able to produce a significantly denser point cloud with more multiple return reflections than CFD for bathymetry. The ability to recover multiple returns by the waveform methods is especially significant, because the additional returns are more probable benthic returns for fluvial environments. The multiple returns can also benefit the classification of benthic layer as the last return of multiple returns are assigned as the seed benthic positions for the region growing classification algorithms. This algorithm is different from the method proposed by Allouis et al. [24] who used NIR returns to estimate the water surface; here the mixed LiDAR signal produced by water surface and water bottom reflections was directly processed through the CWT to extract both surface and benthic locations. One of the challenges for single band bathymetric LiDAR is to recover both the water surface and bottom position from the full waveform. The longer pulse width laser used in the Aquarius system exacerbated the mixture of water surface, water column and benthic returns. In the future we plan to examine our methodology on short pulse width bathymetric full waveform LiDAR systems such as the Riegl VQ 820-G, AHAB Hawkeye III, EAARL, and Optech Titan.

The results of the study also suggest that there is no superior full waveform processing algorithm for all bathymetric situations which agrees with the conclusions of Parrish et al. [10]. ESR performed the best in the Snake River using a NIR water surface, with an R^2 of 0.92 and the lowest Std. of 13 cm. However, the c_G and CWT results for the Snake River with the NIR surface were statistically quite similar to the ESR results. With a green water surface the CWT performed marginally better than ESR with R^2 of 0.92 vs. 0.88, both with a Std. of 14 cm vs. 14 cm. LiDAR for the Blue/Colorado River did not perform nearly as well as the Snake River study due to the significant water turbidity. CWT water depths with either an RTK or green water surface gave the best performance (R^2 of 0.57 and 0.58, respectively). In general the approaches that model expected signal shape (Gaussian and ESR) performed quite poorly for the Blue/Colorado River, suggesting that the water turbidity causes significant distortion to the return waveform shape. Based on this we can safely conclude that CWT is more stable than the other full waveform processing algorithms for shallow water fluvial environments. Both the ESR and CWT showed good bathymetric performance for difference cases, confirming that it is critical for commercial software to include a variety of full waveform processing strategies. However, unfortunately, the optimal processing strategy is not available *a priori*, and therefore a certain level of performance assessment is necessary for users to determine the best processing strategy for their study conditions.

We have also compared water surfaces estimated by both NIR and Green LiDAR returns. There is a definite vertical bias between the two surface estimates. The comparison of the NIR and green water surfaces for the Snake River study showed a maximum mean vertical offset of 45 cm and 33 cm of Std. for ESR. The minimum average of 18 cm of vertical offset and 10 cm Std. are observed for the discrete green water surface. Overall, it appears that the NIR water surface gives slightly better results than using a green surface (for clear water). Turbid water greatly degraded the green water surface performance with large mean error (s_G: 82 cm, c_G: 79 cm, CWT: 72 cm, ESR: 63 cm). The deterioration of water surface performance compared with the clear Snake River indicates that turbidity can skew the return full waveform toward the benthic layer. Mckean et al. [25] suggested that suspended sediment and dissolved organic materials can scatter and absorb incidence laser radiation. They also reported that turbid water exacerbates laser penetration for the EAARL system when turbidity reached 4.5 to 12 NTU. This agree with our results, the Blue/Colorado River presented turbidity ranging from 2 to 12, which negatively impacted Aquarius performance due to substantial water column scattering. It also further confirms that a multiple wavelength LiDAR may be essential for bathymetric applications, especially for turbid water. A leading edge detection method was proposed and tested over these two river conditions; it was found that leading edge detection is effective if more water volume scattering is present (i.e., high turbidity), but waveform-fitting methods are more effective at low turbidity due to the identification of more water surface returns.

6. Conclusions

The objective of the study was to evaluate the performance of a single band full waveform bathymetric LiDAR with different processing algorithms and water surface definitions in two distinct fluvial environments. We proposed a novel full waveform processing algorithm based on a continuous waveform transformation; the detected peaks from CWT are used as candidate seed peaks for both Gaussian and ESR decomposition. The wavelet transformation was assessed in comparison with a more standard approach of using Gaussian decomposition with initial peak estimates from a second derivative analysis. Water depths from each waveform method, along with discrete points produced by the real-time constant fraction discriminator have been compared to field measured water depths. All the methods have been applied to two fluvial environments: the clear and shallow (mostly < 2 m) water of the Snake River, and the turbid and shallow (mostly < 1.5 m) fluvial environment of the Blue/Colorado River.

In a summary, full waveform processing can produce more points than discrete CFD processing to provide better coverage and more multiple returns for better discrimination of benthic returns from water surface returns. However, with all approaches it is difficult to acquire good quality data for turbid water, especially when the water is shallow. The proposed CWT method shows better stability through varying water clarity conditions than the Gaussian or ESR decomposition methods also tested. A single band full waveform bathymetric LiDAR does not appear to be as accurate as a two wavelengths system that recovers the water surface using a NIR laser. However, with an appropriate full waveform processing algorithm, the error in determining the water surface from a single band green LiDAR can be mitigated; these results are encouraging because they seem to indicate that with improved detection of the water surface from the green LiDAR we can expect a single band LiDAR bathymetry system to perform similarly to a two band (NIR and green) bathymetric system. For this to be realized, however, we must successfully extract the water surface from the relatively complex backscatter at the air/water interface, which we were unable to do with the waveform processing algorithms presented. In [23], they proposed a quadrilateral signal to model the effect of water column scattering, and show it to be effective with simulated bathymetric LiDAR data. However, our initial analysis of this methodology has not shown a significant improvement in water surface estimation for the Aquarius datasets examined. Future work will therefore focus on decoupling the water surface and water column scattering at the air/water interface.

Acknowledgments: Partial support for authors one, two and four was provided by a United States National Science Foundation facility grant to the National Center for Airborne Laser Mapping (EAR #1339015). Funding for the Blue/Colorado River data collection was provided by the Office of Naval Research (awards N000141010873 and N000141210737) and a supplement from NSF to the National Center for Airborne Laser Mapping (EAR #1339015). The third author (Hartzell) was supported through an appointment to the Student Research Participation Program at the U.S. Army Cold Regions Research and Engineering Laboratory (CRREL) administered by the Oak Ridge Institute for Science and Education through an interagency agreement between the U.S. Department of Energy and CRREL. The National Park Service granted permission to collect field measurements and conduct remote sensing flights within Grand Teton National Park. The University of Wyoming-National Park Service Research Station provided logistical support.

Author Contributions: Zhigang Pan developed the continuous wavelet algorithm and processed the full waveform data and water surface estimates for the study, interpreted the results, prepared the manuscript and coordinated the revisions of the manuscript. Craig Glennie assisted in the overall study design, interpretation of the results and revisions of the manuscript. Preston Hartzell developed the ESR model, processed the full waveform using this approach, and assisted in manuscript review. Juan Carlos Fernandez-Diaz collected the airborne LiDAR data for the study and reviewed the manuscript. Carl Legleiter and Brandon Overstreet collected all of the in situ observations including field water depth data and turbidity measurements and also assisted in review of the manuscript.

Conflicts of Interest: The authors declare no conflict of interest.

References

1. Wehr, A.; Lohr, U. Airborne laser scanning—An introduction and overview. *ISPRS J. Photogramm. Remote Sens.* **1999**, *54*, 68–82.
2. Baltsavias, E.P. Airborne laser scanning: Existing systems and firms and other resources. *ISPRS J. Photogramm. Remote Sens.* **1999**, *54*, 164–198.
3. Glennie, C.; Brooks, B.; Ericksen, T.; Hauser, D.; Hudnut, K.; Foster, J.; Avery, J. Compact multipurpose mobile laser scanning system—Initial tests and results. *Remote Sens.* **2013**, *5*, 521–538.
4. Williams, K.; Olsen, M.J.; Roe, G.V.; Glennie, C. Synthesis of transportation applications of mobile LiDAR. *Remote Sens.* **2013**, *5*, 4652–4692.
5. Sheridan, R.D.; Popescu, S.C.; Gatziolis, D.; Morgan, C.L.S.; Ku, N.-W. Modeling forest aboveground biomass and volume using airborne LiDAR metrics and forest inventory and analysis data in the Pacific Northwest. *Remote Sens.* **2015**, *7*, 229–255.
6. Fernandez-Diaz, J.; Carter, W.; Shrestha, R.; Glennie, C. Now you see it...Now you don't: Understanding airborne mapping LiDAR collection and data product generation for archaeological research in Mesoamerica. *Remote Sens.* **2014**, *6*, 9951–10001.
7. Kim, H.B.; Sohn, G. Point-based classification of power line corridor scene using random forests. *Photogramm. Eng. Remote Sens.* **2013**, *79*, 821–833.
8. Glennie, C.L.; Hinojosa-Corona, A.; Nissen, E.; Kusari, A.; Oskin, M.E.; Arrowsmith, J.R.; Borsa, A. Optimization of legacy lidar data sets for measuring near-field earthquake displacements. *Geophys. Res. Lett.* **2014**, *41*, 3494–3501.
9. Lakowicz, J.R. *Principles of Fluorescence Spectroscopy*, 3rd ed.; Springer: New York, NY, USA, 2006.
10. Parrish, C.E.; Jeong, I.; Nowak, R.D.; Brent Smith, R. Empirical comparison of full-waveform lidar algorithms: Range extraction and discrimination performance. *Photogramm. Eng. Remote Sens.* **2011**, *77*, 825–838.
11. Mallet, C.; Bretar, F. Full-waveform topographic lidar: State-of-the-art. *ISPRS J. Photogramm. Remote Sens.* **2009**, *64*, 1–16.
12. Irish, J.L.; McClung, J.K.; Lillycrop, W.J.; Chust, G.; Grande, M.; Galparsoro, I.; Uriarte, A.; Borja, Á.; Tuell, G.; Barbor, K.; et al. Airborne lidar bathymetry: The SHOALS system. *Bull. Navig. Assoc.* **2010**, *7695*, 43–54.
13. Richardson, J.J.; Moskal, L.M. Assessing the utility of green LiDAR for characterizing bathymetry of heavily forested narrow streams. *Remote Sens. Lett.* **2014**, *5*, 352–357.
14. Fernandez-Diaz, J.C.; Glennie, C.L.; Carter, W.E.; Shrestha, R.L.; Sartori, M.P.; Singhania, A.; Legleiter, C.J.; Overstreet, B.T. Early results of simultaneous terrain and shallow water bathymetry mapping using a single-wavelength airborne LiDAR sensor. *IEEE J. Sel. Top. Appl. Earth Obs. Remote Sens.* **2014**, *7*, 623–635.
15. Jutzi, B.; Stilla, U. Range determination with waveform recording laser systems using a Wiener Filter. *ISPRS J. Photogramm. Remote Sens.* **2006**, *61*, 95–107.
16. Hofton, M. a.; Minster, J.B.; Blair, J.B. Decomposition of laser altimeter waveforms. *IEEE Trans. Geosci. Remote Sens.* **2000**, *38*, 1989–1996.
17. Wagner, W.; Ullrich, A.; Ducic, V.; Melzer, T.; Studnicka, N. Gaussian decomposition and calibration of a novel small-footprint full-waveform digitising airborne laser scanner. *ISPRS J. Photogramm. Remote Sens.* **2006**, *60*, 100–112.
18. Chauve, A.; Mallet, C.; Bretar, F.; Durrieu, S.; Deseilligny, M.P.; Puech, W. Processing full-waveform lidar data: Modelling raw signals. In Proceedings of ISPRS Workshop on Laser Scanning 2007, Espoo, Finland, 12 September 2007; Volume 36, Part 3/W52, pp. 102–107.
19. Roncat, A.; Bergauer, G.; Pfeifer, N. B-spline deconvolution for differential target cross-section determination in full-waveform laser scanning data. *ISPRS J. Photogramm. Remote Sens.* **2011**, *66*, 418–428.
20. Hartzell, P.J.; Glennie, C.L.; Finnegan, D.C. Empirical waveform decomposition and radiometric calibration of a terrestrial full-waveform laser scanner. *IEEE Trans. Geosci. Remote Sens.* **2014**, *53*, 162–172.

21. Wu, J.; van Aardt, J.A.N.; Asner, G.P. A comparison of signal deconvolution algorithms based on small-footprint LiDAR waveform simulation. *IEEE Trans. Geosci. Remote Sens.* **2011**, *49*, 2402–2414.
22. Guenther, G.C.; Cunningham, A.G.; LaRocque, P.E.; Reid, D.J.; Service, N.O.; Highway, E.; Spring, S. Meeting the accuracy challenge in airborne lidar bathymetry. In Proceedings of EARSeL-SIG-Workshop LIDAR, Dresden/FRG, Germany, 16–17 June, 2000.
23. Abady, L.; Bailly, J.-S.; Baghdadi, N.; Pastol, Y.; Abdallah, H. Assessment of quadrilateral fitting of the water column contribution in Lidar waveforms on bathymetry estimates. *IEEE Geosci. Remote Sens. Lett.* **2014**, *11*, 813–817.
24. Allouis, T.; Bailly, J.S.; Pastol, Y.; Le Roux, C. Comparison of LiDAR waveform processing methods for very shallow water bathymetry using Raman, near-infrared and green signals. *Earth Surf. Process. Landf.* **2010**, *35*, 640–650.
25. McKean, J.; Nagel, D.; Tonina, D.; Bailey, P.; Wright, C.W.; Bohn, C.; Nayegandhi, A. Remote sensing of channels and riparian zones with a narrow-beam aquatic-terrestrial LIDAR. *Remote Sens.* **2009**, *1*, 1065–1096.
26. Pfennigbauer, M.; Ullrich, A.; Steinbacher, F.; Aufleger, M. High-resolution hydrographic airborne laser scanner for surveying inland waters and shallow coastal zones. *Proc. of SPIE* **2011**, *8037*, 803706.
27. Wang, C.; Li, Q.; Liu, Y.; Wu, G.; Liu, P.; Ding, X. A comparison of waveform processing algorithms for single-wavelength LiDAR bathymetry. *ISPRS J. Photogramm. Remote Sens.* **2015**, *101*, 22–35.
28. Persson, Å.; Söderman, U.; Töpel, J.; Ahlberg, S. Visualization and analysis of full-waveform airborne laser scanner data. In Proceedings of International Archives of the Photogrammetry, Remote Sensing and Spatial Information Sciences, Enschede, the Netherlands, 12–14 September, 2005; ISPRS 2005, *36.3/W19*, 103–108.
29. Vetterli, M.; Herley, C. Wavelets and filter banks: Theory and design. *IEEE Trans. Signal Process.* **1992**, *40*, 2207–2232.
30. Heil, C.E.; Walnut, D.F. Continuous and discrete wavelet transforms. *SIAM Rev.* **1989**, *31*, 628–666.
31. Gregoire, J.M.; Dale, D.; van Dover, R.B. A wavelet transform algorithm for peak detection and application to powder x-ray diffraction data. *Rev. Sci. Instrum.* **2011**, *82*, 015105.
32. Oliver, J.; Baxter, R.; Wallace, C. Unsupervised learning using MML. In *Machine Learning*, Proceedings of the Thirteenth International Conference, Bari, Italy, 3–6 July, 1996.
33. Ghilani, C.D. *Adjustment Computations: Spatial Data Analysis*; John Wiley & Sons, Inc.: Hoboken, NJ, USA, 2010; p. 672.
34. Axelsson, P. DEM generation from laser scanner data using adaptive TIN models. *Int. Arch. Photogramm. Remote Sens.* **2000**, *33*, 111–118.
35. Irish, J.L.; Lillycrop, W.J. Scanning laser mapping of the coastal zone: The SHOALS system. *ISPRS J. Photogramm. Remote Sens.* **1999**, *54*, 123–129.
36. Hauser, D.L. Three-Dimensional Accuracy Analysis of a Mapping-Grade Mobile Laser Scanning System. Master's Thesis, University of Houston, Houston, TX, USA, 2013; p. 94.

remote sensing

MDPI

Article

Tropical Forests of Réunion Island Classified from Airborne Full-Waveform LiDAR Measurements

Xiaoxia Shang [1,*], Patrick Chazette [1], Julien Totems [1], Elsa Dieudonné [1], Eric Hamonou [1], Valentin Duflot [2], Dominique Strasberg [3], Olivier Flores [4], Jacques Fournel [3] and Pierre Tulet [2]

[1] Laboratoire des Sciences du Climat et de l'Environnement (LSCE), Commissariat à l'Energie Atomique et aux énergies alternatives—Centre National de la Recherche Scientifique—Université de Versailles Saint-Quentin-en-Yvelines, 91191 Gif sur Yvette Cedex, France; patrick.chazette@lsce.ipsl.fr (P.C.); julien.totems@lsce.ipsl.fr (J.T.); Elsa.Dieudonne@univ-littoral.fr (E.D.); Eric.Hamonou@lsce.ipsl.fr (E.H.)

[2] Laboratoire de l'Atmosphère et Cyclones (LACy), Université de la Réunion, CNRS, Météo-France, 15 Avenue René Cassin, CS 92003, 97744 Saint-Denis Messag, France; valentin.duflot@univ-reunion.fr (V.D.); pierre.tulet@univ-reunion.fr (P.T.)

[3] UMR PVBMT, Peuplements Végétaux et Bioagresseurs en Milieu Tropical, Université de La Réunion, 15 Avenue R. Cassin, CS 92003, 97744 Saint-Denis Messag, France; dominique.strasberg@univ-reunion.fr (D.S.); jacques.fournel@univ-reunion.fr (J.F.)

[4] UMR PVBMT, Peuplements Végétaux et Bioagresseurs en Milieu Tropical, Université de la Réunion, Pôle de Protection des Végétaux, 7, Chemin de l'IRAT, 97410 Saint Pierre, France; olivier.flores@cirad.fr

* Correspondence: xiaoxia.shang@gmail.com; Tel.: +33-1442-75945

Academic Editors: Sangram Ganguly, Compton Tucker, Clement Atzberger and Prasad S. Thenkabail
Received: 3 December 2015 ; Accepted: 29 December 2015 ; Published: 7 January 2016

Abstract: From an unprecedented experiment using airborne measurements performed over the rich forests of Réunion Island, this paper aims to present a methodology for the classification of diverse tropical forest biomes as retrieved from vertical profiles measured using a full-waveform LiDAR. This objective is met through the retrieval of both the canopy height and the Leaf Area Index (LAI), obtained as an integral of the foliage profile. The campaign involved sites ranging from coastal to rain forest, including tropical montane cloud forest, as found on the Bélouve plateau. The mean values of estimated LAI retrieved from the apparent foliage profile are between ~5 and 8 m^2/m^2, and the mean canopy height values are ~15 m for both tropical montane cloud and rain forests. Good agreement is found between LiDAR- and MODIS-derived LAI for moderate LAI (~5 m^2/m^2), but the LAI retrieved from LiDAR is larger than MODIS on thick rain forest sites (~8 against ~6 m^2/m^2 from MODIS). Regarding the characterization of tropical forest biomes, we show that the rain and montane tropical forests can be well distinguished from planted forests by the use of the parameters directly retrieved from LiDAR measurements.

Keywords: tropical forest; airborne LiDAR; canopy height; Leaf Area Index; apparent foliage

1. Introduction

Tropical forest areas are difficult to monitor/to classify using either remote sensing or *in situ* approaches, because of their tremendous heterogeneity and complex structure. The fundamental challenge is thus to acquire information about the forest vegetation structure given the fact that forest vegetation limits the ability to acquire information. Forest horizontal patterns are accessible using passive multispectral sensors [1,2] and hyper-spectral sensors [3–5], but these sensors are not adequate to penetrate beyond the upper canopy layer [6]. Active remote sensing instruments, including LiDAR and radar, have more of a chance to peer through the forest canopy down to the ground level [7]. Radar yields volumetric scattering information in addition to surface scattering observations, but the retrieval of the vegetation vertical structure is not direct, unlike with LiDAR.

Ground-based LiDAR systems, either terrestrial or portable, can accurately estimate canopy structural parameters [8–12]; however, covering large areas with such systems is impractical. Airborne/spaceborne LiDAR technology has been used to rapidly describe forest structure over large areas; whereas the observations of optical remote sensing is often limited by cloud in tropical areas. Several airborne discrete return LiDAR datasets have been acquired over tropical forests and have been successfully used to derive structural characteristics, such as canopy height, canopy cover and aboveground biomass [13–15]. A full description of the forest vertical structure (including canopy top, tree crown base height and understory structures) has also been obtained by airborne full-waveform LiDAR, both with infrared wavelengths [16,17] and ultraviolet wavelengths [18,19]. Recently, the airborne demonstration instrument called the Laser Vegetation Imaging Sensor (LVIS) [20] has shown that a full-waveform infrared LiDAR with a large footprint can reliably extract the vertical structure and Leaf Area Index (LAI) of a tropical rainforest (Costa Rica [21]) as well as a mid-latitude forest (California [22]), even with a dense canopy cover. The LVIS team however acknowledges the need of a broader dataset on multiple tropical biomes to confirm these findings and compare the extracted features.

The overarching goal of this paper is to report on a methodology of classification using a full-waveform ultraviolet airborne LiDAR with a large footprint, of varied tropical forest types on Réunion Island, which is a rich diversity of tropical ecosystems listed as World Heritage by UNESCO. The classification, obtained using LiDAR-derived canopy height and LAI, has distinguished native forest from plantations/exotic forests. The study sites and data collection methods will be described in Section 2, where the main steps of LiDAR processing for forest studies and the sampling strategy will also be presented. The retrieved forest structural and ecological properties will be analyzed in Section 3, along with comparisons to ground-based census and spaceborne observations. The classification of tropical forest sites will be also presented in this section.

2. Materials and Methods

2.1. Study Sites

Réunion Island is a French overseas department located in the Indian Ocean (20°06′52″S, 55°31′57″E; Figure 1). It is a small (2512 km²) tropical volcanic island, which reaches 3070 m in altitude at its highest point (Piton des Neiges). In spite of the transformation of its habitats [23], the island still shelters 100,000 ha of native ecosystems (included in a national park) and is home to the last remnants of intact tropical forests in the Mascarenes archipelago (Réunion, Mauritius, Rodrigues).

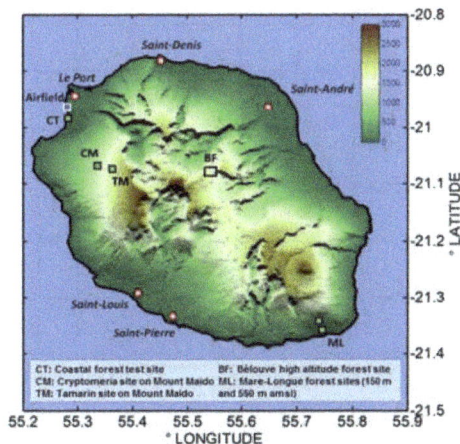

Figure 1. Location of the study sites and topography of Réunion Island.

Seven plots on Réunion Island were used for forest sampling in our study (Figure 1; Table 1). The coastal test site (CT) has only exotic vegetation. The Cryptomeria (CM) and Tamarind (TM) plots, located on Mount Maïdo, and the Bélouve (BF) site, located on a central plateau, are tropical montane cloud forests, which in particular still cover large areas of Réunion Island (60,000 ha), extending from 800 to 1900 m above mean-sea-level (amsl) on the windward side and from 1100 to 2000 m amsl on the leeward side of the island. This dense forest within cultivated forests on moderate slopes is very similar to *Acacia koa* forests in Hawaii. The CM is a monoculture of *Cryptomeria japonica*, an introduced species in the Taxodiaceae family. These trees commonly reach 20 to 25 m in height on the site; they produce a dense canopy, under which light is very scarce. Most acacia stands, like the TM site, are composed of secondary forest and display a monospecific acacia (highland tamarind) canopy with shrubby vegetation in the understory, of which the structure can vary with the intensity of human activities (stock farming in particular). The Mare-Longue sites (three plots dubbed ML-150, -250 and -550, according to their altitude) are located in the National Park of Réunion Island in the former Mare-Longue nature reserve, which shelters the last remnant of lowland tropical rainforest in the Mascarene Islands with around 4000 mm of yearly rainfall. This lowland forest grows on a non-altered basaltic pahoehoe lava flow dated between four and six centuries old [24]. This forest displays the greatest tree species diversity on Réunion Island with an average richness of 40 tree species per hectare [25]. Whereas average tree height remains very low (15 to 20 m), the stem density exceeds 1000 trees/ha (diameter at breast height >10 cm). The most abundant tree species in the sampled plots is *Labourdonnaisia calophylloides* (Sapotaceae), endemic from the Mascarene Islands.

Table 1. Main characteristics of the study sites and associated available LiDAR profiles.

	Forest Sites	Altitude (m·amsl [1])	Dominant Tree Species	Sub-Plot	LiDAR Cover	Laser Shots	Ground Slope > 30°
Exotic Vegetation	Coastal forest (CT)	10	Only exotic vegetation	-	22 ha	2046	0.06%
Tropical Montane Cloud Forests	Cryptomeria (CM)	1230	*Cryptomeria japonica*	40 m × 40 m	10 ha	9646	22.6%
	Tamarind (TM)	1750	*Acacia heterophylla*	40 m × 40 m	10 ha	14,738	12%
	Bélouve (BF)	1600	*Acacia heterophylla*	-	400 ha	48,410	7%
Tropical Lowland Rainforest	Mare-Longue (ML) ML-150	150	*Labourdonnaisia calophylloides*	50 m × 100 m	2.8 ha	9808	2%
	ML-250	250		50 m × 50 m	1.4 ha	1552	0.2%
	ML-550	550		-	0.7 ha	667	16%

[1] amsl: above mean-sea-level.

2.2. Data Collection

The following estimation of forest parameters on Réunion Island was performed in May 2014, combining mainly airborne LiDAR measurements with *in situ* approaches.

2.2.1. Airborne LiDAR and Instrumentation

The LiDAR system used during the campaign is the Ultraviolet LiDAR for Canopy Experiment (ULICE; [19]) developed at Laboratoire des Sciences du Climat et de l'Environnement (LSCE) with the support of CNES (Centre National d'Etudes Spatiales). It was integrated into an autonomous payload flown on an ultra-light aircraft (ULA) shown in Figure 2. The ULICE system characteristics and the airborne payload are given in Table 2. The ultraviolet domain is well suited both for eye safety at low carrier altitude and for precise retrievals over dense forests with little distortion due to multiple scattering (Shang and Chazette, 2014). As recommended by several authors [20,26–29], a large LiDAR footprint is preferred so that the laser can better penetrate the dense tropical forests. Shang and Chazette [26] estimated an optimal laser footprint diameter around 20 m for dense forests, whereas Riaño *et al.* [30] found that the Leaf Area Index (LAI) was better estimated using laser footprints between 7.5 and 12.5 m. The ULICE system was thus modified to obtain a large and controllable sounding area on the ground (approximately a 1-m to 10-m footprint diameter for a flight altitude of ~350 m above the ground level (agl)). The effect of LiDAR footprint size will be studied in Section 2.4.

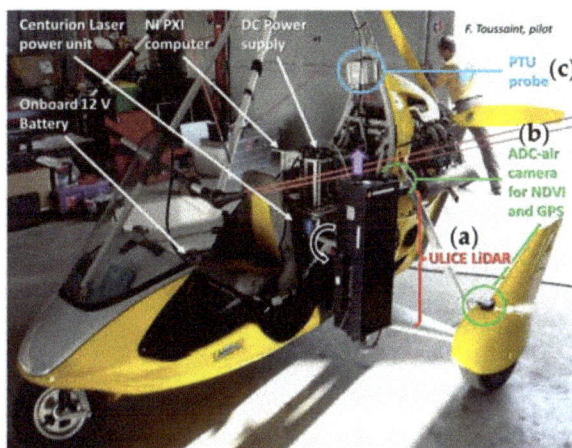

Figure 2. Autonomous payload (~80 kg) implemented on an ultra-light aircraft, including (**a**) the Ultraviolet LiDAR for Canopy Experiment (ULICE). The other instruments are also onboard: (**b**) a Tetracam ADC (Agricultural Digital Camera) air camera is used to get the photosynthesis activity index (Normalized Difference Vegetation Index (NDVI)) images over the forest canopy; the ancillary positioning instrument, called the MTi-G system, consists of a Global Positioning System (5-m accuracy) and an inclinometer (0.7° accuracy); (**c**) a Vaisala PTU-300 pressure/temperature/relative humidity probe is used for altitude correction and control of the tropical high humidity conditions that could affect the transmittance of LiDAR optics.

Table 2. Summary of the ULICE characteristics.

ULICE	Characteristics
Emitter (laser)	Quantel Centurion, diode-pumped, air-cooled, 6 mJ, 8 ns, 100 Hz, 354.7 nm
Output beam	Eyesafe ~40 × 30-mm beam, tunable 0 to 40 mrad divergence with Altechna MoTex Expander (at $1/e^2$)
Receiver	2 channels with different optical densities (OD)
Telescope	Refractive, 150-mm diameter, 280-mm effective focal length
Filtering	No spatial filtering, wideband Thorlabs 355 nm ± 5 nm interference filters for large angular acceptancy (36 mrad)
Field of view	33 mrad for Channel #0, 26 mrad for Channel #1
Detection	Hamamatsu H10721 photo-multiplier tubes. Channel #0: 3.0 OD; Channel #1: 4.0 OD
Data acquisition	12 bits, 200-MHz sampling, 2-channel NI-5124 digitizer, 33-Hz actual profile frequency
Sounding area	Tunable up to ~30 mrad on Channel #0, <22 mrad on Channel #1 (at $1/e^2$)

An ancillary positioning instrument (inclinometer and GPS), an MTi-G system by XSense, is also onboard the ULA. It provides the horizontal geolocation of the ULA with 5-m accuracy and the direction of the laser beam with 0.7° accuracy (*i.e.*, 3.6 m at the ground for a flight altitude of 300 m·agl).With such uncertainties, the study performed at Réunion Island should be statistical, because we cannot distinguish one tree from another. A Tetracam ADC (Agricultural Digital Camera) air camera is also onboard to map the photosynthesis activity index (Normalized Difference Vegetation Index (NDVI)) over the forest canopy to check the scene heterogeneity. However, its images showed high NDVI (>0.65) over all of the observed sites, making it rather irrelevant for the validation of other ecological parameters, such as LAI [31].

2.2.2. Field Data Collection

During two months around the airborne measurements, four representative sub-plots of ~0.2 ha were set up in four forest sites (CM, TM, ML-150 and ML-250 sites, as shown in Table 1), where *in situ* measurements were performed. Within each sub-plot, all trees with diameter at breast height (DBH) >7 cm were identified. The tree top height (TTH) and the DBH were measured using a dendrometer and forestry measuring tapes, respectively. For trees with multiple stems, each significant stem was recorded individually. The uncertainties on the TTH from *in situ* measurements in a dense tropical forest have been evaluated during the experiment in the order of ±4 m (several measurements on the same tree with different operators with a dendrometer). This is due to the difficulty in identifying the tree top among other branch extremities.

2.2.3. Other Data Collections

Digital terrain models (DTM) of 500-m or 5-m resolution for the whole Réunion Island were provided by the Parc National de la Réunion (J.-C. Notter, personal communication). The topography of Réunion Island is given in Figure 1, using the DTM-500 m. The slope of each sampled site was evaluated using the DTM-5 m.

MODIS (Moderate Resolution Imaging Spectroradiometer) Level 3 land products were compared with LiDAR observations. The 8-day LAI products derived from MODIS onboard Terra and Aqua are considered [31]. On 8-day syntheses from May to August 2014, LAI values retrieved on 1-km pixels were averaged after screening for cloud contamination (*i.e.*, values below the median were removed).

2.3. LiDAR Data Processing

From airborne LiDAR measurements, the forest structural and optical parameters are estimated; they are in turn used to evaluate ecological parameters, such as LAI. In this study, three key parameters are estimated from the LiDAR backscatter profiles to characterize the sampled forest sites: the canopy height (CH), the vertical profile of apparent foliage (F_{app}), which informs both the canopy density and the vertical distribution of leaf biomass along the profile, and the LAI, which is linked to the integral of the latter parameter.

2.3.1. Forest Structural Parameter: Canopy Height

The canopy height (CH) parameter is assessed from the full-waveform LiDAR profile using the threshold approach documented in Chazette *et al.* [32] and applied to forest detection by Cuesta *et al.* [18] and Shang and Chazette [19]. CH is estimated as the distance between the first return, at the upper surface of the vegetation, and the last return, which is normally the ground echo. For dense forests, the laser beam cannot always penetrate the leaves and reach the ground, so the last return of the backscattered LiDAR signal is not necessarily the ground echo. Nevertheless, as frequent measurements were performed (33 Hz), allowing some overlap, the ground level can be correctly located thanks to time-integrating signal processing in almost all cases and be the reference to estimate the CH. An example is given in Figure 3. Note that a parasitic echo (undershoot) can be observed beneath the ground echo, which is due to the rebounding, non-linear response of the detector to the strong pulse returned by the ground. The standard deviation of LiDAR-derived CH was assessed to be ~1.5 m when only considering measurement noise and signal processing errors. Shang and Chazette [26] assessed that LiDAR signal distortion due to the surface slope can lead to a relative CH uncertainty of ~5% for a slope of 30° (see Table 1) and a 10-m footprint, as is the case in our present study. As a result, the standard deviation of our LiDAR-derived CH should be of the order of ~2 m. We do not consider geolocation errors in this statistical study.

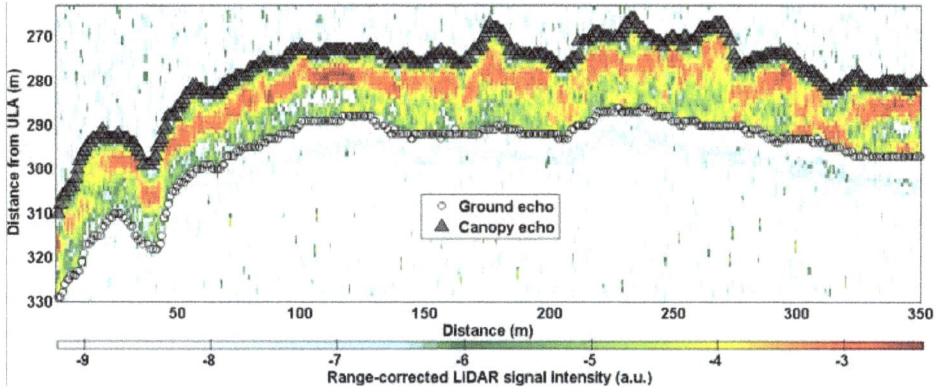

Figure 3. Example of ground echo and canopy top detection from a range-corrected LiDAR signal explained in arbitrary units (a.u.) for a section of flight over the tropical forest of Bélouve (BF). The difference between the ranges of these two points yields the canopy height (CH). Note that the y-axis is not the ground elevation, but the distance from the emitter. ULA, ultra-light aircraft.

2.3.2. Forest Optical Parameters

The range-corrected backscattered airborne LiDAR signal S_v [33,34], taken at a height above ground level (agl) h inside the forest cover, can be expressed by the LiDAR equation [35]:

$$S_v (h) = K \times T_a^2 \times BER \times \alpha_{canopy} (h) \times \exp(-FOT (h)) \tag{1}$$

where K is the instrumental constant and T_a is the atmospheric transmission. The backscatter to extinction ratio BER is a classical parameter used in LiDAR analyses [26,36], which characterizes the probability that an intercepted photon would be backscattered by a scattering layer; it is assumed to be constant for all canopy levels in this study. The canopy extinction coefficient $\alpha_{canopy} (h)$ is defined as the sum of the absorption and scattering coefficients in the canopy. The forest optical thickness (FOT) is defined only in the forest layer between the considered height (h) and the canopy height (CH) and is given by:

$$FOT (h) = \int_h^{CH} \alpha_{canopy} \left(h'\right) \cdot dh' \tag{2}$$

Following the method proposed by Ni-Meister *et al.* [37], we define the transmittance height profile (THP) by taking a ratio of the energy from canopy returns to the total energy, which characterizes the amount of skylight intercepted by vegetation at a given level [38]:

$$THP (h) = \frac{R_v (h)}{R_v (0)} \times \varepsilon \tag{3}$$

with:

$$\varepsilon = \frac{1}{1 + \frac{\rho_v}{\rho_g} \cdot \frac{R_g}{R_v(0)}} \tag{4}$$

where ρ_v and ρ_g are the canopy and ground reflectance, respectively. The integrated range-corrected canopy return $R_v(h)$ (respectively ground return R_g) is defined as the integral of the LiDAR signal from the canopy top CH to height level h (respectively in the equivalent width of the ground echo Δh_{GE}, Δh_{GE} ~4 m):

$$
\left\{
\begin{array}{l}
R_v(h) = \int_h^{CH} S_v(h') \cdot dh' = K \cdot T_a^2 \cdot BER \cdot [1 - exp(-FOT(h))] \\[2ex]
R_g(h) = \int_{-\frac{\Delta h_{GE}}{2}}^{+\frac{\Delta h_{GE}}{2}} S_g(h') \cdot dh' = K \cdot T_a^2 \cdot exp(-FOT(0)) \cdot \frac{\rho_g}{\pi \cdot \Delta h_{GE}} \cdot \int_{-\frac{\Delta h_{GE}}{2}}^{+\frac{\Delta h_{GE}}{2}} g(h') \cdot dh'
\end{array}
\right.
\tag{5}
$$

where the normalized ground echo g(h) is modelled as a Gaussian function [39] and can be calibrated by using the returned laser pulse at nadir over a flat surface.

Thus, as *BER* is equal to $\frac{\rho_v}{\pi}$, THP can be expressed as a function of FOT:

$$
THP(h) = 1 - exp(-FOT(h))
\tag{6}
$$

The ε parameter is usually estimated using a known ratio of canopy and ground reflectance, which was estimated around 2.5 [21] or 2 [38,40] at a 1064 nm wavelength. However, the reflectance values are not available in our study area. Nevertheless, the reflectance ratio can be estimated using only LiDAR measurements as follows. Assuming $FOT(0) >> 1$ (*i.e.*, $\varepsilon \approx 1$), which is realistic for thick tropical forests, as highlighted by Shang and Chazette [26], an initial FOT estimator can be evaluated from Equations (3) to (6), and is given for $h > 0$ by:

$$
\widetilde{FOT_i}(h) \approx -ln\left(1 - \frac{R_v(h)}{R_v(0)}\right)
\tag{7}
$$

This leads to a second assessment:

$$
\widetilde{FOT}(h) = -ln\left(1 - \varepsilon \cdot \frac{R_v(h)}{R_v(0)}\right)
\tag{8}
$$

which is made after a correction of the first order using the ε parameter explained as:

$$
\varepsilon = 1 - exp\left(-\widetilde{FOT}(h_0)\right)
\tag{9}
$$

with h_0 chosen inside the undergrowth layer of the forest (2 to 3 m above the ground level).

Meanwhile, the reflectance ratio (ρ_v/ρ_g) can be determined using the LiDAR signal by:

$$
\frac{\rho_v}{\rho_g} \approx \frac{R_v(0)}{R_g} \cdot \frac{exp\left(-\widetilde{FOT}(h_0)\right)}{1 - exp\left(-\widetilde{FOT}(h_0)\right)}
\tag{10}
$$

A final estimate is obtained after correcting for bias towards high values, which can be assessed very reliably using a simulator of the LiDAR measurements taking inversed profiles for each sampling site as an input. Such an algorithm converges within a relative uncertainty of ~20% after corrections of the bias.

This iterative approach was chosen as an alternative to the one based on ground echo normalization of forest transmittance as proposed by Ni-Meister *et al.* [37]. The reliance of the latter on an accurate estimate of the ground to vegetation reflectance ratio is incompatible with the very diverse and variable forest grounds found at Réunion Island (leaves and debris, soil, lava) and leads to important errors on the retrieved ecological parameters. This inversion process has been applied to each suitable LiDAR profile acquired during the flights above the tropical forests of Réunion Island, in order to characterize the various tropical forest sites.

2.3.3. Forest Ecological Parameters

Several studies have shown that LiDAR is a powerful instrument to retrieve the LAI [41,42]. The LAI can be derived from LiDAR measurements by (e.g., [21]):

$$LAI\,(h) = C \times \int_h^{CH} \frac{F_{app}\,(z)}{G} \cdot dz \tag{11}$$

with the "apparent foliage profile" (F_{app}), which can be identified as the vertical profile of vertical projections of foliage elements, defined by the following equation as in Ni-Meister *et al.* [37]:

$$F_{app}\,(h) = \frac{\text{dln}\left(1 - \varepsilon \cdot \frac{R_v(h)}{R_v(0)}\right)}{\text{d}h} \tag{12}$$

Comparing Equations (2) and (8), we find that the F_{app} is actually the canopy extinction coefficient α_{canopy} in the classical LiDAR equation. We will use F_{app} in the following.

All of this mathematical development can be used whatever the wavelength, but for visible or infrared wavelengths, it may be necessary to consider the multiple scattering effect due to leaves and branches [26]. The multiple scattering enhances the backscatter LiDAR signal and makes the LiDAR signal distorted. It can be taken into account by using a multiple scattering parameter as in Platt [43], Berthier *et al.* [44] and Shang and Chazette [26].

The LAI is an integration of F_{app}. The random orientation of foliage [37] is introduced as $G = 0.5$. Clumping coefficient C has been assessed by Chen *et al.* [45] around 1.58 using the bidirectional reflectances derived from the Polarization and Directionality of the Earth Reflectance (POLDER, [46]) instrument onboard the Advanced Earth Observing Satellite (ADEOS), as well as by He *et al.* [47] using the MODIS BRDF (Bidirectional Reflectance Distribution Function) product, yielding $C = 1.54 \pm 0.05$ over most tropical forests and on Réunion Island. Note that this clumping coefficient is the reciprocal of the clumping index defined as the ratio of the effective LAI and the true LAI in some literature [45,47]. The LAI here calculated is a crude estimate of the true LAI, because it takes into account the contributions of branches and trunks. Tang *et al.* [21] considered that the majority of the backscattered energy measured (93%) was due to leaves, whereas only 7% came from the rest of the tree. Nevertheless, such a value is not justified in their article, and we do not have the capacity to verify it in the current study.

2.4. Sampling Strategy

Spatial sampling is a key parameter when using airborne LiDAR to characterize forest plots. Figure 4 gives an overview of the study sites and examples of LiDAR measurements performed continuously from the ULA. Note that the availability of ground echoes with a good signal-to-noise ratio (SNR) is highly variable with the forest site, depending on the vegetation density or the existence of gaps due to dead trees. We therefore had to adapt the sampling approach or find a representative sample for all of the sites.

Weather conditions with almost daily cloud formation over the tropical forest sites, coupled with trade winds or recirculation currents often exceeding $10\ \text{m} \cdot \text{s}^{-1}$, forced us to revise our initial sampling strategy of forest sites. It was not realistic to expect ground traces to be sufficiently numerous and close to each other to reproduce a 3D vision of forest structures. Nevertheless, we had to check that our samples remained representative.

The horizontal sampling grid during our airborne LiDAR experiment is defined by three independent parameters: the diameter of the laser footprint (d), the sampling along the ground track of the ULA (ΔX) and the sampling along the perpendicular to the ground track (ΔY). The last one must be more specifically defined, as it is based on successive passes of the ULA above the same forest site.

Laser footprint: In order to evaluate the influence of the laser footprint size (d) on the retrieval of the ground echo for a flat surface, three specific flights have been conducted at the same flight altitude over the Tamarind site (TM) with laser footprints at the ground level of 4, 10 and 20 m, respectively. We note no significant difference in the statistical distributions of tree structures between these experiments. Indeed, the treefall gaps help to identify the ground echo when using a sufficiently large laser footprint. A laser footprint of ~10 m associated with a ~350-m flight altitude was therefore considered for the entire sampling campaign. This footprint size is comparable to the overall span of dominant trees, and the ground echoes could be perceived from the optically thinner areas between the trees at each laser shot. Such a value is adequate for a correct assessment of the LAI, as shown by Riaño *et al.* [30], who found that LAI was better estimated using laser footprints between 7.5 and 12.5 m.

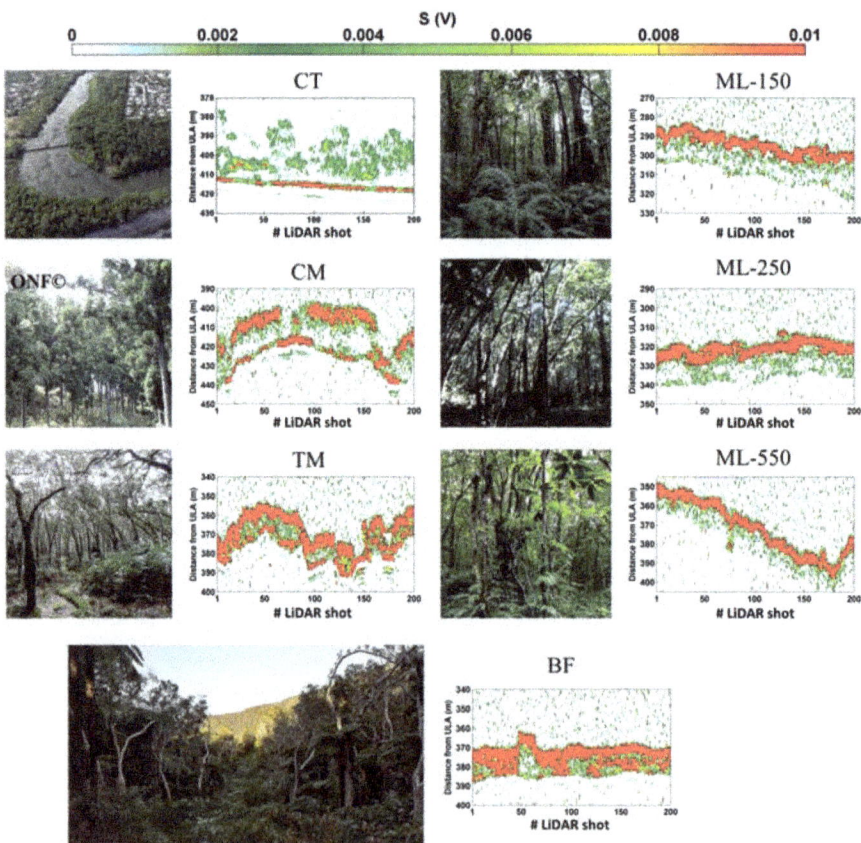

Figure 4. Photos for the 7 studied sites: coastal (CT), Tamarind (TM), Cryptomeria (CM), Bélouve (BF), and Mare-Longue (ML-150, -250 and -550) sites. Examples of continuous LiDAR measurements performed over the sites are also given.

Along-track and cross-track samplings: The influence of the along-track and cross-track sampling distances (ΔX and ΔY) on the horizontal sampling was evaluated, in order to ensure that the sampling was sufficient to accurately retrieve the canopy structure, *i.e.*, the correct CH distribution. After the accumulation of a sufficient number of samples, ΔX and ΔY were found to be log-normally distributed.

On the long flight (~40 km) performed over the Bélouve site (BF), we assessed the mean value and standard deviation of ΔX to be 0.9 ± 0.5 m. Such a value is fully suitable for sampling dense tropical forest from an airborne LiDAR. The sampling distance ΔY between each ground track was around 100-times (respectively 10-times) larger than ΔX in the BF site (respectively other sites), because a flight pattern including too many overpasses over the forest plot is not feasible. To assess the effect of reducing the horizontal sampling frequency, the CH distribution is computed with artificially increased ΔX values ranging from 1 to 100 m, *i.e.*, using only a fraction of the LiDAR shots. The results, presented in Figure 5, show that the CH distribution at $\Delta X = 10$ m is similar to the reference CH distribution, which is the one at the native resolution of $\Delta X \approx 1$ m. Moreover, in Figure 5d, CH histograms appear similar for values of ΔX between 1 and 100 m. Hence, the distance ΔY, which is most difficult to keep during the flights, does not significantly affect the statistical studies performed on the different tropical forest sites. These results on the horizontal sampling also demonstrate the strong homogeneity of the dense tropical cloud forest of Bélouve. Note that similar results are obtained for the tropical rain forest of Mare-Longue.

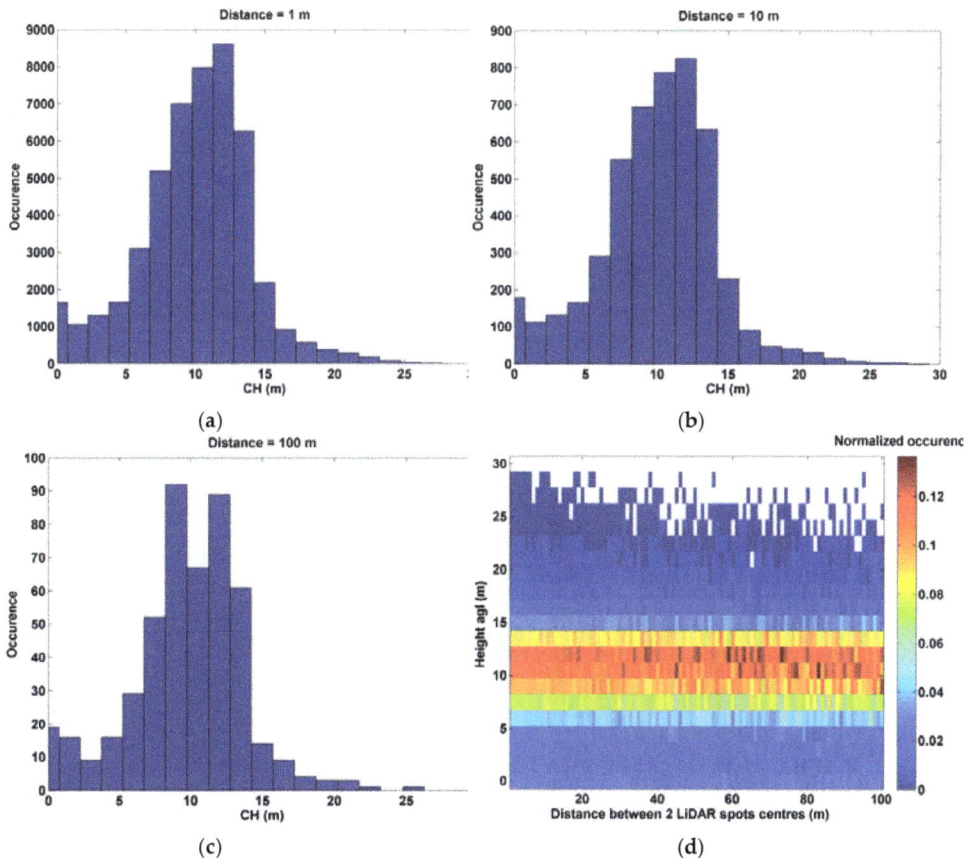

Figure 5. Distribution of the canopy height (CH) computed using a varying fraction of the LiDAR footprints in order to simulate an effective sampling distance between two consecutive footprints along the ULA ground-track of 1 m (**a**); 10 m (**b**) and 100 m (**c**). The two-dimensional representation is given in (**d**).

3. Results and Discussion on Retrieved Tropical Forest Parameters

Each parameter estimated from the airborne LiDAR measurements over the tropical forests of Réunion Island will be analyzed and discussed in this section. The results derived from LiDAR measurements on the various sites will also be compared.

3.1. Canopy Height

3.1.1. LiDAR-Derived Canopy Height

The number of available laser profiles on each sampled site is closely related to weather conditions during the experiment (Table 3). The two last sites of the Mare-Longue area are likely insufficiently characterized because the number of samplings is not enough for a reliable statistic. Nevertheless they are also taken into account in Table 3 where the mean, median and maximum values of retrieved CH are given, together with the standard deviation around the mean value, which represents so-called canopy rugosity. The statistic has been established on ~90% of the LiDAR profiles, when the ground echoes can be well located. The largest CH values are in the same range on all sites (~30 m), except for ML-250 and -550. The mean and median values are similar, which indicates that there is no significant bias due to outliers in the statistics. The standard deviation is larger than 5 m for the coastal site (CT), *i.e.*, large canopy rugosity, pointing out larger differences in terms of tree maturity on these sites. On the Cryptomeria site (CM), LiDAR observations also show large canopy rugosity. That is because the sub-plot of interest, which has uniform tree height (~22 m), is surrounded with low vegetation. As expected, for the Mare-Longue sites, we observe that CH decreases when altitude increases. It is less noticeable elsewhere, because tree species vary between plots.

Table 3. Statistics (mean, median and standard deviation (SD)) for both the canopy height (CH) and the assessment of the LAI on the 7 sites: coastal (CT), Tamarind (TM), Cryptomeria (CM), Bélouve (BF) and Mare-Longue (ML-150, -250 and -550) sites. Profiles with CH < 5 m are not considered. Bold characters highlight the 4 forest sites where *in situ* measurements are available (see Table 1). The maximal CH derived from the LiDAR is also indicated.

	CT	TM	CM	BF	ML-150	ML-250	ML-550
Number of samples	1621	**12,660**	**5790**	42,714	**9639**	**1518**	658
CH (m)							
Mean	16.2	**14.4**	**15.8**	11.4	**16.3**	**15.0**	13.2
Median	16.5	**15.0**	**16.5**	11.3	**17.3**	**15.0**	12.8
SD (rugosity)	5.7	**3.1**	**6.1**	3.1	**3.8**	**3.6**	2.5
Max	29.3	**28.5**	**29.3**	28.5	**30.8**	**24.0**	21.0
LAI (m²/m²)							
Mean	3.5	**4.8**	**5.0**	5.1	**7.8**	**7.5**	6.7
Median	2.8	**4.2**	**4.1**	4.5	**6.7**	**6.7**	5.9
SD	2.7	**2.5**	**3.3**	3.0	**3.9**	**3.9**	3.7
LAI from MODIS (m²/m²)							
Number of pixels (number of valid observations per pixel)		**2 (10)**		4 (14)		**3 (14)**	
Mean	-	**5.4**		5.1		**5.9**	
SD	-	**0.5**		0.4		**0.3**	

Figure 6 gives examples of samplings performed over the TM and CM sites with sampling distances $\Delta X \approx 1$ m and $\Delta Y \approx 10$ m. The presence of both valleys and clear areas observed on the ADC-air vegetation camera image can explain the inhomogeneity in the LiDAR CH measurements.

(a)

(b)

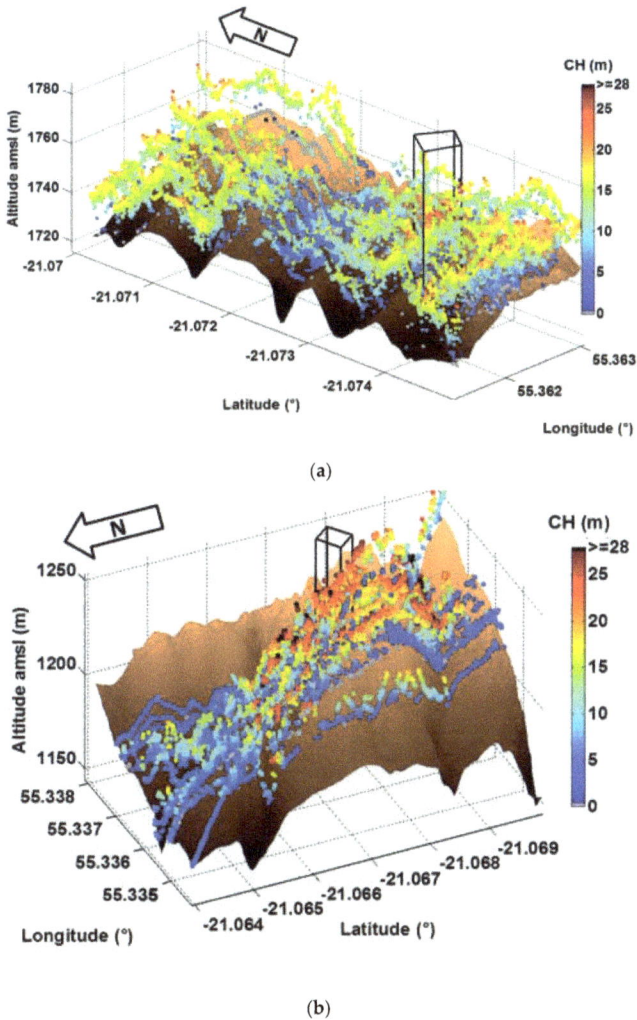

Figure 6. Horizontal samplings performed on the (**a**) Tamarind (TM) and (**b**) Cryptomeria (CM) sites with a laser footprint of 10 m. The horizontal sampling distances along and perpendicular to the ground track are ~1 and ~10 m, respectively. The brown color corresponds to the ground numerical model at a 5-m resolution provided by the Parc National de la Réunion (J.-C. Notter, personal communication). The locations of the sub-plots where *in situ* measurements were performed are highlighted using thick black lines.

3.1.2. Comparison with *in Situ* Measurements

In the four sampled sub-plots, the LiDAR-derived canopy heights (CH) were compared to the tree top heights (TTH) from *in situ* measurements. Their statistical moments and normalized distributions are given in Table 4 and in Figure 7, respectively. Understandably, for *in situ* measurements, the sampling areas were reduced (~0.2 ha), but the sub-plots were chosen so as to be as representative as possible of the extended sites. The agreement is quite good between the two distributions for the Cryptomeria site (CM). This is not the case for the others. With a LiDAR footprint diameter of 10 m, numerous smaller trees are hidden by the higher trees; the LiDAR-derived CH distribution is then biased toward the higher trees, whereas ground measurements can underestimate actual TTH as the tree top may not always be well identified because of the complex canopy. It is thus necessary to consider the apparent foliage profile to better identify the underlying trees.

Table 4. Statistics (mean, median, maximal values and standard deviation (SD)) for both the tree top height (TTH) as measured from the *in situ* census and the canopy height (CH) retrieved by LiDAR measurements in 4 sub-plots of ~0.2 ha: Tamarind (TM), Cryptomeria (CM) and Mare-Longue (ML-150, -250) sites. The values are given for sub-plots well identified in the main sites (Table 1). All trees with diameters at breast height higher than 7 cm are considered in *in situ* measurements.

	Sub-Plot			
	TM	**CM**	**ML-150**	**ML-250**
Location	21°4′25″S, 55°21′44″E	21°4′3″S, 55°20′13″E	21°21′29″S, 55°44′43″E	21°21′2″S, 55°44′39″E
Number of LiDAR profiles	635	245	2371	164
Number of *in situ* values	176	161	170	259
LiDAR CH (m)				
Mean	14.0	22.6	17.9	18.5
Median	13.5	22.5	18.0	18.0
SD	2.2	1.8	2.5	1.8
max	21.8	27.8	30.8	23.3
In Situ TTH (m)				
Mean	7.5	19.5	12.9	11.8
Median	5.8	20.4	11.8	12.1
SD	3.7	3.2	4.8	3.4
max	15	24.1	28.9	19.2

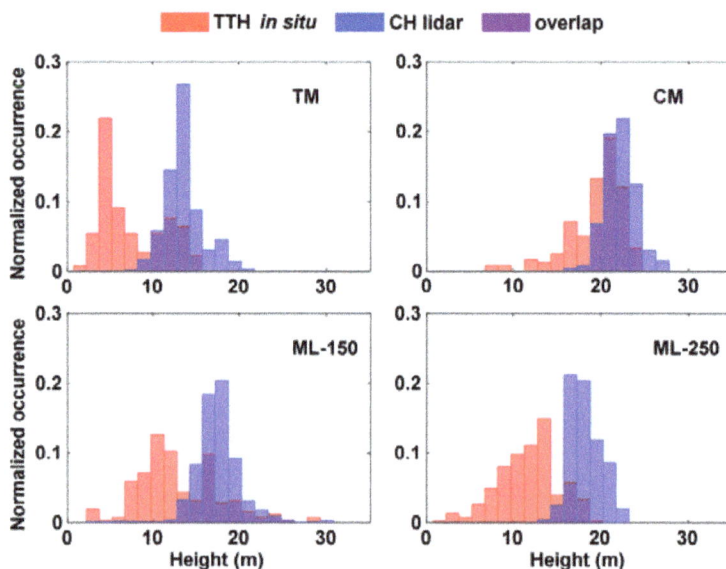

Figure 7. Distributions of the tree top height (TTH) from *in situ* (red) and canopy height (CH) from LiDAR (blue) measurements for the sub-plots of Tamarind site (TM), the Cryptomeria site (CM) and Mare-Longue tropical rain forest sites at 150 m (ML-150) and 250 m (ML-250) amsl. The overlapping parts are in purple. Uncertainty on retrieved heights is 4 m (2 m) for *in situ* (LiDAR) measurements.

In Figure 6, the Tamarind (TM) and Cryptomeria (CM) sub-plots are highlighted. We notice that the TM sub-plot has scarcer vegetation than its surroundings, whereas the CM sub-plot is denser than its surroundings. Images of the ADC-air vegetation camera obtained over Mare-Longue (ML) and Bélouve (BF) present on the contrary a good homogeneity, as well as high NDVI (>0.8), typical of primary tropical forest growing on regular slopes, which is coherent with the results previously discussed in Figure 5. We also found out that LiDAR-derived CHs are comparable between Tables 3 and 4 for the TM and ML-150 sites. Note that for BF and ML-500, it was difficult to access the site and almost impossible to identify the top of a tree from neighbors. Consequently, *in situ* measurements have not been considered as valid for these two sites. For the CM site where trees have about the same maturity, the differences between CH derived from LiDAR and TTH derived from *in situ* measurements are ~3 m, which is included in the standard deviation of CH and TTH as discussed previously (~2 m and ~4 m, respectively). For the others, the discrepancy is larger (>6 m) as for the TM site.

3.2. Understanding Results with Apparent Foliage Profiles

The canopy height (CH) does not provide enough resolved information on the vertical structure of the forest systems, which can be very complex because of the presence of multiple layers of saplings, as well as undergrowth (*i.e.*, tree ferns, ferns, bushes). Hence, it is preferable to consider the vertical profile of the apparent foliage (F_{app}). Such a profile is corrected from the extinction of the upper canopy. Parker *et al.* [48] have shown that F_{app} is an important constraint for energy, water and nutrient flows through forest cover. This is due to the contrasted contributions of the different canopy levels to both photosynthesis and carbon storage [48].

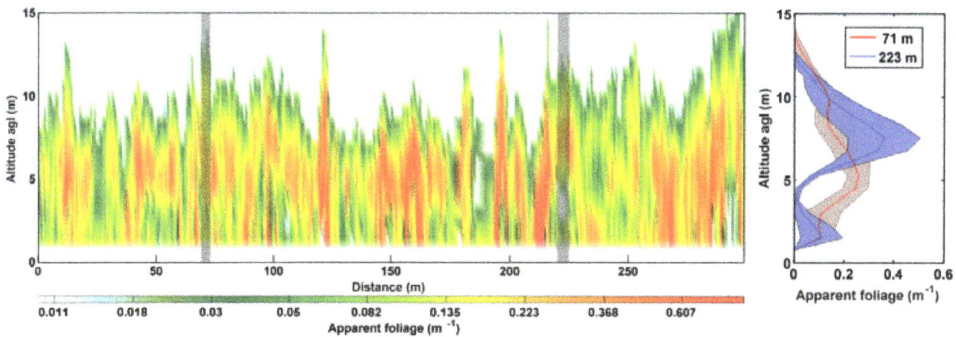

Figure 8. Evolution of the forest vertical profile of the LiDAR-derived apparent foliage (F_{app}) along a transect of the Bélouve site (BF, left panel). Two specific vertical profiles are shown in the right panel, and their locations are highlighted in transparent gray in the left panel.

As an example, we focus here on two different flight segments obtained on the Bélouve (BF) and Tamarind (TM) sites. The first site is very dense with continuous vegetation from ground to the canopy (as seen in the field), whereas the second one is composed of several distinct internal structures. Figure 8 shows the evolution of the apparent foliage as a function of distance along the transect in the BF. Two typical vertical profiles are also given. Certain profiles can show pronounced peaks, which identify the precise position of the tree crown or, on the contrary, smoother shapes due to the likely contribution of branches of nearby trees, lianas and, in the lower part of the profile, important undergrowth. Overall, because of the high density of trees in the tropical montane forest of BF, the LiDAR profiles mainly highlight only one vertical structure with one peak. This is also the case for the Cryptomeria (CM) site, but not for the same reason, because it is an exploited plot and there are generally no overlapping trees. On the contrary, for the TM site, which lies on the slope of Piton Maïdo, the convoluted tree trunks and complex development in response to storm winds lead to the existence of two superimposed layers, as can be seen in Figure 9. The profiles show there is generally an area with a lower density of vegetation between the two layers (between 4 and 8 m· agl), leading to less backscattered signal. This complex structure may be the source of discrepancies between airborne measurements and the *in situ* census made from the ground level. Another interesting and concrete conclusion is that the energy and water vapor fluxes between the forest and the atmosphere are mainly at the crown level of the trees for Tamarinds, even if undergrowth also contributes below ~8 m· agl. In contrast, for the BF site, these fluxes are distributed over the whole vertical forest structure.

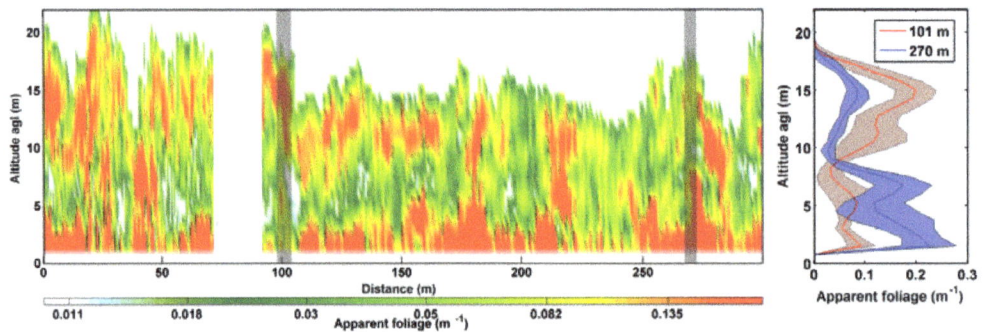

Figure 9. Evolution of the forest vertical profile of the LiDAR-derived apparent foliage (F_{app}) along a transect of the Tamarind site (TM, left panel). Two specific vertical profiles are given in the right panel, and their locations are highlighted in transparent gray in the left panel. The gap between the distances of 70 to 90 m corresponds to LiDAR shots with big pointing angles (the angle between the actual LiDAR line of sight and the nadir direction is larger than 20°), which were not considered in our study.

3.3. Leaf Area Index

LAI is also a key parameter linked to the plant respiration and photosynthesis, as explained by Gower and Norman [49]. It is important for vegetation growth (carbon sequestration) estimation [50]. Such a parameter is also a strong constraint for forest ecosystem modelling. It characterizes the forest interaction surface and exchange efficiency with the atmosphere.

3.3.1. LiDAR-Derived LAI

The LAI has been retrieved from each individual LiDAR profile to complement the characterization of the sampled forest sites. The LAI values are found to be log-normal distributed for all sampled forest sites, with the mean LAI ($\overline{LAI}$) ranging from about 3.5 to 7.8 m^2/m^2 (Table 3). The standard deviation is between 2.5 and 3.9 m^2/m^2, not necessarily correlated with the one of CH (Table 3). The tropical rain forests of Réunion Island (ML-150) have been shown to be associated with the higher mean LAI of 7.8 m^2/m^2. Such a value contrasts with the one retrieved for the tropical montane cloud forest of Bélouve, which is shown to be ~5 m^2/m^2 on average. This difference may be explained in terms of plant nutrient supply between these two forest categories [51]. As an example, the histogram of LAI derived from the LiDAR for the Bélouve site is also presented in Figure 10. The LAI value varies much for the same site from one point to another, likely due to very important difference in terms of nutrient availability in the ground.

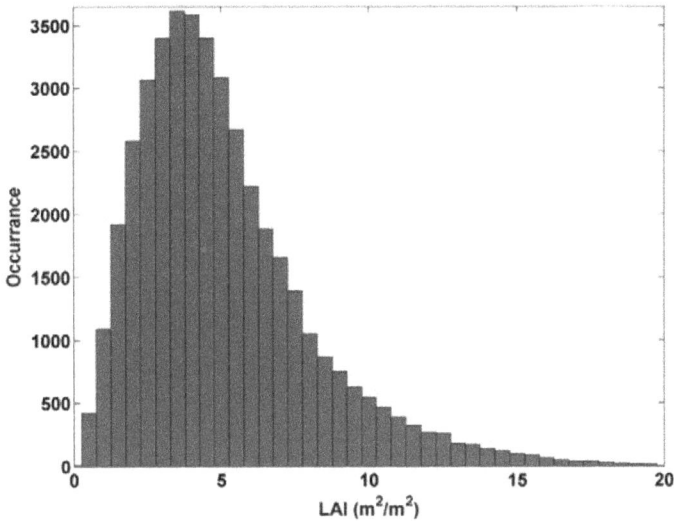

Figure 10. Histogram of the LAI derived from the airborne LiDAR for the tropical mountain rain forests of Bélouve (BF).

3.3.2. Inter-Comparison and Discussion

In this part, we will compare the LiDAR- and MODIS-derived LAI and discuss previous results published in the scientific literature.

The comparison with our observations was possible only on the large sampled areas due to the low spatial resolution of MODIS. Even though the LAI algorithm from satellite observations usually assumes that the ground is flat and does not handle mutual shadows due to the terrain, the agreement is very good on Bélouve (ground almost flat; Table 1) and sites situated on the foothills of the Piton Maïdo (TM and CM, ~5 m^2/m^2; ground with a certain slope; Table 1). LiDAR-derived mean LAIs for ML-150, -250 and -550 (7.8, 7.5 and 6.7 m^2/m^2, respectively) present higher differences with MODIS-derived LAI (5.9 m^2/m^2). The small slopes in these sites (Table 1) are not related to a better agreement, as the sampled surface is too small. We believe that LiDAR measurements offer a better assessment of LAI than MODIS, as this LiDAR metric has already been validated by Tang *et al.* [21] and MODIS-derived LAI saturates at certain value levels because of the nature of the tool [31]. In addition, the clumping of vegetation structure is less considered by the MODIS LAI algorithm; so it is normal that the LAI is underestimated. As previously explained, it is difficult to conclude for both ML-250 and ML-550, because the number of samples obtained over these sites is not significant.

Mature evergreen tropical forests usually have large LAI, more than 4 m^2/m^2, as shown for example in Doughty and Goulden [52] using monthly MODIS observations over tropical forests of Brazil. Cristiano *et al.* [53] also found mean LAI to be larger than 7 m^2/m^2 for native subtropical forests of Argentina, Brazil and Paraguay. Moreover, the mean LAI derived from our LiDAR measurements is within the range of values deduced from the LiDAR- or tower-derived cumulative LAI of Tang *et al.* [21], which give LAI between 5 and 9 m^2/m^2 for secondary and old-growth forests. Asner *et al.* [54] performed a global synthesis of LAI from various ecological and remote sensing studies. The mountain tropical forest of Bélouve, the TM and the CM sites are associated with mean LAI very close to the one compiled by these authors, which found for tropical deciduous and evergreen broadleaves LAI between 3.9 and 4.8 m^2/m^2 on 78 samplings, with a maximal value close to 9 m^2/m^2. Our standard deviations are in the same range as obtained by Asner *et al.* [54], which give values from 0.7 to 4.3 m^2/m^2.

3.4. Classification

Regarding the average values and variability of both CH and $\overline{LAI}$, the seven sites are in two distinct groups (Figure 11): the endemic forests and the planted/exotic forests, with their locations shown in Figure 1. The higher $\overline{LAI}$ are for the three Mare-Longue (ML) sites; the largest value is found for ML-150. The site of Bélouve (BF) with $\overline{LAI}$ = 5.1 appears to be isolated from the other endemic forest sites ($\overline{LAI}$ = 7.8 for ML-150). Indeed, although the ML and BF sites shelter the densest and most native tropical forests of the island, they are located at very different altitudes. The BF forest is characterized by a soil composed of little mineralized organic material, as limited by the temperature conditions. This can contribute to a limitation of nutrient supplies, as shown by Tanner *et al.* [55] for tropical mountain rain forests of Hawaii, Costa Rica and Colombia, which grow mainly on a lava substrate. The previous coarse separation of forest biomes confirms the work of Strasberg *et al.* [23].

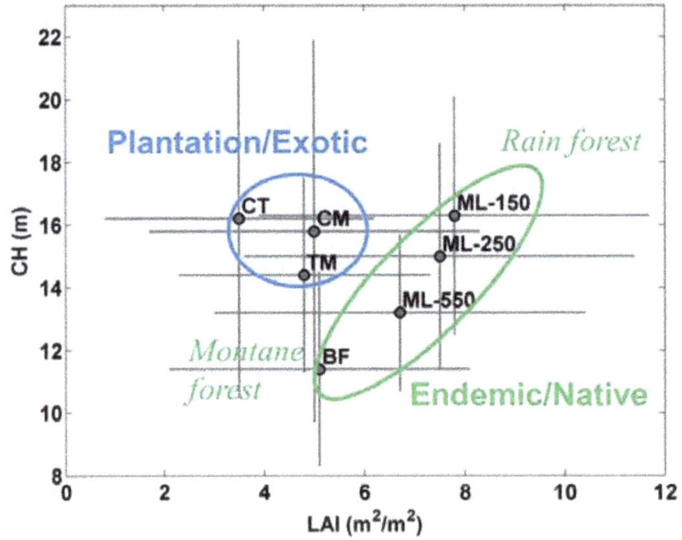

Figure 11. Mean (dot) and standard deviations (line segments) of forest LiDAR-derived LAI against the canopy height (CH) for the seven sites: coastal (CT), Tamarind (TM), Cryptomeria (CM), Bélouve (BF) and Mare-Longue (ML-150, -250 and -550) sites. Measurements with CH < 5 m are not considered.

4. Conclusions

Airborne LiDAR measurements conducted in May 2014 over several tropical forest sites of Réunion Island allow one to clearly identify the different types of coverage thanks to key parameters derived from airborne LiDAR measurements: the canopy height (CH), the forest LAI and the vertical profiles of the apparent foliage (F_{app}), introduced in previous works (e.g., [37]), but evaluated without measuring ground and vegetation reflectance values.

LiDAR-derived CH histograms have been compared to the tree top height (TTH) measured during *in situ* censuses. The CH and TTH statistical values (mean, median, maximal values and standard deviation (SD)) are shown to be in agreement, where it is possible for an operator to distinguish the tree top from the ground (e.g., Cryptomeria site). We derived the LAI from LiDAR measurements alone. The LiDAR- and MODIS-derived LAI are shown to be in good agreement. The LiDAR-derived LAI values are also in the range of usual values regarding previous results published in the scientific literature.

We have shown that the simultaneous use of LiDAR-derived CH and LAI is sufficient to coarsely classify forests of Réunion Island. The endemic and exotic forests are well distinguished. Hence, we can recommend airborne LiDAR measurements as a relevant means for studying forest where ground-based observations are scarce and difficult to obtain. Besides this main result, this study gives new technical insights on the capability of LiDAR to penetrate through dense forest, on the choice of the laser footprint and spatial sampling, taking into account the forest heterogeneity, and on the retrieval of the LAI, in complementarity with the method of Tang *at al.* [21].

Finally, this campaign was an opportunity to compose an original and diverse LiDAR database that will help further works on remote sensing of tropical forests, mainly for the inter-annual evolution of the forest cover of Réunion Island, which is a rich United-European (French) diverse tropical ecosystem listed as World Heritage by UNESCO.

Acknowledgments: Acknowledgments: The experiments have been funded by the Centre National d'Etudes Spatiales (CNES), the Commissariat à l'Energie Atomique et aux Energies Alternatives (CEA) and the Université de la Réunion through the federation Observatoire des Milieux Naturels et des Changements Globaux (OMNCG) of the Observatoire des Sciences de l'Univers de la Réunion (OSU-R). We are also grateful for the support offered by the Direction Générale de l'Armement (DGA) and the Institut Pierre Simon Laplace (IPSL). We thank J.-C. Notter (Parc National de la Réunion) for providing the digital terrain model. We also thank F. Maignan (LSCE), C. Bacour (LSCE) and G. Dedieu (Centre d'Etudes Spatiales de la BIOsphère) for useful discussions and F. Toussaint for planning and performing the ULA flights.

Author Contributions: Author Contributions: All authors contributed significantly to this manuscript. Xiaoxia Shang wrote the manuscript draft with contributions from Patrick Chazette. Xiaoxia Shang and Patrick Chazette analyzed the data. All authors performed the experiments. Patrick Chazette directed the experiments. All authors revised the manuscript draft and provided valuable suggestions for the revision.

Conflicts of Interest: Conflicts of Interest: The authors declare no conflict of interest.

References

1. Franklin, S.E.; Hall, R.J.; Moskal, L.M.; Maudie, A.J.; Lavigne, M.B. Incorporating texture into classification of forest species composition from airborne multispectral images. *Int. J. Remote Sens.* **2000**, *21*, 61–79. [CrossRef]
2. Popescu, S.C.; Wynne, R.H.; Scrivani, J.A. Fusion of small-footprint LiDAR and multispectral data to estimate plot- level volume and biomass in deciduous and pine forests in Virginia, USA. *Soc. Am. For.* **2001**, *50*, 551–565.
3. Etteieb, S.; Louhaichi, M.; Kalaitzidis, C.; Gitas, I.Z. Mediterranean forest mapping using hyper-spectral satellite imagery. *Arab. J. Geosci.* **2012**, *6*, 5017–5032. [CrossRef]
4. Hilton, F.; Armante, R.; August, T.; Barnet, C.; Bouchard, A.; Camy-Peyret, C.; Capelle, V.; Clarisse, L.; Clerbaux, C.; Coheur, P.-F.; *et al.* Hyperspectral earth observation from IASI: Five years of accomplishments. *Bull. Am. Meteorol. Soc.* **2012**, *93*, 347–370. [CrossRef]
5. Huesca, M.; Merino-de-Miguel, S.; Gonzalez-Alonso, F.; Martinez, S.; Miguel Cuevas, J.; Calle, A. Using AHS hyper-spectral images to study forest vegetation recovery after a fire. *Int. J. Remote Sens.* **2013**, *34*, 4025–4048. [CrossRef]
6. Hyde, P.; Dubayah, R.; Walker, W.; Blair, J.B.; Hofton, M.; Hunsaker, C. Mapping forest structure for wildlife habitat analysis using multi-sensor (LiDAR, SAR/InSAR, ETM+, Quickbird) synergy. *Remote Sens. Environ.* **2006**, *102*, 63–73. [CrossRef]
7. Beets, P.N.; Reutebuch, S.; Kimberley, M.O.; Oliver, G.R.; Pearce, S.H.; McGaughey, R.J. Leaf Area Index, biomass carbon and growth rate of radiata pine genetic types and relationships with LiDAR. *Forests* **2011**, *2*, 637–659. [CrossRef]
8. Parker, G.G.; Harding, D.J.; Berger, M.L. A portable LiDAR system for rapid determination of forest canopy structure. *J. Appl. Ecol.* **2004**, *41*, 755–767. [CrossRef]
9. Jupp, D.L.B.; Culvenor, D.S.; Lovell, J.L.; Newnham, G.J.; Strahler, A.H.; Woodcock, C.E. Estimating forest LAI profiles and structural parameters using a ground-based laser called "Echidna". *Tree Phys.* **2009**, *29*, 171–181. [CrossRef] [PubMed]
10. Yang, X.; Strahler, A.H.; Schaaf, C.B.; Jupp, D.L.B.; Yao, T.; Zhao, F.; Wang, Z.; Culvenor, D.S.; Newnham, G.J.; Lovell, J.L.; *et al.* Three-dimensional forest reconstruction and structural parameter retrievals using a terrestrial full-waveform LiDAR instrument (Echidna®). *Remote Sens. Environ.* **2013**, *135*, 36–51. [CrossRef]
11. Greaves, H.E.; Vierling, L.A.; Eitel, J.U.H.; Boelman, N.T.; Magney, T.S.; Prager, C.M.; Griffin, K.L. Estimating aboveground biomass and leaf area of low-stature Arctic shrubs with terrestrial LiDAR. *Remote Sens. Environ.* **2015**, *164*, 26–35. [CrossRef]

12. Dassot, M.; Constant, T.; Fournier, M. The use of terrestrial LiDAR technology in forest science: Application fields, benefits and challenges. *Ann. For. Sci.* **2011**, *68*, 959–974. [CrossRef]
13. Asner, G.P.; Mascaro, J.; Muller-Landau, H.C.; Vieilledent, G.; Vaudry, R.; Rasamoelina, M.; Hall, J.S.; Breugel, M. A universal airborne LiDAR approach for tropical forest carbon mapping. *Oecologia* **2012**, *168*, 1147–1160. [CrossRef] [PubMed]
14. Asner, G.P.; Powell, G.V.N.; Mascaro, J.; Knapp, D.E.; Clark, J.K.; Jacobson, J.; Kennedy-Bowdoin, T.; Balaji, A.; Paez-Acosta, G.; Victoria, E.; *et al.* High-resolution forest carbon stocks and emissions in the Amazon. *Proc. Natl. Acad. Sci. USA* **2010**, *107*, 16738–16742. [CrossRef] [PubMed]
15. Leitold, V.; Keller, M.; Morton, D.C.; Cook, B.D.; Shimabukuro, Y.E. Airborne LiDAR-based estimates of tropical forest structure in complex terrain: Opportunities and trade-offs for REDD+. *Carbon Balance Manag.* **2015**, *10*. [CrossRef] [PubMed]
16. Drake, J.B.; Dubayah, R.O.; Clark, D.B.; Knox, R.G.; Blair, J.B.; Hofton, M.A.; Chazdon, R.L.; Weishampel, J.F.; Prince, S. Estimation of tropical forest structural characteristics using large-footprint LiDAR. *Remote Sens. Environ.* **2002**, *79*, 305–319. [CrossRef]
17. Dubayah, R.O.; Sheldon, S.L.; Clark, D.B.; Hofton, M.A.; Blair, J.B.; Hurtt, G.C.; Chazdon, R.L. Estimation of tropical forest height and biomass dynamics using LiDAR remote sensing at La Selva, Costa Rica. *J. Geophys. Res.* **2010**, *115*. [CrossRef]
18. Cuesta, J.; Chazette, P.; Allouis, T.; Flamant, P.H.; Durrieu, S.; Sanak, J.; Genau, P.; Guyon, D.; Loustau, D.; Flamant, C. Observing the forest canopy with a new ultra-violet compact airborne LiDAR. *Sensors* **2010**, *10*, 7386–7403. [CrossRef] [PubMed]
19. Shang, X.; Chazette, P. Interest of a full-waveform flown UV LiDAR to derive forest vertical structures and aboveground carbon. *Forests* **2014**, *5*, 1454–1480. [CrossRef]
20. Blair, J.B.; Rabine, D.L.; Hofton, M.A. The laser vegetation imaging sensor: A medium-altitude, digitisation-only, airborne laser altimeter for mapping vegetation and topography. *ISPRS J. Photogramm. Remote Sens.* **1999**, *54*, 115–122. [CrossRef]
21. Tang, H.; Dubayah, R.; Swatantran, A.; Hofton, M.; Sheldon, S.; Clark, D.B.; Blair, B. Retrieval of vertical LAI profiles over tropical rain forests using waveform LiDAR at La Selva, Costa Rica. *Remote Sens. Environ.* **2012**, *124*, 242–250. [CrossRef]
22. Tang, H.; Brolly, M.; Zhao, F.; Strahler, A.H.; Schaaf, C.L.; Ganguly, S.; Zhang, G.; Dubayah, R. Deriving and validating Leaf Area Index (LAI) at multiple spatial scales through LiDAR remote sensing: A case study in Sierra National Forest, CA. *Remote Sens. Environ.* **2014**, *143*, 131–141. [CrossRef]
23. Strasberg, D.; Rouget, M.; Richardson, D.M.; Baret, S.; Dupont, J.; Cowling, R.M. An assessment of habitat diversity and transformation on La Réunion Island (Mascarene Islands, Indian Ocean) as a Basis for Identifying Broad-scale Conservation Priorities. *Biodivers. Conserv.* **2005**, *14*, 3015–3032. [CrossRef]
24. Cadet, T.; Figier, J. *Réserve Naturelle de Mare-Longue: Etude Floristique et Ecologique*; Université de la Réunion: Saint-Denis, France, 1989.
25. Strasberg, D. Diversity, size composition and spatial aggregation among trees on a 1-ha rain forest plot at La Réunion. *Biodivers. Conserv.* **1996**, *5*, 825–840. [CrossRef]
26. Shang, X.; Chazette, P. End-to-end simulation for a forest-dedicated full-waveform LiDAR onboard a satellite initialized from airborne ultraviolet LiDAR experiments. *Remote Sens.* **2015**, *7*, 5222–5255. [CrossRef]
27. Lefsky, M.A.; Harding, D.; Cohen, W.B.; Parker, G.; Shugart, H.H. Surface LiDAR remote sensing of basal area and biomass in deciduous forests of eastern Maryland, USA. *Remote Sens. Environ.* **1999**, *67*, 83–98. [CrossRef]
28. Drake, J.B.; Dubayah, R.O.; Knox, R.G.; Clark, D.B.; Blair, J.B. Sensitivity of large-footprint LiDAR to canopy structure and biomass in a neotropical rainforest. *Remote Sens. Environ.* **2002**, *81*, 378–392. [CrossRef]
29. Means, J.E.; Acker, S.A.; Harding, D.J.; Blair, J.B.; Lefsky, M.A.; Cohen, W.B.; Harmon, M.E.; McKee, W.A. Use of large-footprint scanning airborne LiDAR to estimate forest stand characteristics in the western cascades of Oregon. *Remote Sens. Environ.* **1999**, *67*, 298–308. [CrossRef]
30. Riaño, D.; Valladares, F.; Condés, S.; Chuvieco, E. Estimation of Leaf Area Index and covered ground from airborne laser scanner (LiDAR) in two contrasting forests. *Agric. For. Meteorol.* **2004**, *124*, 269–275. [CrossRef]
31. Yang, W.; Tan, B.; Huang, D.; Rautiainen, M.; Shabanov, N.V.; Wang, Y.; Privette, J.L.; Huemmrich, K.F.; Fensholt, R.; Sandholt, I.; *et al.* MODIS Leaf Area Index products: From validation to algorithm improvement. *IEEE Trans. Geosci. Remote Sens.* **2006**, *44*, 1885–1898. [CrossRef]

32. Chazette, P.; Pelon, J.; Mégie, G. Determination by spaceborne backscatter LiDAR of the structural parameters of atmospheric scattering layers. *Appl. Opt.* **2001**, *40*, 3428–3440. [CrossRef] [PubMed]

33. Chazette, P.; Dabas, A.; Sanak, J.; Lardier, M.; Royer, P. French airborne LiDAR measurements for Eyjafjallajökull ash plume survey. *Atmos. Chem. Phys.* **2012**, *12*, 7059–7072. [CrossRef]

34. Chazette, P.; Bocquet, M.; Royer, P.; Winiarek, V.; Raut, J.-C.; Labazuy, P.; Gouhier, M.; Lardier, M.; Cariou, J.-P. Eyjafjallajökull ash concentrations derived from both LiDAR and modeling. *J. Geophys. Res. Atmos.* **2012**, *117*. [CrossRef]

35. Measures, R.M. *Laser Remote Sensing: Fundamentals and Applications*; Wiley, J., Ed.; Krieger Publishing Company: Malabar, FL, USA, 1984.

36. Chazette, P.; David, C.; Lefrère, J.; Godin, S.; Pelon, J.; Mégie, G. Comparative LiDAR study of the optical, geometrical, and dynamical properties of stratospheric post-volcanic aerosols, following the eruptions of El Chichon and Mount Pinatubo. *J. Geophys. Res.* **1995**, *100*. [CrossRef]

37. Ni-Meister, W.; Jupp, D.L.B.; Dubayah, R. Modeling LiDAR waveforms in heterogeneous and discrete canopies. *IEEE Trans. Geosci. Remote Sens.* **2001**, *39*, 1943–1958. [CrossRef]

38. Ahmed, R.; Siqueira, P.; Hensley, S. A study of forest biomass estimates from LiDAR in the northern temperate forests of New England. *Remote Sens. Environ.* **2013**, *130*, 121–135. [CrossRef]

39. Hofton, M.A.; Minster, J.B.; Blair, J.B. Decomposition of laser altimeter waveforms. *IEEE Trans. Geosci. Remote Sens.* **2000**, *38*, 1989–1996. [CrossRef]

40. Lefsky, M.A.; Cohen, W.B.; Acker, S.A.; Parker, G.G.; Spies, T.A.; Harding, D. LiDAR remote sensing of the canopy structure and biophysical properties of Douglas-fir western hemlock forests. *Remote Sens. Environ.* **1999**, *70*, 339–361. [CrossRef]

41. Farid, A.; Goodrich, D.C.; Bryant, R.; Sorooshian, S. Using airborne LiDAR to predict Leaf Area Index in cottonwood trees and refine riparian water-use estimates. *J. Arid Environ.* **2008**, *72*, 1–15. [CrossRef]

42. Solberg, S.; Brunner, A.; Hanssen, K.H.; Lange, H.; Næsset, E.; Rautiainen, M.; Stenberg, P. Mapping LAI in a Norway spruce forest using airborne laser scanning. *Remote Sens. Environ.* **2009**, *113*, 2317–2327. [CrossRef]

43. Platt, C.M.R. Remote sounding of high clouds. III: Monte Carlo calculations of multiple-scattered LiDAR returns. *J. Atmos. Sci.* **1981**, *38*, 156–167. [CrossRef]

44. Berthier, S.; Chazette, P.; Couvert, P.; Pelon, J.; Dulac, F.; Thieuleux, F.; Moulin, C.; Pain, T. Desert dust aerosol columnar properties over ocean and continental Africa from LiDAR in-Space Technology Experiment (LITE) and Meteosat synergy. *J. Geophys. Res.* **2006**, *111*, D21202. [CrossRef]

45. Chen, J.M.; Rich, P.M.; Gower, S.T.; Norman, J.M.; Plummer, S. Leaf Area Index of boreal forests: Theory, techniques, and measurements. *J. Geophys. Res.* **1997**, *102*, 29429–29443. [CrossRef]

46. Leroy, M.; Deuzé, J.L.; Bréon, F.M.; Hautecoeur, O.; Herman, M.; Buriez, J.C.; Tanré, D.; Bouffiès, S.; Chazette, P.; Roujean, J.L. Retrieval of atmospheric properties and surface bidirectional reflectances over land from POLDER/ADEOS. *J. Geophys. Res.* **1997**, *102*, 17023–17037. [CrossRef]

47. He, L.; Chen, J.M.; Pisek, J.; Schaaf, C.B.; Strahler, A.H. Global clumping index map derived from the MODIS BRDF product. *Remote Sens. Environ.* **2012**, *119*, 118–130. [CrossRef]

48. Parker, G.G.; Lefsky, M.A.; Harding, D.J. Light transmittance in forest canopies determined using airborne laser altimetry and in-canopy quantum measurements. *Remote Sens. Environ.* **2001**, *76*, 298–309. [CrossRef]

49. Gower, S.T.; Norman, J.M. Rapid estimation of Leaf Area Index in conifer and broad-leaf plantations. *Ecology* **1991**, *72*, 1896. [CrossRef]

50. Ellsworth, D.S.; Reich, P.B. Canopy structure and vertical patterns of photosynthesis and related leaf traits in a deciduous forest. *Oecologia* **1993**, *96*, 169–178. [CrossRef]

51. Knyazikhin, Y.; Glassy, J.; Privette, J.L.; Tian, Y.; Lotsch, A.; Zhang, Y.; Wang, Y.; Morisette, J.T.; Votava, P.; Myneni, R.B.; *et al.* MODIS Leaf Area Index (LAI) and fraction of photosynthetically active radiation absorbed by vegetation (FPAR) product (MOD15). *Algorithm Theor. Basis Doc.* **1999**, *4*, 1–14.

52. Doughty, C.E.; Goulden, M.L. Seasonal patterns of tropical forest Leaf Area Index and CO_2 exchange. *J. Geophys. Res.* **2008**, *113*, G00B06.

53. Cristiano, P.; Madanes, N.; Campanello, P.; Francescantonio, D.; Rodríguez, S.; Zhang, Y.-J.; Carrasco, L.; Goldstein, G. High NDVI and potential canopy photosynthesis of South American subtropical forests despite seasonal changes in Leaf Area Index and air temperature. *Forests* **2014**, *5*, 287–308. [CrossRef]

54. Asner, G.P.; Scurlock, J.M.O.; Hicke, J. Global synthesis of Leaf Area Index observations: Implications for ecological and remote sensing studies. *Glob. Ecol. Biogeogr.* **2003**, *12*, 191–205. [CrossRef]
55. Tanner, E.V.J.; Vitousek, P.M.; Cuevas, E. Experimental investigation of nutrient limitation of forest growth on wet tropical mountains. *Ecology* **1998**, *79*, 10–22. [CrossRef]

Part 4
Registration of Point Clouds

remote sensing

MDPI

Article

Hierarchical Registration Method for Airborne and Vehicle LiDAR Point Cloud

Liang Cheng [1,2,3,4], Yang Wu [4], Lihua Tong [4], Yanming Chen [1,2,4,*] and Manchun Li [1,2,4,5,*]

[1] Jiangsu Provincial Key Laboratory of Geographic Information Science and Technology,
   Nanjing University, Nanjing 210093, China; E-Mail: lcheng@nju.edu.cn
[2] Collaborative Innovation Center for the South Sea Studies, Nanjing University,
   Nanjing 210093, China
[3] Collaborative Innovation Center of Novel Software Technology and Industrialization,
   Nanjing University, Nanjing 210093, China
[4] Department of Geographic Information Science, Nanjing University, Nanjing 210093, China; E-Mails:
   Wuyang_nju@126.com (Y.W.); buqingyuntlh@sina.cn (L.T.)
[5] Jiangsu Center for Collaborative Innovation in Geographical Information Resource Development and
   Application, Nanjing 210023, China
* Authors to whom correspondence should be addressed; E-Mails: chenyanming@nju.edu.cn (Y.C.);
   limanchun_nju@126.com (M.L.); Tel.: +86-25-8968-1185 (Y.C.); +86-25-8968-0799 (M.L.).

Academic Editors: Juha Hyyppä, Norman Kerle and Prasad S. Thenkabail

Received: 13 July 2015/ Accepted: 15 October 2015 / Published: 23 October 2015

Abstract: A new hierarchical method for the automatic registration of airborne and vehicle light detection and ranging (LiDAR) data is proposed, using three-dimensional (3D) road networks and 3D building contours. Firstly, 3D road networks are extracted from airborne LiDAR data and then registered with vehicle trajectory lines. During the registration of airborne road networks and vehicle trajectory lines, a network matching rate is introduced for the determination of reliable transformation matrix. Then, the RIMM (reversed iterative mathematic morphological) method and a height value accumulation method are employed to extract 3D building contours from airborne and vehicle LiDAR data, respectively. The Rodriguez matrix and collinearity equation are used for the determination of conjugate building contours. Based on this, a rule is defined to determine reliable conjugate contours, which are finally used for the fine registration of airborne and vehicle LiDAR data. The experiments show that the coarse registration method with 3D road networks can contribute to a reliable initial registration result, and the fine registration using 3D building contours obtains a final registration result with high reliability and geometric accuracy.

Keywords: airborne LiDAR; vehicle LiDAR; 3D road network; 3D building contour; point cloud registration

1. Introduction

The advent of the Light Detection and Ranging (LiDAR) system represents a technological breakthrough in acquiring three-dimensional (3D) data of surface objects in a rapid and cost-effective manner [1]. By now, a variety of LiDAR techniques have emerged, including satellite-based laser scanning (SLS), airborne laser scanning (ALS), vehicle laser scanning (VLS), and terrestrial laser scanning (TLS). The airborne LiDAR system, due to its extraordinary capability in gathering highly accurate and dense elevation measurements in large scales, has become an important new technique for data filtering [2], 3D reconstruction [3,4], and object identification [5,6]. The vehicle LiDAR system, mounted on high speed vehicles, can quickly obtain detailed facade information of surface objects with an extremely high point density [7,8]. On one hand, both airborne and vehicle LiDAR

Remote Sens. **2015**, *7*, 13921–13944

are endowed with a large-range scanning area, which makes the integration practical. On the other hand, massive façade information is included in vehicle LiDAR with little top information, while more top information is obtained using airborne LiDAR with little façade information [9]. Considering the similarity and complementarity of airborne and vehicle LiDAR data, the integration of airborne and vehicle LiDAR can be performed to lead to new applications, such as a new presentation of our city, named "point city" (shown in Figure 1). Point cloud registration is a prerequisite to integrate airborne and vehicle LiDAR data.

In general, the point cloud registration methods can be divided into two types: the auxiliary method and the data-based method. Auxiliary data (e.g., spectral image and global positioning system (GPS) data) and artificial targets have been widely used in the auxiliary method to assist the registration of point clouds. In study [10], a method is proposed for the registration of terrestrial LiDAR point clouds based on the local features extracted from the images. Firstly, the characteristic two-dimensional (2D) points based on the scale invariant feature transform (SIFT) feature are extracted from the reflectance images of two scans. Then these 2D points are projected into 3D space by using interpolated range information. Finally, the transformation parameters of point clouds can be estimated from the 3D-2D correspondences. Some studies use GPS data, which can achieve centimeter accuracy, to directly georeference the LiDAR point clouds [11,12]. However, in complex urban and forest environments, GPS receivers frequently lose lock, which would reduce the positioning accuracy of the GPS system. Based on this consideration, after using GPS data for registration, the multi-station registration method which minimizes the surface error of point cloud data is applied [13,14]. As auxiliary data is required in the above work, these methods are limited in their application scope.

The data-based method conducts registration based on the point clouds without using auxiliary data. The iterative closest point (ICP) algorithm is a classic data-based method [15]. Based on the ICP method, many variant approaches appeared later. A flaw of these ICP-based approaches is the calling for a good initial alignment of the point clouds to enable convergence to the local minimum, which is not always easy to access. Another branch of the data-based registration approach lies in the use of the local normal and curvature feature [16–19]. The basic idea of those methods is that the conjugate features are determined based on the changes in geometric curvature and the approximate normal vector of the local surface around each point.

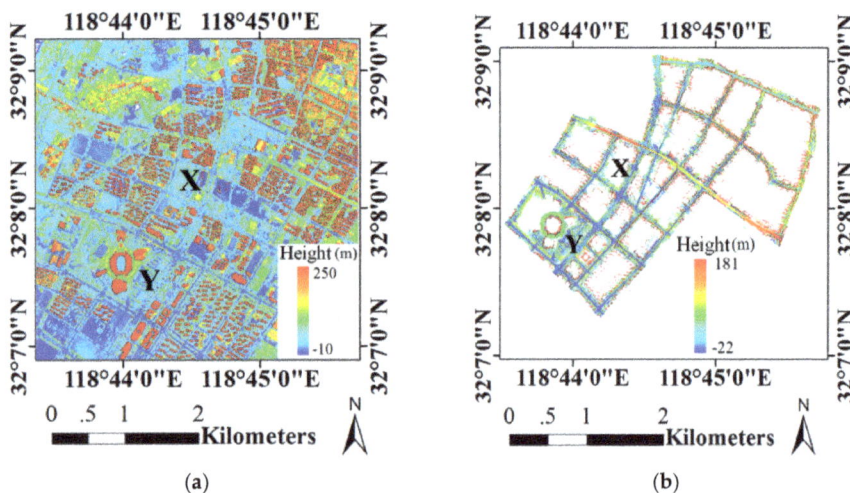

(a)

(b)

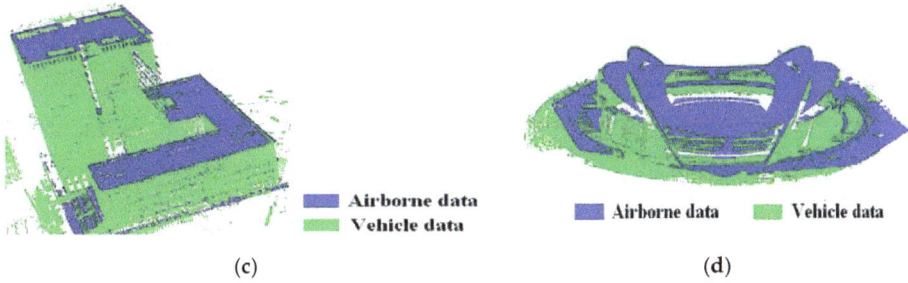

Figure 1. An example of "point city" with registered airborne and vehicle LiDAR points. (**a**) Airborne LiDAR points; (**b**) Vehicle LiDAR points; (**c**) Building X in "point city" (blue points: airborne data, green points: vehicle data); (**d**) Building Y in "point city" (blue points: airborne data, green points: vehicle data).

Besides, some data-based methods conduct registration by using points, lines, planes, and other geometric features extracted from point clouds. (1) Point feature is the basic control feature, and has been widely used in point cloud registration. Barnea and Filin [20] conducted the registration of terrestrial LiDAR data by extracting key points from the panoramic range image. Cheng *et al.* [21] extracted 3D building corners from the intersection of 3D building boundaries from airborne and terrestrial LiDAR data, and the 3D building corners are matched through an automatic iterative process. In [22], a shiftable leading point method is optimized for high accuracy registration of 3D corners; (2) Lines can serve as essential features for the registration of point clouds in areas with abundant man-made structures. Better registration results can be offered with line features than with point features [23]. Jaw and Chuang [24] provided a line-based registration approach, where the matching of 3D line features is performed with angle and distance as constraints. Lee *et al.* [25] extracted 3D line features through the intersection of neighboring planar patches and used them to adjust the discrepancies between overlapping data strips. Recently, Bucksch and Khoshelham [26] extracted skeletons from tree branches and utilized them for the precise registration of point clouds. As line features present geometric evidence of edges, which are quite prominent and extraordinary, the matching of conjugate lines can be performed precisely; (3) Plane features are also used for the registration of point clouds. Teo and Huang [27] proposed a scheme for the registration of airborne and terrestrial LiDAR points using the least squares 3D surface registration technique to minimize the surfaces between two datasets. In von Hansen's work, single plane correspondences are used to find the transformation parameters with some additional assumptions [28]. Zhang *et al.* [29] extracted plane features using a segmentation method and adopted the Rodriguez matrix for the registration work. Brenner *et al.* [30] showed that three plane matches can help to determine the transformation parameters between two point clouds, and their method has been compared with the point-based Normal Distributions Transform method [31]. The results of this comparison showed that the plane-based method tends to be more accurate and the point-based method is conceptually simple and fast. Besides, Wu *et al.* [32] proposed a registration method for airborne and terrestrial LiDAR point clouds based on building profiles, and achieved registration accuracies of 0.15 to 0.5 m in the horizontal direction and 0.20 m in the vertical direction.

However, some technical challenges exist for the integration of airborne and vehicle data due to different perspectives, huge differences of point density, different coverage, and the discrete nature of points [33]. Carlberg *et al.* [34] and Hu *et al.* [35] conducted the registration of airborne and vehicle LiDAR points by using GPS data directly. However, Früh and Zakhor [36] believe it is not reliable enough to use GPS data in urban street canyons. As a result, in their method aerial image was selected as a reference and the Monte Carlo Localization method was applied for the registration. Some studies related to the registration of LiDAR data and imagery also give good inspirations for this topic [37,38]. Among the existing literature, auxiliary data is often used to supplement the registration. In view of the availability and precision of these data, it is essential to get rid of these auxiliary data and seek conjugate features for the automatic registration. Based on the above, it is necessary to find reliable matching primitives (e.g., road and wall) from the LiDAR point cloud.

In complex urban environments, the GPS receiver of the vehicle LiDAR platform frequently loses lock, which would reduce the measurement accuracy of the whole system, and cause the drift of the local scanning point cloud. The vehicle LiDAR data used in this study has been corrected through a topographic map of 1:500. Thus, the coordinate of vehicle data has been converted to a local coordinate system, while the initial airborne data is in the WGS-84 coordinate frame. In this study, a hierarchical registration method is proposed, including coarse registration with 3D road networks and fine registration with 3D building contours.

2. Method

Based on the above consideration, a hierarchical registration method (Figure 2), including coarse registration with 3D road networks and fine registration with 3D building contours, is used for the registration of airborne and vehicle LiDAR points.

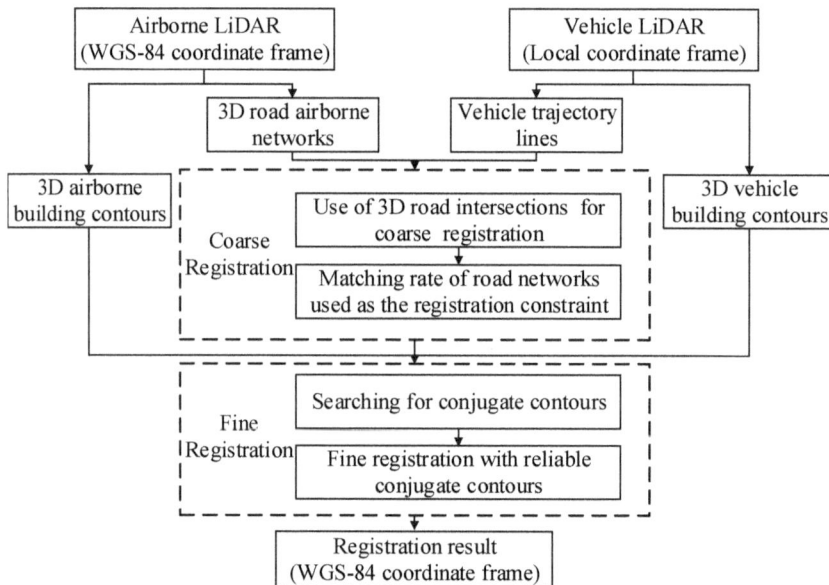

Figure 2. A flowchart of the proposed registration method.

2.1. Coarse Registration with 3D Road Networks

Two sets of 3D road networks, which contain road links and intersections, are used for the coarse registration of large-range LiDAR data. Three-dimensional airborne road networks are extracted using an existing method [39] from airborne LiDAR data. Existing vehicle trajectory lines provided along with vehicle LiDAR points are used as 3D vehicle road networks. Both airborne and vehicle road networks are used for the coarse registration. Considering that the coarse registration just offers a rough registration relationship, road networks with high positioning precision are not necessary.

2.1.1. 3D Road Networks from Airborne LiDAR

A method, proposed in literature [39], for automatic detection and vectorization of roads from airborne LiDAR data, is used in this section. A digital surface model (DSM) is generated by using the morphology algorithm and intensity values are used for the detection of road points. Due to the homogenous and consistent nature of roads, a local point density and a minimum bounding rectangle are introduced to finalize the detection. On this basis, the road centerline is extracted by the convolution of the binary road image with a complex-valued disk named the Phase Coded Disk (PCD). Both the width and direction of the road at the centerline are obtained from the convolution. The direction of the road facilitates the successful vectorization of the magnitude image by using the described tracing algorithm. The vectorization of any classified road networks captured in a binary image can be achieved using the PCD method.

After obtaining 2D airborne road networks, a 3D buffer zone along the networks, forming a cylinder, is constructed. The buffer area is usually several times the average spacing of the point cloud, and the elevations of the points inside the cylinder volume are averaged to obtain 3D road networks.

2.1.2. Coarse Registration with 3D Road Networks

In this section, an intersection point registration method with 3D road networks as constraints, is introduced for the coarse registration. In this method, the 3D intersection points, derived from airborne and vehicle road networks, respectively, are used for calculating the transformation matrices. Additionally, the road networks of airborne and vehicle LiDAR data are regarded as constraints to evaluate the reliability of the matrices.

As shown in Figure 3a,b, the 3D intersection points of airborne and vehicle road networks are $PA = \{PA_i, i = 0,1,2,\ldots,\mu\}$ and $PB = \{PB_i, i = 0,1,2,\ldots,v\}$, with their corresponding road networks being $RA = \{RA_i, i = 0,1,2,\ldots,m\}$ and $RB = \{RB_i, i = 0,1,2,\ldots,n\}$, respectively. The coarse registration of airborne and vehicle LiDAR with road networks is conducted as follows:

(1)   Selecting intersections.

Choose three pairs of points from intersection sets *PA*, *PB* randomly and calculate the translation matrix T_1 and rotation matrix R_1.

(2)   Road network transformation.

Use T_1 and R_1 to convert the road networks *RB*. The converted road network $RC = \{RC_i, i = 0,1,2,\ldots,n\}$ is obtained.

(3)   Calculating the match rate of 3D road networks.

Determine the matching rate of road networks in *RA* and *RC*. In Figure 3c, R_1 and R_2 refer to road segments selected from road networks *RA* and *RC*, respectively. Starting from *Endpoint NA_0* of *Segment R_1*, draw 3D section planes at an interval θ. The section plane is drawn as a circle with its radius set according to the positioning precision of extracted road networks. If precise road networks are provided, a small radius is set and *vice versa*. As for each node point (e.g., NA_0 and NA_1), if it intersects with the other road segments, record the reciprocal of the distance between the foot point and intersection point as the matching rate $pme = \begin{cases} 1/d_i, & d_i > 0.1 \\ 10, & d_i \leq 0.1 \end{cases}$, in which d_i represents the distance of nodes NA_i and NB_i. Otherwise, $pme = 0$. If several roads intersect with a section plane leading to $pme_1, pme_2, \ldots, pme_n$, then the largest matching rate is regarded as its matching rate. As for road segments R_1 and R_2, the matching rate is $lme = \sum_{i=1}^{k} \frac{1}{d_i}$, where k is the number of nodes in R_1.

The final matching rate between two road networks is $tme = \sum_{j=1}^{m} lme_j$, where m is the number of road segments in *RA*.

(4)   Determination of optimal transformation.

Repeat (1)–(3) until the matching rate reaches a certain threshold or all intersection pairs have been iteratively selected from *PA* and *PB*. The transformation matrix with the highest matching rate

is selected as the optimal transformation matrix. All points of *PB* are transformed and recorded as
$PC = \{PC_i, i = 0,1,2,\ldots,v\}$.

(5) Coarse registration.

The successfully matched points from *PA* and *PC* are selected as $MA = \{MA_i, i = 0,1,2,\ldots,m\}$
and $MC = \{MC_i, i = 0,1,2,\ldots,n\}$, respectively. The least mean square (LMS) method is used to calculate
the registration relationships between set *MA* and set *MC*. The acquired registration transformational
matrices *R* and *T* are used to transform the LiDAR points. Finally, the coarse registration is finished.

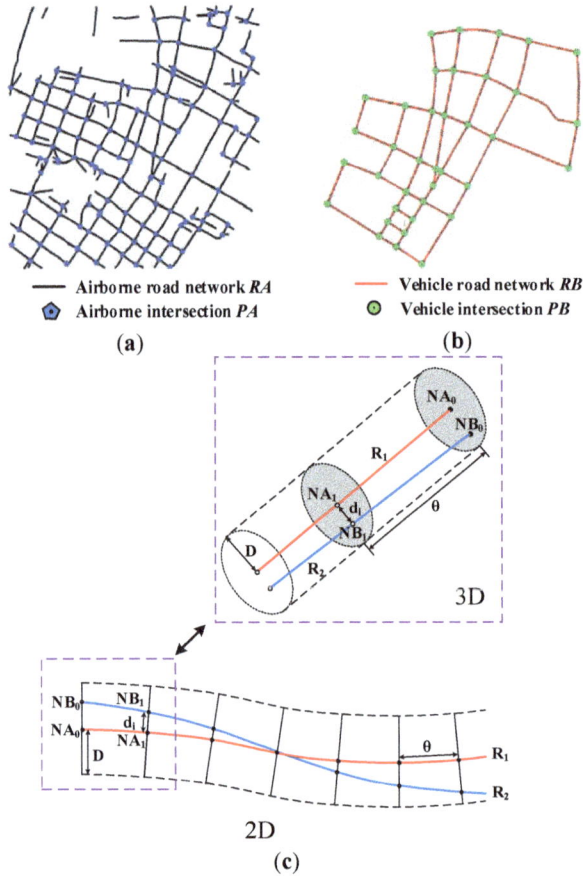

Figure 3. Coarse registration of 3D road networks. (**a**) Airborne road networks and intersections; (**b**)
Vehicle road networks and intersections; (**c**) Matching rate of two single road segments from 3D road
networks.

2.2. Fine Registration with 3D Building Contours

In this section, on the basis of the above coarse registration, a locally fine registration work is
done with 3D building contours. Coarse registration with road networks can provide a rough
registration result of points; nevertheless, due to the positioning precision of extracted road networks,
the coarse registration of point clouds is not so satisfactory. In addition, it is difficult to achieve good
registration results by applying a single registration relationship for large-scale datasets.

2.2.1. 2D Building Contours from Vehicle LiDAR

In this section, the 2D building contours are extracted from vehicle LiDAR data based on two steps, including elevation difference filtering and height value accumulation.

(1) Elevation difference filtering.

Compared with other surface features, obvious elevation differences exist along building contours. Thus, the building contours can be extracted by setting an elevation difference threshold. Here a 2D regular grid is constructed with the point cloud projected on it. The elevation differences of points inside each grid are analyzed and the grids with larger elevation differences than the preset threshold are retained.

(2) Height value accumulation.

Disturbed by the trees and some other objects, the extraction result only from using the elevation difference is not satisfactory. As shown in Figure 4a, building contours of both stage-style and herringbone roofs are in a certain elevation range. Exact building contours can be acquired by obtaining the corresponding elevation range. For this purpose, a further filtering procedure with height value accumulation is proposed.

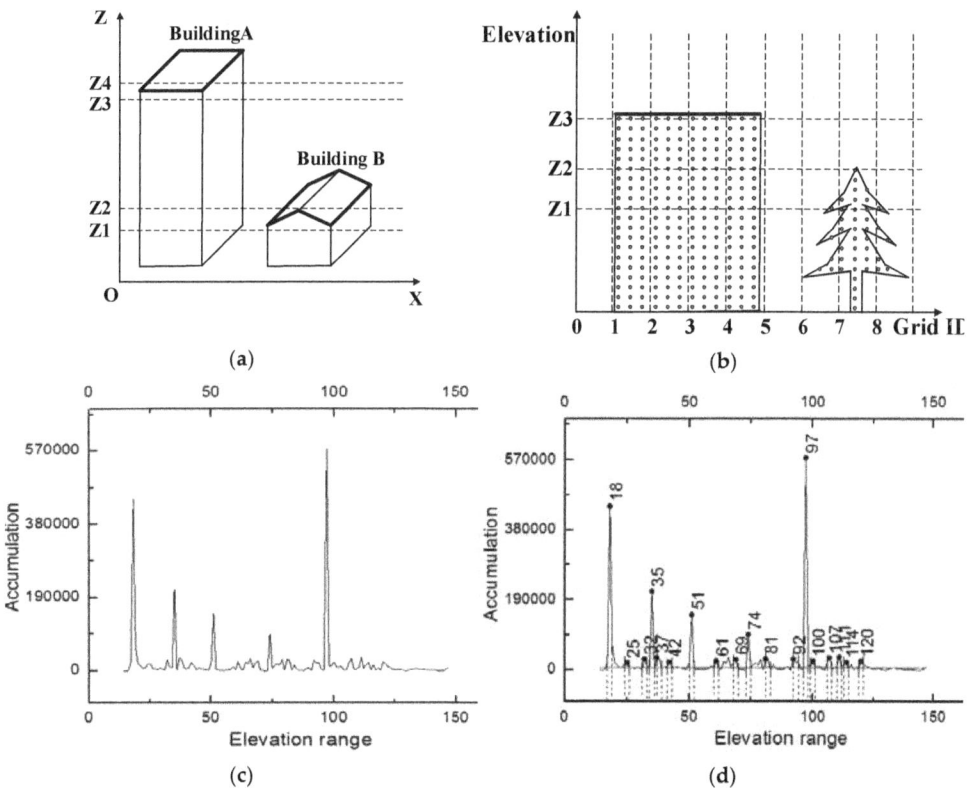

Figure 4. Theory of height value accumulation. (**a**) A sample of building roofs; (**b**) A sample of height value accumulation; (**c**) Height histogram; (**d**) Extraction of elevation range of building contours.

In Figure 4b, LiDAR points of a building facade and a tree are presented. As we can see, the building contour points are divided into four grids and the tree points are divided into three grids. The grids corresponding to the building have consistent highest point (Z3), while that of the tree vary.

More importantly, grids corresponding to buildings obviously have a large number of points. If the point number of grids corresponding to each elevation range is added, the range which contains buildings' highest points will stand out as peaks. According to this, buildings can be extracted from the elevation range. Details are shown as follows:

Step 1: Projecting points to 2D grids.

Construct 2D regular grids and divide the projected points. Then, select the highest point Z_i and calculate the point number N_i of each grid.

Step 2: Calculating the elevation range.

Calculate the elevation range (Z_{min}, Z_{max}) with the highest point and lowest point in the point cloud. Set a small interval as Z_s and divide the elevation range, getting the set $S = \{S_j, j = 1,2,...,n\}$, where

$n = (Z_{max}-Z_{min})/Z_s$.

Step 3: Height value accumulation.

For all the grids, if the highest point Z_i is within the interval S_j; then the accumulation value Acc_j is recalculated as $Acc_j = Acc_j + N_i$, and the height histogram is shown in Figure 4c.

Step 4: Obtaining the elevation range interval of buildings.

The median of the height value accumulation in Figure 4c is set as a threshold, and the data which is smaller than the threshold is eliminated. The partial derivative is calculated, and the peak interval is obtained. Finally, the elevation range interval of the buildings (Figure 4d) is used to extract building contours.

(3)   2D contour extraction.

As for the extracted grids, the points inside them are determined. A random sample consensus (RANSAC) algorithm [40] is used to fit a plane. Additionally, the 2D building contours are obtained through the projection of the plane.

2.2.2. 2D Building Contours from Airborne LiDAR

In this section, the 2D building contours are extracted from airborne LiDAR data by using the RIMM algorithm [9].

Airborne LiDAR points provide rich and abundant top information of buildings, from which reliable architecture area can be extracted. Here, building areas are extracted using the RIMM algorithm. For this method, an opening operation is first conducted with a window larger than all the buildings in a particular area. On this basis, the opening operation is iterated by gradually decreasing the window at a fixed step length. Furthermore, the surface features can be extracted by fitting the size of the corresponding window. The elevation differences between two consecutive iterations are compared, and parts with elevation differences exceeding the minimum building height are regarded as buildings. Binary images are generated using the extracted building areas. Hough transformation is conducted for obtaining the 2D airborne building contours.

2.2.3. Extraction of 3D Building Contours

The aim of this section is to extract the 3D building contours. As building contours are extracted from 2D grids, isolated 3D contours may be extracted as a 2D single contour, as shown in Figure 5a. Thus, a 3D contour extraction method with point elevation is introduced [21]. Details are shown as follows.

(1)   Projection and division of points.

A plane perpendicular to the XY plane is constructed, and point clouds inside the contour grids are projected onto this plane (Figure 5b). The projected points are divided into blocks with a small interval, which is usually slightly larger than the average point spacing.

(2) Points clustering.

The elevation and gradient difference of neighboring blocks are calculated according to the highest point in each block. If the differences are small enough, the neighboring blocks are clustered; otherwise, a new cluster is generated.

(3) 3D contour fitting.

The RANSAC approach is used to fit points of each cluster and 3D building contours are obtained.

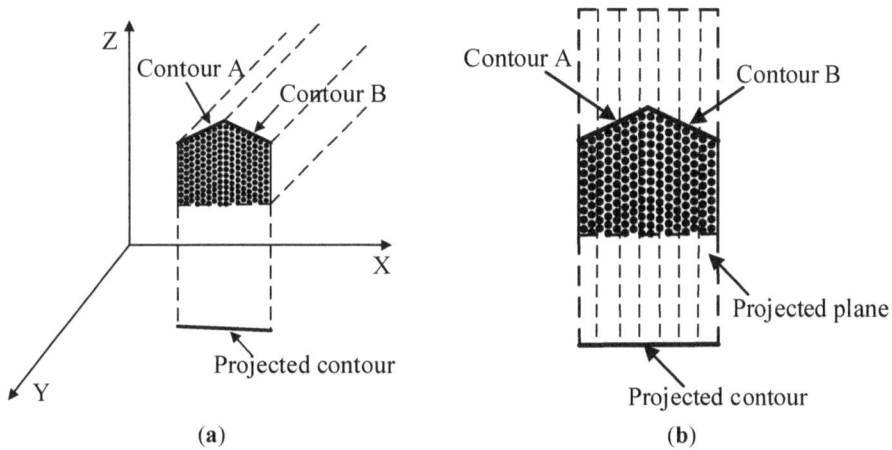

Figure 5. Contour segmentation using point elevation. (**a**) A sample of a projected 2D contour; (**b**) A sample of 3D contour segments.

2.2.4. Fine Registration with 3D Building Contours

Fine registration with 3D building contours consists of two steps: searching for conjugate contours and fine registration with reliable conjugate contours. Firstly, two pairs of conjugate contours are used to acquire a registration relationship of airborne and vehicle contours, and all conjugate contours can be obtained after that. On this basis, a rule is defined to determine reliable conjugate contours, which are finally used for the fine registration of airborne and vehicle LiDAR.

The airborne and vehicle building contours are recorded as $LA = \{LA_i, i = 1,2,\ldots,m\}$ and $LB = \{LB_i, i = 1,2,\ldots,n\}$, respectively. The fine registration with building contours are as follows:

(A) Searching for conjugate contours

(1) *Selection of two pairs of conjugate contours.*

After the coarse registration with 3D road networks, a rough registration relationship between two datasets is obtained. Therefore, further fine registration with building contours should be conducted within a searching space. Randomly select two conjugate contours from LA and LB respectively, recorded as LA_1, LA_2, LB_1, and LB_2. The following conditions should be met: (a)

$$\begin{cases} lAng \leq Thre_{ang} \\ lDist \leq Thre_{dist} \\ lDif \leq Thre_{dif} \end{cases}, \text{ where } lAng = \arccos(\frac{\vec{l_1} \cdot \vec{l_2}}{|\vec{l_1}| \cdot |\vec{l_2}|}) \text{ is the angle between the straight lines in which the}$$

two contours are located, $lDist = \left| \dfrac{(\vec{l_1} \times \vec{l_2}) \cdot \overrightarrow{P_1 P_2}}{|\vec{l_1} \times \vec{l_2}|} \right|$ is the distance between the straight lines in which

the two contours are located, $lDif = ||\vec{l_1}| - |\vec{l_2}||$ is the length difference of the two contours. $\vec{l_1}$ and $\vec{l_2}$ refer to the directions of the two contours. (b) LA_1 and LB_1 are neither parallel nor coplanar, nor are LA_2 and LB_2. (c) The angles and distances between LA_1 and LB_1 equal those between LA_1 and LB_1.

(2) *Calculation of rotation matrix.*

Use two pairs of conjugate segments to calculate the rotation matrix. Here a vector-based transformation model [29] is adopted to calculate the rotation matrix R by using two pairs of conjugate segments. The elements in rotation matrix R are determined by three rotation angles.

Supposing there is an antisymmetric matrix $S = \begin{bmatrix} 0 & -\gamma & -\beta \\ \gamma & 0 & -\alpha \\ \beta & \alpha & 0 \end{bmatrix}$ where α, β, γ are independent.

Rotation matrix R can be expressed using the elements of S as

$$R = \frac{1}{\Delta} \begin{bmatrix} 1 + \alpha^2 - \beta^2 - \gamma^2 & -2\gamma - 2\alpha\beta & 2\alpha\gamma - 2\beta \\ 2\gamma - 2\alpha\beta & 1 - \alpha^2 + \beta^2 - \gamma^2 & -2\alpha - 2\beta\gamma \\ 2\alpha\gamma + 2\beta & 2\alpha - 2\beta\gamma & 1 - \alpha^2 - \beta^2 + \gamma^2 \end{bmatrix} \tag{1}$$

where $\Delta = 1 + \alpha^2 + \beta^2 + \gamma^2$; R is the Rodrigues matrix. The characteristics of the antisymmetry matrix and Rodrigues matrix are:

$$\begin{cases} 1)\ S^T = -S \\ 2)\ R = (I - S)^{-1}(I + S) \end{cases} \tag{2}$$

where I is a 3×3 unit matrix and $RR^T = I$. In this study, spatial vectors are first obtained with the 3D building contours. Assuming that the unit vectors corresponding to two pairs of conjugate contours are v and w, we get equation:

$$v = Rw \tag{3}$$

From (2) of Equation (2), Equation (3) should be modified as follows:

$$L = Sb \tag{4}$$

where $L = v - w$; and $b = v + w$. After S is substituted in Equation (4), the following form can be written:

$$\begin{bmatrix} 0 & -b_z & -b_y \\ -b_z & 0 & b_x \\ b_y & b_x & 0 \end{bmatrix} \begin{bmatrix} \alpha \\ \beta \\ \gamma \end{bmatrix} = \begin{bmatrix} L_x \\ L_y \\ L_z \end{bmatrix} \tag{5}$$

where (b_x, b_y, b_z) are the vector components of b, and (L_x, L_y, L_z) are those of L. Then the parameter vector x can be approximated by using the following linear least squares estimation.

$$x = (M^T M)^{-1} M^T L \tag{6}$$

where M is the coefficient matrix of Equation (5), and rotation matrix R should be calculated by Equations (1) and (6).

(3) *Calculation of translation matrix.*

After obtaining the rotation matrix R, the transition matrix can be obtained through collinearity equations. As shown in Figure 6, ab is a contour in LA, $c'd'$ is the corresponding contour of LB.

Theoretically, ab and $c'd'$ should be collinear, meaning that the actual position of $c'd'$ is cd, in

which $\begin{pmatrix} X_c \\ Y_c \\ Z_c \end{pmatrix} = \begin{pmatrix} X_{c'} \\ Y_{c'} \\ Z_{c'} \end{pmatrix} + \begin{pmatrix} d_x \\ d_y \\ d_z \end{pmatrix}, \begin{pmatrix} X_d \\ Y_d \\ Z_d \end{pmatrix} = \begin{pmatrix} X_{d'} \\ Y_{d'} \\ Z_{d'} \end{pmatrix} + \begin{pmatrix} d_x \\ d_y \\ d_z \end{pmatrix}$. According to the collinearity equation, the formula

$$\frac{X_c - X_a}{X_b - X_a} = \frac{Y_c - Y_a}{Y_b - Y_a} = \frac{Z_c - Z_a}{Z_b - Z_a}$$

can be procured for each pair of contours. After derivation,

$$\frac{X_d - X_a}{X_b - X_a} = \frac{Y_d - Y_a}{Y_b - Y_a} = \frac{Z_d - Z_a}{Z_b - Z_a}$$

$$\begin{cases} (Y_b - Y_a)d_x + (X_a - X_b)d_y = (X_b - X_a)(Y_{c'} - Y_a) - (Y_b - Y_a)(X_{c'} - X_a) \\ (Z_b - Z_a)d_y + (Y_a - Y_b)d_z = (Y_b - Y_a)(Z_{c'} - Z_a) - (Z_b - Z_a)(Y_{c'} - Y_a) \end{cases}$$

$$\begin{cases} (Y_b - Y_a)d_x + (X_a - X_b)d_y = (X_b - X_a)(Y_{d'} - Y_a) - (Y_b - Y_a)(X_{d'} - X_a) \\ (Z_b - Z_a)d_y + (Y_a - Y_b)d_z = (Y_b - Y_a)(Z_{d'} - Z_a) - (Z_b - Z_a)(Y_{d'} - Y_a) \end{cases}$$

is obtained. With two pairs of conjugate

contours, a least square method can be used to solve the equations with d_x, d_y, d_z as the variables.

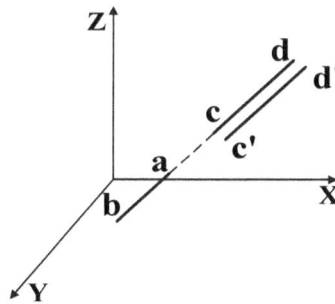

Figure 6. A sample of collinear lines.

(4) *Determination of all conjugate contours.*

Transforming all contours by using the obtained matrix. If the conditions in $\begin{cases} lAng \leq Thre_{ang} \\ lDist \leq Thre_{dist} \\ lDif \leq Thre_{dif} \end{cases}$

are satisfied, the contours are considered as conjugate contours.

(B) Fine registration with reliable conjugate contours

Fine registration with reliable conjugate contours is carried out to obtain a more precise transformation matrix. In the above searching procedure for conjugate contours, only two pairs of conjugate contours are used. Thus, the registration accuracy can be enhanced by using more contours.

(1) *Selection of reliable conjugate contours.*

If the building rooftops have eaves, there may be obvious errors with airborne and vehicle building contours. It is inappropriate to utilize all the contours for the fine registration. Here the angles and distances between conjugate contours are analyzed and the obvious outliers are discarded. Details are as follows:

Step 1: *Selecting contours by angle.*

As for all conjugate contours of LA and LB, calculate their angles as $\alpha = \{\alpha_i, i = 0,1,2,\ldots,k\}$. Group the angles in α and calculate the total number of conjugate contours in each group. If the group with the largest number of contours reaches a certain percent of the total contours, the contours in the group are considered as reliable ones. Otherwise, join other groups of segment pairs until the proportion of matched pairs is more than a certain threshold. Here the percent threshold is set according to the overall precision of extracted building contours.

Step 2: *Selecting contours by distance.*

Just as in Step 1, distances between conjugate contours are also used for selecting the remaining contours.

(2) *Fine registration.*

As for all the reliable conjugate contours, use the vector-based transformation model [29] and collinearity equation for calculating the transformation matrix. The fine transformation matrix with maximum matching rate can be obtained until all reliable conjugate contours have been iteratively selected. Finally, the fine registration of airborne and vehicle LiDAR data can be achieved.

2.3. Summary of Threshold Parameters

As a few parameters are used in this study, a summarization on the setting of the key thresholds is shown in Table 1. The setting basis of these thresholds includes three types: data source, calculation, and empiric. The term "data source" means that a threshold is set according to the real data. The term "calculation" indicates that a threshold can be calculated automatically in the method. If this method is applied in some other experimental data, the "data source" and "calculation" thresholds are easy to automatically determine, which cannot limit the applicability of the proposed method. The term "empiric" means that the thresholds are set empirically.

The calculation formulas are described briefly as follows. In the coarse registration with road networks, the radius of the circles is set to 1 m (*i.e.*, $W_{min}/5$, where $W_{min} = 5$ m is the minimum width of road) for obtaining 3D road networks.

In the extraction of 3D building contours, the size of blocks is one to two times the average point spacing. In this study, we take 100% as the maximum roof slope. The elevation difference of neighboring blocks is set to 2 m (*i.e.*, $2 \times D \times i$, where $D = 1$ m is the block size and $i = 100\%$ is the maximum roof slope). The distance threshold (*Thre_dif*) is set to 5 m according to the width of a lane.

Table 1. Key parameters used in the proposed method.

Method		Parameter	Scale	Setting Basis
Coarse registration with road networks	Extraction of three dimensional (3D) road networks	Radius of small circle	1 m·W/4	Calculation
	Determination of matching rate	Interval of 3D section planes	1 m	Empiric
		The radius of section plane	60 m	Data source
Fine registration with building contours	Extraction of two dimensional (2D) building contours from vehicle LiDAR	2D regular grid	1 m × 1 m	Data source
		Elevation difference	15 m	Data source
		Elevation interval Z_s	4–5 times the average point spacing	Empiric
	Extraction of 3D building contours	Elevation difference	$2 \times D \times I$	Calculation
		Angle difference	20°	Empiric
	Fine registration	Angle threshold *Thre_ang*	5°	Empiric
		Distance threshold *Thre_dist*	5 m (width of a lane)	Calculation
		Length difference *Thre_dif*	10 m	Empiric

3. Experiments and Analysis

3.1. Experimental Data

The experimental area is located around Olympic Sports Center, Nanjing, China (32.0°N, 118.7°E), with a total area of 4000 m × 4000 m. The experimental data includes airborne LiDAR points (Figure 7a), vehicle LiDAR points (Figure 7b), and the trajectory path of vehicle LiDAR (Figure 7c). The SSW (Shou shi/Si wei) mobile mapping system (360° scanning cope, surveying range 2–300 m, reflectance 80%, and point frequency 200,000 points/s) invented by the Chinese Academy of Surveying and Mapping is used to capture the vehicle LiDAR data and a topographic map of 1:500 is used for the correction. The whole data is about 30 G. The total amount of vehicle points is nearly one billion. The average point spacing of airborne LiDAR points is about 0.5 m, with a horizontal precision of 30 cm and vertical precision of 15 cm. The total amount of airborne points is about 78 million. As shown in Figure 7a,b, the whole area is used for the coarse registration and *Area A* is selected for the locally fine registration, and *Area B* is used for evaluating the registration accuracy.

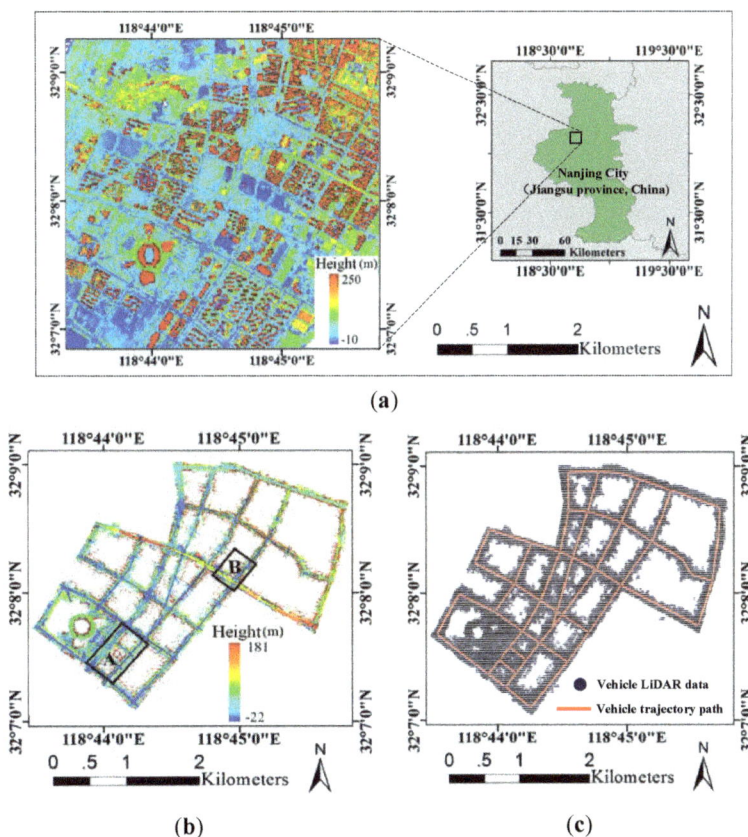

Figure 7. Experimental data. (**a**) Airborne LiDAR data; (**b**) Vehicle LiDAR data; (**c**) Vehicle trajectory path.

3.2. Coarse Registration with 3D Road Networks

Figure 8a shows the extracted 3D airborne road networks using the method mentioned in Section 2.1.1. As we can see, there are 131 road intersections in the airborne road network (black lines).

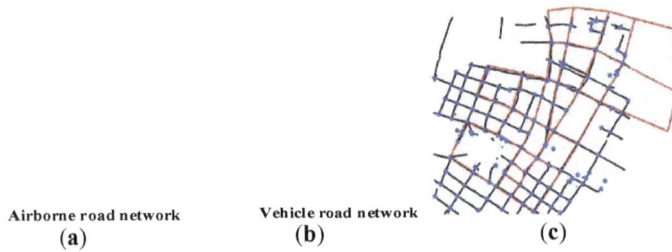

Airborne road network   Vehicle road network
(a)   (b)   (c)

Figure 8. Coarse registration with road networks. (**a**) Airborne road network (black lines); (**b**) Vehicle road network (red lines); (**c**) Registered road networks.

In Figure 8b, 37 road intersections are seen in vehicle trajectory path (red lines). The actual number of conjugate intersections is 30. The coarse registration with 3D road networks is conducted, in which the radius of the section circle is 80 m, and the interval θ is set 1 m. The calculated matching rate is 4012.5 and the coarse registration result is shown in Figure 8c, where 30 pairs of intersections are successfully matched. As we can see, airborne and vehicle road networks are well matched. Little distortion (*i.e.*, geometric scaling) is seen in the matched networks, which presents that little rotation is seen along the Z axis. This coarse registration can offer a rough transformation relationship for the two datasets.

3.3. Fine Registration with 3D Building Contours

3.3.1. Extraction of 3D Building Contours

Figure 9a presents local vehicle LiDAR points around *Area A*. The points are projected to regular grids (1 m × 1 m) and the elevation threshold is set to 15 m for the filtering. In Figure 9b, after elevation filtering, most of the building contours are retained with other non-building grids eliminated. Nevertheless, as we can see along certain building contours (e.g., *SA*, *SB* and *SC*), the contour grids are of large width, which may lead to the imprecision of vectorized contours. The proposed high value accumulation is used for a further filtering. In Figure 9c, 18 peaks are acquired, corresponding to 18 elevation ranges. Using these elevation ranges, more precise contours are obtained (Figure 9d). Compared with Figure 9b, the width of contour grids is obviously lessened, especially in *Area SA, SB* and *SC*. Finally, with Hough transform, 42 building contours are obtained from vehicle LiDAR data.

As for airborne LiDAR in Figure 9e, the RIMM algorithm is used to extract the buildings. The thresholds are set as follows: the size of original morphological window, 106 m; the size of window reduction in each step, 10 m; height difference, 3 m; roughness, 1.6. The extracted building areas are shown in Figure 9f. Through the extraction of 3D building contours, 562 building contours are obtained (Figure 9g).

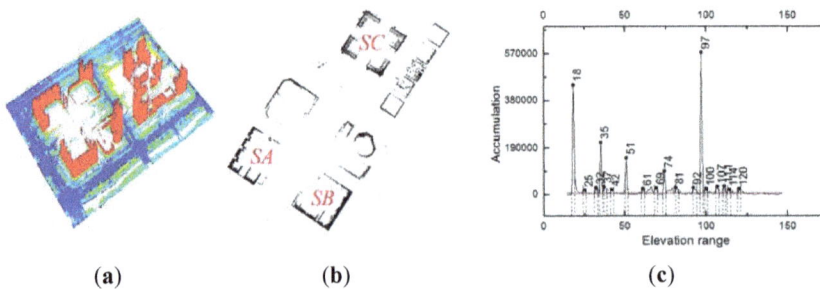

(a)   (b)   (c)

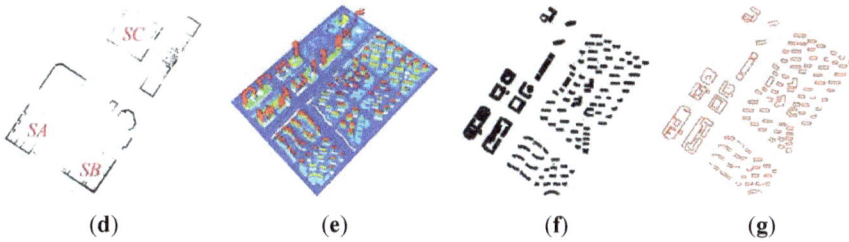

Figure 9. Building contour extraction of *Area A*. (**a**) Vehicle LiDAR data; (**b**) Elevation difference filtering; (**c**) Height value accumulation; (**d**) Vehicle building contours; (**e**) Airborne LiDAR data; (**f**) RIMM for buildings; (**g**) Airborne building contours.

3.3.2. Fine Registration with 3D Building Contours

Figure 10 is the fine registration of airborne and vehicle LiDAR around *Area A*. Figure 10a shows the result after coarse registration, in which the black segments refer to airborne contours and red segments refer to vehicle contours. After searching the conjugate building contours, Figure 10b is obtained, where 15 pairs of building contours are successfully matched. A further fine registration is conducted in Figure 10c. Compared with Figure 10b, almost all the building contours match well, especially in *FA*, *FB* and *FC*.

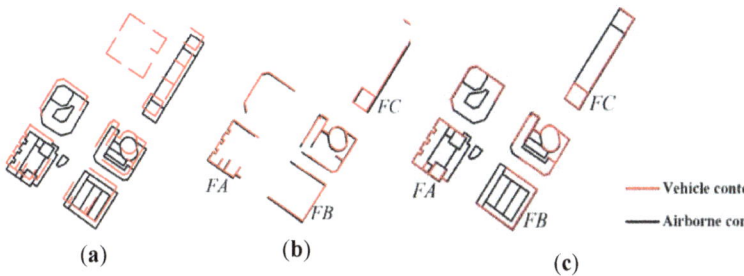

Figure 10. Fine registration with building contours in *Area A*. (**a**) Result of coarse registration; (**b**) Result after using the conjugate contours; (**c**) Result of fine registration by only using conjugate contours in (**b**), and all airborne contours (black lines) are retained for visualization.

3.4. Result and Analysis

3.4.1. Visual Evaluation

Figure 11 shows the final registration result around *Area A*, where the blue points refer to airborne LiDAR points and green ones refer to vehicle LiDAR points.

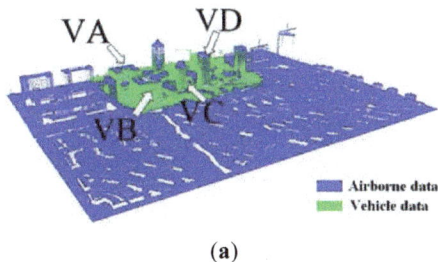

(**a**)

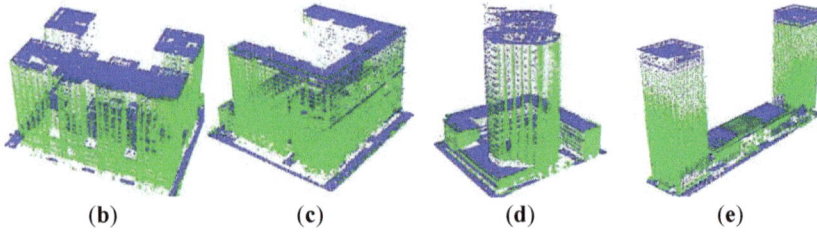

Figure 11. Registration result of *Area A* using the proposed method. (**a**) Registration result in *Area A*; (**b**) Details of *VA*; (**c**) Details of *VB*; (**d**) Details of *VC*; (**e**) Details of *VD*.

The overall registration result of airborne and vehicle points is presented in Figure 11a. Figure 11b–e are the four details of *VA*, *VB*, *VC* and *VD* in Figure 11a, in which the green points and blue points match well along the building contours. With these integrated airborne and vehicle points, a multi-view all-round information of surface features is obtained. Further quantitative evaluations are conducted with building contours and ground points for the horizontal precision and vertical precision, respectively.

3.4.2. Evaluation on Horizontal Accuracy with Building Contours

An evaluation by using building contours is conducted to evaluate the horizontal accuracy of the registration result. Here the building contours are extracted by manual means from registered airborne and vehicle LiDAR, respectively. To make a quantitative assessment of these registration results, the transect distance and angle between the contours are calculated. The calculation method of transect distance is described as follows. One contour is taken as a baseline, and lines perpendicular to this baseline are then constructed with an interval. Thus, the distance between the baseline and corresponding contour is a transect distance. The average transect distance between two conjugate contours is the distance between two contours.

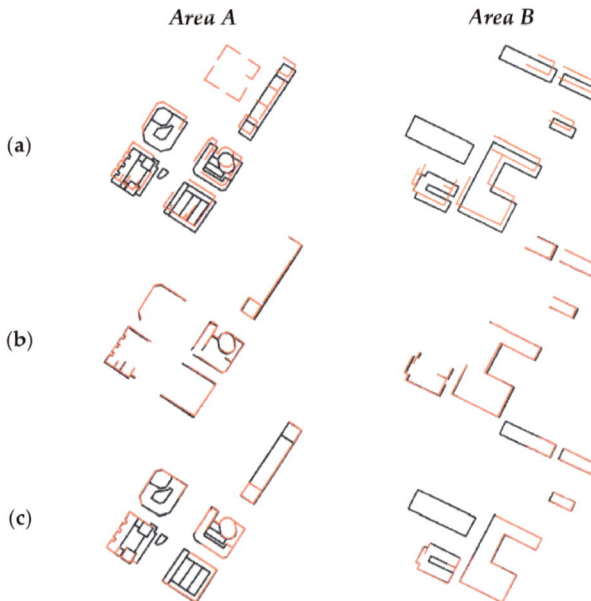

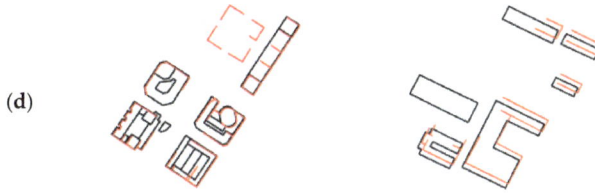

(d)

Figure 12. Evaluation on horizontal accuracy using building contours of *Area A* (left) and *Area B* (right). (**a**) Result of coarse registration; (**b**) Result by using the conjugate contours; (**c**) Result of fine registration (all airborne contours are retained for visualization); (**d**) ICP refined result.

Figure 12a–d are the digitized building contours in *Area A* (left) and *Area B* (right), in which black and red segments are building contours extracted from airborne and vehicle LiDAR data, respectively. Figure 12a shows the digitized building contours after coarse registration with 3D road networks, where obvious offset is seen. As shown in Table 2, the average and maximum distance of *Area A* is 17.44 m and 21.03 m, respectively, with the average and maximum angle offset being 0.95° and 1.7°. The average and maximum distance of *Area B* is 7.43 m and 12.68 m, respectively, with the average and maximum angle offset being 0.88° and 2.1°. Figure 12b shows the registration result after using two pairs of building contours. As we can see, the digitized contours match well and only small offsets are seen among the contours. Through statistics, the average and maximum distance of *Area A* is 2.53 m and 4.30 m, respectively, with the average and maximum angle offset being 0.82° and 1.6°. The average and maximum distance of *Area B* is 1.59 m and 3.79 m, respectively, with the average and maximum angle offset being 0.69° and 1.3°. The horizontal registration accuracy has been greatly improved by using building contours. Figure 12c is the final registration result with reliable conjugate building contours. The building contours digitized from airborne and vehicle LiDAR data match even better than Figure 12b. In Table 2, the average and maximum distance of *Area A* is 0.73 m and 1.90 m, respectively, with the average and maximum angle offset being 0.32° and 1.2°. The average and maximum distance of *Area B* is 0.63 m and 1.73 m, respectively, with the average and maximum angle offset being 0.48° and 1.1°. Figure 12d is the registration result by directly using the ICP method after the coarse registration. The average and maximum distance of *Area A* is 1.52 m and 2.55 m, respectively, with the average and maximum angle offset being 0.47° and 1.5°. However, the average and maximum distance of *Area B* is 5.27 m and 11.23 m, respectively, with the average and maximum angle offset being 0.72° and 1.9°.

Table 2. Horizontal accuracy of the registration result.

Method	Transect Distance (m)				Line Angle (°)			
	Average		Max		Average		Max	
	A	*B*	*A*	*B*	*A*	*B*	*A*	*B*
Coarse registration	17.44	7.43	21.03	12.68	0.95	0.88	1.7	2.1
Searching result	2.53	1.59	4.30	3.79	0.82	0.69	1.6	1.3
Fine registration	0.73	0.63	1.90	1.73	0.32	0.48	1.2	1.1
ICP refined result	1.52	5.27	2.55	11.23	0.47	0.72	1.5	1.9

According to the above, we can see that the fine registration horizontal accuracy reaches the meter level, which is relatively high. Compared with the registration result using only two pairs of building contours, the fine registration with reliable conjugate building contours can achieve better results. For horizontal accuracy, the fine registration results are better than the ICP results both in *Area A* and *B*.

3.4.3. Evaluation on Vertical Accuracy with Common Ground Points

In order to evaluate the registration accuracy along the vertical direction, some common ground points are selected. As shown in Figure 13, the ground points (white circles) are evenly distributed in *Area A* and *Area B*, respectively. The elevation of each ground point is obtained by averaging the *Z* values of its neighboring points. The elevation differences between airborne and vehicle ground points are shown in Table 3.

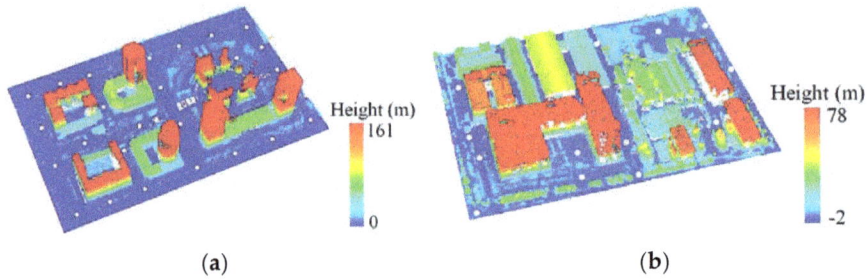

Figure 13. Location of common ground points. (**a**) Common ground points of *Area A*; (**b**) Common ground points of *Area B*.

After coarse registration with road networks, the Average error, Max error, and RMSE error of common ground points in *Area A* are 0.92 m, 1.17 m, and 0.97 m, respectively. The Average error, Max error, and RMSE error of common ground points in *Area B* are 1.08 m, 1.33 m, and 0.84 m, respectively. As we can see, in spite of dozens of meters of errors along the horizontal direction, the errors along the vertical direction reached the decimeter level after coarse registration. The reason lies in the fact that the experimental area is flat in general, which leads to similar elevations of road intersections. If the experiment is carried out in areas with undulating terrains, the errors will be larger. Though relatively high registration accuracy has been achieved with road networks, the registration is further improved after fine registration using building contours as shown in Table 3. After searching the conjugate contours, the Average error, Max error, and RMSE error in *Area A* are 0.46 m, 0.63 m, and 0.50 m. In *Area B*, the Average error, Max error, and RMSE error of searching result are 0.59 m, 0.92 m, and 0.68 m. Based on this, after fine registration, the Average error, Max error, and RMSE error in *Area A* are 0.39 m, 0.50 m, and 0.42 m, respectively. The Average error, Max error, and RMSE error in *Area B* are 0.43 m, 0.75 m, and 0.36 m, respectively. The ICP result is performed based on the points of *Area A* and *B* after coarse registration. In *Area A*, the Average error, Max error, and RMSE error of the ICP result are 0.37 m, 0.61 m, and 0.28 m, respectively. In *Area B*, the Average error, Max error, and RMSE error of the ICP result are 0.46 m, 0.72 m, and 0.21 m, respectively.

As we can see, high registration accuracy is obtained using the proposed method. However, the vertical accuracies of ICP registration results are better than the fine registration results in *Area A* and *Area B*.

Table 3. Vertical accuracy of the registration result.

Method	Average Error (m)	Max Error (m)	RMSE (m)
	A/B	A/B	A/B
Coarse registration	0.92/1.08	1.17/1.33	0.97/0.84
Searching result	0.46/0.59	0.63/0.92	0.50/0.68
Fine registration	0.39/0.43	0.50/0.75	0.42/0.36
ICP result	0.37/0.46	0.61/0.72	0.28/0.21

4. Discussion

The registration accuracy of the airborne and vehicle LiDAR point cloud not only depends on the registration algorithm, but also relies on the precision of the laser scanning system and the scope of the experimental area. The whole experimental area in the study is about 4000 m × 4000 m and the average point spacing of the airborne LiDAR data is about 0.50 m. In this circumstance, it is difficult to achieve extremely high registration accuracy for a large scene dataset with the same parameters. In study [32], the registration results of airborne and terrestrial LiDAR data reached a horizontal accuracy of 0.15 to 0.5 m and a vertical accuracy of 0.20 m. Thus, a hierarchical registration strategy, which contains coarse and fine registration procedures, is used to meet the needs of different registration accuracies.

The registration result in Figure 11 and the accuracy evaluation in Tables 2 and 3 demonstrate the relatively good performance of the proposed method in an urban area. For comparison, the ICP registration method is performed over two small experimental areas (*Area A* and *B*). The ICP method is conducted based on the coarse registered point clouds with a good initial state. In addition, due to the massive LiDAR data (about 30 G), the ICP method could not be directly applied to the whole area. The horizontal accuracy of the fine registration results in *Area A* and *B* are 0.39 m and 0.43 m, respectively. The horizontal accuracy of the ICP results in *Area A* and *B* are 1.42 m and 5.27 m, respectively. According to the experimental results, the horizontal accuracy of the proposed method is better than the ICP method. However, the vertical accuracies of the ICP registration result are slightly better than the fine registration result both in *Area A* and *B*. The ICP method is performed mainly based on the common ground points of airborne and vehicle LiDAR data, which is prone to lead a horizontal shift. Thus, the ICP method would achieve a registration result with good vertical accuracy and unstable horizontal accuracy.

Although the proposed registration scheme can obtain good results in the urban area, there are some limitations. Firstly, the proposed registration method is hard to deal with in non-urban regions, which contain almost no geometric buildings. Besides, if the road networks of a city are following a regular grid, the coarse registration would not work due to the strong similarity of road networks. In addition, if most of the buildings in the experimental area have significant eaves, there may be obvious errors between airborne and vehicle building contours. As a result, few conjugate building contours can be extracted, and it would be difficult to achieve the ideal registration result with fine registration.

5. Conclusions

This paper proposes a hierarchical registration approach for airborne and vehicle LiDAR data. The keys of this approach lie in a coarse registration method with 3D road networks and a fine registration method with 3D building contours. In the coarse registration procedure, the extracted airborne road networks are registered with vehicle trajectory lines based on the road network matching rate. In the fine registration procedure, the matched conjugate contours are obtained using the Rodriguez matrix and collinearity equation. Finally, the fine registration is conducted by using reliable conjugate contours. Through the experiment, the following conclusions can be reached:

(1) The coarse registration method with 3D road networks can provide a rough transformation matrix for a long-range registration task. With the coarse registration, further fine registration can be done within a small searching space, thus effectively avoiding local optimal result and greatly reducing the calculation amount.

(2) Three-dimensional building contours present high positioning precision and the fine registration method with 3D building contours can achieve a relatively good registration result. The fine registration result achieves an accuracy of 0.73 m in the horizontal direction and 0.39 m in the vertical direction.

In future work, the proposed method will be applied to different experimental areas to test its robustness and effectiveness. However, it is still difficult to achieve ideal results for the registration of those buildings with eaves. Therefore, we will perform a future study on eaves by using the center points of building roofs or other auxiliary data.

Acknowledgments: This work is supported by the National Natural Science Foundation of China (Grant No. 41501456, 41371017). Sincere thanks are given for the comments and contributions of the anonymous reviewers and members of the editorial team.

Remote Sens. **2015**, *7*, 13921–13944

Author Contributions: Liang Cheng proposes the core idea of hierarchical registration method and designs the technical framework of the paper. Yang Wu implements the registration method of airborne and vehicle LiDAR data. Lihua Tong implements the method of extraction of road networks from airborne and terrestrial LiDAR data. Yanming Chen implements the method of extraction of building contours, and is responsible for the implementation of the comparative experiments. Manchun Li is responsible for experimental analysis and discussion.

Conflicts of Interest: The authors declare no conflict of interest.

References

1. Wu, B.; Yu, B.; Yue, W.; Shu, S.; Tan, W.; Hu, C.; Huang, Y.; Wu, J.; Liu, H. A voxel-based method for automated identification and morphological parameters estimation of individual street trees from mobile laser scanning data. *Remote Sens.* **2013**, *5*, 584–611.
2. Meng, X.; Currit, N.; Zhao, K. Ground filtering algorithms for airborne LiDAR data: A review of critical issues. *Remote Sens.* **2010**, *2*, 833–860.
3. Cheng, L.; Gong, J.; Li, M.; Liu, Y. 3D building model reconstruction from multi-view aerial imagery and LiDAR data. *Photogramm. Eng. Remote Sens.* **2011**, *77*, 125–139.
4. Cheng, L.; Wu, Y.; Wang, Y.; Zhong, L.; Chen, Y.; Li, M. Three-dimensional reconstruction of large multilayer interchange bridge using airborne LiDAR data. *IEEE J. Sel. Top. Appl. Earth Obs. Remote Sens.* **2014**, *8*, 691–707.
5. Kent, R.; Lindsell, J.A.; Laurin, G.V.; Valentini, R.; Coomes, D.A. Airborne LiDAR detects selectively logged tropical forest even in an advanced stage of recovery. *Remote Sens.* **2015**, *7*, 8348–8367.
6. Cheng, L.; Tong, L.; Wang, Y.; Li, M. Extraction of urban power lines from vehicle-borne LiDAR data. *Remote Sens.* **2014**, *6*, 3302–3320.
7. Lehtomäki, M.; Jaakkola, A.; Hyyppä, J.; Kukko, A.; Kaartinen, H. Detection of vertical pole-like objects in a road environment using vehicle-based laser scanning data. *Remote Sens.* **2010**, *2*, 641–664.
8. Yang, B.; Wei, Z.; Li, Q.; Li, J. Automated extraction of street-scene objects from mobile LiDAR point clouds. *Int. J. Remote Sens.* **2012**, *33*, 5839–5861.
9. Cheng, L.; Zhao, W.; Han, P.; Zhang, W.; Shan, J.; Liu, Y.; Li, M. Building region derivation from LiDAR data using a reversed iterative mathematic morphological algorithm. *Opt. Commun.* **2012**, *286*, 244–250.
10. Weinmann, M.; Weinmann, M.; Hinz, S.; Jutzi, B. Fast and automatic image-based registration of TLS data. *ISPRS-J. Photogramm. Remote Sens.* **2011**, *66*, S62–S70.
11. Böhm, J.; Haala, N. Efficient integration of aerial and terrestrial laser data for virtual city modeling using lasermaps. In Proceedings of ISPRS WG III/3, III/4, V/3 Workshop on Laser Scanning, Enschede, the Netherlands, 12–14 September 2005; pp. 12–14.
12. Hohenthal, J.; Alho, P.; Hyyppä, J.; Hyyppä, H. Laser scanning applications in fluvial studies. *Prog. Phys. Geogr.* **2011**, *35*, 782–809.
13. Bremer, M.; Sass, O. Combining airborne and terrestrial laser scanning for quantifying erosion and deposition by a debris flow event. *Geomorphology* **2012**, *138*, 49–60.
14. Heckman, T.; Bimböse, M.; Krautblatter, M.; Haas, F.; Becht, M.; Morche, D. From geotechnical analysis to quantification and modelling using LiDAR data: A study on rockfall in the Reintal catchment, Bavarian Alps, Germany. *Earth Surf. Process. Landf.* **2012**, *37*, 119–133.
15. Besl, P.J.; McKay, N.D. A method for registration of 3-D shapes. *IEEE Trans. Pattern Anal. Mach. Intell.* **1992**, *14*, 239–256.
16. Chen, H.; Bhanu, B. 3D free-form object recognition in range images using local surface patches. *Pattern Recognit. Lett.* **2007**, *28*, 1252–1262.
17. He, B.; Lin, Z.; Li, Y.F. An automatic registration algorithm for the scattered point clouds based on the curvature feature. *Opt. Laser Technol.* **2012**, *46*, 53–60.
18. Makadia, A.; Patterson, A.I. Daniilidis, K. Fully automatic registration of 3D point clouds. In Proceedings of IEEE Computer Society Conference on Computer Vision and Pattern Recognition, New York, NY, USA, 17–22 June 2006; pp. 1297–1304.
19. Bae, K.; Lichti, D. A method for automated registration of unorganized point clouds. *ISPRS J. Photogramm. Remote Sens.* **2008**, *63*, 36–54.
20. Barnea, S.; Filin, S. Keypoint based autonomous registration of terrestrial laser point-clouds. *ISPRS-J. Photogramm. Remote Sens.* **2008**, *63*, 19–35.

21. Cheng, L.; Tong, L.; Li, M.; Liu, Y. Semi-automatic registration of airborne and terrestrial laser scanning data using building corner matching with boundaries as reliability check. *Remote Sens.* **2013**, *5*, 6260–6283.

22. Cheng, L.; Tong, L.; Wu, Y.; Chen, Y.; Li, M. Shiftable leading point method for high accuracy registration of airborne and terrestrial LiDAR data. *Remote Sens.* **2015**, *7*, 1915–1936.

23. Eo, Y.D.; Pyeon, M.W.; Kim, S.W.; Kim, J.R. Coregistration of terrestrial LiDAR points by adaptive scale-invariant feature transformation with constrained geometry. *Autom. Constr.* **2012**, *25*, 49–58.

24. Jaw, J.J.; Chuang, T.Y. Registration of ground-based LiDAR point clouds by means of 3D line features. *J. Chin. Inst. Eng.* **2008**, *31*, 1031–1045.

25. Lee, J.; Yu, K.; Kim, Y.; Habib, A.F. Adjustment of discrepancies between LiDAR data strips using linear features. *IEEE Geosci. Remote Sens. Lett.* **2007**, *4*, 475–479.

26. Bucksch, A.; Khoshelham, K. Localized registration of point clouds of botanic trees. *IEEE Geosci. Remote Sens. Lett.* **2013**, *10*, 631–635.

27. Teo, T.A.; Huang, S.H. Surface-based registration of airborne and terrestrial mobile LiDAR point clouds. *Remote Sens.* **2014**, *6*, 12686–12707.

28. Von Hansen, W. Robust automatic marker-free registration of terrestrial scan data. In Proceedings of the Photogrammetric Computer Vision, Bonn, Germany, 20–22 September 2006; pp. 105–110.

29. Zhang, D.; Huang, T.; Li, G.; Jiang, M. Robust algorithm for registration of building point clouds using planar patches. *J. Sur. Eng.* **2011**, *138*, 31–36.

30. Brenner, C.; Dold, C.; Ripperda, N. Coarse orientation of terrestrial laser scans in urban environments. *ISPRS-J. Photogramm. Remote Sens.* **2008**, *63*, 4–18.

31. Biber, P.; Straßer, W. The normal distributions transform: A new approach to laser scan matching. In Proceedings of the 2003 IEEE/RSJ International Conference on Intelligent Robots and Systems, Las Vegas, NV, USA, 27–31 October 2003; pp. 2743–2748.

32. Wu, H.; Scaioni, M.; Li, H.; Li, N.; Lu, M.; Liu, C. Feature-constrained registration of building point clouds acquired by terrestrial and airborne laser scanners. *Int. J. Appl. Remote Sens.* **2014**, *8*, 083587.

33. Hansen, W.; Gross, H.; Thoennessen, U. Line-based registration of terrestrial and aerial LiDAR data. In Proceedings of International Archives of the Photogrammetry, Remote sensing, and Spatial Information Sciences, Beijing, China, 3–11 July 2008; pp. 161–166.

34. Carlberg, M.; Andrews, J.; Gao, P.; Zakhor, A. Fast surface reconstruction and segmentation with ground-based and airborne LiDAR range data. In Proceedings of the Fourth International Symposium on 3D Data Processing, Visualization and Transmission (3DPVT), Atlanta, GA, USA, 18–20 June 2008.

35. Hu, J.; You, S.; Neumann, U. Approaches to large-scale urban modelling. *IEEE Comput. Graph. Appl.* **2003**, *23*, 62–69.

36. Früh, C.; Zakhor, A. Constructing 3D city models by merging aerial and ground views. *IEEE Comput. Graph. Appl.* **2003**, *23*, 52–61.

37. Li, N.; Huang, X.; Zhang, F.; Wang, L. Registration of aerial imagery and LiDAR data in desert area using the centroids of bushes as control information. *Photogramm. Eng. Remote Sens.* **2013**, *79*, 743–752.

38. Parmehr, E.G.; Fraser, C.S.; Zhang, C.; Leach, J. Automatic registration of optical imagery with 3D LiDAR data using statistical similarity. *ISPRS-J. Photogramm. Remote Sens.* **2014**, *88*, 28–40.

39. Clode, S.; Rottensteiner, F.; Kootsookos, P.J.; Zelniker, E.E. Detection and vectorisation of roads from LiDAR data. *Photogramm. Eng. Remote Sens.* **2007**, *73*, 517–535.

40. Fischler, M.A.; Bolles, R.C. Random sample consensus: A paradigm for model fitting with applications to image analysis and automated cartography. *Commun. ACM* **1981**, *24*, 381–395.

remote sensing

MDPI

Article

Surface-Based Registration of Airborne and Terrestrial Mobile LiDAR Point Clouds

Tee-Ann Teo * and Shih-Han Huang

Department of Civil Engineering, National Chiao Tung University, Hsinchu 30010, Taiwan;
E-Mail: tateo@mail.nctu.edu.tw
* Author to whom correspondence should be addressed; E-Mail: tateo@mail.nctu.edu.tw;
  Tel.: +886-3-571-2121 (ext. 54929); Fax: +886-3-571-6257.

External Editors: Gonzalo Pajares Martinsanz and Prasad S. Thenkabail
Received: 9 October 2014; in revised form: 8 December 2014 / Accepted: 15 December 2014 / Published: 17 December 2014

Abstract: Light Detection and Ranging (LiDAR) is an active sensor that can effectively acquire a large number of three-dimensional (3-D) points. LiDAR systems can be equipped on different platforms for different applications, but to integrate the data, point cloud registration is needed to improve geometric consistency. The registration of airborne and terrestrial mobile LiDAR is a challenging task because the point densities and scanning directions differ. We proposed a scheme for the registration of airborne and terrestrial mobile LiDAR using the least squares 3-D surface registration technique to minimize the surfaces between two datasets. To analyze the effect of point density in registration, the simulation data simulated different conditions and estimated the theoretical errors. The test data were the point clouds of the airborne LiDAR system (ALS) and the mobile LiDAR system (MLS), which were acquired by Optech ALTM 3070 and Lynx, respectively. The resulting simulation analysis indicated that the accuracy of registration improved as the density increased. For the test dataset, the registration error of mobile LiDAR between different trajectories improved from 40 cm to 4 cm, and the registration error between ALS and MLS improved from 84 cm to 4 cm. These results indicate that the proposed methods can obtain 5 cm accuracy between ALS and MLS.

Keywords: LiDAR; point clouds; least squares surface matching; registration

1. Introduction

Light detection and ranging (LiDAR) systems are currently common tools to acquire three-dimensional (3-D) surface information. This technology integrates a laser scanner, a Global Positioning System (GPS), and an inertial navigation system (INS) and thus can effectively obtain 3-D surface models. Different platforms, such as aircraft and land-based vehicles can be equipped with LiDAR systems, which can be generally classified into two categories: airborne and terrestrial. Airborne LiDAR acquires data from the air to the ground to obtain the 3-D points on building rooftops and object surfaces, while terrestrial LiDAR usually acquires the 3-D points on building façades and object surfaces. Because terrestrial LiDAR cannot easily acquire 3-D points from building roofs, airborne LiDAR can be incorporated to provide building roof information. Hence, the integration of airborne LiDAR and terrestrial LiDAR is needed to form a complete dataset for 3-D buildings.

Point cloud registration is a procedure to eliminate the inconsistency between different point clouds acquired from different platforms. Point cloud data acquired by different platforms have different characteristics according to scanning distance, scanning rate, and scanning direction. For example, the scanning distance and beam divergence angle of airborne LiDAR is larger than ground-based LiDAR and, consequently, the point density of airborne LiDAR is lower than ground-based LiDAR. In addition, the scan direction of airborne LiDAR and mobile LiDAR are different, and the acquired 3-D points partially overlap on the object surface. Because the scanning range of airborne LiDAR is longer than ground-based LiDAR, the scanning area of airborne LiDAR is usually larger than ground-based LiDAR. Hence, the registration of airborne and mobile LiDAR is a challenging topic in data co-registration.

Data registration is a procedure to transform a dataset from its own coordinate system to another system. It can be classified into 2-D data and 3-D data registration. For example, image registration is the most common 2-D data registration, and 3-D point cloud registration is one of the 3-D data registrations. The 3-D data registration includes three types of control features: control point, control line, and control surface.

The control points represent a set of 3-D point features in different datasets. This feature is widely used in registration because the control point is the basic control feature. Iterative Closest Point (ICP) [1–4] is acquired through point features. The ICP algorithm selects the two closest points as a conjugate pair and then calculates the transformation parameters to minimize the mean square error iteratively until the distance between the point pair is less than the threshold. Rusinkiewicz and Levoy [5] analyzed the original ICP and improved the performance and precision; the new ICP can register more complex models. The registration precision indicates the geometrical difference between two systems. Barnea and Filin [6] transferred the 3-D point clouds to 2-D panoramic range images and extracted the registration key points to improve the computational time of conjugate points selection.

The second control feature, control line, is a linear feature consisting of a set of 3-D line features in different datasets. This type of line feature mainly occurs in man-made objects such as buildings. Linear features cannot be extracted directly from a LiDAR point cloud and are usually intersected by two planes. The reliable linear features can be used as control entities to calculate the transformation parameters. Habib et al. [7] used line features to register LiDAR point clouds and image data. For the image, the control lines were extracted manually; for the LiDAR point cloud, the line features were intersected by two near planes. Jaw and Chuang [8] also proposed a line-based method to register terrestrial LiDAR point cloud scanned from different stations.

The third control feature, control surface, is suitable for LiDAR registration because the LiDAR systems provide an abundance of 3-D surface features. Rosenholm and Torlegard [9] used digital elevation models (DEM) as reference data in absolute orientation of the stereo model from stereopairs. Gruen and Akca [10] used least squares 3-D surface matching (LS3D) to minimize the 3-D distance between the reference data and model data. Akca [11] also used LS3D to register point clouds by their geometry and spectrum characteristics. This LS3D method has been applied to many applications, such as surface registration for land deformation [12].

Multi-strips or multi-stations LiDAR registration is a standard process before delivering LiDAR data [13]. LiDAR systems are available on several different platforms, such as airborne LiDAR systems (ALS), terrestrial static LiDAR systems, and terrestrial mobile LiDAR systems. In this study, the terrestrial LiDAR system (TLS) and mobile LiDAR system (MLS) refer to terrestrial static and mobile LiDAR systems, respectively. The registration of LiDAR data can be classified into four categories: registration of multi-strip ALS, registration of multi-station TLS, registration of multi-strip MLS, and registration of different platforms.

Multi-strip ALS not only enlarges acquisition areas but also improves the point density in the overlapped area. The registration of ALS includes two mathematics models: system-driven models and data-driven methods [14]. The system-driven approach considers the physical sensor model of ALS and usually requires the trajectories of ALS. In the contrast, the data-driven approach does not require physical orientation parameters; it minimizes the Euclidean distance and models the discrepancy between strips using actual LiDAR points. The geometric features of ALS registration can be a signalized target, control line, or control surface. To avoid the effect of irregular points caused by trees, one possibility is to use ground points to calculate the transformation coefficients. In addition, LiDAR intensity can be also be integrated in registration.

The registration of multi-stations TLS combines the partially scanned objects to obtain a complete scene. Because the platform of TLS is fixed during scanning, the point clouds of TLS are treated as a rigid body, and the 3-D similarity transformation model is usually adopted in the registration of TLS. The registration of TLS can be classified into two categories: Range-based registration and image-based registration. Similar to ALS, range-based registration uses 3-D points to extract geometric features, including signalized target and non-signalized natural targets. Signalized targets such as spherical targets are suitable for an area without man-made objects, and non-signalized natural targets such as control lines and control planes can be extracted from man-made objects. The image-based registration interpolates the 3-D points into a panorama image using the LiDAR intensity. The feature points can then be extracted for registration by image processing techniques such as Scale-Invariant Feature Transform (SIFT) and Speeded Up Robust Features (SURF).

The mobile terrestrial LiDAR systems collect and perform mapping from a moving vehicle on a road. The aim of MLS registration is to register the LiDAR points from different trajectories; therefore, to obtain larger street sections, MLS usually acquires data from direct and reverse lanes using a scanning mechanism similar to ALS. MLS uses an odometer, GPS, and INS to determine the position and orientation of LiDAR sensors for direct georeferencing; however, the GPS condition in urban corridors usually affects the positioning accuracy of MLS. To compare the contribution of GPS positioning error to the overall accuracy between ALS and MLS, the scanning distance of MLS (range < 200 m) is shorter than ALS (range > 1000 m); consequently, the effect of GPS on MLS is much larger than on ALS systems.

Because ALS and TLS acquire data from different viewpoints, the integration of these systems is beneficial to obtain complete data for several applications. Several researchers have investigated possible ALS and TLS registration methods. In work by Böhm and Haala [15], who used ICP methods, TLS provided the geometric information of vertical walls while ALS provided the geometric information of roof-tops for city modelling. Gressin et al. [4] also applied ICP in multi-platform LiDAR registration and integrated different types of point features into the tie points selection, including point features for registration such as eigenvalues, eigenvectores, dimensionality features, and entropy from neighborhood points. Von Hansen et al. [16] extracted the linear features from the building boundaries and then applied control lines. Boulaassal et al. [17] extracted the 3-D vectors of buildings from ALS, TLS, and MLS separately and then registered all the extracted 3-D vectors by linear feature for a detailed 3-D building model; they combined the vector data rather than point clouds. Cheng et al. [18] extracted the 3-D building corners from the intersection of 3-D building boundaries from ALS and TLS and then applied the 3-D building corners. Wu et al. [19] combined the control lines and planes extracted from buildings for the registration of ALS and TLS.

Most previous studies focused on ALS and TLS registration; relatively few discussed the registration of ALS and MLS. The challenge of ALS and MLS registration is to obtain reliable control entities from these two different systems. Because ALS and MLS acquire an abundance of 3-D surface points, the 3-D surface features such as road and wall surfaces can be utilized in ALS and MLS registration.

The airborne and terrestrial mobile LiDAR systems acquire data efficiently. The objective of this study was to co-register the point clouds acquired by airborne LiDAR and terrestrial mobile LiDAR and use these complementary data to improve the point coverage in urban areas. The point clouds scanned from two platforms can be located at difference coordinate systems, and the point clouds must first be registered to remove the error between the point cloud data from the two platforms. This study proposes a scheme to register airborne and terrestrial point clouds by surface features and discusses the effect between different point densities.

The terrestrial mobile data use an odometer, GPS, and INS to determine the position and orientation of LiDAR sensors for direct georeferencing; however, the GPS condition in urban corridors usually affects the positioning accuracy for terrestrial mobile LiDAR. Because the GPS condition of airborne LiDAR is better than mobile LiDAR, we assumed that the airborne LiDAR have been georeferenced to a world coordinate system. The terrestrial mobile LiDAR is then transformed to the coordinate system of airborne LiDAR to improve the accuracy of mobile LiDAR in urban corridors.

2. The Proposed Scheme

The framework consists of two major parts: (1) registration of multi-strips terrestrial mobile LiDAR data and (2) registration of airborne and terrestrial mobile LiDAR. First, we co-registered the multi-strips terrestrial LiDARs to enlarge the coverage of the dataset. The registered terrestrial LiDARs were then transformed to the ALS coordinate system. We used the least squares 3-D surface matching (LS3D) algorithm to minimize the Euclidean surface distance between the airborne and terrestrial mobile LiDAR point clouds. The registration features in this study are 3-D planes, and the applied transformation model is a 3-D similarity transformation. The steps of LS3D are (1) extracting planar features; (2) matching criterion; (3) mathematical modeling; (4) solving transformation parameters; and (5) applying the parameters to the model data.

2.1. Planar Feature Extraction

LiDAR point clouds are composed of a large number of irregular points. To improve the computational performance, the irregular points must be structuralized into organized points. In this study, we used a voxel structure to structuralize the airborne and terrestrial mobile LiDAR. The boundaries of the voxel structure were calculated from maximum and minimum values of all points, and the grid size depended on the point density. Both ALS and MLS used the same boundaries and grid size, and therefore we can search the corresponding points from two different systems effectively. After the data structuralization, all the points were indexed into a 3-D grid, and the points in a voxel were selected to calculate the planar feature (Figure 1).

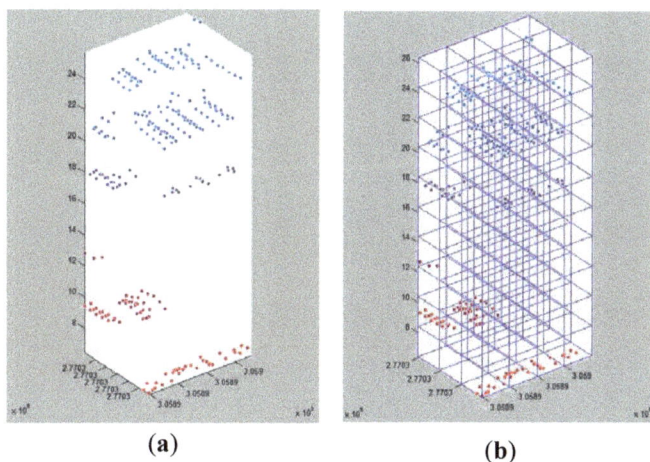

(a) **(b)**

Figure 1. An example of structured points: (**a**) irregular points, (**b**) voxel of points.

3-D planes were used as the control entity for registration; therefore, we used principal component analysis (PCA) [20,21] to analyze and calculate the plane features. The points inside each voxel are used for PCA calculation. The covariance matrix of points was calculated using Equation (1), in which (x_i, y_i, z_i) represent the ith point in the voxel, and $(\bar{x}, \bar{y}, \bar{z})$ are the mean of the points in a voxel. The eigenvalues ($\lambda_1 > \lambda_2 > \lambda_3$) and eigenvectors ($S_1$, S_2, S_3) of covariance matrix Mc can be extracted by Equation (2). When flatly distributed points are analyzed, the first and second eigenvalues are similar and the third eigenvalue is smaller than the other eigenvalues ($\lambda_1 \approx \lambda_2 > \lambda_3$). We defined λ_k by Equation (3) [22] to extract the planar features. When λ_k is smaller than a predefined threshold, the points in the voxel can be considered as a plane and the normal vector equal to the corresponding eigenvector. Otherwise, the points in a voxel are the less identifiable points (Figure 2). Figure 2 shows an example of normal vector extraction. Figure 2a shows a number of voxels after structuralization. We select a voxel from Figure 2a and show the points inside the voxel in Figure 2b. Figure 3b shows the extracted plane and corresponding normal vector.

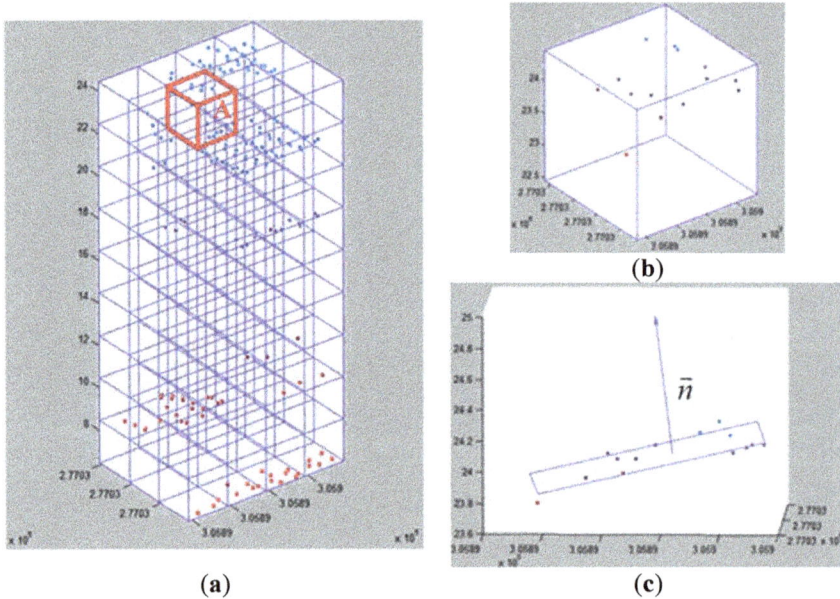

(b)

(a)   **(c)**

Figure 2. Illustration of points to normal vectors: (**a**) voxels, (**b**) points in voxels A, (**c**) normal vector of plane in voxel A.

$$M_c = \begin{bmatrix} x_1-\overline{x} & x_2-\overline{x} & \cdots & x_n-\overline{x} \\ y_1-\overline{y} & y_2-\overline{y} & \cdots & y_n-\overline{y} \\ z_1-\overline{z} & z_2-\overline{z} & \cdots & z_n-\overline{z} \end{bmatrix} \times \begin{bmatrix} x_1-\overline{x} & y_1-\overline{y} & z_1-\overline{z} \\ x_2-\overline{x} & y_2-\overline{y} & z_2-\overline{z} \\ \vdots & \vdots & \vdots \\ x_n-\overline{x} & y_n-\overline{y} & z_n-\overline{z} \end{bmatrix} \tag{1}$$

$$M_c = \begin{bmatrix} s_1 & s_2 & s_3 \end{bmatrix} \begin{bmatrix} \lambda_1 & 0 & 0 \\ 0 & \lambda_2 & 0 \\ 0 & 0 & \lambda_3 \end{bmatrix} \begin{bmatrix} s_1 & s_2 & s_3 \end{bmatrix}^T \tag{2}$$

$$\lambda_k = \frac{\lambda_3}{\lambda_1 + \lambda_2 + \lambda_3} \tag{3}$$

To summarize the process of planar object extraction, the extraction of 3D surface features from irregular points include the following steps: (1) generating voxel structure for irregular points; (2) removing voxels that contain less than 5 LiDAR points; (3) calculating eigenvalues from points inside the voxels; and (4) extracting planar object based on parameter λ_k. The extracted planes could be located on walls, roofs, and road surfaces in any direction. The control planes do not have to follow the same direction to obtain transformation coefficients; on the contrary, the air-to-ground LiDAR registration requires different plane directions to avoid singular problems. The plane equation for

236

each voxel, Equation (4), is suitable to represent a plane in any directions. The plane coefficients are calculated from normal vector and mean points from Equation (5):

$$Ax + By + Cz + D = 0 \qquad (4)$$

$$A = n_x; B = n_y; C = n_z$$
$$D = -n_x \bar{x} - n_y \bar{y} - n_z \bar{z} \qquad (5)$$

where x, y, z are plane coordinates; $(\bar{x}, \bar{y}, \bar{z})$ aremean points of a plane; A, B, C, D are coefficients of a plane; and n_x, n_y, n_z are normal vectors.

2.2. Matching Criterion

After planar feature extraction, we used the extracted planes to search the corresponding planes between the two LiDAR systems. The plane-matching criteria included distance and angle thresholds (Figure 3). If the Euclidean distance of mean points between ALS and MLS was smaller than the distance thresholds and the normal vectors between ALS and MLS had similar orientation, the planes from ALS and MLS were treated as a conjugate plane pair. All selected conjugate planes were used in 3-D surface minimization. The angle can be calculated using Equation (6).

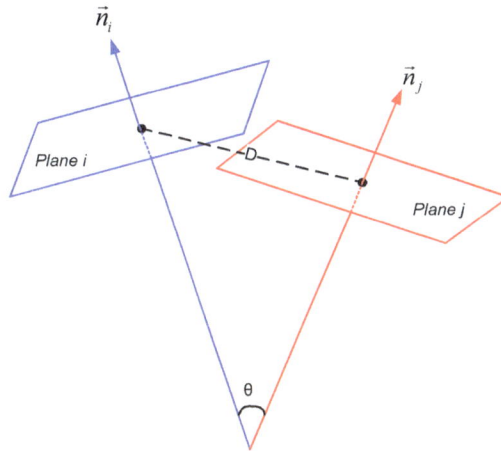

Figure 3. Illustration of angle and distance criteria.

$$\theta_{ij} = \cos^{-1}\left(\frac{\vec{n_i} \bullet \vec{n_j}}{|\vec{n_i}| \cdot |\vec{n_j}|}\right) \qquad (6)$$

where: n_i is the normal vector of plane i; n_j is the normal vector of plane j; and θ_{ij} is the angle between normal vector n_i and n_j.

The conjugate planes selection from ALS and transformed MLS's planes is based on these two criteria: (1) the distance of mean points between ALS and transformed MLS's is smaller than the predefined distance threshold; and (2) the intersection angle of normal vectors between ALS and transformed MLS's is smaller than the predefined angle threshold. After the determination of the unknown parameters, we update the transformed MLS's plane by calculate the parameters and refine the automatic conjugate planes. The threshold selection is based on the data itself. In the first three iterations, we use large thresholds to handle the large differences between ALS and transformed MLS. After three iterations, the thresholds are determined by standard deviation of distance and intersection angle between ALS's plane and transformed MLS's plane. Below shows the pseudo codes for selection of thresholds:

Pseudo Codes for Selection of Thresholds
1   **while** (iteration < 20) do
2   **if** iteration < 4 then
3   Threshold_distance = 1m;Threshold_angle = 15deg;
4   **else**
5   **if** std(distance) > 0.10m & std(angle) > 5deg then
6   Threshold_distance = 2 × std(distance);Threshold_angle=2 × std(angle);
7   **else**
8   Threshold_distance = 0.10m;Threshold_angle = 5deg;
9   **end if**
10   Find plane pair and calculate transformation parameters
11   **while** stopping criterion is satisfied do
12   exit loop
13   **end while**
14   Transform MLS's plane using calculated parameters
15   Calculate the distance between ALS and transformed MLS's planes
16   Calculate the intersection angle between ALS and transformed MLS's planes
17   Iteration = iteration + 1
18   **end while**

2.3. Least Squares 3-D Surface Matching (LS3D)

The LS3D algorithm, developed by Gruen and Akca [10], minimizes the 3-D distance between surfaces, while the ICP algorithm minimizes the Euclidean distances between points. Compared to LS3D, ICP requires relatively higher iteration numbers [23] while LS3D quickly converges to an optimal solution. LS3D assumes that two surfaces are created from the same object by different processes. In this study, one surface acquired by ALS is called the template surface $f(x, y, z)$, while the other surface from MLS is called the search surface $g(x, y, z)$. If the error function $e(x, y, z)$ is zero, these two surfaces should be the same, and all the surfaces in the template surface can correspond to the surfaces in the search surface, as represented by Equation (7).

In reality, the two surfaces are not equal. We used error function $e(x, y, z)$ to describe the inconsistency between the two conjugate surfaces; hence, Equation (7) can be rewritten as Equation (8). To minimize the error function $e(x, y, z)$, the coordinate system of the MLS (x_0, y_0 z_0) was subjected to a general 3-D translation, scaling, and rotation transformation (the so called "3-D similarity transformation") used to minimize the integrated squared error function between these two conjugate surfaces over a well-defined common spatial domain. The transformation parameters of similarity transformation included a translation vector (t_x, t_y, t_z), three rotation angles (ω,φ,κ), and one scale factor (m) (see Equation (9)). The rotation angle is counterclockwise. The detail of rotation matrix can be found at [24].These parameters were used to minimize the errors between these two conjugate surfaces. The aim of LS3D is to determine these 7 parameters using conjugate planes between ALS and MLS.

$$f(x, y, z) = g(x, y, z) \tag{7}$$

$$f(x, y, z) - e(x, y, z) = g(x, y, z) \tag{8}$$

$$x = t_x + s(r_{11}x_0 + r_{12}y_0 + r_{13}z_0)$$
$$y = t_y + s(r_{21}x_0 + r_{22}y_0 + r_{23}z_0)$$
$$z = t_z + s(r_{31}x_0 + r_{32}y_0 + r_{33}z_0)$$

(9)

where $f(x, y, z)$ and $g(x, y, z)$ are the template and search surfaces; $e(x, y, z)$ is error vector; (x, y, z) are the coordinate systems of ALS-derived surfaces; (x_0, y_0, z_0) are the coordinate systems of MLS- derived surfaces; t_x, t_y, and t_z are the three translation parameters along three axes; $r_{11} \sim r_{33}$ are elements of the rotation matrix formed by three rotation angles ω, φ and κ around three axes; and s is the scale factor we assume is close to 1.

To perform least squares estimation, Equation (8) should be linearized by Taylor expansion, ignoring the higher-order terms, resulting in Equation (10). The template surfaces $f(x, y, z)$ and search surfaces $g_0(x, y, z)$ are planar surface patches, represented by Equations (11) and (12).

$$f(x, y, z) - e(x, y, z) = g_0(x, y, z) + \frac{\partial g_0(x, y, z)}{\partial x} dx + \frac{\partial g_0(x, y, z)}{\partial y} dy + \frac{\partial g_0(x, y, z)}{\partial z} dz$$
$$= g_0(x, y, z) + g_x dx + g_y dy + g_z dz$$

(10)

$$f(x, y, z) = A_f x + B_f y + C_f z + D_f$$

(11)

$$g_0(x, y, z) = A_g x + B_g y + C_g z + D_g$$

(12)

where $g_0(x, y, z)$ is the initial approximation of search surfaces; g_x, g_y, g_z are numeric first derivatives of $g(x, y, z)$; dx, dy, dz are the differentiation terms; A_f, B_f, C_f, D_f are coefficients of a target plane; and A_g, B_g, C_g, D_g are coefficients of a search plane.

In the linearized Equation (10), elements dx, dy, and dz can be combined with 3-D similarity parameters in Equation (9). Equation (13) shows the differentiation terms of dx, dy, and dz. The numerical derivatives g_x, g_y, and g_z can be derived from plane equations as Equation (14).

$$dx = \frac{\partial x}{\partial p_i} dp_i = dt_x + a_{10}ds + a_{11}d\omega + a_{12}d\phi + a_{13}d\kappa;$$

$$dy = \frac{\partial y}{\partial p_i} dp_i = dt_y + a_{20}ds + a_{21}d\omega + a_{22}d\phi + a_{23}d\kappa;$$

(13)

$$dz = \frac{\partial z}{\partial p_i} dp_i = dt_z + a_{30}ds + a_{31}d\omega + a_{32}d\phi + a_{33}d\kappa$$

$$g_x = \frac{A_g}{\sqrt{A_g^2 + B_g^2 + C_g^2}}; g_y = \frac{B_g}{\sqrt{A_g^2 + B_g^2 + C_g^2}}; g_z = \frac{C_g}{\sqrt{A_g^2 + B_g^2 + C_g^2}}$$

(14)

where $p_i \in (t_x, t_y, t_z, s, \omega, \varphi, \kappa)$; and $a_{10} \sim a_{33}$ are the coefficient elements.

The final observation is Equation (15), which is derived from Equations (13) and (14). The least squares adjustment algorithm was applied to minimize the errors. We iteratively minimized the sum of squared errors between MLS and ALS surface using the LS3D approach. Because the observation equation is a nonlinear function, we measured three initial registration points between ALS and MLS to obtain initial approximations. The initial approximations were determinate by a noniterative approach [25], which is able to provide stable initial parameters because the manual registration points are less than four points.

In Equation (15), the $f(x, y, z)$ and $g_0(x, y, z)$ are ALS and MLS planes, respectively. The initial values of unknown parameters for Equation (15) are calculated by three initial registration points between ALS and MLS. We apply these initial parameters for converting MLS's planes to ALS's coordinate systems. Then, we use distance and angle thresholds to find a large number of conjugate planes between ALS and The transformed MLS's planes. As the observation equations are larger than the unknown parameters, the parameters of Equation (15) are calculated by least squares adjustment through an iterative process [10]. A more in depth description of LS3D details regarding the parameter determinations can be found in [26].

$$
\begin{aligned}
-e\left(x, y, z\right) = {} & g_x dt_x + g_y dt_y + g_z dt_z + \left(g_x a_{10} + g_y a_{20} + g_z a_{30}\right) dm \\
& + \left(g_x a_{11} + g_y a_{21} + g_z a_{31}\right) d\omega + \left(g_x a_{12} + g_y a_{22} + g_z a_{32}\right) d\phi \\
& + \left(g_x a_{13} + g_y a_{23} + g_z a_{33}\right) d\kappa - \left[f\left(x, y, z\right) - g_0\left(x, y, z\right)\right]
\end{aligned}
\tag{15}
$$

2.4. Airborne and Terrestrial Mobile LiDARs Registration

In this study, the ALS data were transformed into a world coordinate system by differential GPS (DGPS) and strip adjustment [13]. When a GPS outage occurs in an urban corridor, the direct-georeferencing of MLS can only rely on the Inertial Measurement Unit (IMU) and speedometer; consequently, the MLS point clouds may contain systematic errors. The ALS data were therefore treated as reference data, and the MLS data were transformed into the coordinate system of ALS data.

MLS data usually include forward and reverse trajectories of a road and are used to obtain additional 3-D points to describe the road environment. There are two possibilities to co-register the forward MLS, reverse MLS and ALS. The first approach co-registers these multi-trajectory MLS data before the registration of ALS and MLS; the second approach performs the registration for forward and reverse MLS separately. Considering that the similarity between multi-trajectories is higher than the similarity between MLS and ALS, the first approach may derive more control features from multi-trajectories MLS for data registration. Furthermore, the registration of multi-trajectories may enlarge the MLS coverage for the registration of ALS and MLS. This study therefore adopted the first approach to co-register the multi-trajectory MLS using least square surfaces matching.

3. Experimental Results

The test data include airborne LiDAR and terrestrial mobile LiDAR data. The test area is about 190 m by 900 m. The airborne point cloud data were scanned by Optech ALTM30/70 using 7 flightlines. The total number of points is 2,774,371, and the average point density is about 15 points/m². The terrestrial mobile LiDAR was scanned by Optech Lynx, and the length of the road is about 900 m. The MLS data include the two different trajectories, left-to-right and right-to-left. The total number of points is about 79,170,000, and the average point spacing in the horizontal plane is about 5 cm. Figure 4 shows the ALS digital surface model (DSM), MLS DSM, and reference orthophoto of test area. Only the overlapped are used in registration. Figure 5 compares the profiles of ALS and MLS in the test area. The buildings beside the road and trees along streets in the urban corridor cause GPS signal occlusion, which significantly degrades the navigation performance; therefore, the MLS data need post-processing (i.e., point registration) to obtain precise locations. The parameters used are listed in Table 1.

(a)

(b)

(c)

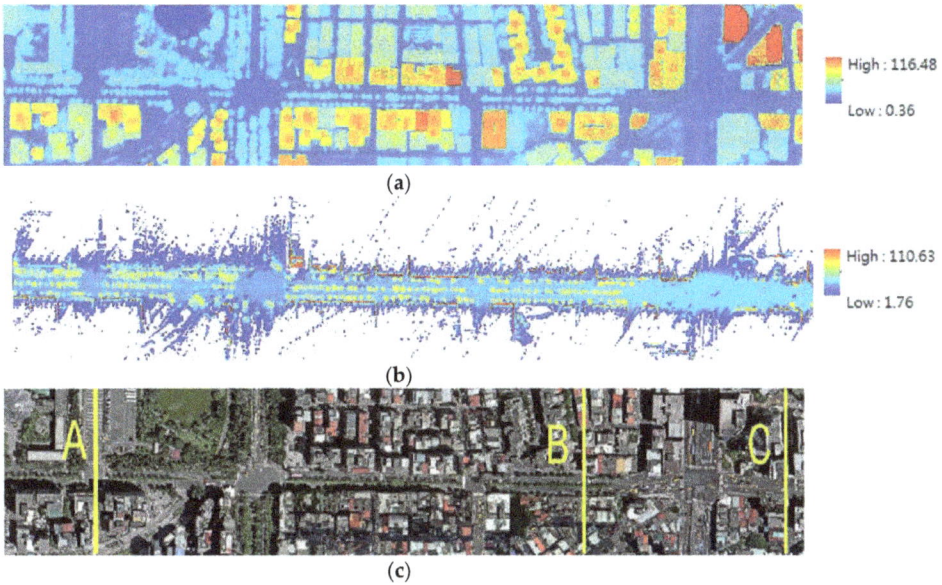

Figure 4. Test data: (**a**) ALS colored by elevation, (**b**) MLS colored by elevation, (**c**) reference orthophoto.

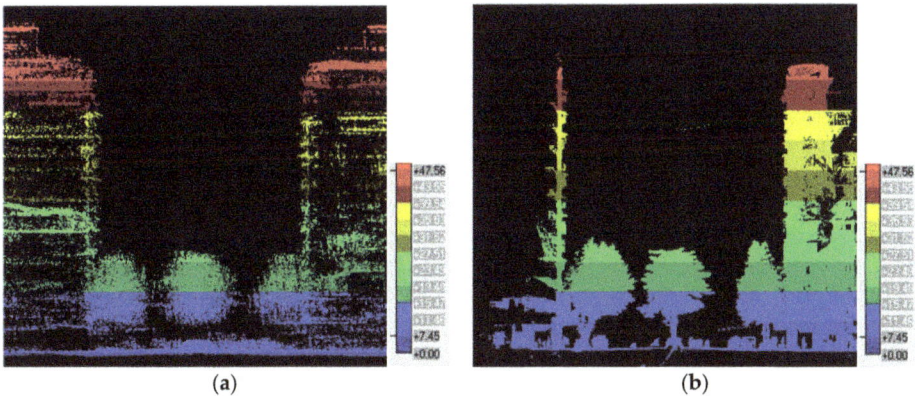

(a)

(b)

Figure 5. Profiles of ALS and corrected MLS data: (**a**) profile of ALS colored by elevation, (**b**) profile of MLS colored by elevation.

Table 1. The parameters setting and descriptions.

Parameter	Descriptions	Parameter Setting
Voxel size	The voxel size is a parameter to structuralize irregular points to regular 3-D voxels. This parameter is related to point density. A lower point density needs larger voxel size to aggregate more points in a 3-D voxel.	1 m
Number of point in a voxel	This parameter defines the minimum number of point in a voxel. If the number of point is larger than this threshold, these points are used to calculate the plane equation. Any 3 points can determine a plane, so we used more than 3 points in a voxel to determine the plane parameters.	5 points

Table 1. *Cont.*

λ_k	When λ_k is smaller than a predefined threshold, the points in the voxel is considered as a plane. We observed the data set to define this empirical parameter.	0.2
Intersection angle	If the intersection angle of two planes is smaller than the angle threshold, these two planes are treated as a conjugate plane pair. We observed the data set to define this empirical parameter.	Fifteen degrees for the first 3 iterations. After 3 iterations, this parameter is 2 × std (intersection angle of plane pair after registration). (see Section 2.2)
Distance	If the Euclidean distance of mean points between two planes is smaller than the distance threshold, these two planes are treated as a conjugate plane pair. This parameter is defined by voxel size and pre-alignment quality.	One meter for the first 3 iterations. After 3 iterations, this parameter is 2 × std (distance of plane pair after registration). (see Section 2.2)
Maximum iteration	The matching process terminates when the iteration number exceeds predefined maximum number of iteration. For good data configuration, the convergence of LS3D is relatively faster than ICP approach.	20
Stopping criterion	The matching process meets the optimal solution when the corrections of transformation parameters are smaller than the predefined thresholds. The small thresholds are selected to ensure the quality.	$\lvert dt_x \rvert < 0.001$ m; $\lvert dt_y \rvert < 0.001$ m; $\lvert dt_z \rvert < 0.001$ m $\lvert S \rvert < 0.0001$; $\lvert d\omega \rvert < 0.001$ deg; $\lvert d\varphi \rvert < 0.001$ deg; $\lvert d\kappa \rvert < 0.001$ deg

The experiments in this study used both simulation and real data to analyze the performance of point registration. The validation experiments were carried out in three parts. First, the 3-D points with different point densities and standard errors were simulated; second, the relative accuracy between forward and reverse of MLS data was examined; and third, the errors between MLS and ALS from the proposed methods were checked.

3.1. Simulation Data Registration

First, we used a simulation dataset to verify the precision of registration at different densities. The registration precision indicates the geometrical difference between two systems. In this experiment, the point-density ratio indicated the ratio between target points and search points and was used to simulate different point densities between LiDAR system 1 and system 2. We simulated 3-D points distributed on a prismatic building model shaped like a 50 m × 50 m × 50 m box. The point densities of simulated system 1 were 100, 90, 80, 70, 60, 50, 40, 30, 20, 10 points/m², while the point density of simulated system 2 was 100 points/m²; therefore, the point-density ratios were 1/1, 1/2, 1/3, 1/4, 1/5, 1/6, 1/7, 1/8, 1/9, 1/10. The noise levels (random error) of simulated point were 0.10 m and 0.05 m for system 1 and 2, respectively. The 3-D transformation parameters were predefined using the maximum transformation parameters (tx, ty, tz, omega, phi, kappa) in Table 1. One hundred data sets are simulated for each point-density ratio. After the data simulation, we used LS3D to solve the transformation parameters and applied the parameters to the simulated system 2 data. Every transformed point can be compared with the original point. We used the differences along X, Y, and Z axes as a precision index (Figure 6). The simulation results indicate that accuracy of registration improved as the density increased. The distance error may preserve 5 cm precision when the point-density ratio is reduced to 1/10.

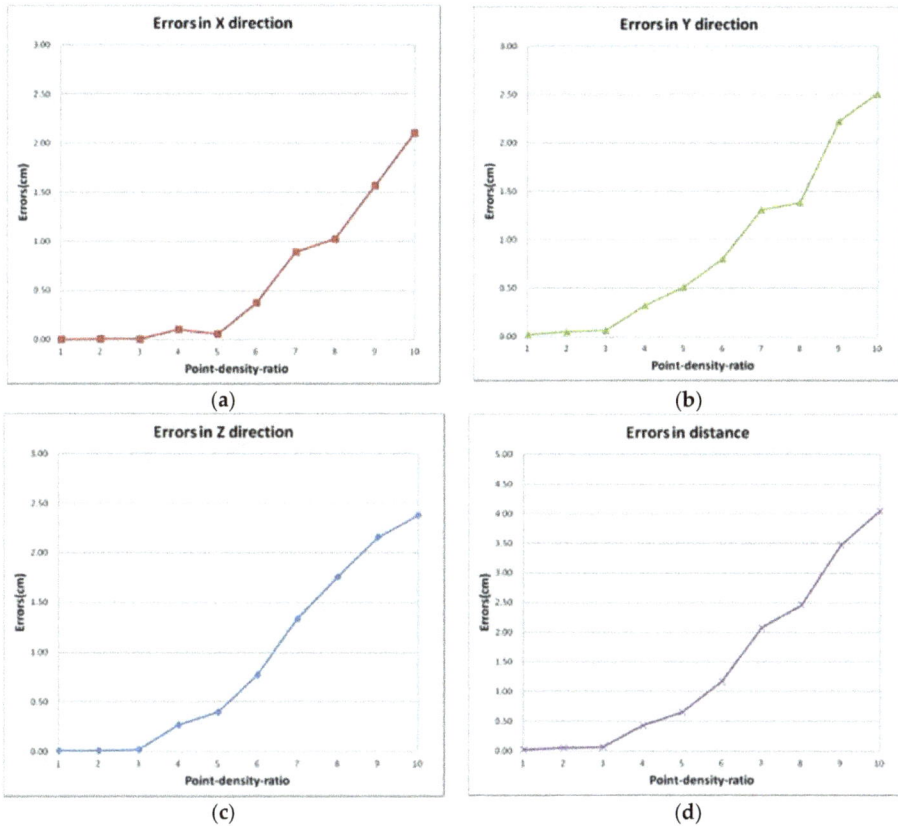

Figure 6. Simulation results: (**a**) Position difference along X-axis. (**b**) Position difference along Y-axis. (**c**) Position difference along Z-axis. (**d**) Distance between transformed point and original point.

3.2. Terrestrial Mobile LiDARs Registration

Because the points of MLS number about 79 million, we reduced the points to accelerate the computation by using a 1 m 2D grid to remove the nonoverlapped points. We apply initial translation (tx, ty, tz) in point reduction. It can avoid large difference between ALS and MLS. Besides, we use 2D grid to accelerate the process of point reduction. In other words, all the MLS points are projected to 2D horizontal plane and elevation is not considered in this stage. If the points from both trajectories appeared in same cell, then that cell was marked as an overlapped cell and all the points in this overlapped cell were preserved for registration. The grid size was estimated by the inconsistency of points from two different trajectories. The smaller grid size may remove the overlapped points incorrectly due to the problem of direct georeferencing. After the nonoverlapped point reduction, the number of points was reduced to 44 million, and the compression rate was about 44%.

To obtain high point density in urban environments in this study, the maximum vehicle speed of MLS was <60 km/hr, and arbitrary lane-changing was not allowed, resulting in a continuous and smooth driving path. Considering the road length and data volume, we split the data into several road segments. In addition, the error behaviors for certain road segments were considered to be systematic errors, and a 3-D similarity transformation was considered to compensate the geometric inconsistency between two road segments. The length of the road section was 25 m, with a 3 m overlap. Each trajectory had 36 segments, and the total number of road segments was 72 in this stage. We used the forward trajectory (from left to right) as reference data. The registration coefficients converted the points from reverse trajectory to reference data. Because the LS3D algorithm requires

243

initial transformation parameters, we manually measured three registration points in the first segment and calculated the initial transformation parameters. The precise registration transformation parameters were then determined by LS3D algorithm. We assumed that the actual road path was continuous and smooth; hence, the initial parameters for the next road segment could be obtained from the previous road segment, and therefore only three registration points were needed to begin the TLS data registration.

The accuracy assessments in this section include three parts: analysis of derived transformation parameters, point distance between road marks from two trajectories, and visualization of MLS data before and after registration.

For these road segments, we calculated 36 sets of transformation parameters (Table 2). The scale coefficient is very close to 1. It is ranged from 0.9998 to 1.0004. To compare the translation and rotation parameters, the standard deviation of translation parameters (i.e., 12.6 cm to 40.2 cm) was much larger than rotation angles (i.e., 1.7 cm to 3.5 cm). In addition, the error in vertical direction was also larger than horizontal direction. The error in vertical direction was larger than 1 m when the GPS solution was not available, similar to error behavior in an urban corridor [27]. In other words, the rotation angles were more consistent than the translation parameters. This phenomenon is referred to as the positioning errors of GPS outage. Figure 7 plots the translation parameters at different road sections and indicates the continuity of translation parameters trajectories.

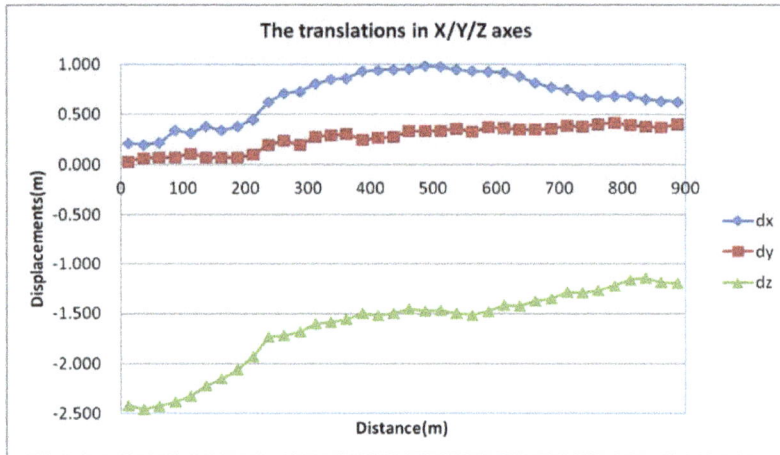

Figure 7. The translations in X/Y/Z directions (MLS/MLS registration).

Table 2. The summary of transformation parameters for MLS/MLS registration. .

	tx (m)	ty (m)	tz (m)	Omega	Phi	Kappa
Mean	0.689	0.267	−1.639	0.003deg (0.1cm@25m)	−0.001deg (0.04cm@25m)	0.043deg (1.9cm@25m)
Std	0.247	0.126	0.402	0.080deg (3.5cm@25m)	0.038deg (1.7cm@25m)	0.057deg (2.5cm@25m)
Min	0.200	0.030	−2.457	−0.130deg (−5.7cm@25m)	−0.088deg (−3.8cm@25m)	−0.090deg(−3.9cm@25m)
Max	0.984	0.417	−1.148	0.195deg (8.5cm@25m)	0.087deg (3.8cm@25m)	0.198 deg(8.6cm@25m)

To evaluate the accuracy of registration, we manually measured 36 well-defined points to compare the point distance before and after registration. Because the point spacing of MLS is about 5 cm, we could identify the corner point of the road marks from the LiDAR intensity. We use lines intersection to determine the corner of road mark. Figure 8 is an example of a check-point located on a marked pedestrian "zebra" crossing. The total number of independent check-points (CP) is 36. The error vectors of check point before and after registration are shown as Figure 9.

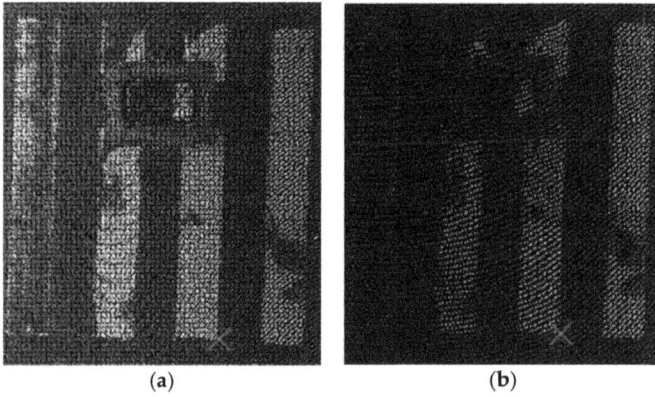

Figure 8. Illustration of independent check-point between trajectories: (**a**) check point in reference strip, (**b**) check point in registered strip.

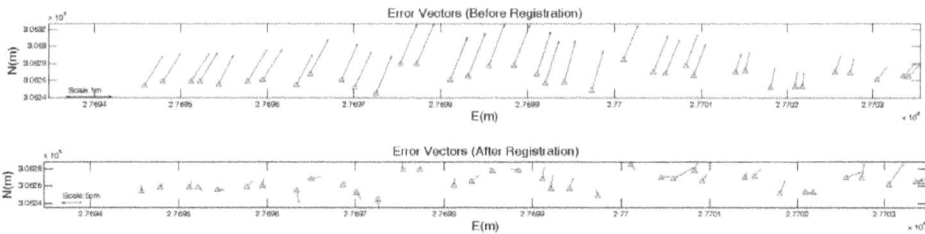

Figure 9. Error vectors of check points before and after registration.

Table 3 shows the results of check points before and after registration. The mean errors before registration ranged from −1.634 m to 0.719 m. The mean error in vertical direction is larger than horizontal directions. After the registration, the mean errors were significantly reduced to 0.002 m to 0.025 m. The standard deviations of check point before registration ranged from 0.114 m to 0.401 m and fell to <5 cm after registration.

Table 3. Statistics of the independent check points.

		Max (m)	Min (m)	Mean (m)	Std (m)
dY	Before	1.017	0.295	0.719	0.228
	After	0.042	0.000	0.025	0.043
dX	Before	0.460	0.073	0.293	0.114
	After	0.056	0.003	0.024	0.043
dZ	Before	2.469	1.159	-1.634	0.401
	After	0.069	0.000	0.002	0.028

After the numerical analysis, we also selected some profiles to visually compare MLS before and after registration. Figure 10 shows three profiles of the road point cloud before and after registration. The width of the profile is 1.5 m, and the different colors indicate points from different trajectories. In the figure, the discrepancy was removed after registration.

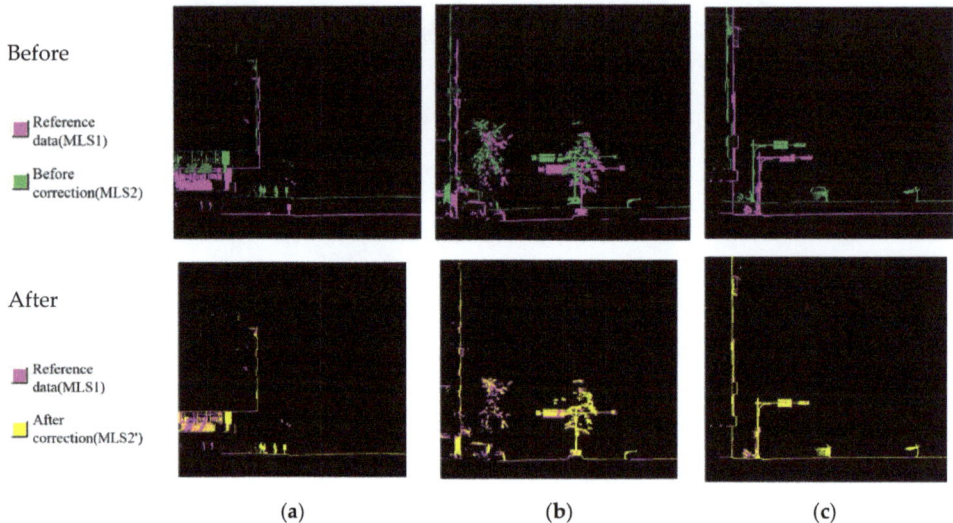

Before

Reference
data(MLS1)

Before
correction(MLS2)

After

Reference
data(MLS1)

After
correction(MLS2')

(a)   (b)   (c)

Figure 10. Profiles of MLS before and after registration: (**a**) profile A, (**b**) profile B, (**c**) profile C.

3.3. Airborne LiDAR and Terrestrial Mobile LiDAR Registration

Once the MLS data from different trajectories were well co-registered, we performed the LS3D registration for ALS and MLS data. The spacing of voxel for ALS and MLS was 1 m. The total number of points should be more than 5 points for normal angle calculation; the number of voxel for ALS and MLS were 40,907 and 218,212 respectively. The registration features included surfaces of roads, walls, and other objects. We use elevation and azimuth angles to discuss the distribution of normal vector. The elevation angle is a vertical angle from horizontal plane to normal vector; the azimuth angle is a horizontal angle from north to horizontal projected normal vector. Elevation and azimuth angles of normal vectors (Figure 11) show that the elevation angles of the road and roof points were mostly close to 90 degrees, while the elevation angles of the walls were mostly close 0 degrees. The number of normal vectors from the road and roofs were larger than the normal vectors from walls (Figure 11a), and the total number of elevation angles <10 degrees was 768 (about 2%). These are important control elements for registration. The MLS contain more normal vectors from roads and walls (Figure 11b), and the azimuths of the normal vectors are distributed on different directions (Figure 11c,d). Therefore, the controlling capability of control surfaces is well-distributed to cover all directions.

The statistics of transformation parameters for ALS/MLS registration is shown as Table 4. The mean errors in translation parameters were ranged from −0.885m to −0.220m. It was larger than the rotation parameters. It means the errors were mostly conducted by GPS positioning error. The standard deviation of translation parameters (3.6 cm to 9 cm) was also larger than rotation angles (i.e., 0.75~3.5 cm). Figure 12 shows the translation at different distances.

The accuracy assessments included two parts: plane distance between surfaces from ALS and MLS, and visualization of ALS and MLS data before and after registration. Because measuring check-points between ALS and MLS is difficult, we used the check plane in the accuracy assessment of ALS and MLS registration. We manually selected the 74 conjugate planes from ALS and MLS; a least squares plane fitting was then applied to estimate the optimal plane equation. The perpendicular distance between two planes was calculated as the quality index of ALS and MLS registration (Table 5). The data registration improved the mean error datum of ALS and MLS. In addition, the precision of registration was improved from 0.847 m between ALS and MLS from −0.979 m to −0.02 m. The mean error was mainly caused by the GPS/IMU signals and different to 0.033 m. A comparison of the error vectors of perpendicular distance between plane before and after registration shows that the systematic error was compensated by LS3D after registration (Figure 13). Hence, the LS3D was able to automatically improve the consistency between ALS and MLS.

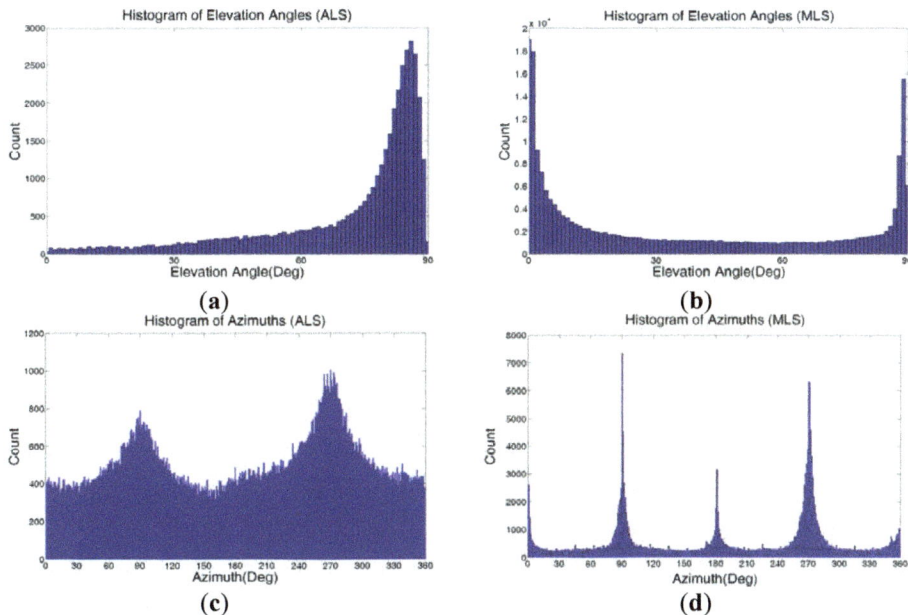

Figure 11. Histogram of elevation and azimuth angles: (**a**) elevation angles of ALS, (**b**) elevation angles of MLS, (**c**) azimuth angles of ALS, (**d**) azimuth angles of MLS.

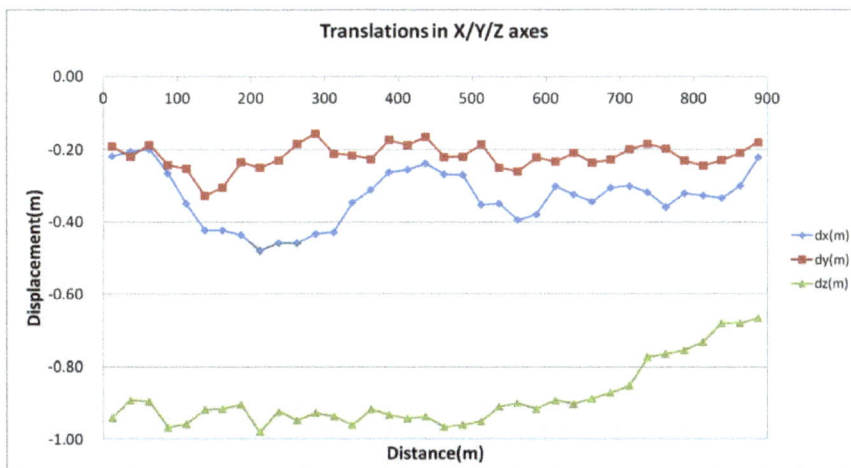

Figure 12. The translations in X/Y/Z directions (ALS/MLS registration).

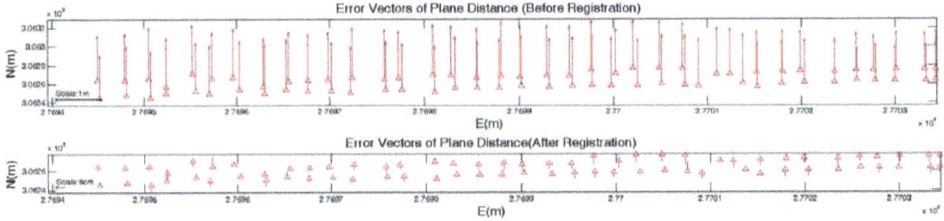

Figure 13. Error vectors of plane distance before and after registration.

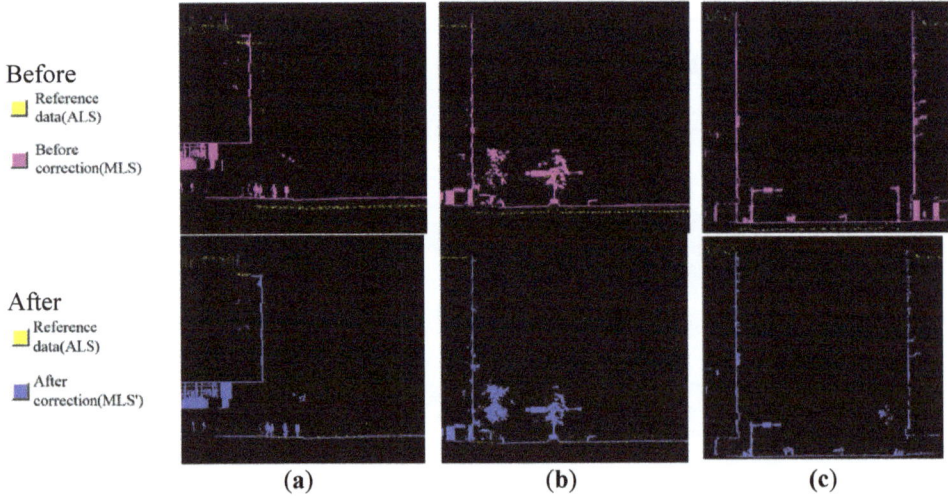

(a)	**(b)**	**(c)**

Figure 14. The profiles of the airborne LiDAR and terrestrial mobile LiDAR point cloud: (a) profile A, (b) profile B, (c) profile C.

Table 4. The summary of transformation parameters for ALS/MLS registration.

	tx (m)	ty (m)	tz (m)	Omega	Phi	Kappa
Mean	−0.333	−0.220	−0.885	0.001 deg (0.1cm@25m)	−0.008 deg (0.4cm@25m)	0.011 deg (0.5cm@25m)
Std	0.076	0.036	0.088	0.016 deg (0.75cm@25m)	0.025 deg (1.1cm@25m)	0.080 deg (3.5cm@25m)
Min	−0.480	−0.328	−0.980	−0.041 deg (−1.8cm@25m)	−0.038 deg (−1.6cm@25m)	−0.154 deg (−6.7cm@25m)
Max	−0.199	−0.156	−0.667	0.0335 deg (1.4cm@25m)	0.077 deg (3.3cm@25m)	0.218 deg (9.5cm@25m)

Table. 5. The statistic of the check planes.

	Before Registration	After Registration
Mean (m)	−0.979	−0.020
Std (m)	0.847	0.033
Max (m)	0.929	0.068
Min (m)	0.581	−0.045

Because the conjugate planes were mostly located on road surfaces, we selected some profiles to evaluate the continuity of walls. Three profiles of the road point cloud before and after registration (Figure 14) show that the width of the profile is 1.5 m, and different colors indicate points from ALS

Remote Sens. **2014**, *6*, 12686–12707

and MLS. The discrepancy was removed after registration, indicating that the continuity between the building roof and wall was improved after registration.

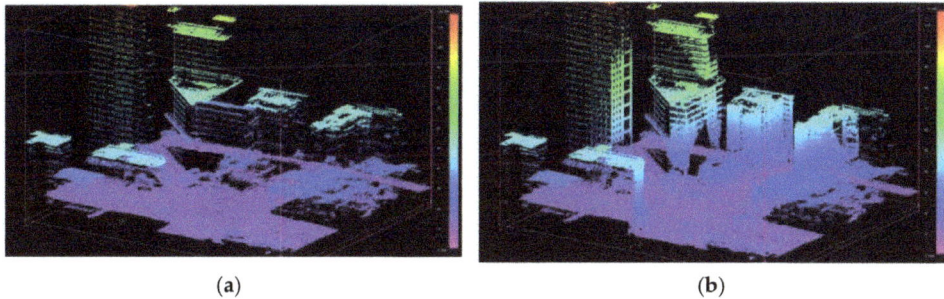

(a) (b)

Figure 15. The results of ALS and MLS registration: (**a**) ALS only (colored by height), (**b**) registration of ALS and MLS (colored by height).

The perspective view of registration results (Figure 15a,b) shows that the proposed method worked well for the registration of ALS and MLS. In addition, the combination of ALS and MLS may provide a more complete scene in urban environments. For example, the MLS may improve the wall features for ALS.

4. Conclusions

This study proposed a scheme to co-register the 3-D point clouds scanned from airborne and terrestrial vehicle platforms to increase the details of urban scenes. In addition, the data co-registration georeferences the uncorrected points from MLS to world coordinates of ALS. The proposed method utilized least squares 3-D surface matching to minimize the surfaces between different systems. The conjugate surfaces were established by computing the angle and the distance between features and can be treated as a control entity to minimum the surface distance between two systems. The registrations include two major parts: (1) registration between multi-trajectories MLS, and (2) registration of ALS and MLS. The experimental results indicated that the proposed method may improve the geometric consistency between ALS and MLS. Under conditions of poor GPS reception for MLS, the maximum error between trajectories was improved from 1.6 m to 0.02 m, and the standard errors also improved from 0.4 m to 0.05 m. We used the ALS data as reference data, and the MLS data were transformed into the coordinate system of ALS data. The geometric consistency between MLT and TLS may reach 0.05 m. This model can be applied to point cloud registration from different platforms. Notices that the preconditions of the proposed method are: (1) the coordinate system of ALS and MLS is similar and the change between these two systems is relatively low. (2) the proposed scheme needs control planes in different directions to improve the controlling capability. Both horizontal and vertical planes are needed for the proposed methods. In this study, the initial parameters for least squares matching were calculated from the manual measured tie points. The further improvement of this study will be the automatic tie point selection. Besides, we use equal-grid voxels to obtain planar feature, however, the grid size is an important issue in planar feature extraction. The future work will adopt optimal neighborhood radius selection [28] in planar feature extraction.

Acknowledgments: The authors gratefully acknowledge the financial support provided by the Ministry of Science and Technology of Taiwan (NSC 101-2628-E-009-019-MY3). The authors would also like to thank the Taipei City Government and Taiwan Instrument Co., Ltd. Taiwan for providing the test data.

Author Contributions: Tee-Ann Teo provided the overall conception of this research, designs the methodologies and experiments, and wrote the majority of the manuscript; Shih-Han Huang contributed to the implementation of proposed algorithms, conducts the experiments and performs the data analyses.

Remote Sens. **2014**, 6, 12686–12707

Conflicts of Interest: The authors declare no conflict of interest.

Reference

1. Besl, P.J.; McKay, N.D. A method for registration of 3-D shape. *Proc. SPIE* **1992**, *1611*, 239–256.
2. Yang, C.; Medioni, G. Object modelling by registration of multiple range images. *Image Vision Comput.* **1992**, *10*, 145–155.
3. Zhang, Z. Iterative point matching for registration of free-form curves. *Int. J. Comput. Vision* **1992**, *13*, 119–152.
4. Gressin, A.; Mallet, C.; David, N. Improving 3D LiDAR Point Cloud Registration Using Optimal Neighborhood Knowledge. Available online: http://www.isprs-ann-photogramm-remote-sens-spatial-inf-sci.net/I-3/111/2012/isprsannals-I-3-111-2012.pdf (accessed on 9 October 2014).
5. Rusinkiewicz, S.; Levoy, M. Efficient variants of the ICP algorithm. In Proceedings of 3-D Digital Imaging and Modeling, Quebec City, Canada, 28 May–1 June 2001; pp. 145–152.
6. Barnea, S.; Filin, S. Keypoint based autonomous registration of terrestrial laser point-clouds. *ISPRS J. Photogramm. Remote Sens.* **2008**, *63*, 19–35.
7. Habib, A.; Ghanma, M.; Morgan, M.; Al-Ruzouq, R. Photogrammetric and LiDAR data registration using linear features. *Photogramm. Eng. Remote Sens.* **2005**, *71*, 699–707.
8. Jaw, J.J.; Chuang, T.Y. Registration of ground-based LiDAR point clouds by means of 3D line features. *J. Chin. Inst. Eng.* **2008**, *31*, 1031–1045.
9. Rosenholm, D.; Torlegard, K. Three-dimensional absolute orientation of stereo models using digital elevation models. *Photogramm. Eng. Remote Sens.* **1988**, *54*, 1385–1389.
10. Gruen, A.; Akca, D. Least squares 3D surface and curve matching. *ISPRS J. Photogramm. Remote Sens.* **2005**, *59*, 151–174.
11. Akca, D. Matching of 3D surfaces and their intensities. *ISPRS J. Photogramm. Remote Sens.* **2007**, *62*, 112–121.
12. Monserrat, O.; Crosetto, M. Deformation measurement using terrestrial laser scanning data and least squares 3D surface matching. *ISPRS J. Photogramm. Remote Sens.* **2008**, *63*, 142–154.
13. USGS LiDAR Guidelines and Base Specification. Available online: http://lidar.cr.usgs.gov/USG-SNGP%20Lidar%20Guidelines%20and%20Base%20Specification%20v13 (ILMF).pdf (accessed on 7 September 2014).
14. Habib, A.F.; Kersting, A.P.; Bang, K.I.; Zhai, R.; Al-Durgham, M. A strip adjustment procedure to mitigate the impact of inaccurate mounting parameters in parallel LiDAR strips. *Photogramm. Rec.* **2009**, *24*, 171–195.
15. Böhm, J.; Haala, N. Efficient integration of aerial and terrestrial laser data for virtual city modeling using LASERMAPs. In Proceedings of ISPRS Workshop Laser Scanning 2005, Enschede, The Netherlands, 12–14 September 2005.
16. Von Hansen, W.; Gross, H.; Thoennessen, U. Line-Based Registration of Terrestrial and Airborne LiDAR Data. Available online: http://www.isprs.org/proceedings/XXXVII/congress/3_pdf/24.pdf (accessed on 9 October 2014).
17. Boulaassal, H.; Landes, T.; Grussenmeyer, P. Reconstruction of 3D Vector Models of Buildings by Combination of ALS, TLS and VLS data. Available online: http://www.int-arch-photogramm-remote-sens-spatial-inf-sci.net/XXXVIII-5-W16/239/2011/isprsarchives-XXXVIII-5-W16-239-2011.pdf (accessed on 9 October 2014).
18. Cheng, L.; Tong, L.; Li, M.; Liu, Y. Semi-automatic registration of airborne and terrestrial laser scanning data using building corner matching with boundaries as reliability check. *Remote Sens.* **2013**, *5*, 6260–6283.
19. Wu, H.; Marco, S.; Li, H.; Li, N.; Lu, M.; Liu, C. Feature-constrained registration of building point clouds acquired by terrestrial and airborne laser scanners. *J. Appl. Remote Sens.* **2014**, *8*, doi:10.1117/1.JRS.8.083587.
20. Pauly, M.; Gross, M.; Kobbelt, L.P. Efficient Simplification of Point-Sampled Surfaces. Available online: http://www-i8.informatik.rwth-aachen.de/publication/122/p_Pau021.pdf (accessed on 9 October 2014).
21. Lagüela, S.; Armesto, J.; Arias, P.; Zakhor, A. Automatic Procedure for the Registration of Thermographic Images with Point Clouds. Available online: http://www-video.eecs.berkeley.edu/papers/slaguela/isprsarchives-XXXIX-B5-211-2012_published.pdf (accessed on 9 October 2014).
22. Pauly, M.; Gross, M.; Kobbelt, L.P. Efficient Simplification of Point-Sampled Surfaces. Available online: https://dl.acm.org/purchase.cfm?id=602123&CFID=463596832&CFTOKEN=12462253 (accessed on 9 October 2014).

Remote Sens. **2014**, *6*, 12686–12707

23. Pottmann, H.; Leopoldseder, S.; Hofer, M. Registration without ICP. *Comput. Vision Image Und.* **2004**, *95*, 54–71.

24. Ghilani, C.D. *Adjustment Computations: Spatial Data Analysis*, 5th ed.; John Wiley & Sons: New Jersey, NJ, USA, 2010.

25. Han, J.Y. A Noniterative Approach for the quick alignment of multistation unregistered LiDAR point clouds. *IEEE Geosci. Remote Sens. Lett.* **2010**, *7*, 727–730.

26. Akca, D. *Least Squares 3D Surface Matching*; ETH Zurich: Zürich, Switzerland, 2007.

27. Han, J.Y.; Chen, C.S.; Lo, C.T. Time-variant registration of point clouds acquired by a mobile mapping system. *IEEE Geosci. Remote Sens. Lett.* **2014**, *11*, 196–199.

28. Demantké, J.; Mallet, C.; David, N.; Vallet, B. Dimensionality based Scale Selection in 3D LiDAR Point Cloud. Availablie online: http://www.int-arch-photogramm-remote-sens-spatial-inf-sci.net/XXXVIII-5-W12/97/2011/isprsarchives-XXXVIII-5-W12-97-2011.pdf (accessed on 9 October 2014).

remote sensing

MDPI

Article
Multi-Feature Registration of Point Clouds

Tzu-Yi Chuang and Jen-Jer Jaw *

Department of Civil Engineering, National Taiwan University, Taipei City, Taiwan 10617; jtychuang@ntu.edu.tw
* Correspondence: jejaw@ntu.edu.tw; Tel.: +886-2-3366-4276

Academic Editors: Jie Shan, Juha Hyyppä, Richard Gloaguen and Prasad S. Thenkabail
Received: 23 October 2016; Accepted: 12 March 2017; Published: 16 March 2017

Abstract: Light detection and ranging (LiDAR) has become a mainstream technique for rapid acquisition of 3-D geometry. Current LiDAR platforms can be mainly categorized into spaceborne LiDAR system (SLS), airborne LiDAR system (ALS), mobile LiDAR system (MLS), and terrestrial LiDAR system (TLS). Point cloud registration between different scans of the same platform or different platforms is essential for establishing a complete scene description and improving geometric consistency. The discrepancies in data characteristics should be manipulated properly for precise transformation estimation. This paper proposes a multi-feature registration scheme suitable for utilizing point, line, and plane features extracted from raw point clouds to realize the registrations of scans acquired within the same LIDAR system or across the different platforms. By exploiting the full geometric strength of the features, different features are used exclusively or combined with others. The uncertainty of feature observations is also considered within the proposed method, in which the registration of multiple scans can be simultaneously achieved. The simulated test with an ideal geometry and data simplification was performed to assess the contribution of different features towards point cloud registration in a very essential fashion. On the other hand, three real cases of registration between LIDAR scans from single platform and between those acquired by different platforms were demonstrated to validate the effectiveness of the proposed method. In light of the experimental results, it was found that the proposed model with simultaneous and weighted adjustment rendered satisfactory registration results and showed that not only features inherited in the scene can be more exploited to increase the robustness and reliability for transformation estimation, but also the weak geometry of poorly overlapping scans can be better treated than utilizing only one single type of feature. The registration errors of multiple scans in all tests were all less than point interval or positional error, whichever dominating, of the LiDAR data.

Keywords: LiDAR; multiple features; registration; simultaneous adjustment; cross-platform

1. Introduction

Light detection and ranging (LiDAR) has been an effective technique for obtaining dense and accurate 3-D point clouds. Related applications have emerged in many fields because of advances made in the evolutions of data acquisition and processing [1,2]. Different LiDAR platforms are designed to operate with different scanning distances, scanning rates, and scanning directions. Therefore, point clouds generated from different platforms may vary in density, accuracy, coordinate systems, and scenic extent. Airborne LiDAR system (ALS) renders point clouds covering explicit top appearance of objects (Figure 1a), while mobiles LiDAR system (MLS) and terrestrial LiDAR system (TLS) provide façades and object surfaces (Figure 1b). To make use of the complementary information and generate a complete dataset of 3-D scenes (Figure 1c,d), data integration of different platforms or scans is indispensable. Even if the global navigation satellite systems (GNSS) and inertial navigation systems (INS) are available to the ALS and MLS, problems caused by position and orientation system, mounting errors, and shadowed effect warrant a rectification step for quality products. Thus, registration can be regarded as a pretreatment in terrestrial LiDAR data processing to transfer point clouds collected from different scans onto a suitable reference coordinate system, and also utilized as a refinement procedure to adjust small misalignments either between adjacent strips of airborne or neighboring mobile LiDAR data or between the data collected through two different platforms. A proper registration methodology, being able to tackle the inconsistency between different point clouds, appears crucial to numerous point cloud applications.

Indeed, considerable solutions for point cloud registration have been proposed in the literature. Existing methods can be classified into several types, in which surface- and feature-based approaches are commonly employed and yield high-quality results. The iterative closest point (ICP) algorithm developed by [3,4] is the most commonly used surface-based algorithm. Many follow-up improvements [5–8] have promoted ICP efficiency. In addition, other related studies, such as the iterative closest patch [9], the iterative closest projected point [10], and the least-squares surface matching method (LS3D) [11,12], have also proven their effectiveness in point cloud registration. On the other hand, numerous feature-based techniques have been developed based on geometric primitives, such as points, lines [13–16], planes [13,17–20], curves and surface [21], and specific objects [22]. Methods manipulating image techniques for laser scans [23–26], and refining the 3D fusion based on the correlation of orthographic depth maps [27] have also been studied for registration purposes. Wu et al. [28] employed points, lines, and planes to register lunar topographic models obtained by using different sensors. Han [29] and Han and Jaw [30] applied hybrid geometric features to solve a 3-D similarity transformation between two reference frames. An extended work of Chuang and Jaw [31] aimed at finding multiple feature correspondences with a side effect of sequentially gaining 3-D transformation parameters, facilitating the level of automatic data processing. Furthermore, the ideas of 2D feature descriptors, such as scale-invariant feature transform (SIFT) [32], smallest univalue segment assimilating nucleus (SUSAN) [33], and Harris [34] operators, were also extended as Thrift [35] and Harris 3D [36] to work with 3-D fusion and similarity search applications [37–42], but the current studies mostly undertook the evaluation of shape retrieval [43] or 3D object recognition [44] using techniques such as the shape histograms [45], scale-space signature vector descriptor [46], and local surface descriptors based on salient features [47].

Studies with a focus on cross-based registration can be found as in [15,48] which registered terrestrial and airborne LiDAR data based on line features and combinations of line and plane features extracted from buildings, respectively. Teo and Huang [49] and Cheng et al. [50] proposed a surface-based and a point-based registration, respectively, for airborne and ground-based datasets. However, the inadequate overlapping area of the point clouds of the ALS and MLS or TLS as the scenarios demonstrated in Figure 1a,b raises challenges for acquiring sufficient conjugate features and thus is apt to result in unqualified transformation estimation. In addition, the majority of post-processing methods perform pair-wise registration rather than simultaneous registration for consecutive scans, and the uncertainty of related observations has rarely been considered during the transformation estimation. The registration errors tend to accumulate as running each pair transformation sequentially. On the other hand, the ignorance of varying levels of feature observations errors would lead to the bias of the estimation.

It also drew our attention to the fact that the existing studies of feature-based point cloud registration employing line and plane features either utilized only partial geometric information [13,28] or needed to derive necessary geometric components with alternative ways [29,30]. In this study, we focused on the employment of features versatile in types as well as both complete and straightforward in geometric trajectory and quality information. The main contribution of this paper is the rigorous mathematical and statistical models of the multi-feature transformation capable of carrying out global registration for point cloud registration purposes. Considering the point, line, and plane features, the transformation model was driven by the existing models to improve the exploitation of features on both geometric and weighting aspects and to offer a simultaneous adjustment for global point cloud registration. Each type of feature can be exclusively used or combined with the others to resolve the 3-D similarity transformation. Since the multiple features with full geometric constraints are exploited, a better quality of geometric network is more likely to be achieved. To correctly evaluate the random effect of feature observations derived from the different quality of point cloud data, the variance–covariance matrix of the observations is considered based on the fidelity of data sources. Thus, the transformation parameters among multiple datasets are simultaneously determined via a weighted least-squares adjustment.

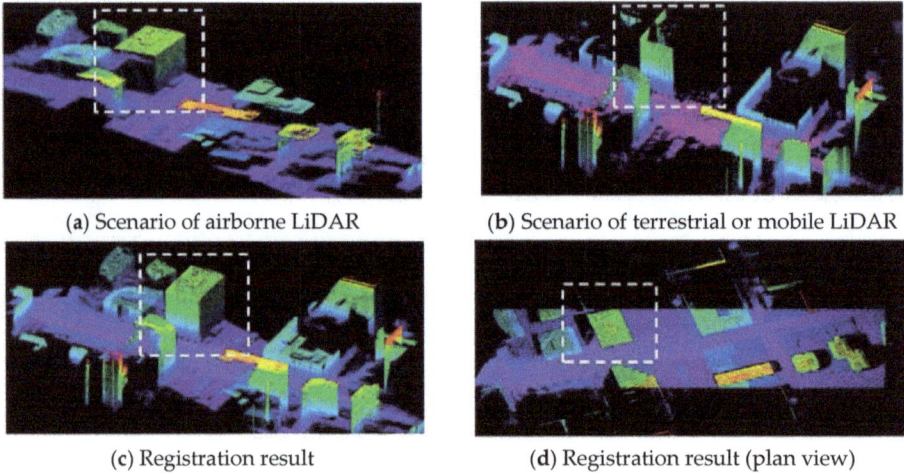

(**a**) Scenario of airborne LiDAR

(**b**) Scenario of terrestrial or mobile LiDAR

(**c**) Registration result

(**d**) Registration result (plan view)

Figure 1. The demonstration of cross-platform point cloud registration.

The workflow of the multi-feature registration involves three steps comprising the feature extraction, feature matching, and transformation estimation. Although considerable researches on extracting point, line [51], and plane [49,52] features from point clouds have been reported, fully automated feature extraction is still a relevant research topic. We leveraged a rapid plane extraction [53] coupled with a line extractor [51] to acquire accurate and evenly-distributed features and then performed the multiple feature matching approach called RSTG [31] to confirm the conjugate features and to derive approximations for the nonlinear transformation estimation model. Nevertheless, the authors placed a focus on transformation estimation modeling and left aside the feature extraction and feature matching with a brief address referred to the existing literature. Two equivalent formulae for line and plane features, respectively, using different parametric presentation were proposed in the transformation model. The detailed methodology of how the transformation models were established and implemented is given in the following paragraph. The terms "point feature" stands for "corner or point-like feature", while "line feature" employs only straight line feature throughout the paragraphs that follow.

2. Concepts and Methodology

The proposed multi-feature transformation model integrates point-, line-, and plane-based 3-D similarity transformations to tackle various LiDAR point cloud registrations. Seven parameters, including one scale parameter (s), three elements of a translation vector ($[t_X \ t_Y \ t_Z]^T$), and three angular elements of rotation matrix ($R\{\omega, \varphi, \kappa\}$), are considered for the transformation of scan pairs. The feasible parametric forms of observations used in this paper are the coordinates of points ($P(X_i, Y_i, Z_i)$), two endpoints ($L_e\{X_{i,1}, Y_{i,1}, Z_{i,1}, X_{i,2}, Y_{i,2}, Z_{i,2}\}$) as well as six-parameter ($L_p\{t_i, u_i, v_i, X_{0i}, Y_{0i}, Z_{0i}\}$) or four-parameter ($L_p\{d_i, e_i, 1, p_i, q_i, 0\}$) forms of lines, and two parametric forms of planes ($Pl_p\{\theta_i, \vartheta_i, \rho_i\}$ and $Pl_n\{a_i, b_i, c_i\}$). The different parametric forms for the same type of feature were proposed to fit to the need of feature collection or measurement. In fact, they are mathematically equivalent. Note that the geometric uncertainty, namely the variance–covariance matrix, of these feature observations has been integrated into the transformation estimation in this study.

2.1. Transformation Models

At the methodological level, the spatial transformation of point clouds can be established by the correspondence between conjugate points, the co-trajectory relation of conjugate lines, or conjugate planes. Equation (1) presents the Helmert transformation formula for two datasets on a point-to-point correspondence. To balance the transformation equation for solving seven parameters while providing a sufficient geometric configuration, at least three non-collinear corresponding point pairs are required for a non-singular solution. Furthermore, Equation (1) is employed to derive line- and plane-based transformations in the following paragraphs.

$$
F_{P_i} = \begin{bmatrix} X_{(2)i} \\ Y_{(2)i} \\ Z_{(2)i} \end{bmatrix} - sR \begin{bmatrix} X_{(1)i} \\ Y_{(1)i} \\ Z_{(1)i} \end{bmatrix} - \begin{bmatrix} t_X \\ t_Y \\ t_Z \end{bmatrix} = \begin{bmatrix} 0 \\ 0 \\ 0 \end{bmatrix}
\tag{1}
$$

with

$$
\begin{aligned}
R &= \begin{bmatrix} m_{11} & m_{12} & m_{13} \\ m_{21} & m_{22} & m_{23} \\ m_{31} & m_{32} & m_{33} \end{bmatrix} \\
&= \begin{bmatrix} \cos(\kappa)\cos(\varphi) & \sin(\kappa)\cos(\omega) + \cos(\kappa)\sin(\varphi)\sin(\omega) & \sin(\kappa)\sin(\omega) - \cos(\kappa)\sin(\varphi)\cos(\omega) \\ -\sin(\kappa)\cos(\varphi) & \cos(\kappa)\cos(\omega) - \sin(\kappa)\sin(\varphi)\sin(\omega) & \cos(\kappa)\sin(\omega) + \sin(\kappa)\sin(\varphi)\cos(\omega) \\ \sin(\varphi) & -\cos(\varphi)\sin(\omega) & \cos(\varphi)\cos(\omega) \end{bmatrix}
\end{aligned}
$$

where s is the scale parameter; $[t_X \ t_Y \ t_Z]^T$ is translation vector; R is a rotation matrix; and ω, φ and κ are the sequential rotation angles. $\left(X_{(1)i}, Y_{(1)i}, Z_{(1)i} \right)$ and $\left(X_{(2)i}, Y_{(2)i}, Z_{(2)i} \right)$ represent the coordinates of the ith conjugate point in systems 1 and 2, respectively.

2.1.1. Transformation for Line Features

Two types of 3-D line expressions commonly employed in data acquisition were considered. The transformation model for the two-endpoint type of the line features (Figure 2a) is established by collinear endpoints of the line features with their conjugate counterparts when transformed to another coordinate system. The collinearity property for one endpoint may present three alternative forms, one of which is formulized in Equation (2).

$$
F_{L_e} = \begin{cases} u_{(2)i}\left(X_{o(2)i} - X_{(1')i,j} \right) - t_{(2)i}\left(Y_{o(2)i} - Y_{(1')i,j} \right) = 0 \\ v_{(2)i}\left(Y_{o(2)i} - Y_{(1')i,j} \right) - u_{(2)i}\left(Z_{o(2)i} - Z_{(1')i,j} \right) = 0 \end{cases}
\tag{2}
$$

where i denotes the ith line; $j = \{1,2\}$; $\left(X_{(1')i,j}, Y_{(1')i,j}, Z_{(1')i,j} \right)$ shows the transformed jth endpoint $\left(X_{(1)i,j}, Y_{(1)i,j}, Z_{(1)i,j} \right)$ by Equation (1) from coordinate system 1 to system 2; for $\left(X_{o(2)i}, Y_{o(2)i}, Z_{o(2)i} \right)$, either endpoint of $\left(X_{(2)i,j}, Y_{(2)i,j}, Z_{(2)i,j} \right)$ represents the reference point of the ith conjugate line in coordinate system 2; and $\left[t_{(2)i} \ u_{(2)i} \ v_{(2)i} \right]^T$, derived by subtracting the corresponding coordinate component of the first end-point from the second one, represents the dicovt of the ith line in coordinate system 2.

As revealed in Equation (2), one pair of 3-D line correspondences contributes four equations (two for each endpoint) to the transformation. Similar to the two-endpoint model, the condition for the transformation model based on four-parameter line features can be established by constraining the direction vector and penetration point (which is determined by selecting the X–Y, Y–Z, or X–Z plane, called penetration plane, that supports the best intersection for the 3-D line) of the 3-D line to attain collinearity with their conjugate correspondents when transformed to another coordinate system (Figure 2b).

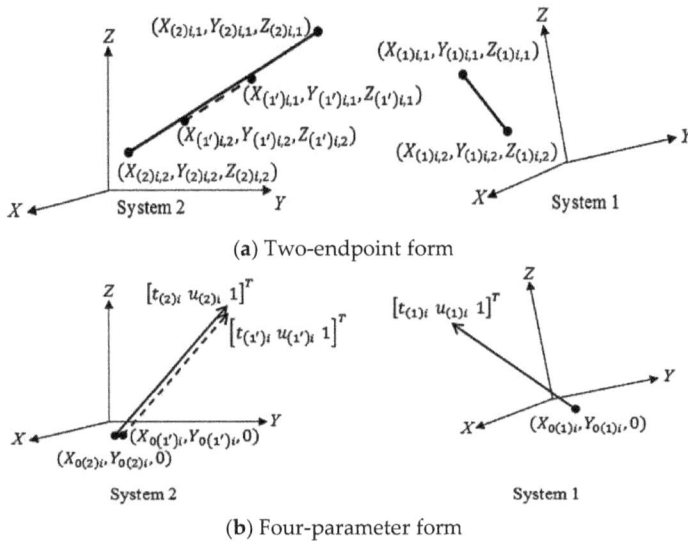

(a) Two-endpoint form

(b) Four-parameter form

Figure 2. Two-endpoint and four-parameter forms of the line-based transformation.

The right-hand side of Equation (3) shows the four independent parameters of the 3-D line when choosing X-Y intersection plane and they can also be derived from the six-parameter set. The mathematical formula of transformation based on four-parameter lines can be established in Equation (4).

$$
\begin{bmatrix} X_i \\ Y_i \\ Z_i \end{bmatrix} = \begin{bmatrix} X_{0i} \\ Y_{0i} \\ Z_{0i} \end{bmatrix} + k_i \begin{bmatrix} t_i \\ u_i \\ v_i \end{bmatrix} = \begin{bmatrix} p_i \\ q_i \\ 0 \end{bmatrix} + z_i \begin{bmatrix} d_i \\ e_i \\ 1 \end{bmatrix} \tag{3}
$$

where (X_{0i}, Y_{0i}, Z_{0i}) and $\begin{bmatrix} t_i & u_i & v_i \end{bmatrix}^T$ represent the reference point and direction vector of the ith line, respectively; $(p_i, q_i, 0)$ is the penetration point on the X–Y plane; $\begin{bmatrix} d_i & e_i & 1 \end{bmatrix}^T$ is the reductive direction vector; and k_i and z_i are the scaling variables.

$$
F_{L_p}^+ = \begin{cases} u_{(2)i}\left(X_{0(2)i} - X_{0(1')i}\right) - t_{(2)i}\left(Y_{0(2)i} - Y_{0(1')i}\right) = 0 \\ v_{(2)i}\left(Y_{0(2)i} - Y_{0(1')i}\right) - u_{(2)i}\left(Z_{0(2)i} - Z_{0(1')i}\right) = 0 \\ u_{(2)i}t_{(1')i} - t_{(2)i}u_{(1')i} = 0 \\ v_{(2)i}u_{(1')i} - u_{(2)i}v_{(1')i} = 0 \end{cases} \tag{4}
$$

where $\left(X_{0(1')i}, Y_{0(1')i}, Z_{0(1')i}\right)$ denotes the transformed reference point (six-parameter type) or penetration point $\left(X_{0(1)i}, Y_{0(1)i}, Z_{0(1)i}\right)$ of the ith line by Equation (1) from coordinate system 1 to system 2; $\left(X_{0(2)i}, \, Y_{0(2)i}, \, Z_{0(2)i} \right)$ is the reference point or penetration point of the *i*th conjugate line in coordinate system 2; and $\left[t_{(1')i} \, u_{(1')i} \, v_{(1')i} \right]^{T}$ represents the transformed direction vector $\left[t_{(1)i} \, u_{(1)i} \, v_{(1)i} \right]^{T}$ of the *i*th line by Equation (5) from coordinate system 1 to system 2.

$$\begin{bmatrix} t_{(1')i} \\ u_{(1')i} \\ v_{(1')i} \end{bmatrix} = R \begin{bmatrix} t_{(1)i} \\ u_{(1)i} \\ v_{(1)i} \end{bmatrix} \tag{5}$$

The six positional variables in Equation (4) may vary with the different penetration planes. Moreover, the conjugate lines from different coordinate systems may have varied penetration planes. Table 1 shows the appropriate use of parameters according to the facing condition.

Table 1. Variable setting in the four-parameter model of line-based transformation.

Penetration Plane	Parameter	Constant
X–Y	$X_{0i} = p_i, \, Y_{0i} = q_i, \, t_i = d_i, \, u_i = e_i$	$Z_{0i} = 0, \, v_i = 1$
Y–Z	$Y_{0i} = p_i, \, Z_{0i} = q_i, \, u_i = d_i, \, v_i = e_i$	$X_{0i} = 0, \, t_i = 1$
X–Z	$X_{0i} = p_i, \, Z_{0i} = q_i, \, t_i = d_i, \, v_i = e_i$	$Y_{0i} = 0, \, u_i = 1$

The dual representation forms of the line-based transformation model given in Equations (2) and (4) are mathematically equivalent and suggest that one pair of conjugate lines offer four equations. At least two non-coplanar conjugate 3-D line pairs must be used to solve the transformation parameters. Furthermore, the line trajectory correspondence underpinning the 3-D line transformation allows flexible manipulation of feature measurements compared with the point-to-point approach.

2.1.2. Transformation for Plane Features

With the similar manipulation mentioned in the line-based model, two equivalent formulae using three independent parameters for plane-based transformation model were formed. A plane can be formulated in a normal vector form by considering the dot product of the normal vector and the point vector of the plane as one (Equation (6)) (zero if the plane includes the origin), or it can be formulated in a polar form by utilizing horizontal angle (ϑ) and zenith angle (θ) along with the projection distance (ρ) from the origin (Equation (8)). Notice that the normal vector employed in Equation (6) is the normalized normal vector of the plane divided by the projection distance. The transformation models are expressed in Equations (7) and (9) and both forms are depicted in Figure 3.

$$Pl_n\{a_i, \, b_i, \, c_i\} = a_i X + b_i Y + c_i Z = \begin{bmatrix} a_i \\ b_i \\ c_i \end{bmatrix} \cdot \begin{bmatrix} X \\ Y \\ Z \end{bmatrix} = 1 \tag{6}$$

$$F_{pl_n} = \begin{cases} a_{(1)i} - s \left(a_{(2)i}m_{11} + b_{(2)i}m_{21} + c_{(2)i}m_{31} \right) / \left(1 - a_{(2)i}t_X - b_{(2)i}t_Y - c_{(2)i}t_Z \right) = 0 \\ b_{(1)i} - s \left(a_{(2)i}m_{12} + b_{(2)i}m_{22} + c_{(2)i}m_{32} \right) / \left(1 - a_{(2)i}t_X - b_{(2)i}t_Y - c_{(2)i}t_Z \right) = 0 \\ c_{(1)i} - s \left(a_{(2)i}m_{13} + b_{(2)i}m_{23} + c_{(2)i}m_{33} \right) / \left(1 - a_{(2)i}t_X - b_{(2)i}t_Y - c_{(2)i}t_Z \right) = 0 \end{cases} \tag{7}$$

where *i* denotes the *i*th plane; and $\left[a_{(1)i} \, b_{(1)i} \, c_{(1)i} \right]^{T}$ and $\left[a_{(2)i} \, b_{(2)i} \, c_{(2)i} \right]^{T}$ stand for the normal vectors of the *i*th conjugate plane in coordinate systems 1 and 2, respectively.

$$sin\theta_i cos\vartheta_i X + sin\theta_i sin\vartheta_i Y + cos\theta_i Z = \rho_i \tag{8}$$

$$F^+_{Pl_p} = \begin{cases} sin\theta_{(1)i}cos\vartheta_{(1)i}/\rho_{(1)i} - F_1/F_4 = 0 \\ sin\theta_{(1)i}sin\vartheta_{(1)i}/\rho_{(1)i} - F_2/F_4 = 0 \\ cos\theta_{(1)i}/\rho_{(1)i} - F_3/F_4 = 0 \end{cases} \tag{9}$$

with

$$F_1 = s\left(sin\theta_{(2)i}cos\vartheta_{(2)i}m_{11} + sin\theta_{(2)i}sin\vartheta_{(2)i}m_{21} + cos\theta_{(2)i}m_{31}\right);$$

$$F_2 = s\left(sin\theta_{(2)i}cos\vartheta_{(2)i}m_{21} + sin\theta_{(2)i}sin\vartheta_{(2)i}m_{22} + cos\theta_{(2)i}m_{23}\right);$$

$$F_3 = s\left(sin\theta_{(2)i}cos\vartheta_{(2)i}m_{31} + sin\theta_{(2)i}sin\vartheta_{(2)i}m_{32} + cos\theta_{(2)i}m_{33}\right);$$

$$F_4 = \rho_{(2)i} - sin\theta_{(2)i}cos\vartheta_{(2)i}t_X - sin\theta_{(2)i}sin\vartheta_{(2)i}t_Y - cos\theta_{(2)i}t_Z;$$

where $(\theta_{(1)i}, \vartheta_{(1)i}, \rho_{(1)i})$ and $(\theta_{(2)i}, \vartheta_{(2)i}, \rho_{(2)i})$ represent the parameter sets of the ith conjugate planes in coordinate systems 1 and 2, respectively. In fact, with $sin\theta_i cos\vartheta_i/\rho_i = a_i$, $sin\theta_i sin\vartheta_i/\rho_i = b_i$, and $cos\theta_i/\rho_i = c_i$, Equations (7) and (9) are mathematically equivalent.

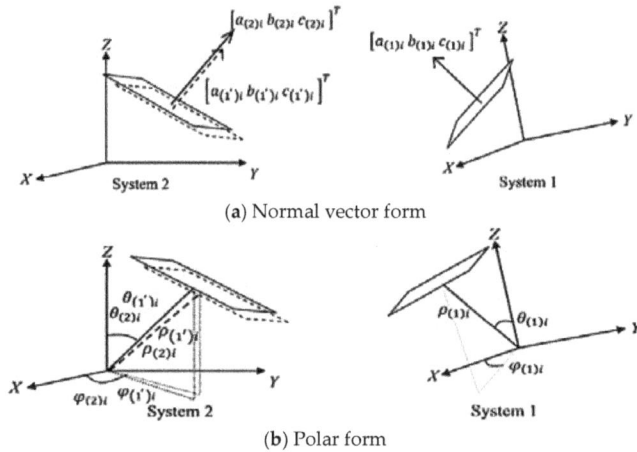

Figure 3. Normal vector and polar forms of the plane-based transformation.

2.2. Contributions of Features toward the Transformation Model

The characteristics of geometric constraints provided by these three types of observations vary from one type to another. A single point only contributes to translational (or locative) information, whereas unknown rotation and scale must be solved by points that span the area of interest. By contrast, a line feature provides the co-trajectory constraint along the line. This feature leaves the degrees of freedom of one scale, one translation, and one rotation parameter. A plane feature results in singularities in one scale, one rotation, and two translation parameters on the plane. Thus, line-based and plane-based approaches require features that supply complementary constraints for an entire solution. The application of different primitives is extremely appealing for the support of effective geometric constraints and can resolve situations that are considered difficult when the traditional point-based approaches are used.

Table 2 summarizes the transformation modes for the three primitives, where the numbers of independent pair correspondences for points, lines, and planes are notated by n_{Pt}, n_L, and n_{Pl}, respectively, and n_s is the number of scans. The effective number of equations generated by each feature is identical to the number of independent parameters, i.e., three, four, and three for a pair correspondence of point, line, and plane, respectively. The third column in Table 2 shows that the line features outperform the others in terms of redundancy.

Table 2. The configurations of the transformation models.

Mode	Number of Equations	Redundancy	Number of Features for a Minimum Solution as $n_s = 2$
Point-based	$3n_{Pt}$	$3n_{Pt} - 7(n_s - 1)$	3
Line-based	$4n_L$	$4n_L - 7(n_s - 1)$	2
Plane-based	$3n_{Pl}$	$3n_{Pl} - 7(n_s - 1)$	4

When dealing with two scans, three point feature pairs with nine equations or two non-coplanar line feature pairs with eight equations are able to solve seven parameters of transformation. Nevertheless, according to Equations (7) and (9), it simply states that one conjugate plane pair will produce three equations which convey all the transformation parameters. However, four pairs of independent and well-distributed conjugate planes are the minimum requirements to solve seven parameters because planes, even if spatially isolated, may not be so complementary as an effective response to the transformation parameters. This statement can be understood by considering that three planes mutually perpendicular to each other contain only the information of three translation elements and three rotation angles even they supply nine equations. The scale factor cannot be fixed until a fourth plane is added.

Moreover, the intersected feature may gain geometric significance together with good quality of positional location at the expense of failing to completely represent the geometric property of the original data used for the intersection. An obvious example is that a corner point resulting from intersecting three neighboring planes brings forth only three equations for transformation, whereas these three planes are able to solve six parameters of 3-D rigid-body transformation. Besides the distinctive geometric characters elaborated by the individual primitive, appreciating the feature combinations realized by one point + one line, one point + two planes, one line + two planes, or two points + one plane in a minimum solution for seven-parameter similarity transformation of two scans is practically meaningful. Such configurations are arranged and analyzed in the experiments.

2.3. Simultaneous Adjustment Model for Global Registration

As mentioned before, the registration process for multiple scans often follows a pair-wise approach and then combines all results for the entire datasets. Consequently, registration errors between neighboring point clouds would gradually accumulate and result in a larger impact on those data distant from the reference one. In contrast, a global adjustment approach that simultaneously addresses the registration of all involved scans would minimize errors among all datasets. In such a simultaneously global registration manner, each feature observation in a scan is associated with only one residual, instead of receiving a residual as it is conducted by passing the pairs if the feature appears in more than two scans.

The Gauss–Helmert model of the least-squares adjustment is used to cope with diversified uncertainty treatment of registration tasks. Equation (10) expresses the mathematical form of the simultaneous adjustment model. Once a data frame is chosen as the reference, all other frames are co-registered by conducting the multi-feature transformation.

$$w_{j \times 1} = A_{j \times [7(n_s-1)]} \xi_{[7(n_s-1)] \times 1} + B_{j \times k}(y_{k \times 1} + e_{k \times 1}), \; e \sim \left(0, \Sigma = \sigma_0^2 W^{-1}\right) \tag{10}$$

with

$$\begin{cases} j = \sum_{i=1}^{n_{m_{Pt}}} (3(n_{Point,i} - 1)) + \sum_{i=1}^{n_{m_L}} (4(n_{Line,i} - 1)) + \sum_{i=1}^{n_{m_{Pl}}} (3(n_{Plane,i} - 1)) \\ k = 3\sum_{i=1}^{n_{m_{Pt}}} n_{Point,i} + n_l \sum_{i=1}^{n_{m_L}} n_{Line,i} + 3\sum_{i=1}^{n_{m_{PL}}} n_{Plane,i} \end{cases}$$

where Equations (2) and (7) are used for line and plane features in this formula. y, e, w, and ξ denote the observation vector, error vector, discrepancy vector, and incremental transformation parameter vector, respectively; W is the weight matrix; A and B are the partial derivative coefficient matrices with respect to unknown parameters and observations, respectively; σ_0^2 is the variance component; $n_{m_{Pt}}$,

n_{m_L}, and $n_{m_{PL}}$ are the numbers of total matched mates of points, lines, and planes, respectively; $n_{Point,i}$, $n_{Line,i}$, and $n_{Plane,i}$ represent the numbers of the conjugate features of ith matched mate; $n_l = 6$ based on Equation (2), otherwise $n_l = 4$ based on Equation (4); and n_s is the number of all scans. On the other hand, if Equations (4) and (9) are employed, the $F_{L_{ei}}$ and $F_{Pl_{ni}}$ in matrices A and B should be replaced with $F_{L_{pi}}^{+}$ and $F_{Pl_{pi}}^{+}$, and the $\left\{ L_{e(1),i}, L_{e(2),i} \right\}$ and $\left\{ Pl_{n(1),i}, Pl_{n(2),i} \right\}$ are replaced with $\left\{ L_{p(1),i}, L_{p(2),i} \right\}$ and $\left\{ Pl_{p(1),i}, Pl_{p(2),i} \right\}$, respectively.

According to the Equation (10), the convergent least-squares solutions of the estimated parameter vector $\hat{\xi}$, residual vector $\tilde{e}$, posterior variance component $\hat{\sigma}_0^2$, and posterior variance–covariance matrix of the estimated parameter $\hat{\Sigma}_{\hat{\xi}}$ can be computed through Equations (11) to (14), where r represents the redundancy.

$$\hat{\xi} = \left(A^T \left(BW^{-1}B^T \right)^{-1} A \right)^{-1} A^T \left(BW^{-1}B^T \right)^{-1} w \tag{11}$$

$$\tilde{e} = W^{-1}B^T \left(BW^{-1}B^T \right)^{-1} (w - A\hat{\xi}) \tag{12}$$

$$\hat{\sigma}_0^2 = \tilde{e}^T W \tilde{e} / r \tag{13}$$

$$\hat{\Sigma}_{\hat{\xi}} = \hat{\sigma}_0^2 \left(A^T \left(BW^{-1}B^T \right)^{-1} A \right)^{-1} \tag{14}$$

with

$$W = \sigma_0^2 \begin{bmatrix} \Sigma_{Point}^{-1} & \Sigma\Sigma_{Point, Line}^{-1} & \Sigma\Sigma_{Point,Plane}^{-1} \\ \Sigma\Sigma_{Point, Line}^{-1} & \Sigma_{Line}^{-1} & \Sigma\Sigma_{Line,Plane}^{-1} \\ \Sigma\Sigma_{Point,Plane}^{-1} & \Sigma\Sigma_{Line,Plane}^{-1} & \Sigma_{Plane}^{-1} \end{bmatrix};$$

The uncertainty of the observations, resulting from feature extraction, is considered and propagated based on how the feature parameters, namely the parametric forms as previously introduced, are acquired to form the variance–covariance matrix (Σ) of the observed features. The $\Sigma\Sigma$ indicates the covariance between two different feature observations if needed. The quantities revealed in the variance–covariance matrix present the feature quality that often results from the measuring errors and discrepancy between the real scene and the hypothesized scene. Basing the observation uncertainty on the fidelity of the data source is considered an optimal way of assessing and modeling weights in this work. The weight matrix (W), which is derived from the variance–covariance matrix of the observations, is then fed into the adjustment model.

3. Experiments and Analyses

The effectiveness of the registration quality provided by the individual and combined features was validated using the simulated data. The configurations of simulated data were designed to put more concerns on validating mathematical models in solving the transformation parameters with varied features, different levels of data errors, minimal and nearly minimal observations, thus rendered so ideal as well as fundamental as compared to the practical data that often involved much complicated scenes with feature distribution and quality demanding cautious treatment.

Real scenes with feature appearances collected by TLS, ALS and MLS that dealt with both single- and cross-platform registrations were subsequently demonstrated to reveal the feasibility and effectiveness of the proposed method. To fairly evaluate the merits of the proposed registration scheme in handling multiple scans, we compared our results with those obtained by the ICP algorithm initiated by Besl and McKay [3] and the line-based method [51]. These methods were coded in Matlab and realized on a Windows system.

The customary way of assessing transformation and registration quality is to utilize the root mean square error (RMSE) based on the check points with RMSE_P = $\sqrt{(\text{RMSE\_X})^2 + (\text{RMSE\_Y})^2 + (\text{RMSE\_Z})^2}$. Under the mode of multi-feature transformation, the positional discrepancies of the conjugate points, geometric relationships quantified by angles and distances for both conjugate lines and planes are considered as registration quality indicators. Apart from the RMSE criterion, a distance indicator $Q_{distance}$ in Equation (15), the average of the spatial distances of all the conjugate line segments and plane patches, is used to evaluate positional agreement. Similarly, an angular indicator Q_{angle} in Equation (16), the average angle derived from all the conjugate line segments and plane patches, is used for evaluating geometric similarity.

$$Q_{distance} = 0.5 \left(\sum_{i=1}^{m_{line}} D_{L_i} / m_{line} + \sum_{i=1}^{m_{plane}} D_{Pl_i} / m_{plane} \right) \tag{15}$$

$$Q_{angle} = 0.5 \left(\sum_{i=1}^{m_{line}} A_{L_i} / m_{line} + \sum_{i=1}^{m_{plane}} A_{Pl_i} / m_{plane} \right). \tag{16}$$

where D_{L_i} is the average distance of each middle point of the ith conjugate line segments to its line counterpart along the direction perpendicular to the line, and D_{Pl_i} is the average distance of each centroid of the ith conjugate planes to its plane counterpart along the normal direction. m_{line} and m_{plane} are the total numbers of independent conjugate line and plane pairs, respectively. A_{L_i} is the angle of the ith conjugate line pair computed by the direction vectors. A_{Pl_i} is the angle of the ith conjugate plane pair computed by the normal vectors.

The positional accuracy of each laser point is a function of range, scanning angles, and positioning/pose observations, if applicable. The proposed quality indicators reflect the registration accuracy upon the reference data. The point interval of a point cloud intimates the minimum discrepancy of the raw quality considering the discrete and blind data nature. Thus, we intuitively included the point spacing of the point cloud into quality analysis for the real datasets, particularly if it prevails over the point error. In addition to employing check features, we also introduced the registered features into Equation (15) to derive internal accuracy (IA), which has the same meaning as the mean square error (MSE) index of the ICP algorithm [3]. The index resulting from transformation adjustment would not only be used to assess the registration quality but also suggest whether the internal accuracy and the external one derived from check features are consistent.

3.1. Simulated Data and Evaluation

To simplify the scene geometry and concentrate on validating the mathematical models, simulated data with a size of approximately 10 m × 10 m × 10 m were generated in this experiment. Figure 4a shows a point cloud forming a cube, in which 6 planes, 12 lines, and 8 points can be extracted as observations. Four layers of check points, denoted as rhombuses in a total of 400 points, as shown in Figure 4b, were densely arranged and situated inside the cube for exclusively evaluating the registration quality. The point cloud was transformed using a pre-defined set of similarity transformation parameters to represent the data viewed from another station. To simulate the random errors of terrestrial scans, both the simulated and transformed point clouds were contaminated with a noise of zero mean and σ cm standard deviation, where $\sigma \in \{0.5, 1.5, 2.5, 3.5, 4.5\}$ in each coordinate component. In this case, the points and lines directly measured from the point clouds were considered the coarse features as they (points and endpoints) were determined with the same positional quality as that of the original point. On the other hand, features extracted by fitting process or by intersecting neighboring fitted planes were deemed the fine observations with better accuracy than original point cloud.

First, the influence of the coarse features versus fine features on the transformation quality was observed as well as the registration effectiveness derived from individually and unitedly as the observations when σ = 1.5 cm was added. This exhibited that transformation can be solved by either applying a single type of features or the combination of multiple features. Figure 5 reveals the registration quality. To assess the quality of estimated transformation parameters, the solved parameters were used to transform the 400 check points from one station to another and computed the RMSE_P.

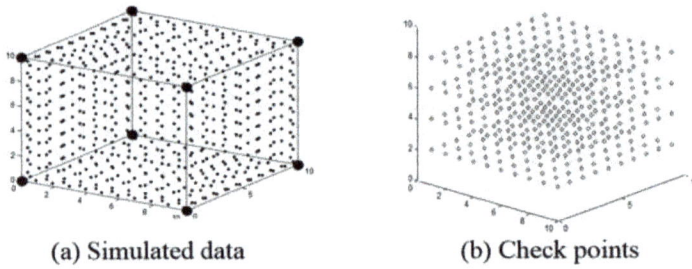

(a) Simulated data   (b) Check points

Figure 4. Illustrations of experimental configurations.

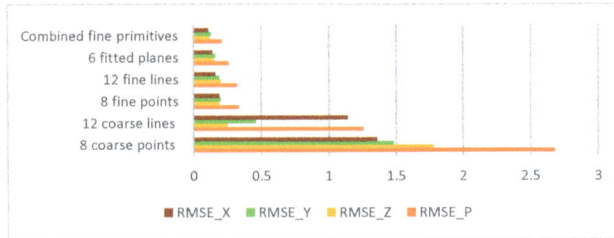

Figure 5. Quality assessment.

In light of Figure 5, applying fine features only in this case achieves registration quality ranging from 0.23 cm to 0.35 cm under the 1.5 cm standard deviation in each coordinate component of the point clouds, which is much better than the results of applying coarse features only, ranging from 1.26 to 2.68 cm. Using fine features even supplies transformation parameter estimation with sub-point accuracy of point clouds in this case. Obviously, combining all features also results in satisfactory registration quality. Further, to hypothesize the TLS data acquisition with reasonable error range, Figure 6 shows the RMSE_P obtained from each observation set when the standard deviation of the point clouds is set as $\sigma \in \{0.5\ cm, 1.5\ cm, 2.5\ cm, 3.5\ cm, 4.5\ cm\}$, respectively. As anticipated, combining all types of fine features resulted in the best RMSE_P under each data quality. As the deviation increased, namely worsened observations, the advantage of the feature integration became more obvious. It can be seen in Figure 6 that the error trend of employing feature combination, benefiting from the accurate observations and high redundancy (83 in this case), grew moderately as compared to those utilizing single feature type.

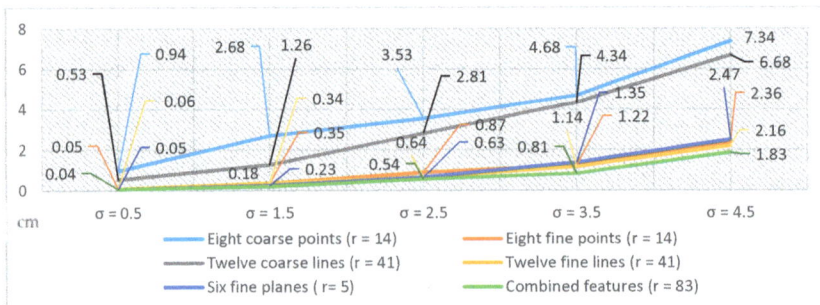

Figure 6. Quality assessment (*r* indicates the redundancy).

Further, nine minimal or nearly minimal configurations of observations for transformation estimation were investigated to show the complementariness among observed points, lines, and planes. The distribution of the features is illustrated in Figure 7a–i. Specifically, Figure 7a–c was designed to analyze the different benefits gained from an intersected point versus three planes that determined the point. Figure 7d–i was configured to verify the feasibility of multiple feature integration by laying out the feature combinations under the condition of only a few or nearly minimum measurements for estimating the transformation parameters. The evaluations of transformation quality in relation to all the cases are presented in Figure 8, where *r* indicates the redundancy.

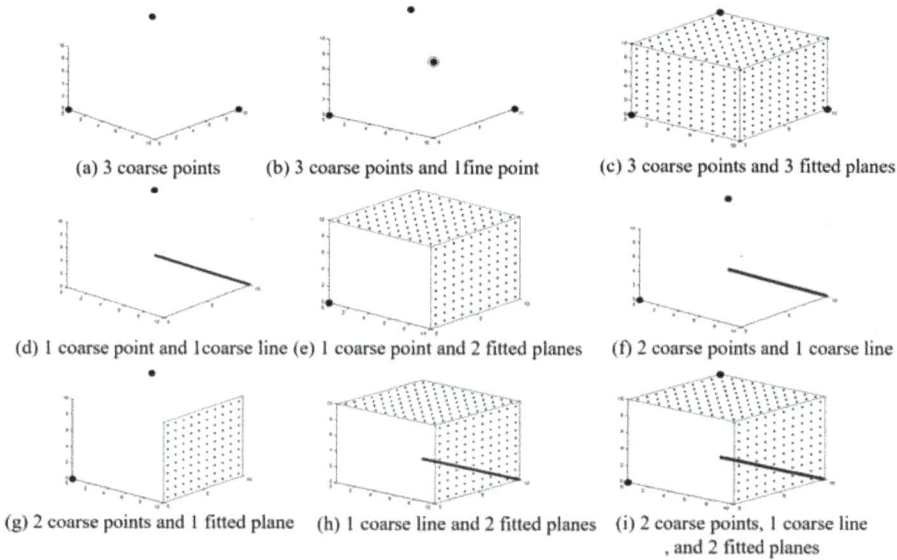

(a) 3 coarse points   (b) 3 coarse points and 1 fine point   (c) 3 coarse points and 3 fitted planes

(d) 1 coarse point and 1coarse line   (e) 1 coarse point and 2 fitted planes   (f) 2 coarse points and 1 coarse line

(g) 2 coarse points and 1 fitted plane   (h) 1 coarse line and 2 fitted planes   (i) 2 coarse points, 1 coarse line , and 2 fitted planes

Figure 7. Illustrations of feature combinations.

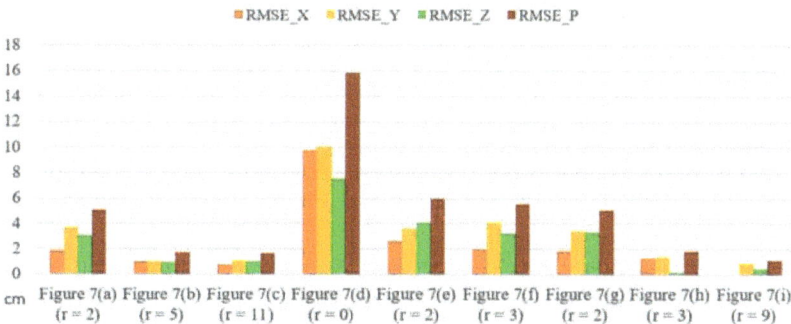

Figure 8. Assessment of the transformation quality of feature combinations.

Again, transformations in all nearly minimal solution configurations were resolved. From the comparison of transformation results of Figure 7b versus Figure 7c, both sharing the three coarse points as shown in Figure 7a, it showed that when the three fitted planes, deemed as fine features, were added, the transformation accuracy was slightly increased with a lower RMSE_P (0.07 cm) than adding the fine point intersected by these three planes. This effect gave a numerical example to further add to what was stated in Section 2.2. In the configurations of Figure 7d–h, the transformation estimations using the united features under the minimum measurement condition were demonstrated. It should be noted that the specified number of each feature was insufficient to solve the transformation alone in these cases. Considering the RMSE results in Figure 8, the combination of one point and one line (Figure 7d) resulted in dramatic transformation errors because of its poor distribution and zero redundancy. Transformation errors were reduced with better-distributed features, as those in Figure 7a,e–h. On the other hand, the configuration of Figure 7i which combined the well-situated two points, one line, and two planes, gained the best transformation quality among all the cases.

Considering points are relatively likely to be obstructed between scans than other features as well as the distribution of planes in urban scenes is often monotonous, combinations of these feature primitives would bring more adaptability to conquer distribution restrictions frequently confronted in practical cases. The simulated experiments demonstrated the flexibility of feature combinations and quantified the merits of solving the point cloud registration in a multi-feature manner.

3.2. Real LiDAR Data

To validate the advantage of the proposed registration scheme on real scenes, three cases consisting of the registration of terrestrial point clouds, the registration between terrestrial and airborne LiDAR point clouds, and the registration between mobile and airborne LiDAR point clouds were conducted. The first case demonstrated the routine terrestrial point cloud registration, while the second and third cases tackled the point clouds with different scanning directions, distances, point densities, accuracy, and having rare overlapping areas which is unfavorable for most surface-based approaches, e.g., ICP [3] or LS3D [11].

To evaluate the effectiveness of the registration quality, the registration results were compared with those derived from the ICP algorithm [3] and the line-based transformation model [51]. The comparison with the ICP highlighted the essential differences of the algorithm in terms of operational procedures (e.g., calculating strategy and with/without the need of manual intervention) and the adaptability for cross-based LiDAR platforms. On the other hand, the comparison with the line-based transformation was to illustrate the advantages of feature integration on the operational flexibility and accuracy improvement. Moreover, the last case demonstrated both the feasibility and working efficiency of the proposed method in an urban area with larger extent as compared to the other two cases. Notably, the experiments were performed on the point clouds without any data pre-processing including noise removal, downsampling, or interpolation.

3.2.1. Feature Extraction and Matching

As fully automated point cloud extraction is still a developing research topic, the proposed registration scheme may couple with existing feature extraction approaches. From a standpoint of point cloud registration, to efficiently acquire qualified and well-distributed features is more important than acquiring complete features of the point cloud. In this study, we leveraged a rapid extractor [51,53] to acquire line and plane features and subsequently intersected neighboring lines or planes for points as long as an intersection condition is met [53]. Afterward, a 3D multiple feature matching approach [31] was implemented to obtain corresponding features and the approximations of similarity transformation between two data frames. The results were then introduced into the proposed model for rigorous transformation estimation. Figure 9 shows the workflow of feature acquisition in this study.

Conceptually, the extractor conducted a coarse-to-fine strategy to increase working efficiency. The point clouds were first converted into a range image and a 2.5-D grid data with a reasonable grid size according to the data volume, and the correspondence of coordinates was preserved. Although the process deteriorated the quality of the original data, the computational complexity was reduced. For line features, Canny edge detection and Hough transformation were implemented in the range image for detecting straight image lines. The image coordinates of line endpoints were then referred back to the original point cloud and used to collect candidate points for 3-D line fitting. On the other hand, the normal vectors of each 2.5-D grid cell were computed and clustered to find the main directions of planes in the point cloud. The information was introduced to the plane Hough transformation as constraints to reduce the searching space, so the computational burden for the plane extraction can be eased. Subsequently, lines or planes that met the intersection condition were used to generate point features [53]. Taking lines as an example, the extreme points within the group of points which described the line were first computed. Then, the extreme points of the line, being the endpoints, were used to compute the distances with the extreme points of other lines. If the shortest distance between the two lines was smaller than a tolerance (pre-defined by point space of data), the two lines were regarded as neighbors and used to intersect point feature. With the aid of this condition, virtual and unreliable intersected points, along with unnecessary calculations, were avoided. Besides, the quality measures of the resulting features were provided in the form of a variance–covariance matrix. Indeed, even though the coarse-to-fine procedure, which reveals approximate areas where features are possibly located and then aims at these areas to precisely extract the features from the raw point cloud, promotes the working efficiency and encourages the level of automation, the feature extraction is still the most computationally expensive step in this study.

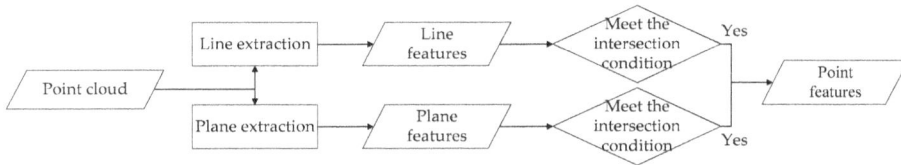

Figure 9. The structure for feature acquisition.

Afterward, the extracted features were introduced to the multiple feature matching approach, called RSTG [31], to confirm the feature correspondence and to derive the initial values for the transformation estimation. The RSTG is the abbreviation of the matching approach which comprises four steps, namely rotation alignment, scale estimation, translation alignment, and geometric check. It applied a hypothesis-and-test process to retrieve the feature correspondence and to recover the Helmert transformation in an order of rotation, scale, then translation. The rotation matrix was determined by leveraging the vector geometry [29] and singular value decomposition, whereas the scale factor was calculated by a distance ratio between two coordinate systems. After that, the translation vector was derived from a line-based similarity transformation model [51] based on a collinearity constraint. The approach got rid of the point-to-point correspondence of the Helmert transformation, leading to a more flexible process. Details of the extraction and matching processes can be referred to [31,51,53].

3.2.2. Registration of Terrestrial Point Clouds

In this case, five successive scans, which cover a volume of 26 m (length) × 18 m (width) × 10 m (height), collected by Trimble (Mensi) GS200 describing the main gate of National Taiwan University in Taipei, Taiwan, were used. The nominal positional accuracy of 4 shots as reported by the Trimble manufacturer was up to 2.5 mm at 25 m range. The details of this dataset and the point clouds are shown in Table 3 and Figure 10, respectively.

The number of acquired features in all scans was 95 features in total, consisting of 17 points, 57 lines, and 21 planes. They were passed on to the RSTG matcher. Figure 11a illustrates these features superimposed on the point clouds. The extracted lines were highlighted in red lines, the extracted points were marked in blue rhombuses, and the points belonging to the same plane were coded by the same color. Figure 11b shows the feature correspondences, where the corresponding planes are in blue.

Table 3. Information of the terrestrial point clouds.

	Scan 1	Scan 2	Scan 3	Scan 4	Scan 5
Number of points	529,837	624,229	1,432,191	1,105,417	635,438
Avg. scanning distance (m)	26.0	29.7	21.9	22.6	28.2
Density (pts./m^2)	8726	1190	2475	2383	1409
Avg. point spacing (cm)	1.1	2.9	2.0	2.0	2.7
Nominal positional accuracy (mm)	2.6	3.0	2.2	2.3	2.8

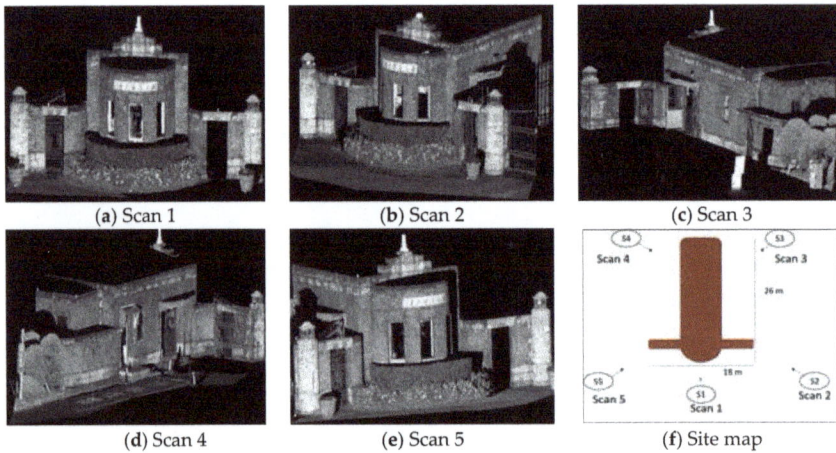

(a) Scan 1   (b) Scan 2   (c) Scan 3

(d) Scan 4   (e) Scan 5   (f) Site map

Figure 10. Illustration of the terrestrial scans.

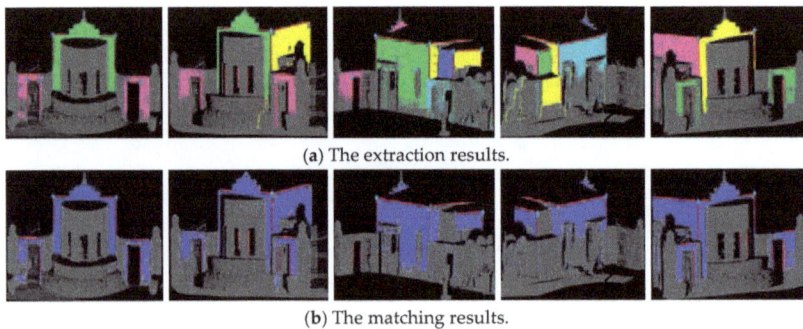

(a) The extraction results.

(b) The matching results.

Figure 11. The feature extraction and matching results.

The RSTG approach successfully matched three points, 17 lines, and seven planes among all overlapping scan pairs. The feature correspondences and the approximations of transformation parameters were delivered to the process of transformation estimation. The registration of the terrestrial LiDAR point clouds was performed by the proposed simultaneous adjustment model of the multi-feature transformation. In addition, the approximations were applied to the ICP algorithm and the line-based transformation. To evaluate the registration quality, there were 24 check points acquired by best-fit spherical markers as well as 18 check lines and six check planes measured evenly and manually serving as independent check features.

The overall quality of registration indicated in Table 4 revealed that the positional and distance discrepancies were about 1 cm, and the angle deviation was about 0.3 degrees. The positional discrepancy RMSE_P derived from check points was quite close to the distance discrepancy ($Q_{distance}$) calculated by check lines and planes. The angle deviation also represented similar spatial discrepancy as the other two indicators when considering the employed line features within 3 m length. This suggested that these three quality indicators were effective and consistent.

Table 4. The quality assessment of the registration results.

	MSE (cm)	IA-$Q_{distance}$/std. (cm)	Q_{angle}/std. (deg.)	$Q_{distance}$/std. (cm)	RMSE_P (cm)
ICP	0.01	-	0.48/0.04	1.30/0.23	1.36
Line-based	-	0.87/0.13	0.32/0.02	1.09/0.18	1.17
Proposed method	-	0.75/0.12	0.31/0.02	1.04/0.15	1.03

In light of Table 3, the quality and density of the point cloud data were high. As a result, the three approaches yielded similar registration performances at the same quantitative level. Both the internal accuracy (IA) and the external accuracy derived from the check features reported satisfactory registration. The proposed multi-feature method slightly outperformed the baseline implementation of the ICP and line-based transformation methods based on the same configuration: the same initial values and one-time execution. It is well known that the ICP method has numerous variants which apply other techniques, such as limiting the maximum distance in the X, Y and Z directions or filtering points containing large residuals, to reduce the number of inadequacy points because points far from the data center do not contribute important information about the scene and some of them could even be considered as noise. The afore-mentioned process would trade between the points in the vertical and the horizontal surfaces to avoid a final misalignment if the point number of one of the surfaces greatly dominates the other. Indeed, the registration quality can be improved by adjusting parameters and executing the calculation repeatedly. However, our study was interested in comparing the fundamental methodologies without ad-hoc parameter tuning and repeated manual intervention to assess the feasibility and effectiveness of these algorithms.

The results in Table 4 validated that integrating multiple features not only offered better operational flexibility, but also gained better registration quality than just exploiting line features. The visual inspections of the registration results of the proposed method can be found in Figure 12. The point clouds of each scan are drawn with different colors, where the discrepancy in overlapping areas is nearly undetectable and the facades collected by different scans are registered precisely.

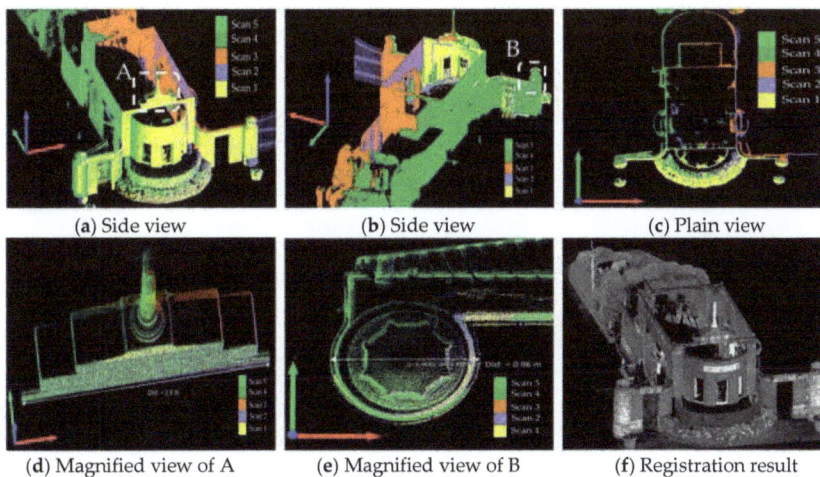

(a) Side view	(b) Side view	(c) Plain view
(d) Magnified view of A	(e) Magnified view of B	(f) Registration result

Figure 12. The registration results by the proposed method.

The registration working process is nearly fully automated. In this case, the execution time for feature acquisition was about 3.24 min, and the matching and estimation process was about 41 and 2.7 s, respectively. The working time of the extraction and matching processes highly relates to the amount and complexity of point cloud data. Once the corresponding features are determined, the transformation estimation will be done in few seconds.

3.2.3. Registration of Terrestrial and Airborne Point Clouds

A complete scene of man-made structures that often appears in urban areas was considered in this case. A total of seven successive scans of terrestrial LiDAR point clouds were collected by the Optech ILRIS-3-D system. As displayed in Figure 13a–g, the scans cover the facades of a campus library building with a size of approximately 94 m (length) × 75 m (width) × 40 m (height). The nominal positional accuracy of a single shot for ILRIS-3-D is 8 mm at a range of 100 m. In addition, the airborne point cloud of the same building as shown in Figure 13h was collected via Optech ALTM 3070 at a flying height of 450 m with a reported elevation accuracy of up to 15 cm at a range of 1200 m and horizontal accuracy better than $(1/2000) \times$ flying height, which is approximately 23 cm in this case. Table 5 presents the information of all point clouds, in which scan 6 rendered the largest number of points and the highest density due to the large coverage of the target and the perpendicular scanning to the building facade.

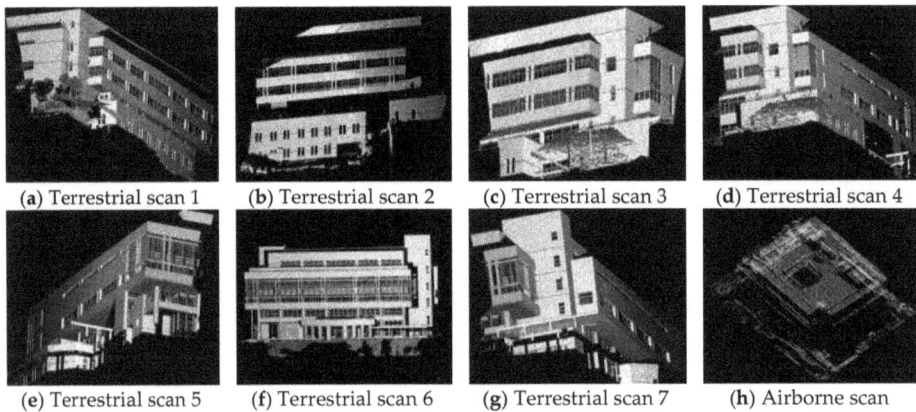

(**a**) Terrestrial scan 1   (**b**) Terrestrial scan 2   (**c**) Terrestrial scan 3   (**d**) Terrestrial scan 4

(**e**) Terrestrial scan 5   (**f**) Terrestrial scan 6   (**g**) Terrestrial scan 7   (**h**) Airborne scan

Figure 13. Terrestrial and airborne LiDAR point clouds.

Table 5. Information of the point clouds.

Scan	Number of Points	Avg. Scanning Distance (m)	Avg. Point Spacing (cm)	Nominal Positional Accuracy (mm)
Terrestrial scan 1	543,001	37.366	9.0	2.9
Terrestrial scan 2	574,007	37.716	6.7	3.0
Terrestrial scan 3	565,367	37.831	5.5	3.0
Terrestrial scan 4	511,254	37.601	9.1	3.0
Terrestrial scan 5	484,352	33.779	7.3	2.7
Terrestrial scan 6	1,745,490	94.370	4.6	7.5
Terrestrial scan 7	728,603	36.145	7.7	2.9
Airborne scan	37,678	450	44.7	237

The extracted features were superimposed onto the point clouds, as illustrated by colors, following a similar coding as previous cases, in Figure 14. There were 297 well distributed features extracted within the scene. The numbers of extracted features are listed in Table 6.

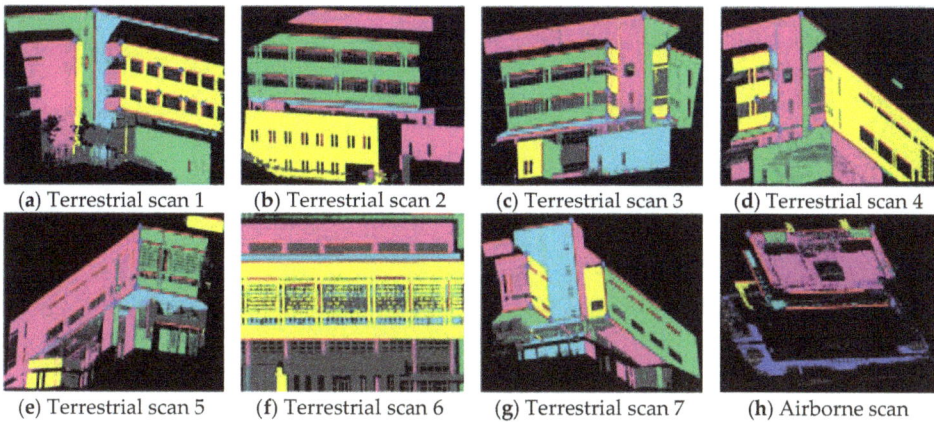

(a) Terrestrial scan 1   (b) Terrestrial scan 2   (c) Terrestrial scan 3   (d) Terrestrial scan 4

(e) Terrestrial scan 5   (f) Terrestrial scan 6   (g) Terrestrial scan 7   (h) Airborne scan

Figure 14. The feature extraction results.

Table 6. Numbers of extracted features in each scan.

	T-1	T-2	T-3	T-4	T-5	T-6	T-7	A-1	Total
Point	11	1	9	8	4	0	7	3	43
Line	33	23	35	29	23	15	22	13	193
Plane	7	6	11	10	7	5	9	6	61
Total	51	30	55	47	34	20	38	22	297

As seen in Figure 15, each terrestrial scan rarely overlaps the airborne point cloud, and no single feature can fulfill the minimum requirement of the transformation estimation due to the poor overlap geometry in this case. Therefore, both the matching and transformation tasks were alternatively turned to process TLS first and then match and transform TLS to ALS. Again, the RSTG approach was applied to find the corresponding features of seven terrestrial scans and to render the approximations simultaneously. The successfully matched features among all overlapping scan pairs comprised nine points, 39 lines, and 15 planes, as shown in Figure 15a–g. The corresponding features were colored with the blue rhombus points, red lines, and blue planes, respectively. The correspondences along with the variance–covariance matrix of features and the approximations of transformation parameters were then introduced into the multi-feature transformation. However, the corresponding features between terrestrial scan 1 and scan 7 were not sufficient to construct stable transformation leading to an open-loop registration in this case. With the same basis of the initial transformation values, the ICP and line-based methods were performed to align all scans onto the reference frame (scan 6) accordingly.

Apart from the observed features, there were 20 lines and six planes, coded by green color in Figure 15, manually measured serving as independent check features to evaluate registration quality of the terrestrial registration. With the larger scanning distances and distance variations than the previous case, the significance of scale factor was studied in this case. Table 7 shows the estimated seven transformation parameters using the proposed simultaneous adjustment, whereas Table 8 reveals the six parameter results when the scale factor was fixed as 1. Further, the quantitative quality assessment of the terrestrial registration results is displayed in Table 9.

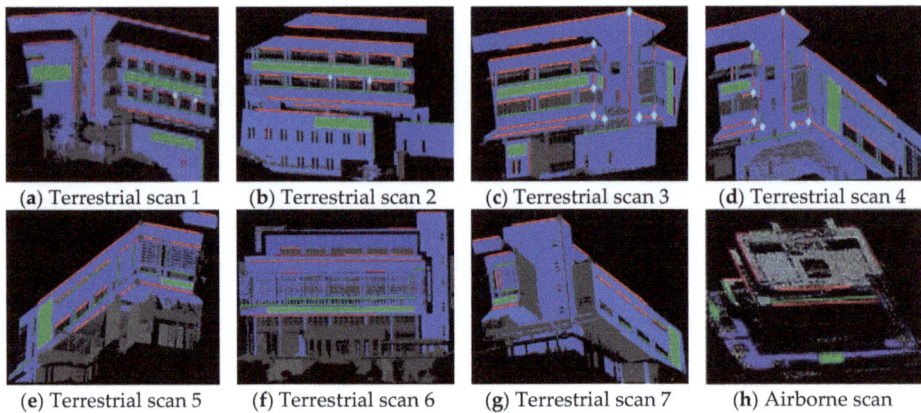

(a) Terrestrial scan 1   (b) Terrestrial scan 2   (c) Terrestrial scan 3   (d) Terrestrial scan 4

(e) Terrestrial scan 5   (f) Terrestrial scan 6   (g) Terrestrial scan 7   (h) Airborne scan

Figure 15. The feature matching results.

Table 7. The estimated transformation parameters.

Parameters	S	ω (rad)	φ (rad)	κ (rad)	T_X (m)	T_Y (m)	T_Z (m)	$\hat{\sigma}_0$
Scan 1-6	1.0151	−3.4493	2.8075	−1.6508	−94.612	53.321	−18.798	
Scan 2-6	1.0098	−3.2476	2.9891	−1.8388	−71.119	34.458	−12.529	
Scan 3-6	0.9987	−3.1344	9.4287	−15.905	0.10513	0.0277	0.3087	0.855
Scan 4-6	0.9975	18.248	−12.14	2.3005	−68.693	126.67	−53.181	
Scan 5-6	1.002	−3.0229	−3.559	−14.151	−152.29	88.494	−13.467	
Scan 7-6	1.0012	2.8475	15.355	−7.9346	−93.779	53.096	−17.995	

Table 8. The estimated transformation parameters.

Parameters	S	ω (rad)	φ (rad)	κ (rad)	T_X (m)	T_Y (m)	T_Z (m)	$\hat{\sigma}_0$
Scan 1-6	1.0	−3.4493	2.8075	−1.6508	−94.614	53.323	−18.798	
Scan 2-6	1.0	−3.2476	2.9891	−1.8388	−71.118	34.456	−12.530	
Scan 3-6	1.0	−3.1344	9.4287	−15.905	0.105	0.028	0.310	0.894
Scan 4-6	1.0	18.248	−12.14	2.3005	−68.692	126.67	−53.180	
Scan 5-6	1.0	−3.0229	−3.559	−14.151	−152.290	88.495	−13.466	
Scan 7-6	1.0	2.8475	15.355	−7.9346	−93.779	53.097	−17.995	

Table 9. The quality assessment of the terrestrial point cloud registration.

	MSE (cm)	IA-$Q_{distance}$/std. (cm)	Q_{angle}/std. (deg.)	$Q_{distance}$/std. (cm)
ICP	3.31	-	9.30/1.31	6.82/2.42
Line-based	-	1.97/0.23	0.64/0.40	2.04/0.27
Proposed method	-	1.35/0.11	0.42/0.31	1.61/0.21
Proposed method ($S = 1$)	-	1.61/0.47	0.44/0.39	1.73/0.64

In the six-parameter case, the scale factor was fixed as 1, ignoring this effect as most rigid body-based methods would do. Considering Tables 7–9, the values of posterior variance $\hat{\sigma}_0$, internal accuracy, Q_{angle}, and $Q_{distance}$ report that taking the scale factor into account can obtain better registration results. The scale may bear certain systematic and random errors in this case, suggesting that solving seven parameters should render more comprehensive applications. However, the solved scale parameter needs also to be verified through significance test or based on prior information, if any, as not to accept a distorted solution. For example, the scale factor provided upon RSTG can be a very good approximation to be referred to. In Table 9, the proposed method yielded the best registration result, about 0.4 degrees in angle deviation and 1.6 cm in distance discrepancy, which is smaller than the average point spacing of the raw point clouds. Both the line-based and the proposed methods kept the consistency between the internal and external accuracy while the ICP did not. The MSE estimates resulted from each registration pair of the ICP nevertheless reached up to 3.3 cm, which indicated high internal consistency, even though the external accuracy was quite poor compared to either the proposed or line-based methods. This would suggest that the computation of the ICP might converge toward a local minimum. In addition, it could be understood that the ICP was affected by noise and its sequential pair-wise registration scheme made the cumulative errors grow significantly while dealing with open-loop scans.

Figure 16 shows the distance and angular discrepancies at each scan pair. It reveals that the misalignment errors of the line-based and proposed methods were spread more uniformly among successive point clouds, while the discrepancies of the ICP were accumulated. Figure 17 illustrates the visual inspections of the registration result, where the point clouds of each scan are depicted in different colors. With the better feature distribution and redundancy (186 in this case), this terrestrial experiment highlighted the merits of the proposed simultaneous least-squares adjustment and the multiple feature integration in dealing with the global registration of successive scans.

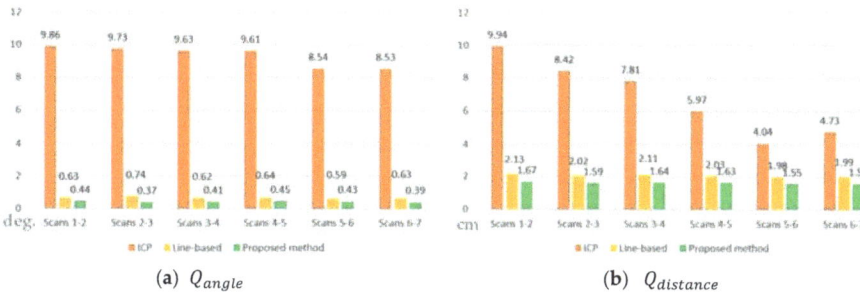

(a) Q_{angle} (b) $Q_{distance}$

Figure 16. The discrepancies in each scan pair.

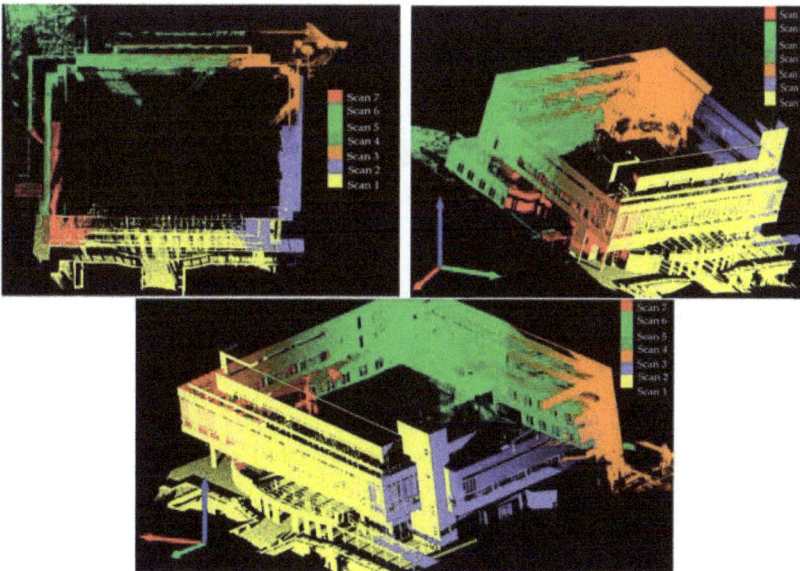

Figure 17. The terrestrial registration results reviewing from different directions.

After accomplishing the terrestrial registration, all features had been transformed onto the same coordinate system. Again, the RSTG approach was implemented to the terrestrial and airborne feature datasets for finding the conjugate features. There were nine lines and one plane correspondences found between the local coordinate and the global coordinate systems. It should be noted that the bulk of these data involved large geometric discrepancy raising the challenge of finding feature correspondences and the unavailability of reliable conjugate points. Nevertheless, it can be seen in Figure 15 that the normal vectors of plane features either in terrestrial or airborne scans are of less variation in geometric distribution. That is, most of the normal vectors of planes point to similar directions. It is intrinsically due to the specific scanning axis of LiDAR platform which would restrict the usage of surface-based approaches and plane features in cross-platform point cloud registration.

After feature extraction and matching, the TLS data were registered onto the ALS one by the three methods, and the registration quality was evaluated by 6 check lines and 1 check plane which were manually extracted from the point clouds. Figure 18 shows the matched features (red lines and blue plane) and check features (green).

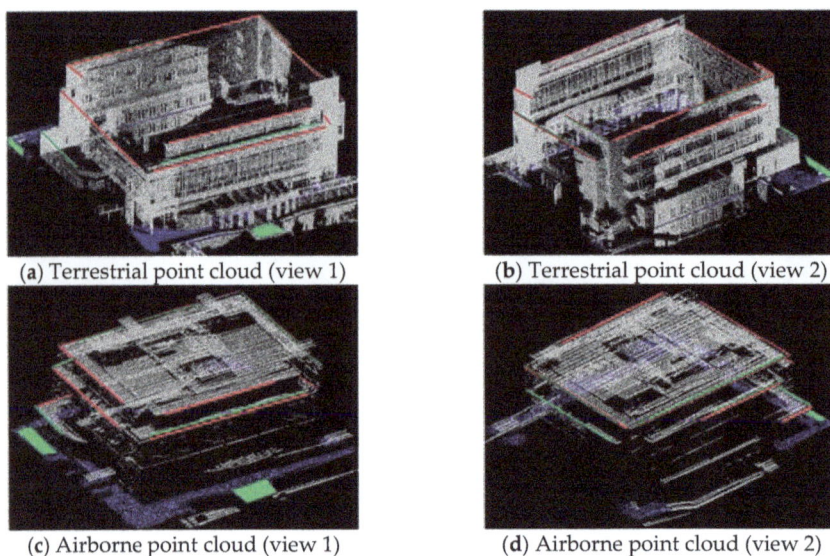

(a) Terrestrial point cloud (view 1)

(b) Terrestrial point cloud (view 2)

(c) Airborne point cloud (view 1)

(d) Airborne point cloud (view 2)

Figure 18. The feature correspondences and check features.

As shown in Table 10, the check features point out that the registration quality of the proposed method is about 12 cm in distance discrepancy, slightly better than that of the line-based method. Apparently, considering the nature of the data characters, the dominant registered errors would inherit from the airborne point cloud. On the other hand, the ICP algorithm spoiled the registration in this case because the terrestrial and airborne datasets barely contained overlapping areas along the boundaries of the roof. The insufficient overlapping surfaces and the uneven data quality led the ICP method to converge to a local minimum. Indeed, the uneven data quality also directly affected the effectiveness of feature extraction and thus resulted in unequal accuracy of feature observations. The proposed multi-feature transformation did introduce variance–covariance matrices of observations into the adjustment process based on the fidelity of the data quality. If the uncertainty of features was not considered, namely unweighted, the registration quality deteriorated and the positional quality dropped almost 2 cm as compared to weighted result in this case. Therefore, by assigning appropriate weights and adjusting the feature observations accordingly would lead to the better transformation estimations.

Figure 19 shows the visual inspections on the registered point clouds. It can be seen that the point spacing between airborne and terrestrial scans was extremely distinct. However, the junctions and the profiles in Figure 19b outlined the build boundary quite consistently. In addition, it is worth mentioning that the quantity of distance discrepancy listed in Table 10 is far less than the average point spacing (44 cm) of the airborne point cloud.

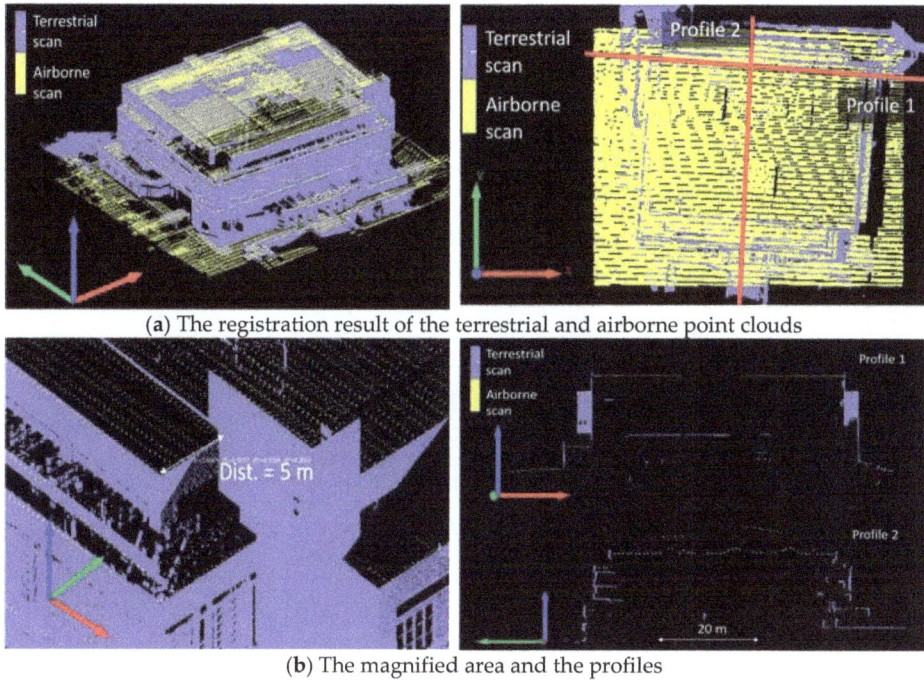

(a) The registration result of the terrestrial and airborne point clouds

(b) The magnified area and the profiles

Figure 19. The visual inspections on the registered point clouds.

Table 10. The quality assessment of the airborne and terrestrial point cloud registration.

	IA-$Q_{distance}$/std. (cm)	Q_{angle}/std. (deg.)	$Q_{distance}$/std. (cm)
ICP	-	-	-
Line-based	11.13/1.13	2.03/0.07	12.48/1.32
Proposed method (unweighted)	12.73/3.41	2.06/0.04	13.62/1.79
Proposed method (weighted)	11.12/1.11	2.02/0.03	11.23/1.24

The registration between TLS and ALS manipulated only the line and plane features since the conjugate points were not available in this case. As a result, the internal accuracy of the line-based and the proposed methods appeared very similar but the external accuracy indicated that the proposed method was slightly better than the line-based method. The registration quality of line-based method may be influenced by the check plane which was away from the effective area of line observations. Even though line features were abundant, their reliability could be inferior to plane features considering their high uncertainty in the discontinuity area of point clouds [54]. Therefore, integrating multiple features into registration tasks would offer better solutions, especially for those scans rendering poor overlapping geometry. The total execution time of this case was about 4.5 min for feature extraction, 76.8 s for matching, and 2.4 s for transformation estimation.

3.2.4. Registration of Mobile and Airborne Point clouds

In this case, we applied the proposed method to the point cloud registration between mobile LiDAR system (MLS) and airborne LiDAR system (ALS). Regarding ALS and MLS, the exterior orientation of the sensor is typically known, so point cloud registration serves as a means of producing a best-fit surface through an adjustment process that compensates for small misalignments between adjacent data sets. Figure 20 shows the mobile and airborne scans of an urban area in Toronto, Canada. The mobile point clouds were collected utilizing Optech LYNX system. The absolute accuracy was 5 cm at 10 m range [55]. The airborne point cloud was collected by Optech ALTM 3100 at a flying height of 385 m with a reported elevation accuracy of up to 15 cm at a range of 1200 m [55] and approximately 19.2 cm horizontal accuracy in this case. The overlap area between the two data was about 412 m in length and 162 m in width, as shown in Figure 20a. The information of point clouds can be found in Table 11.

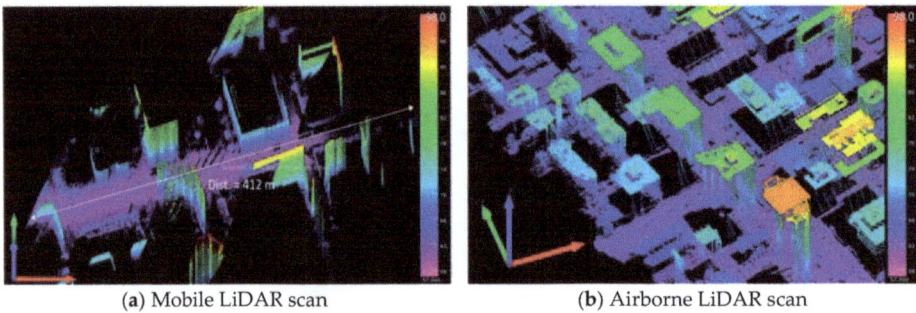

(a) Mobile LiDAR scan   (b) Airborne LiDAR scan

Figure 20. Mobile and airborne LiDAR data.

Table 11. Information of the point clouds.

Scan	Number of Points	Avg. scanning Distance (m)	Avg. Point Spacing (cm)	Nominal Positional Accuracy (cm)
Mobile scan	2,911,811	47.7	3	24
Airborne scan	766,488	385	40	19.8

In order to reduce the computational complexity, we trimmed the airborne point cloud according to the mobile one and left the overlap area for feature extraction and matching. As illustrated in Figure 21a, there were two points, nine lines, and one plane retrieved from the overlap area, and even-distributed eight lines and one plane were manually collected as check features for quality assessment.

The quality indicators in Table 12 show that the remaining discrepancies are 8.7 cm in distance and 0.01 degrees in geometric similarity. Nevertheless, the visual registration result in Figure 21b demonstrates the detailed roof and façade of building structures, where the purple and the yellow points represented the airborne LiDAR and mobile LDAR data, respectively.

Table 12. The quality assessment of the airborne and mobile point cloud registration.

	IA-$Q_{distance}$/std. (cm)	Q_{angle}/std. (deg.)	$Q_{distance}$/std. (cm)
Proposed method	6.33/1.01	0.017/0.002	8.72/1.18

Apart from the quantity evaluation, three profiles (Figure 21c) were selected to visually inspect the junctions between MLS and ALS after registration. Profile 1 displayed the road point cloud and showed the consistency in plain area; and Profiles 2 and 3 demonstrated the roof and façade of the same building in two directions. The geometric agreement appeared not only on solid structure objects, but also on the tree-crown. Figure 21d exhibits the integrated point cloud result colored by height, which benefits further geometric exploration. The feature extraction process took about 7.5 min since the space extent of this point cloud data was larger than previous cases. The operation time for the feature matching and transformation was around 3.12 s.

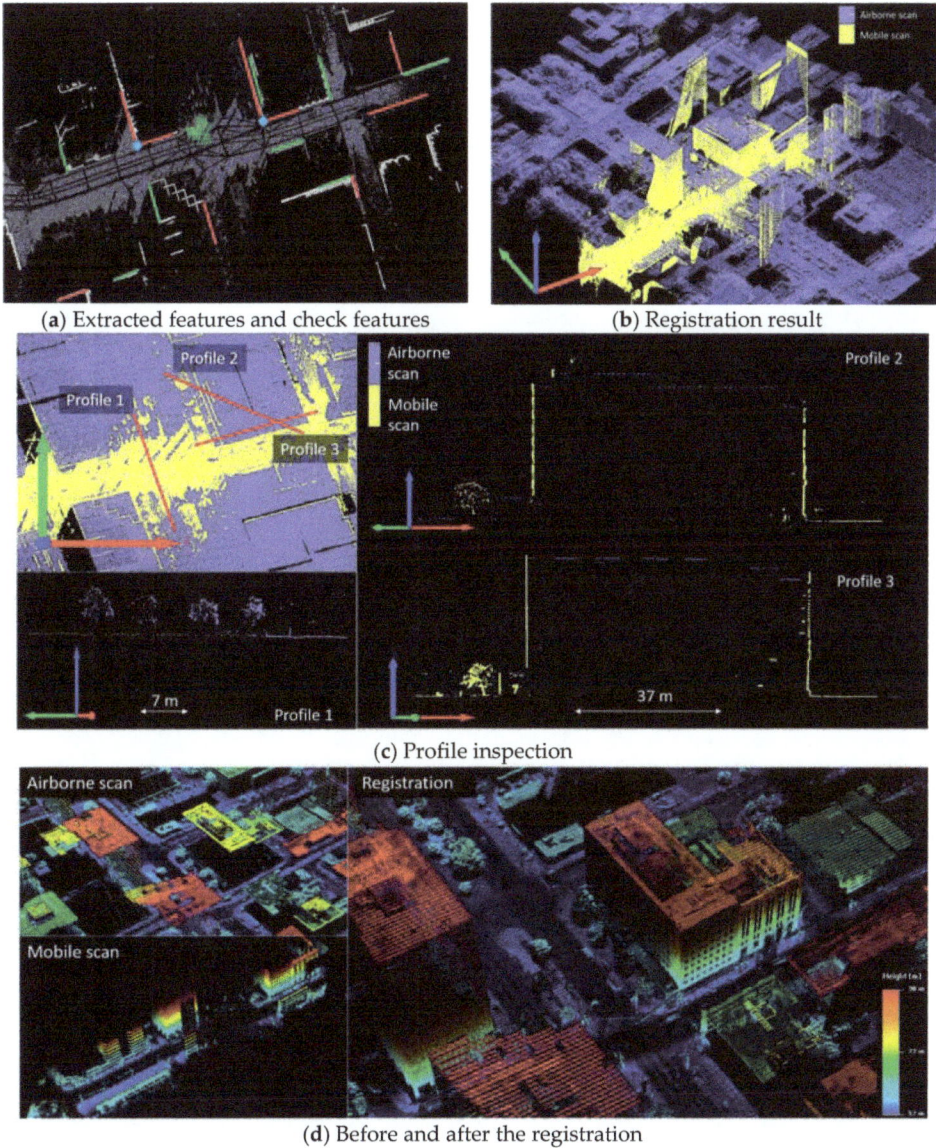

(**a**) Extracted features and check features (**b**) Registration result

(**c**) Profile inspection

(**d**) Before and after the registration

Figure 21. Cross-platform LiDAR registration.

4. Conclusions and Further Work

In this paper, the authors presented a multi-feature transformation model, exploiting the full geometric properties and random information of points, lines, and planes, for the registration of LiDAR point clouds. Simulated and real datasets were used to validate the robustness and effectiveness of the proposed method. The essential behavior of individual features through coarse versus fine measurements, single type of feature versus combined features, different levels of point cloud errors, and minimal and nearly minimal configuration of combined features applicable to the transformation was realized and evaluated both qualitatively and quantitatively by simulated data. The experimental results using real scene datasets, on the other hand, showed that the multi-feature transformation obtained better registration quality compared with the line-based and the surface-based ICP methods in the first two cases with features inherited in the scene. This highlights the advantages of integrating multiple features in a simultaneous registration scheme and suggests a comprehensive transformation model for dealing with scenes involving complicated structures rich in line and plane features. Practically speaking, by combining the existing feature extraction and feature matching mechanism, the multi-feature transformation can be considered an automated and efficient technique for achieving accurate global registration. Under current computing configuration, it consumed about four and six minutes for all processes in Sections 3.2.2 and 3.2.3 , respectively, with feature extraction taking about 80% of total computation time. In addition, this paper investigated into the effectiveness using different feature primitives to exploit the advantages of feature combinations toward registration purposes within the same platform and across different platforms. Furthermore, Case 3, with much larger extent of operation area as compared to the other two cases, successfully demonstrated the applicability of the proposed method tackling the cross-platform registration between ALS and MLS point clouds with satisfactory result.

To increase the working flexibility and feasibility, future improvements following the proposed method will explore other feature types and integrate surface- and feature-based approaches to where geometric features may serve as constraints to gain registration benefit in an effective way.

Acknowledgments: The authors express the most sincere thanks to the Ministry of Science and Technology (formerly the National Science Council), Taiwan, ROC for granting this study through projects NSC 100-2221-E-002-216, 98-2221-E-002-177-MY2, and 97-2221-E-002-194. In addition, this publication would not be possible without the constructive suggestions from three reviewers, which are greatly appreciated.

Author Contributions: Jen-Jer Jaw provided the research conception and design and analysis of the experiments and revised and finalised the manuscript. Tzu-Yi Chuang contributed to drafting the manuscript and the implementation of proposed algorithms and the analysis of experiments.

Conflicts of Interest: The authors declare no conflict of interest.

References

1. Shan, J.; Toth, C. *Topographic Laser Ranging and Scanning: Principles and Processing*; CRC Press: Boca Raton, FL, USA, 2008; p. 590.
2. Vosselman, G.; Maas, H.G. *Airborne and Terrestrial Laser Scanning*; CRC Press: Boca Raton, FL, USA, 2010; p. 320.
3. Besl, P.J.; McKay, N.D. A method for registration of 3-D shape. *IEEE Trans. Pattern Anal. Mach. Intell.* **1992**, *14*, 239–256. [CrossRef]
4. Chen, Y.; Medioni, G. Object modeling by registration of multiple range images. *Image Vis. Comput.* **1992**, *10*, 145–155. [CrossRef]
5. Mitra, N.J.; Gelfand, N.; Pottmann, H.; Guibas, L. Registration of point cloud data from a geometric optimization perspective. In Proceedings of the 2004 Eurographics Symposium on Geometry Processing, Nice, France, 8–10 July 2004.
6. Makadia, A.; Patterson, A.; Daniilidis, K. Fully automatic registration of 3-D point clouds. In Proceedings of the IEEE Computer Society Conference on Computer Vision and Pattern Recognition, New York, NY, USA, 17–22 June 2006; pp. 1297–1304.

7. Bae, K.-H.; Lichti, D.D. Automated registration of unorganised point clouds from terrestrial laser scanners. In Proceedings of the XXth ISPRS Congress: Geo-Imagery Bridging Continents, Istanbul, Turkey, 12–23 July 2004; pp. 222–227.

8. Bae, K.H.; Lichti, D.D. A method for automated registration of unorganised point clouds. *ISPRS J. Photogramm. Remote Sens.* **2008**, *63*, 36–54. [CrossRef]

9. Habib, A.; Bang, K.I.; Kersting, A.P.; Chow, J. Alternative methodologies for LiDAR system calibration. *Remote Sens.* **2010**, *2*, 874–907. [CrossRef]

10. Al-Durgham, M.; Habib, A. A framework for the registration and segmentation of heterogeneous LiDAR data. *Photogramm. Eng. Remote Sens.* **2013**, *79*, 135–145. [CrossRef]

11. Gruen, A.; Akca, D. Least squares 3-D surface and curve matching. *ISPRS J. Photogramm. Remote Sens.* **2005**, *59*, 151–174. [CrossRef]

12. Akca, D. Co-registration of surfaces by 3-D least squares matching. *Photogramm. Eng. Remote Sens.* **2010**, *76*, 307–318. [CrossRef]

13. Stamos, I.; Leordeanu, M. Automated feature-based range registration of urban scenes of large scale. In Proceedings of the IEEE Computer Society Conference on Computer Vision and Pattern Recognition, Madison, WI, USA, 18–20 June 2003; pp. 555–561.

14. Habib, A.; Mwafag, G.; Michel, M.; Al-Ruzouq, R. Photogrammetric and LiDAR data registration using linear features. *Photogramm. Eng. Remote Sens.* **2005**, *71*, 699–707. [CrossRef]

15. Von Hansen, W.; Gross, H.; Thoennessen, U. Line-based registration of terrestrial and aerial LiDAR data. In Proceedings of the International Society for Photogrammetry and Remote Sensing Congress, Beijing, China, 3–11 July 2008; pp. 161–166.

16. Al-Durgham, M.; Habib, A. Association-matrix-based sample consensus approach for automated registration of terrestrial laser scans using linear features. *Photogramm. Eng. Remote Sens.* **2014**, *80*, 1029–1039. [CrossRef]

17. Von Hansen, W. Registration of Agia Sanmarina LiDAR Data Using Surface Elements. In Proceedings of the ISPRS Workshop on Laser Scanning 2007 and SilviLaser 2007, Espoo, Finland, 12–14 September 2007; pp. 93–97.

18. Brenner, C.; Dold, C. Automatic relative orientation of terrestrial laser scans using planar structures and angle constraints. In Proceedings of the ISPRS Workshop on Laser Scanning 2007 and SilviLaser 2007, Espoo, Finland, 12–14 September 2007; pp. 84–89.

19. Dold, C.; Brenner, C. Analysis of Score Functions for the Automatic Registration of Terrestrial Laser Scans. In Proceedings of the International Archives of Photogrammetry. Remote Sensing and Spatial Information Sciences, Beijing, China, 3–11 July 2008; pp. 417–422.

20. Lichti, D.D.; Chow, J.C.K. Inner constraints for planar features. *Photogramm. Rec.* **2013**, *28*, 74–85. [CrossRef]

21. Zhang, Z. Iterative point matching for registration of free-form curves and surfaces. *Int. J. Comput. Vis.* **1994**, *13*, 119–152. [CrossRef]

22. Rabbani, T.; Dijkman, S.; Heuvel, F.; Vosselman, G. An integrated approach for modelling and global registration of point clouds. *ISPRS J. Photogramm. Remote Sens.* **2007**, *61*, 355–370. [CrossRef]

23. Barnea, S.; Filin, S. Keypoint based autonomous registration of terrestrial laser point-clouds. *ISPRS J. Photogramm. Remote Sens.* **2008**, *63*, 19–35. [CrossRef]

24. Al-Manasir, K.; Fraser, C.S. Registration of terrestrial laser scanner data using imagery. *Photogramm. Rec.* **2006**, *21*, 255–268. [CrossRef]

25. Wendt, A. A concept for feature based data registration by simultaneous consideration of laser scanner data and photogrammetric images. *ISPRS J. Photogramm. Remote Sens.* **2007**, *62*, 122–134. [CrossRef]

26. Han, J.Y.; Perng, N.H.; Chen, H.J. LiDAR point cloud registration by image detection technique. *IEEE Geosci. Remote Sens. Lett.* **2013**, *10*, 746–750. [CrossRef]

27. Wendel, A.; Hoppe, C.; Bischof, H.; Leberl, F. Automatic fusion of partial reconstructions. *ISPRS Ann. Photogramm. Remote Sens. Spat. Inf. Sci.* **2012**, *I-3*, 81–86. [CrossRef]

28. Wu, B.; Guo, J.; Hu, H.; Li, Z.; Chen, Y. Co-registration of lunar topographic models derived from Chang'E-1, SELENE, and LRO laser altimeter data based on a novel surface matching method. *Earth Planet. Sci. Lett.* **2013**, *364*, 68–84. [CrossRef]

29. Han, J.Y. A non-iterative approach for the quick alignment of multistation unregistered LiDAR point clouds. *IEEE Geosci. Remote Sens. Lett.* **2010**, *7*, 727–730. [CrossRef]

30. Han, J.Y.; Jaw, J.J. Solving a similarity transformation between two reference frames using hybrid geometric control features. *J. Chin. Inst. Eng.* **2013**, *36*, 304–313. [CrossRef]

31. Chuang, T.Y.; Jaw, J.J. Automated 3d feature matching. *Photogramm. Rec.* **2015**, *30*, 8–29. [CrossRef]

32. Lowe, D.G. Object recognition from local scale-invariant features. In Proceedings of the 7th IEEE International Conference on Computer Vision, Kerkyra, Greece, 20–27 September 1999; pp. 1150–1157.

33. Smith, S.M. A new class of corner finder. In Proceedings of the 3rd British Machine Vision Conference, Leeds, UK, 22–24 September 1992; pp. 139–148.

34. Harris, C.; Stephens, M. A combined corner and edge detector. In Proceedings of the Fourth Alvey Vision Conference, Manchester, UK, 31 August–September 1988.

35. Flint, A.; Dick, A.; van den Hengel, A. Thrift: Local 3D structure recognition. In Proceedings of the 9th Biennial Conference of the Australian Pattern Recognition Society on Digital Image Computing Techniques and Applications, Glenelg, Australia, 3–5 December 2007; pp. 182–188.

36. Sipiran, I.; Bustos, B. Harris 3D: A robust extension of the Harris operator for interest point detection on 3D meshes. *Vis. Comput.* **2011**, *27*, 963–976. [CrossRef]

37. Hänsch, R.; Weber, T.; Hellwich, O. Comparison of 3D interest point detectors and descriptors for point cloud fusion. ISPRS Annals of Photogrammetry. *Remote Sens. Spat. Inf. Sci.* **2014**, *II-3*, 57–64.

38. Ankerst, M.; Kastenmuller, G.; Kriegel, H.P.; Seidl, T. 3-D shape histograms for similarity search and classification in spatial databases. In Proceedings of the Symposium on Large Spatial Databases, Hong Kong, China, 20–23 July 1999; pp. 207–226.

39. Heczko, M.; Keim, D.; Saupe, D.; Vranic, D.V. Methods for similarity search of 3D objects. *Datenbank-Spektrum* **2002**, *2*, 54–63.

40. Chen, D.Y.; Tian, X.P.; Shen, Y.T.; Ouhyoung, M. On visual similarity based 3d model retrieval. *Comput. Graph. Forum* **2003**, *22*, 223–232. [CrossRef]

41. Chua, C.S.; Jarvis, R. Point signatures: A new representation for 3D object recognition. *Int. J. Comput. Vis.* **1997**, *25*, 63–85. [CrossRef]

42. Belongie, S.; Malik, J.; Puzicha, J. Matching shapes. In Proceedings of the 8th IEEE International Conference on Computer Vision, Vancouver, BC, Canada, 7–14 July 2001; Volume 1, pp. 454–461.

43. Bronstein, A.M.; Bronstein, M.M.; Bustos, B.; Castellani, U.; Crisani, M.; Falcidieno, B.; Guibas, L.J.; Kokkinos, I.; Murino, V.; Ovsjanikov, M.; et al. SHREC'10 track: Feature detection and description. In Proceedings of the 3rd Eurographics Conference on 3D Object Retrieval, EG 3DOR'10, Norrköping, Sweden, 2 May 2010; pp. 79–86.

44. Salti, S.; Tombari, F.; Di Stefano, L. A performance evaluation of 3D keypoint detectors. In Proceedings of the 2011 International Conference on 3D Imaging, Modeling, Processing, Visualization and Transmission (3DIMPVT), Hangzhou, China, 16–19 May 2011; pp. 236–243.

45. Gelfand, N.; Mitra, N.J.; Guibas, L.J.; Pottmann, H. Robust global registration. In Proceedings of the 3rd Eurographics Symposium on Geometry Processing, Vienna, Austria, 4–6 July 2005; pp. 197–206.

46. Li, X.; Guskov, I. Multi-scale features for approximate alignment of point-based surfaces. In Proceedings of the 3rd Eurographics Symposium on Geometry Processing, Vienna, Austria, 4–6 July 2005; pp. 217–226.

47. Gal, R.; Cohen-Or, D. Salient geometric features for partial shape matching and similarity. *ACM Trans. Graph.* **2006**, *25*, 130–150. [CrossRef]

48. Wu, H.; Marco, S.; Li, H.; Li, N.; Lu, M.; Liu, C. Feature-constrained registration of building point clouds acquired by terrestrial and airborne laser scanners. *J. Appl. Remote Sens.* **2014**, *8*, 083587. [CrossRef]

49. Teo, T.A.; Huang, S.H. Surface-based registration of airborne and terrestrial mobile LiDAR point clouds. *Remote Sens.* **2014**, *6*, 12686–12707. [CrossRef]

50. Cheng, L.; Tong, L.; Wu, Y.; Chen, Y.; Li, M. Shiftable leading point method for high accuracy registration of airborne and terrestrial LiDAR data. *Remote Sens.* **2015**, *7*, 1915–1936. [CrossRef]

51. Jaw, J.J.; Chuang, T.Y. Registration of lidar point clouds by means of 3-d line features. *J. Chin. Inst. Eng.* **2008**, *31*, 1031–1045. [CrossRef]

52. Wang, M.; Tseng, Y.H. Automatic segmentation of lidar data into coplanar point clusters using an octree-based split-and-merge algorithm. *Photogramm. Eng. Remote Sens.* **2010**, *76*, 407–420. [CrossRef]

53. Chuang, T.Y. Feature-Based Registration of LiDAR Point Clouds. Ph.D. Thesis, National Taiwan University, Taipei, Taiwan, 2012.

54. Briese, C. Breakline Modelling from Airborne Laser Scanner Data. Ph.D. Thesis, Institute of Photogrammetry and Remote Sensing, Vienna University of Technology, Vienna, Austria, 2004.

55. Teledyne Optech. Available online: http://www.teledyneoptech.com (accessed on 14 March 2017).

Part 5
Trees and Terrain

remote sensing

MDPI

Article

An Easy-to-Use Airborne LiDAR Data Filtering Method Based on Cloth Simulation

Wuming Zhang [1], Jianbo Qi [1,*], Peng Wan [1], Hongtao Wang [2], Donghui Xie [1], Xiaoyan Wang [1] and Guangjian Yan [1]

[1] State Key Laboratory of Remote Sensing Science, Beijing Key Laboratory of Environmental Remote Sensing and Digital City, School of Geography, Beijing Normal University, Beijing 100875, China; wumingz@bnu.edu.cn (W.Z.); wanpeng@mail.bnu.edu.cn (P.W.); xiedonghui@bnu.edu.cn (D.X.); xinxin1594@aliyun.com (X.W.); gjyan@bnu.edu.cn (G.Y.)
[2] School of Surveying and Land Information Engineering, Henan Polytechnic University, Jiaozuo 454003, China; wht_31@hpu.edu.cn
* Correspondence: qijb@mail.bnu.edu.cn; Tel.: +86-10-5880-9246

Academic Editors: Jie Shan, Juha Hyyppä, Lars T. Waser and Prasad S. Thenkabail
Received: 13 March 2016; Accepted: 3 June 2016; Published: 15 June 2016

Abstract: Separating point clouds into ground and non-ground measurements is an essential step to generate digital terrain models (DTMs) from airborne LiDAR (light detection and ranging) data. However, most filtering algorithms need to carefully set up a number of complicated parameters to achieve high accuracy. In this paper, we present a new filtering method which only needs a few easy-to-set integer and Boolean parameters. Within the proposed approach, a LiDAR point cloud is inverted, and then a rigid cloth is used to cover the inverted surface. By analyzing the interactions between the cloth nodes and the corresponding LiDAR points, the locations of the cloth nodes can be determined to generate an approximation of the ground surface. Finally, the ground points can be extracted from the LiDAR point cloud by comparing the original LiDAR points and the generated surface. Benchmark datasets provided by ISPRS (International Society for Photogrammetry and Remote Sensing) working Group III/3 are used to validate the proposed filtering method, and the experimental results yield an average total error of 4.58%, which is comparable with most of the state-of-the-art filtering algorithms. The proposed easy-to-use filtering method may help the users without much experience to use LiDAR data and related technology in their own applications more easily.

Keywords: LiDAR point cloud; ground filtering algorithm; cloth simulation

1. Introduction

High-resolution digital terrain models (DTMs) are critical for flood simulation, landslide monitoring, road design, land-cover classification, and forest management [1]. Light detection and ranging (LiDAR) technology, which is an efficient way to collect three-dimensional point clouds over a large area, has been widely used to produce DTMs. To generate DTMs, ground and non-ground measurements have to be separated from the LiDAR point clouds, which is a filtering process. Consequently, various types of filtering algorithms have been proposed to automatically extract ground points from LiDAR point clouds. However, developing an automatic and easy-to-use filtering algorithm that is universally applicable for various landscapes is still a challenge.

Many ground filtering algorithms have been proposed during previous decades, and these filtering methods can be mainly categorized as slope-based methods, mathematical morphology-based methods, and surface-based methods. The common assumption of slope-based algorithms is that the change in the slope of terrain is usually gradual in a neighborhood, while the change in slope between buildings or trees and the ground is very large. Based on this assumption, Vosselman [2] developed a slope-based filtering algorithm by comparing slopes between a LiDAR point and its neighbors. To improve the calculation efficiency, Shan and Aparajithan [3] calculated the slopes between neighbor points along a scan line in a specified direction, which was extended to multidirectional scan lines by Meng *et al.* [4]. Acquiring an optimal slope threshold that can be applied to terrain with different topographic features is difficult with these methods. To overcome this limitation, various automatic threshold definitions have been studied, e.g., adaptive filters [5,6] and dual-direction filters [7]. Nonetheless, their results suggested that slope-based algorithms were not guaranteed to function well in complex terrain, as the filtering accuracy decreased with increasingly steeper slopes [8].

Another type of filtering method uses mathematical morphology to remove non-ground LiDAR points. Selecting an optimal window size is critical for these filtering methods [9]. A small window size can efficiently filter out small objects but preserve larger buildings in ground points. On the other hand, a large window size tends to smooth terrain details such as mountain peaks, ridges and cliffs. To solve this problem, Zhang *et al.* [10] developed a progressive morphological filter to remove non-ground measurements by comparing the elevation differences of original and morphologically opened surfaces with increasing window sizes. However, poor ground extraction results may occur because the terrain slope is assumed to be a constant value in the whole processing area. To overcome this constant slope constraint, Chen *et al.* [11] extended this algorithm by defining a set of tunable parameters to describe the local terrain topography. Other improved algorithms that are based on mathematical morphology can be found in [12–15]. The advantage of mathematical morphology-based methods is that they are conceptually simple and can be easily implemented. The accuracy of morphological based approach is also relatively good. However, additional priori knowledge of the study area is usually required to define a suitable window size because local operators are used [16].

Previous algorithms separated ground and non-ground measurements by removing non-ground points from LiDAR datasets. In contrast to these algorithms, surface-based methods gradually approximate the ground surface by iteratively selecting ground measurements from the original dataset, and the core of this type of filtering method is to create a surface that approximates the bare earth. Axelsson [17] proposed an adaptive triangulated irregular network (TIN) filtering algorithm that gradually densified a sparse TIN that was generated from the selected seed points. In this algorithm, two important threshold parameters need to be carefully set: one is the distance of a candidate point to the TIN facet, and the other is the angle between the TIN facet and the line that connects the candidate point with the facet's closest vertex. These parameters are constant values for the entire study area in the adaptive TIN filter, which makes it difficult to detect ground points around break lines and steep terrain. Recently, Zhang and Lin [18] improved this algorithm by embedding smoothness-constrained segmentation to handle surfaces with discontinuities. Another typical surface-based filtering algorithm was developed by Kraus and Pfeifer [19], who used a weighted linear least-squares interpolation to identify ground points from LiDAR data. This algorithm was first used to remove tree measurements and generate DTMs in forest areas and was subsequently extended to process LiDAR points in urban areas by incorporating a hierarchical approach [20]. By using this filtering algorithm, ground measurements can be successfully detected on flat terrain, but the filtering results are less reliable on terrain with steep slopes and large variability. To cope with this problem, multi-resolution hierarchical filtering methods have been proposed to identify ground LiDAR points based on point residuals from a thin plate spline-interpolated surface [16,21,22]. Recently, Hui *et al.* [23] proposed an improved filtering algorithm which combines the traditional morphological filtering algorithm and multi-level interpolation filtering algorithm. It can achieve promising results in both of urban areas and rural areas.

Another special surface-based filtering algorithm was proposed by Elmqvist [24], who employed active shape models to approximate the ground surface. In this algorithm, an energy function is designed as a weighted combination of internal forces from the shape of the contour and external forces from the LiDAR point clouds. Minimizing this energy function determines the ground surface, which behaves like a membrane that sticks to the lowest points. This algorithm is a new idea to model ground surface from LiDAR data, but the optimization only achieve an global optimum solution, which does not guarantee to get all the local optimums. Thus, some local details may be ignored. Meanwhile, this algorithm performs relatively poorly in complex areas as reported in [25]. They may also fail to effectively model terrain with steep slopes and large variability because they are based on the assumption that the terrain is a smooth surface. Furthermore, another challenge of these methods is how to increase the efficiency when the accuracy is fixed [18].

The use of the aforementioned filtering algorithms has proven to be successful, but the performance of these algorithms changes according to the topographic features of the area, and the filtering results are usually unreliable in complex cityscapes and very steep areas. In addition, the implementation of these filtering methods requires a number of suitable parameters to achieve satisfactory results, which are difficult to determine because the optimal filter parameters vary from landscape to landscape, so these filtering methods are not easy to use by users without much experience. To cope with these problems, this paper proposes a novel filtering algorithm which is capable of approximating the ground surface with a few parameters. Different from other algorithms, the proposed method filters the ground points by simulating a physical process that an virtual cloth drops down to an inverted (upside-down) point cloud. Compared to existing filtering algorithms, the proposed filtering method has some advantages: (1) few parameters are used in the proposed algorithm, and these parameters are easy to understand and set; (2) the proposed algorithm can be applied to various landscapes without determining elaborate filtering parameters; and (3) this method works on raw LiDAR data.

The remainder of this paper is organized as follows. A new ground filtering algorithm is proposed in Section 2. Section 3 presents the experimental results, and the proposed algorithm is discussed in Section 4. Finally, Section 5 concludes this paper.

2. Method

Our method is based on the simulation of a simple physical process. Imagine a piece of cloth is placed above a terrain, and then this cloth drops because of gravity. Assuming that the cloth is soft enough to stick to the surface, the final shape of the cloth is the DSM (digital surface model). However, if the terrain is firstly turned upside down and the cloth is defined with rigidness, then the final shape of the cloth is the DTM. To simulate this physical process, we employ a technique that is called cloth simulation [26]. Based on this technique, we developed our cloth simulation filtering (CSF) algorithm to extract ground points from LiDAR points. The overview of the proposed algorithm is illustrated in Figure 1. First, the original point cloud is turned upside down, and then a cloth drops to the inverted surface from above. By analyzing the interactions between the nodes of the cloth and the corresponding LiDAR points, the final shape of the cloth can be determined and used as a base to classify the original points into ground and non-ground parts.

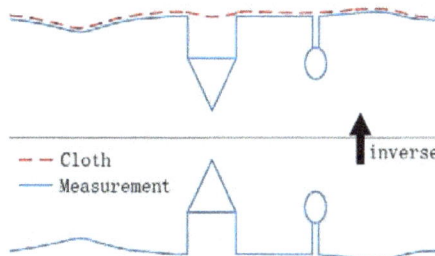

Figure 1. Overview of the cloth simulation algorithm.

2.1. Fundamental of the Cloth Simulation

Cloth simulation is a term of 3D computer graphics. It is also called cloth modeling, which is used for simulating cloth within a computer program. During cloth simulation, the cloth can be modeled as a grid that consists of particles with mass and interconnections, called a Mass-Spring Model [27]. Figure 2 shows the structure of the grid model. A particle on the node of the grid has no size but is assigned with a constant mass. The positions of the particles in three-dimensional space determine the position and shape of the cloth. In this model, the interconnection between particles is modeled as a "virtual spring", which connects two particles and obeys Hooke's law. To fully describe the characteristics of the cloth, three types of springs have been defined: shear spring, traction spring and flexion spring. A detailed description about the functions of these different springs can be found in [27].

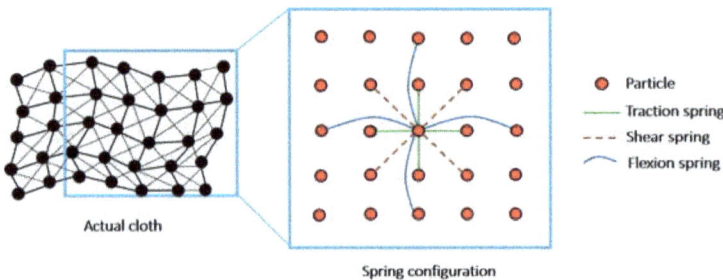

Actual cloth

Spring configuration

- ● Particle
- —— Traction spring
- --- Shear spring
- ⌒ Flexion spring

Figure 2. Schematic illustration of mass-spring model. Each circle indicates a particle and each line represents a spring.

To simulate the shape of the cloth at a specific time, the positions of all of the particles in the 3D space are computed. The position and velocity of a particle are determined by the forces that act upon it. According to Newton's second law, the relationship between position and forces is determined by Equation (1):

$$m\frac{\partial X(t)}{\partial t^2} = F_{ext}(X,t) + F_{int}(X,t) \tag{1}$$

where X means the position of a particle at time t; $F_{ext}(X,t)$ stands for the external force, which consists of gravity and collision forces that are produced by obstacles when a particle meets some objects in the direction of its movement; and $F_{int}(X,t)$ stands for the internal forces (produced by interconnections) of a particle at position X and time t. Because both the internal and external forces vary with time t, Equation (1) is usually solved by a numerical integration (e.g., Euler method) in the conventional implementation of cloth simulation.

2.2. Modification of the Cloth Simulation

When applying the cloth simulation to LiDAR point filtering, a number of modifications have been made to make this algorithm adaptable to point cloud filtering. First, the movement of a particle is constrained to be in vertical direction, so the collision detection can be implemented by comparing the height values of the particle and the terrain (e.g., when the position of a particle is below or equal to the terrain, the particle intersects with the terrain). Second, when a particle reaches the "right position", *i.e.*, the ground, this particle is set as unmovable. Third, the forces are divided into two discrete steps to achieve simplicity and relatively high performance. Usually, the position of a particle is determined by the net force of the external and internal forces. In this modified cloth simulation, we first compute the displacement of a particle from gravity (the particle is set as unmovable when it reaches the ground, so the collision force can be omitted) and then modify the position of this particle according to the internal forces. This process is illustrated in Figure 3.

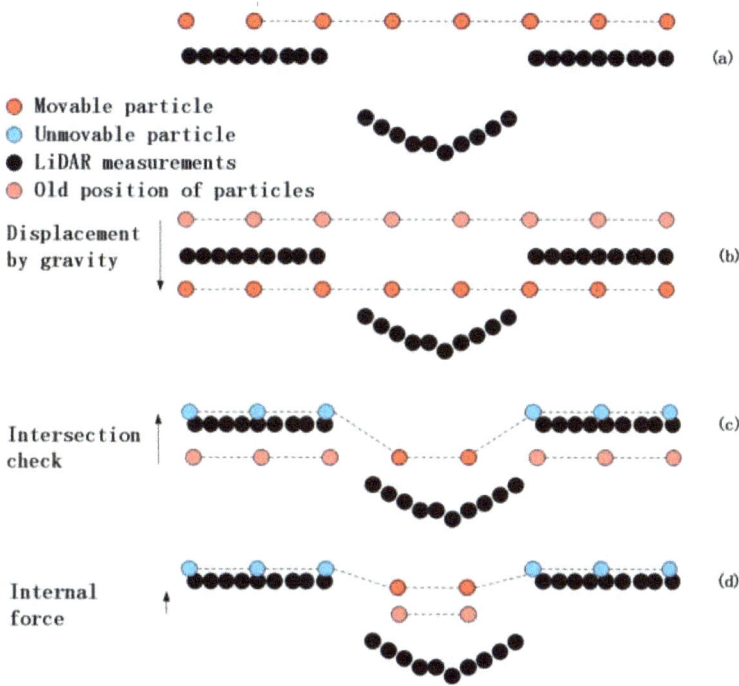

Figure 3. Main Steps in CSF: (**a**) Initial state. A cloth is place above the inverted LiDAR measurements; (**b**) The displacement of each particle is calculated under the influence of gravity. Thus, some particles may appear under the ground measurements; (**c**) Intersection check. For those who are under the ground, they are moved on the ground and set as unmovable; (**d**) Considering internal forces. The movable particles are moved according to forces produced by neighbour particles.

2.3. Implementation of CSF

As described above, the forces that act on a particle are considered as two discrete steps. This modification was inspired by [28]. First, we calculate the displacement of each particle only from gravity, *i.e.*, solve Equation (1) with internal forces equal to zero. Then, the explicit integration form of this equation is

$$X(t + \Delta t) = 2X(t) - X(t - \Delta t) + \frac{G}{m}\Delta t^2 \tag{2}$$

where m is the mass of the particle (usually, m is set to 1) and Δt is the time step. This equation is very simple to solve. Given the time step and initial position, the current position can be calculated directly because G is a constant.

To constrain the displacement of particles in the void areas of the inverted surface, we consider the internal forces at the second step after the particles have been moved by gravity. Because of internal forces, particles will try to stay in the grid and return to the initial position. Instead of considering neighbors of each particle one by one, we simply traverse all the springs. For each spring, we compare the height difference between the two particles which form this spring. Thus, the 2-dimensional (2-D) problem are abstracted as a one-dimensional (1-D) problem, which is illustrated in Figure 4.

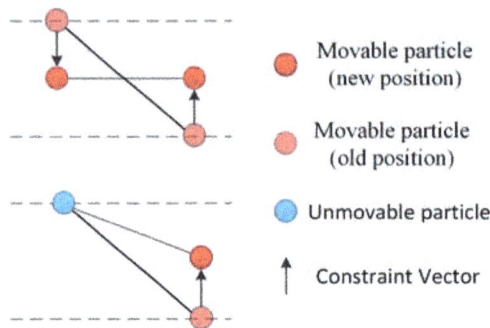

Figure 4. Constraint between particles.

As we have restricted the movement directions of the particles, two particles with different height values will try to move to the same horizon plane (cloth grid is horizontally placed at the beginning). If both connected particles are movable, we move them by the same amount in the opposite direction. If one of them is unmovable, then the other will be moved. Otherwise, if these two particles have the same height value, neither of them will be moved. Thus, the displacement (vector) of each particle can be calculated by the following equation:

$$\vec{d} = \frac{1}{2}b(\vec{p_i} - \vec{p_0}) \cdot \vec{n} \tag{3}$$

where $\vec{d}$ represents the displacement vector of a particle; b equals to 1 when the particle is movable, otherwise it equals to 0. $\vec{p_0}$ is the position of current particle that is ready to be moved. $\vec{p_i}$ is the position neighboring particle that connects with p_0; and $\vec{n}$ is a normalized vector that points to vertical direction, $\vec{n} = (0,0,1)^T$. This movement process can be repeated; we set a parameter rigidness (RI) to represent the repeated times. This parameterization process is shown in Figure 5. If RI is set to 1, the movable particle is just moved only once, and the displacement is half of the vertical distance (VD) between the two particles. If the RI is set to 2, the movable particle will be moved twice, the total displacement is $3/4$VD. Finally, if RI is set to 3, the movable particle will be moved three times and the total displacement is $7/8$VD. The value of 3 is enough to produce a very hard cloth. Thus, we constrain the rigidness to values of 1, 2 and 3. The larger the rigidness is, the more rigidly the cloth will behave.

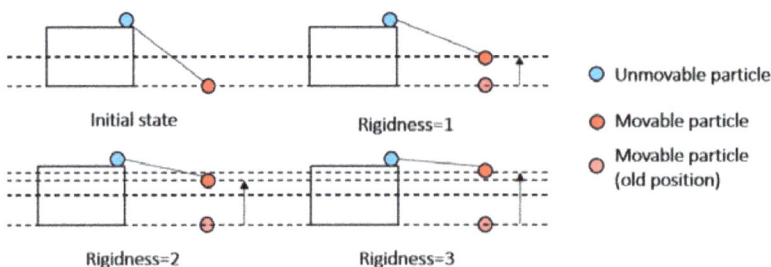

Figure 5. Parameterization of rigidness.

The main implementation procedures of CSF are described as follows. First, we project the cloth particles and LiDAR points into the same horizontal plane and then find a nearest LiDAR point (named corresponding point, CP) for each cloth particle in this 2D plane. An intersection height value (IHV) is defined to record the height value (before projection) of CP. This value represents the lowest position that a particle can reach (*i.e.*, if the particle reaches the lowest position that is defined by this value, it cannot move forward anymore). During each iteration, we compare the current height value (CHV) of a particle with IHV; if CHV is equal or lower than IHV, we move the particle back to the position of IHV and make the particle unmovable.

An approximation of the real terrain is obtained after the simulation, and then the distances between the original LiDAR points and simulated particles are calculated by using a cloud-to-cloud distance computation algorithm [29]. LiDAR points with distances that are less than a threshold h_{cc} are classified as BE (bare earth), while the remaining points are OBJ (objects).

The procedure of the proposed filtering algorithm is presented as follows:

1. Automatic or manual outliers handling using some third party software (such as cloudcompare).
2. Inverting the original LiDAR point cloud.
3. Initiating cloth grid. Determining number of particles according to the user defined grid resolution (GR). The initial position of cloth is usually set above the highest point.
4. Projecting all the LiDAR points and grid particles to a horizontal plane and finding the CP for each grid particle in this plane. Then recording the IHV.
5. For each grid particle, calculating the position affected by gravity if this particle is movable, and comparing the height of this cloth particle with IHV. If the height of particle is equal to or less than IHV, then this particle is placed at the height of IHV and is set as "unmovable".
6. For each grid particle, calculating the displacement of each particle affected by internal forces.
7. Repeating (5)–(6). The simulation process will terminate when the maximum height variation (M_HV) of all particles is small enough or when it exceeds the maximum iteration number which is specified by the user.
8. Computing the cloud to cloud distance between the grid particles and LiDAR point cloud.
9. Differentiating ground from non-ground points. For each LiDAR points, if the distance to the simulated particles is smaller than h_{cc}, this point is classified as BE, otherwise it is classified as OBJ.

2.4. Post-Processing

For steep slopes, this algorithm may yield relatively large errors because the simulated cloth is above the steep slopes and does not fit with the ground measurements very well due to the internal constraints among particles, which is illustrated in Figure 6. Some ground measurements around steep slopes are mistakenly classified as OBJ. This problem can be solved by a post-processing method that smoothes the margins of steep slopes. This post-processing method finds an unmovable particle in the four adjacent neighborhoods of each movable particle and compares the height values of CPs. If the height difference is within a threshold (h_{cp}), the movable particle is moved to the ground and set as unmovable. For example, for point D in Figure 6, we find that point A is the unmovable particle from the four adjacent neighbors of D. Then, we compare the height values between C and B (the CPs for D and A, respectively). If the height difference is less than h_{cp}, then this candidate point D is moved to C and is set as unmovable. We repeat this procedure until all the movable particles are properly handled (either set as unmovable or kept movable).

To implement post-processing, all the movable particles should be traversed, if we scan the cloth grid row by row, the results may be affected by this particular scan direction. Thus, we first build up sets of strongly connected components (SCCs) and each SCC contains a set of connected movable particles. In a SCC, it usually contains two kinds of particles, those particles that have at least one unmovable neighbor are marked as type M1, and the others are marked as M2 (see Figure 7). Using M1 as initial seeds, we perform the breath-first traversal for the SCC, with which movable particles are handled one by one from 1 to 18 in Figure 7. This process guarantees that the post-processing are performed from edge to center, regardless of scan direction.

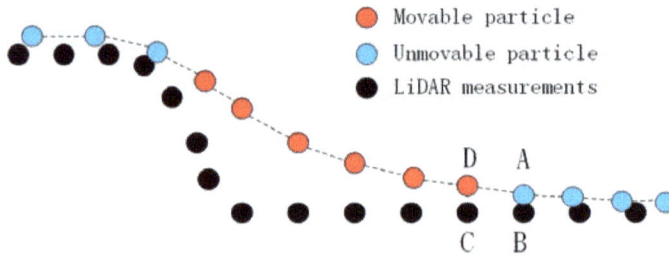

Figure 6. Post-processing of the steep slope.

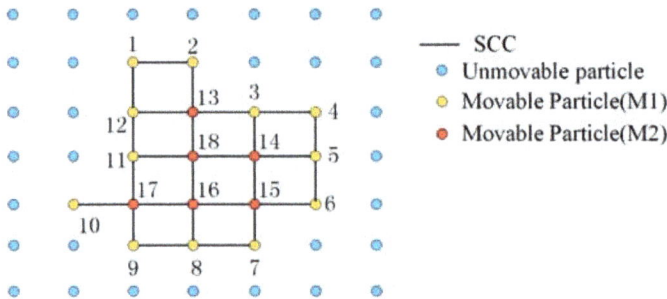

Figure 7. Illustration of strongly connected component (SCC). Movable particles are handled from 1 to 18.

2.5. Parameters

CSF mainly consists of four user-defined parameters: grid resolution (GR), which represents the horizontal distance between two neighboring particles; time step (dT), which controls the displacement of particles from gravity during each iteration; rigidness (RI), which controls the rigidness of the cloth; and an optional parameter steep slope fit factor (ST), which indicates whether the post-processing of handling steep slopes is required or not.

In addition to these user-defined parameters, two threshold parameters have been used in this algorithm to aid the identification of ground points. The first is a distance threshold (h_{cc}) that governs the final classification of the LiDAR points as BE and OBJ based on the distances to the cloth grid. This parameter is set as a fixed value of 0.5 m. Another threshold parameter is the height difference (h_{cp}), which is used during post-processing to determine whether a movable particle should be moved to the ground or not. This parameter is set to 0.3 m for all of the datasets.

3. Experiment and Results

3.1. Validation of the Filtering

This method was first tested by datasets that were provided by the International Society for Photogrammetry and Remote Sensing (ISPRS) Working Group III/3 to quantitatively test the performance of different filters and identify directions for future research [30]. In these datasets, fifteen samples with different characteristics were selected to test the performance of the proposed CSF algorithm, which are shown in Table 1. The reference datasets were generated by manually filtering the LiDAR datasets, and each point in the samples was classified as BE or OBJ.

Table 1. Characteristics of all samples [15].

Environment	Site	Sample	Features
Urban	1	11	Mixture of vegetation and buildings on hillside
		12	Buildings on hillside
	2	21	Large buildings and bridge
		22	Irregularly shaped buildings
		23	Large, irregularly shaped buildings
		24	Steep slopes
	3	31	Complex buildings
	4	41	Data gaps
		42	Railway station with trains
Rural	5	51	Mixture of vegetation and buildings on hillside
		52	Buildings on hillside
		53	Large buildings and bridge
		54	Irregularly shaped buildings
	6	61	Large, irregularly shaped buildings
	7	71	Steep slopes

According to the implementation of CSF, when the original LiDAR point cloud is turned upside down, the objects above ground will appear below the ground measurements. Then, the complexity of object measurements (such as rooftops) seldom influences the simulation process. Based on this feature, we visually classified the samples into different groups according to the properties of the topography. These properties indicate the existence of steep slopes or terraced slopes. If the terrain is very flat and has no steep or terraced slopes, RI is set to a relatively large value ($RI = 3$), and no post-processing is needed ($ST = false$). If steep slopes exist (e.g., river bank, ditch, and terrace), a medium soft cloth ($RI = 2$) and post-processing ($ST = true$) are needed. When handling very steep slopes, we need post-processing ($ST = true$) and a very soft cloth ($RI = 1$). Thus, the fifteen samples are classified into three groups, each sharing the same set of parameters, which are illustrated in Table 2.

Table 2. Parameters for each group of samples ($dT = 0.65$, $GR = 0.5$).

Group	Feature	Parameters	Samples
I	Flat terrain or gentle slope, no steep slopes	RI = 3 ST = false	21, 31, 42, 51, 54
II	With steep or terraced slopes (e.g., river bank, ditch, terrace)	RI = 2 ST = true	11, 12, 22, 23, 24, 41
III	High and steep slopes (e.g., pit, cliff)	RI = 1 ST = true	52, 53, 61, 71

The main parameters that control the results and vary with the scene type were RI and ST, which reduce the complexity and improve the usability of this algorithm. For dT and GR, we set them as fixed values of 0.65 and 0.5, respectively. These two values are universally applicable to all of the reference datasets according to our tests. The influence of these two parameters on the results is discussed in Section 4.2.

To evaluate the performance of this algorithm, the type I (T.I), type II (T.II) and total errors (T.E.) for all the fifteen samples were calculated. The type I error is the number of BE points that are incorrectly classified as OBJ divided by the true number of BE points; the type II error is the number of OBJ points that that incorrectly classified as BE points divided by the true number of OBJ points, and the total error is the number of mistakenly classified points divided by the total points. Besides, the Cohen's Kappa coefficient [31], which measures the overall agreement between two judges [15,21], is also calculated in this study. The calculations of T.I, T.II, T.E. and Kappa coefficient can refer to Hu *et al.* [32].

The errors and the Kappa coefficients are shown in Table 3. The results show that CSF has relatively good results for samp21, samp31, samp42, samp51 and samp54 respecting total error and kappa coefficient. All these samples belong to group I, which indicates that the most suitable types for our method is urban areas. For group II and group III, the total errors are relatively large (especially for samp11) compared to group I, which shows that CSF performs relatively poorly in complex regions similar to other filtering algorithms [21]. Overall, our method is not sensitive to the type and distribution of object above ground because the LiDAR point cloud is inverted and the shape of the terrain mostly determines the filtering accuracy. However, in high relief areas with very steep slopes (e.g., pit, cliff) and low rise buildings, our method perform worst, because when a soft cloth fit with the terrain, it may also reach the rooftops of low rise buildings.

Table 3. Errors and Kappa coefficients for all samples.

Samples	T.I(%)	T.II(%)	T.E.(%)	Kappa(%)
samp11	7.23	18.44	12.01	75.17
samp12	1.15	4.9	2.97	94.04
samp21	3.89	1.78	3.42	90.47
samp22	1.29	25.9	8.94	77.72
samp23	3.52	6.21	4.79	90.38
samp24	1.03	7.73	2.87	92.68
samp31	0.96	2.38	1.61	96.75
samp41	1.48	8.78	5.14	89.73
samp42	3.28	0.87	1.58	96.18
samp51	2.67	4.57	3.08	91.13
samp52	1.01	28.79	3.93	77.05
samp53	3.85	37.08	5.2	46.86
samp54	3.79	2.64	3.18	93.61
samp61	0.87	18.94	1.49	78.1
samp71	1.61	37.85	5.71	68.03

Figure 8 shows the original dataset, reference DTM, produced DTM and the space distribution of Type I and Type II errors of some representative samples (samp31, samp11 and samp51). Compared with the reference DTM, the produced DTM has successfully preserved the main terrain shape and also the microtopography, especially in mine field (Figure 8k). It can also be seen that the error points mainly exist on the edges of objects for samp11 and samp31, in which some object measurements with low height values may be classified as BE and some ground measurements with relatively high height values may also be classified as OBJ. samp11 has a large group of error points (type II) that almost classifies a whole building as ground measurements, it occurs because the building is located on a slope and the roof is nearly connected to the ground, causing the building to be treated as ground in the post-processing step. For samp53 (the same as samp52 and samp61), the very soft cloth (group III) makes some small vegetation points mistakenly classified as BE. Besides, the total number of OBJ points is much smaller compared to BE points. These jointly cause a large Type II error. However, if a harder cloth is used it may yield the opposite error (BE points around the steep slopes may be identified as OBJ). Thus, adjusting the parameters is necessary to balance type I and type II errors. If a large number of low objects exists above the ground, the cloth should be harder (RI should be set larger), which will guarantee that fewer object measurements are mistakenly classified as ground objects. Among the samples which have very poor Type II error, samp22 and samp71 are caused by the incorrectly identification of bridges, which is classified as BE under CSF. The details will be discussed in Section 4.4.

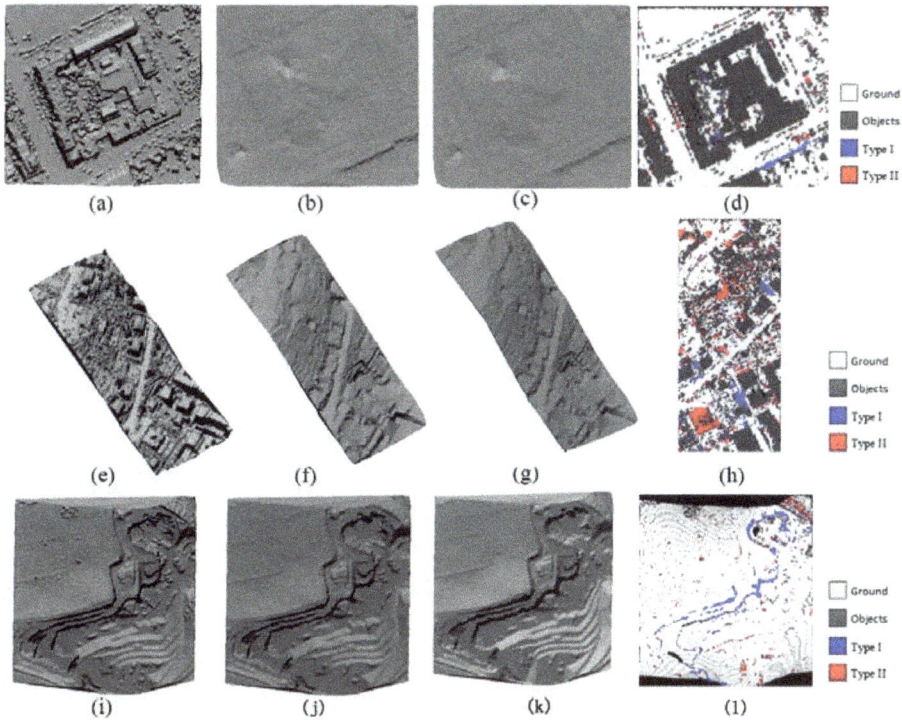

Figure 8. Results of each group (choose samp31, samp11, and samp53 as representatives): (**the first column**) are original datasets; (**the second column**) are the DTMs that are generated from the reference data of samp31, samp11, and samp53; (**the third column**) are the DTMs that are produced from the CSF algorithm; (**the last column**) are the spatial distributions of the type I and type II errors.

To quantitatively analyze the accuracy of the CSF algorithm, we compared the total error and kappa coefficient with some existing top algorithms. The total errors and Kappa coefficients of these algorithms and our algorithm are shown in Tables 4 and 5. On the whole, the accuracy of our method is close to some top filtering algorithms, except for samples 22 and 71. The total errors for sample 22 and 71 are relatively high, which are also mainly caused by the bridge.

Table 4. Total error compared to other reported algorithms (%).

Samples	Axelsson (1999)	Elmqvist (2000)	Pfeifer (2001)	Mongus (2012)	Li (2013)	Chen (2013)	Pingel (2013)	Zhang (2013)	Hu (2014)	Mongus (2014)	Hui (2016)	CSF
samp11	10.76	22.4	17.35	11.01	12.85	13.01	8.28	18.49	8.31	7.5	13.34	12.01
samp12	3.25	8.18	4.5	5.17	3.74	3.38	2.92	5.92	2.58	2.55	3.5	2.97
samp21	4.25	8.53	2.57	1.98	2.55	1.34	1.1	4.95	0.95	1.23	2.21	3.42
samp22	3.63	8.93	6.71	6.56	4.06	4.67	3.35	14.18	3.23	2.83	5.41	8.94
samp23	4.00	12.28	8.22	5.83	6.16	5.24	4.61	12.06	4.42	4.34	5.11	4.79
samp24	4.42	13.83	8.64	7.98	5.67	6.29	3.52	20.26	3.80	3.58	7.47	2.87
samp31	4.78	5.34	1.8	3.34	2.47	1.11	0.91	2.32	0.90	0.97	1.33	1.61
samp41	13.91	8.76	10.75	3.71	6.71	5.58	5.91	20.44	5.91	3.18	10.6	5.14
samp42	1.62	3.68	2.64	5.72	3.06	1.72	1.48	3.94	0.73	1.35	1.92	1.58
samp51	2.72	21.31	3.71	2.59	3.92	1.64	1.43	5.31	2.04	2.73	4.88	3.08
samp52	3.07	57.95	19.64	7.11	15.43	4.18	3.82	12.98	2.52	3.11	6.56	3.93
samp53	8.91	48.45	12.6	8.52	11.71	7.29	2.43	5.58	2.74	2.19	7.47	5.2
samp54	3.23	21.26	5.47	6.73	3.93	3.09	2.27	6.4	2.35	2.16	4.16	3.18
samp61	2.08	35.87	6.91	4.85	5.81	1.81	0.86	16.13	0.84	0.96	2.33	1.49
samp71	1.63	34.22	8.85	3.14	4.58	1.33	1.65	10.44	1.50	2.49	3.73	5.71
Avg.	4.82	20.73	8.02	5.62	6.18	4.11	2.97	10.63	2.85	2.74	5.33	4.39
Std.	3.44	15.92	5.09	2.39	3.84	3.06	2.00	6.01	2.03	1.64	3.23	2.76

Table 5. Kappa coefficient compared to other reported algorithms (%).

Samples	Axelsson (1999)	Elmqvist (2000)	Pfeifer (2001)	Chen (2013)	Pingel (2013)	Hu (2014)	Hui (2016)	CSF
samp11	78.48	56.68	66.09	74.12	83.12	82.97	72.92	75.17
samp12	93.51	83.66	91	93.23	94.15	94.83	93.00	94.04
samp21	86.34	77.4	92.51	96.1	96.77	97.23	93.35	90.47
samp22	91.33	80.3	84.68	89.03	92.21	92.04	87.58	77.72
samp23	91.97	75.59	83.59	89.49	90.73	91.14	89.74	90.38
samp24	88.5	54.13	78.43	84.53	91.13	90.39	81.93	92.68
samp31	90.43	89.31	96.37	97.76	98.17	98.19	97.33	96.75
samp41	72.21	82.46	78.51	88.83	88.18	88.18	78.78	89.73
samp42	96.15	90.86	93.67	95.81	96.48	98.25	95.38	96.18
samp51	91.68	52.74	89.61	95.17	95.76	93.9	85.06	91.13
samp52	83.63	9.36	41.02	78.91	81.04	86.24	69.51	77.05
samp53	39.13	7.05	30.83	46.69	68.12	66.43	41.84	46.86
samp54	93.52	55.88	88.93	93.9	95.44	95.28	91.63	93.61
samp61	74.52	10.31	47.09	77.36	87.22	86.76	67.82	78.1
samp71	91.44	26.26	75.27	93.19	91.81	92.59	79.86	68.03
Avg.	84.19	56.8	75.84	86.27	90.02	90.29	81.72	83.86
Std.	13.9	29.18	19.87	12.72	7.58	7.74	13.95	13.12

3.2. Testing with Dense Point Cloud

The datasets from ISPRS were obtained many years ago, as the development of LiDAR technology, the density of collected point cloud are continuously arising. Thus, we tested the performance of CSF with datasets that have more dense points with average point distance equal to 0.6 m–0.8 m, the information of these datasets are shown in Table 6. By comparing with the true ground obtained by standard industrial semi-automatic software, the accuracy of CSF is evaluated. The results are shown in Table 7. It can be seen that CSF has achieved a relatively high accuracy for all the datasets from urban to rural areas. However, A large T.I error also has been noted for dataset 3, this is because ground measurements is very sparse in this area. For dataset 4, the error mainly occurs around steep slopes, since this area contains large number of steep slopes.

Table 6. Characteristics of testing datasets.

Dataset	Type	Point Number	Scope	Features
1	Urban	1559933	1 km × 1 km	Flat terrain, large and dense buildings, high vegetation coverage
2	Urban	1522256	1 km × 1 km	Flat terrain with dense bungalow areas
3	Rural	2093506	2 km × 1 km	dense vegetation coverage
4	Rural	1418228	0.5 km × 0.5 km	Large number of steep slopes

Table 7. Accuracy evaluation with true ground measurements.

Dataset	T.I(%)	T.II(%)	T.E.(%)
1	0.72	13.36	6.84
2	5.29	9.29	7.84
3	36.09	1.84	5.49
4	8.57	22.61	14.09

To analyse the details of CSF, we presented the generated DTM and the cross section of each dataset. Since CSF will first invert the original point cloud, thus large buildings will produce large holes. However, when applying a relative hard cloth, this hole can be crossed. Then large buildings can be removed (see Figure 9). If post-processing is enabled ($ST = true$), the cloth can fit the ground very well. Through this way, microtopography (e.g., low-lying areas) in urban areas can be preserved (Figure 10). In mountain areas, CSF performs relatively poorly, especially in dense vegetation areas where ground measurements are usually sparse. If the cloth is too soft, many object measurements may be mistakenly classified as BE. Otherwise, ground measurements may be classified as OBJ due to the hilly topography (see Figure 11). For areas with large number of steep slopes, cloth should be more soft and post-process is also needed. Figure 12 shows a typical area with large number of steep slopes, it can be seen that the main skeleton of terrain has been preserved well. However, some small houses may be missed (red circle in Figure 12c).

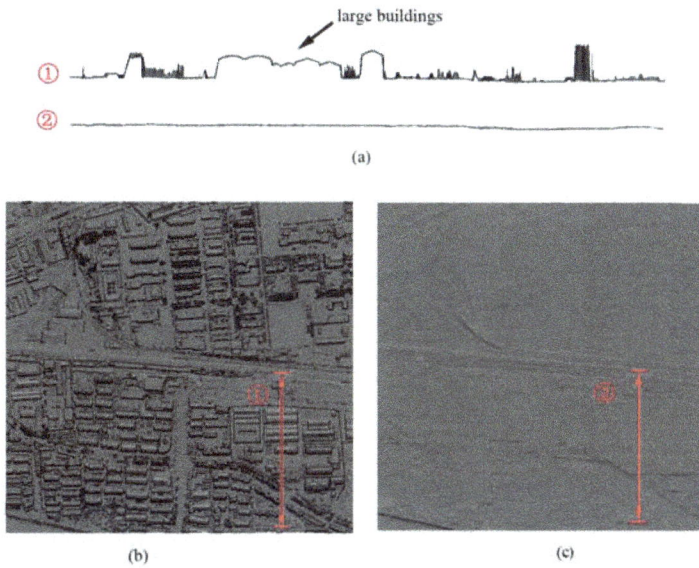

Figure 9. Removal of large buildings in urban area: (**a**) Cross sections from (**b**) and (**c**); (**b**) Dataset 1; (**c**) Produced DTM. In this dataset, it contains a number of connected large low buildings (see the cross section), when the cloth is relatively hard, it will not drop into this large hole, then these buildings can be removed.

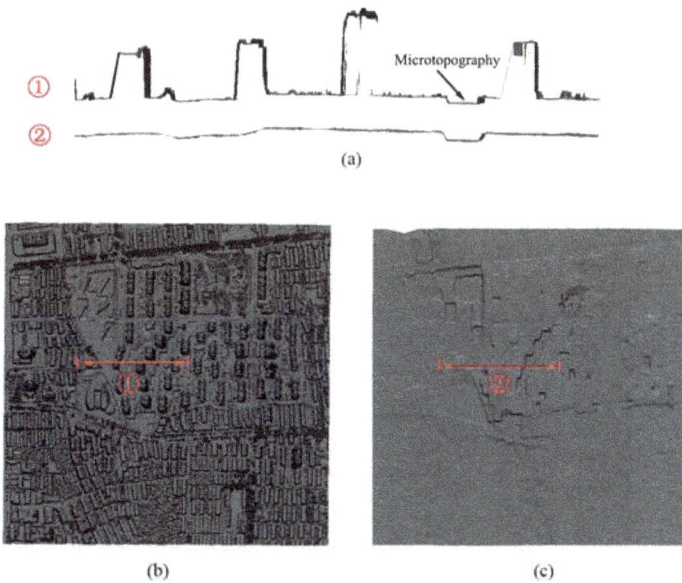

Figure 10. Preservation of microtopography: (**a**) Cross sections from (**b**) and (**c**); (**b**) Dataset 2; (**c**) Produced DTM. When post-processing is enabled, the cloth can stick to the surface more closely, some small steep slopes can be preserved.

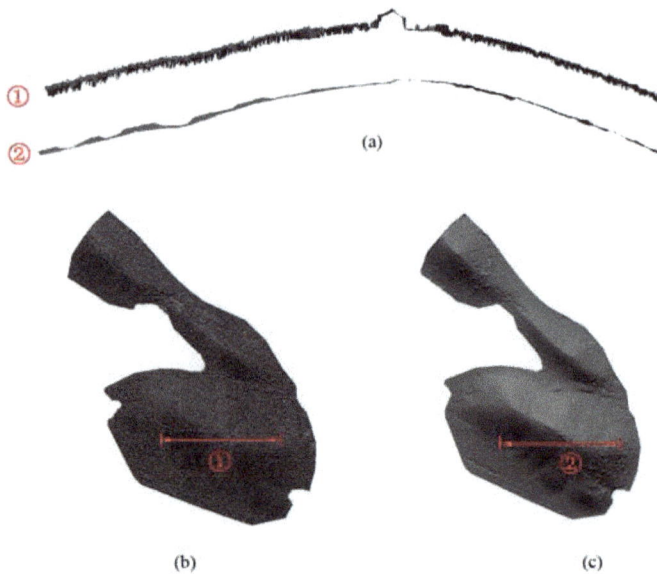

Figure 11. Sparse ground measurements: (**a**) Cross sections from (**b**) and (**c**); (**b**) Dataset 3; (**c**) Produced DTM. In some hilly topography areas, some parts of the cloth may not sticks to the ground well, this will cause classification errors (BE may be treated as OBJ).

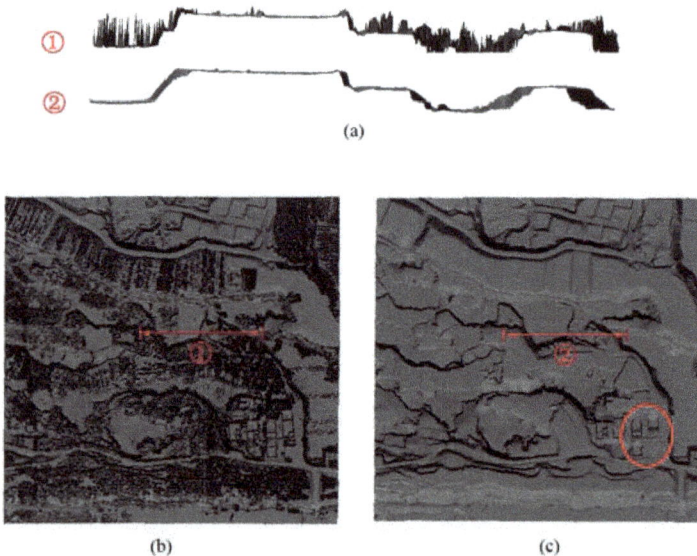

Figure 12. Preservation of steep slopes: (**a**) Cross sections from (**b**) and (**c**); (**b**) Dataset 4; (**c**) Produced DTM. In this dataset, the main objects are vegetation and contains large number of ground measurements, the cloth can be very soft to maximumly fit the terrain shape with less consideration of T.II error. Companied with post-processing, large steep slopes can be preserved well.

4. Discussion

4.1. Accuracy

Among all the reported algorithms, the Axelsson's algorithm had been implemented in a commercial software package called Terrasolid [33] and the Pfeifer's algorithm was implemented into a commercially available software package called SCOP++ from the German company Inpho GmbH [34]. The overall performance of our algorithm also showed high accuracy and stability, as both the mean total error (4.39) and standard deviation (2.76) of all the samples are relatively low compared to all of the other algorithms. This result is inspirational and demonstrates that our algorithm can be adapted to various environments and achieves relatively high accuracy.

4.2. Parameter Setting

Usually, we only modify RI and ST for different groups of samples and set dT and GR as fixed values. These universally applicable parameters (dT and GR) were determined by a number of tests. Theoretically, smaller dT value would make behavior of simulated cloth more like a true cloth, but it dramatically increases the computing time. To quantitatively evaluate the influence of dT, we tested all of the samples with different time steps (from 0.4 to 1.5 with steps of 0.05; 0.4 was chosen because the value would take too much time to compute when it was smaller than 0.4). The total errors of each group and the mean total error of all the samples, which depend on the time step, are illustrated in Figure 13. This figure indicates that the total error increases after an initial decline for all groups, and all of them achieve the lowest total error around the 0.65 time step. When dT is small, the displacement of particle in each step is also small. As we have set a maximum iteration number of 500, if dT is too small, the cloth may not reach the LiDAR measurements or fit with the terrain well after simulation process. Thus, very small dT may produce large error. On the other hand, When the dT is too large, the simulated cloth may stick to rooftops, this also increases the total error. Thus, 0.65 was chosen as the value of dT because it produces relatively good results for all of the samples and it can be applied to many situations without adjustments.

The grid resolution (GR) parameter in the simulation process has strong relationship with simulation time because it determines how many cloth particles are created for a specific dataset. Figure 14 shows the total errors at different GR values. It can be seen that the accuracies of group I and group III are relatively stable than group II because group II usually have complicated terrain shape and buildings (e.g., areas with terraced slopes and low rise buildings). However, almost all samples get the highest accuracy around 0.5, which was then been used as a fixed value.

Except the parameters above, there are two threshold parameters: h_{cc} and h_{cp}. h_{cc} governs the final classification which separates LiDAR measurements into BE or OBJ. Most particles will stick to ground after simulation. And OBJ measurements (e.g., buildings and trees) are usually taller than 0.5 m. Thus, we set h_{cc} as 0.5, which is also a fixed value. The influences of h_{cc} on total errors are illustrated in Figure 15. It shows that this value has limited impact on total errors. As for the parameter h_{cp}, it is used in the post-processing to decide whether a movable particle should be moved to ground according to its neighbors. We simply set this parameter to 0.3 m, which indicates the height difference between two adjacent ground measurements is usually less than 0.3 m on a flat terrain. Since this parameter is only used when post-process is enabled, and it also only influence the movable particles over steep slopes, the influence is also limited.

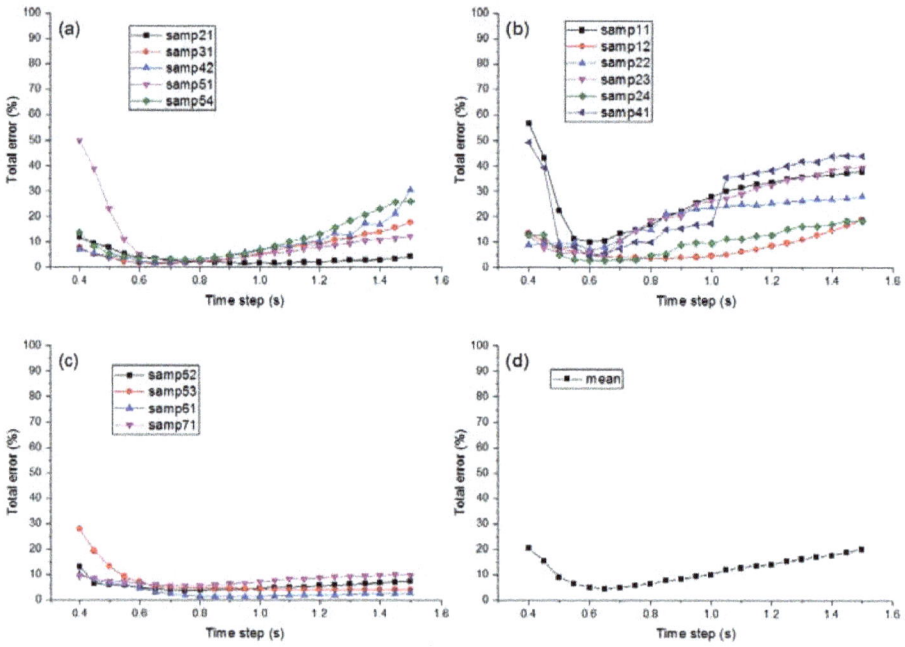

Figure 13. Total errors for each time step: Group I (**a**); Group II (**b**); Group III (**c**); Mean (**d**).

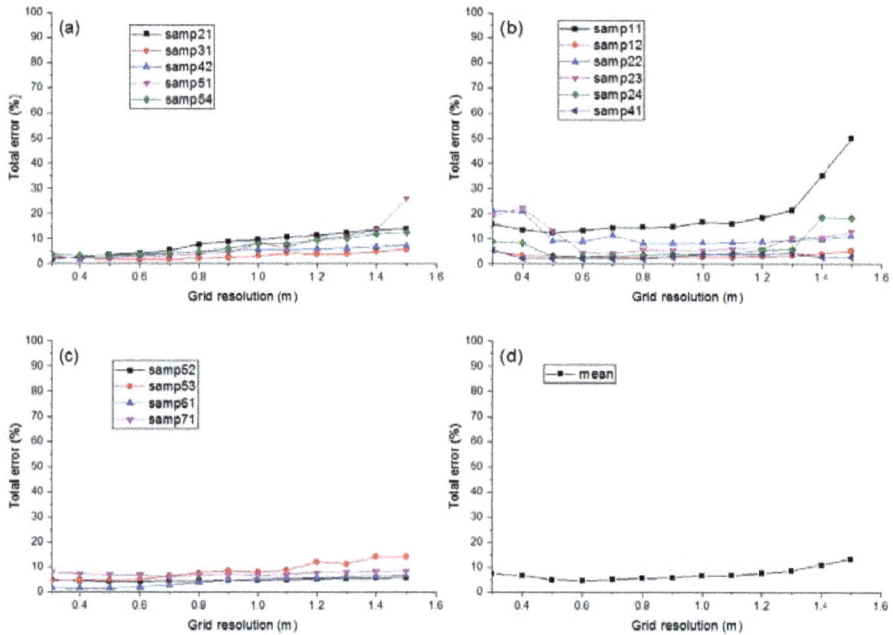

Figure 14. Total errors for each grid resolution: Group I (**a**); Group II (**b**); Group III (**c**); Mean (**d**).

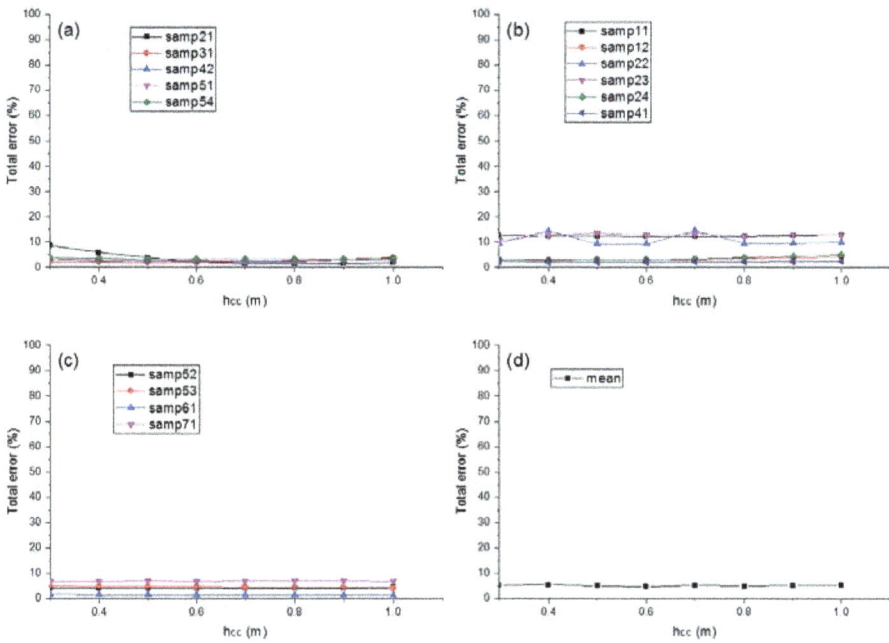

Figure 15. The influences of h_{cc} on total errors: Group I (**a**); Group II (**b**); Group III (**c**); Mean (**d**).

During the simulation, CSF will terminate when the M_HV is less than a threshold or the maximum iterations reach a user specified value. Usually, this user-defined value is set to 500. However in most cases, CSF will end according to the former criterial. Figure 16 shows the trend of M_HV and A_HV (average height variation) of a typical scene. It can be seen that both M_HV and A_HV decreased to a very low value around 100 iterations. Actually, M_HV is 0.0033 when iteration number equal to 150. That means the CSF will give a satisfied result with a relatively low iteration times.

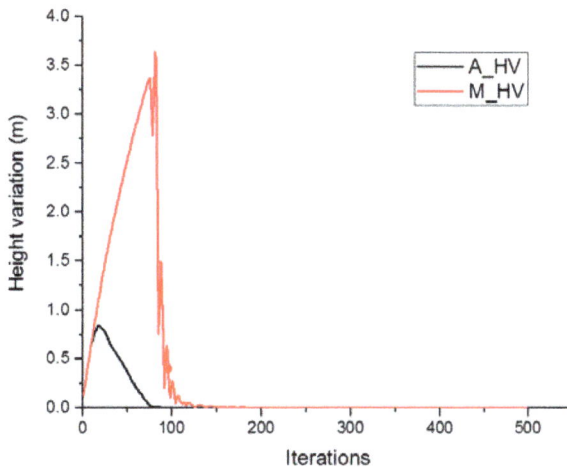

Figure 16. Maximum height variation (M_HV) and average height variation (A_HV).

From the discussion above, the parameter settings in the CSF algorithm are relatively simple and intuitive. Only two parameters (RI and ST) must be determined through visual judgment by the user. A rough estimation is enough to determine these parameter values. Usually, RI can be set to 1, 2 or 3 according to the features of the terrain, they are applied to areas with high steep slopes, terraced slopes and gentle slopes, respectively. ST is set to "true" or "false". "true" means that there exists steep slopes and post-processing is needed. "false" means post-processing is not needed (See details in Section 4.3).

4.3. Steep Slopes

For group II and group III, which contain many steep or terraced slopes in the scene, the slope is an important factor that influences the accuracy of the CSF algorithm. The simulated cloth will lie over the slope but usually cannot stick to the ground perfectly; at the edge of the slope, some distances will appear between the cloth and the ground. If this distance exceeds h_{cc}, the ground measurements will be mistakenly classified as OBJ. A direct method to mitigate this problem is to set the rigidness to a lower value, but some low objects may be classified as BE as a result. To balance these two types of errors, a post-processing method for the margin area is proposed in this study, as mentioned in Section 2.3. Thus, we can use a relatively hard cloth and post-processing to remove lower objects and correctly handle steep slope areas. A typical result is shown in Figure 17, in which the cloth properly sticks to the terrain near the steep margin after the post-processing procedure.

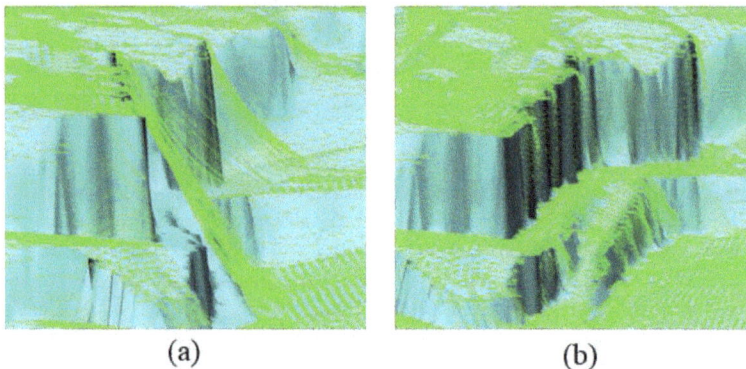

(a)   (b)

Figure 17. Simulated cloth over an area with steep slopes: (**a**) simulated cloth before post-processing; (**b**) simulated cloth after post-processing.

4.4. Bridge

When handling steep slopes, an exception is made for bridges, which are defined as OBJ in the ISPRS reference datasets, but the adjacent road is treated as BE. Usually, this scene will show a gentle slope along the direction of the road but will show an abrupt elevation change in the direction of the river (Figure 18). During the post-processing procedure, movable particles over the bridge may be set as unmovable particles and stick to the bridge because the height difference between two CPs is very small along the direction of the road. Thus, the bridge will usually be classified as BE through post-processing in our algorithm.

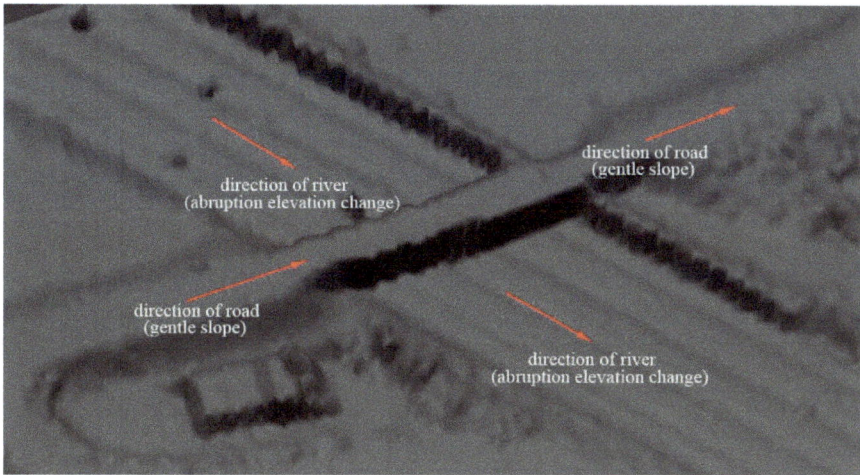

Figure 18. Illustration of a bridge: height variation along road is much less than that in direction of river.

4.5. Outlier Processing

As described in Section 2.2, the cloth particles will stop moving as soon as they reach the IHV; if some outliers exist under the ground measurements, then some particles will be obscured, and the cloth is usually propped up by these outliers. This phenomenon can increase the errors around the outliers because of the rigidness of the cloth. Thus, the outliers should be removed by some statistical filters before applying the CSF algorithm to the point cloud. Among the fifteen samples that were used in this study, sample 41 is an exception that has a large area of outliers [11] under the ground, which cannot be easily removed by a statistical approach. If the outliers could be removed either automatically or manually before applying the CSF algorithm, the results can be satisfactory. The total error before removing the outliers was 5.14, which reduced to 1.63 after removing the outliers.

5. Conclusions

This research proposes a novel filtering method named CSF based on a physical process. It utilizes the nature of cloth and modifies the physical process of cloth simulation to adapt to point cloud filtering. Compared to conventional filtering algorithms, the parameters are less numerous and are easy to set. Regardless of the complexity of ground objects, the samples were divided into three categories according to the shape of the terrain. Few parameters are needed, and these parameters hardly changed among the three sample categories; only an integer parameter rigidness and a Boolean parameter ST are required to be set by the user. These three groups of parameters exhibit relatively high accuracies for all fifteen samples of the ISPRS benchmark datasets. Another benefit of the CSF algorithm is that the simulated cloth can be directly treated as the final generated DTM for some circumstances, which avoids the interpolation of ground points, and can also recover areas of missing data. Moreover, we have released our software to the public [35], and we will also release our source code to the research community. We hope that the proposed novel physical process simulation-based ground filtering algorithm could help promote the scientific, government, and the public's use of LiDAR data and technology to the applications of flood simulation, landslide monitoring, road design, land-cover classification, and forest management.

However, the CSF algorithm has limitations as well. Because we have modified the physical processes of particle movement into two discrete steps, the particles may stick to roofs and some OBJ points may be mistakenly classified as BE when dealing with very large low buildings. This process usually produces some isolated points at the centers of roofs, an noise filtering can help to mitigate this problem. Additionally, the CSF algorithm cannot distinguish objects that are connected to the ground (e.g., bridge). In the future, we will try to use the geometry information of LiDAR points or combine optical images (such as multispectral images) to clearly distinguish bridges from roads.

Acknowledgments: This work was supported by the National Basic Research Program of China (973 Program) Grant No. 2013CB733402 and the CAS Key Laboratory of Lunar and Deep Space Exploration through Grant No. YQSYS-HT-140630-1. This work was also supported by the National Natural Science Foundation of China Grant No. 41171265, 41331171 and 40801131, and was partially supported by China Scholarship Council (CSC). Special thanks to LiDAR 2015 conference (http://LiDAR2015.org/) for providing the testing data, and Daniel, the maintainer of Cloudcompare (http://www.cloudcompare.org/), who gave us much valuable advice and helped to optimize the source code.

Author Contributions: Wuming Zhang had the original idea that uses cloth simulation to filter LiDAR data. Jianbo Qi refined and implemented this idea, and wrote this manualscript. Peng Wan helps to refine some parameters settings, and developed cloudcompare plugin. Hongtao Wang summarized some existing filtering algorithms and give much useful advice to promote this method. Donghui Xie and Xiaoyan Wang helped to process some LiDAR data, and tested this method with them. Guangjian Yan helped to review this article and give much advice.

Conflicts of Interest: The authors declare no conflict of interest.

References

1.   Kobler, A.; Pfeifer, N.; Ogrinc, P.; Todorovski, L.; Oštir, K.; Džeroski, S. Repetitive interpolation: A robust algorithm for DTM generation from Aerial Laser Scanner Data in forested terrain. *Remote Sens. Environ.* **2007**, *108*, 9–23.

2.   Vosselman, G. Slope based filtering of laser altimetry data. *Int. Arch. Photogramm. Remote Sens.* **2000**, *33*, 935–942.

3.   Shan, J.; Aparajithan, S. Urban DEM generation from raw lidar data. *Photogramm. Eng. Remote Sens.* **2005**, *71*, 217–226.

4.   Meng, X.; Wang, L.; Silván-Cárdenas, J.L.; Currit, N. A multi-directional ground filtering algorithm for airborne LIDAR. *ISPRS J. Photogramm. Remote Sens.* **2009**, *64*, 117–124.

5.   Sithole, G. Filtering of laser altimetry data using a slope adaptive filter. *Int. Arch. Photogramm. Remote Sens. Spat. Inf. Sci.* **2001**, *34*, 203–210.

6.   Susaki, J. Adaptive slope filtering of airborne LiDAR data in urban areas for digital terrain model (DTM) generation. *Remote Sens.* **2012**, *4*, 1804–1819.

7.   Wang, C.K.; Tseng, Y.H. Dem generation from airborne LiDAR data by an adaptive dual-directional slope filter. In Proceedings of the ISPRS Commission VII Mid-Term Symposium 100 Years ISPRS—Advancing Remote Sensing Science, Vienna, Austria, 5–7 July 2010.

8.   Liu, X. Airborne LiDAR for DEM generation: Some critical issues. *Progress Phys. Geogr.* **2008**, *32*, 31–49.

9.   Sithole, G.; Vosselman, G. Filtering of airborne laser scanner data based on segmented point clouds. *Int. Arch. Photogramm. Remote Sens. Spat. Inf. Sci.* **2005**, *36*, W19.

10. Zhang, K.; Chen, S.C.; Whitman, D.; Shyu, M.L.; Yan, J.; Zhang, C. A progressive morphological filter for removing nonground measurements from airborne LIDAR data. *IEEE Trans. Geosci. Remote Sens.* **2003**, *41*, 872–882.

11. Chen, Q.; Gong, P.; Baldocchi, D.; Xie, G. Filtering airborne laser scanning data with morphological methods. *Photogramm. Eng. Remote Sens.* **2007**, *73*, 175–185.

12. Li, Y. Filtering Airborne LIDAR Data by AN Improved Morphological Method Based on Multi-Gradient Analysis. *ISPRS Int. Arch. Photogramm. Remote Sens. Spat. Inf. Sci.* **2013**, *XL-1/W1*, 191–194.

13. Li, Y.; Yong, B.; Wu, H.; An, R.; Xu, H. An Improved Top-Hat Filter with Sloped Brim for Extracting Ground Points from Airborne Lidar Point Clouds. *Remote Sens.* **2014**, *6*, 12885–12908.

14. Mongus, D.; Lukač, N.; Žalik, B. Ground and building extraction from LiDAR data based on differential morphological profiles and locally fitted surfaces. *ISPRS J. Photogramm. Remote Sens.* **2014**, *93*, 145–156.

15. Pingel, T.J.; Clarke, K.C.; McBride, W.A. An improved simple morphological filter for the terrain classification of airborne LIDAR data. *ISPRS J. Photogramm. Remote Sens.* **2013**, *77*, 21–30.

16. Mongus, D.; Žalik, B. Parameter-free ground filtering of LiDAR data for automatic DTM generation. *ISPRS J. Photogramm. Remote Sens.* **2012**, *67*, 1–12.

17. Axelsson, P. DEM generation from laser scanner data using adaptive TIN models. *Int. Arch. Photogramm. Remote Sens.* **2000**, *33*, 111–118.

18. Zhang, J.; Lin, X. Filtering airborne LiDAR data by embedding smoothness-constrained segmentation in progressive TIN densification. *ISPRS J. Photogramm. Remote Sens.* **2013**, *81*, 44–59.

19. Kraus, K.; Pfeifer, N. Determination of terrain models in wooded areas with airborne laser scanner data. *ISPRS J. Photogramm. Remote Sens.* **1998**, *53*, 193–203.

20. Pfeifer, N.; Reiter, T.; Briese, C.; Rieger, W. Interpolation of high quality ground models from laser scanner data in forested areas. *Int. Arch. Photogramm. Remote Sens.* **1999**, *32*, 31–36.

21. Chen, C.; Li, Y.; Li, W.; Dai, H. A multiresolution hierarchical classification algorithm for filtering airborne LiDAR data. *ISPRS J. Photogramm. Remote Sens.* **2013**, *82*, 1–9.

22. Su, W.; Sun, Z.; Zhong, R.; Huang, J.; Li, M.; Zhu, J.; Zhang, K.; Wu, H.; Zhu, D. A new hierarchical moving curve-fitting algorithm for filtering lidar data for automatic DTM generation. *Int. J. Remote Sens.* **2015**, *36*, 3616–3635.

23. Hui, Z.; Hu, Y.; Yevenyo, Y.; Yu, X. An Improved Morphological Algorithm for Filtering Airborne LiDAR Point Cloud Based on Multi-Level Kriging Interpolation. *Remote Sens.* **2016**, *8*, 35.

24. Elmqvist, M. Automatic Ground Modelling Using Laser Radar Data. Master's Thesis, Linköping University, Linköping, Sweden, 2000.

25. Guan, H.; Li, J.; Yu, Y.; Zhong, L.; Ji, Z. DEM generation from lidar data in wooded mountain areas by cross-section-plane analysis. *Int. J. Remote Sens.* **2014**, *35*, 927–948.

26. Weil, J. The synthesis of cloth objects. *ACM Siggraph Comput. Graph.* **1986**, *20*, 49–54.

27. Provot, X. *Deformation Constraints in a Mass-Spring Model to Describe Rigid Cloth Behaviour*; Graphics Interface; Canadian Information Processing Society: Mississauga, ON, Canada, 1995.

28. Mosegaards Cloth Simulation Coding Tutorial. Available online: http://cg.alexandra.dk/?p=147 (accessed on 8 June 2016).

29. Girardeau-Montaut, D. *Cloud Compare-Open Source Project*; OpenSource Project: Grenoble, France, 2011.

30. Sithole, G.; Vosselman, G. Experimental comparison of filter algorithms for bare-Earth extraction from airborne laser scanning point clouds. *ISPRS J. Photogramm. Remote Sens.* **2004**, *59*, 85–101.

31. Cohen, J. A coefficient of agreement for nominal scales. *Educ. Psychol. Measur.* **1960**, *20*, 37–46.

32. Hu, H.; Ding, Y.; Zhu, Q.; Wu, B.; Lin, H.; Du, Z.; Zhang, Y.; Zhang, Y. An adaptive surface filter for airborne laser scanning point clouds by means of regularization and bending energy. *ISPRS J. Photogramm. Remote Sens.* **2014**, *92*, 98–111.

33. Terrasolid, Ltd. *TerraScan User's Guide*; Terrasolid, Ltd.: Helsinki, Finland, 2010.

34. Pfeifer, N.; Stadler, P.; Briese, C. Derivation of digital terrain models in the SCOP++ environment. In Proceedings of the OEEPE Workshop on Airborne Laserscanning and Interferometric SAR for Detailed Digital Terrain Models, Stockholm, Sweden, 1–3 March 2001; Volume 3612.

35. CSF Software and Introduction. Available online: http://ramm.bnu.edu.cn/researchers/wumingzhang/english/default_contributions.htm (accessed on 8 June 2016).

remote sensing

MDPI

Article

Extracting Canopy Surface Texture from Airborne Laser Scanning Data for the Supervised and Unsupervised Prediction of Area-Based Forest Characteristics

Mikko T. Niemi [1] and Jari Vauhkonen [1,2,*]

[1] Department of Forest Sciences, University of Helsinki, P.O. Box 27, FI-00014 Helsinki, Finland; mikko.t.niemi@helsinki.fi

[2] School of Forest Sciences, University of Eastern Finland, P.O. Box 111, FI-80101 Joensuu, Finland

* Correspondence: jari.vauhkonen@uef.fi; Tel.: +358-50-442-4432

Academic Editors: Jie Shan, Juha Hyyppä, Nicolas Baghdadi and Prasad S. Thenkabail

Received: 31 May 2016; Accepted: 5 July 2016; Published: 9 July 2016

Abstract: Area-based analyses of airborne laser scanning (ALS) data are an established approach to obtain wall-to-wall predictions of forest characteristics for vast areas. The analyses of sparse data in particular are based on the height value distributions, which do not produce optimal information on the horizontal forest structure. We evaluated the complementary potential of features quantifying the textural variation of ALS-based canopy height models (CHMs) for both supervised (linear regression) and unsupervised (k-Means clustering) analyses. Based on a comprehensive literature review, we identified a total of four texture analysis methods that produced rotation-invariant features of different order and scale. The CHMs and the textural features were derived from practical sparse-density, leaf-off ALS data originally acquired for ground elevation modeling. The features were extracted from a circular window of 254 m^2 and related with boreal forest characteristics observed from altogether 155 field sample plots. Features based on gray-level histograms, distribution of forest patches, and gray-level co-occurrence matrices were related with plot volume, basal area, and mean diameter with coefficients of determination (R^2) of up to 0.63–0.70, whereas features that measured the uniformity of local binary patterns of the CHMs performed poorer. Overall, the textural features compared favorably with benchmark features based on the point data, indicating that the textural features contain additional information useful for the prediction of forest characteristics. Due to the developed processing routines for raster data, the CHM features may potentially be extracted with a lower computational burden, which promotes their use for applications such as pre-stratification or guiding the field plot sampling based solely on ALS data.

Keywords: forest inventory; Light Detection And Ranging (LiDAR); surface modeling; Inverse Distance Weighting (IDW) interpolation; image texture anisotropy

1. Introduction

Airborne laser scanning (ALS; also referred to in some instances as "airborne scanning LiDAR") has become a routinely operated technique for wall-to-wall prediction and mapping of various forest characteristics [1]. Due to straightforward implementation with ALS data acquired broadly for ground elevation modeling [2–4], so-called area-based approaches are the most established prediction techniques. By area-based approaches, we fundamentally refer to the two-stage procedure [5] in which: (i) models to predict the forest attributes of interest for the individual areas of interest (AOIs) are fit based on a set of training field plots; and (ii) the resulting models are applied to all the AOIs of the entire inventory area to produce wall-to-wall predictions. Operational implementations are elaborated upon in [6–8].

To develop predictive relationships between ALS and field training data, relevant features need to be extracted. The sparse pulse densities (typically $< 1 \text{ m}^{-2}$ according to [6,7]) allow extracting features related to the distributions of height values or proportions of certain types of echoes [5,9], which do not account for the full horizontal information available in the data. It is possible to extract structural and volumetric features from sparse, area-based data [10–12], whereas applying notably higher pulse densities has allowed understory- [13–15] or tree-level descriptions based on canopy density and height models [16–18]. Canopy height models (CHMs) are rasterized images representing interpolated difference in elevations between the top of the vegetation and ground level. Importantly, the sparse pulse density of ALS data constrains the CHM pixel size and therefore affects the accuracies of the subsequent forest attribute predictions. However, some researchers have related the CHM texture derived even from sparse ALS data to the properties of the growing stock such as the spatial pattern of the trees [19,20].

Texture analysis is a well-established field of image analysis with early forest applications for optical data acquired by varying sensors [21–27]. Additionally, Tuominen and Pekkarinen [28] provide a comprehensive overview of parameters affecting the prediction accuracies of most essential forest attributes based on very high-resolution image data in forest conditions closely corresponding to those of the present study. Textural features derived from the aerial images were used as predictor variables with ALS point-based features in [29,30]. Even though ALS-based CHMs were readily suitable for texture analyses, those did not appear until about a decade after the introduction of the ALS-based CHM products [31]. CHMs derived from high-density ALS data were analyzed for their texture in [32–34], and those were derived using practical sparse pulse densities in [19,20,35]. The applications are, to date, related to predicting species, diversity, and the spatial arrangement of the trees, whereas the relationships between the textural features derived from the CHMs and forest attributes are not comprehensively known like in the case of aerial images [28].

In all ALS studies cited in the previous paragraph, the characterization of the image texture was based on computing a gray-level co-occurrence matrix (GLCM) and a set of descriptive features following the principles presented by Haralick et al., in 1973 [36]. However, there are several alternative approaches to quantify image texture. The use of Gabor or wavelet filtering produced good results compared to using the GLCM features with other remotely sensed materials besides ALS data [37]. Autocorrelation functions or (semi-)variograms [38] were tested with optical images [39–41] and with ALS data [42]. Patch or landscape features [43] were proposed as coarse measures of CHM texture [35]. More detailed approaches, such as analyzing local microstructures [44] or geometrically [45–47] or topologically [48] driven partitioning of image patches, have been proposed as powerful texture discriminators. To the best of our knowledge, however, the latter approaches are tested for built material classification only. Thus, the suitability of these methods and sensitivity of their parameters for remotely sensed vegetation analyses cannot be deduced based on earlier studies.

In summary, various texture analyses can be identified in the literature to improve the assessments of forest growing stock attributes, but analyses considering aspects other than GLCM-based textural features, especially in the case of ALS-based CHMs, are lacking. Even though conventional point-based ALS features most likely produce the best prediction outcome due to their potential to depict multilayered vertical strata [49,50], the textural features may be expected to improve the predictions by allowing a more detailed description of the horizontal forest structure [19,20]. Beside the use of textural features in supervised learning approaches, the favorable computational properties of the CHMs [51] may open up possibilities for inventories, in which ALS data have been acquired but no field reference data yet exist [52]. In those cases, the texture features may be used for pre-stratification or guiding a later field inventory to cover the variation of forest attributes within the area, which is challenging but important [53–55]. Thus, testing textural features for both the supervised and unsupervised learning approaches is well reasoned.

The purpose of this study is thus to compare texture analysis methods on CHMs derived from practically available sparse, leaf-off ALS data. Appropriate methods for extracting the textural features were identified based on a comprehensive literature review. The features were related to selected forest biophysical properties measured from field reference plots with an area of 254 m^2. The performance of these features was tested in: (i) a supervised prediction of the total stem volume, basal area, and mean diameter; and (ii) an unsupervised classification of the forest area for a simulated sampling of the field plots. For benchmarking purposes, the extracted features were compared to the most common point-based ALS predictor variables in the supervised predictions.

2. Materials and Methods

2.1. Experimental Data

2.1.1. Study Area

The study area is located in the southern boreal forest zone in Evo, Finland (61.19°N, 25.11°E). The area of altogether approximately 2000 ha is a part of a state-owned forest. The elevation is typically 125–145 m above sea level and mineral soils with gentle slopes prevail. The forest stands vary from natural to intensively managed in terms of their silvicultural status. The area is dominated by Scots pine (*Pinus sylvestris* L.) and Norway spruce (*Picea abies* [L.] H. Karst.), which altogether constitute approximately 84% of the growing tree stock, with minor proportions of deciduous trees mainly occurring below the dominant canopy.

2.1.2. ALS Data

The ALS data used in the study were acquired by the National Land Survey of Finland as a part of their data-acquisition campaign for creating a nationwide ground elevation model for Finland. The data were downloaded from a file service [56], from which they are available for free and with extensive permissions of use. The data were acquired in two separate campaigns: on 7 May 2012 with Optech ALTM Gemini scanner operated from a flying altitude of 1830 m, and on 13 May 2012 with Leica ALS50 scanner operated from 2200 m. In both the acquisitions, a scanning angle of $\pm20°$, a ground footprint of approximately 50 cm, and otherwise similar scanning parameters were applied to yield a nominal pulse density of 0.7–0.8 m^{-2}. Both the scanner systems recorded up to four echoes per each emitted pulse. The data provider had detected and classified the ground level, on which the normalization of the vegetation height values was based, using an adaptive filtering algorithm [57] implemented in TerraScan software (Terrasolid Ltd., Helsinki, Finland). As the data are meant specifically for ground elevation modeling, we assumed the accuracy of this classification to be appropriate for our purposes. Following similar data acquisition and processing principles, the standard errors of terrain elevation values were found to be in order of 15 cm [58].

2.1.3. Field Measurements

The field sample was selected based on auxiliary information from the ALS data [59]. Altogether, five ALS features were extracted to quantify height and canopy cover of the dominant vegetation, and density, diversity, and proportion of understory trees and shrub layer, determined as below 60% of the maximum height and below 5 m [59], respectively. These data were clustered, applying an unsupervised *k*-Means algorithm with equal weights for each feature to obtain an ALS-based stratification of forest structure. The strategy was found efficient to distribute the sample across the spatial, size, and age distributions of the tree stock.

A field measurement protocol based on circular plots with the radius of 9 m was applied on a total of 155 plots. The species and diameter-at-breast height (DBH) were measured for each tree with a DBH ⩾ 5 cm. For each tree species of the plot, a tree with a DBH corresponding to the median tree was determined in the field and measured for height. These attributes were used as the median tree attributes per plot and species. The similarity of close surroundings of each circular reference plot was quantified by establishing an additional field plot 16 m from the original plot center in each cardinal direction ($0°$, $90°$, $180°$, and $270°$). In these satellite plots, only DBHs were measured from tally trees selected employing a basal area factor of $2\,m^2a^{-1}$.

The positions of the field plot centers were measured with a Trimble® Nomad® 900G Global Positioning System and Global Navigation Satellite System receiver with an external antenna and battery. At least 100 signals were collected from the center of each plot, and its position was computed from these observations. The receiver was connected to a GSM phone, which provided a connection to a reference station for real-time differential correction to help the positioning in the field. A differential post-processing was finally applied using Trimble®Pathfinder®Office and a local, permanent base station.

2.1.4. Field Plot Attributes Derived for the Evaluation

Plot-level forest attributes were compiled and aggregated from the tree-level measurements. Plot basal area (G) was computed by summing the DBH measurements. The variations in the within-plot DBH distribution and G among the main plot and the satellite plots were quantified using the coefficient of variation (CV), i.e., the ratio of the standard deviation to the mean of the specific values. The missing tree heights were predicted by calibrating the parameters of Näslund's height curve [60] using the species-specific median tree diameter and height estimates [61]. The basal-area-weighted DBH (D_{gM}) and height (H_{gM}) were computed based on all the trees of a plot. The volumes of the individual trees were predicted, employing the DBH and height as predictors of species-specific equations [62]. The volume models for birch were used for all deciduous trees. The volumes (V) were summed to the plot level and scaled per ha. Descriptive statistics of the 155 reference plots measured are presented in Table 1.

Table 1. Central total and species-specific characteristics of the reference data, aggregated to the plot-level.

Attribute	Range	Mean	SD *
V, m^3ha^{-1}	7.0–673.0	184.0	107.0
G, m^2ha^{-1}	1.3–48.0	19.8	9.0
D_{gM}, cm	8.4–42.7	22.9	6.9
H_{gM}, m	7.9–32.3	19.0	4.7
V_{pine}, m^3ha^{-1}	0–360.0	101.0	89.0
G_{pine}, m^2ha^{-1}	0–33.1	10.6	8.7
D_{gM_pine}, cm	5.3–50.0	24.3	9.0
H_{gM_pine}, m	3.5–32.0	19.4	5.3
V_{spruce}, m^3ha^{-1}	0–464.0	53.8	87.9
G_{spruce}, m^2ha^{-1}	0–36.3	5.7	7.7
D_{gM_spruce}, cm	5.0–40.0	15.9	9.8
H_{gM_spruce}, m	4.0–32.5	13.7	7.6
$V_{deciduous}, m^3ha^{-1}$	0–228.0	29.0	43.2
$G_{deciduous}, m^2ha^{-1}$	0–24.9	3.5	4.9
$D_{gM\_deciduous}$, cm	5.0–52.0	16.3	9.3
$H_{gM\_deciduous}$, m	6.0–33.0	16.2	6.2

* Standard deviation.

2.2. Canopy Height Model (CHM) Generation

CHM rasters were generated based on the first-of-many and only echoes of the normalized ALS data, aiming to obtain the main information from the data while retaining most generalization abilities over sensors that record a different number of echo categories [6]. The low pulse density of ALS data was observed to restrict the realism of the CHM, as a requirement for a smaller pixel size would have meant that fewer CHM values were observed and more were interpolated. Consequently, different pixel sizes were tested and an inverse-distance weighting (IDW) technique [63] was applied to fill the pixels without observations. The IDW algorithm was selected due to its simplicity and controllability based on only few parameters (see further notes in Section 4). The proportions of the observed CHM values with the considered pixel sizes are shown on Table 2.

Table 2. The proportion of observed canopy height model pixel values with the considered pixel size in the reference plots.

Pixel Size	Mean	SD *
0.5 m	27%	11%
0.75 m	49%	13%
1.0 m	59%	12%
1.5 m	93%	6%
2.0 m	97%	3%
3.0 m	100%	2%

* Standard deviation.

Using the six different pixel sizes (Table 2), the CHM values for all the pixels containing at least one ALS echo were initially filled by selecting the maximum height value of observations inside the pixel. A non-filled value $\hat{z}$ at pixel S_0 was interpolated using the observed CHM values z at pixels S_i (Equation (1)), where λ_i is the weight calculated according to Equation (2) based on an interpolation parameter α and d_{0i} as the distance between S_0 and S_i:

$$\hat{z}(S_0) = \sum_{i=1}^{n} \lambda_i z(S_i)$$ (1)

$$\lambda_i = d_{0i}^{-\alpha} / \sum_{i=1}^{n} d_{0i}^{-\alpha}$$ (2)

The maximum distance between S_0 and S_i was set to 3 m, i.e., observations located further than 3 m from the center of an empty pixel had no effect to the interpolation of the specific CHM value. Values 0.1, 0.5, 1, 2, 3, ... , 9, and 10 were tested for α. Figure 1 shows examples of CHMs derived using different pixel sizes and $\alpha = 5$, which was used in most of the analyses.

2.3. Textural Feature Extraction from the CHM

Image texture analysis has been a subject of very extensive research since the 1970s—for a review, see, e.g., Chapter 2 in [64]. This makes an exhaustive comparison of all the published methods impossible. The properties of the algorithms therefore were analyzed by relating earlier literature to our research objectives, i.e., to model forest attributes aggregated from field plots with a size of 254 m^2 based on the extracted textural features. Several aspects such as invariance and robustness of the extraction method relative to the properties of the analyzed images were considered when choosing the algorithms [65].

Texture analyses are typically labeled as: (i) statistical; (ii) geometrical; (iii) model-based; or (iv) signal processing [65,66], in which the discrimination between different textures is based on: (i) comparing the statistics computed over different (sub-)regions; (ii–iii) determining geometric or image-model-based primitives that compose the texture; or (iv) filtering the image in the spatial or frequency domain to be able to compute the frequency components of the image signal. Approaches (ii) and (iii) were deemed inappropriate for our purposes due to the requirement to assume a geometrical or image model behind the process that generated the texture, i.e., to assume the spatial and size variation in the trees of the plots. Approach (iv) was found to be sensitive to a particular frequency and orientation, and even if the invariance to these factors was achieved by a multichannel filtering approach [64], the method appeared as overly complicated for our purposes. Hence, all the methods included in the present comparison could be labeled as statistical texture analyses. The number of these methods was further reduced by a requirement to obtain rotation-invariant features, i.e., not having to assume that the surfaces analyzed are captured from the same viewpoint.

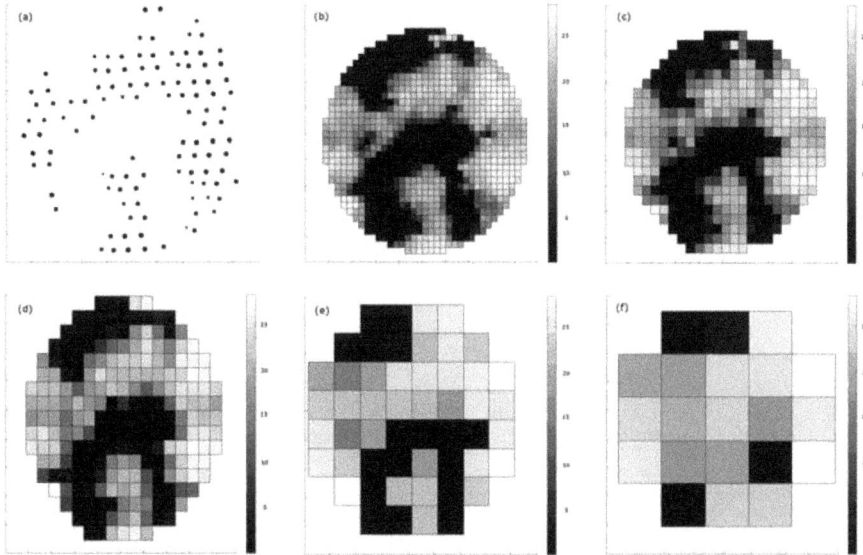

Figure 1. (**a**) Points with height values >5 m above ground level extracted from an example plot. Canopy height models interpolated from the point data using pixel sizes of: (**b**) 0.5 m; (**c**) 0.75 m; (**d**) 1.0 m; (**e**) 2.0 m; and (**f**) 3.0 m. In (**a**), the sizes of the dots are scaled according to the height values, whereas the tick marks of each sub-plot correspond to a horizontal distance of 2 m.

The methods selected for the analyses are listed and characterized according to the order and scale considered (Table 3). "Order" determines the immediate neighborhood of pixels considered as an individual observation, whereas "scale" indicates the spatial domain in which the features are computed from the individual observations. First-order statistics are based on individual pixels, whereas second-order statistics also account for the spatial co-occurrence of the pixels [66]. The features considered here are extracted from the area of full plots ("global" scale) or the plots re-scaled using the GLCMs ("macro" scale) or detailed local neighborhoods ("micro" scale). The features were extracted from a circular window with a size of 254 m^2 corresponding to the reference plots with 9 m radii. In addition, a 30 m × 30 m square window centered to the location of the plots were tested as an alternative computation unit in particular to evaluate the prediction of the CV of G among the satellite plots.

First-order statistics computed in the global scale are assumed to be neutral to the rotations of the image, whereas the use of second-order statistics (based on co-occurrence of two pixels) requires considering the rotation. In the GLCM approach [36], the rotation invariance is obtained by averaging the features extracted from the GLCMs over all directions. In addition to these second-order, macro-scale statistics, detailed micro-scale statistics were derived by computing gray-scale changes in local neighborhoods using the local binary pattern (LBP) approach [44]. In LBP, the rotation invariance is obtained by computing occurrence statistics for each individual rotation pattern of the neighborhood considered [44].

Since the use of higher-than-second-order statistics or other scales is rare in the literature, we believe that the results obtained by the selected methods (Table 3) can be extended to other texture quantification methods characterized by the same order and scale. The feature groups considered in the study are listed as follows.

Table 3. Textural feature groups considered.

Group	Order	Scale	Image Data Source	Features and Their Abbreviations
HIST	1st	global	Gray-level histogram	Mean (CHM_{mean}), Standard deviation (CHM_{std}), Maximum (CHM_{max})
PATCH	2nd	global	Thresholded image [a]	Proportion of field plot total area covered by tree patches (TP_t_%), Number of tree patches per hectare (TP_t_den), Average area of tree patches (TP_t_avg), Standard deviation of the area of tree patches (TP_t_std), Average number of pixels in 4-neighborhood belonging to the same tree patch (TP_t_N)
GLCM	2nd	macro	Grey-level co-occurrence matrix [b]	Angular second moment (ASM), Contrast (CON), Correlation (COR), Sum of squares: variance (VAR), Inverse difference moment (IDM), Sum average (SAVG), Sum variance (SVAR), Sum entropy (SENT), Entropy (ENT), Difference variance (DVAR), Difference entropy (DENT), 2 × Information measure of correlation (IMC1, IMC2)
LBP	2nd	micro	Local binary pattern (LBP) [c]	Uniformity of a LBP computed as percentages of at most 0 (U_0), 2 (U_2) or 4 (U_4) bitwise transitions in the binary code of a circular neighborhood

[a] In the abbreviations of the features, subscript *t* is either 5 m or *ad*, indicating the use of a constant (5 m) or adaptive ($ad = h_{max}-h_{std}$) height threshold, respectively; [b] All features were computed using a lag of 1, 2, 3, 4, or 5 pixels; [c] All features were computed using a lag of 1, 2, or 3 pixels.

2.3.1. HIST Features

The following first-order, general statistics were derived from the histograms of CHM pixel values: mean, standard deviation, and maximum (Table 3). These features explain vegetation height and its variation directly.

2.3.2. PATCH Features

The patch metrics are considered as coarse measures of texture that eventually produce information on the co-occurrence of groups of pixels obtained by thresholding an image. The spatial pattern of trees was modeled by thresholding CHMs for representing ground and tree patches [19,20]. In [19], a threshold of 5 m was used to assign all CHM values above that height as tree patches. In [20], an adaptive difference between the maximum height (h_{max}) and the standard deviation of the height values (h_{std}) was reported as the best canopy threshold for determining the spatial pattern of trees. Both the aforementioned constant and adaptive thresholds were tested in this study, and the patch features described in Table 3 were extracted from the thresholded CHMs.

2.3.3. GLCM Features

Textural features were computed from the GLCMs according to [36] by varying the lag parameter (i.e., the spatial interval between co-occurring pixel values) from 1 to 5. To get rotation invariant features, angular GLCMs were computed for four different offsets (0°, 45°, 90°, and 135°), and the mean values of the four angular textural features were analyzed. One out of the original 14 features [36], namely "Maximal Correlation Coefficient", was not included in our analyses due to known computational instability problems associated with this feature. The derived features and their abbreviations are listed in Table 3. For more detailed descriptions, see [36].

2.3.4. LBP Features

The rotation invariant textural features based on the LBPs [44,67,68] are computed for every pixel by comparing the CHM value with its local circular neighborhood. The values are re-coded depending on the neighborhood: if the value of the pixel is equal or smaller than the value of its neighboring pixel, the local binary is coded as 1 and otherwise as 0. A single LBP of a pixel thus consists of the eight binary values derived from its circular neighborhood. Following [44], in which the uniformity of the image was measured by counting the number of bitwise 0/1 changes in the binary codes, we used the percentages of LBP patterns having uniformity of at most 0, 2, and 4 (abbreviated as U_0, U_2 and U_4, respectively) as the extracted LBP features for the reference plots. All the LBP features were computed using the lags of 1, 2, and 3 CHM pixels.

2.3.5. Conventional Point-Based ALS Features

For benchmarking purposes, the most common point-based ALS predictor variables [5,9], i.e., the maximum, the mean, and the standard deviation of the height values; proportion of echoes above 2 m vegetation threshold; the 5th, 10th, 20th, . . . , 90th, and 95th percentiles; and the corresponding proportional densities of the ALS-based canopy height distribution were calculated according to [69] (pp. 502–503). The features were computed separately based on the first and last pulse data, i.e., "only" or "first of many" and "only" or "last of many" echoes, respectively, of up to 4 laser echoes recorded per pulse.

2.4. Supervised Prediction of Forest Attributes

The performance of the extracted features was first assessed in supervised prediction of forest attributes based on the observations made from the field-measured reference plots. The strength of the relationship between the CHM features and various forest inventory attributes was quantified using the coefficient of determination (R^2, [70]). The main attention was focused on central attributes related to the properties of the forest growing stock, i.e., total stem volume (V), basal area (G), and basal-area weighted mean diameter (D_{gM}). Due to the earlier reported potential to link the textural features with the diversity of the tree size and vegetation structure [34,71], we additionally analyzed the relationships between the textural features and the number of understory trees and variation of the DBH and G observed from within the 9 m plots and among the satellite plots, respectively.

The applicability of the extracted textural features and the benchmark variables was tested for predicting V, G, and D_{gM} attributes of the sample plots, as these were found to be most related to the textural features in the initial tests. To normalize the potentially non-linear relationships between the forest attributes and the ALS features, three most common transformations (square, square root, and natural logarithm) of all the textural and point-based features were included as candidate predictors in addition to the absolute values in a simple linear regression (LR) analysis. The final predictors of the LR models were selected by inserting one of the predictor candidates at a time into each model and iteratively appending the feature that produced the best prediction accuracy. The LR models were formed applying a leave-out-one-plot cross-validation, i.e., excluding each plot at the time from the model fitting data and predicting for the excluded plot. The prediction accuracy was measured by root mean square error (RMSE) and relative root mean square error (RMSE%):

$$RMSE = \sqrt{\frac{\sum_{i=1}^{n}(\hat{y}_i - y_i)^2}{n}} \tag{3}$$

$$RMSE\% = 100 * \frac{RMSE}{\bar{y}} \tag{4}$$

where y_i is the observed value of variable y on plot i, $\hat{y}_i$ is the predicted value of variable y on plot i, $\bar{y}$ is the mean of the observed values, and n is the number of reference plots.

2.5. Unsupervised Classification of the Forested Area

Second, the textural features were evaluated for an unsupervised classification of the forested area, subsequently prioritizing the plots to be measured for field reference data. Earlier studies have suggested that the information in the ALS data may be condensed to a few metrics [72,73], the partitioning of which will provide a stratification corresponding closely to the structural complexity observed in the field [59,74,75]. In this study, a similar partitioning was carried out using the textural features, and the applicability of the obtained information was demonstrated by prioritizing the field plots to be measured for predicting plot V using other features.

The data were stratified using an unsupervised *k*-Means algorithm [76] implemented in R statistical computing environment [77]. The algorithm partitions the observations into *k* clusters such that the sum of squares from the observations to the assigned cluster centers is minimized. The clustering was based on the standard Euclidean distance and initialized by altogether 10,000 random solutions for the initial cluster centers to minimize the randomness in the final result. To use the standard Euclidean distance, the selected variables were required to have low inter-correlation to minimize redundancy of the information and thereby ensure an equal weight of each variable in the unsupervised classification. For this reason, the unsupervised classification was based on altogether six textural features hand-picked to capture the main variation in the forest structural characteristics but to have as low inter-correlation as possible. To include all the ALS features with the same importance in the clustering, the original feature values were normalized between 0 and 1. The normalized feature values were obtained according to Equation (5):

$$r_{ij} = \frac{q_{ij} - q_i^{min}}{q_i^{max} - q_i^{min}} \tag{5}$$

where r_{ij} and q_{ij} are the normalized and original values of the *j*th observation of feature *i* and q_i^{min} and q_i^{max} are the smallest and largest values of *i* among all plots.

Since it was found problematic to determine a fixed *k*, i.e., to fix the expected value of the structural classes, we determined the final partitioning according to the persistence of the clusters on an interval of different *k* values. Based on initial tests in the data studied, the clustering was repeated by gradually increasing the value of *k* from 2 to 7, using an R package *clue* [78] to manage the resulting ensemble of the *k*-Means partitions. During the iterations, it was recorded whether the individual cells persisted in a cluster or shifted from one cluster to another one. As a result, each grid cell was labeled with an identifier describing its path along the clustering hierarchy. Therefore, as opposed to results with a fixed *k*, the final partitioning was based on the composition of the sub-clusters formed during the clustering. More details on the applied methodology are provided by [59].

The applicability of the information obtained by the stratification was demonstrated by prioritizing the sample plots to be measured for field reference data. The accuracy of the predictions depends on how extensively the reference data represent the full range of the phenomenon to be modeled [52]. We assume that the textural variation extracted from the CHMs is related to the real-world variation of the forest attributes and therefore indicative of which plots include the extreme variation that must be captured in the modeling data to produce realistic predictions. This hypothesis was tested by simulating the sampling of a reduced number of field reference plots for predicting V (cf. [52–54]), i.e., applying Probability Proportional to Size (PPS) sampling for the selection of the first-stage sample units [79].

The non-linear relationship between aggregated stem volume (V) and a product of ALS-point-based mean height (H) and canopy cover (CC) was modeled as:

$$V = a + b\,(H \times CC)^c \tag{6}$$

where a, b, and c were model parameters solved using the *nls* function of R [77]. The plots to be included in the modeling data were prioritized according to the distance to their respective cluster center. The distribution of the data to the different clusters was assumed to indicate different forest structure types and the distance to the cluster center to indicate whether a plot represented an extreme (high distance) or a typical observation of the cluster. The plots were ordered according to the decreasing distance and inserted one by one to the data used for fitting Equation (6). The RMSE of Equation (6) was recorded after including each plot.

3. Results

3.1. Effects of the CHM Parameters to the Textural Features

The spatial interpolation of unknown CHM values was based on the premise that CHM values of two pixels are related to each other, and the similarity is inversely related to the distance between their locations. In the IDW method, this was affected by the parameter α: the larger the value, the less weight was given for observations located far from the unknown CHM pixel. Figure 2 shows how the R^2 of various textural features and total stem volume were affected, when the values of α were 0.1, 0.5, 1, 2 . . . 9, and 10 with the pixel size of either 0.5 m or 1.0 m. The rest of the analyses were performed using $\alpha = 5$, as with the pixel size of 0.5 m the highest correlations were obtained with that particular parameter α value, and no obvious improvements by other parameter values were discovered with the pixel size of 1.0 m.

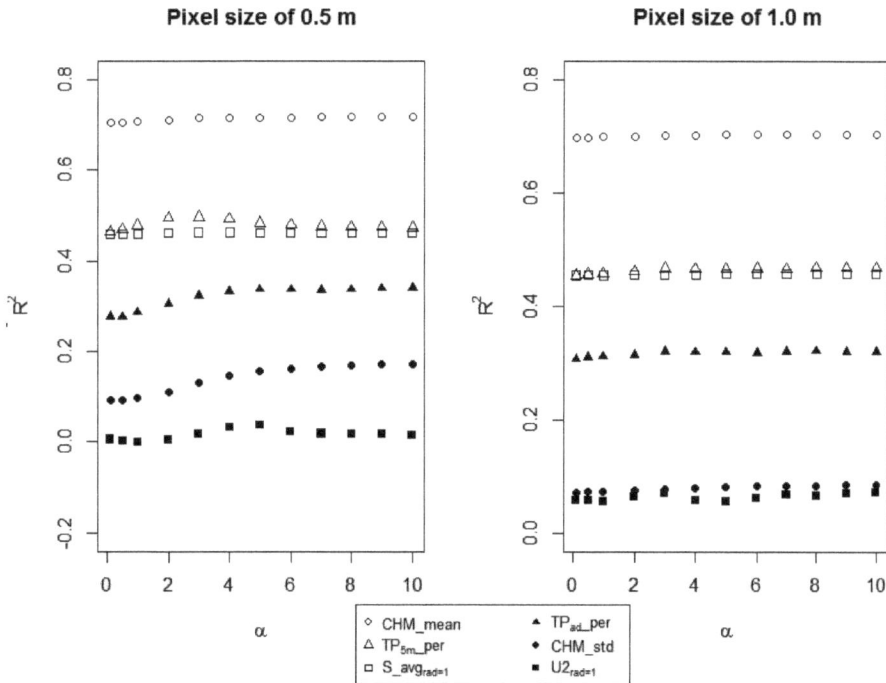

Figure 2. The R^2 of selected textural features and total stem volume with the pixel sizes of 0.5 m and 1.0 m. The interpolation parameter (α) was set to 0.1, 0.5, 1, 2 . . . , 9 and 10.

The pixel size affected the textural features and subsequently the prediction of forest attributes. To find the optimal pixel size, we evaluated R^2 of the selected textural features and the forest attributes with varying pixel sizes (Figure 3). Several important textural features had the strongest correlations with stem volume and basal area when the pixel size was 0.5–1.0 m. No significant differences between these levels could be observed based on a visual analysis of Figure 3, but since a higher number of pixel values were observed on the latter (Table 2), the pixel size was set to 1.0 m in the following analysis.

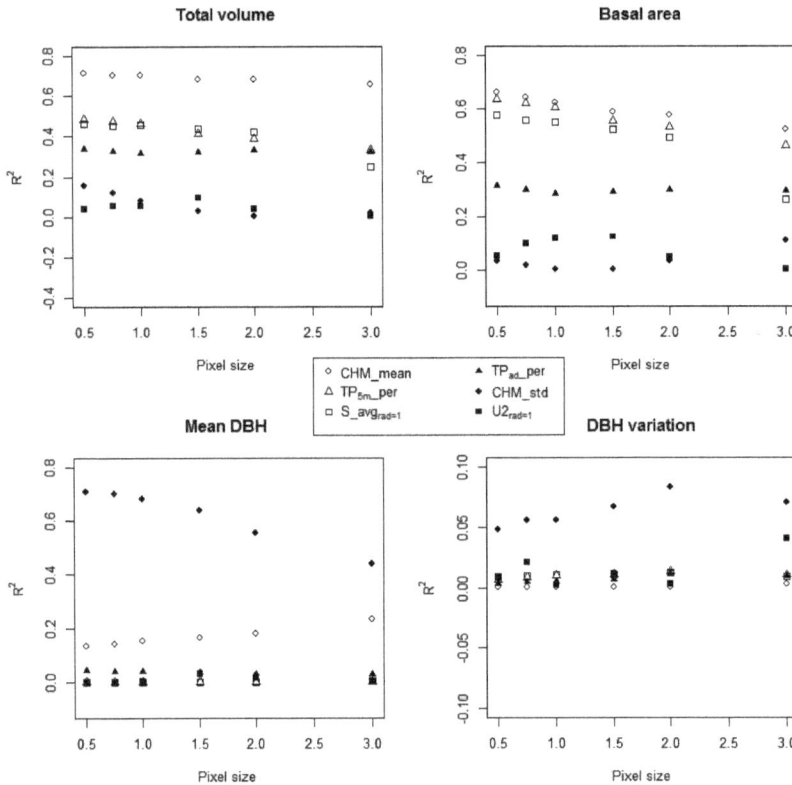

Figure 3. The effect of the pixel size to the coefficients of determination (R^2) of selected textural features and plot-level forest attributes (total stem volume, basal area, mean diameter (DBH), and DBH variation). The interpolation parameter α was set to 5.

3.2. The Degree of Determination between the Textural Features and Forest Attributes

The R^2 of selected textural features and the most common structural forest attributes are presented in Table 4, while the R^2 of all features are presented as Supplementary Data. For benchmarking purposes, selected point-based features are also shown. None of the ALS-based features correlated well with CV(DBH) or the number of understory trees on the field plots. The observed R^2 of the textural features and CV(G) on the large plots were more promising, as some of the GLCM features explained approximately one third of the variation in CV(G). Nevertheless, obtaining approximately the same degree of determination between the textural features extracted from the 9 m plots and CV(G) of the satellite plots does not support the causality of this observation. Thus, the rest of the analyses are focused on the accuracies of the tree-based forest attributes (i.e., V, G, and D_{gM}).

The extracted textural features were fairly insensitive to the computation parameters used. The patch features explained the variation of V and G more clearly when the constant threshold (5 m) was used instead of the adaptive threshold. Some GLCM-derived features such as SAVG correlated well with V and G (respective R^2 of 0.46 and 0.55 regardless of the lag) but always poorly with the D_{gM}. The lag parameter of the GLCM features affected only a few of the relationships between the GLCM features and forest attributes. In particular, when a larger GLCM lag value (from 1 to 5) was used for calculating the CON feature, the stronger R^2 was observed with the D_{gM} (from 0.01 to 0.22). LBP-based CHM uniformity features U_0 and U_2 were also dependent on G having the dyadic R^2 up to 0.20.

Table 4. The coefficients of determination (R^2) of selected features (17 textural features and three point-based features) and three forest attributes with the pixel size of 1.0 m and interpolation parameter of 5.

Feature	V	G	D_{gM}
CHM_{mean}	0.70	0.63	0.15
CHM_{std}	0.08	0.00	0.68
CHM_{max}	0.22	0.07	0.69
$TP_{5m\_}\%$	0.47	0.61	0.00
$TP_{5m\_}N$	0.35	0.43	0.06
$TP_{ad\_}\%$	0.32	0.28	0.04
$TP_{ad\_}N$	0.17	0.14	0.19
$CON_{lag = 1}$	0.02	0.01	0.01
$CON_{lag = 3}$	0.00	0.00	0.11
$CON_{lag = 5}$	0.00	0.01	0.22
$SAVG_{lag = 1}$	0.46	0.55	0.00
$SAVG_{lag = 3}$	0.46	0.55	0.00
$SAVG_{lag = 5}$	0.45	0.55	0.00
$U_{0,lag = 1}$	0.13	0.18	0.00
$U_{0,lag = 1}$	0.14	0.20	0.01
$U_{2,lag = 1}$	0.06	0.12	0.00
$U_{2,lag = 2}$	0.08	0.13	0.03
H_mean_{FP}	0.43	0.22	0.51
$VegR_{FP}$	0.38	0.58	0.00
$H70_{FP}$	0.23	0.07	0.74

Among all ALS-based variables, CHM_{mean} was the best feature for predicting V and G with respective R^2 of 0.70 and 0.63 (Table 4). In addition, $TP_{5m\_}\%$ predicted the respective forest attributes better than any single point-based feature (R^2 of 0.47 and 0.61). H_mean_{FP} and $VegR_{FP}$ were the best point-based features for predicting V and G with R^2 of 0.43 and 0.58, respectively. In addition, based on graphical comparisons with the point-based mean height features, CHM_{mean} had a stronger but also more non-linear relationship with V and G. A square root transformation was required to fix the linearity of the relationship and improved the R^2 of both CHM and point-based features. Conversely, the point-based features were better for predicting D_{gM} (the best feature $H70_{FP}$ had an R^2 of 0.74 with D_{gM}) compared to the best textural features (CHM_{max} and CHM_{std} with R^2 of 0.69 and 0.68 with D_{gM}, respectively).

3.3. Prediction Accuracies Based on the Linear Regression Models

The CHM features and the point-based features were applied for predicting V, G, and D_{gM} using LR. The best prediction accuracies of V (RMSE% of 25.0%) and G (RMSE% of 20.9%) were achieved by combining both textural and point-based features, whereas an optimal prediction model of D_{gM} (RMSE% of 12.9%) was based only on point-based features (Table 5).

When all ALS-based predictor features were applied for the LR model of V, as an example, CHM_{mean}, $TP_{ad\_}\%$, $U_{0,lag = 1}$, $VegR_{LP}$, H_mean_{LP}, $SVAR_{lag = 1}$, $H10_{LP}$, TP_{ad}_std, and TP_{ad}_avg were the first 10 predictor features selected to the model (Table 5). A similar combination of the different feature groups was used for modeling G, whereas the D_{gM} models were more frequently based on the point features. The results suggest that combining different feature groups improves the prediction accuracies. Considering the interactions between the features in a feature selection procedure also results in a slightly different ranking of the features than the correlation analyses above.

Table 5. The plot-level prediction accuracy of total stem volume, basal area and mean diameter using different sets of predictor features and their transformations in linear regression analyses. Altogether, 10 predictors are listed from each feature group in the order they were selected (footnote). The predictors included in the models were the first 1, 2, 3, or 5 predictors mentioned.

Predictors	V, m^3ha^{-1}		G, m^2ha^{-1}		D_{gM}, cm	
	RMSE	RMSE%	RMSE	RMSE%	RMSE	RMSE%
CHM Textural Features [a]						
The best feature	59.8	32.6	5.6	28.3	3.7	16.3
Set of 2 features	54.3	29.6	5.2	26.2	3.4	15.0
Set of 3 features	52.3	28.5	4.9	24.6	3.3	14.6
Set of 5 features	50.6	27.6	4.8	23.9	3.1	13.7
Point-based Features [b]						
The best feature	81.6	44.4	5.7	29.0	3.3	14.4
Set of 2 features	57.4	31.2	4.7	23.8	3.0	13.2
Set of 3 features	55.2	30.0	4.7	23.5	3.0	13.0
Set of 5 features	53.0	28.9	4.6	23.0	2.8	12.2
CHM Textural Features and Point-based Features Combined [c]						
The best feature	59.8	32.6	5.6	28.3	3.3	14.4
Set of 2 features	54.3	29.6	4.9	24.5	3.0	13.2
Set of 3 features	52.3	28.5	4.7	23.5	2.9	12.5
Set of 5 features	49.3	26.8	4.4	21.9	2.7	11.8

[a] $V = f(CHM_{mean} + TP_{ad}\_\%^2 + CHM_{mean}^2 + \sqrt{U}_{0,lag = 1} + SVAR_{lag = 1} + TP_{5m}\_std^2 + TP_{ad}\_std^2 + TP_{ad}\_avg^2 + COR_{lag = 1} + \sqrt{TP_{ad}\_std})$; $G = f(CHM_{mean} + TP_{5m}\_\%^2 + TP_{ad}\_\% + CHM_{max}^2 + COR_{lag = 1}^2 + \sqrt{U}_{0,lag = 1} + COR_{lag = 1} + TP_{5m}\_std + U_{0,lag = 2}^2 + log(IMC2_{lag = 1}))$; $D = f(CHM_{max}^2 + CON_{lag = 5} + IMC1_{lag = 1}^2 + CHM_{mean}^2 + \sqrt{U}_{0,lag = 1} + \sqrt{U}_{0,lag = 2} + log(SVAR) + \sqrt{CHM_{mean}} + TP_{ad}\_\%^2 + log(TP_{ad}\_avg))$; [b] $V = f(H\_mean_{FP}^2 + VegR_{FP}^2 + H10_{FP} + H\_mean_{LP}^2 + H10_{LP}^2 + VegR_{LP}^2 + D05_{LP}^2 + H\_std_{LP}^2 + H30_{FP} + log(H05_{LP}))$; $G = f(VegR_{FP}^2 + H\_mean_{LP}^2 + I110_{LP}^2 + \sqrt{VegR_{FP}} + log(H\_mean_{FP}) + VegR_{LP}^2 + H50_{LP}^2 + D05_{FP}^2 + H30_{FP} + log(D90_{FP}) + H\_max_{FP})$; $D = f(H70_{FP}^2 + H60_{LP}^2 + H20_{FP} + \sqrt{H10_{LP}} + \sqrt{H30_{FP}} + H60_{FP} + log(D90_{FP}) + D95_{LP}^2 + H90_{LP}^2 + H90_{FP}^2)$; [c] $V = f(CHM_{mean} + TP_{ad}\_\%^2 + CHM_{mean}^2 + \sqrt{U}_{0,lag = 1} + VegR_{LP}^2 + H\_mean_{LP}^2 + SVAR_{lag = 1} + H10_{LP}^2 + TP_{ad}\_std + TP_{ad}\_avg^2)$; $G = f(CHM_{mean} + VegR_{FP}^2 + H\_mean_{LP}^2 + TP_{ad}\_den^2 + VegR_{LP}^2 + IDM_{lag = 3}^2 + H10_{FP}^2 + \sqrt{U}_{0,lag = 1} + TP_{5m}\_N^2 + H05_{LP}^2)$; $D = f(H70_{FP}^2 + H60_{LP}^2 + log(IMC2_{lag = 1}) + H10_{LP} + TP_{ad}\_\%^2 + log(TP_{ad}\_avg) + \sqrt{H20_{FP}} + \sqrt{H30_{FP}} + H05_{FP}^2 + D50_{LP}^2)$.

3.4. Unsupervised Classification

The partitioning of the 155 plots to 2–7 clusters according to the selected features (CHM_{mean}, CHM_{std}, TP_{ad}_avg, $ASM_{lag = 1}$, $IMC2_{lag = 1}$, and $U_{2,lag = 2}$) resulted in altogether 27 sub-clusters in the clustering hierarchy. Excluding clusters formed by single plots, the composition of the remaining nine sub-clusters ordered according to CHM_{mean} and labeled as A–I are shown in Supplementary Data (Figure S1) in terms of the textural feature values that generated the clusters.

When examined against the field measurements, the clustering could be linked with forest size-related attributes, particularly the aggregated stem volume, basal area, and mean diameter (Supplementary Data, Figure S2). Cluster A included the most of the youngest stands, but it was not completely uniform because of considerable variation in diameter distribution. Maturity and stocking increased from cluster B to E. Cluster F included the most stocked stands in the data, and F–I were the most prominent stands in terms of maturity for harvesting. Except for Cluster A, the clusters could not be linked to dominant species or the heterogeneity in the within-plot diameter or between-plot basal-area distributions.

The example of using the cluster dispersion obtained solely from the ALS data as a criterion to prioritize the selection of the sample plots for model fitting is illustrated in Figure 4. The RMSE of V based on applying Equation (6) to the full data considerably varied when inserting the initial plots (Figure 4b). The left column of Figure 4 suggests plots with the highest priorities in terms of the cluster dispersion are needed to capture the correct form of the relationship, whereas adding plots with lower priorities does not add information on the relationship. The RMSE stabilized after including 70 plots with the highest priority (Figure 4b).

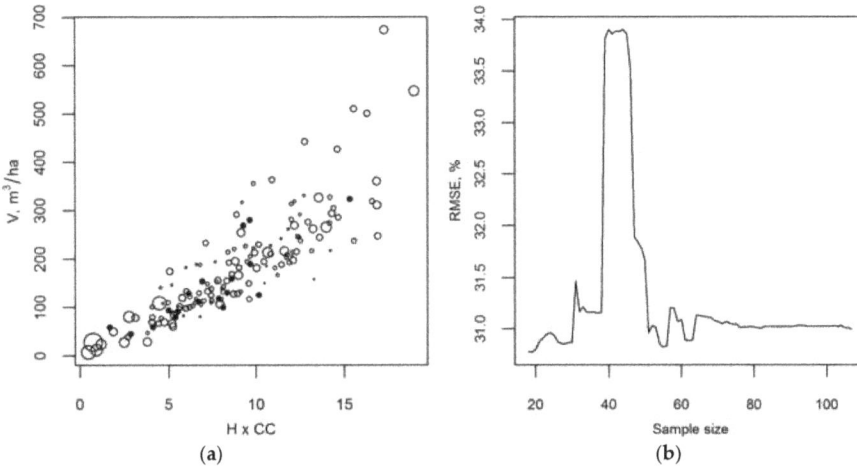

(a) (b)

Figure 4. (**a**) The relationship between the airborne laser scanning estimate of mean height (H) × canopy cover (CC) and plot volume (V) in the data studied. The filled and open circles correspond to clusters including single or several plots, respectively, and the size of the open circles indicates the dispersion of the plot from the respective cluster center; (**b**) The development of the root-mean-squared-error, when the prediction of V is based on Equation (6) and an increasing number of sample plots for model fitting, selected in the order of the decreasing dispersion.

4. Discussion

Unlike in earlier studies based on ALS, we performed a comprehensive comparison of methods for extracting and quantifying the textural variation present in the CHMs derived from a boreal forest area. Although earlier studies had related the textural variation of aerial images to size [28] and diversity [71] attributes of the vegetation and explained the effects of the parameters affecting the textural features extracted from aerial images [28], a new study was justified since the CHMs interpolated from the height values differ considerably from the spectral response recorded by the aerial images. The CHMs directly reflect the size and structure variation in the forest and were expected to thus readily improve the forest attribute prediction, eluding problems related to variations in geometry and radiometry of the spectral images [80]. Further, a considerably wider range of textural features was included in our analyses compared to [28]. The results are based on practical data, which are broadly available due to the frequent acquisitions of such data for ground elevation modeling. The results of this study may be seen as a continuation of studies promoting the usability of such data for various forest inventory applications [2–4,12,59,81].

While an exhaustive comparison of the texture quantification methods was found to be impossible due to the multitude of the developed methods, we attempted to select the most important ones based on a literature review of the properties of the algorithms. It could be argued whether the selection of the methods covered by our analyses was representative. However, our problem of relating the extracted features to the forest attributes aggregated from field plots with an area of 254 m^2 placed some constraints on the methods. A particular emphasis was placed on the rotation invariance, which removed the requirement to run the training procedure applying the same rotation. On the other hand, the use of the isotropic features may have caused a loss of directionality from these features, which could have been an important characteristic when classifying the texture.

Assuming an underlying spatial pattern or geometric model of the trees would have enabled the use of additional approaches to quantify the texture. However, there is no evidence that the spatial pattern of the trees could be correctly classified based on remotely sensed data [20]. Even if such a classification were useful for forestry applications, departures from the assumptions could cause severe error trends [82]. The inherent spatial autocorrelation of the data further prevented the use of some texture quantification approaches. Given these challenges, it was well reasoned to focus on the statistical texture analyses and the selected combinations of order and scale (Table 3). Note that similar considerations on the feasibility of texture analyses and roles of assumptions and spatial scales were already discussed based on optical satellite images [83].

The generation of CHMs from the low-density ALS data affected the textural features derived using the selected methods. A part of the textural variation in the CHMs originates from the choice of whether to interpolate pixel values or increase the pixel size in order to use more observed height values. The use of small pixel sizes required a number of pixels to be interpolated, whereas increasing the pixel size would have allowed more pixel values to be based on real height observations. On the other hand, the degree of detail of the CHM reduced according to the pixel size. Nevertheless, Figure 1 indicates that the vertical and horizontal variation in the forest structure, as captured by the ALS point data, can be reproduced and analyzed as textural variation in the CHMs. An obvious solution to add the level of detail of the CHMs would be to increase the point density of the data. Data with a higher density might improve the texture analyses, but also allow the use of more efficient analysis methods such as detecting individual tree tops and using them as predictors [18].

The sparse point density led to, at best, a large CHM pixel size. Increasing the pixel size resulted in an increase of the CHM values observed, but a loss of information in terms of the detail. At small pixel sizes (0.5 or 1.0 m), the R^2 of textural features and forest attributes were stable, whereas at larger pixel sizes their R^2 degraded. The IDW interpolation parameter α only had a minor impact on the textural features computed in the study. Notably, many other techniques could have been employed in the interpolation of the CHM rasters. Acknowledging that the effects of interpolation to the quality of the CHMs should also be further studied, we selected to fill the CHMs using IDW interpolation due to its simplicity and controllability based on only few parameters. It would further have been possible to fill the CHMs using more than only or first echoes per emitted pulse, which has become popular especially when the detail of the CHM must enable individual tree detection [18]. Even though such alternative may be recommendable for practical purposes, the results obtained here using only first echo data are more likely reproducible with sensors that record a different number of echo categories [6].

The performance of the textural features derived using the selected pixel sizes (0.5 or 1.0 m) was moderately good for predicting V, G, and D_{gM}. Among the tested feature sources, those based on the HIST, PATCH, and GLCM clearly outperformed those based on LBP. The HIST features derived from the grey-level histograms directly characterized the forest canopy height based on the CHMs. CHM_{mean} was clearly the best individual feature for predicting stem volume and basal area among all the ALS-based features. CHM_{std} and CHM_{max} were almost as good as the point-based features to explain the variation of mean DBH. This is slightly surprising, since the CHM features should overall include less information compared to the height features computed from the entire point cloud. The reasons for the more stable relationship with the forest attributes may be related to the smoothing due to the CHM interpolation step. A similar result was reported in [4] and the potential to improve forest variables this way should be further examined. Overall, our results indicate that the description of the forest structure may be considerably simplified using the CHM_{mean} as a substitute (or complement) for the point-based features. For improved accuracies, one should combine the CHM height features with the density features derived from the point data.

Among other textural feature groups, the PATCH features were useful in the prediction of the volume and basal area. The features derived using adaptive CHM threshold appeared as important predictors of the volume, whereas the basal area was better modeled when a constant threshold was applied. Among the GLCM features, $SAVG_{lag = 1}$ was the best individual feature, while $SVAR_{lag = 1}$ was most often selected to the LR models from the set of other predictor candidates. Conversely, the uniformity features derived from the LBPs did not stand out with respect to any application considered. This is also slightly surprising, since LBPs should fundamentally detect more detailed microstructures than the aforementioned techniques. Analyzing the individual LBPs directly using more complex machine learning methods might provide better results. The poor performance of the method may also be related to the low resolution of the data analyzed. Nevertheless, the result may also indicate that texture quantification methods performing well in man-made material classifications may underperform in the case of vegetation with considerable natural variations.

Interestingly, the textural features sometimes outperformed point-based features regardless of which echo types were used for the computation. The use of the area-based approach with point-based canopy height and density features is a common and established practice for predicting forest inventory attributes from the ALS data [5,8]. Mainly features based on the vertical distribution of the ALS point cloud are used, however, for which reason these analyses optimally characterize only the vertical canopy conditions. The textural features of the present study were computed from the interpolated CHMs, aiming to improve the characterization of horizontal canopy conditions in ALS forest inventories. From a forest management planning perspective, the horizontal information in terms of the stem density and clustering is related to the timing of thinnings, and the inclusion of the textural features is proposed to improve these decisions [19]. Although Ozdemir and Donoghue [34] and Wood et al. [71] indicated a potential to link the textural features with the diversity of the tree size and vegetation structure, respectively, our results did not support this. The textural features were poorly correlated with the coefficients of variation of the DBH and G values as well as with the number of understory trees. The features mainly described the properties of the trees in the dominant canopy such as the total stem volume, basal area, and mean diameter. Direct measures of the understory are difficult to obtain due to transmission losses occurring in the upper canopy [84] (see also [50]) and no correlations between the textural features quantifying the top of the canopy and the understory were observed here. Although a slight correlation between the textural features and CV(G) computed from the large plots was observed, a similar correlation was obtained when the textural features were computed from a smaller window. A better solution for quantifying the within-stand variation in a practical prediction based on the grid cells could be to analyze the variation between the neighboring cells. A higher number of cells would also likely be needed to make solid conclusions.

In addition to the linear regression analyses, the textural features were demonstrated in an unsupervised classification using the well-known *k*-Means approach. The approach provided a data-driven partitioning of the feature space to the given number of clusters. As there are no objective criteria for determining the number of clusters [74], we followed an earlier example on determining this number [59]. By initially experimenting with different numbers of clusters, we found that 4–7 structural classes could be separated based on our study area and data. There are five classes in both the site type and development stage classification system applying to the studied forest area. However, not all classes are likely present and the data may not discriminate between all these properties. Therefore, the applied split and merge criteria resulted in a reasonable number of clusters.

Notably, the results obtained here are a re-clustering of the entire study area [59]. Whether applied to the full study area, it is reasonable to expect more clusters to be found similar to [59]. Nevertheless, the main purpose of this analysis was to determine if the developed features could be used for pre-stratification of the inventory area based on ALS data, as proposed in the introduction section. It was found that the dispersion of the clusters derived in an unsupervised mode could be used as an indicator for prioritizing the plots to be measured as the sample to form the reference data for the wall-to-wall models. We acknowledge that these are indicative results and the generalizability of this conclusion should be verified in more extensive data, allowing a split to training and validation data. For example, the textural features derived show a preliminary potential for streamlining the sampling design of the field plots to be used for later inventories, which should be further studied.

Practical inventory projects of large areas will most likely need to increasingly account for computation costs related to extracting the features. We see a considerable potential of the features developed here related specifically to their computational feasibility. For example, when designing the sampling protocol to obtain the field reference data for an ALS-based forest inventory, the sizes and orientations of potential plot locations should not be fixed but instead allowed to vary in order to design a field sample capturing the essential variation of the forest. Whether guiding the field plot sampling requires extracting features from, e.g., moving windows of multiple scales, the related computations are likely much more efficient when based on CHMs rather than underlying point data due to the efficient processing routines developed for raster data. Specifically, the time complexity related to CHM-based features reduces from the number of points to the number of pixels of the CHM, which appears to be a considerable factor in large inventory areas. Relating the computation costs of extracting various features to the added value obtained could be a potentially interesting future topic, since at least we are not aware of studies related to the time and space complexity of the ALS algorithms.

5. Conclusions

CHMs interpolated from the ALS data were found to reflect some degree of textural variation that was useful for modeling the underlying forest attributes, particularly plot volume and basal area. Applications based on supervised (linear regression) and unsupervised (k-Means clustering) learning of forest structure were demonstrated. The latter indicates the utility of the derived features for improving the sampling of the field plots for forest inventories, which should be further verified with separate validation data. Among the features considered, the statistical, patch, and GLCM features outperformed those based on point data, indicating that improved information is contained in the textural features. Features based on LBP were less useful for the same purpose. Although the tested features were selected based on a comprehensive review of potential methods to quantify image texture, we acknowledge that many more textural feature extraction techniques could be considered, especially if the requirement to produce rotation-invariant features was relaxed. In all, even the sparse-density data include potential to develop features that quantify very different aspects of the data, which should be employed together to improve analyses of vertical and horizontal forest structure.

Supplementary Materials: The following are available online at www.mdpi.com/2072-4292/8/7/582/s1, Table S1: Coefficients of determination (R^2) between 87 CHM textural features and various forest attributes; Table S2: Coefficients of determination (R^2) between 52 point-based features and various forest attributes; Figure S1: The distribution of the textural features CHM_{mean}, CHM_{std}, $TP_{ad\_}avg$, $ASM_{lag = 1}$, $IMC2_{lag = 1}$ and $U_{2,lag = 2}$ in the sub-clusters formed by the k-Means algorithm; Figure S2: The distribution of the field-measured forest structural attributes according to the sub-clusters formed based on the ALS data.

Acknowledgments: The contribution of Mikko T. Niemi was enabled by the funding of Metsämiesten Säätiö Foundation. The field measurements, preparation of this study, and open access publishing costs were financed by the Research Funds of the University of Helsinki. The study is a contribution to Academy of Finland, project 272195. The authors would like to thank Joni Imponen and Mika Salmi for their contributions related to the field measurements and Markus Holopainen and Mikko Vastaranta for discussions related to the study.

Author Contributions: Jari Vauhkonen conceived and designed the experiments; Mikko T. Niemi computed the plot-level forest attributes, implemented the CHM generation and textural feature extraction, and analyzed the data with respect to the supervised prediction of forest attributes; Jari Vauhkonen analyzed the data with respect to the unsupervised classification method; and Mikko T. Niemi and Jari Vauhkonen wrote the paper.

Conflicts of Interest: The authors declare no conflict of interest.

References

1. Maltamo, M.; Næsset, E.; Vauhkonen, J. *Forestry Applications of Airborne Laser Scanning—Concepts and Case Studies*, 1st ed.; Springer: Dordrecht, The Netherlands, 2014.

2. Nord-Larsen, T.; Schumacher, J. Estimation of forest resources from a country wide laser scanning survey and national forest inventory data. *Remote Sens. Environ.* **2012**, *119*, 148–157. [CrossRef]

3. Villikka, M.; Packalén, P.; Maltamo, M. The suitability of leaf-off airborne laser scanning data in an area-based forest inventory of coniferous and deciduous trees. *Silva Fenn.* **2012**, *46*, 99–110. [CrossRef]

4. Kankare, V.; Vauhkonen, J.; Holopainen, M.; Vastaranta, M.; Hyyppä, J.; Hyyppä, H.; Alho, P. Sparse density, leaf-off airborne laser scanning data in aboveground biomass component prediction. *Forests* **2015**, *6*, 1839–1857. [CrossRef]

5. Næsset, E. Predicting forest stand characteristics with airborne scanning laser using a practical two-stage procedure and field data. *Remote Sens. Environ.* **2002**, *80*, 88–99. [CrossRef]

6. Næsset, E. Area-based inventory in Norway—From innovation to an operational reality. In *Forestry Applications of Airborne Laser Scanning—Concepts and Case Studies*, 1st ed.; Maltamo, M., Næsset, E., Vauhkonen, J., Eds.; Springer: Dordrecht, The Netherlands, 2014; pp. 215–240.

7. Maltamo, M.; Packalen, P. Species-specific management inventory in Finland. In *Forestry Applications of Airborne Laser Scanning—Concepts and Case Studies, Managing Forest Ecosystems*, 1st ed.; Maltamo, M., Næsset, E., Vauhkonen, J., Eds.; Springer: Dordrecht, The Netherlands, 2014; pp. 241–252.

8. White, J.C.; Wulder, M.A.; Varhola, A.; Vastaranta, M.; Coops, N.C.; Cook, B.D.; Pitt, D.; Woods, M. *A Best Practices Guide for Generating Forest Inventory Attributes from Airborne Laser Scanning Data Using an Area-Based Approach (Version 2.0)*; Information Report FI-X-010; The Canadian Wood Fibre Centre: Victoria, BC, Canada, 2013.

9. Magnussen, S.; Boudewyn, P. Derivations of stand heights from airborne laser scanner data with canopy-based quantile estimators. *Can. J. For. Res.* **1998**, *28*, 1016–1031. [CrossRef]

10. Vauhkonen, J.; Seppänen, A.; Packalén, P.; Tokola, T. Improving species-specific plot volume estimates based on airborne laser scanning and image data using alpha shape metrics and balanced field data. *Remote Sens. Environ.* **2012**, *124*, 534–541. [CrossRef]

11. Vauhkonen, J.; Næsset, E.; Gobakken, T. Deriving airborne laser scanning based computational canopy volume for forest biomass and allometry studies. *ISPRS J. Photogramm. Remote Sens.* **2014**, *96*, 57–66. [CrossRef]

12. Vauhkonen, J.; Holopainen, M.; Kankare, V.; Vastaranta, M.; Viitala, R. Geometrically explicit description of forest canopy based on 3D triangulations of airborne laser scanning data. *Remote Sens. Environ.* **2016**, *173*, 248–257. [CrossRef]

13. Vehmas, M.; Packalén, P.; Maltamo, M.; Eerikäinen, K. Using airborne laser scanning data for detecting canopy gaps and their understory type in mature boreal forest. *Ann. For. Sci.* **2011**, *68*, 825–835. [CrossRef]

14. Bouvier, M.; Durrieu, S.; Fournier, R.A.; Renaud, J. Generalizing predictive models of forest inventory attributes using an area-based approach with airborne LiDAR data. *Remote Sens. Environ.* **2015**, *156*, 322–334. [CrossRef]

15. Latifi, H.; Heurich, M.; Hartig, F.; Müller, J.; Krzystek, P.; Jehl, H.; Dech, S. Estimating over-and understorey canopy density of temperate mixed stands by airborne LiDAR data. *Forestry* **2016**, *89*, 69–81. [CrossRef]

16. Lee, A.C.; Lucas, R.M. A LiDAR-derived canopy density model for tree stem and crown mapping in Australian forests. *Remote Sens. Environ.* **2007**, *111*, 493–518. [CrossRef]

17. Ferraz, A.; Mallet, C.; Jacquemoud, S.; Rito Gonçalves, G.; Tomé, M.; Soares, P.; Gomes Pereira, L.; Bretar, F. Canopy density model: A new ALS-derived product to generate multilayer crown cover maps. *IEEE Trans. Geosci. Remote Sens.* **2015**, *53*, 6776–6790. [CrossRef]

18. Hyyppä, J.; Yu, X.; Hyyppä, H.; Vastaranta, M.; Holopainen, M.; Kukko, A.; Kaartinen, H.; Jaakkola, A.; Vaaja, M.; Koskinen, J.; Alho, P. Advances in forest inventory using airborne laser scanning. *Remote Sens.* **2012**, *4*, 1190–1207. [CrossRef]

19. Pippuri, I.; Kallio, E.; Maltamo, M.; Peltola, H.; Packalén, P. Exploring horizontal area-based metrics to discriminate the spatial pattern of trees and need for first thinning using airborne laser scanning. *Forestry* **2012**, *85*, 305–314. [CrossRef]

20. Packalén, P.; Vauhkonen, J.; Kallio, E.; Peuhkurinen, J.; Pitkänen, J.; Pippuri, I.; Strunk, J.; Maltamo, M. Predicting the spatial pattern of trees by airborne laser scanning. *Int. J. Remote Sens.* **2013**, *34*, 5154–5165. [CrossRef]

21. Cohen, W.B.; Spies, T.A. Estimating structural attributes of Douglas-fir/western hemlock forest stands from Landsat and SPOT imagery. *Remote Sens. Environ.* **1992**, *41*, 1–17. [CrossRef]

22. Franklin, S.E.; Waring, R.H.; McCreight, R.W.; Cohen, W.B.; Fiorella, M. Aerial and satellite sensor detection and classification of western spruce budworm defoliation in a subalpine forest. *Can. J. Remote Sens.* **1995**, *21*, 299–308. [CrossRef]

23. Wulder, M.A.; Franklin, S.E.; Lavigne, M.B. High spatial resolution optical image texture for improved estimation of forest stand leaf area index. *Can. J. Remote Sens.* **1996**, *22*, 441–449. [CrossRef]

24. Franklin, S.E.; Hall, R.J.; Moskal, L.M.; Maudie, A.J.; Lavigne, M.B. Incorporating texture into classification of forest species composition from airborne multispectral images. *Int. J. Remote Sens.* **2000**, *21*, 61–79. [CrossRef]

25. Franklin, S.E.; Wulder, M.A.; Gerylo, G.R. Texture analysis of IKONOS panchromatic data for Douglas-fir forest age class separability in British Columbia. *Int. J. Remote Sens.* **2001**, *22*, 2627–2632. [CrossRef]

26. Muinonen, E.; Maltamo, M.; Hyppänen, H.; Vainikainen, V. Forest stand characteristics estimation using a most similar neighbor approach and image spatial structure information. *Remote Sens. Environ.* **2001**, *78*, 223–228. [CrossRef]

27. Anttila, P. Nonparametric estimation of stand volume using spectral and spatial features of aerial photographs and old inventory data. *Can. J. For. Res.* **2002**, *32*, 1849–1857. [CrossRef]

28. Tuominen, S.; Pekkarinen, A. Performance of different spectral and textural aerial photograph features in multi-source inventory. *Remote Sens. Environ.* **2005**, *94*, 256–268. [CrossRef]

29. Maltamo, M.; Malinen, J.; Packalén, P.; Suvanto, A.; Kangas, J. Nonparametric estimation of stem volume using airborne laser scanning, aerial photography, and stand-register data. *Can. J. For. Res.* **2006**, *36*, 426–436. [CrossRef]

30. Packalén, P.; Maltamo, M. Predicting the plot volume by tree species using airborne laser scanning and aerial photographs. *For. Sci.* **2006**, *52*, 611–622.

31. Hyyppä, J.; Inkinen, M. Detecting and estimating attributes for single trees using laser scanner. *Photogramm. J. Finl.* **1999**, *16*, 27–42.

32. Vauhkonen, J.; Tokola, T.; Maltamo, M.; Packalén, P. Effects of pulse density on predicting characteristics of individual trees of Scandinavian commercial species using alpha shape metrics based on airborne laser scanning data. *Can. J. Remote Sens.* **2008**, *34*, S441–S459. [CrossRef]

33. Heinzel, J.; Koch, B. Investigating multiple data sources for tree species classification in temperate forest and use for single tree delineation. *Int. J. Appl. Earth Obs. Geoinf.* **2012**, *18*, 101–110. [CrossRef]

34. Ozdemir, I.; Donoghue, D.N.M. Modelling tree size diversity from airborne laser scanning using canopy height models with image texture measures. *For. Ecol. Manag.* **2013**, *295*, 28–37. [CrossRef]

35. Pippuri, I.; Suvanto, A.; Maltamo, M.; Korhonen, K.T.; Pitkänen, J.; Packalen, P. Classification of forest land attributes using multi-source resolution data. *Int. J. Appl. Earth Obs. Geoinf.* **2016**, *44*, 11–22. [CrossRef]

36. Haralick, R.M.; Shanmugam, K.; Dinstein, J. Textural features for image classification. *IEEE Trans. Syst. Man Cybernetics* **1973**, *3*, 610–621. [CrossRef]

37. Ruiz, L.A.; Fdez-Sarría, A.; Recio, J.A. Texture feature extraction for classification of remote sensing data using wavelet decomposition: A comparative study. In Proceedings of the 20th ISPRS Conference, Istanbul, Turkey, 12–23 July 2004.

38. Cressie, N. *Statistics for Spatial Data: Wiley Series in Probability and Statistics*; John Wiley & Sons: Hoboken, NJ, USA, 1993; pp. 105–209.

39. St-Onge, B.A.; Cavayas, F. Estimating forest stand structure from high resolution imagery using the directional variogram. *Int. J. Remote Sens.* **1995**, *16*, 1999–2021. [CrossRef]

40. St-Onge, B.A.; Cavayas, F. Automated forest structure mapping from high resolution imagery based on directional semivariogram estimates. *Remote Sens. Environ.* **1997**, *61*, 82–95. [CrossRef]

41. Wulder, M.; Boots, B. Local spatial autocorrelation characteristics of remotely sensed imagery assessed with the Getis statistic. *Int. J. Remote Sens.* **1998**, *19*, 2223–2231. [CrossRef]

42. Wallerman, J.; Holmgren, J. Estimating field-plot data of forest stands using airborne laser scanning and SPOT HRG data. *Remote Sens. Environ.* **2007**, *110*, 501–508. [CrossRef]

43. Riitters, K.H.; O'Neill, R.V.; Hunsaker, C.T.; Wickham, J.D.; Yankee, D.H.; Timmins, S.P.; Jones, K.B.; Jackson, B.L. A factor analysis of landscape pattern and structure metrics. *Landsc. Ecol.* **1995**, *10*, 23–39. [CrossRef]

44. Ojala, T.; Pietikäinen, M.; Mäenpää, T. Multiresolution gray-scale and rotation invariant texture classification with local binary patterns. *IEEE Trans. Patt. Anal. Mach. Intell.* **2002**, *24*, 971–987. [CrossRef]

45. Varma, M.; Zisserman, A. A statistical approach to texture classification from single images. *Int. J. Comput. Vis.* **2005**, *62*, 61–81. [CrossRef]
46. Varma, M.; Zisserman, A. A statistical approach to material classification using image patch exemplars. *IEEE Trans. Patt. Anal. Mach. Intell.* **2009**, *31*, 2032–2047. [CrossRef]
47. Crosier, M.; Griffin, L.D. Using basic image features for texture classification. *Int. J. Comput. Vis.* **2010**, *88*, 447–460. [CrossRef]
48. Perea, J.A.; Carlsson, G.A. Klein-Bottle-Based dictionary for texture representation. *Int. J. Comput. Vis.* **2014**, *107*, 75–97. [CrossRef]
49. Zimble, D.A.; Evans, D.L.; Carlson, G.C.; Parker, R.C.; Grado, S.C.; Gerard, P.D. Characterizing vertical forest structure using small-footprint airborne LiDAR. *Remote Sens. Environ.* **2003**, *87*, 171–182. [CrossRef]
50. Maltamo, M.; Packalén, P.; Yu, X.; Eerikäinen, K.; Hyyppä, J.; Pitkänen, J. Identifying and quantifying structural characteristics of heterogeneous boreal forests using laser scanner data. *For. Ecol. Manag.* **2005**, *216*, 41–50. [CrossRef]
51. Liu, X. Airborne LiDAR for DEM generation: Some critical issues. *Progr. Phys. Geogr.* **2008**, *32*, 31–49.
52. Maltamo, M.; Bollandsås, O.M.; Næsset, E.; Gobakken, T.; Packalén, P. Different plot selection strategies for field training data in ALS-assisted forest inventory. *Forestry* **2011**, *84*, 23–31. [CrossRef]
53. Dalponte, M.; Martinez, C.; Rodeghiero, M.; Gianelle, D. The role of ground reference data collection in the prediction of stem volume with LiDAR data in mountain areas. *ISPRS J. Photogramm. Remote Sens.* **2011**, *66*, 787–797. [CrossRef]
54. Gobakken, T.; Korhonen, L.; Næsset, E. Laser-assisted selection of field plots for an area-based forest inventory. *Silva Fenn.* **2013**. [CrossRef]
55. Maltamo, M.; Ørka, H.O.; Bollandsås, O.M.; Gobakken, T.; Næsset, E. Using pre-classification to improve the accuracy of species-specific forest attribute estimates from airborne laser scanner data and aerial images. *Scand. J. For. Res.* **2015**, *30*, 336–345. [CrossRef]
56. National Land Survey of Finland. *File Service of Open Data*. Available online: https://tiedostopalvelu.maanmittauslaitos.fi/tp/kartta?lang=en (accessed on 12 May 2016).
57. Axelsson, P. DEM generation from laserscanner data using adaptive TIN models. *Int. Arch. Photogramm. Remote Sens.* **2000**, *33*, 16–22.
58. Vilhomaa, J.; Laaksonen, H. Valtakunnallinen laserkeilaus—Testityöstä tuotantoon (in Finnish for "National laser scanning—From testing to production"). *Photogramm. J. Finl.* **2011**, *22*, 82–91.
59. Vauhkonen, J.; Imponen, J. Unsupervised classification of airborne laser scanning data to locate potential wildlife habitats for forest management planning. *Forestry* **2016**. [CrossRef]
60. Näslund, M. Skogsförsöksanstaltens gallringsförsök i tallskog (in Swedish for "Forestry Institute's thinning trial in a pine forest"). *Meddelanden* **1936**, *29*, 169.
61. Siipilehto, J. Improving the accuracy of predicted basal-area diameter distribution in advanced stands by determining stem number. *Silva Fenn.* **1999**, *33*, 281–301. [CrossRef]
62. Laasasenaho, J. Taper curve volume functions for pine, spruce and birch. *Comm. Inst. For. Fenn.* **1982**, *108*, 1–74.
63. Lu, G.Y.; Wong, D.W. An adaptive inverse-distance weighting spatial interpolation technique. *Comput. Geosci.* **2008**, *34*, 1044–1055. [CrossRef]
64. Wu, J. Rotation Invariant Classification of 3D Surface Texture Using photometric Stereo. Ph.D. Thesis, Heriot-Watt University, Edinburgh, UK, 2003.
65. Ojala, T.; Pietikäinen, M. Texture Classification. In *CVonline—Compendium of Computer Vision*; Fisher, R.B., Ed.; Available online: http://homepages.inf.ed.ac.uk/rbf/CVonline/LOCAL_COPIES/OJALA1/texclas.htm (accessed on 12 May 2016).
66. Tuceryan, M.; Vain, A.K. Texture analysis. In *Handbook of Pattern Recognition and Vision*; Chen, C.H., Pau, L.F., Wang, P.S.P., Eds.; World Scientific: Singapore, Singapore, 1993; pp. 235–276.
67. Ahonen, T.; Pietikäinen, M. Face description with local binary patterns: Application to face recognition. *IEEE Trans. Patt. Anal. Mach. Intell.* **2006**, *28*, 2037–2041. [CrossRef]
68. Guo, Z.; Zhang, L.; Zhang, D. A completed modeling of local binary pattern operator for texture classification. *IEEE Trans. Image Proc.* **2010**, *19*, 1657–1663.
69. Korhonen, L.; Peuhkurinen, J.; Malinen, J.; Suvanto, A.; Maltamo, M.; Packalén, P.; Kangas, J. The use of airborne laser scanning to estimate sawlog volumes. *Forestry* **2008**, *81*, 499–510.

70. Nagelkerke, N.J.D. A note on a general definition of the coefficient of determination. *Biometrika* **1991**, *78*, 691–692. [CrossRef]
71. Wood, E.M.; Pidgeon, A.M.; Radeloff, V.C.; Keuler, N.S. Image texture as a remotely sensed measure of vegetation structure. *Remote Sens. Environ.* **2012**, *121*, 516–526. [CrossRef]
72. Kane, V.R.; McGaughey, R.J.; Bakker, J.D.; Gersonde, R.F.; Lutz, J.A.; Franklin, J.F. Comparisons between field-and LiDAR-based measures of stand structural complexity. *Can. J. For. Res.* **2010**, *40*, 761–773. [CrossRef]
73. Leiterer, R.; Furrer, R.; Schaepman, M.E.; Morsdorf, F. Forest canopy-structure characterization: A data-driven approach. *For. Ecol. Manag.* **2015**, *358*, 48–61. [CrossRef]
74. Pascual, C.; García-Abril, A.; García-Montero, L.G.; Martín-Fernández, S.; Cohen, W.B. Object-based semi-automatic approach for forest structure characterization using lidar data in heterogeneous *Pinus sylvestris* stands. *For. Ecol. Manag.* **2008**, *255*, 3677–3685. [CrossRef]
75. Thompson, S.D.; Nelson, T.A.; Giesbrecht, I.; Frazer, G.; Saunders, S.C. Data-driven regionalization of forested and non-forested ecosystems in coastal British Columbia with LiDAR and RapidEye imagery. *Appl. Geogr.* **2016**, *69*, 35–50. [CrossRef]
76. Hartigan, J.A.; Wong, M.A. A k-means clustering algorithm. *Appl. Stat.* **1979**, *28*, 100–108. [CrossRef]
77. R Core Team. *R: A Language and Environment for Statistical Computing*; R Foundation for Statistical Computing: Vienna, Austria, 2016.
78. Hornik, K. A clue for cluster ensembles. *J. Stat. Softw.* **2005**, *14*, 1–25. [CrossRef]
79. Pesonen, A.; Leino, O.; Maltamo, M.; Kangas, A. Comparison of field sampling methods for assessing coarse woody debris and use of airborne laser scanning as auxiliary information. *For. Ecol. Manag.* **2009**, *257*, 1532–1541. [CrossRef]
80. Mäkinen, A.; Korpela, I.; Tokola, T.; Kangas, A. Effects of imaging conditions on crown diameter measurements from high-resolution aerial images. *Can. J. For. Res.* **2006**, *36*, 1206–1217. [CrossRef]
81. Niemi, M.; Vastaranta, M.; Peuhkurinen, J.; Holopainen, M. Forest inventory attribute prediction using airborne laser scanning in low-productive forestry-drained boreal peatlands. *Silva Fenn.* **2015**, *49*. [CrossRef]
82. Vauhkonen, J.; Mehtätalo, L. Matching remotely sensed and field-measured tree size distributions. *Can. J. For. Res.* **2015**, *45*, 353–363. [CrossRef]
83. Woodcock, C.E.; Strahler, A.H. The factor of scale in remote sensing. *Remote Sens. Environ.* **1987**, *21*, 311–332. [CrossRef]
84. Korpela, I.; Hovi, A.; Morsdorf, F. Understory trees in airborne LiDAR data—Selective mapping due to transmission losses and echo-triggering mechanisms. *Remote Sens. Environ.* **2012**, *119*, 92–104. [CrossRef]

![remote sensing logo] *remote sensing*

MDPI

Article

Detection and Segmentation of Small Trees in the Forest-Tundra Ecotone Using Airborne Laser Scanning

Marius Hauglin * and Erik Næsset

Norwegian University of Life Sciences, Department of Ecology and Natural Resource Management, P.O. Box 5003, N-1432 Ås, Norway; erik.naesset@nmbu.no
* Correspondence: marius.hauglin@nmbu.no; Tel.: +47-672-316-89

Academic Editors: Jie Shan, Juha Hyyppä, Lars T. Waser and Prasad S. Thenkabail
Received: 10 December 2015; Accepted: 4 May 2016; Published: 11 May 2016

Abstract: Due to expected climate change and increased focus on forests as a potential carbon sink, it is of interest to map and monitor even marginal forests where trees exist close to their tolerance limits, such as small pioneer trees in the forest-tundra ecotone. Such small trees might indicate tree line migrations and expansion of the forests into treeless areas. Airborne laser scanning (ALS) has been suggested and tested as a tool for this purpose and in the present study a novel procedure for identification and segmentation of small trees is proposed. The study was carried out in the Rollag municipality in southeastern Norway, where ALS data and field measurements of individual trees were acquired. The point density of the ALS data was eight points per m^2, and the field tree heights ranged from 0.04 to 6.3 m, with a mean of 1.4 m. The proposed method is based on an allometric model relating field-measured tree height to crown diameter, and another model relating field-measured tree height to ALS-derived height. These models are calibrated with local field data. Using these simple models, every positive above-ground height derived from the ALS data can be related to a crown diameter, and by assuming a circular crown shape, this crown diameter can be extended to a crown segment. Applying this model to all ALS echoes with a positive above-ground height value yields an initial map of possible circular crown segments. The final crown segments were then derived by applying a set of simple rules to this initial "map" of segments. The resulting segments were validated by comparison with field-measured crown segments. Overall, 46% of the field-measured trees were successfully detected. The detection rate increased with tree size. For trees with height >3 m the detection rate was 80%. The relatively large detection errors were partly due to the inherent limitations in the ALS data; a substantial fraction of the smaller trees was hit by no or just a few laser pulses. This prevents reliable detection of changes at an individual tree level, but monitoring changes on an area level could be a possible application of the method. The results further showed that some variation must be expected when the method is used for repeated measurements, but no significant differences in the mean number of segmented trees were found over an intensively measured test area of 11.4 ha.

Keywords: airborne laser scanning; treeline; monitoring

---

1. Introduction

Airborne laser scanning (ALS) is used today as a tool for forest applications, both for research purposes as well as in operational settings. Productive forest has in many places been the main target, but ALS can also be applied in other types of forest. The very frontiers of the forests have in many places gradually expanded into alpine areas [1], and this expansion is believed to be caused by several factors, with reduced grazing by domestic livestock and climate changes as two dominating causes.

Expansion of forests into areas such as the forest-tundra ecotone will influence carbon sequestration, but will also in many places have a direct effect on the climate through the so-called "albedo-effect". The darker-colored trees will reflect less of the solar radiation than bare ground, especially in the winter when the ground is covered with snow. The net effect of this phenomenon is warming [2].

It is therefore of interest to map and monitor possible changes taking place in the vegetation structure of the forest-tundra ecotone, such as the appearance of pioneer trees and the migration of the tree line. ALS has through several studies been proposed and tested as a tool for this task [3–8]. Several of these studies investigate and document the potential for discriminating between echoes reflected from trees and echoes reflected from other objects using metrics derived from ALS data, such as the height above the modeled terrain surface, the backscatter intensity of the echoes and the properties of the spatial distribution of echoes. There is, however, a need to apply this knowledge and further develop methods to derive quantitative properties such as tree numbers, crown coverage or tree size distributions to enable the establishment of efficient monitoring methodologies. A segmentation and identification of single trees from the ALS data would be one possible way of deriving such properties. Numerous studies have already proposed and tested methods to derive single-tree segments from ALS data [9–13]. An introduction to tree segmentation and an overview of these methods can be found in Koch *et al.* [14]. Common to all of these studies are that they are focused on mature forests, with an emphasis on trees considerably larger than those typically found in the forest-tundra ecotone. We did not consider any of the described methods to be directly applicable to the task of deriving single-tree information for smaller trees in the forest-tundra ecotone because with an average point density of, for example, 5–10 points per m^2, the number of echoes from each individual tree will typically range from one single echo up to less than 100 in most cases.

Many of the existing segmentation methods involve interpolation of the ALS point cloud to a raster or to a three-dimensional voxel space [14]. However, such methods typically assume a choice of a fixed pixel—or voxel—size. This pixel size will be closely linked to the range of tree sizes which can be detected. A large pixel size will smooth out the information inherent in the ALS point cloud and therefore make the detection of small trees harder, whereas smaller pixels will likely cause over-segmentation of echoes from larger trees. With these existing methods one is, in practice, faced with a choice of detecting trees within a limited size range, through the choice of a fixed pixel size. The chosen pixel size and the level of smoothing applied will also determine the spatial extent of each segmented tree crown. The extent of the segmented tree crowns is typically represented by pixels, which could limit the ability to accurately represent the crown of small trees. We wanted a method that could detect trees ranging from small to medium in size, and rather than modifying any of the existing methods, we developed a simple and novel segmentation procedure. Thus, the proposed segmentation procedure was specifically tailored to the detection of small trees, with as few as only one laser echo. It should be noted that its area of application could be wider than just small trees in the forest-tundra ecotone, including, for example, the monitoring of seedlings in forest stands planted after final fellings in managed boreal forests, or the detection and monitoring of small trees in afforested areas in the tropics.

The objectives of the present study were (1) to develop a procedure for automatic detection and segmentation of small trees using ALS data and (2) to assess the accuracy of the method by comparing the results with field reference data. We consider small trees in this context to be trees with heights up to 7 m and crown diameters up to 6 m. We further wanted to assess the suitability for monitoring purposes by testing the stability of the method across two separate ALS acquisitions for the same study area.

2. Materials and Methods

2.1. Study Area

The study area is located in the Rollag municipality in southeastern Norway (60°0′N 9°01′E, 910–950 m above sea level) and is constituted by a rectangle of 200 × 600 m centered on a mountain ridge. The data materials used were from registrations in the tree line which, at this location, is around

900–940 m above sea level (Figure 1). The main tree species are downy birch (*Betula pubescens* ssp *czerepanovii*), Scots pine (*Pinus sylvestris* L.) and Norway spruce (*Picea abies* (L.) Karst.).

Figure 1. Picture of the landscape and vegetation in the study area.

2.2. Field Data

Field registrations from 472 positioned trees ranging from 0.04 to 6.3 m in height were used in the present study (Table 1). The field work was conducted during the summer of 2012. Within the study area, 40 points were systematically laid out, and at each point up to 16 trees were selected according to the so-called point-centered quarter sampling method sampling procedure [15]: At each point a circular area with a radius of 25 m was sectioned into four quadrants along the cardinal directions N–S and E–W. Using four height classes (0–1, 1–2, 2–3 and >3 m), the tree nearest to the center point in each class and quadrant was selected, giving a maximum of four sample trees per quadrant. The purpose of the procedure was to establish a consistent method to sample trees across the entire range of tree heights found in the area.

Table 1. Summary of the field-measured trees.

	Height (m)	Mean Crown Diameter (m)	n
	min–max *(mean)*	min–max *(mean)*	
Deciduous trees [a]	0.04–6.30 *(1.73)*	0.03–5.50 *(1.37)*	193
Pine	0.06–1.99 *(0.47)*	0.03–1.35 *(0.37)*	83
Spruce	0.05–3.08 *(1.49)*	0.07–3.35 *(1.28)*	196
All	0.04–6.30 *(1.41)*	0.03–5.50 *(1.15)*	472

[a] Mainly birch.

Each of the sample trees were positioned with real-time differential Global Navigation Satellite Systems, with an expected accuracy of 3–4 cm. For each tree, the species, height and horizontal crown diameter in the N–S and E–W directions were recorded. The height and crown diameters were recorded with a measuring tape, with the heights of the highest trees recorded with a Haglöfs Vertex III hypsometer.

The horizontal extent—or crown projection—of all field-measured trees in the data material was defined as an ellipse created from the tree position and the two perpendicular crown diameter measurements. These field-measured *crown segments* were used when extracting ALS echoes from individual trees, and as a field reference in the validation.

2.3. ALS Data

Two sets of ALS data were used in the present study. The first set of ALS data was acquired in July 2006 with an Optech ALTM 3100 laser scanner mounted on a fixed-wing aircraft. These data were acquired in two overlapping flight lines, which means that parts of the study area were covered by ALS data from both flight lines. This dataset was used to test the stability of the proposed segmentation procedure (further described in Section 2.4).

The second set of ALS data was acquired in August 2012 with a Leica ALS70 laser scanner mounted on a fixed-wing aircraft. This dataset corresponded in time with the field registrations and was used to develop and test the segmentation procedure. The two datasets are denoted ALS2006 and ALS2012 throughout this paper. Further details of the ALS2006 and ALS2012 datasets are given in Table 2.

Table 2. Specifications of the two ALS datasets. Mean echo and pulse density calculated from the data, information in the other fields from the data vendor.

Dataset	ALS2006	ALS2012
Sensor	Optech ALTM3100	Leica ALS70
Scan frequency (Hz)	70	-
Pulse frequency (kHz)	100	154.4
Flying speed (m/s)	75	69
Mean flying altitude (a.g.l.)	800 m	1800 m
Mean point density (echoes per m^2)	8	15
Mean pulse density (pulses per m^2)	8	15
Footprint diameter (m)	0.21	0.27
Vertical accuracy (m)	0.10	0.12
Planimetric accuracy (m)	0.13	0.20
Maximum iteration angle (degrees)	9	7.5
Maximum iteration distance (m)	1	1.9

First return echoes were used from the ALS2006 dataset, and all returns from the ALS2012 dataset. Note that—with respect to return categories—the difference between the two datasets was small, since most pulses yield only a single echo from the generally low vegetation in the study area. We found that approximately 97% of the echoes in the ALS2012 dataset were, in fact, single returns. The positional accuracy of the laser echoes was expected to be in the range of 0.1—0.2 m for both sensors, according to Ussyshkin and Smith [16] and the Leica ALS70 product brochure [17].

The ALS echoes were, for both datasets, classified into ground and non-ground using the Terrascan software, following the triangular irregular network (TIN) densification algorithm described by Axelsson [18]. Control parameters for the ground classification, the so-called "maximum iteration angle" and "maximum iteration distance", were set to 9 degrees and 1 m for the first ALS dataset. In the second dataset, values of 7.5 degrees and 1.9 m were used (see [19]). Above-ground heights were calculated for all echoes, as the distance between the TIN and the recorded ellipsoidal heights.

A selection of the ALS and field data is visualized in Figure 2.

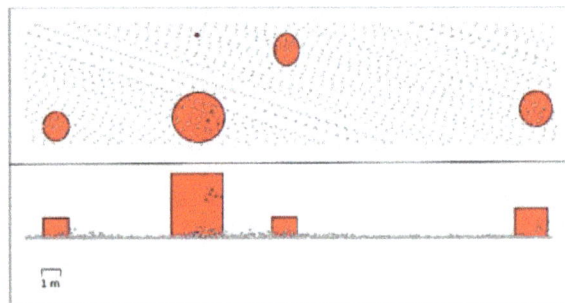

Figure 2. Visualization of five field-measured trees, and the ALS echoes from ALS2012 in the corresponding area. Viewed from above (**upper figure**) and from the side (**lower figure**). The field-measured height and crown extent is colored red, and the ALS echoes are colored from grey to black. The highest echoes are colored black. Note that the terrain height has been subtracted from the ALS echo heights (see text for details), and that only a sample of trees was measured.

2.4. Calculations and Analysis

The field dataset was split in to a modeling dataset consisting of four of the 40 sample locations (locations #10, #20, #30 and #40), and a validation dataset consisting of the remaining 36 locations. This resulted in a set of 39 trees for modeling, and a set of 433 trees for validation. Note that the validation dataset was used to validate the whole segmentation process, whereas the modeling data were used for the two models described in the following. The rationale for this split of the data was to have a sufficient number of trees for the two allometric models, but with an emphasis on the validation of the whole segmentation process. We considered the chosen number of trees in the modeling dataset to be sufficient for the two simple linear models. A summary of the model and validation datasets is given in Table 3.

Table 3. Number of trees in the modeling and validation datasets.

Height Class	Model Dataset	Validation Dataset
0–1 m	18 *(46%)*	202 *(47%)*
1–2 m	11 *(28%)*	119 *(27%)*
2–3 m	6 *(16%)*	61 *(14%)*
>3 m	4 *(10%)*	51 *(12%)*
all	39	433

The proposed method is based on two models, one allometric model relating field-measured tree height to crown diameter, and one model relating field-measured tree height to the above-ground height of the ALS echoes.

2.4.1. Height–Crown Diameter Model

The field registrations in the modeling data were used to fit a non-intercept linear regression model relating crown diameter to tree height

$$\hat{cd} = \beta_a \cdot h + \epsilon_a \tag{1}$$

where $\hat{cd}$ is the crown diameter defined as the mean of the two perpendicular crown diameter measurements, h is the field-measured tree height, β_a is the parameter to be estimated and ϵ_a is an error term, expected to be normally distributed with mean zero. Other model forms and transformations of the variables were tested, but did not result in substantially better models with this data. The simplest linear model was used in the present study, and other models are not further documented. A non-intercept model was chosen in order to ensure positive predictions of cd for all $h > 0$, and to satisfy the condition that $cd = 0$ when $h = 0$. The coefficient of determination for the non-intercept model was calculated as the square of the Pearson's correlation of the fitted and observed values.

The relationship between height and crown diameter might vary between species, but since species information cannot easily be obtained from ALS data we could not use species-specific models for predictions. The species information were therefore not used when we fitted the model given by Equation (1).

2.4.2. ALS Echo Height–Field-Measured Height Model

The crown segments formed from the field-measured crown diameters were used to extract echoes from the ALS2012 dataset for each tree. All echoes inside the crown segment were assigned to the tree for which the segment was created. We did not introduce specific procedures for handling overlapping crowns, which means that a single echo could theoretically be assigned to more than one tree. It further means that any given echo assigned to a tree could have been reflected from an overlapping part of another tree. In the modeling data the maximum above-ground height of the echoes assigned to a tree was denoted $hmax_{ALS}$ and related to the field-measured tree height through a linear regression model

$$\hat{h} = \beta_{b0} + \beta_{b1} \cdot hmax_{ALS} + \epsilon_b \tag{2}$$

where $\hat{h}$ is the estimated tree height, β_{b0} and β_{b1} are parameters to be estimated and ϵ_b is an error term.

2.4.3. Model Fit and Validation

Model fit was assessed by inspecting the coefficient of determination, and the models were further validated through a leave-one-out cross-validation procedure. Root mean squared error (RMSE) was computed as

$$\text{RMSE} = \sqrt{\frac{\sum_{i=1}^{n}(x_i - \hat{x}_i)^2}{n}} \tag{3}$$

where n is the number of trees, x_i is the reference value of the ith tree and $\hat{x}_i$ is the corresponding value predicted by the model constructed from the remaining n-1 trees in the cross-validation. We denoted RMSE as a percentage of the mean reference value as RMSE%.

2.4.4. Crown Segments

Using the two described models (Equations (1) and (2)), every individual echo with a positive above-ground height was related to a circular crown segment, positioned with the given echo in its center (Figures 3 and 4). Equation (2) was used to estimate a tree height from the above-ground height of the ALS echo, and then this tree height was used in Equation (1) to get an estimated crown width, and thereby produce an initial circular crown segment. Thus, at this initial stage all echoes with a positive above-ground height were related to a positioned crown segment. The final crown segments were then determined by applying a set of simple rules to the initial crown segments:

- All echoes that fell within the circular crown segment of an echo higher above ground (*i.e.*, a larger segment as per the height-crown diameter model) were assumed to belong to the larger segment, and such smaller segments were therefore removed before the subsequent steps. This procedure was carried out according to segment size, so that echoes within larger segments were removed first.
- In the next step, overlapping segments were identified and, based on the degree of overlap, the two underlying echoes were either assumed to be from different trees and the corresponding segments kept separate, or they were assumed to be reflected from the same tree. If the latter was true, the smaller segment was merged with the larger. Segments with an overlapping part of the two radii of more than s times the smaller radius were merged. In the case of merging two segments, the lowest echo was added as a new vertex in the largest segment (Figures 3 and 4). In the present study we tested values of s between 0.05 and 0.85.

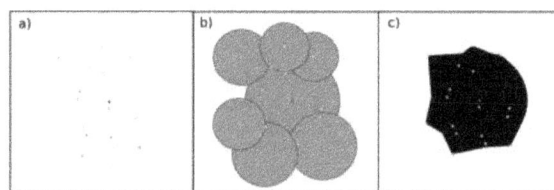

Figure 3. Graphical representation of steps in the segmentation procedure: (**a**) Laser echoes viewed from above, darker color indicates echoes higher above ground; (**b**) Each echo is associated with a circular segment. Note that the segment of the echo highest above ground is created first, and echoes inside this segment are treated as reflected from this segment; (**c**) Overlapping circular segments are merged based on the degree of overlap (see text for details), with the echoes of the smaller segments added as vertices in the larger segment, forming the final segment (shown in black).

The procedure described in this section was implemented in the programming language R [20] and C++ as a fully automated algorithm.

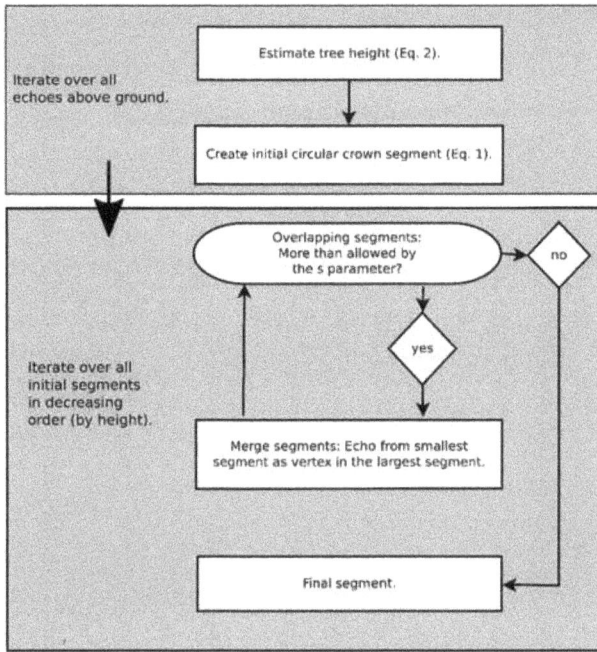

Figure 4. Flow diagram showing an outline of the segmentation process.

2.4.5. Validation

The ALS-derived crown segments were compared to the field-measured crown segments, and the number of field-measured segments matched by an ALS-derived segment was noted. A "match" in this context is not universally defined, so we had to rely on a set of criteria to decide if two segments matched up or not. The set of criteria we used basically defines the degree of similarity in terms of size and position that is required to label a field-measured segment as matched by an ALS-derived segment, thus regarding the tree as detected and correctly segmented.

An ALS-derived segment was defined to be successfully matched with a field tree by using the following procedure:

Each field-measured tree was linked to an ALS-derived tree if the position of the ALS-derived tree was inside the field-measured crown segment. If more than one ALS-derived tree was inside the field-measured crown segment, the ALS-derived tree with the smallest planar distance to the field tree position was used. The location of the highest ALS echo within the ALS-derived segment was used as the ALS-derived tree position.

Linked pairs of trees with large differences in height were excluded. This was done by fitting a regression model $\hat{h} \sim h$, and excluding all pairs of linked trees with a height difference larger than two times the standard error of the model.

This procedure corresponds to the procedure used in a comparison of segmentation algorithms by Vauhkonen *et al.* [21].

We calculated detection rates for the individual height classes described in Section 2.2 as the number of correctly segmented field trees in the particular height class to the total number field trees in that class.

2.4.6. Stability—Number of Segmented Trees

Due to the design of the field work, we could not calculate the commission error. We did, however, test the stability of the number of ALS-segmented trees between two separate ALS acquisitions within a given area. In other words, we tested—for different height classes—if a similar number of trees would be segmented from a separate, second acquisition of ALS data. One aim of the current project was to develop a method that was suitable for monitoring, or repeated measurements. In such a monitoring approach it is desirable to have ALS-derived metrics which are stable, *i.e.*, which vary little due to properties of the scanning and segmentation process itself. Thus, as much as possible of the variation between similar metrics derived from two separate acquisitions should ideally stem from actual changes in the vegetation.

The ALS2006 dataset was used, with two acquisitions carried out on the same day. The same sensor and flight parameters were used for the two acquisitions. The study area was divided into hexagonal cells of 200 m^2 and after application of the proposed segmentation algorithm, the number of segmented trees from within each cell was counted. This was done separately with data from each of the two ALS acquisitions. Note that we, in the segmentation procedure here, used the existing models derived by Equations (1) and (2) and the field and ALS data from 2012.

Differences in the number of segmented trees between the two acquisitions were tested by fitting linear mixed effects models, described further in this section. The use of this approach was motivated by the ability to incorporate assumptions about spatial correlation in the test procedure. We asserted that the observations might be spatially correlated, which would violate the assumption of independent observations in statistical tests such as a paired t-test. Following the approach described by Zuur *et al.* [22] and Pinheiro and Bates [23], we tested the difference between the mean number of trees in each height class by fitting a linear mixed effects model:

$$y_i = x_i\beta + b_i + \epsilon_i, \quad i = 1, \ldots, M \qquad (4)$$

where y_i is a vector with the number of segmented trees in cell i and x_i is a vector of the corresponding acquisitions as factors. We derived the number of trees from two different acquisitions, so y_i and x_i will be vectors of length two. Then β is a vector of the regression model parameters (fixed effects), b_i is a vector of random effects allowed to differ for each cell, ϵ_i is an error vector and M is the number of cells. In this model framework it is assumed that

$$b_i \sim N\left(0, \sigma_b^2\right), \quad \epsilon_i \sim N\left(0, \sigma^2 I\right) \qquad (5)$$

where $\sigma_b{}^2$ and σ^2 are the within-cell and between-cell variance, respectively. I denotes an identity matrix. The *lme* function from the nlme package [24] in the statistical software R was used to fit the models. The t-statistic and the corresponding p-value for the slope in this model should be identical to the values obtained from a comparison of the two acquisitions using a paired t-test [25]. We verified this for all the models in this study by performing paired t-tests using the *t.test* function in R, and comparing the results with the slope statistics in the output from the model-fitting using the *lme* function. The model given by Equation (4) can thus be used as a comparison of the two acquisitions. As described by Zuur *et al.* [22], assumptions about spatial autocorrelation between subjects can be introduced in a linear mixed effects model by replacing I with a matrix V, such that

$$\epsilon_i \sim N\left(0, \sigma^2 V\right) \qquad (6)$$

with V depending on the given correlation structure. Since the data from the two acquisitions could exhibit spatially-dependent variation, we tested if incorporating assumptions about spatial autocorrelation led to models which differed from the models without such assumptions.

We fitted separate models with spherical and Gaussian correlation structures [23]. Each of these models was then compared to the model with assumed uncorrelated errors, given by Equations (4) and (5). Since this model is nested within the models with assumptions about correlated errors, a likelihood ratio test could be used for the comparisons [23].

3. Results

3.1. Regression Models

Two linear regression models were fit to the modeling dataset (Equations (1) and (2)). Both models showed good fit, with R^2 values of 0.86 and 0.72 (Table 4 and Figure 5). A leave-one-out cross-validation of the two regression models resulted in RMSE% of 28.5% and 44.7% for the height model (Equation (2)) and crown diameter model (Equation (1)), respectively (Table 4).

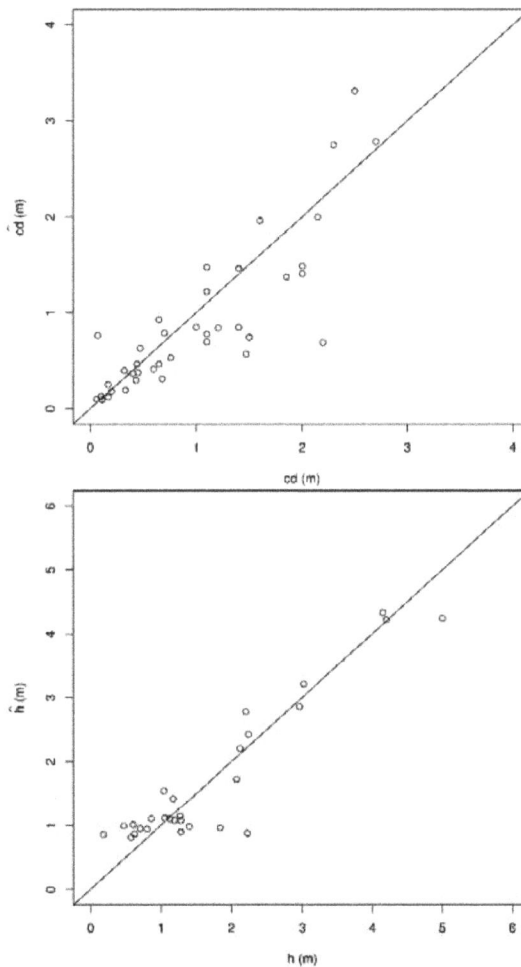

Figure 5. Observed *versus* predicted values for the two linear models: Height–crown diameter model given in Equation (1) (**top**), and ALS height–field height given in Equation (2) (**bottom**).

Table 4. Regression model variables, parameter values and goodness of fit. RMSE from leave-on-out cross-validation.

Dependent Variable	Independent Variable	Parameter Values [a]		R^2	RMSE (RMSE%)
h	$hmax_{ALS}$	β_{b0}: 0.8030	β_{b1}: 0.9590	0.86	0.49 (28.5)
cd	h	β_a: 0.6621		0.72	0.45 (44.7)

[a] Significance level for all parameters: $p < 0.000$.

3.2. Detection and Segmentation

Following the procedure described in Section 2.4, individual crown segments were formed from the ALS echoes for the entire study area (Figures 6 and 7). We report results for the segmentation procedure with parameter $s = 0.2$. Other values of s gave only minor changes in the results, and this is further discussed in Section 4. The ALS-derived segments were compared to the field-measured crown segments (Figure 4). A successful match, *i.e.*, a correctly segmented tree, was registered using the criteria given in Section 2.4. Overall, 46.2% of the trees were successfully segmented. Detection rates ranged from 15.8% to 80.4% for the individual height classes (see Section 2.2), with the detection rate increasing with tree size (Table 5).

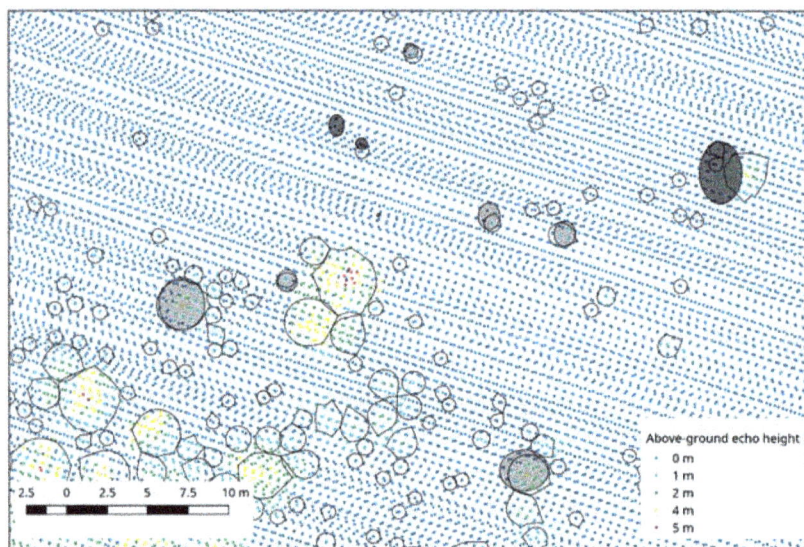

Figure 6. Single-tree segments from the described procedure (hollow segments) and field-measured crown ellipses of detected (light grey) and undetected (dark grey) trees. Note that only a sample of the trees was measured in the field. ALS echoes are colored according to the above-ground height.

Table 5. Detection rates with field-measured segments as reference.

Height Class	Correctly Segmented Trees (%)
0–1 m	15.8
1–2 m	68.1
2–3 m	75.5
>3 m	80.4
All	46.2

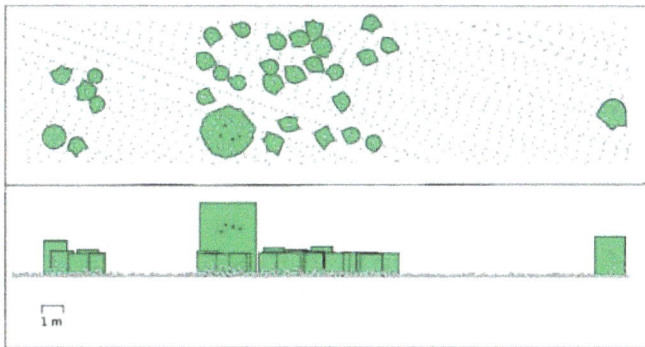

Figure 7. Visualization of single-tree segments and ALS echoes from ALS2012, covering the same area as Figure 2. Viewed from above (**upper figure**) and from the side (**lower figure**). The segments extent and estimated tree heights are colored green, and the ALS echoes are colored from grey to black. The highest echoes are colored black. Note that the terrain height has been subtracted from the ALS echo heights (see Section 2.3 for details).

3.3. Stability

Comparison of the linear mixed effects models with and without assumptions about spatial autocorrelation revealed that no differences could be found between the models, with *p*-values for the likelihood ratio tests ranging from 0.15 to 0.99. This suggests that spatial autocorrelation does not have a major influence in this case, and that using an ordinary paired *t*-test is sufficient for testing the differences between the number of trees derived from the two acquisitions in each 200 m^2 cell.

These tests for differences between the mean number of segmented trees from the two acquisitions resulted in *p*-values ranging from 0.27 to 0.65, and it is evident from this result that no significant differences could be detected through these tests. It should be noted that this range included *p*-values from all comparisons, for the slope coefficients in the mixed models in which assumptions about spatial autocorrelation were included, as well as ordinary paired *t*-tests. Overall, this shows that for the 569 cells—with a total area of 11.4 ha—there were no significant differences between the mean number of segmented trees derived from the two same-day ALS acquisitions.

Some variation between the number of segmented trees from the two acquisitions was, however, observed, with the highest variation for the smaller trees (Table 6). Overall, the mean number of segmented trees in each cell was 52.0 and 51.7 for the two acquisitions. The mean of the differences between the number of segmented trees in each cell for the two acquisitions was 0.31 with a standard deviation of 10.20 (Table 6). It is, from this, evident that the number of segmented trees varies for individual cells. This variation is smaller if one considers only larger trees (Table 6).

Table 6. Comparison of the number of segmented trees within each 200 m^2 cell derived using two separate ALS acquisitions (569 cells of 200 m^2). Mean difference between the two acquisitions and the standard deviation for the difference (sd).

| Height Class | Number of Trees | | | |
| | Acquisition 1 | Acquisition 2 | Difference [a] | |
	Min–Max (*Mean*)	Min–Max (*Mean*)	Mean	sd
0–1 m	0–139 *(37.8)*	0–170 *(37.7)*	−0.38	10.04
1–2 m	0–30 *(4.6)*	0–28 *(4.5)*	−0.11	2.21
2–3 m	0–30 *(3.7)*	0–32 *(3.8)*	*0.08*	1.82
>3 m	0–31 *(6.3)*	0–30 *(6.3)*	0.04	1.27
All	1–147 *(51.7)*	1–181 *(52.0)*	0.31	10.20

[a] No significant differences were observed, see text for details.

4. Discussion

Over all the four height classes, 46.2% of the trees were correctly segmented by the proposed method. A higher proportion of the larger trees were correctly segmented and the detection rate decreased with the decreasing tree size. There are few directly comparable studies, since previous studies typically targeted trees that are larger than in the present study. Vauhkonen *et al.* [21] compared six single-tree detection algorithms in different types of mature forests. The average detection rates for the different forest types reported by Vauhkonen *et al.* varied between 54% and 91%. These detection rates were, however, calculated using the plot-wise total number of tree segments in relation to the number of field-measured trees. They cannot be directly compared to the detection rates in the present study, which were calculated based on the linking field and ALS-derived trees. Vauhkonen *et al.* report a corresponding number called "treetop candidates linked to field trees", and this varied between 42% and 60%. The mean tree diameter at the different sites in the data material used by Vauhkonen *et al.* was 18.5–35.8 cm, and the ALS point density was 1.5–30 per m^2. The detection rates observed for trees with $h < 1$ m in the present study were considerably lower than the rates reported by Vauhkonen *et al.* [21]. These trees are, however, much smaller than any tree in the data material used in that study. For the trees with $h > 1$ m, the proportion of detected field trees in the present study seems to be within the range reported in Vauhkonen *et al.* [21], as well as in recent studies by Liu *et al.* [26], and Mongus and Zalik [27].

The advantage of the proposed method is that it is simple, and can be implemented using models developed from a limited number of sample trees. A disadvantage is the need for these sample trees, as well as the detection errors discussed in the following.

From the current study it is evident that direct detection of all individual trees by ALS is not possible when the properties of the ALS data are similar to those of the current study. Omission errors will occur, caused by two different factors: firstly, due to limitations in the data material itself; ALS can be viewed as measuring distances from the aircraft to the ground, with the distance to only some particular spots on the ground being acquired. In the present study the average density of laser pulses was eight pulses per m^2, and given the footprint size together with the non-uniform spatial distribution of the laser pulses, this means that some trees will not be hit at all. The fraction of the trees that is not hit by any laser pulse will depend on a range of factors, such as the pulse density, the pulse footprint size and the degree of unevenness in the spatial distribution of the pulses on the ground. It is, however, clear that the size of a tree directly affects the probability for it to be hit by a laser pulse, so smaller trees are less likely to be hit than larger trees. The trees which are not hit by any laser pulses cannot be directly detected using the ALS data, and it is hence a definite limit to direct detection of individual trees inherent in the data material itself. There is, under such conditions, not enough information in the data material to directly detect all trees, and omission errors are unavoidable if all trees, even the smallest ones, are considered. The chance of being hit by a laser pulse increases with the tree size, but even an echo reflected from the tree is in itself not sufficient to ensure a successful detection. To be able to separate it from the surrounding terrain, the echo must have a positive above-ground height.

The second cause of omission errors is in the segmentation procedure, which in some cases will fail to produce a segment that matches that of the tree on the ground. This can be seen in Figure 6, at the rightmost field-measured tree. The ALS-derived segment is, in this case, not similar enough to be considered a correct segmentation. The reason can be measurement errors, neighboring trees or other factors affecting how the echoes are being reflected from that particular tree. Segments from multiple trees can also be erroneously merged, and thus lead to omissions. How the procedure merges segments is controlled by the *s* parameter, and this is discussed later in this section.

All echoes with a positive above-ground height will, however, not be reflected from trees, which is one out of two types of commission errors. Objects such as rocks, hummocks and bushes may all result in positive above-ground heights, and thus result in falsely detected trees. The intensity value of the echoes could, however, hold some information that can be used to distinguish between trees and the surrounding terrain and vegetation, and some studies have found a positive contribution from the intensity values when classifying tree and non-tree echoes [7].

The second type of commission error is over-segmentation, which means detecting several trees from the echoes reflected from a single tree. Due to the design of the field work in which only a sample of the trees was measured, we were not able to fully assess these commission errors. The parameter s controls how the segmentation procedure merges initially overlapping segments, and the number of segments will increase as s increases. So with $s = 1$ all initial segments are kept as separate segments, whereas with $s = 0$ all overlapping segments are merged, *i.e.*, no final segments overlap. Since tree crowns sometimes do overlap, a reasonable value of s should be somewhere between the two extremes. The detection rates in the present study varied, however, very little for the different tested values of s. This can be attributed to several factors, first of all that changing the value of s will only affect segments that initially overlapped. Furthermore, it is clear that the presence of small segments at the edge of larger segments does not have a large influence on either detection or omission errors. So even with some small segments overlapping, the larger segment will still be connected with the field tree. These small segments will, on the other hand, directly influence the commission errors, and this is the type of error we were unable to control in the present study. Further research is needed in order to find an optimal value for s, and to fully assess the commission errors. It should be noted that the study area, as well as other transition zones between forest and alpine areas, is, in parts, sparsely populated with trees. In more dense forests, a larger proportion of the trees will have overlapping crowns, which will affect the performance of the proposed segmentation procedure, as well as the optimal value of the s parameter.

The initial processing of ALS data for most applications related to forests or trees involves a choice of algorithms and corresponding parameter values. In the current study, this involves the echo classification as well as the computation of the above-ground heights. This choice of algorithm and parameter values will most likely influence further use of the data in, for example, single tree detection. The widely used classification algorithm based on the principles described by Axelsson [18,28] is used in the present study. This algorithm requires parameter values for "iteration angle" and "iteration distance". The effect of the iteration angle on echoes reflected from small trees in the forest-tundra ecotone was investigated by Næsset [3]. In that study, an increase in omission errors and a decrease in commission errors were observed when the iteration angle was increased from six to 12 degrees. A tree was, in that study, regarded as detected if it yielded at least one echo with a positive above-ground height, and the terms omission and commission error refer to that definition. A conclusive suggestion on an optimal iteration angle was, however, not given based on those results. The use of a model-chain as in the proposed algorithm will cause errors to propagate and add up through the chain. Errors in the allometric model given by Equation (2) will, for example, affect the results from applying the model given in Equation (1), and finally the resulting single-tree segments.

When evaluating single-tree detection algorithms, the obtained tree segments will deviate from the field measurements. The process of choosing and defining the detection criteria will inevitably involve subjectivity. The choice of detection criteria will affect the detection rates, and the effect of detection criteria should be incorporated in the evaluation of segmentation algorithms.

In the case of a change assessment in which an identical detection method is applied at two points in time, omission and commission errors should theoretically be of less consequence. Given that these errors occur with the same magnitude in each of the two segmentations, actual changes on the ground between two ALS acquisitions should lead to corresponding differences in the two sets of segmented trees. The stability of the ALS-derived variables plays a role in this case. The amount of variation that is due to the scanning and the segmentation process itself will determine the magnitude of the vegetation changes that can be reliably detected for a given area. The results from the present study indicate that for the proposed segmentation procedure, some variation must be expected for smaller areas. The magnitude of the changes that can be reliably detected using the proposed method could be further investigated. The spatial distribution of the laser echoes on the ground and in the vegetation will differ from one acquisition to another. We assessed, in the present study, the influence of these differences on the resulting single-tree segments by using two separate acquisitions from the same sensor. The use of different sensors in multi-temporal data acquisition will further contribute to differences between the two sets of data. Expected effects and possible calibration methods to mitigate these could be subject to further research.

5. Conclusions

Moderate detection rates were observed when using the proposed segmentation algorithm. Overall, 46.2% of the trees were segmented correctly. The detection rates were higher for larger trees, and conversely, lower for smaller trees. The high proportion of undetected trees was partly due to limitations in the data material itself; some trees were not hit by any laser pulses at all. No significant differences between the number of segmented trees derived from two separate ALS acquisitions were found in the present study. This indicates that it can be suitable for monitoring purposes. Even though the magnitude of the detection errors prevents the detection of changes at an individual tree level, the method might potentially be used to detect changes at an area level. The use of the proposed method for area-based monitoring and change detection in the forest-tundra ecotone could be subject to further research. The proposed method could also be suitable for detection and monitoring of small trees in other biomes, such as seedlings in boreal forest or regeneration in tropical forests. This could be further investigated.

Acknowledgments: We wish to thank Eirik Næsset Ramtvedt for participating in the field work. Terratec AS and Blom ASA acquired and pre-processed the ALS data. We will also like to thank the anonymous reviewers for valuable comments and suggestions. The project was funded by the Norwegian Research Council through grant #184636/S30.

Author Contributions: E.N. planned the acquisition of ALS data, planned and conducted the acquisition of field data and revised the manuscript. M.H. conceived and conducted the study and wrote the paper.

Conflicts of Interest: The authors declare no conflict of interest.

Abbreviations

The following abbreviations are used in this manuscript:

ALS	Airborne laser scanning
TIN	Triangular irregular network
RMSE	Root mean squared error
sd	Standard deviation

References

1. Holtmeier, F.-K.; Broll, G. Treeline advance—Driving processes and adverse factors. *Landsc. Online* **2007**, *1*, 1–32. [CrossRef]
2. De Wit, H.A.; Bryn, A.; Hofgaard, A.; Karstensen, J.; Kvalevåg, M.M.; Peters, G.P. Climate warming feedback from mountain birch forest expansion: Reduced albedo dominates carbon uptake. *Glob. Chang. Biol.* **2014**, *20*, 2344–2355. [CrossRef] [PubMed]
3. Næsset, E. Influence of terrain model smoothing and flight and sensor configurations on detection of small pioneer trees in the boreal–alpine transition zone utilizing height metrics derived from airborne scanning lasers. *Remote Sens. Environ.* **2009**, *113*, 2210–2223. [CrossRef]
4. Næsset, E.; Nelson, R. Using airborne laser scanning to monitor tree migration in the boreal–alpine transition zone. *Remote Sens. Environ.* **2007**, *110*, 357–369. [CrossRef]
5. Reese, H.; Nystrom, M.; Nordkvist, K.; Olsson, H. Combining airborne laser scanning data and optical satellite data for classification of alpine vegetation. *Int. J. Appl. Earth Obs. Geoinf.* **2014**, *27*, 81–90. [CrossRef]
6. Rees, W.G. Characterisation of Arctic treelines by LiDAR and multispectral imagery. *Polar Rec.* **2007**, *43*, 345–352. [CrossRef]
7. Stumberg, N.; Ørka, H.O.; Bollandsås, O.M.; Gobakken, T.; Næsset, E. Classifying tree and nontree echoes from airborne laser scanning in the forest–tundra ecotone. *Can. J. Remote Sens.* **2012**, *38*, 655–666. [CrossRef]

8. Thieme, N.; Martin Bollandsås, O.; Gobakken, T.; Næsset, E. Detection of small single trees in the forest–tundra ecotone using height values from airborne laser scanning. *Can. J. Remote Sens.* **2011**, *37*, 264–274. [CrossRef]

9. Chang, A.; Eo, Y.; Kim, Y.; Kim, Y. Identification of individual tree crowns from LiDAR data using a circle fitting algorithm with local maxima and minima filtering. *Remote Sens. Lett.* **2013**, *4*, 29–37. [CrossRef]

10. Hyyppä, J.; Kelle, O.; Lehikoinen, M.; Inkinen, M. A segmentation-based method to retrieve stem volume estimates from 3-D tree height models produced by laser scanners. *IEEE Trans. Geosci. Remote Sens.* **2001**, *39*, 969–975. [CrossRef]

11. Persson, A.; Holmgren, J.; Söderman, U. Detecting and measuring individual trees using an airborne laser scanner. *Photogramm. Eng. Remote Sens.* **2002**, *68*, 925–932.

12. Reitberger, J.; Schnörr, C.; Krzystek, P.; Stilla, U. 3D segmentation of single trees exploiting full waveform LIDAR data. *ISPRS J. Photogramm. Remote Sens.* **2009**, *64*, 561–574. [CrossRef]

13. Solberg, S.; Næsset, E.; Bollandsås, O.M. Single tree segmentation using airborne laser scanner data in a structurally heterogeneous spruce forest. *Photogramm. Eng. Remote Sens.* **2006**, *72*, 1369–1378. [CrossRef]

14. Koch, B.; Kattenborn, T.; Straub, C.; Vauhkonen, J. Segmentation of Forest to Tree Objects. In *Forestry Applications of Airborne Laser Scanning*; Maltamo, M., Næsset, E., Vauhkonen, J., Eds.; Springer: Dordrecht, The Netherlands, 2014; Volume 27, pp. 89–112.

15. Cottam, G.; Curtis, J.T. The use of distance measures in phytosociological sampling. *Ecology* **1956**, *37*, 451–460. [CrossRef]

16. Ussyshkin, R.V.; Smith, B. Performance analysis of ALTM 3100EA: Instrument specifications and accuracy of LiDAR data. In Proceedings of the ISPRS Conference, Commission I Symposium, Paris, France, 4–6 May 2006.

17. Leica ALS70-Airborne Laser Scanners—Performance for Diverse Applications. Available online: http://leica-geosystems.com/products/airborne-systems/lidar/leica-als70-airborne-laser-scanner (accessed on 10 December 2015).

18. Axelsson, P. DEM generation from laser scanner data using adaptive TIN models. *Int. Arch. Photogramm. Remote Sens.* **2000**, *33*, 110–117.

19. Terrasolid Ltd. *TerraScan User's Guide*; Terrasolid Ltd.: Helsinki, Finland, 2011.

20. R Development Core Team. *R: A Language and Environment for Statistical Computing*; R Foundation for Statistical Computing: Vienna, Austria, 2011.

21. Vauhkonen, J.; Ene, L.; Gupta, S.; Heinzel, J.; Holmgren, J.; Pitkanen, J.; Solberg, S.; Wang, Y.; Weinacker, H.; Hauglin, K.M.; *et al.* Comparative testing of single-tree detection algorithms under different types of forest. *Forestry* **2012**, *85*, 27–40. [CrossRef]

22. Zuur, A.F.; Ieno, E.N.; Walker, N.; Saveliev, A.A.; Smith, G.M. *Mixed Effects Models and Extensions in Ecology with R*; Statistics for Biology and Health; Springer New York: New York, NY, USA, 2009.

23. Pinheiro, J.; Bates, D. *Mixed-Effects Models in S and S-PLUS*; Springer: New York, NY, USA, 2013.

24. Pinheiro, J.; Bates, D.; DebRoy, S.; Sarkar, D.; R Core Team. NLME: Linear and Nonlinear Mixed Effects Models. Available online: https://CRAN.R-project.org/package=nlme (accessed on 3 March 2016).

25. Nyberg, J.S. The Paired T-Test: Does PROC MIXED Produce the Same Results as PROC TTEST? In Proceedings of the PharmaSUG Conference, San Diego, CA, USA, 23–26 May 2004; Volume 5.

26. Liu, T.; Im, J.; Quackenbush, L.J. A novel transferable individual tree crown delineation model based on Fishing Net Dragging and boundary classification. *ISPRS-J. Photogramm. Remote Sens.* **2015**, *110*, 34–47. [CrossRef]

27. Mongus, D.; Žalik, B. An efficient approach to 3D single tree-crown delineation in LiDAR data. *ISPRS J. Photogramm. Remote Sens.* **2015**, *108*, 219–233. [CrossRef]

28. Axelsson, P. Processing of laser scanner data—Algorithms and applications. *ISPRS J. Photogramm. Remote Sens.* **1999**, *54*, 138–147. [CrossRef]

![remote sensing logo] *remote sensing*

MDPI

Article

Deep-Learning-Based Classification for DTM Extraction from ALS Point Cloud

Xiangyun Hu [1,2] and Yi Yuan [1,*]

[1] School of Remote Sensing and Information Engineering, 129 Luoyu Road, Wuhan University, Wuhan 430079, China; huxy@whu.edu.cn
[2] Collaborative Innovation Center of Geospatial Technology, Wuhan University, Wuhan 430079, China
*   Correspondence: yuan_yi@whu.edu.cn; Tel.: +86-138-8605-6791

Academic Editors: Jie Shan, Juha Hyyppä, Lars T. Waser, Xiaofeng Li and Prasad S. Thenkabail
Received: 20 May 2016; Accepted: 29 August 2016; Published: 5 September 2016

Abstract: Airborne laser scanning (ALS) point cloud data are suitable for digital terrain model (DTM) extraction given its high accuracy in elevation. Existing filtering algorithms that eliminate non-ground points mostly depend on terrain feature assumptions or representations; these assumptions result in errors when the scene is complex. This paper proposes a new method for ground point extraction based on deep learning using deep convolutional neural networks (CNN). For every point with spatial context, the neighboring points within a window are extracted and transformed into an image. Then, the classification of a point can be treated as the classification of an image; the point-to-image transformation is carefully crafted by considering the height information in the neighborhood area. After being trained on approximately 17 million labeled ALS points, the deep CNN model can learn how a human operator recognizes a point as a ground point or not. The model performs better than typical existing algorithms in terms of error rate, indicating the significant potential of deep-learning-based methods in feature extraction from a point cloud.

Keywords: deep learning; convolutional neural network (CNN); digital terrain model (DTM); ALS; ground point classification

1. Introduction

In recent decades, airborne laser scanning (ALS) has become more important in the process of digital terrain model (DTM) production [1]. ALS can provide a description of a surface on a terrain with high accuracy and density. However, ALS also records the information of non-terrain objects, such as buildings and trees. Thus, the ALS filtration is important in the processing of ALS data. Given the various non-ground objects on the surface and the lack of topology among the points, filtering of the ALS point cloud can be difficult and troublesome. In fact, point filtering often occupies approximately 80% of the workload of ALS data processing in DTM production. The algorithms of ALS filtration can be divided into three categories based on their characteristics, as follows:

(1) Slope-based methods. The kernel foundation of these methods considers that two adjacent points are likely to belong to different categories if they have a mutation in height [2,3]. Slope-based methods are fast and easy to implement. Their shortcoming is their dependency on different thresholds in different terrains.

(2) Mathematical morphology-based methods. These methods are composed of a series of 3D morphological operations on the ALS points. The results of morphological methods heavily rely on the filter window size. Small windows can only filter small non-ground objects, such as telegraph poles or small cars. By contrast, large windows often filter several ground points and make the results of filtration smooth. Zhang [4] proposed progressive morphological filters,

which can filter large non-ground objects with ground points preserved by varying the filter window size, to overcome this problem.

(3) Progressive triangular irregular network (TIN)-based method. Axelsson [5] proposed the iterative TIN; this network has been used in some business software. The TIN selects the coarse lowest points as ground points and builds a triangulated surface from them. Then, the TIN adds new points to the triangular surface under many constrains for slope and distance. However, the method is easily affected by negative outliers; these outliers draw the triangular surface downward.

(4) Surface-based methods. These methods maintain a surface model of the ground based on the interpolation of ground points [6–9]. However, these methods are sensitive to input parameters and negative outliers.

Other recent algorithms try to use optimization to obtain accurate classification. For instance, semi-global filtering (SGF) [10] employs a novel energy function balanced by adaptive ground saliency to adapt to steep slopes, discontinuous terrains, and complex objects. Then, the SGF uses semi-global optimization to determine labels by minimizing the energy.

Although the existing methods have done well in ALS filtration, they still need much human labor to generate DTM based on the filtration results. We want to make full use of the existing ALS and responding DTM by learning a deep neural network from a big amount of the existing data. Neural networks has been used in pattern recognition and classification for a long time [11,12]. The deep convolutional neural networks (CNN) [13] are inspired by biological vision systems; these networks have recently shown their ability to extract high-level representations through compositions of low-level features [14]. In the present study, we propose a new filtering algorithm based on deep CNN. First, training samples are obtained from many labeled points. Each image is generated from the point and its neighboring points; the image can be a positive or negative training sample depending on the label. Second, a deep CNN model is trained using the labeled data. Images generated from points are treated as input of the deep CNN model. Then, the input will be processed by several components being comprised of a convolution layer, a batch Normalization layer, an activation layer and a pooling layer after some components. At last, the results of the last pooling layer will be connected to subsequent three fully-connected layers, the last fully-connected layer will produce the probability for the input to be a ground point or a non-ground point. Detailed construction of deep CNN model can be seen in Section 2.2. The deep CNN model can learn the important feature of the input automatically from the huge training data, which usually work better than hand-craft features. Finally, each point is mapped to an image to classify a raw ALS point cloud; this image is classified as an image belonging to a ground point or is not used by the trained CNN model.

This paper is organized as follows: Section 2 describes the proposed method. Section 3 presents the ALS filtration results and analysis. Section 3.3 compares the proposed method with other methods. Section 4 concludes this study and identifies several aspects for improvement.

2. Methods

The workflow of our approach for filtering is shown in Figure 1. ALS filtering means to find and delete all non-ground points from ALS data. We treat filtering as a binary classification problem to classify all the points of ALS data as ground points or non-ground points. The major steps include the calculation of context information for each point from the neighboring points in a window, the transformation of the information of the window into an image, and the training and classification based on the images using the CNN model. Training sample points are selected from a large number of point clouds with different terrain complexities. A deep CNN model is trained from the labeled images.

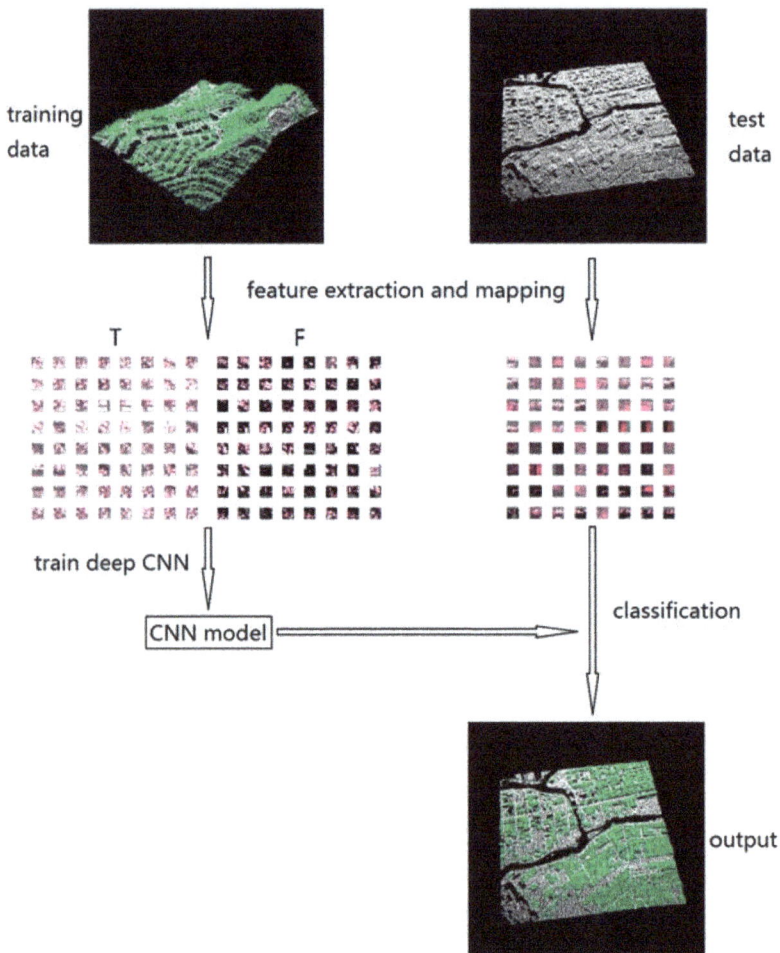

Figure 1. Workflow of the proposed approach. "T" means the samples of ground points and "F" means the samples of non-ground points.

2.1. Information Extraction and Image Generation

For each ALS point (P_i), its surrounding points within its "square window" are divided into many cells. The "square window" means a square in (x, y) spatial coordinates. It is a two dimensional window. In the method, the size of "square window" is 96 m × 96 m, which is divided into 128 × 128 cells. Each "square window" extracted for a point can be transferred to a 128 × 128 image by mapping each cell to a pixel with red, blue, and green colors. For each cell, the maximum (Z_{max}), minimum (Z_{min}), and mean (Z_{mean}) of the height among all points within the cell are obtained. Then, the difference values between Z_{max}, Z_{min}, and Z_{mean} and the height (Z_i) of point (P_i) are transferred to three integers within 0 to 255 following Equations (1) and (2); these integers would be the red, green, and blue values of the corresponding pixel in the image transformed from the cells, as follows:

$$
\begin{aligned}
F_{red} &= \lfloor 255 * Sigmoid(Z_{max} - Z_i) - 0.5 \rfloor \\
F_{green} &= \lfloor 255 * Sigmoid(Z_{min} - Z_i) - 0.5 \rfloor \\
F_{blue} &= \lfloor 255 * Sigmoid(Z_{mean} - Z_i) - 0.5 \rfloor
\end{aligned}
\tag{1}
$$

The sigmoid function is expressed in Equation (2), as follows:

$$Sigmoid(x) = (1 + e^{-x})^{-1} \tag{2}$$

An example for the point-to-image transformation is shown in Figure 2.

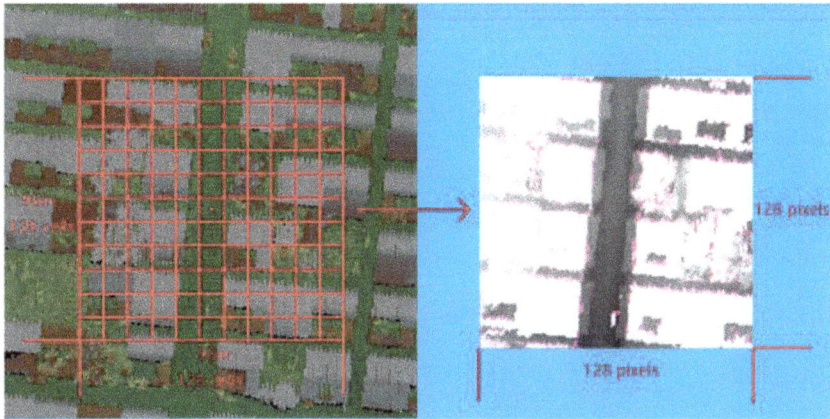

Figure 2. Point-to-image transformation (source: own study in the "FugroViewer").

2.2. Convolutional Neural Network

CNNs have been the focus of considerable attention for a wide range of vision-related [15–18], audio-related [19], or language-related [20] tasks. The existing best-performing models [21–23] on ImageNet ILSVRC have all been based on deep CNNs since 2012. CNNs are designed to process data that come in the form of multiple arrays, such as 1D arrays for signals like language and 2D arrays for images or audio spectrograms. CNNs have four key ideas, namely, local connections, shared weights, pooling, and use of many layers [24]. More detailed description of CNN can be found in [25].

The architecture of the CNN model used in our approach is shown in Figure 3.

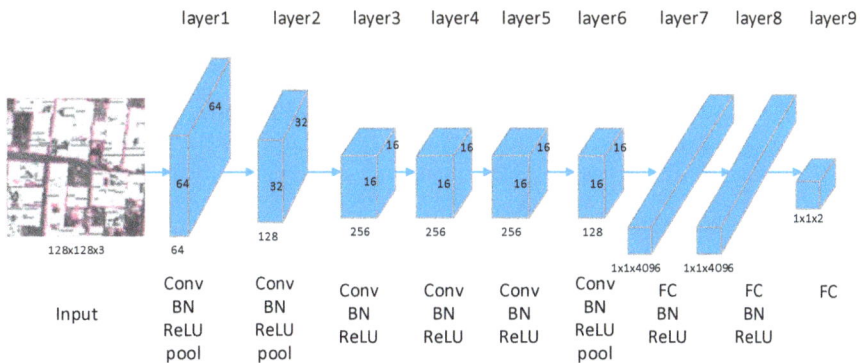

Figure 3. The architecture of the proposed deep CNN.

Deep CNN model is comprised of 6 kinds of layers, the size of layers can be defined as $width \times height \times depth$ in which $width \times height$ describes the spatial size and $depth$ refers to the number of channels of its feature maps. The detailed explanation of the layers in Figure 3 can be seen below:

(1) An input layer is denoted as *Input* here. The input layer contains the input data for the network. The input of the deep CNN model is a three-channel (red, green, blue) 128 × 128 image generated from an ALS points.

(2) A convolution layer is denoted as *Conv* here. The convolution layer is the core building block of a convolutional network that performs most of the computational heavy lifting. Convolutional layers convolve the input image or feature maps with a learnable linear filter, which have a small receptive field (local connections) but extend through the full depth of the input volume. The output feature maps represent the responses of each filter on the input image or feature maps. Each filter is replicated across the entire visual field and the replicated unit share the same weights and bias (shared weights), which allows for features to be detected regardless of their position in the visual field. As a result, the network learns filters that activate when they see a specific type of feature at some spatial position in the input. In our model, all of the convolution layers use the same sized 3 × 3 convolution kernel.

(3) A batch normalization layer is denoted as *BN* here. Given that the deep CNN often has a large number of parameters, taking care to prevent overfitting is necessary, particularly when the number of training samples is relatively small. Batch normalization [26] normalizes the data in each mini-batch, rather than merely performing normalization once at the beginning, using the following equation:

$$y = \frac{x - \mu}{\sqrt{\sigma^2 + \varepsilon}} \gamma + \beta \tag{3}$$

The input of *BN* is normalized to zero mean and unit variance and then linearly transformed. During training, μ and σ^2 are the mean and variance of the input mini-batch. During testing, μ and σ^2 are the average statistics calculated from the training data. γ and β are learned parameters which scale and shift the normalized value. ε is a constant added to the mini-batch variance for numerical stability.

Batch normalization can significantly reduce overfitting, allow higher learning rates and accelerate the training for deep network.

(1) A rectified linear units layer is denoted as *ReLU* here. Activation layers are neuron layers that apply nonlinear activations on input neurons. They increase the nonlinear properties of the decision function and of the overall network without affecting the receptive fields of the convolution layer. Rectified linear units (ReLU) proposed by Nair and Hinton in 2010 [27] is the most popular activation function. ReLU can be trained faster than typical smoother nonlinear functions and allows the training of a deep supervised network without unsupervised pretraining. The function of ReLU can be demonstrated as $f(x) = max(0, x)$.

(2) A pooling layer is denoted as *Pooling* here. Pooling layers are nonlinear downsampling layers that achieve maximum or average values in each sub-region of input image or feature maps. The intuition is that once a feature has been found, its exact location is not as important as its rough location relative to other features. Pooling layers increase the robustness of translation and reduce the number of network parameters.

(3) A fully-connected layer is denoted as *FC* here. After several convolutional and max pooling layers, high-level reasoning in the neural network is performed via fully-connected layers. A fully-connected layer takes all neurons in the previous layer and connects it to every single neuron it has. Fully-connected layers are not spatially located anymore, thereby making them suitable for classification rather than location or semantic segmentation.

A *BN* and a *ReLU* are applied after every *conv* layer and the first two *FC* layers. Thus, layers 1 to 6 are composed of *Conv* → *BN* → *ReLU* and layers 7 and 8 are composed of *FC* → *BN* → *ReLU*. Pooling layers are applied after layers 1, 2, and 6. The output of the last *FC* layer is fed to a 2-way softmax, which produces a distribution over the 2 class labels. Our network maximizes the multinomial logistic regression objective.

To train the model of the deep CNN, over 150 million parameters need to be learned. Two measures are taken to avoid overfitting: huge amount of training data and batch normalization layers which are proven to be effective.

3. Experimental Analysis

3.1. Experimental Data

A total of 17,280,000 labeled points are sampled evenly from 900 airborne ALS datasets in south China to be used to train a general model tested in variety types of terrains to evaluate the proposed approach. Each dataset has an area size of 500 m by 500 m and an average density of 4 points/m^2. Moreover, 40 scenes outside the training areas and the International Society for Photogrammetry and Remote Sensing (ISPRS) benchmark datasets provided by the ISPRS Commission III/WG2 [28] are classified to validate the trained CNN model. The 40 scenes have the same area size as the training data and approximately 40 million points. All training and testing ground truths are produced by a procedure of DTM production, including automatic filtering by TerraScan software and post manual editing. Examples of the training dataset and feature maps of the training samples are shown in Figures 4 and 5, respectively.

It is easy to see from Figure 5 that as the ground points usually being lower than their surrounding points while non-ground points more probable being higher than their surrounding points, most of the F_{red}, F_{green} and F_{blue} calculated from surrounding cells of ground points by Equation (1) are much bigger than non-ground points, which causes that the feature images of ground points are much brighter than non-ground points.

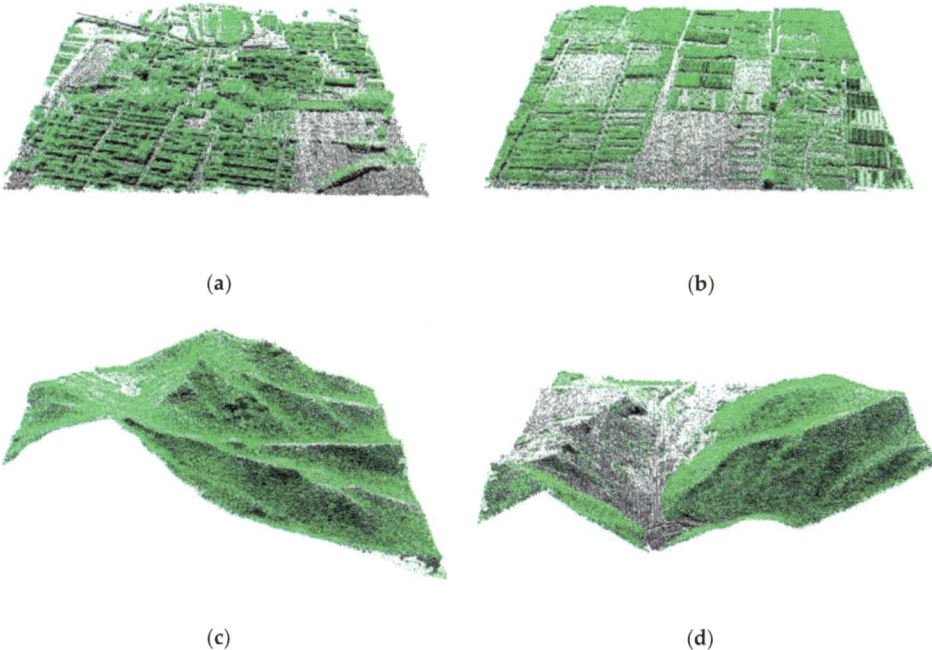

(a)

(b)

(c)

(d)

Figure 4. Four examples of the training ALS point clouds with different terrain features: (**a**,**b**) flat terrain with buildings and farmland; (**c**,**d**): mountainous terrain. White denotes the ground points, and green denotes the non-ground points.

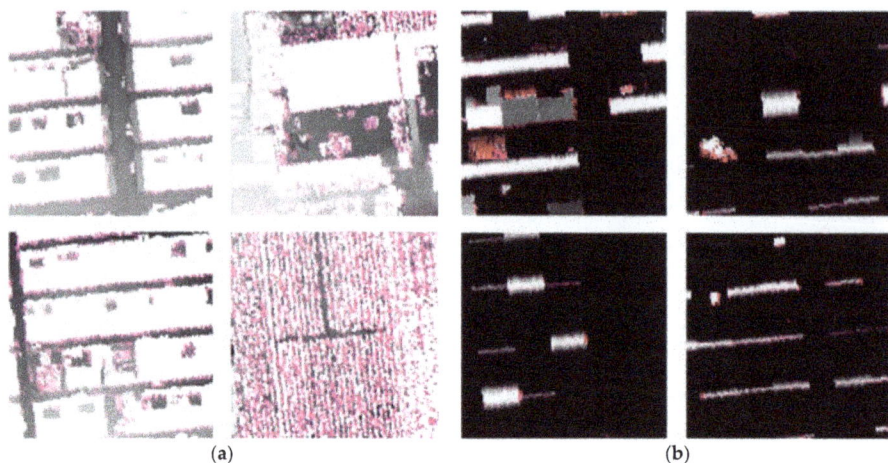

Figure 5. Training samples of the feature images corresponding to: (**a**) ground points; and (**b**) non-ground points.

3.2. Training

Batch gradient descent with a batch size of 256 examples, momentum of 0.9, and weight decay of 0.0005 to estimate the CNN parameters is used for the training. To find a local minimum of a function, gradient descent takes steps proportional to the negative of the gradient (or of the approximate gradient) of the function at the current point. In batch gradient descent, the gradient is approximately estimated by the mini-batch in each iteration.

The loss function of the CNN model can be calculated as:

$$L = -\frac{1}{m} \left[\sum_{i=1}^{m} \sum_{j=1}^{k} 1\{y^{(i)} = j\} \log \frac{e^{w_j^T x^{(i)}}}{\sum_{l=1}^{k} e^{w_l^T x^{(i)}}} \right], \tag{4}$$

where m is the size of batch 256, and k is the number of classes (in here k = 2 because there are 2 classes, ground points and non-ground points), w is the parameters of the model, x is the output of the upper layer and for the first hidden layer, and x is the input layer. $y^{(i)}$ is the label of training sample i. The value of $1\{y^{(i)} = j\}$ equals 1 while $y^{(i)} = j$ and 0 otherwise.

The update rule for weight w was:

$$v_{t+1} := 0.9 \cdot v_t - 0.0005 \cdot \varepsilon \cdot W_t - \varepsilon \cdot \left\langle \frac{\partial L}{\partial w} \Big|_{W_t} \right\rangle_{D_t}$$

$$W_{t+1} := W_t + v_{t+1} \tag{5}$$

where t is the iteration index, mini-batch D_t is the m training samples which will be used to estimate the gradient in this iteration, v is the momentum variable, ε is the learning rate, and $\left\langle \frac{\partial L}{\partial w} \Big|_{W_t} \right\rangle_{D_t}$ is the average over the t th batch D_t of the derivative of the objective loss function with respect to W, evaluated at W_t [21].

The training of the CNN model takes approximately three weeks on a PC with Intel i7-4790 CPU, 32 GB RAM, and a NIVIDIA GTX TitanX GPU.

3.3. Results and Comparison with Other Filtering Algorithms

We compare the deep CNN model with the popular commercial software TerraSolid TerraScan, Mongus's parameter-free ground filtering algorithm in 2012 [1], SGF [10], Axelsson's algorithm, and Mongus's connected operators-based algorithm in 2014 [29] on the ISPRS benchmark dataset. TerraScan uses the TIN-based filtering method; this software produces a significantly low average total error when a set of tunable parameters of the data is processed using the algorithm. The classification that CNN used for this test over two datasets is the one trained by the 900 airborne ALS datasets in south China to challenge the versatility of the CNN.

The filtering accuracy is measured based on the Type I error, which is the percentage of rejected bare ground points; Type II error, which is the percentage of accepted non-ground points; and total error, which is the overall probability of points being incorrectly classified. The results are shown in Table 1 and Figures 6–8. The classification using deep CNN model takes approximately 200 s on a test ALS dataset with a million points using a computer with an i7-4790 CPU and a TitanX GPU.

We also compare deep CNN model with TerraScan on 40 cases from the test ALS data with various terrain complexities in the aspects of both error rates and root mean square error of DTM. The comparison of error rates is shown in Figure 7 and comparison of root mean square error (RMSE) between the generated DTM with the ground truths is shown in Figure 8.

Table 1. Comparison of deep CNN model and other methods on the ISPRS dataset.

	Type I Error (%)	Type II Error (%)	Total Error (%)
TerraScan	11.05	4.52	7.61
Mongus 2012	3.49	9.39	5.62
SGF	5.25	4.46	4.85
Axelsson	5.55	7.46	4.82
Mongus 2014	2.68	12.79	4.41
Deep CNN	0.67	2.262	1.22

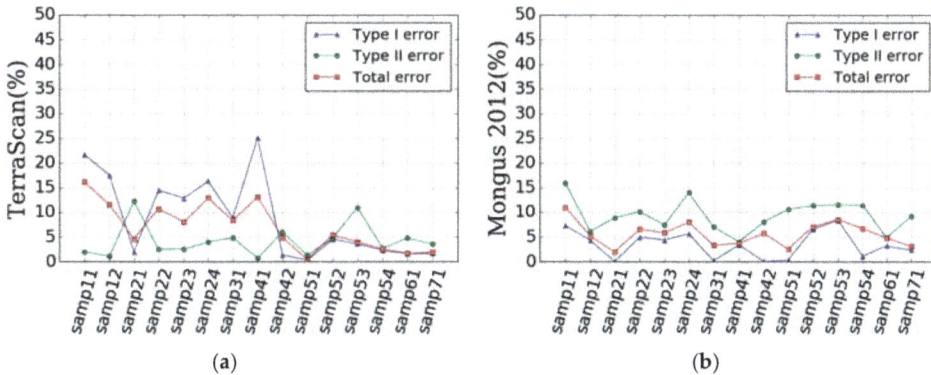

(a)   (b)

Figure 6. *Cont.*

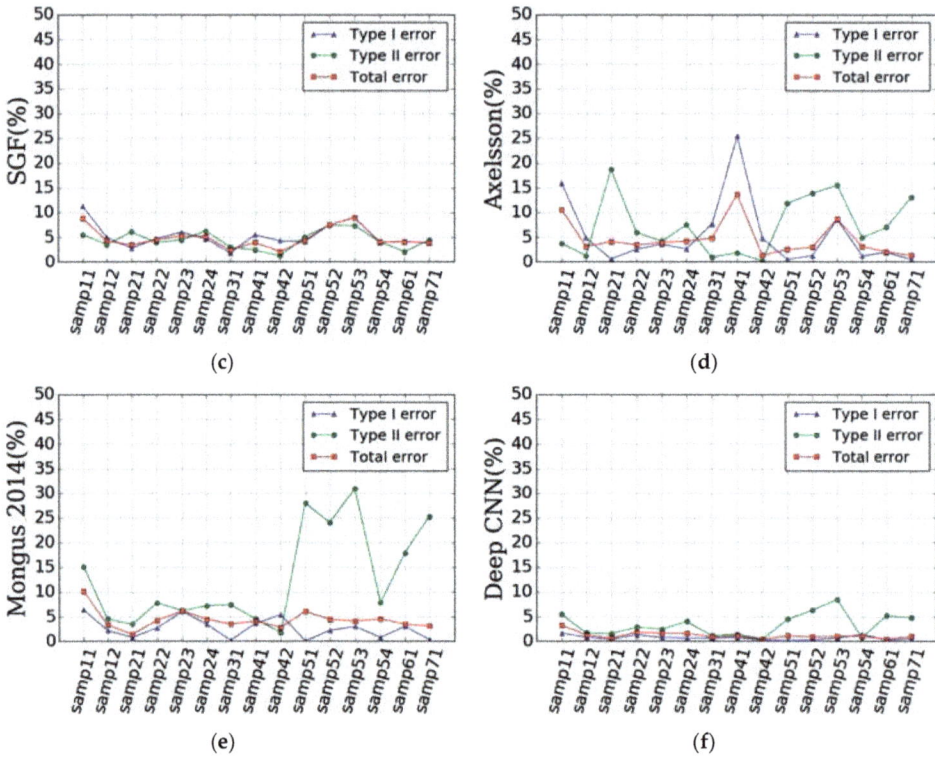

Figure 6. Detailed comparison with other methods and the proposed algorithm across 15 samples in the ISPRS dataset: (**a**) error rates of Terrasan, (**b**) error rates of Mongus 2012, (**c**) error rates of SGF, (**d**) error rates of Axelsson, (**e**) error rates of Mongus 2014, (**f**) error rates of Deep CNN.

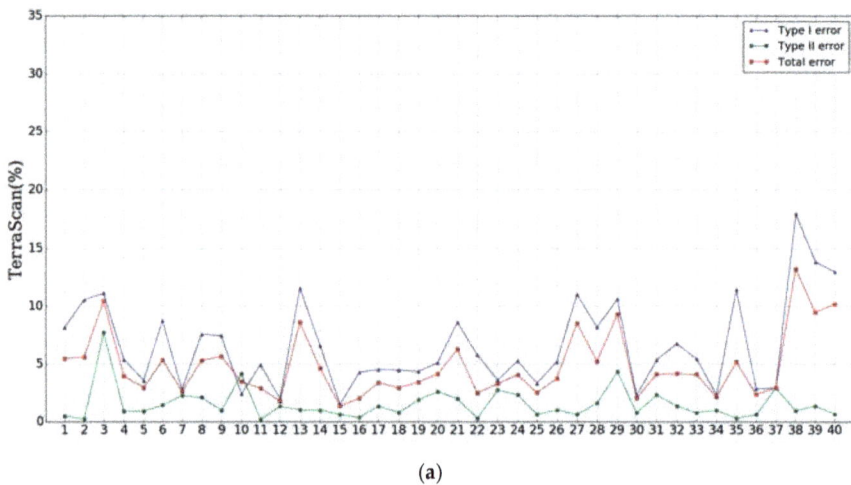

(**a**)

Figure 7. *Cont.*

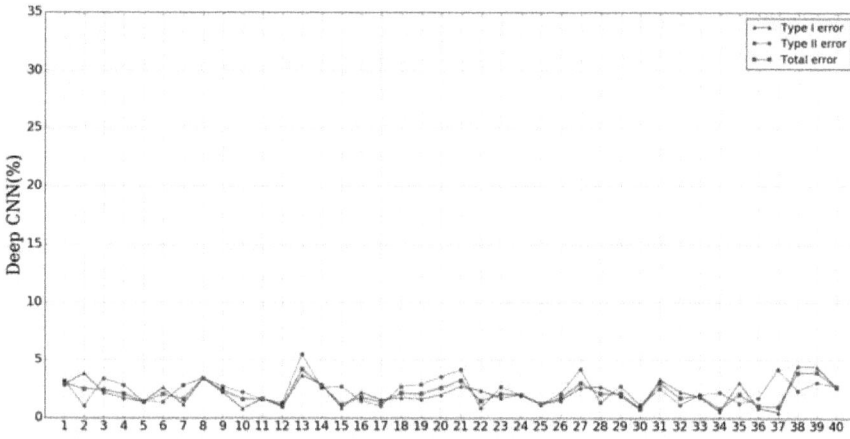

(**b**)

Figure 7. Error rate of TerraScan (**a**) and Deep CNN (**b**) in 40 test cases.

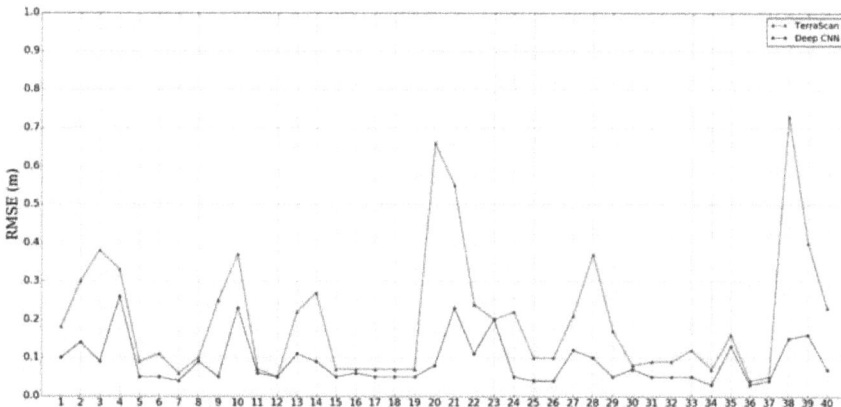

Figure 8. Comparison of root mean square error (RMSE) between the generated DTM with the ground truths.

The comparison of total error over 40 various complex terrains can be seen in Table 2 below and the detailed comparison of several examples with different terrains are shown in Figures 9–12.

Table 2. Comparison of total error over 40 various complex terrains between TerraScan and deep CNN model.

Error	TerraScan	Deep CNN
type I	10.5%	3.6%
type II	1.4%	2.2%
total	6.3%	2.9%

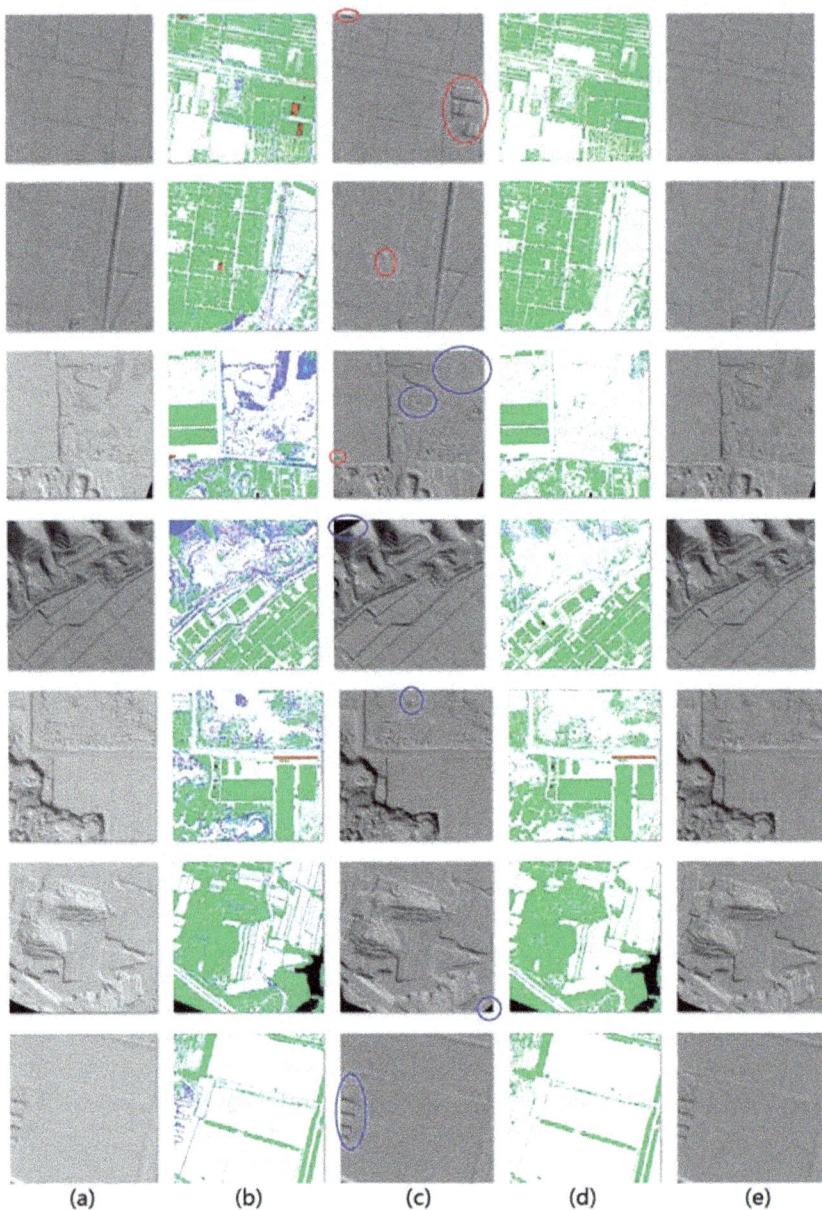

Figure 9. Comparison of the proposed method and TerraScan on the detailed difference of the DTM. Column (**a**) is the ground truth TIN-rendered gray image of the test data. Columns (b,d) are the results of filtration by TerraScan and Deep CNN, respectively; the white points denote correctly classified ground points, the green points denote correctly classified non-ground points, the red points denote accepted non-ground points, and the blue points denote rejected ground points. Columns (**c**) and (**e**) are TIN-rendered DTM extracted from the results of filtration by TerraScan and deep CNN model, respectively. In Column (**c**), blue ellipses denote type I error and red ellipses denote type II error.

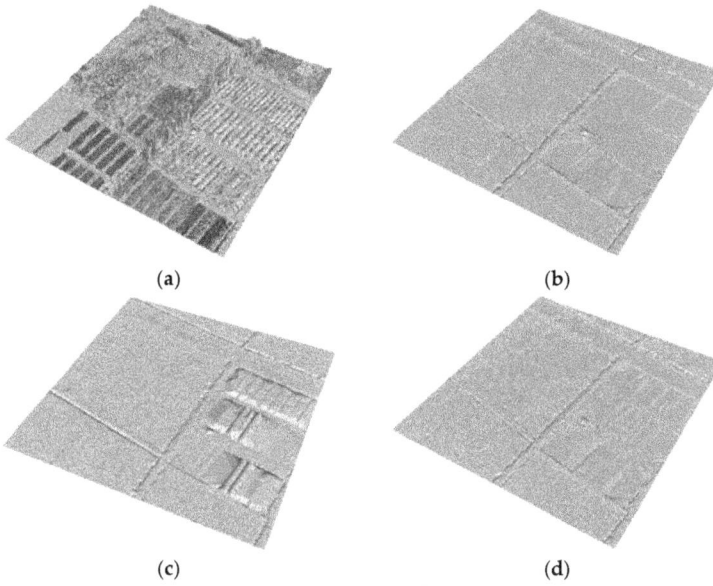

Figure 10. Comparison of the proposed method and TerraScan on the details of the plain area: (a) raw ALS data; (b) ground truth; (c) result of TerraScan; and (d) result of deep CNN model.

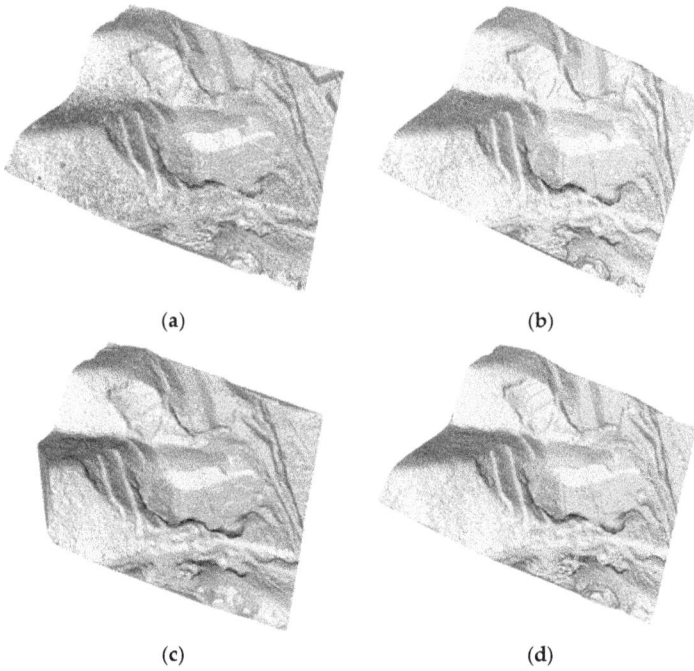

Figure 11. Comparison of the proposed method and TerraScan on the details of the mountain area: (a) raw ALS data; (b) ground truth; (c) result of TerraScan; and (d) result of deep CNN model.

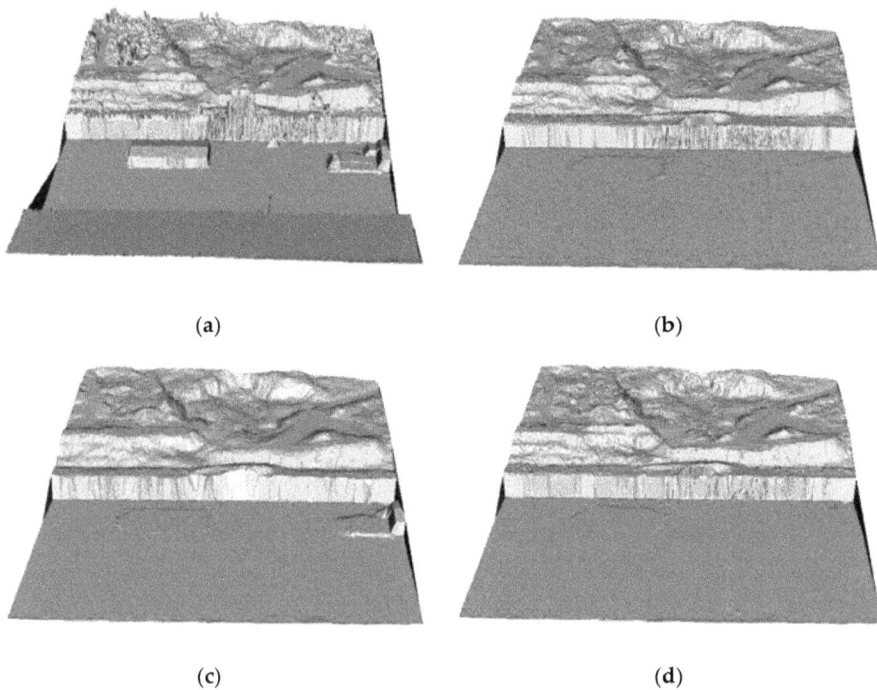

(a)

(b)

(c)

(d)

Figure 12. Comparison of the proposed method and TerraScan on the details of the complex area: (**a**) raw ALS data; (**b**) ground truth; (**c**) result of TerraScan; and (**d**) result of deep CNN model.

The tests strongly indicate that the CNN model produces significantly low Type I error; this result indicates only a slightly tedious manual post-editing for DTM production because removing the non-ground points (Type II error) is usually easier than finding the incorrectly rejected ground points. The proposed CNN-based classification can generate high-quality DTM, particularly to retain subtle, micro, and steep terrains; existing handcrafted algorithms may produce more Type I error.

Figures 10 and 12 show that deep CNN model does well in some big scale non-ground situations which are hard for TerraScan such as buildings and farmlands. Figure 11 shows that deep CNN model conserve the terrain feature well even in the mountains which often has little ground points caused by the shield of trees. The ground hit density in mountain area of Figure 11 is about 1–2 ground points per square meter while the ground hit density in plain area in the same data is about 4–6 ground points per square meter.

However, there are still some cases that deep CNN model cannot deal with very well, as shown in Figure 13.

The CNN model generates some wrong results (type II error) as shown in Figure 13. This is mainly because these wrong non-ground points belong to very low (down to 10 cm) man-made structures, which are too close to the ground. Modifying points-image transformation to represent shape and structure information may improve the accuracy in the similar cases.

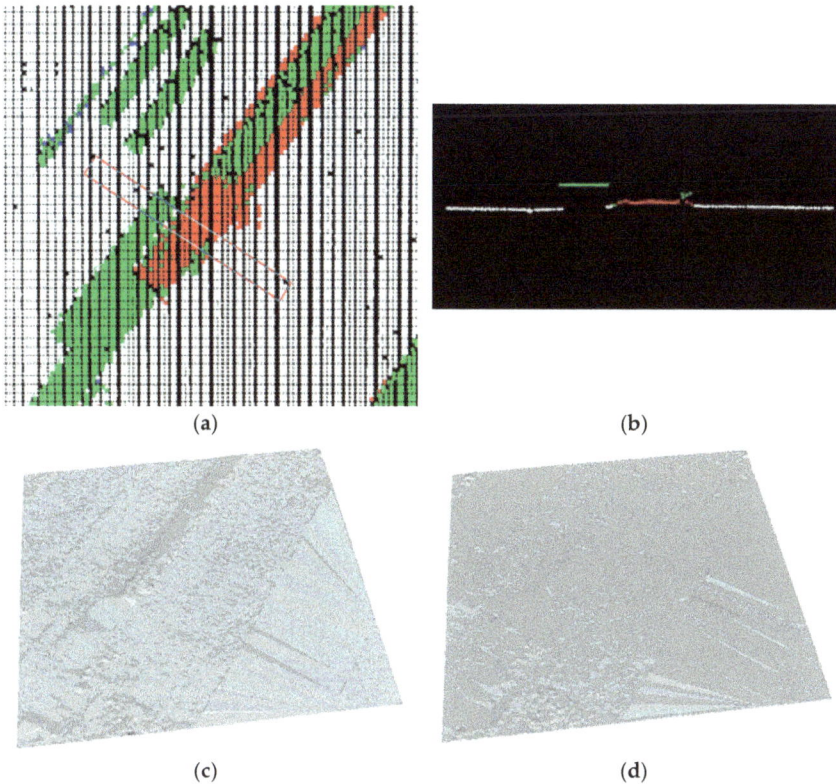

Figure 13. (a) Area that deep CNN model accept many wrong ground points, where the white points denote correctly classified ground points, the green points denote correctly classified non-ground points, the red points denote accepted non-ground points, and the blue points denote rejected ground points; (b) the profile of that area; (c) DTM of that area by deep CNN model; and (d) ground truth DTM of that area. The root mean square error (RMSE) between the DTM by deep CNN model with the ground truth DTM in this section shown below is 0.1 m.

4. Conclusions

To the best of our knowledge, this study is the first one that reports using deep CNN-based classification for DTM extraction from ALS data. Relative elevation differences between each point and its surrounding points are extracted and transformed into an image representing the point feature. Then, the deep CNN model is used to train and classify the images. Each point to be processed can be classified as a ground or non-ground point by the trained deep CNN. A total of 40 ALS point clouds with 40 million points and the ISPRS benchmark dataset with various scene complexities and terrain types are tested using one CNN trained by 17 million labeled points. The results show the high accuracy of the proposed method. The developed method provides a general framework for ALS point cloud classification.

However, the drawback of deep-learning-based methods is that they usually require large labeled data and powerful computational resources. Future work should focus on better point-image transformation and making more compact classification through the deep CNN model to improve the training and classification. Future work should also perform tests on larger datasets.

Acknowledgments: This study was partially supported by the research funding from Guangdong Province (2013B090400008) and Guangzhou City (201508020054) of China. The authors thank Guangzhou Jiantong Surveying, Mapping, and Geographic Information Technology Ltd. for providing the data used in this research project.

Author Contributions: Xiangyun Hu originally proposed using deep CNN for the classification-based filtering and transforming the point context to an image; he also guided the algorithm design and revised the paper. Yi Yuan conducted the detailed algorithm design, implementation, and experimental study. He also wrote the paper.

Conflicts of Interest: The authors declare no conflict of interest.

References

1. Mongus, D.; Žalik, B. Parameter-free ground filtering of LiDAR data for automatic DTM generation. *ISPRS J. Photogramm. Remote Sens.* **2012**, *67*, 1–12. [CrossRef]
2. Vosselman, G. Slope based filtering of laser altimetry data. *Int. Arch. Photogram. Remote Sens.* **2000**, *33*, 935–942.
3. Brzank, A.; Heipke, C. Classification of Lidar Data into water and land points in coastal areas. *Int. Arch. Photogram. Remote Sens. Spat. Inform. Sci.* **2006**, *36*, 197–202.
4. Zhang, K.; Chen, S.-C.; Whitman, D.; Shyu, M.-L.; Yan, J.; Zhang, C. A progressive morphological filter for removing non-ground measurements from airborne LiDAR data. *IEEE Trans. Geosci. Remote Sens.* **2003**, *41*, 872–882. [CrossRef]
5. Axelsson, P. DEM generation from laser scanner data using adaptive TIN models. *Proc. Int. Arch. Photogramm. Remote Sens.* **2000**, *33*, 110–117.
6. Kraus, K.; Pfeifer, N. Determination of terrain models in wooded areas with airborne laser scanner data. *ISPRS J. Photogramm. Remote Sens.* **1998**, *53*, 193–203. [CrossRef]
7. Kamiński, W. M-Estimation in the ALS cloud point filtration used for DTM creation. In *GIS FOR GEOSCIENTISTS*; Hrvatski Informatički Zbor-GIS Forum: Zagreb, Croatia; University of Silesia: Katowice, Poland, 2012; p. 50.
8. Błaszczak-Bąk, W.; Janowski, A.; Kamiński, W.; Rapiński, J. Application of the Msplit method for filtering airborne laser scanning datasets to estimate digital terrain models. *Int. J. Remote Sens.* **2015**, *36*, 2421–2437. [CrossRef]
9. Błaszczak-Bąk, W.; Janowski, A.; Kamiński, W.; Rapiński, J. ALS Data Filtration with Fuzzy Logic. *J. Indian Soc. Remote Sens.* **2011**, *39*, 591–597.
10. Hu, X.; Ye, L.; Pang, S.; Shan, J. Semi-Global Filtering of Airborne LiDAR Data for Fast Extraction of Digital Terrain Models. *Remote Sens.* **2015**, *7*, 10996–11015. [CrossRef]
11. Kubik, T.; Paluszynski, W.; Netzel, P. *Classification of Raster Images Using Neural Networks and Statistical Classification Methods*; University of Wroclaw: Wroclaw, Poland, 2008.
12. Meng, L. Application of neural network in cartographic pattern recognition. In Proceedings of the 16th International Cartographic Conference, Cologne, Germnay, 3–9 May 1993; Volume 1, pp. 192–202.
13. LeCun, Y.; Boser, B.; Denker, J.S.; Henderson, D.; Howard, R.E.; Hubbard, W.; Jackel, L.D. Backpropagation applied to handwritten zip code recognition. *Neural Comput.* **1989**, *1*, 541–551. [CrossRef]
14. Girshick, R.; Donahue, J.; Darrell, T.; Malik, J. Rich feature hierarchies for accurate object detection and semantic segmentation. In Proceedings of the 2014 IEEE Conference on Computer Vision and Pattern Recognition (CVPR), Columbus, OH, USA, 24–27 June 2014; pp. 580–587.
15. Tompson, J.; Goroshin, R.; Jain, A.; LeCun, Y.; Bregler, C. Efficient object localization using convolutional networks. In Proceedings of the IEEE Conference on Computer Vision and Pattern Recognition, Boston, MA, USA, 7–12 June 2015; pp. 648–656.
16. Taigman, Y.; Yang, M.; Ranzato, M.; Wolf, L. Deepface: Closing the gap to human-level performance in face verification. In Proceedings of the Conference on Computer Vision and Pattern Recognition, Columbus, OH, USA, 24–27 June 2014.
17. He, K.; Zhang, X.; Ren, S.; Sun, J. Deep residual learning for image recognition. *arXiv Preprint*, 2015. arXiv:1512.03385.

18. Sermanet, P.; Eigen, D.; Zhang, X.; Mathieu, M.; Fergus, R.; Lecun, Y. Overfeat: Integrated recognition, localization and detection using convolutional networks. In Proceedings of the International Conference on Learning Representations, Banff, AB, Canada, 14–16 April 2014.
19. Waibel, A.; Hanazawa, T.; Hinton, G.E.; Shikano, K.; Lang, K. Phoneme recognition using time-delay neural networks. *IEEE Trans. Acoust. Speech Signal Process.* **1989**, *37*, 328–339. [CrossRef]
20. Collobert, R.; Weston, J.; Bottou, L.; Karlen, M.; Kavukcuoglu, K.; Kuksa, P. Natural language processing (almost) from scratch. *J. Mach. Learn. Res.* **2011**, *12*, 2493–2537.
21. Krizhevsky, A.; Sutskever, I.; Hinton, G. ImageNet classification with deep convolutional neural networks. In Proceedings of the Advances in Neural Information Processing Systems 2012, Lake Tahoe, NV, USA, 3–8 December 2012.
22. Szegedy, C.; Liu, W.; Jia, Y.; Sermanet, P.; Reed, S.; Anguelov, D.; Erhan, D.; Vanhoucke, V.; Rabinovich, A. Going deeper with convolutions. In Proceedings of the IEEE Conference on Computer Vision and Pattern Recognition, Columbus, OH, USA, 24–27 June 2014.
23. Simonyan, K.; Zisserman, A. Very deep convolutional networks for large-scale image recognition. In Proceedings of the International Conference on Learning Representations, Banff, Canada, 16 April 2014.
24. LeCun, Y.; Yoshua, B.; Geoffrey, H. Deep learning. *Nature* **2015**, *521*, 436–444. [CrossRef]
25. Ian, G.; Yoshua, B.; Aaron, C. *Deep Learning*; MIT Press: Cambridge, MA, USA, 2016.
26. Ioffe, S.; Szegedy, C. Batch normalization: Accelerating deep network training by reducing internal covariate shift. In Proceedings of the International Conference on Machine Learning (ICML), Lille, France, 6–11 July 2015.
27. Nair, V.; Hinton, G.E. Rectified linear units improve restricted boltzmann machines. In Proceedings of the 27th International Conference on Machine Learning, Haifa, Israel, 21–24 June 2010.
28. WG III/2: Point Cloud Processing. Available online: http://www.commission3.isprs.org/wg2/ (accessed on 1 September 2016).
29. Mongus, D.; Zalik, B. Computationally efficient method for the generation of a digital terrain model from airborne LiDAR data using connected operators. *IEEE J. Sel. Top. Appl. Remote Sens.* **2014**, *7*, 340–351. [CrossRef]

remote sensing

MDPI

Article

Detecting Terrain Stoniness From Airborne Laser Scanning Data [†]

Paavo Nevalainen [1,*], Maarit Middleton [2], Raimo Sutinen [2], Jukka Heikkonen [1] and Tapio Pahikkala [1]

[1] Department of Information Technology, University of Turku, FI-20014 Turku, Finland; jukhei@utu.fi (J.H.); Tapio.Pahikkala@utu.fi (T.P.)
[2] Geological Survey of Finland, P.O. Box 77, Lähteentie 2, 96101 Rovaniemi, Finland; Maarit.Middleton@gtk.fi (M.M.); Raimo.Sutinen@gtk.fi (R.S.)
* Correspondence: ptneva@utu.fi; Tel.: +358-40-351-8236
† This paper is an extended version of our paper published in Paavo Nevalainen, Ilkka Kaate, Tapio Pahikkala, Raimo Sutinen, Maarit Middleton, and Jukka Heikkonen. Detecting stony areas based on ground surface curvature distribution. In The 5th International Conference on Image Processing Theory, Tools and Applications, 2015.

Academic Editors: Jie Shan, Juha Hyyppä, Lars T. Waser and Prasad S. Thenkabail
Received: 29 June 2016; Accepted: 17 August 2016; Published: 31 August 2016

Abstract: Three methods to estimate the presence of ground surface stones from publicly available Airborne Laser Scanning (ALS) point clouds are presented. The first method approximates the local curvature by local linear multi-scale fitting, and the second method uses Discrete-Differential Gaussian curvature based on the ground surface triangulation. The third baseline method applies Laplace filtering to Digital Elevation Model (DEM) in a 2 m regular grid data. All methods produce an approximate Gaussian curvature distribution which is then vectorized and classified by logistic regression. Two training data sets consisted of 88 and 674 polygons of mass-flow deposits, respectively. The locality of the polygon samples is a sparse canopy boreal forest, where the density of ALS ground returns is sufficiently high to reveal information about terrain micro-topography. The surface stoniness of each polygon sample was categorized for supervised learning by expert observation on the site. The leave-pair-out (L2O) cross-validation of the local linear fit method results in the area under curve $AUC = 0.74$ and $AUC = 0.85$ on two data sets, respectively. This performance can be expected to suit real world applications such as detecting coarse-grained sediments for infrastructure construction. A wall-to-wall predictor based on the study was demonstrated.

Keywords: aerial laser scan; point cloud; digital elevation model; logistic regression; stoniness; natural resources; micro-topography; Gaussian curvature

1. Introduction

There is an increased attention towards classification of the small scale patterns of terrain surface. Recognition of micro-topography may help in arctic infrastructure planning [1], terrain trafficability prediction [2], in hydraulic modeling [3], and in detecting geomorphologic features like in [3,4], and terrain analysis and modelling.

In Finland, a nationwide airborne light detection and ranging (LiDAR) mapping program has provided the means for detecting ground objects with the ground return density $\rho \approx 0.8$ m^{-2}. Since one needs at least one point per stone, and to define the stone radius one needs at least 4 points per stone, this leads to an absolute theoretical detection limit of stone radius $r_{min} = 0.6...1.2$ m. The real limit is naturally somewhat higher. The actual stone sizes fall into this critical range (as discussed in Section 2.2) making the stoniness detection a difficult problem.

One aspect of the ground surface is the presence of stones and boulders, which can be characterized by the stone coverage and by stone size distribution. Mass-flow deposits are recognized by irregular distribution of boulders and stones on their surface. Mass-flow deposits may have regional significance in aggregate production if they occur in fields as they do in the Kemijärvi region in Finland. Mass-flow sediments are often moderately sorted sediments with low fine grained fraction (clay and silt content, <0.006 mm) being less than 12 % [5]. In addition they contain boulders and stones in their sediments which may be crushed for aggregates. Therefore, they are potential aggregates for infrastructure construction. Mass-flow deposits can be detected in a two-step process: First candidate polygons are found by analyzing geomorphological features in a process which can be automated, then surface stone detection based on airborne LiDAR data is performed to limit the set of candidates. Various other geomorphological features like paleolandslides [6], fluvial point bars [7], neotectonic faults [3] and Pulju moraines [8] in Finland and several other types of glacial landforms elsewhere (see summary in [9]) have already been mapped using LiDAR data.

The intent of this paper is to document various methods, which analyze airborne laser scanning data (ALS) or digital elevation model (DEM) to detect stony areas. Our hypothesis is that a direct approach may be able to detect a signal of a target feature like stoniness better than methods using DEM. This is because DEM is a general smoothed representation of the ground surface for generic purposes [10]. This paper focuses on binary classification of the stoniness of sample areas. The approach results in a classifier, which is subjected to 20 m × 20 m point cloud patches to produce a binary mask about stoniness covering whole Northern Finland. Stoniness is just one example of micro-topographic features, which could be detected from public ALS data. Even the positive samples of the data sets focus on stony mass-flow deposits, algorithms are developed for general stoniness detection, which can be later targeted to various specific purposes depending on the available teaching data. It is our hope that the research community finds our results and methods useful in the future.

This paper is an expansion of [11], which studied only one polygon set *data2014* using curvature estimation based on local linear fit (LLC). In comparison to [11], this paper uses an additional data set *data2015*, additional public DEM data format and two additional methods: local curvature estimation based on triangulated surface model computed from LiDAR (LTC) and Laplace filtering of a DEM grid (DEC). LTC uses triangulated irregular network (TIN) produced by a solid angle filtering (SAF). An overview of the relation of computational methods and various data formats can be seen in Figure 1.

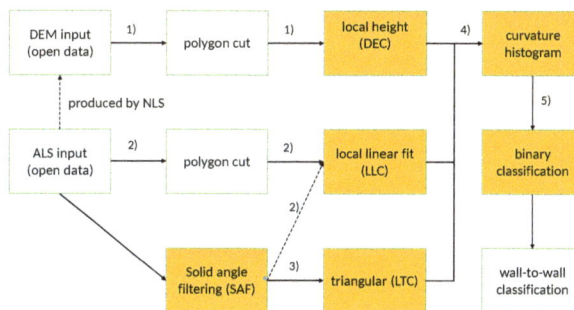

Figure 1. The process flow, methods covered in this paper are highlighted. Data formats: (1) 2 m raster; (2) point cloud; (3) task-specific TIN model; (4) curvature value sets; (5) sample vectors. LLC can optionally use either original point cloud (2) or vertex points (3) produced by SAF TIN model. Wall-to-wall classification is a possibility provided by the resulting binary classifier.

The structure of the paper is as follows: a summary of current research applicable to micro-topographic feature detection is in Section 1. Data sets and data formats are explained in Section 2.2. The solid angle filtering (SAF) can be used by two presented methods, it has been described in Section 2.4. Three methods are documented in Sections 2.5–2.7. Methods are compared in Section 3. Possible improvements are discussed in Section 4 and further application areas considered in Section 5.

Current Research

A good presentation of general-purpose ALS classification is in [12]. Our work relates to some of the contour and TIN -based ground filtering algorithms mentioned in [13], since all of our methods either directly or indirectly use or produce a tailor-made ground model. Methods described in [13] are usually more generic accommodating to infrastructure signatures etc. It is possible, that methods described in our paper have to be combined with existing generic ground model algorithms, where an assembly of methods would use e.g., a voting arrangement at the approximity of constructed environment.

Solid angle filtering (SAF) in Section 2.4 resembles the despike algorithm presented in [14]. Two problems are mentioned in [14]: Unnecessary corner removals (rounding off the vertices of e.g., block-like structures) and effects of negative blunders (false points dramatically below the real surface level). Our routine was specifically designed to eliminate these problems. SAF can also be used in canopy removal. An interesting new technique in limiting the ground return points is min-cut based segmentation of k-nearest neighbors graph (k-NNG) [15]. The graph is fast to compute with space partitioning, and it could have served as a basis for stoniness analysis directly e.g., by fast local principal components analysis (PCA) and local normal estimation with vector voting procedure, as in [16]. The literature focuses mostly on laser clouds of technological environment, where the problem of eliminating the canopy (noise) and finding the ground returns (a smooth technical surface) are not combined. Our experiments with local normal approximation and vector voting were inferior to results presented in this paper. There is great potential in local analysis based on k-NNG, though.

There seems to be no research concerning the application of ALS data to stoniness detection in forest areas. Usually target areas have no tree cover [17], objects are elongated (walls, ditches, archaeological road tracks, etc.) [17,18] and often multi-source data like photogrammetry or wide-spectrum tools are used. curbstones which separate the pavement and road in [18]. Their data has the sample density $\rho = 5/\,m^2$ which produces geometric error of size 0.3 m which is larger than the observed shapes (curbstones) and thus not practical. Effects of foliage and woody debris are discussed in [19]. They mention that even a high-density ALS campaign is not able to get a dense sampling of the ground surface in a non-boreal forest (Pennsylvania, U.S.). They reported ground return ratio is 40% with the ground sample density $\rho = 4/\,m^2$, which is much higher than $\rho \approx 0.8/\,m^2$ in our study. The distribution of the local ground sample density was not reported in [19] but is probably much higher than in our case.

DEM in Figure 1) is a standard data type used by geographic information systems (GIS). Many implementations and heuristics exist (see e.g., [20]) to form DEM from ALS format *.las* defined by [21]. Usually, the smallest raster grid size is dictated by the sample density and in this case DEM grid size $\delta = 2\,m$ is possible, and $\delta = 1$ m already suffers from numerical instability and noise.

A rare reference to DEM based detection of a relatively small local ground feature (cave openings) in forest circumstances is presented in [22]. In that paper the target usually is at least partially exposed from canopy and the cave opening is more than 5 m in diameter. On the other hand, the forest canopy was denser than at our site in general. Another application is detecting karst depressions, where slope histograms [23] and local sink depth [24] were used to detect karst depressions. There are similarities with our study, e.g., application of several computational steps and tuning of critical parameters (e.g., the depression depth limit in [23]), although the horizontal micro-topology feature size is much larger than in our study (diameter of doline depressions is 10–200 m vs. 1.5–6 m diameter of stones in our study). The vertical height differences are at the same range, 0.5–1.5 m in our study and in [23,24], though. A similar study of [25] uses higher density LiDAR data with $\rho = 30\,\text{m}^{-2}$ to detect karst depressions of size 26 m and more. The vertical height difference (depth) was considerably larger than in in [23,24]. The high density point cloud and a carefully designed multi-step process results in quantitative analysis of sinkholes in [25], unlike in our study, where the stoniness likelihood of a binary classifier is the only output.

One reference [19] lists several alternative LiDAR based DEM features, which could be used in stone detection, too. These include fractal dimension, curvature eigenvectors, and analyzing variograms generated locally over multiple scales. Some of the features are common in GIS software, but most should be implemented for stoniness detection.

Hough method adapted to finding hemispherical objects is considerably slower than previous ones, although there is a recent publication about various possible optimizations, see e.g., [26]. These optimizations are mainly about better spatial partitioning.

Minimum description length (MDL) is presented in [27] with an application to detect planes from the point cloud. The approach is very basic, but can be modified to detect spherical gaps rather easily. MDL formalism can provide a choice between two hypotheses: *a plain spot/a spot with a stone*. Currently, there is no cloud point set with individual stones tagged to train a method based on MDL. MDL formalism could have been used without such an annotated data set, but we left this approach for further study. In addition, probably at least 4..8 returns per stone is needed and thus a higher ground return density than is currently available.

This paper presents two methods based on ALS data and one method using DEM and acting as a baseline method. The DEM method was designed according to the following considerations: It has to be easy to integrate to GIS and it would start from a DEM raster file, then generate one or many texture features for the segmentation phase. The possible texture features for this approach are the following:

- local height difference, see Laplace filtering Section 2.7. This feature was chosen as the baseline method since it is a typical and straightforward GIS technique for a problem like stoniness detection.
- various roughness measures, e.g., rugosity (related trigonometrically to the average slope), local curvature, standard deviation of slope, standard deviation of curvature, mount leveling metric (opposite to a pit fill metric mentioned in [19]).
- multiscale curvature presented in [28]. It is used for dividing the point cloud to ground and non-ground returns, but could be modified to bring both texture information and curvature distribution information. The latter could then be used for the stoninesss prediction like in this study. The methods, possibly excluding interpolation based on TIN, seem to be numerically more costly than our approach.

Possible GIS -integrated texture segmentation methods would be heavily influenced on the choices made above. Most of the features listed are standard tools in GIS systems or can be implemented by minimal coding. An example is application of the so called mount leveling metric to stoniness detection, which would require negating the height parameter at one procedure.

Terrain roughness studied in [19] is a concept which is close to stoniness. Authors mention that the point density increase from $\rho = 0.7/\text{m}^2$ to $\rho = 10/\text{m}^2$ did not improve the terrain roughness observations considerably. This is understandable since the vertical error of the surface signal is at the same range as the average nearest point distance of the latter data set. The paper states that algorithms producing the terrain roughness feature have importance to success. This led us to experiment with various new algorithms.

Point cloud features based on neighborhoods of variable size are experimented with in [29]. Many texture recognition problems are sensitive to the raster scale used, thus we tested a combination of many scales, too. According to [16], curvature estimation on triangulated surfaces can be divided to three main approaches:

- surface fitting methods: a parametric surface is fitted to data. Our local linear fit LLC falls on this category, yet does not necessarily require triangularization as a preliminary step.
- total curvature methods: curvature approximant is derived as a function of location. Our local triangular curvature LTC is of this category of methods.
- curve fitting methods.

LLC has a performance bottleneck in local linear fit procedure described in Section 2.5. This problem has been addressed recently in [30], where an algebraic-symbolic method is used to solve a set of total least squares problems with Gaussian error distribution in a parallelizable and efficient way. That method would require modification and experimentations with e.g., a gamma distributed error term due to asymmetric vegetation and canopy returns.

The vector voting method presented in [16] decreases noise and achieves good approximative surface normals for symmetrically noisy data sets of point clouds of technological targets. Our target cloud has asymmetrical noise (vegetation returns are always above the ground), and returns under the ground (e.g., reflection errors) are extremely rare. Usually vector voting methods are used in image processing. They are based on triangular neighborhood and any similarity measure between vertices, focusing signal to fewer points and making it sometimes easier to detect. Neighborhood voting possibilities are being discussed int Section 4.

General references of available curvature tensor approximation methods in case of triangulated surfaces are [31,32]. A derivation of Gaussian curvature κ_G and mean curvature κ_H is in [33]:

$$\kappa_G = \kappa_1 \kappa_2 \tag{1}$$
$$\kappa_H = (\kappa_1 + \kappa_2)/2, \tag{2}$$

where κ_l, $l \in \{1,2\}$ are the two eigenvalues of the curvature tensor. Perhaps the best theoretical overview of general concepts involved in curvature approximation on discrete surfaces based on discrete differential geometry (DDG) is [34].

We experimented with methods which can produce both mean and Gaussian curvatures, giving access to curvature eigenvalues and eigenvectors. Our experiments failed since the mean curvature κ_H seems to be very noise-sensitive to compute and would require a special noise filtering post-processing step. Difficulties in estimating the mean curvature from a noisy data have been widely noted, see e.g., [29].

In comparison to previous references, this paper is an independent study based on the following facts: point cloud density is low relative to the field objects of interest (stones), ratio of ground returns amongst the point cloud is high providing relatively even coverage of the ground, a direct approach without texture methods based on regular grids was preferred, individual stones are not tagged in the test data, and the methods are for a single focused application. Furthermore, we wanted to avoid complexities of segmentation-based filtering described in [35] and the method parameters had to be tunable by cross-validation approach.

2. Materials and Methods

Test data is presented in Section 2.2. It is available online, details are at the end of this paper.

Figure 1 presents the process flow of stone detection. Two data sources at left are introduced in Section 2.2, tested methods (DEC, LLC, LTC) are detailed in Sections 2.5–2.7. The vectorization of samples varies depending on the method in question, details can be found in Section 2.8. The solid angle filtering SAF of Section 2.4 is a necessary preprocessing step for LTC, but could be used also before LLC for computational gain.

2.1. Study Area

The study area is a rectangle of 1080 km^2 located in the Kemijärvi municipality, in Finnish Lapland, see Figure 2. The 675 sample polygons cover approx. 10.7 km^2 of the area. Mass-flow sediment fields such as the Kemijärvi field, have regional significance for aggregate production as there is an abundance of closely spaced mass-flow formations within a relatively short distance from the road network.

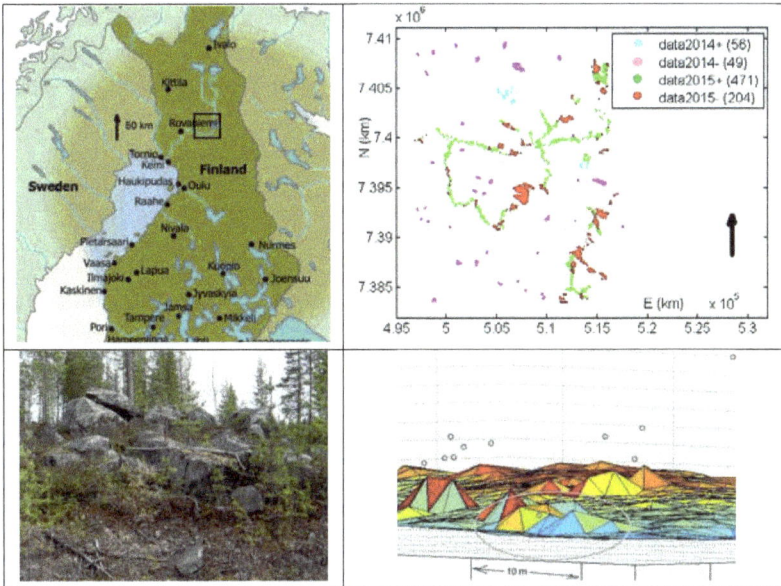

Figure 2. Upper left: The site near Kemijärvi Finland. The research area covered by 120 open data *.las* files covering 1080 km^2. **Upper right**: the relative location of sample polygons. Amount of sample sets in parenthesis. **Lower left**: A view of a sample site in boreal forest. **Lower right**: approximately the same view as at lower left after solid angle filtering (see Section 2.4) of the point cloud. The stone formation has been circled. Location is at UTM map T5212C3, polygon 11240.

2.2. Materials

Table 1 gives a short summary of the data sets: the first set *data2014* is rather small with positive samples occupied by large boulders. The second set *data2015* has an imbalance of many positive samples with smaller stones vs. fewer negative samples. Data sets are depicted in Figure 2. The acquisition of data sets differ: the classification of *data2014* was based on cumulated field photographs and the land survey annotations of the general topographic map (stone landmarks). There is no stone size distribution data available for *data2014*, though. The set *data2015* was classified and the approximative stone size and coverage statistics recorded by a geology expert. The second data set seems to present more difficult classification task—the areas are more varied and stone size probably smaller than in the first set. The advantage of having two data sets of different origin is that the resilience and generality of the methods can be better asserted.

Table 1. Some characteristics of the two data sets.

Data Set	Stony Samples	Area km^2	Non-Stony Samples	Area km^2	Acquisition
data2014	56	1.7	49	1.7	cumulated observations
data2015	471	4.7	204	6.0	field campaign

The data preprocessing consists of the following three steps:

1. All hummocky landforms (i.e., hills) with a convex topographic form were delineated from the ALS derived digital elevation model and its tilt derivative with an Object-Based Image Analysis algorithm developed in eCognition software, see [1]. This step produced *data2014* and *data2015* polygon sets (see Table 1 and Figure 2).
2. A $10\,m \times 10\,m$ space partitioning grid was used to cut both the point cloud (ALS) and DEM to polygon samples.
3. Point cloud was cut to 2 m height from initial approximate ground level. The mode of heights in $2\,m \times 2\,m$ partitions was used as the ground level.

ALS LiDAR data was produced by National Land Survey (NLS) (NLS laser data: http://www.maanmittauslaitos.fi/en/maps-5) in fall 2012 with a Leica ALS50-II laser scanner (Leica Geosystems, St. Gallen, Switzerland), the flight altitude was 2000 m. Last-return data has approx. 0.09–0.1 m vertical ground resolution and average footprint of 0.6 m. ALS data has several additional information fields per cloud point, see e.g., [21]. We used only x-y-z components of the data. Approximately 25% of the data are canopy returns, the rest is ground returns. Reflection errors causing outlier points occur approximately once per 0.5×10^6 returns.

DEM data is 2 m regular grid data available from NLS. It is nationwide data aimed for general purposes (geoengineering, construction industry). Its vertical accuracy is 0.3–1.0 m std. Both ALS and DEM data were cut to polygon samples by using $10\,m \times 10\,m$ space partitioning slots. See two example polygon shapes in Figure 3. The further processing focused only to the point cloud limited by each polygon sample.

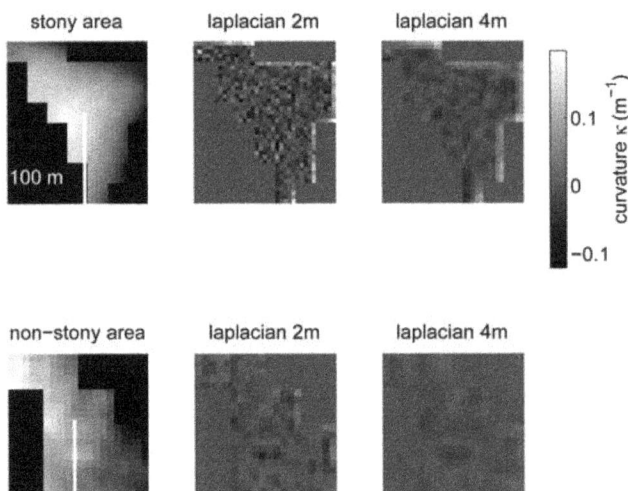

Figure 3. A stony (upper row) and a non-stony (lower row) sample polygon. Original polygons are approximated by $10\,m \times 10\,m$ batches. The ground height (DEM 2 m) and its Laplace discrete operator signals with 2 m and 4 m radius are depicted. The border noise has been removed from actual analysis. The 100 m scale is aligned to North.

Stones are bare and the vegetation is thin due to high Northern latitudes. The point cloud on this site has approx. 25 % returns to forest canopy and approx. 75 % ground returns, so the ground signal is rather strong. The reflection errors were extremely rare, approx. 1 per 10^6 returns. Together the sample sets represent rather well the Kemijärvi study area.

2.3. Materials online

Sample data sets including some polygon point clouds, DEM data of two map pages, sample vectors, some field images, SAF algorithm document, a short description of the data set and the problem are available at: http://users.utu.fi/ptneva/ALS/.

2.4. Solid Angle Filtering (SAF)

The proposed solid angle filtering is a novel method to produce an alternative TIN DEM sensitive to stones and boulders on the ground. Filtering starts by forming an initial TIN either from a full point cloud or after an industry standard preliminary canopy and tree point elimination. The cut was made 2 meters above the local mode of the point cloud height. The 2D projection of TIN satisfies Delaunay condition at all times during the iterative process of point elimination. The prominent 'pikes' in the intermediate TIN are removed in random order while the Delaunay triangulation is updated correspondingly. The implementation requires a dynamical Delaunay algorithm, which facilitates incremental removal of points. We used an industry standard approach described in [36] with $O(k)$ computational complexity per removed point, where k stands for the average number of nearest neighbors of a point.

A second iterative phase removes 'pits' in a similar fashion. The prominence of pikes and pits is measured by solid angle Ω_k, which is the spatial angle of the surrounding ground when viewed from a TIN vertex point p_k. Appendix A provides the technical definition of computing Ω_k.

Each state of TIN is achieved by dropping one point which fails the following inequality:

$$\Omega_{min} \leq \Omega_k \leq \Omega_{max}, \tag{3}$$

where solid angle limits $\Omega_{min} = 1.80$ sr (steradians) and $\Omega_{max} = 12.35$ sr correspond to solid angles of two spherical cones with opening angles 89 $^\circ$ and 330 $^\circ$, respectively. The choice affects the prediction performance: if both limits are close to a planar situation of $\Omega \approx 2\pi$, there is a loss of points. If there are no limitations ($\Omega_{min} \equiv 0$, $\Omega_{max} = 4\pi$), data is dominated by the noise from canopy and tree trunks. The solid angle limits were defined by maximizing the Kolmogorov-Smirnov (K-S) test [37] difference using 95 % confidence limit. Figure 4 depicts the difference between solid angle distributions at positive and negative sample sets at the choice we made. A pike at approx. 2.5 sr indicates stones.

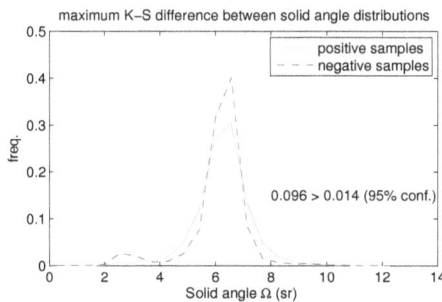

Figure 4. The solid angle distribution of positive and negative samples among the *data2015* data set. Averages can be distinguished well but variation among samples is high.

The resulting ground surface triangularization (see lower left part of Figure 2) resembles the end product of the 3D alpha shape algorithms [38], when alpha shape radius is chosen suitably. It produces an alternative TIN model which hopefully contains a signal needed in stone detection. In this paper this method is used as a preprocessing step for LTC method (see Section 2.6) and for LLC method (see Section 2.5, and Figure 1).

2.5. Curvature Estimation Based on Local Linear Fit (LLC)

LLC method is based on a curvature approximation method described in [33]. The method requires surface normals to be available at the triangle vertices. LLC provides these normals by a local linear fit to the point cloud at a regular horizontal grid. Since the resulting curvature function is not continuous at the border of the triangles, a voting procedure is needed to choose a suitable value for each grid point.

Finding the local planes is a similar task to finding the moving total least squares (MTLS) model in [39]. The differences are the following:

- a space partitioning approach is used instead of a radial kernel function to select the participant points. This is because ground surface can be conveniently space partitioned horizontally unlike in [39], where the point cloud can have all kinds of surface orientations.
- the point set is not from constructed environment. Canopy returns create a 3D point cloud, thus the loss function cannot be symmetrical, but must penalize points below the approximate local ground plane.

The LLC process has 6 steps, which are expounded in Appendix B. Step 1 is cutting the foliage dominated part of the point cloud, step 2 approximates ground with local linear planes at regular grid points. Step 3 spans the grid with triangles avoiding spots with missing data. Step 4 defines the curvature within each triangle. Step 5 combines the curvature values of the neighboring triangles to each grid point. Step 6 is about forming a histogram over the whole grid of the sample polygon.

LLC is a multi-scale method like [28]. Steps 1 through 6 are repeated with differing grid lengths $\delta_j, j \in [1,6]$ of the grid, see Table 2. There is a potential danger for overfitting, so the qualities of grid sizes are discussed here from that point of view.

Table 2. Square grid sizes used in local linear fit of LLC method.

Grid Version	1	2	3	4	5	6
Grid constant δ_m (m)	1.25	2.0	3.0	4.0	5.0	6.0

The smallest grid size $\delta_1 = 1.25$ m has approx. 85 % of the grid slots with only 1 to 3 points as shown in the left part of Figure 5 and so it represents a practical low limit of the local planar fit. A practical upper limit is $\delta_6 = 6$ m because only the largest boulders get registered on this scale. Such large boulders are few, as shown in the right part of Figure 5.

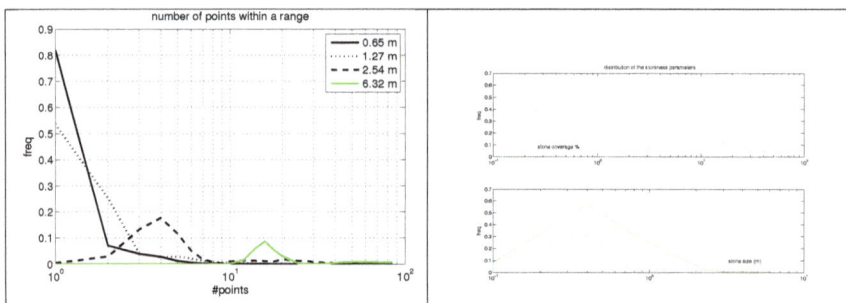

Figure 5. Approximative properties of *data2015* data set. Similar qualities of *data2014* are not available. **Left**: The number of stones at a spatial partition when the partitioning range (the grid size δ) changes. A sensible approximation of e.g., local ground inclination is possible only when there are at least 3 points per grid square. **Right**: The difference between positive and negative samples is mainly in stone size distribution. The practical detection limit in size is approx. 1.0 m.

2.6. Local Curvature Based on Ground Triangularization (LTC)

The input is a triangulated surface model generated by SAF method described in Section 2.4 and Figure 1. LTC produces the curvature estimates directly to the vertex points, so local linearization step 3 of the LLC method of Section 2.5 is not needed. Vectorization (step 4 of LLC) has to be done, but only once, since there are no grids nor multi-grids in this method. The idea is to calculate the value κ_k of a Gaussian curvature operator at each ground point p_k based on the Gauss-Bonnett theorem as described in [31].

The right side of the Figure A1 depicts a point p_k and its neighboring points p_i and p_j. The neighboring triangles T_k of a vertex point k define a so-called spherical excess, which is the difference between the sum of triangular surface angles β_{ikj} and the full planar angle 2π. Now, one can write an estimator for the Gaussian curvature κ_k at point p_k based solely on local triangularization formed by the Delaunay process:

$$\beta_{ikj} = \operatorname{acos}(p_i - p_k, p_j - p_k)$$

$$A_k = \sum_{t \in T_k} A_t / 3 \tag{4}$$

$$\kappa_k \approx \left(2\pi - \sum_{(i,k,j) \in T_k} \beta_{ikj}\right) / A_k, \tag{5}$$

where β_{ikj} is the angle at vertex k in a surface triangle $t = (i, k, j) \in T_k$, $\operatorname{acos}(.,.)$ is the angle between two vectors defined in Equation (A1) in Appendix A, and A_k is a characteristic surface area associated with the vertex k. The characteristic area has been defined approximately by taking one third of the area A_t of each adjoining triangle t. There are locally more stable but also more complicated ways to calculate A_k, see e.g., [31,32]. The choice made in Equation (4) causes noise because the area is approximate but seems to allow effective histogram vectors.

2.7. Curvature Based on Filtering DEM by a Modified Discrete Laplace Operator (DEC)

The third method is traditional, fast and easy to implement in the GIS framework and thus provides a convenient baseline for the previous two methods. Local height difference is converted to local curvature approximant. Curvature histograms are then vectorized as in previous methods.

DEM data with a regular grid with the grid size $\delta = 2.0$ m was utilized. Data is publicly available over most of Finland. The discrete 2D Laplace operator with radius $r_{horiz} = 2.0$ m is well suited for detecting bumpy features like stones at the grid detection limit. It simply returns the difference between the average height of 4 surrounding grid points and the height of the center point. A modified Laplacian filter with $r_{horiz} = 4.0$ m (length of two grid squares) was used to estimate the local height difference on the larger scale, see Figure 6. A postprocessing transformation by Equations (7) and (8) was applied to produce correspondence to Gaussian geometric curvature κ_k at point k. A geometric justification for the transformation is depicted in Figure 6. A stone is assumed to be a perfect spherical gap with perfect horizontal surrounding plane. The mean curvature κ_H can be approximated from the observed local height difference of Equation (6) by using the geometric relation Equation (7), see Figure 6. $\bar{z}(r_{horiz})$ is the average height at the perimeter of horizontal radius r. The local height difference $\bar{Z}$ is the key signal produced by Laplacian filter. Gaussian curvature is approximately the square of the mean curvature, when perfect sphericality is assumed, see Equation (8). The sign of the Gaussian curvature approximant κ_{Gk} at point c_k can be decided on the sign of the height difference $\bar{Z}$ at vertex k. The index k of the vertex point p_k is omitted for brevity. Equation (7) comes from rectangular triangle in Figure 6.

$$\bar{Z} = z - \bar{z}(r) \qquad \text{local height difference} \tag{6}$$

$$1/\kappa_H^2 \approx r_{horiz}^2 + (1/\kappa_H - \bar{Z})^2 \qquad \text{approximate mean curvature condition} \tag{7}$$

$$\kappa_H \approx 2\bar{Z}/(\bar{Z}^2 - r_{horiz}^2) \qquad \text{mean curvature solved from Equation (7)}$$

$$\kappa_G \approx -\operatorname{sign}(\bar{Z})\kappa_H^2 \qquad \text{approximate Gaussian curvature} \tag{8}$$

Figure 6 shows grid squares of size $2\,\text{m} \times 2\,\text{m}$. Points A are used to calculate the average height $\bar{z}(2.0\,\text{m})$ and points B the average height $\bar{z}(4.0\,\text{m})$.

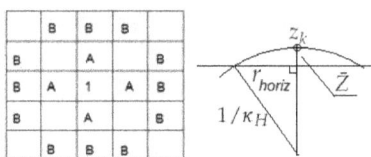

Figure 6. Left: The Laplace difference operator returns the height difference between the center point (1) and the average of points A. The modified Laplace difference operator does the same but using points B. These two kernels define each an average circumferential height difference $\bar{Z}$. **Right**: The geometric relation between $\bar{Z}$ and approximate mean curvature κ_H. Horizontal line represents average ground level at the circumference.

Many sample polygons are relatively small. The above described difference operator produces numerical boundary disturbance, see Figure 3. This can be countered by limiting the perimeter average $\bar{z}(r)$ only to the part inside the polygon, and then removing the boundary pixels from the histogram summation.

The next step is to build the sample histograms as with other methods. Histogram vectors from two filters are concatenated to produce a sample vector of a polygon. The details of forming the histogram are given in Section 2.8.

2.8. Vectorization

All three methods produce histograms of Gaussian curvature $\kappa_G = \pm 1/r^2$, where r is the local characteristic curvature radius and the curvature sign has been chosen in Equation (8) so that potential stone tops have negative curvature and "pit bottoms" have positive curvature. An ideally planar spot k has curvature radius $r_k \equiv \infty$ and curvature $\kappa_G \equiv 0$. Recall that minimum detectable stone radius is approx. $r_{min} = 0.6...1.2$ m, which leads to a Gaussian curvature interval of $\kappa_G \in = [-1.8, 1.8]$ m^{-2}. This range was spanned by histogram bins. LLC and DEC are rather insensitive to bin choice, so a common ad hoc choice was made for these methods, see Table 3. The LTC method proved sensitive to bin choices, so the values were derived using a subset of 10 positive and 10 negative samples in leave-pair-out cross-validation. This set was excluded from later performance measurements.

Table 3. Curvature histogram bins.

Method	Positive Half of the Bin Values
LLC and DEM	0.010, 0.030, 0.060, 0.13, 0.25, 0.50, 1.0, 2.0
LTC	0.031, 0.12, 0.25, 0.44 ,0.71, 1.13, 1.8

The histogram creates a vector representation $\mathbf{x}_i$, $i = 1..n$ for all sample polygons i. The LTC method produces one histogram vector, the DEC method produces two vectors (for $r = 2$ and $r = 4$ m) and LLC produces 6 vectors (for 6 different grids), which are then concatenated to form the final sample vector $\mathbf{x}_i$.

Figure 7 provides a summary of average curvature distributions produced by each of the three methods. The planar situation with $\kappa_G \equiv 0$ is the most common. Occurrences with characteristic radius $r < 1$ m are very rare. Useful information is contained within the range $\kappa_G \in [-1, 1]$ m^{-2}. With LLC method, grid size $\delta = 2$ m is able to detect greater curvatures and grid size $\delta = 5$ m is the last useful grid size. DEM is remarkably similar to 2 m LLC grid, which was to be expected.

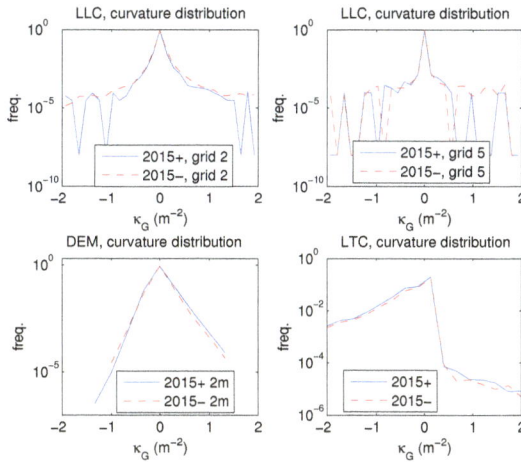

Figure 7. Curvature distributions produced by each method. **Upper left**: LLC and grid size 2m. **Upper right**: LLC and grid size 4 m. Larger grid size results in narrow band around $\kappa = 0$. **Lower left**: DEM curvatures are characterized by kurtosis. **Lower right**: the LTC distribution.

Negative samples (dashed lines) have about the same amount of exactly planar samples, but are a little bit more slightly curved ($|\kappa_G| \approx 0.4$ m^2) samples. This probably results from the basic ground curvature distribution. If the presented three methods were to be adapted elsewhere, the changing background curvature spectrum may result in changes in the prediction performance.

The LTC method is able to detect the negative curvature around the stone causing the curvature distribution to be asymmetric. Unfortunately, this method is also very noise-sensitive reducing its performance.

Each method requires more testing and especially test data with known stone properties (relative coverage of the surface, radius and height distribution, individually located and labeled stones).

2.9. Logistic Regression

The label vector $y_i \in \{-1, 1\}$, $i \in D = 1...n$ was acquired by field campaign done by a geology expert. D is the index set of the full sample set, size $n = |D|$ varies depending e.g., on the different sensitivity to sparse point cloud of each method. The sample vectors $x_i \in \mathbb{R}^d$ are produced by histogram vectorization described in Section 2.8. Dimensionality d varies depending on the method. We use the affine form $z_i = (1, x_i)$ to shorten the following treatise. Vectors $\{z_i\}_{i \in D}$ are also standardized before solving the regression problem.

This is a qualitative response problem, so logistic regression was chosen to predict a label $\hat{y}$ from a given sample vector x. The prediction coefficient $\beta \in \mathbb{R}^{d+1}$ is tuned by usual maximum likelihood approach to optimal value $\beta^*_{D'}$ using a sample set $\{(z_i, y_i)\}_{i \in D'}$, $D' \subset D$ where D' is the training set used:

$$f(z_i, \beta) = Pr(y_i = 1 \mid z_i) = (1 + \exp[-\beta \cdot z_i])^{-1}$$
$$\hat{y}^{(D')}(z) = \begin{cases} 1 & \text{when } f(z, \beta^*_{D'}) \geq 1/2 \\ -1 & \text{otherwise} \end{cases} \tag{9}$$

The area under curve (AUC) performance measure is natural in this application area, where cost functions do exist but are not exactly known. The sample set *data2014* is rather small and the sample set *data2015* has imbalanced samples of positive and negative cases and the n-fold cross-validation may produce too optimistic estimates, as mentioned in [40]. It is recommended in [40] in this case:

- to perform a leave-pair-out (L2O) test over all possible positive-negative label pairs P, and
- to measure L2O area under curve AUC by using the Heaviside function $H(.)$ for summation.

$$
\begin{aligned}
P &= \{(i,j)\,|\,y_i = 1,\, y_j = -1,\, i,j \in D\} &\text{all possible (+,–) pairs} \\
AUC &= \sum_{(i,j)\in P} H(\hat{y}_{ij}(\mathbf{z}_i) - \hat{y}_{ij}(\mathbf{z}_j))/|P| &\text{leave-pair-out AUC} \\
\hat{y}_{ij}(\mathbf{z}) &= \hat{y}^{(D\setminus\{i,j\})}(\mathbf{z}) &\text{prediction without a pair } i,j \text{ based on } \mathbf{z} \\
H(\Delta\hat{y}) &= (1 + \text{sign}(\Delta\hat{y}))/2 &\text{Heaviside over the prediction difference}
\end{aligned}
\tag{10}
$$

2.10. Method Parameters and Design Choices

Some parameters were optimized by nested cross-validation or K-S test and thus settled. Some parameters are ad-hoc built-in parameters, values of which are chosen mostly during the coding process. These should be taken into consideration if the methods are utilized under slightly different conditions. A list of the open parameters, their potential range and some of the discrete design choices available, follows. Table 4 is a summary of the treatise.

Table 4. Effective method parameters, a summary.

Method	Parameters	Binary Choices
SAF	2	0
LLC	3–15	63
LTC	0	1
DEC	2	1

LLC: There are 11 non-zero shape parameters of the planar distance weight function $g(.)$ presented in [11]. The validation of the choices made will be a separate publication. There are some minor design choices, an example is Equation (B2): one can use either median or mean rule in composing curvature from surrounding triangle vertices, and results do not change noticeably. Median rule was used to reduce occasional outliers. Another example is the 6 grid sizes in Table 2. The number of possible subsets of grids to be used equals $2^6 - 1 = 63$.

LTC: There is a design choice of using the local surface area $A_t/3$ in Equation (4) or a more complex definition given in [32]. This is listed as one binary choice in Table 4.

DEC: There is a binary choice of either choosing Laplacian filter signal $\tilde{Z}$ or the Gaussian approximant κ_k of Equation (8) based on the signal $\tilde{Z}$.

2.11. General Wall-to-Wall Prediction

Methods presented in Sections 2.4–2.9 were applied only to given polygon areas, since teaching is possible only where the response value is known. But after the parameters of predictor have been settled, the area to be inspected can be a generic one. As a demonstration and speed test, we applied methods to a 1080 km^2 area divided to 20 m $\times$ 20 m pixels with approx. 320 points from a point cloud of density $\rho = 0.8\,\text{m}^{-2}$. Pixels have 6 m overlapping margins to increase the sample area to 32 m $\times$ 32 m (approx. 820 points) to avoid partially populated histograms, which would not be recognized correctly by the classifier. See Figure 8 for the DEC method wall-to-wall result.

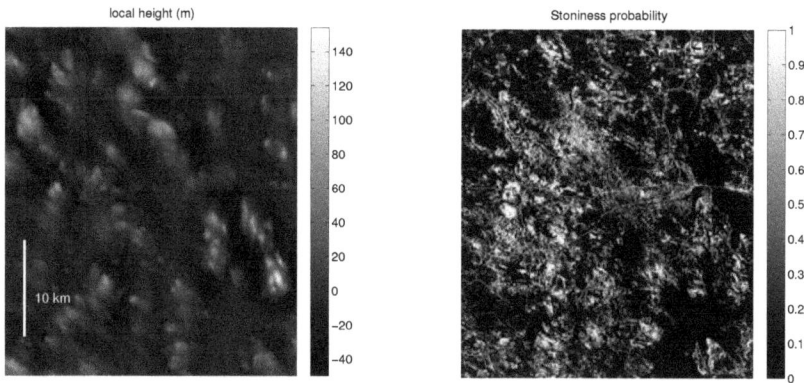

Figure 8. Left: The local height from DEM files, 30 km × 36 km area depicted. The scale is oriented northwards. The general location of the rectangle can be seen in upper left part of the Figure 2. **Right**: Stoniness probability by DEC method. The scale is probabilty of having stones on a particular pixel. Roads and waterways are classified as stony areas. LLC and LTC methods are much less sensitive to roads and constructed details.

3. Results

Binary classification of stoniness was done by logarithmic regression over curvature histogram vectors cumulated over each sample polygon area. Three methods were used; they differ on how the curvature approximants were produced(Table 5):

- Local linear fitting (LLC) divides the polygon into 6 different grids. Each grid square is fit by a plane approximating the local ground height of the center of the plane and the plane orientation. Curvatures are computed from these center points and their orientation normals.
- Curvature from DEM (DEC) uses traditional DEM data. Curvatures are approximated by the observed local height difference delivered by a modified discrete Laplace operator.
- Curvature by local triangulation (LTC) has a TIN computed by SAF method of Section 2.4. The curvature is then computed triangle by triangle as in LLC.

Area under curve AUC [41] was measured using both data sets and all three methods, see Section 2.9. The AUC measure describes the discriminative power of a predictor over various possible cutoff point choices. A proper cutoff depends on the costs involved and is not known at the moment, justifying the accommodation of AUC. Leave-pair-out variant of AUC was used, see considerations for this choice in Section 2.9.

Table 5. Leave-pair-out AUC results based on three methods used: digital elevation model, local linear fit and local triangular curvature for two polygon sample sets.

Data Set	DEC	LLC	LTC
data2014	0.85	0.82	0.79
data2015	0.68	0.77	0.66

LLC proved best for *data2015*. This data set is large and perhaps more representative of the locality, and the performance $AUC = 0.77$ can be considered adequate for practical application such as pre-selecting possible gravel deposit sites for infrastructure construction. Its performance is also on par with many hard natural resource prediction tasks based on open data, see e.g., [2].

Data set *data2014* is somewhat exceptional, since it contains larger boulders and seems to be an easy prediction task for wide array of methods. Both DEC and LLC performed well. The same holds to several other tested methods which have not been included to this report, e.g., neighborhood voting based on solid angle values.

DEC performance is mediocre with *data2015* set because the average stone size depicted in Figure 5 right side is actually below the theoretical detection limit of a regular 2 m grid. Established DEM computation routines are a trade-off of many general-purpose goals and some of the stone signal seems to be lost in the case of the *data2015* set.

LLC, eventhough it was cross-validated with *data2014*, performed adequately here. There were high hopes about local curvature based on triangularization (LTC), but it performed the worst. This is because LTC computes the curvature directly from a TIN, and the process produces a lot of noise. LTC has been included in this report mainly because the method is fast to compute and there seems to be potential to reduce noise in the future by neighborhood voting methods.

The processing speed (see Table 6) has a linear dependence with the area analyzed. This is because the analysis is done by space partitions of constant point cloud sample size n. All the steps have a linear $O(n)$ time complexity except the Delaunay triangulation in SAF. The point removal phase in SAF is of $O(kn)$ complexity, where $k \approx 5...20$ is the amount of nearest neighboring points. The experiments were run on a desktop computer with Intel Core i5-3470 CPU (3.20 GHz) running Ubuntu Linux 14.10. LLC implementation requires several intermediary file operations, which makes it slow. All implementations are experimental prototypes and many speed improvements are still possible.

Table 6. Analysis speed computed from average of two runs over the data set *data2015*.

Analysis Speed	DEC	LLC	LTC
km^2/h	200	0.5	4.0

4. Discussion

A traditional approach for terrain micro-topography classification is to use DEM model as a basis for a wide array of texture methods. The low end has a simple texture feature computation followed by segmentation tuned manually by an expert. The high end has several texture features extracted, and preferably at least two DEM models of different grid size as a basis for analysis.

This paper presents a way to use the existing ALS LiDAR material to construct an alternative task-specific terrain surface representation which hopefully contains more information e.g., concerning the presence of stones. All methods presented are conceptually simple, although documenting and coding LLC and LTC brings up a multitude of details and ad hoc choices, see e.g., the number of method parameters in Table 4. Each method has potential for further improvement by a more thorough parameter tuning. SAF and LLC have enough method parameters that tuning by new field campaign data at more southern boreal forests could succeed. More southern boreal forest provides a challenge since the ratio of the ground returns is only 30%–60% instead of 70% at our Northern test site.

LLC and DEC perform well enough to be practically usable. The direct utilization of ALS data seems to work on this site.

Because LTC is based on a TIN model, there is available additional geometrical information like mean curvature, curvature eigenvalues and eigenvectors etc. for more complex micro-topographic features. If it is possible to reduce the noise and keep the computation costs at the current low level, a combination of these features could be a basis for fruitful multi-layer texture analysis.

Many terrain micro-topography classification tasks e.g., registering post-glacial landslides, karst depresssion detection and fault lines detection e.g., can be done with DEM and by texture methods, but there may be a need to add stoniness or curvature related features to improve the classification. The current data sets do not provide accurate quantitative information about stoniness for regression methods. The wall-to-wall stoniness result with 20 m $\times$ 20 m pixels produced by current binary classification (see Figure 8) can be utilized as an additional feature in other prediction problems in the future. The wall-to-wall pixel size must be increased when the ground return ratio decreases in the southern dense forests.

Three individual methods or a combination of them can be modified to produce estimates of the relative stone coverage and stone size distribution. This step can be taken only if data sets come available with individual stones and their properties tagged out. Furthermore, more sophisticated probabilistic and minimum description length (MDL) based methods would then be possible. The stone coverage and size distribution information in Figure 5 is approximative only, so current data sets cannot be used for development of quantitative stoniness models.

It is hard to estimate how far to southern forests the three methods can be extended. The ground return density varies with boreal forests, and detection results from spatially rather sparse accessible spots should be somehow extended to nearby areas using other available public data and Machine Learning methods. This line of research requires specific field test sets, though.

There are several other micro-topological problems, e.g., classifying and detecting undergrowth, marshland types, military structures, unauthorized inhabitation and geomorphology e.g., frost phenomena. Some of these require comparison of two snapshots from ultra-light vehicle (UAV) scans, and e.g., a combination of SAF and MDL might perform well in this scenario. We believe SAF has wide adaptivity to several purposes by tuning its 2 parameters (minimum and maximum spatial angle allowed) and MDL can be built to detect a specific shape (a hemisphere in cases of stones, a box, a prism or a cylinder in case of other applications). As mentioned before, MDL would require denser point clouds with $\rho \approx 1.6...3.2\,\text{m}^{-2}$.

5. Conclusions and Future Research

Results in Section 3 show that LLC performs better than DEC but is numerically much more expensive. LLC seems to be robust and useful when the computation costs can be amortized over several subsequential analyses. LTC performed worst but there is room for improvement as discussed in Section 4. Both LLC and DEC are ready to be applied to industrial purposes after prototyping implementations are upgraded to production code.

Current results are bound to stoniness of mass-flow deposits what comes to teaching data, but each method should work in generic stoniness detection, if such a need arises and general teaching data sets become available.

Using direct ALS information either as an alternative data source or supplementary one may help solving a variety of micro-topography detection problems better in the future. The research efforts will be focused on the following topics:

- Extending the analysis to more dense forests, where stoniness detection occurs only at benevolent cicumstances (forest openings, sparse canopy, hilltops). In this environment the acquired stoniness signal has to be combined to a wide array of open data features to extend prediction to unobservable areas. The corresponding field campaigns will be more elaborate.
- Taking into account the stone coverage and size distribution. It is likely that a multi-grid method like LLC might perform well in this prediction task (given suitable teaching data), whereas DEC may be restricted by the general purpose nature of DEM and its modest grid size.
- Topography and vegetation classification of marshlands. Marshlands have similar high ground return ratio as the current case site. SAF can be tuned by cross-validation to produce a tailored TIN and an improved LTC method with added curvature properties (mean curvature, curvature eigenvectors) could detect various micro-topographic marshland features. It is our assumption that the histogram approach would work also with marshland classification, given a suitable teaching polygon quality produced in field campaigns.
- Using min-cut based segmentation of k-NNG graph of ALS data as described in [15] instead of simple Delaunay triangulation. One has to modify the algorithm to include neighborhood voting to reduce noise. This could be a fruitful approach, since it could suit to 3D analysis of forest tree species, providing more motivation for the implementation.

- Utilizing all relevant LiDAR attribute fields, like return intensity, return number, the scan angle etc. (see [21]).

Acknowledgments: Department of Information Technology of University of Turku covered the travel and publication costs. English language was revised by Kent Middleton.

Author Contributions: Maarit Middleton designed and conceived the sample polygon sets; Raimo Sutinen conceived the research problem; Paavo Nevalainen designed the methods and analyzed the data; Paavo Nevalainen wrote the paper; Jukka Heikkonen contributed to the paper; Tapio Pahikkala contributed to the performance analysis.

Conflicts of Interest: The authors declare no conflict of interest.

Abbreviations

The following abbreviations are used in this manuscript:

ALS	Aerial laser scan
AUC	area under curve
DDG	Discrete differential geometry
DEC	Curvature based on DEM
DEM	Digital elevation model
DTM	Digital terrain model
GIS	Geographic information system
k-NNG	k-nearest neihgbors graph
K-S	Kolmorogov-Smirnov test
L2O	Leave-pair-out
LiDAR	Light detection and ranging
LLC	curvature based on local linear fit
LTC	local curvature based on ground triangulation
MDL	Minimum description length principle
MTLS	Moving total least squares
NLS	National Land Survey of Finland
PCA	principal components analysis
SAF	Solid angle filtering
TIN	triangulated irregular network
UAV	Ultra-light vehicle

Appendix A

The computation of a solid angle Ω_k takes place at the approximity of a point p_k, namely at the set of the adjoining triangles T_k, see detail A) of Figure A1. Detail B) presents a tetrahedron defined by points p_k, p_i, p_j, p_l, where $p_i, p_j \in T_k \backslash p_k$ are from the outskirt of the triangle set T_k and p_l is an arbitrary point directly below point p_k. There are several ways to implement the solid angle calculation, a formula based on a classical l'Huillier's theorem [42] is presented here:

$$\mathrm{acos}(\mathbf{a}, \mathbf{b}) = \cos^{-1}(\mathbf{a}^0 \cdot \mathbf{b}^0) \qquad \text{angle between two vectors} \qquad (A1)$$

$$\mathbf{x}^0 := \mathbf{x}/\|\mathbf{x}\|_2 \qquad \text{vector normalization}$$

$$\alpha_i = \mathrm{acos}(p_l - p_k, p_j - p_k) \qquad \text{compartment angles } i, j, k$$

$$\alpha_j = \mathrm{acos}(p_i - p_k, p_l - p_k)$$

$$\alpha_l = \mathrm{acos}(p_j - p_k, p_i - p_k)$$

$$\alpha_0 = \alpha_i + \alpha_j + \alpha_l \qquad \text{basic product term}$$

$$\omega_{ilj} = 4\tan^{-1}\sqrt{\tan\tfrac{\alpha_0}{4}\Pi_{v \in \{i,j,l\}}\tan\tfrac{\alpha_0 - 2\alpha_v}{4}} \qquad \text{compartment angle} \qquad (A2)$$

$$\Omega_k = \Sigma_{ilj \in T_k}\omega_{ilj} \qquad \text{solid angle at point } p_k \qquad (A3)$$

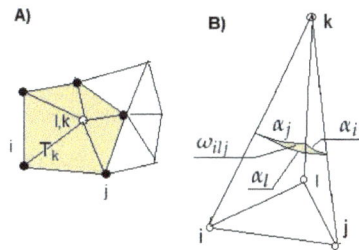

Figure A1. Solid angle filtering. (**A**) The set of adjoining triangles T_k of a point p_k seen from above; (**B**) A compartment *ijl* of the vertex point p_k presented in detail. A solid angle Ω_k is a sum of compartment angles ω_{ilj} of Equation (A2). Point p_l is an arbitrary point directly below the vertex point p_k.

Equation (A2) approaches so called Heron's planar trigonometric formula [42] when angles $\alpha_i, \ldots$ approach zero. The practical implementation requires a combination of space partitioning to a manageable point cloud patches of approx. 700...2000 points and a 2D Delaunay triangulation with an efficient point removal method. We use a batch version (python *scipy.spatial.delaunay*) with our own industry standard routine for deletion. This combination is simple to implement and excels in practise as [36] mentions, even though there exists faster incremental deletion arrangements with 2..3 times slower construction phase.

Appendix B

Step 1: Data can be preprocessed in three different ways before the LLC step. Alternatives are listed here in the order of increasing computational efficiency and decreasing amount of points:

(a) raw 3D ALS data
(b) same as (a) with tree and foliage returns cut from approx. 2 m height from approximative ground level
(c) TIN model produced e.g., by solid angle filtering of Section 2.4

The local linear fitting step 2 finds similar ground model per each alternative, only the speed of convergence varies. All three alternatives seem to result in about the same quality when measured by predictor performance.

Step 2: The fitting of the plane resembles a normal regression problem with an ad hoc nonlinear loss function, which penalizes residuals below the plane to force the ground fit. By applying the fitting process to planes of varying sizes one gets an assembly of plane orientations and plane centers. The neighboring planes can then be used to approximate local curvature. Since sample density is low, some of the plane fits cannot be performed. Therefore, it is numerically more resilient to use triangulation over neighboring planes and define curvature over each triangle using formulation developed in [33]. Another approach would be to produce a triangular mesh and estimate curvature based on it, as in [31,32]. There is a large filtering effect in this approach, since vertex normals depend on surrounding vertices. Local linear fit seemed to pick up the stoniness signal better, especially since we used multi-scale grids.

The grid division has been depicted in Figure B1. At each grid slot, one has to find the best fitting plane $\mathcal{P}(p, \mathbf{n})$, where $p \in \mathbf{R}^3$ is the center point of the plane $\mathcal{P}$ and $\mathbf{n}$ its normal. The initial state for the plane is: $\mathcal{P}_0 = \mathcal{P}(\text{lowest point of local sample}, (0,0,1)^T)\,\delta$.

The plane $\mathcal{P}$ represents a good local ground linearization provided that the weight function $g(l)$ of the orthogonal distance l penalizes heavily points below the approximated ground level. The details of weight function have been published in [11]. The practical considerations in selecting the weight function shape are at rapid and guaranteed convergence, whereas the influence to prediction performance comes from the actual ground returns and their geometry.

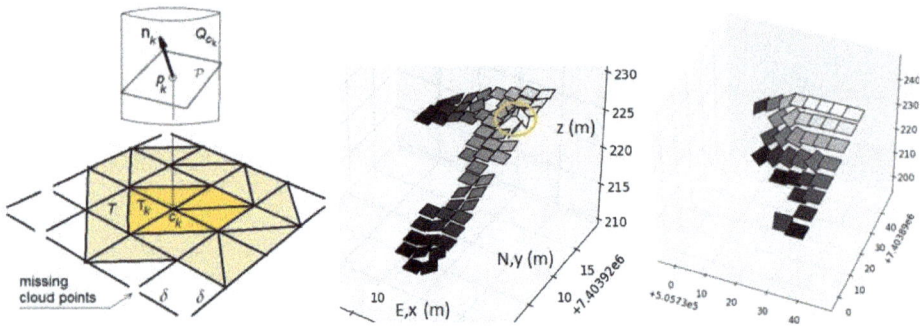

Figure B1. Left: An individual local plane $\mathcal{P}(p_k, \mathbf{n}_k)$ at grid point c_k and its parameters (local plane center point p_k and normal $\mathbf{n}_k$). A triangulation T of the grid avoids squares with incomplete data. A local cloud point set Q_{c_k} and neighboring triangles $T_k \subset T$ of a grid slot c_k are also depicted. **Center**: a stone revealed by two adjacent tilted planes. This stone provides a signal with the grid size $\delta = 2$ m. Note the amount of missing planes due to a lack of cloud points. **Right**: The grid of size $\delta = 4$ m at the same spot. The stone does not appear, local variation has disappeared but the grid is almost full approximating the sample polygon shape.

The optimal fit at each grid slot concerns now coordinate components $\mathbf{w}^T = (p_z, n_x, n_y)$ of p and $\mathbf{n}$. The local plane $\mathcal{P}$ is found by a numerical minimization:

$$\mathbf{w}^* = \operatorname*{argmin}_{\mathbf{w}} \sum_{q_j \in Q_c} g[(q_j - p) \cdot \mathbf{n}] \tag{B1}$$

Step 3: Triangulation is based on the grid centers, where the surface normal is known. Because of relatively low point density, some grid locations are bound to have no points and thus have to be omitted from triangulation, see Figure B1. The triangularization T outlined in Figure B1 is generated randomly, see Figure B1. The end result dictates the adjoining triangle sets $T_k \subset T$ of each grid point k. The size $|T_k|$ of the adjoining triangle sets varies depending on how dense or sparse point cloud is nearby point k: $1 \leq |T_k| \leq 8$.

Step 4: The curvature is approximated on each vertex of each triangle as in [33]. There are several other similar formulations e.g., using rectangular grids, but those are not so suitable in the presence of sparse and missing cloud points. The end result is set of candidate curvatures $\kappa_{tk}, t \in T_k$ per each grid point p_k.

Step 5: Now the task is to combine the final curvature approximant at a grid point p_k by taking a median of values available at the adjoining vertices of all surrounding triangles:

$$\kappa(p_k) = \operatorname*{median}_{t \in T_k} \kappa_{tk} \tag{B2}$$

Step 6: We used the normalized histograms of $\mathbf{h} = \text{hist}_{k \in K} \kappa(p_k)$, where K is the set of grid centers and histogram operator hist(.) and its properties are documented in Section 2.8.

References

1. Middleton, M.; Nevalainen, P.; Schnur, P.; Hyvnen, T.; Sutinen, R. Pattern recognition of mass-flow deposits from airborne LiDAR. In Proceedings of the 36th EARSeL Symposium, Bonn, Germany, 20–24 June 2016.
2. Pohjankukka, J.; Nevalainen, P.; Pahikkala, T.; Hyvönen, E.; Middleton, M.; Hänninen, P.; Ala-Ilomäki, J.; Heikkonen, J. Predicting water permeability of the soil based on open data. In Proceedings of the 10th Artificial Intelligence Applications and Innovations (AIAI 2014), Rhodes, Greece, 19–21 September 2014

3. Palmu, J.P.; Ojala, A.; Ruskeeniemi, T.; Sutinen, R.; Mattila, J. LiDAR DEM detection and classification of postglacial faults and seismically-induced. *GFF* **2015**, *137*, 344–352.

4. Johnson, M.D.; Fredin, O.; Ojala, A.; Peterson, G. Unraveling Scandinavian geomorphology: The LiDAR revolution. *GFF* **2015**, *137*, 245–251.

5. Sutinen, R.; Middleton, M.; Liwata, P.; Piekari, M.; Hyvönen, E. Sediment anisotropy coincides with moraine ridge trend in south-central Finnish Lapland. *Boreas* **2009**, *38*,638–646.

6. Sutinen, R.; Hyvönen, E.; Kukkonen, I. LiDAR detection of paleolandslides in the vicinity of the Suasselk posglacial fault, Finnish Lapland. *Int. J. Appl. Earth Obs. Geoinf. A* **2014**, *27*, 91–99.

7. Alho, P.; Vaaja, M.; Kukko, A.; Kasvi, E.; Kurkela, M.; Hyyppä, J.; Hyyppä, H.; Kaartinen, H. Mobile laser scanning in fluvial geomorphology: mapping and change detection of point bars. *Z. Geomorphol. Suppl.* **2011**, *55*, 31–50.

8. Sutinen, R.; Hyvönen, E.; Middleton, M.; Ruskeeniemi, T. Airborne LiDAR detection of postglacial faults and Pulju moraine in Palojärvi, Finnish Lapland. *Glob. Planet. Chang.* **2014**, *115*, 24–32.

9. Gallay, M. *Geomorphological Techniques*, 1st ed.; Chapter Section 2.1.4: Direct Acquisition of Data: Airborne Laser Scanning; British Society for Geomorphology: London, UK, 2012.

10. Li, Z.; Zhu, C.; Gold, C. *Digital Terrain Modeling*; CRC Press, Roca Baton, FL, USA, 2004.

11. Nevalainen, P.; Kaate, I.; Pahikkala, T.; Sutinen, R.; Middleton, M.; Heikkonen, J. Detecting stony areas based on ground surface curvature distribution. In Proceedings of the 5th International Conference Image Processing, Theory, Tools and Applications IPTA2015, Orléans, France, 10–13 November 2015.

12. Waldhauser, C.; Hochreiter, R.; Otepka, J.; Pfeifer, N.; Ghuffar, S.; Korzeniowska, K.; Wagner, G. Automated Classification of Airborne Laser Scanning Point Clouds. *Solv. Comput. Expens. Eng. Probl. Proc. Math. Stat.* **2014**, *97*, 269–292.

13. Meng, X.; Currit, N.; Zhao, K. Ground Filtering Algorithms for Airborne LiDAR Data: A Review of Critical Issues. *Remote Sens.* **2010**, *2*, 833–860.

14. Haugerud, R.; Harding, D. Some algorithms for virtual deforestation (VDF) of LiDAR topographic survey data. *Int. Arch. Photogramm. Remote Sens.* **2001**, *34*, 219–226.

15. Golovinskiy, A.; Funkhouser, T. Min-Cut Based Segmentation of Point Clouds. In Proceedings of the IEEE Workshop on Search in 3D and Video (S3DV) at ICCV, Kyoto, Japan, 27 September–4 October 2009.

16. Page, D.L.; Sun, Y.; Koschan, A.F.; Paik, J.; Abidi, M.A. Normal Vector Voting: Crease Detection and Curvature Estimation on Large, Noisy Meshes. *Graph. Model.* **2002** *64*, 199–229.

17. Johnson, K.M.; Ouimet, W.B. Rediscovering the lost archaeological landscape of southern New. *J. Archaeol. Sci.* **2014**, *43*, 9–20.

18. Vosselman, G.; Zhou, L. Detection of curbstones in airborne laser scanning data. ISPRS, 2009, pp. 111–116.

19. Brubaker, K.M.; Myers, W.L.; Drohan, P.J.; Miller, D.A.; Boyer, E.W. The Use of LiDAR Terrain Data in Characterizing Surface Roughness and Microtopography. *Appl. Environ. Soil Sci.* **2013**, *2013*, 1–13.

20. Hengl, T.; Reuter, H. (Eds.) *Geomorphometry: Concepts, Software, Applications*; Elsevier: Amsterdam, The Netherland, 2008; Volume 33, p. 772.

21. ASPRS. *LAS Specification Version 1.4-R13*; Technical Report; The American Society for Photogrammetry & Remote Sensing(ASPRS): Bethesda, ML, USA, 2013.

22. Weishampel, J.F.; Hightower, J.N.; Chase, A.F.; Chase, D.Z.; Patrick, R.A. Detection and morphologic analysis of potential below-canopy cave openings in the karst landscape around the Maya polity of Caracol using airborne LiDAR. *J. Cave Karst Stud.* **2011**, *73*, 187–196.

23. Telbisz, T.; Látos, T.; Deák, M.; Székely, B.; Koma, Z.; Standovár, T. The advantage of lidar digital terrain models in doline morphometry copared to topogrpahic map based datasets—Aggtelek karst (Hungary) as an example. *Acta Carsol.* **2016**, *45*, 5–18.

24. de Carvalho Júnior, O.A.; Guimaraes, R.F.; Montgomery, D.R.; Gillespie, A.R.; Gomes, R.A.T.; de Souza Martins, É.; Silva, N.C. Karst Depression Detection Using ASTER, ALOS/PRISM and SRTM-Derived Digital Elevation Models in the BambuiGroup, Brazil. *Remote Sens.* **2014**, *6*, 330–351.

25. Kobal, M.; Bertoncelj, I.; Pirotti, F.; Kutnar, L. LiDAR processing for defining sinkhole characteristics under dense forest cover: A case study in the Dinaric mountains. *Int. Arch. Photogramm. Remote Sens. Spat. Inf. Sci.* **2014**, *XL-7*, 113–118.

26. Hulík, R.; Španěl, M.; Materna, Z.; Smrž, P. Continuous Plane Detection in Point-cloud Data Based on 3D Hough Transform. *J. Vis. Commun. Image Represent.* **2013**, *25*, 86–97.

27. Yang, M.Y.; Förstner, W. *Plane Detection in Point Cloud Data*; Technical Report TR-IGG-P-2010-01; University of Bonn, Bonn, Germany, 2010.

28. Evans, J.S.; Hudak, A.T. A multiscale curvature algorithm for classifying discrete return lidar in forested environments. *IEEE Trans. Geosci. Remote Sens.* **2007**, *45*, 1029–1038.

29. Pauly, M.; Keiser, R.; Gross, M.; Zrich, E. Multi-scale Feature Extraction on Point-sampled Surfaces. *Comput. Graph. Forum* **2003**, *22*, 281–290.

30. Palancz, B.; Awange, J.; Lovas, T.; Fukuda, Y. Algebraic method to speed up robust algorithms: Example of laser-scanned point clouds. *Surv. Rev.* **2016**, 1–11, doi:10.1080/00396265.2016.1183939.

31. Meyer, M.; Desbrun, M.; Schröder, P.; Barr, A.H. Chapter Discrete Differential-Geometry Operators for Triangulated 2-Manifolds. *Visualization and Mathematics III*; Springer Berlin Heidelberg: Berlin, Heidelberg, 2003; Volume 7, pp. 35–57.

32. Mesmoudi, M.M.; Floriani, L.D.; Magillo, P. *Discrete Curvature Estimation Methods for Triangulated Surfaces*; Lecture Notes in Computer Science; Springer: New York, NY, USA, 2010; Volume 7346, pp. 28–42.

33. Theisel, H.; Rössl, C.; Zayer, R.; Seidel, H.P. Normal Based Estimation of the Curvature Tensor for Triangular Meshes. In Proceedings of the 12th Pacific Conference on (PG2004) Computer Graphics and Applications, Seoul, Korea, 6–8 October 2004; pp. 288–297.

34. Crane, K.; de Goes, F.; Desbrun, M.; Schröder, P. *Digital Geometry Processing with Discrete Exterior Calculus*; ACM SIGGRAPH 2013 Courses; ACM: New York, NY, USA, 2013.

35. Lin, X.; Zhang, J. Segmentation-Based Filtering of Airborne LiDAR Point Clouds by Progressive Densification of Terrain Segments. *Remote Sens.* **2014**, *6*, 1294–1326.

36. Devillers, O. On Deletion in Delaunay Triangulations. *Int. J. Comp. Geom. Appl.* **2002**, *12*, 193–205.

37. Massey, F. The Kolmogorov-Smirnov Test for Goodness of Fit. *J. Am. Stat. Assoc.* **1956**, *46*, 66–78.

38. Bernardini, F.; Mittleman, J.; Rushmeier, H.; Silva, C.; Taubin, G. The ball-pivoting algorithm for surface reconstruction. *IEEE Trans. Visual. Comput. Graph.* **1999**, *5*, 349–359.

39. Pauly, M.; Keiser, R.; Kobbelt, L.P.; Gross, M. Shape Modeling with Point-sampled Geometry. *ACM Trans. Graph.* **2003**, *22*, 641–650.

40. Airola, A.; Pahikkala, T.; Waegeman, W.; Baets, B.D.; Salakoski, T. An experimental comparison of cross-validation techniques for estimating the area under the ROC curve. *Comput. Stat. Data Anal.* **2010**, *55*, 1824–1844.

41. Fawcett, T. An Introduction to ROC Analysis. *Pattern Recognit. Lett.* **2006**, *27*, 861–874.

42. Zwillinger, D. *CRC Standard Mathematical Tables and Formulae*, 32th ed.; CRC Press, Boca Raton, FL, USA, 2011.

Part 6
Building Extraction

![remote sensing logo] *remote sensing*

MDPI

Article

Edge Detection and Feature Line Tracing in 3D-Point Clouds by Analyzing Geometric Properties of Neighborhoods

Huan Ni [1], Xiangguo Lin [2], Xiaogang Ning [2] and Jixian Zhang [1,*]

[1] School of Resource and Environmental Sciences, Wuhan University, No. 129 Luoyu Road, Wuhan 430079, China; nih2015@yeah.net
[2] Chinese Academy of Surveying and Mapping, No. 28 Lianhuachixi Road, Beijing 100830, China; linxiangguo@gmail.com (X.L.); ningxg@casm.ac.cn (X.N.)
* Correspondence: zhangjx@casm.ac.cn; Tel.: +86-10-6388-1816

Academic Editors: Jie Shan, Juha Hyyppä, Gonzalo Pajares Martinsanz and Prasad S. Thenkabail
Received: 10 March 2016; Accepted: 24 August 2016; Published: 1 September 2016

Abstract: This paper presents an automated and effective method for detecting 3D edges and tracing feature lines from 3D-point clouds. This method is named Analysis of Geometric Properties of Neighborhoods (AGPN), and it includes two main steps: edge detection and feature line tracing. In the edge detection step, AGPN analyzes geometric properties of each query point's neighborhood, and then combines RANdom SAmple Consensus (RANSAC) and angular gap metric to detect edges. In the feature line tracing step, feature lines are traced by a hybrid method based on region growing and model fitting in the detected edges. Our approach is experimentally validated on complex man-made objects and large-scale urban scenes with millions of points. Comparative studies with state-of-the-art methods demonstrate that our method obtains a promising, reliable, and high performance in detecting edges and tracing feature lines in 3D-point clouds. Moreover, AGPN is insensitive to the point density of the input data.

Keywords: 3D edge; Edge detection; Feature line tracing; RANdom SAmple Consensus (RANSAC); Angular gap

1. Introduction

1.1. Problem Statement

Feature extraction in 2D-images, one of the most important topics in the fields of image analysis and computer vision, has been studied for years [1]. Edges and feature lines are considered as important features in various urban scenes covering a vast number of man-made objects. Generally, once edges are detected, a further step will be done for tracing feature lines from the detected edges. For edge detection in images, edges have been well defined, such as "the boundary element between two regions of different homogeneous luminance" [2] or "large or sudden changes in some image attribute, usually the brightness" [3]. An extensive review of the established edge detection methods can be found in the literature [1]. Apart from the surveyed methods, many outstanding methods have been proposed such as the revised Canny operator [4] and Edison operator [5].

In recent years, benefiting from the advances in sensor technology for both airborne and ground-based laser scanning, dense 3D-point clouds have become increasingly common [6]. Therefore, edge detection and feature line extraction in 3D-point clouds have become a novel research topic. However, in the fields of remote sensing and photogrammetry, the approaches have just scratched the surface, and the definition of 3D edges has not yet been confirmed though feature line extraction has long been a major issue in the 3D city modeling works.

In addition, the established edge detection methods defined for images cannot be applied directly to 3D-point clouds. The main reasons for this are given below:

(1)   The data representation is different. An image is considered as a matrix, whereas a 3D-point cloud is an unorganized and irregularly distributed [7] scattered point set.

(2)   The presented information type is different. An image contains cryptic spatial information, and abundant spectral information. Comparatively, a 3D-point cloud contains explicit spatial information, and the reflected intensity at times [8].

(3)   The spatial neighborhood is different. An image is arranged as a grid-like pattern, and the neighborhood of a pixel can easily be determined. However, a 3D-point cloud is unorganized, and the neighborhood of a point is more complex than that of a pixel in an image. Generally, in 3D-point clouds, there are three types of neighborhoods: spherical neighborhood, cylindrical neighborhood, and k-closest neighbors based neighborhood [9]. The three types of neighborhoods are based on different search methods, and change of the search method alters the neighborhood correspondingly.

To address the aforementioned problems, an automated and effective method is proposed to detect edges and trace feature lines from 3D-point clouds. Prior to presenting our method, the definition of 3D edges is given herein. Traditionally, 2D edges in an image are defined as the following two types [10]:

(1)   Gray level edges, which are often associated with abrupt changes in average gray level.

(2)   Texture edges, which are the abrupt "coarseness" changes between adjacent regions contained the same texture at different scales, or the abrupt "directionality" changes between the directional textures in adjacent regions.

Then, the definition of 2D edges is extended by the literatures [1–4,11]. Specially, the literature [1] defines 2D edges as one-dimensional discontinuities in the intensity surface of the underlying scene. However, the intensity or spectral information cannot describe the complete geometric properties in 3D-point clouds. According to the definitions in images and the characteristics of 3D-point clouds, we visibly define 3D edges as 3D discontinuities of the geometric properties in the underlying 3D-scene. Mathematically, we define 3D edges as the following two types (see Figure 1):

(1)   Boundary elements, which are often associate with an abrupt angular gap in the shape formed by their neighborhoods. The details are presented in Section 2.2. Boundary elements are the edges belonging to roof contours, façade outlines, height jump lines [12], and other types of surface's contours. Specially, the surface is a 3D-plane or a curve surface.

(2)   Fold edges, which are the abrupt "directionality" changes between the normal directions in adjacent surfaces. Generally, two curve or planar intersected surfaces exist in the neighborhood of a fold edge. The details are presented in Section 2.2. Fold edges are the edges belonging to plane intersection lines [13], sharp feature line [14], breaklines [15], and other types of intersections between different surfaces.

Edge detection in 3D-point clouds is similar to 2D image processing. The usual aim of edge detection is to locate edges belonging to boundaries of objects of interest [3]. Most edge detection techniques consist of two stages [11]: (1) converting an original image into features; and (2) assigning points to edges or non-edges. Similarly, our proposed edge detection procedure first computes angular gap feature for all the points in a 3D-point cloud, then assigns the points to edges or non-edges. Sometimes, there is a further stage called edge linking, in which the detected edges are examined once again for instance to obtain closed contours [11]. For 3D-point clouds, the edge linking procedure may relate to a special application such as line segment extraction or building reconstruction. In this paper, we present a feature line tracing procedure for linking the detected edges to feature lines, such as boundaries and intersection lines. The feature lines might be straight or curve, however, must be smooth.

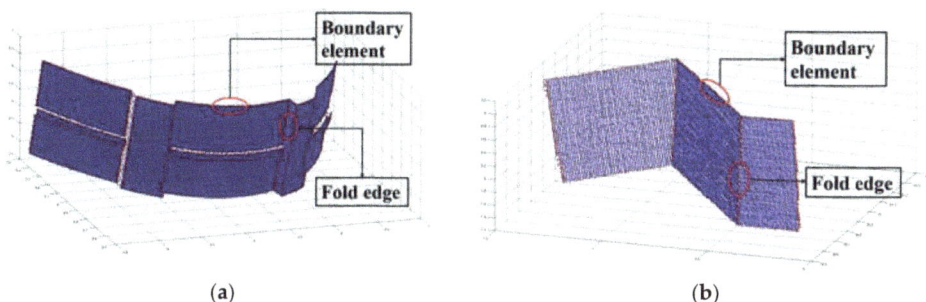

Figure 1. The definition of two types of edges.

To detect the defined 3D edges and trace the feature lines from 3D-point clouds, we propose an Analysis of Geometric Properties of Neighborhoods (AGPN) method. AGPN first analyzes geometric properties of each query point's neighborhood in spatial domain, and then detects 3D edges based on RANSAC and angular gap metric. Finally, to trace feature lines from the 3D edges, a hybrid method based on region growing and model fitting is proposed. With the proposed AGPN method, we can detect the two defined 3D edges—boundary elements and fold edges from 3D-point clouds directly without extra image processing or point cloud preprocessing (e.g., segmentation or object recognition).

1.2. Related Work

In the fields of remote sensing, photogrammetry, and computer vision, edge detection and feature line extraction from airborne or terrestrial laser scanning data, has long been one of the major issues. Specifically, feature line extraction from 3D-point clouds, as a substep in 3D city modeling works, has been a major research topic for years [12,13,15–35]. Research efforts for 3D edge detection and feature line extraction from laser scanning data can be categorized into two groups: (1) direct methods [12,13,15–26], first recognize the building points from a 3D-point cloud, next, segment or cluster building points into planes, finally, detect edges and extract feature lines i.e., plane outlines and plane intersection lines [13]; (2) indirect methods [26–35], first convert a 3D-point cloud into an image, or register with corresponding images, next, detect 2D edges in the images, and then project the 2D edges back into the 3D-point cloud to obtain 3D edges.

The above-noted approaches can detect edges on regular feature lines such as plane intersection lines [13] or breaklines [15,16], roof or wall outlines [15,17] in buildings or piecewise planar objects. We mainly review the aforementioned direct methods herein, because the indirect methods might cause loss of spatial geometric information when a 3D-point cloud is converted into an image. To extract roof boundaries or façade outlines, a boundary estimation method is required in the direct methods. At beginning, 3D-point clouds without a high point density bring challenges to this task [18]. The literature [12] extracts roof boundaries named height jump lines by using segmented ground plans. With the improvement of the point density of 3D-point clouds, two groups of boundary estimation methods are proposed. The first group of methods [17,19,20] extract boundaries based on triangulation. In this case, a Triangular Irregular Network (TIN) for the planar segments is generated first, and then long TIN edges appear only at the outer boundary (segment outlines) or inner boundary (holes). Boundary points are just the end points of the long TIN edges. The second group of methods [21,22] extract boundaries based on convex hull algorithm. In this case, the convex hull algorithm is modified by local convex hull testing to deal with complex boundaries. Then the points which are close enough to the query local convex boundary are picked up and treated as boundary points. In comparison with boundary extraction, the extraction for plane intersection lines is more convenient in the field of photogrammetry. After all the planar segments in a 3D-point cloud have been determined, the topologic relations among the planar segments are computed and represented by an adjacency matrix [15], then all pairs of adjacent planar segments are intersected to extract plane intersection lines [12,13,15,23], and thus, the edges on the intersection lines can be easily detected.

The aforementioned methods can, however, detect edges and extract feature lines in only regular buildings or piecewise planar objects. Furthermore, the literature [6] reviews the drawbacks of this work in the literature [13], some of which are common in state-of-the-art methods using 3D laser scanning data. The drawbacks include: (1) these methods are incapable of fitting a small and narrow plane in a noisy point cloud; (2) these methods may generate unexpected lines at non-planar surfaces when the data become complex.

Some approaches [14,36–40] employ surface meshes or point-based surfaces, which can detect 3D edges or extract feature lines from some irregular objects and more complex surfaces. However, these methods are only applied to a small-scale 3D-point cloud with a single object. Specially, an angular gap based method [39,40], whose variations are widely used in building reconstruction work [41], fan clouds work [42] and the Point Cloud Library (PCL) [43], has been proposed. In this paper, the angular gap based method is utilized as a criterion in our proposed method.

2. Methodology

2.1. Overview

To detect 3D edges and trace feature lines, a two-step strategy based on AGPN, is illustrated in Figure 2. In the diagram shown in Figure 2, the upper part marked by blue dotted lines is the first step, and the lower part marked by red dotted lines is the second step. The first step, detailed in Section 2.2, detects or locates edges in 3D-point clouds based on RANSAC, normal optimization, and angular gap computation. The second step, detailed in Section 2.3, traces feature lines from the detected edges based on neighborhood refinement and growing criterion determination.

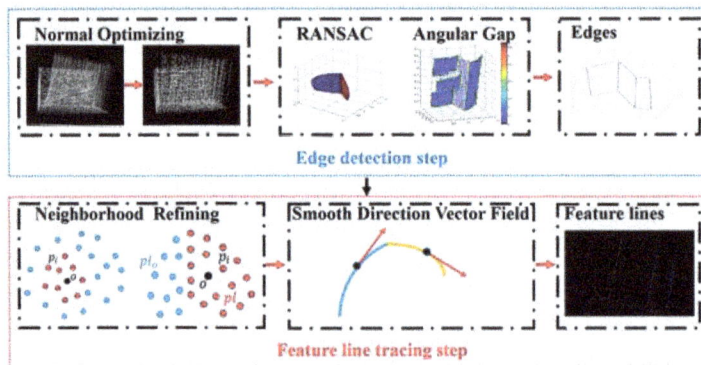

Figure 2. Overview of the proposed AGPN.

The pseudo code of our proposed AGPN method and its parameters are shown in Appendix A.

2.2. Edge Detecton

2.2.1. Geometric Property Analysis

The inherent property of being an edge is that it is the local neighborhood of a point rather than the point itself [44]. Based on this principle, we design an algorithm to determine whether a point is an edge or not by analyzing geometric properties of the point's neighborhood. In the neighborhood of a boundary element, there is only one curve or planar surface. In the neighborhood of a fold edge, there are two or more intersected surfaces. We first present the flowchart of the edge detection step, and then explain why the defined edges can be detected.

The flowchart of the edge detection step in our proposed AGPN is shown in Figure 3. Let o denote an unlabeled point. Our method first searches the nearest neighbor point set P of o based on distance. Next, the point set P is fitted into a local plane, pl, by the RANSAC algorithm, and then P is divided into inliers (on the fitted plane pl) and outliers. The point o will be labeled as non-edge if it does not belong to the inliers. Otherwise, o and the inliers are connected to construct a number of spatial vectors, from which angular gap will be calculated based on the optimized normal detailed in Section 2.2.4. If a substantial angular gap exists between the constructed spatial vectors on pl, o will be labeled as edge. Otherwise, o will be labeled as non-edge. Edges will be detected after all the points in a 3D-point cloud are labeled.

It is noteworthy that we use the angular gap method rather than a modified convex hull boundary detection algorithm, which can be widely found in some references, because the inliers may be a subset of a surface plane model. Moreover, a *kd*-tree is used to determine the nearest neighbor point set P of each point.

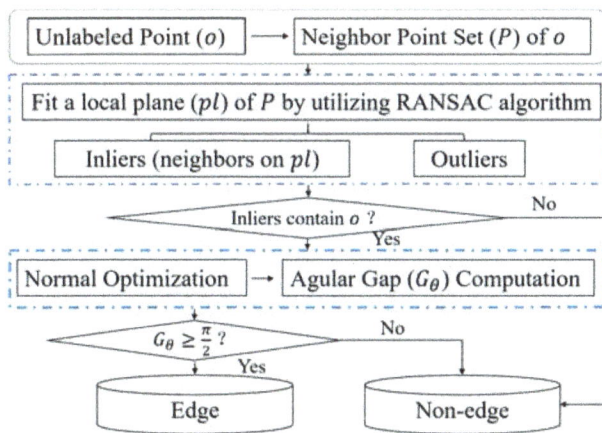

Figure 3. Flowchart for the edge detection step in AGPN.

As shown in Figure 3, the plane model, calculated by the RANSAC algorithm, fits a local surface in the nearest neighbor point set P of the unlabeled point o (see Figure 4). When o is on a plane outline or plane intersection line, the plane model is impeccable. When o is on a curve surface boundary or intersection of different curve surfaces, the plane model is also the most direct and effective geometric model for approximating the local smooth surface, because the neighborhood of o is a local small area of the surface.

Based on the plane model fitted by the RANSAC algorithm, the nearest neighbor point set P are divided into inliers and outliers (see Figure 4). The inliers are in the fitted plane and the outliers are outside. It can be found that, by the RANSAC algorithm, a best plane model can be found in P. If a point is a boundary element, the inliers are on the fitted plane, and the outliers are noise. If a point is a fold edge, the inliers are on the fitted plane lying on one of the intersecting surfaces in the point's neighborhood (see Figure 4).

The edge detection procedure in AGPN can detect both boundary elements and fold edges. The detection of the two types of edges is detailed in Sections 2.2.2 and 2.2.3, respectively.

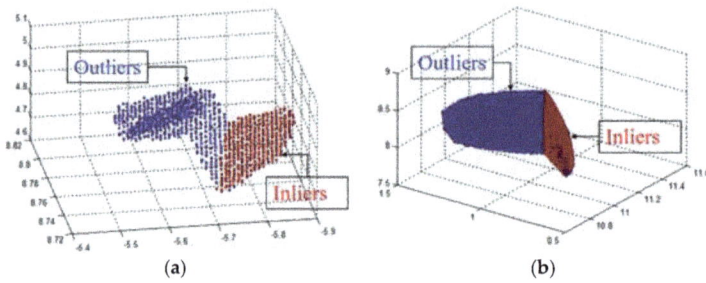

Figure 4. Local plane (rendered in red) fitted by the RANSAC algorithm in the nearest neighbor point set *P*. (**a**,**b**) show two types of neighbor point sets respectively. There are three planes in (**a**) and two planes in (**b**).

2.2.2. Boundary Element Detection

As shown in Figure 5, there is only one surface in the neighborhood of an unlabeled point *o*. *K* points depicted in blue and red are the nearest neighbors of *o*, obtained using a *kd*-tree. Points p_i $(i = 1 \cdots N_r)$ rendered in red are inliers extracted by the RANSAC algorithm. These inliers are on the fitted plane *pl* of the local surface.

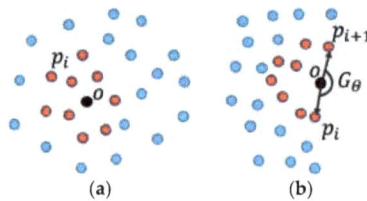

Figure 5. Distribution of the nearest neighbors of an unlabeled point on a surface: (**a**) the neighborhood of an interior point; (**b**) the neighborhood of a point on a boundary.

If *o* is on a surface boundary, there will be a substantial angular gap G_θ (see Figure 5b) between vectors $\overrightarrow{op_i}$ $(i = 1 \cdots N_r)$ on the local plane *pl*. If *o* is an interior point (see Figure 5a), the distribution of the angles between vectors $\overrightarrow{op_i}$ $(i = 1 \cdots N_r)$ will be consecutive, and there will certainly be no substantial angular gap.

Notably, *o* will be labelled as noise or an isolated point if it is an outlier of the local plane *pl*.

2.2.3. Fold Edge Detection

As shown in Figure 6, the neighborhood of an unlabeled point *o* includes two intersecting surfaces. *K* points depicted in blue and red are the nearest neighbors of *o*. Points p_i $(i = 1 \cdots N_r)$ rendered in red are inliers extracted by the RANSAC algorithm. The extracted inliers are on the fitted local plane *pl* lying on one of the intersecting surfaces.

If *o* is a fold edge point, there will be a substantial angular gap G_θ (see Figure 6b) between vectors $\overrightarrow{op_i}$ $(i = 1 \cdots N_r)$ on the local plane *pl*. The local plane *pl* is on one of the intersecting surfaces. If *o* is an interior point on one of the intersecting surfaces (see Figure 6a), the distribution of the angles between vectors $\overrightarrow{op_i}$ $(i = 1 \cdots N_r)$ will be consecutive, and there will certainly be no substantial angular gap.

Figure 6c shows a special case, that is, the point densities of the intersecting surfaces are considerably different. One of the intersecting surfaces contains sparse points, while the other intersecting surface includes much dense points. If *o* is an interior point on the surface with sparse points, the inliers extracted by the RANSAC algorithm will be on the other surface with dense points.

Then, once the angular gaps are computed from these inliers, there will be a substantial angular gap, resulting in *o* mislabeled as an edge. Fortunately, *o* is an outlier of the fitted local plane *pl*, and therefore, it can be rejected by determining whether or not it is an inlier of the fitted local plane *pl*.

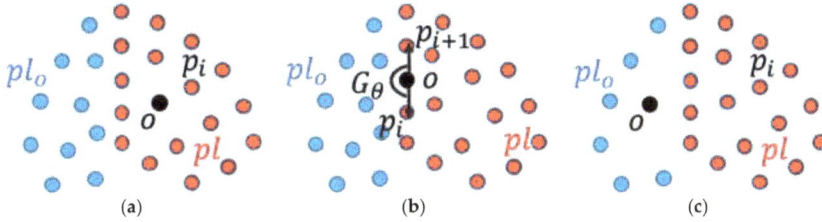

Figure 6. Distribution of the nearest neighbors of an unlabeled point on a surface intersecting structure: (a) the neighborhood of an interior point on one of the intersecting surfaces; (b) the neighborhood of a fold edge point; (c) the neighborhood of a point on the two intersecting surfaces with different point densities.

In addition, if the number of inliers is less than three, the inliers will not be able to fit a plane. In this case, *o* will be labeled as noise or a local extreme, such as the vertex of a circular cone.

2.2.4. Normal Optimization

Generally, the normal of a 3D point is computed by a covariance matrix created from the nearest neighbors of the 3D point [45], which is called PCA-Normal herein. The PCA-Normal is the normal of the tangent plane of a 3D point. In our method, the normal is required to be orthogonal to the local fitted plane *pl*. When we detect boundary elements, only one surface exists in the neighborhood, and *pl* is on the surface. When we detect fold edges, two or more intersecting surfaces exist, and *pl* is on one of the intersecting surfaces. However, the PCA-Normal is orthogonal to the tangent plane, and hence it cannot meet the aforementioned requirement.

The reasons for this are that the PCA algorithm cannot detect any of the intersected planes in a complex neighborhood. Fortunately, the RANSAC algorithm can solve this problem, and deal with noise as well. As shown in Figure 4, the RANSAC algorithm can detect a proper plane in either of the two complex neighborhoods.

In this study, we fit a local plane in the neighborhood to estimate the normal $\vec{n}$ of an unlabeled point *o*. The procedure first searches for the neighbors of *o*. Next, to eliminate the influence of outliers or noises, a local plane *pl* is fitted using the RANSAC algorithm. Then, $\vec{n}$ is the normal of the local fitted plane *pl*. Note that the normal $\vec{n}$ used in this study is named RANSAC-Normal.

2.2.5. Angular Gap Computation

To compute the angular gap, G_θ, a spatial coordinate frame is constructed based on the local plane *pl* and its normal vector $\vec{n}$. The axes of this frame are composed of two perpendicular vectors $\vec{u}$, $\vec{v}$ on *pl* and the normal vector $\vec{n}$. The angular gap G_θ is computed by Equations (1)–(4):

$$d_i^u = \vec{op_i} \odot \vec{u} \tag{1}$$

$$d_i^v = \vec{op_i} \odot \vec{v} \tag{2}$$

$$\theta_i = arc\tan\left(d_i^u / d_i^v\right) \tag{3}$$

$$G_\theta = \max\left(\theta_{i+1} - \theta_i\right) \ (i = 1 \cdots N_r - 1) \tag{4}$$

where $\vec{op_i}$ $(i = 1 \cdots N_r)$ is the vector connecting the unlabeled point *o* to an inlier p_i extracted by the RANSAC algorithm. The distribution of G_θ is shown in Figure 7.

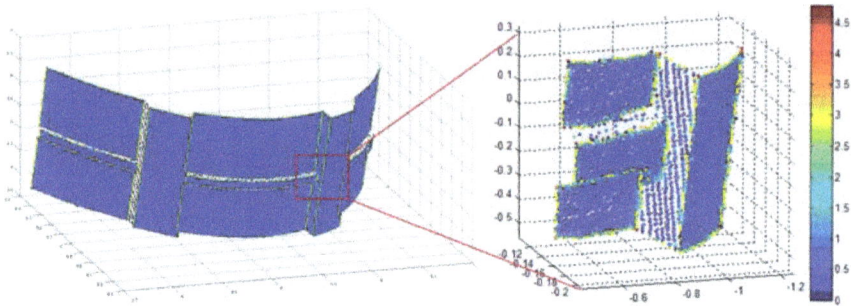

Figure 7. Distribution of G_θ , each point is colored according to its value of G_θ .

The pseudo code of the edge detection step in AGPN and its parameters are shown in Appendix B.

2.3. Feature Line Tracing

Once edges have been detected, a further step will trace the detected edges into segments. Each segment is a point list, in which all the points belong to the same feature line. The proposed feature line tracing method connects edges with similar principle directions and splits edges with abrupt directionality changes. Thus, the traced feature lines are curve or straight, however, must be smooth.

The proposed feature line tracing is a hybrid method based on region growing and model-fitting algorithms. The model-fitting algorithm estimates 3D line parameters in each point's neighborhood by the RANSAC algorithm. The directional parameters of the fitted 3D line denote the principle direction of the edge point. The region growing algorithm clusters the detected edges into segments based on the following two redefined growing criteria related to the refined neighborhood and the principle direction.

The proposed feature line tracing procedure includes two essential steps: neighborhood refinement and growing criterion definition.

2.3.1. Neighborhood Refinement

Compared to an image, a 3D-point cloud is an unorganized and scattered point set, which brings great challenges to neighborhood searching of edge points. Our method first obtains the nearest neighbors of a query point by using the *kd*-tree algorithm. Next, a straight line model is fitted by the RANSAC algorithm, and then, the nearest neighbors are divided into inliers and outliers. The inliers containing the query point are the refined nearest neighbors. Otherwise, the outliers are processed iteratively by the RANSAC algorithm until the updated inliers contain the query point. Therefore, an adaptive neighborhood is designed for each query point.

2.3.2. Growing Criterion Definition

Two growing criteria given by [46] are redefined in this study.

a   Proximity of points. Only points that are near one of the points in the current segment can be added to the stack of the segment. For feature line tracing, this proximity of edge points can be implemented by the aforementioned neighborhood refinement.

b   Smooth direction vector field. Only points that have a similar principal direction with the current tracing segment can be added to the stack of the current segment. In this paper, a line model is first fitted from the refined neighborhood by the RANSAC algorithm, and then the principal direction of the current point is defined as the direction of the fitted line.

In addition, a larger proportion of inliers to all nearest neighbors implies a higher possibility of the presentence of feature lines. Due to the region growing procedure being irreversible, an edge point with greater linearity should be grown first to ensure a better tracing result. Therefore, edge points are first sorted by their proportion values, and then grown in order.

The general region growing procedure for linear feature segmentation is sensitive to the size of the established local neighborhoods and the location of seed points. However, the refined neighborhood and the aforementioned sorted edge points can overcome the problem. It is validated that the proposed feature line tracing procedure can distinguish spatially-adjacent, collinear/coaxial lines (see Figure 8).

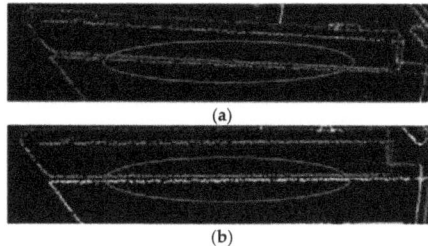

(a)

(b)

Figure 8. Feature line tracing, (**a**) feature line segments generated by region growing method; (**b**) feature line segments generated by the proposed feature line tracing method. The traced segments are marked by different colors.

Three parameters are used in the feature line tracing step, that is, the number of nearest neighbors K^2 for kd-tree algorithm, distance threshold d_r^2 for the RANSAC line model estimation, and smooth direction threshold sm_thr. The traced feature lines are shown in Figure 8b.

3. Experiments and Analysis

The proposed algorithms were implemented in C++ using the PCL. There are five parameters in the proposed AGPN (detailed in Appendix A). In this study, the point spacing of the input data affects the performance of our proposed AGPN most. Moreover, the performance is also affected by the distance thresholds d_r^1 and d_r^2 related to the point spacing. Therefore, we describe the testing data in Section 3.1 and discuss parameters tuning in Section 3.3. The point spacing is measured by the open source software CloudCompare [47].

To quantitatively evaluate the performance of our AGPN, four measures are defined in Section 3.2. We further compare the proposed AGPN with two representative algorithms: the boundary estimation method presented in the PCL and the edge points clustering algorithm presented in the literature [13]. The comparative studies are presented in Section 3.7.

3.1. Testing Data

The open datasets available from the homepage of 3D ToolKit [48] are employed. The datasets are recorded by a Riegl VZ400 scanner in the Bremen city center. The point density of a single 3D scan is uneven.

From the open datasets, we selected two testing sites, i.e., Site 1 and Site 2. The two sites contain a large number of complex man-made objects. Table 1 shows their detailed information, including the number of points, maximum point spacing, minimum point spacing and the parameters used in our experiments.

The point spacings in the two testing sites are quite different. In Site 1, the point density is uneven. In the neighborhood of the scanner, the average point spacing is 0.001 m, and in the areas away from the scanner, the average point spacing is 0.15 m. The variation tendency of the point density is consecutive with the variation of the distances to the scanner. In Site 2, except for the window areas, the average point spacing is 0.005 m. In the window areas, the point density decreases rapidly, and the average point spacing turns to 0.01 m. The variation tendency of the point density is piecewise.

Table 1. Data description and parameter settings for the two testing sites.

	Number of Points	Maximum Point Spacing	Minimum Point Spacing	Parameters				
				K^1	d_r^1	K^2	d_r^2	sm_thr
Site 1	14040449	0.15	0.001	200	0.01	15	0.01	0.2
Site 2	4411599	0.01	0.005	200	0.005	15	0.005	0.2

3.2. Evaluation Metrics

To test the performance of the proposed AGPN, we quantitatively evaluate the results of the edge detection step and the feature line tracing step, respectively, by four measures: p_{dc}(correctness rate of the edge detection step), p_{mj}(mislabeled rate of the edge detection step), p_{dct}(correctness rate of the feature line tracing step), and p_{mjt}(mislabeled rate of the feature line tracing step). To compute p_{dc} and p_{mj} for evaluating the 3D edge results, we count the number of the feature lines which the detected edges belong to. The feature line is curve or straight, however, must be smooth. The four measures are defined as follows:

$$p_{dc} = \frac{N_{dc}}{N_{gc}} \tag{5}$$

$$p_{mj} = \frac{N_{mj}}{N_{gc}} \tag{6}$$

$$p_{dct} = \frac{N_{tc}}{N_{dc} + N_{mj}} \tag{7}$$

$$p_{mjt} = \begin{cases} \frac{N_{mjt}}{N_{dc}+N_{mj}} & if \ N_{mjt} \leq N_{dc} + N_{mj} \\ \frac{N_{mjt}}{N_{mjt}+N_{tc}} & Otherwise \end{cases} \tag{8}$$

where N_{dc} is the number of true positive feature lines contained in the detected edges, N_{tc} is the number of correctly traced feature lines in the feature line tracing step, N_{gc} is the total number of feature lines in the ground truth, N_{mj} is the number of mislabeled feature lines contained in the detected edges, and N_{mjt} is the number of incorrectly traced feature lines in the feature line tracing step. Larger values of p_{dc} and p_{dct}, and smaller values of p_{mj} and p_{mjt} are more desirable.

3.3. Parameter Tuning

One of the major strengths of the proposed AGPN is that the extracted 3D edges in the edge detection step are sufficiently subtle with the default value ($\frac{\pi}{2}$) of the angular gap (G_θ).

Table 1 lists the parameters used in the proposed AGPN and their empirical values. We maintained $K^1 = 200$ and $K^2 = 15$ for Site 1 and Site 2 because both K^1 and K^2 have little influence on the quality of the results.

In addition, because the feature line tracing parameters are mainly dependent on the requirements of users, we only analyze the influence on the application of 3D line segment extraction. Under this condition, to ensure the sufficient quality of the result, sm_thr is set to 0.2.

We selected a number of subsets in Site 1 with the same point spacing (0.01 m) to analyze the sensitivity of d_r^1 and d_r^2. As shown in Figure 9, the x-axis denotes d_r^1 and d_r^2, the y-axis depicts the average values of the correctness and mislabeled rates. It can be seen that when the value of d_r^1 and d_r^2 are close to the point spacing of 0.01 m, the edge detection step and the feature line tracing step both achieve good results. If we set d_r^1 smaller than the point spacing, the mislabeled rate p_{mj} will arise. Although the correctness rate p_{dc} also arises, it is much slighter than p_{mj}. According to this figure, if the point spacings of the input data are the same, a reasonable configuration is obtained with d_r^1 and d_r^2 equal to the point spacing of the input data. Otherwise, we can set the parameters according to the variation tendency of the point density. For example, if the variation tendency of the point density is piecewise, we set d_r^1 smaller than the average point spacing in the edge detection step, and thus all the edges can be ensured to be detected, next, we utilize the feature line tracing step to trace all the edges, then filter the traced segments with small number of points.

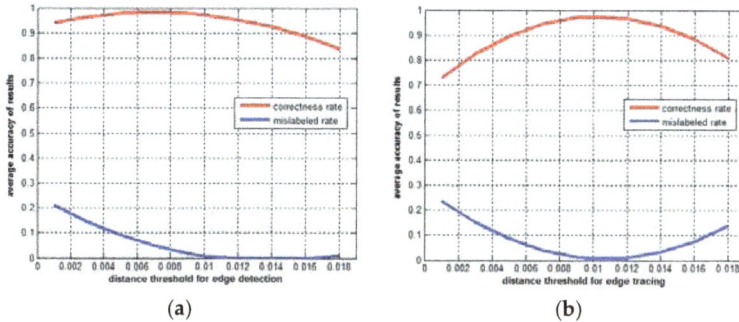

(a)　　　　　　　　　　　　　　　(b)

Figure 9. Average values of correctness and mislabeled rates for different values of d_r^1 and d_r^2 : (a) results of the edge detection step with different values of ; and (b) results of the feature line tracing step with different values of d_r^2.

3.4. Normal Estimation

To demonstrate the feasibility of our RANSAC-Normal, we compare it with the PCA-Normal. The comparison is shown in Figure 10. When a point is a fold edge, there are two intersecting surfaces orthogonal to each other in its neighborhood. The PCA-Normal (see Figure 10a) is orthogonal to the tangent plane rather than one of the intersecting planes, while the RANSAC-Normal (see Figure 10b) is orthogonal to one of the intersecting planes. If we construct a coordinate frame based on the computed RANSAC-Normal, the other two axes $\vec{v}$ and $\vec{u}$ of the coordinate frame are located in one of the intersecting planes. Therefore, the RANSAC-Normal reaches the requirement of our method (see Section 2.2.4).

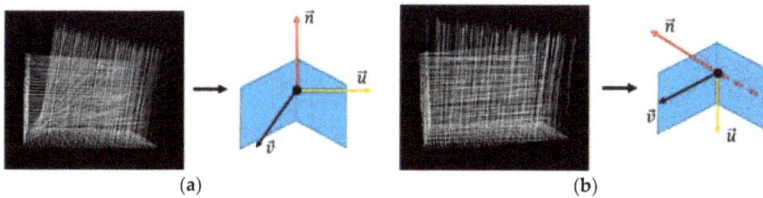

(a)　　　　　　　　　　　　　　　(b)

Figure 10. Normal estimation of the neighborhood with two intersecting planes: (a) PCA-Normal $\vec{n}$; (b) RANSAC-Normal $\vec{n}$ estimated by our method.

3.5. Influence of Point Density

The point density of the input data mainly determines the details of a complex object, such as a chimney, step, or window. To analyze the influence of the input density on the results of the edge detection step and the feature line tracing step, we down-sampled a point cloud data of a window (see Figure 11a) with different point densities.

As shown in Figure 11c, the horizontal axis denotes the number of points in the down-sampled data. The light blue bar represents the ratio of the contained feature lines in the original data of a window to all feature lines in the window (denoted by r_1). When the data is down-sampled into different point densities, the value of r_1 decreases with the decreasing of point density. We can find that the ratio of feature lines in the detected edges to all feature lines (denoted by r_2) and the ratio of traced feature lines to all feature lines (denoted by r_3) are close to the ratio r_1. For example, when the data is down-sampled to 6673 points, only the sixty percent of the original feature lines are contained, the values of r_1, r_2 and r_3 are close to 0.6 simultaneously, and both the correctness rates of the edge detection step and the feature line tracing step are close to 1.0. This indicates that as long as a feature line is contained in the data, our method can detect the edges and trace the feature line segment. Therefore, the experimental results demonstrate that the correctness rates of the edge detection step and the feature line tracing step are not influenced by the point density.

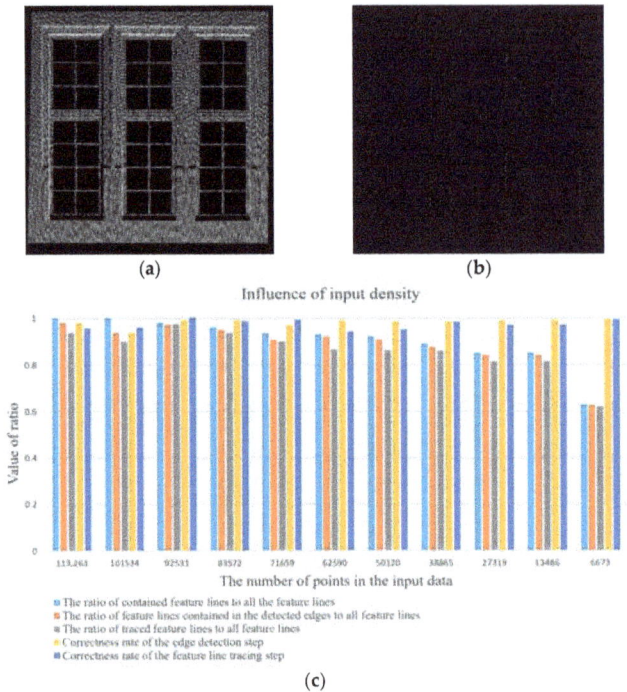

(a)   (b)

Influence of input density

The number of points in the input data

▪ The ratio of contained feature lines to all the feature lines
▪ The ratio of feature lines contained in the detected edges to all feature lines
▪ The ratio of traced feature lines to all feature lines
▪ Correctness rate of the edge detection step
▪ Correctness rate of the feature line tracing step

(c)

Figure 11. (**a**) Original input data without down-sampling; (**b**) the input data down-sampled to 6673 points; (**c**) correctness rates under different densities.

3.6. Results

With the discussed values of the parameters used in the proposed AGPN, the overall performance is evaluated on the aforementioned two testing sites.

To test the capability of the proposed AGPN detecting the defined types of edges presented in Section 1.1, a small scene in Site 2 is selected. The selected small scene and its details are shown in Figure 12a. Four surfaces exist in this scene, i.e., one curve surface rendered by red color and three planar surfaces rendered by blue, green, and yellow colors respectively. There are fold edges belonging to the intersection curves [49] and lines [13], and boundary elements belonging to the plane outlines. Three parts of the edges are marked by white rectangles. Specifically, the fold edges belonging to an intersection curve are the intersetions of the red curve surface and the green planar surface. The results of our proposed AGPN are shown in Figure 12b,c. Figure 12b shows the detected edges. We can find that all the defined edges are detected in the scene of interest. Figure 12c shows the traced feature line segments rendered by different colors. Specifically, the purple and orange segments correspond to the two intersection curves of the scene respectively, and other segments correspond to the intersection lines and outlines.

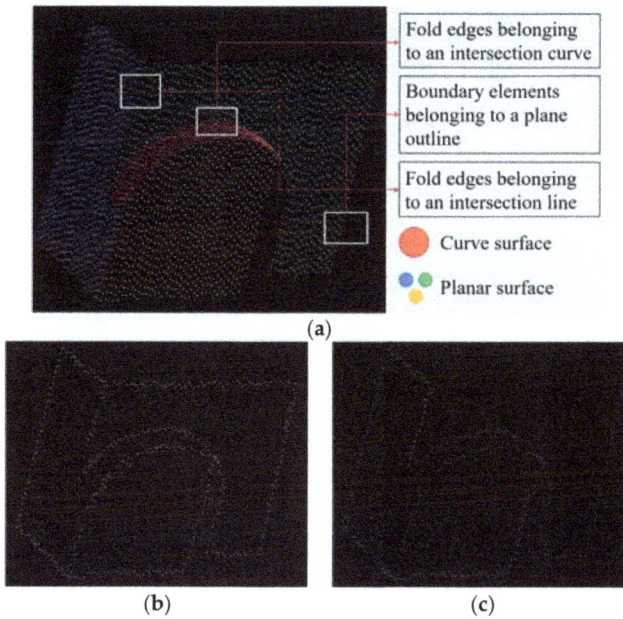

Figure 12. The results of a small area, (**a**) shows the four surfaces, and three kinds of edges in this area; (**b**) shows the edges detected by our method; (**c**) shows the feature line segments traced by our method.

In Site 1, the variation tendency of the point density is consecutive with the variation of the distances to the scanner. Therefore, we set the thresholds of d_r^1 and d_r^2 to the average point spacing 0.01. In Site 2, the variation tendency of the point density is piecewise, i.e., the point spacing in window areas (0.01) is much larger than that in non-window areas (0.005). According to the discussion in Section 3.3, we set the thresholds of d_r^1 and d_r^2 to 0.005 (the average point spacing in non-window areas). The parameters and their values used in the proposed algorithms are listed in Table 1. To clearly demonstrate the performance of the proposed method, the 3D edge detection results are overlaid on the original point cloud.

The results of Sites 1 and 2 are shown in Figures 13 and 14, and the accuracy evaluation results are listed in Table 2. In Table 2, the correctness rates p_{dc} and p_{dct} of Site 2 are higher than that of Site 1. The reason is that the variation range in Site 1 is larger than that in Site 2. According to the analysis in Section 3.3, when the values of d_r^1. and d_r^2 are unequal to the point spacing, the correctness rates will decrease. We can also find that the mislabeled rate p_{mj} of Site 1 is lower than that of Site 2. A close-up visual inspection shows that there are many windows in the second site data, the misjudgments mainly arise in the areas of windows. The reason is that the point spacings in the areas of windows are much larger than the ones in non-window areas. Moreover, the difference of point spacings between windows and non-window areas in Site 2 is larger than that in Site 1. According to the analysis in Section 3.3, when the parameter of distance threshold is smaller than the point spacing, the misjudgments will arise. Furthermore, the tendencies of the correctness rate p_{dc} and the mislabeled rate p_{mj} (in Figure 9a) validates that the mislabeled rates of the edge detection step in Table 2 is reasonable.

In Figures 13d–f and 14d–f, all the defined types of 3D edges are detected by our proposed method, and some details in the results of Sites 1 and 2 are presented. We can find that most of the edges in the window areas or walls with textures belong to intersection curves or intersection lines. Specially, most of the intersected planes or curve surfaces are narrow, which are difficult to be extracted by segmentation or clustering procedures for 3D-point clouds.

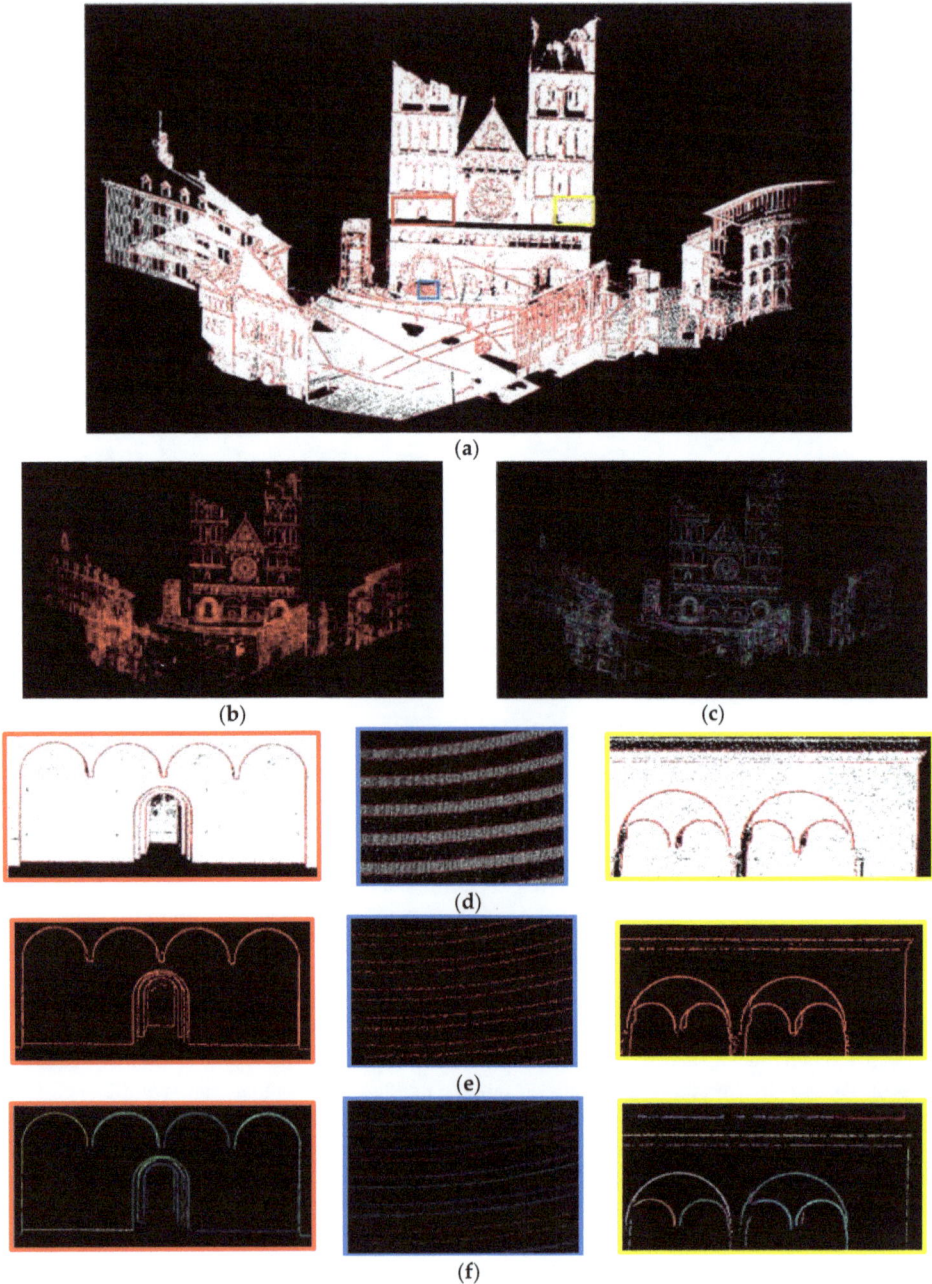

Figure 13. Results of Site 1: (**a**) edge detection result overlaid on the original input data; (**b**) edges; (**c**) traced feature line segments depicted in different colors; (**d–f**) details of edges and traced feature line segments demarcated by different colored outlines corresponding to (**a**).

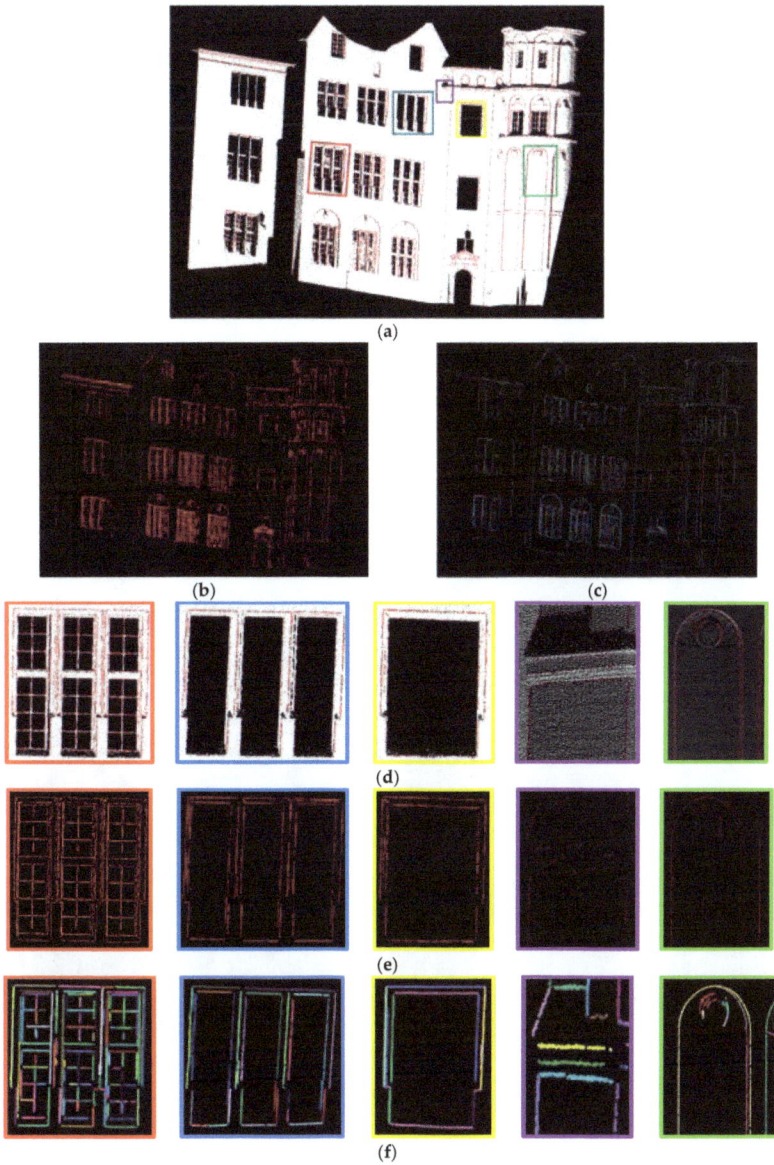

Figure 14. Results of Site 2: (**a**) edge detection result overlaid on the original input data; (**b**) edges; (**c**) traced feature line segments depicted in different colors; (**d–f**) details of edges and traced feature line segments, demarcated by different colored outlines corresponding to (**a**).

Table 2. Accuracy results of the proposed algorithms.

	Edge Detection		Feature Line Tracing	
	p_{dc} (%)	p_{mj} (%)	p_{dct} (%)	p_{mjt} (%)
Site 1	95.6	3.4	86.7	10.1
Site 2	98.1	5.2	87.9	5.3

3.7. Comparative Studies

In our comparison, we do not include all types of the existing methods. This paper compares the proposed AGPN method to "direct methods" because AGPN processes 3D-point clouds directly. In the proposed AGPN, there are two steps, i.e., edge detection step and feature line tracing step. We compare the existing methods with the two steps individually.

A boundary estimation method is presented in PCL, and it uses an angle criterion which is similar to the literature [41]. Comparative results of the boundary estimation method and our edge detection method are shown in Figure 15. Three subsets are selected from the testing datasets. The comparative results show that the proposed edge detection method can detect all types of edges irrespective of how complex the details of the objects are. However, the boundary estimation method in PCL is incapable of detecting fold edges. Table 3 lists the comparative accuracies of our edge detection method and the boundary estimation method.

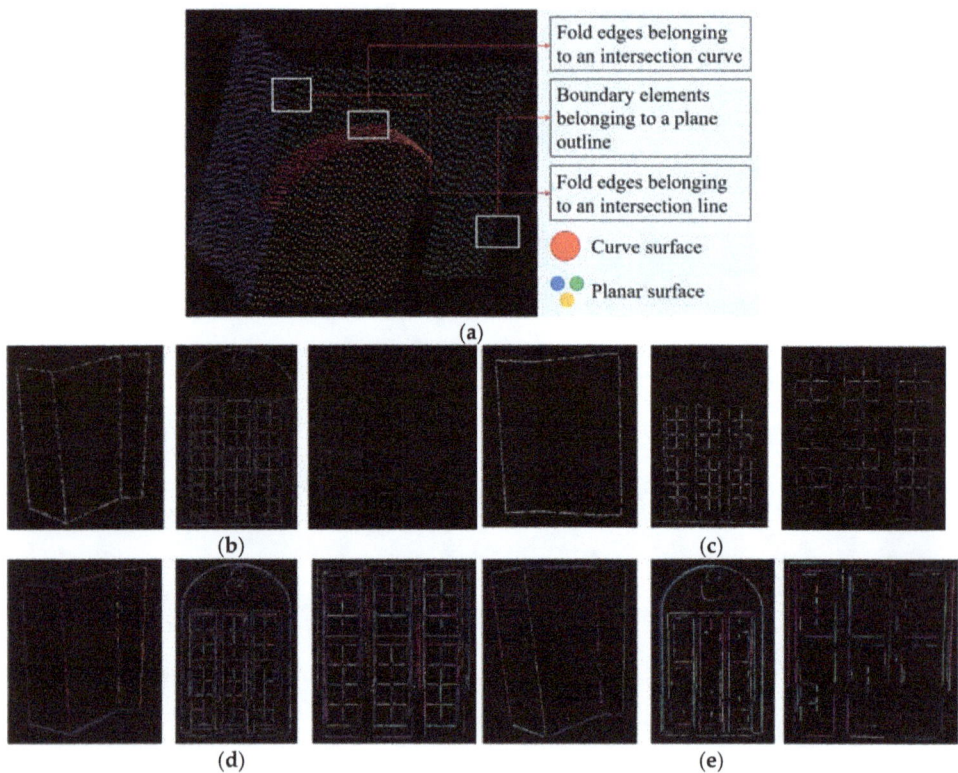

Figure 15. Comparison of the results of the AGPN and existing methods: (**a**) original input data; (**b**) edges detected by our edge detection method; (**c**) edges detected by PCL, where the results are optimal, obtained by training ten sets of parameters; (**d**) feature line segments traced by our method; (**e**) feature line segments traced by the method in the literature [13].

The reason is that the boundary estimation method detects 3D edges using only angle criterion, which is insufficient for detecting fold edges from unorganized point clouds. Besides, the boundary estimation method detects surface boundaries using the PCA-Normal, which ignores the influence of outliers and noises. Therefore, the boundary estimation method is incapable of dealing with an edge with a complex neighborhood.

Table 3. Comparative accuracy results.

	Our edge Detection Method		PCL Method		Our feature Line Tracing Method		Edge Points Clustered by [13]	
	p_{dc} (%)	p_{mj} (%)	p_{dc} (%)	p_{mj} (%)	p_{dct} (%)	p_{mjt} (%)	p_{dct} (%)	p_{mjt} (%)
Data1	100.0	0.0	80.0	0.0	100.0	0.0	100.0	0.0
Data2	88.3	6.3	52.6	2.7	90.3	19.6	50.6	3.4
Data3	86.8	5.2	60.3	5.5	88.9	14.1	31.0	4.4

An edge point clustering method has been proposed by the literature [13]. Comparative results of our feature line tracing method and the algorithm in the literature [13] are shown in Figure 15, where different traced feature line segments are rendered in different colors. Comparative accuracy results of our method and the method in the literature [13] are listed in Table 3. It can be seen that our feature line tracing method performs very well, and the tracing quality of it is obviously better than that of the algorithm in the literature [13].

The reason is that the literature [13] determines the principal direction of each edge point by analyzing a self-correlation matrix constructed by nearest neighbors, but ignoring outliers among the nearest neighbors. When several feature lines intersect, the neighborhood will not be a linear structure, resulting in the edges in this intersecting area will be missed.

4. Conclusions

In this paper, we propose an automated and effective method named AGPN, which detects 3D edges and traces feature lines from 3D-point clouds. There are two main steps in our proposed AGPN: edge detection and feature line tracing. In the first step, AGPN detects edges from 3D-point clouds. This step first analyzes the geometric properties of each query point's neighborhood, and then, combines RANSAC algorithm and angular gap criterion to label the query point as edge or non-edge. In the second step, AGPN traces feature lines from the detected 3D edges. This step is a hybrid method based on region growing and model fitting. In this step, we refine the neighborhood of each query point and redefine two growing criteria to overcome the uncertainties of the region growing procedure.

The contributions of the proposed AGPN method include: (1) image processing or point cloud preprocessing such as segmentation and object recognition are not needed, thereby reducing the complexity of the 3D-point cloud process; (2) the proposed edge detection method in AGPN can detect all the defined types of 3D edges and be insensitive to noise; (3) the feature line tracing step in AGPN is used to distinguish spatially-adjacent, collinear/coaxial lines in complex neighborhoods.

The experimental results show that our proposed AGPN can detect all the defined types of edges irrespective of how complex the details of the objects are. The feature line tracing step can trace feature lines in a complex neighborhood with several intersected or parallel lines. In comparison with state-of-the-art methods, the proposed AGPN can obtain superior results qualitatively and quantitatively. Moreover, we analyze the uncertainties of the results of our proposed method from two aspects, i.e., parameters tuning and influence of point density. In the analysis of the parameters tuning, we present the variation tendency of the correctness rate and the mislabeled rate with different parameters. Then, we achieve a good result according to the point spacing of the input data and the variation tendency. In the analysis of the influence of input density, the experimental results demonstrate that the two steps in our method are not influenced by the input density, though the details of a complex object mainly depend on it.

However, a limitation exists in that over-segmentation arises when the parameters of the feature line tracing procedure are strict or there is a gap in a long feature line. In the future work, we will attempt to solve this problem. In addition, it would be interesting to apply the proposed algorithms to object recognition in large-scale 3D-point clouds.

Acknowledgments: This research was funded by: (1) the General Program sponsored by the National Natural Science Foundations of China (NSFC) under Grant 41371405; (2) the Foundation for Remote Sensing Young Talents by the National Remote Sensing Center of China; and (3) the Basic Research Fund of the Chinese Academy of Surveying and Mapping under Grant 777161103.

Author Contributions: All of the authors contributed extensively to the work presented in this paper. Huan Ni proposed the method, implemented all the algorithms in the method and wrote this manuscript. Xiangguo Lin discussed the structure of the manuscript, named the method with Huan Ni, revised the manuscript, and supplied computers for experiments. Xiaogang Ning revised the manuscript. Jixian Zhang improved the manuscript, supported this study and protected the research process.

Conflicts of Interest: The authors declare no conflict of interest.

Appendix A

There are five parameters in our proposed AGPN. Specifically, the first two parameters are the number of nearest neighbors K^1 and the distance threshold d_r^1 for the RANSAC plane model estimation in the edge detection step. The last three parameters are smooth direction threshold sm_thr, the number of nearest neighbors K^2 for kd-tree, distance threshold d_r^2 for the RANSAC line model estimation in the feature line tracing step. The following pseudo codes are the details.

AGPN

Input: Point cloud = $\{P\}$, parameter K^1, d_r^1, d_r^2, K^2, sm_thr.
1: *Edge points* $\{E\} \leftarrow \varnothing$
2: *Feature line segments* $\{FS\} \leftarrow \varnothing$
3: *Edge detection step* $\{E\} \leftarrow \Xi\left(K^1, d_r^1\right)$
4: *Feature line tracing step* $\{FS\} \leftarrow \Theta\left(E, K^2, d_r^2, sm\_thr\right)$
5: **Return** *Feature line segments* $\{FS\}$

Appendix B

The edge detection step in AGPN uses two parameters. Specifically, the first is the number of nearest neighbors K^1. The second is the distance threshold d_r^1 for the RANSAC model. Furthermore, there is a default threshold $\frac{\pi}{2}$ for the angular gap (G_θ). The following pseudo codes are the details.

3D Edge Detection

Input: Point cloud = $\{P\}$, parameters K^1, d_r^1.
1: *Edge points* $\{E\} \leftarrow \varnothing$
2: **For** $i = 0$ *to* size $(\{P\})$ **do**
3:     *Current neighbors* $\{N_c\} \leftarrow \varnothing$
4:     *Find nearest neighbors of current point* $\{N_c\} \leftarrow \varphi\left(P, i, K^1\right)$
5:     *Current normal vector* $n_c \leftarrow \{0,0,0\}$
6:     *Current inlier list* $\{I_c\} \leftarrow \varnothing$
7:     *Compute current inlier list* $\{I_c\}$ *using RANSAC* $\{I_c\} \leftarrow \Omega\left(N_c, d_r^1\right)$
8:     *Normal optimization* $n_c \leftarrow \phi\left(I_c\right)$
9:     **If** $P_i \notin I_c$ || size (I_c) <3 **then**
10:        Continue
11:    **End If**
12:    *The first axis* $u \leftarrow \{0,0,0\}$, *the second axis* $v \leftarrow \{0,0,0\}$
13:    *Construct coordinate frame* $u, v \leftarrow \Gamma\left(n_c\right)$
14:    *Compute angular gap* $G_\theta \leftarrow \Lambda\left(i, I_c, u, v\right)$
15:    **If** $G_\theta >= \frac{\pi}{2}$ **then**
16:        $\{E\} \leftarrow \{E\} \cup i$
17:    **End If**
18: **End For**
19: **Return** *Edge points* $\{E\}$

References

1. Ando, S. Image field categorization and edge/corner detection from gradient covariance. *IEEE Trans. Pattern Anal. Mach. Intell.* **2000**, *2*, 179–190. [CrossRef]
2. Frei, W.; Chen, C.C. Fast boundary detection: A generalization and a new algorithm. *IEEE Trans. Comput.* **1977**, *10*, 988–998. [CrossRef]
3. Shanmugam, K.S.; Dickey, F.M.; Green, J.A. An optimal frequency domain filter for edge detection in digital pictures. *IEEE Trans. Pattern Anal. Mach. Intell.* **1979**, *1*, 37–49. [CrossRef] [PubMed]
4. McIlhagga, W. The Canny edge detector revisited. *Int. J. Comput Vision* **2011**, *91*, 251–261. [CrossRef]
5. Meer, P.; Georgescu, B. Edge detection with embedded confidence. *IEEE Trans. Pattern Anal. Mach. Intell.* **2001**, *12*, 1351–1365. [CrossRef]
6. Lin, Y.B.; Wang, C.; Cheng, J. Line segment extraction for large scale unorganized point clouds. *ISPRS J. Photogramm. Remote Sens.* **2015**, *102*, 172–183. [CrossRef]
7. Guan, H.Y.; Li, J.; Cao, S.; Yu, Y. Use of mobile LiDAR in road information inventory: A review. *Int. J. Image Data Fusion.* **2016**, *3*, 219–242. [CrossRef]
8. Zhang, J.X.; Lin, X.G. Advances in fusion of optical imagery and LiDAR point cloud applied to photogrammetry and remote sensing. *Int. J. Image Data Fusion.* **2016**. [CrossRef]
9. Weinmann, M.; Jutzi, B.; Hinz, S.; Mallet, C. Semantic point cloud interpretation based on optimal neighborhoods, relevant features and efficient classifiers. *ISPRS J. Photogramm. Remote Sens.* **2015**, *105*, 286–304. [CrossRef]
10. Rosenfeld, A.; Thurston, M. Edge and curve detection for visual scene analysis. *IEEE Trans. Computers.* **1971**, *20*, 562–569. [CrossRef]
11. Vanderheijden, F. Edge and line feature-extraction based on covariance-models. *IEEE Trans. Pattern Anal. Mach. Intell.* **1995**, *17*, 16–33. [CrossRef]
12. Vosselman, G.; Dijkman, S. 3D building model reconstruction from point clouds and ground plans. *Int. Arch. Photogramm. Remote Sens.* **2001**, *34*, 22–24.
13. Borges, P.; Zlot, R.; Bosse, M.; Nuske, S.; Tews, A. Vision-based localization using an edge map extracted from 3D laser range data. In Proceedings of the International Conference on Robotics and Automation (ICRA), Anchorage, AK, USA, 3–7 May 2010.
14. Demarsin, K.; Vanderstraeten, D.; Volodine, T.; Roose, D. Detection of closed sharp edges in point clouds using normal estimation and graph theory. *Comput. Aided Des.* **2007**, *39*, 276–283. [CrossRef]
15. Sampath, A.; Shan, J. Segmentation and reconstruction of polyhedral building roofs from aerial lidar point clouds. *IEEE Geosci. Remote Sens.* **2010**, *48*, 1554–1568. [CrossRef]
16. Sampath, A.; Shan, J. Clustering based planar roof extraction from LiDAR data. In Proceedings of the American Society for Photogrammetry Remote Sensing Annual Conference, Reno, NV, USA, 1–5 May 2006.
17. Pu, S.; Vosselman, G. Knowledge based reconstruction of building models from terrestrial laser scanning data. *ISPRS J. Photogramm. Remote Sens.* **2009**, *64*, 575–584. [CrossRef]
18. Overby, J.; Bodum, L.; Kjems, E.; Iisoe, P.M. Automatic 3D building reconstruction from airborne laser scanning and cadastral data using Hough transform. *Int. Arch. Photogramm. Remote Sens.* **2004**, *35*, 296–301.
19. Pu, S.; Vosselman, G. Extracting windows from terrestrial laser scanning. In Proceedings of the ISPRS Workshop Laser Scanning and Silvi Laser 2007, Espoo, Finland, 12–14 September 2007; pp. 320–325.
20. Boulaassal, H.; Landes, T.; Grussenmeyer, P. Automatic extraction of planar clusters and their contours on building facades recorded by terrestrial laser scanner. *Int. J. Archit Comput.* **2009**, *7*, 1–20. [CrossRef]
21. Sampath, A.; Shan, J. Building boundary tracing and regularization from airborne Lidar point clouds. *Photogramm. Eng. Remote Sens.* **2007**, *7*, 805–812. [CrossRef]
22. Wang, J.; Shan, J. Segmentation of LiDAR point clouds for building extraction. In Proceedings of the American Society for Photogrammetry Remote Sensing Annual Conference, Baltimore, MD, USA, 9–13 March 2009.
23. Nizar, A.; Filin, S.; Doytsher, Y. Reconstruction of buildings from airborne laser scanning data. In Proceedings of the American Society for Photogrammetry Remote Sensing Annual Conference, Reno, NV, USA, 1–6 May 2006.
24. Peternell, M.; Steiner, T. Reconstruction of piecewise planar objects from point clouds. *Comput. Aided Des.* **2004**, *36*, 333–342. [CrossRef]
25. Lafarge, F.; Mallet, C. Creating large-scale city models from 3D-point clouds: A robust approach with hybrid representation. *Int. J. Comput. Vision* **2012**, *1*, 69–85. [CrossRef]

26. Awrangjeb, M. Using point cloud data to identify, trace, and regularize the outlines of buildings. *Int. J. Remote. Sens.* **2016**, *3*, 551–579. [CrossRef]

27. Poullis, C. A framework for automatic modeling from point cloud data. *IEEE Trans. Pattern Anal. Mach. Intell.* **2013**, *11*, 2563–2575. [CrossRef] [PubMed]

28. Heo, J.; Jeong, S.; Park, H.K.; Jung, J.; Han, S.; Hong, S.; Sohn, H.-G. Productive high-complexity 3D city modeling with point clouds collected from terrestrial LiDAR. *Comput. Environ. Urban.* **2013**, *41*, 26–38. [CrossRef]

29. Rottensteiner, F.; Briese, C. Automatic generation of building models from LiDAR data and the integration of aerial images. In Proceedings of the International Society for Photogrammetry and Remote Sensing, Dresden, Germany, 8–10 October 2003; Volume 34, pp. 174–180.

30. Alharthy, A.; Bethel, J. Heuristic filtering and 3d feature extraction from LIDAR data. In Proceedings of the ISPRS Commission III, Graz, Austria, 9–13 September 2002.

31. Alharthy, A.; Bethel, J. Detailed building reconstruction from airborne laser data using a moving surface method. *Int. Arch. Photogramm. Remote Sens.* **2004**, *35*, 213–218.

32. Forlani, G.; Nardinocchi, C.; Scaiono, M.; Zingaretti, P. Complete classification of raw LIDAR and 3D reconstruction of buildings. *Pattern Anal. Appl.* **2006**, *8*, 357–374. [CrossRef]

33. Brenner, C. Building reconstruction from images and laser scanning. *Int. J. Appl. Earth Obs. Geoinf.* **2005**, *6*, 187–198. [CrossRef]

34. Li, H.; Zhong, C.; Hu, X.G. New methodologies for precise building boundary extraction from LiDAR data and high resolution image. *Sensor Rev.* **2013**, *2*, 157–165. [CrossRef]

35. Li, Y.; Wu, H.; An, R. An improved building boundary extraction algorithm based on fusion of optical imagery and LIDAR data. *Optik* **2013**, *124*, 5357–5362. [CrossRef]

36. Hildebrandt, K.; Polthier, K.; Wardetzky, M. Smooth feature lines on surface meshes. In Proceedings of the 3rd Eurographics Symposium on Geometry Processing, Vienna, Austria, 4–6 July 2005.

37. Altantsetseg, E.; Muraki, Y.; Matsuyama, K.; Konno, K. Feature line extraction from unorganized noisy point clouds using truncated Fourier series. *Visual Comput.* **2013**, *29*, 617–626. [CrossRef]

38. Huang, J.; Menq, C. Automatic data segmentation for geometric feature extraction from unorganized 3-D coordinate points. *IEEE Trans. Robot. Autom.* **2001**, *17*, 268–279. [CrossRef]

39. Gumhold, S.; Wang, X.; Macleod, R. Feature extraction from point clouds. In Proceedings of the 10th International Meshing Roundtable, Sandia National Laboratory, Newport Beach, CA, USA, 7–10 Octerber 2001.

40. Linsen, L.; Prautzsch, H. Local versus global triangulations. In Proceedings of the EUROGRAPHICS 2001, Manchester, UK, 2–3 September 2001.

41. Truong-Hong, L.; Laefer, D.F.; Hinks, T. Combining an angle criterion with voxelization and the flying voxel method in reconstructing building models from LiDAR data. *Comput. Aided Civ. Inf.* **2013**, *28*, 112–129. [CrossRef]

42. Linsen, L.; Prautzsch, H. Fan clouds—An alternative to meshes. In Proceedings of the 11th International Workshop on Theoretical Foundations of Computer Vision, Dagstuhl Castle, Germany, 7–12 April 2002.

43. PCL-The Point Cloud Library. 2012. Available online: http://pointclouds.org/ (accessed on 10 March 2016).

44. Bendels, G.H.; Schnabel, R.; Klein, R. Detecting holes in point set surfaces. *J. WSCG* **2006**, *14*, 89–96.

45. Zhang, J.X.; Lin, X.G.; Ning, X.G. SVM-based classification of segmented airborne LiDAR point clouds in urban area. *Remote Sens.* **2013**, *5*, 3749–3775. [CrossRef]

46. Vosselman, G.; Gorte, B.G.H.; Sithole, G.; Tabbani, T. Recognize structure in laser scanning point clouds. *Int. Arch. Photogramm. Remote Sens.* **2004**, *46*, 33–38.

47. CloudCompare-3D Point Cloud and Mesh Processing Software. 2013. Available online: http://www.danielgm.net/cc/ (accessed on 10 March 2016).

48. DTK-The 3D ToolKit. 2011. Available online: http://slam6d.sourceforge.net (accessed on 10 March 2016).

49. Wikipedia, the Free Encyclopedia. Available online: https://en.wikipedia.org/wiki/Intersection_curve (accessed on 5 May 2016).

remote sensing

MDPI

Article

Fast and Accurate Plane Segmentation of Airborne LiDAR Point Cloud Using Cross-Line Elements

Teng Wu [1], Xiangyun Hu [1,2,*] and Lizhi Ye [1]

[1] School of Remote Sensing and Information Engineering, 129 Luoyu Road, Wuhan University,
     Wuhan 430079, China; whurswuteng@whu.edu.cn (T.W.); ye_lizhi@whu.edu.cn (L.Y.)
[2] Collaborative Innovation Center of Geospatial Technology, Wuhan University, Wuhan 430079, China
*   Correspondence: huxy@whu.edu.cn; Tel.: +86-27-6877-1528; Fax: +86-27-6877-8010

Academic Editors: Juha Hyyppä, Nicolas Baghdadi and Prasad S. Thenkabail
Received: 25 February 2016; Accepted: 27 April 2016; Published: 5 May 2016

Abstract: Plane segmentation is an important step in feature extraction and 3D modeling from light detection and ranging (LiDAR) point cloud. The accuracy and speed of plane segmentation are two issues difficult to balance, particularly when dealing with a massive point cloud with millions of points. A fast and easy-to-implement algorithm of plane segmentation based on cross-line element growth (CLEG) is proposed in this study. The point cloud is converted into grid data. The points are segmented into line segments with the Douglas-Peucker algorithm. Each point is then assigned to a cross-line element (CLE) obtained by segmenting the points in the cross-directions. A CLE determines one plane, and this is the rationale of the algorithm. CLE growth and point growth are combined after selecting the seed CLE to obtain the segmented facets. The CLEG algorithm is validated by comparing it with popular methods, such as RANSAC, 3D Hough transformation, principal component analysis (PCA), iterative PCA, and a state-of-the-art global optimization-based algorithm. Experiments indicate that the CLEG algorithm runs much faster than the other algorithms. The method can produce accurate segmentation at a speed of 6 s per 3 million points. The proposed method also exhibits good accuracy.

Keywords: cross-line elements; plane segmentation; airborne LiDAR point cloud; line segmentation; fast segmentation

1. Introduction

To segment a light detection and ranging (LiDAR) point cloud is to partition the points into different groups with homogeneous properties, such as height, density, and normality. Using plane segmentation to extract facets from a point cloud is important in object classification, building extraction, and roof reconstruction. The main methods of plane segmentation are generally categorized as edge detection, profile line analysis, point clustering, model fitting, region growth and optimization.

Edge detection methods [1,2] convert a point cloud into a digital surface model (DSM). Edge detection of the raster DSM is then implemented for segmentation, the quality of which depends on the edge detection operator.

Methods based on profile line analysis employ scan line analysis to identify planes [3]. Proper selection of the scan line direction is essential in these methods [4]. The profiles in one or more directions are utilized to segment the data in order to detect man-made structures (*i.e.*, bridges and buildings) from the LiDAR point cloud [5–7]. These methods are usually effective and fast. However, using profile information for accurate plane segmentation remains insufficiently explored. The algorithm design, quality and performance assessment compared with existing methods need to be comprehensively investigated.

Methods based on point clustering, including octree-based clustering [8,9], K-means clustering [10–12], fuzzy clustering [13,14] and mean shift [15–17], cluster the point cloud into point groups by using similarity measurements, such as distance between points and point density. These methods can produce stable results but may lead to over-segmentation or under-segmentation because of the improper clustering algorithm setup (e.g., parameters of the kernel width and the minimum point number of a valid region in mean shift segmentation) [18].

Methods based on model fitting attempt to solve the plane equation by fitting local points with the presupposed model. Random sample consensus (RANSAC) [19], Hough transform [20], and tensor voting [21] are popular algorithms in this category. RANSAC can outperform methods based on normal vector consistency and outline segmentation [22]. Normal driven RANSAC is an accelerated version of the original RANSAC [23]. The limitation of RANSAC is that the neighborhood of points located on the same plane is not fully considered. The algorithm selects planes with the maximum number of support points in each iteration, which may not be correct. Several improved algorithms have been developed for these problems [24,25]. 3D Hough transform is a voting-based algorithm of plane extraction in 3D Hough space (θ, ϕ, ρ). The disadvantage of this method is that the voting operation in the 3D Hough space is usually slow; the same problem is encountered in selecting support points [26]. Many methods (e.g., random Hough transformation) have been proposed to speed up Hough transformation [27]. Tensor voting obtains 3D normal vector field based on discrete points, by which the maximum tendency is utilized to extract characteristic regions [28,29]. The drawback of the tensor voting method is the dependency on selecting the parameter of the range of influence [28].

Methods based on region growth select seed points or regions as the original patches and cluster the points subordinated to the same patch [30–34]. These methods can also be integrated with model fitting methods. These methods ensure that the points on the same plane are in the neighborhood; they are faster than model fitting methods when the point number is large [35]. The normal vectors of points in the region of growth can be computed through principal component analysis (PCA). The region of growth similar to the image region of growth is then utilized to extract planes [36]. An iterative PCA is developed to estimate local planarity [37]. Region growth methods usually rely on the choice of seed points. The computation of the normal vectors becomes unstable when noise points exist or the supporting points are not properly selected. In addition, these methods may lead to over-segmentation or under-segmentation in the surface intersection region and noisy areas [38].

Optimization-based methods are inspired by image segmentation that uses a graph to represent data elements (e.g., pixels or super pixels) with connected nodes. The segmentation can be modeled as an optimization problem to determine the best graph cut [39–41]. The frequently used graph cut algorithms are minimum spanning tree [42], normalized cut [43,44] and Graphcuts [38]. Other optimization methods, such as level set, are also utilized to segment planes [45]. A recent study has shown that using Graphcuts to optimize the initial segmentation [38] significantly improves the initial over-segmentation and eliminates the cross-planes. The limitation of this method is that the result relies on the initial segmentation, and the speed is low because of its iterative optimization operation [46].

Developing a fast, accurate, and easy-to-implement segmentation algorithm is still necessary to address the various scenarios involving massive point numbers, noisy and complex object contexts.

This paper presents a new segmentation method based on cross-line elements growth (CLEG). This method combines profile analysis, model fitting and region growth. The point cloud is converted into a grid index data structure. The Douglas-Peucker algorithm [47] is subsequently utilized in four directions to extract the cross-line elements (CLEs). CLE can determine a plane, and this is the rationale of the proposed method. The final facets can be obtained after selecting the seed CLEs and combining CLE growth and point growth. Comparison of CLEG with other popular methods, such as RANSAC [19], 3D Hough transformation [27], PCA [36], iterative PCA [37] and a state-of-the-art global optimization-based algorithm [38], shows that the proposed algorithm runs much faster than them and produces stable and accurate results. The remainder of the paper is structured as follows.

Section 2 formulates the proposed segmentation method. Section 3 describes the test data and presents the experimental analysis. Section 4 provides the conclusion.

2. Plane Segmentation Using Cross-Line Elements

In general, a good plane segmentation algorithm has to address some key issues: (1) how to accurately measure local planarity with proper selection of support points for these measurements; (2) how to properly group all spatially adjacent points belonging to one facet; (3) how to efficiently deal with large-scale data. The existing methods, such as model fitting, clustering, region growth and global optimization, have more or less room to improve in these aspects, as presented in the introduction. In this study, aiming at solving these problems, a cross-line element growth (CLEG) method is proposed to segment point cloud accurately and efficiently.

The workflow of the CLEG algorithm is shown in Figure 1; the red lines in the segmentation result are the seed CLEs, and the white points are the gross noise points.

The pseudo-code is listed to describe the principle of the algorithm:

```
CLEG(points, label)
Grids=StoringPointsInGrid(points);
directions = horizon, vertical, upper right, lower right;
for each direction
        LineSegmentation=DouglasPeucker(Grids);
end for
for each grid
    if CLE crossing the grid is stable
        Add grid to seeds;
    end if
end for
Sort(seeds);
for each seed
    if not labeled
        GetPlanFunction(CLE);
        CLEbasedgrowth(label);
        Pointbasedgrowth(label);
    end if
end for
End
```

A CLE is defined as two cross-lines at a cross-point in two directions. In one direction, the cross-line is determined by two planes, *i.e.*, the candidate plane and one special plane (e.g., ZOY plane, plane 1, plane 2, and ZOX plane in Figure 1). The directions of the cross-lines are relative to the equation of the candidate plane. The cross-line is determined by the candidate plane and ZOY plane, for example; Equation (1) is the function of the candidate plane, and Equation (2) is the function of ZOY plane.

$$a \cdot x + b \cdot y + c \cdot z + d = 0 \tag{1}$$

$$y = e \tag{2}$$

The direction vector of the cross-line is then $(c, 0, -a)$. Similarly, the direction vector of the cross-line determined by the candidate plane and the plane 1 is $(-c, c, b - a)$; the direction vector of the cross-line determined by the candidate plane and the plane 2 is $(c, c, -a - b)$; and the direction vector of the cross-line determined by the candidate plane and ZOX plane is $(0, c, -b)$. Therefore, a CLE can determine the plane model.

In 3D space, two intersected straight lines passing the cross-point determine the plane model. In other words, a point and the normal vector formed by the intersected lines are the basic elements in plane detection, which is the rationale of CLEG-based plane segmentation. The CLEG algorithm has the following advantages: (1) the CLEs can be easily and quickly extracted in the profile space; (2) a CLE contains rich information, such as rough plane model and facet size; which can further help in finding better seeds and measurements for the growth of CLEs and points; (3) pre-segmenting the point cloud into CLEs eliminates the problem of selecting support points in clustering and model fitting methods [25], which leads to a more accurate and stable segmentation; and (4) the CLE extraction and growth operation are efficient in terms of computational cost, thereby making it suitable for use when dealing with a massive number of points.

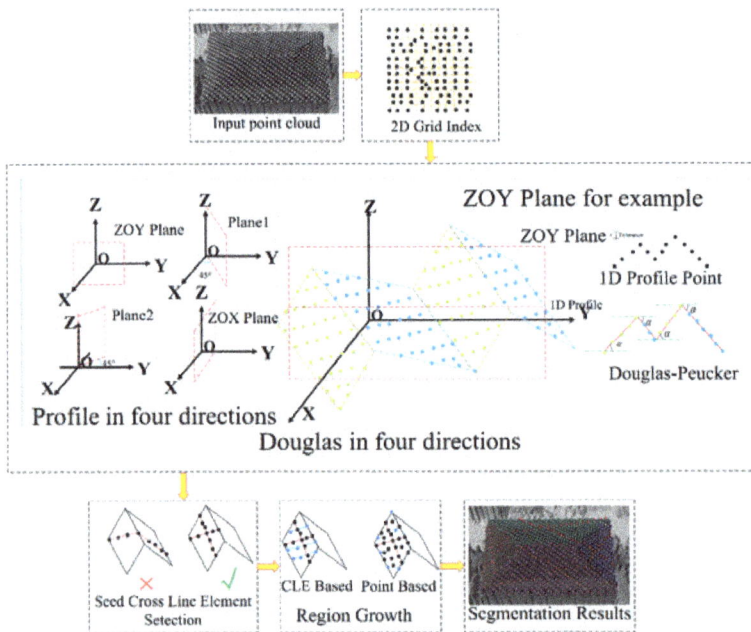

Figure 1. Workflow of plane segmentation using cross-line elements.

2.1. Line Segmentation

The seed CLE is derived by first converting the point cloud into a grid index data structure based on ground sample distance (GSD), which can be obtained from the average point density. The grid index data structure is utilized to improve the efficiency of data inquiry. More than one point may exist in each grid. Some grids may also be null, as shown in Figure 1 (*i.e.*, 2D Grid Index).

An extended line segmentation of scanning line segmentation [7] is employed to segment the profiles in four directions (*i.e.*, vertical, horizontal, upper right, and lower right). The angle between the split line segment and the horizon direction is calculated by using the Douglas-Peucker algorithm (Figure 1) [47]. The tolerance is ε. The difference between the original Douglas-Peucker algorithm and the proposed method is that the angles between the line segments and the horizontal plane are calculated simultaneously (denoted by α in Figure 1). The angles are important in the subsequent steps in seed selection and growth. The length and angle of each grid in each direction is then obtained, as shown in Figure 2. The black points are the uncolored points because more than one point may exist in one grid, and only the highest point is colored. Each grid is crossed by line segments and defined as a cross-point after using the Douglas-Peucker algorithm in four directions.

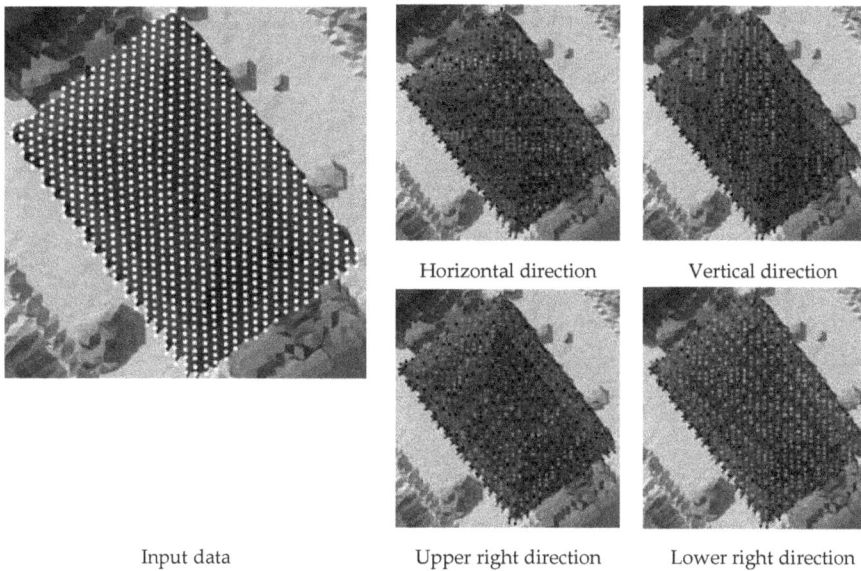

	Horizontal direction	Vertical direction

Input data	Upper right direction	Lower right direction

Figure 2. Line segmentation in four directions.

The length of a line segment becomes relevant to the surface roughness of the region after line segmentation. The lines are much longer on large planes (e.g., ground and roof) and shorter in regions with a significant height difference (e.g., tree area). A valid CLE is defined as the cross-line whose length is longer than threshold *l* at a cross-point in two directions. All the lines crossing the cross-point may be longer than the threshold. The two longer lines indicate the principal directions. The facets are obtained by using the CLEs to select the seed and region growth.

2.2. Selection of Seed CLEs for Growth

A coarse-to-fine strategy is employed to extract prior large planes and guarantee the segmentation quality and stability. The seed CLE is selected based on estimations of the plane property. The seed CLE should satisfy the following conditions.

1. Each line of the seed CLE is longer than the minimum length threshold *l*.
2. The cross-point of the seed CLE should not be the end points of the line segments to ensure the stability of the seed CLE. A false seed CLE is shown in Figure 1. The red cross denotes the false selection of the seed cross-line element.
3. The variance between the angle (*i.e.*, α in Figure 1) of the cross-point and those of the neighbor points should be small. Figure 3 shows red lines, which denote the seed CLE and the red point, which represents the cross-point. The ZOY plane is the segmentation direction and nb_1 nb_2 ... nb_8 are the neighbors of the cross-point. The $\alpha_0, \alpha_1 ... \alpha_9$ variance should be smaller than the threshold and should extend to the four directions to ensure the stability of the seed CLE. Several false seed CLEs could be found in the tree areas if the condition is not applied. The variance in the rough areas can be large because the angles can vary significantly even if the lines of CLE are longer than *l*.

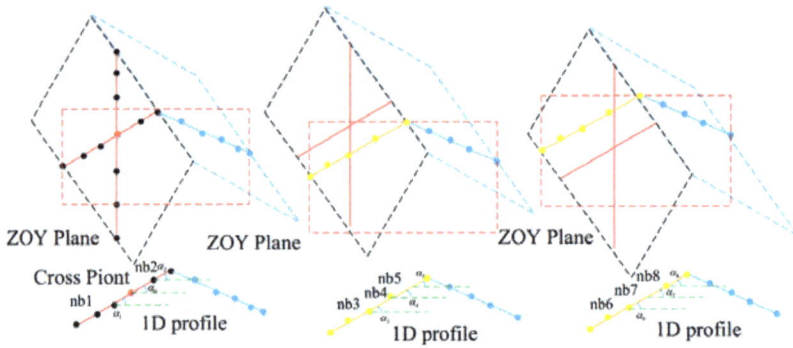

Figure 3. Seed CLE and neighbors of the cross-point.

The cross-points that meet the aforementioned conditions are sorted by using the length of the CLE. The seed cross-points of the CLE are then processed in order.

The points on the CLE may not be on the same plane when the seed CLE is selected. Figure 4 shows the CLE, which is represented by red lines. The CLE should be checked as valid.

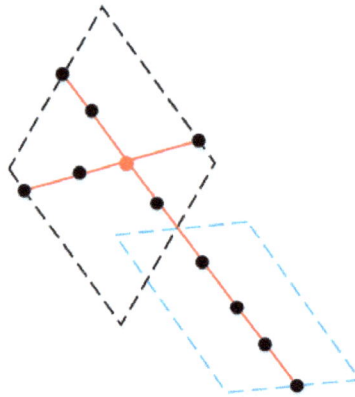

Figure 4. Points on CLE may not be on a plane.

The characteristic of the line intersect with the plane indicates that the angles between the parallel lines and the plane are equal. Accordingly, α_1, α_2, α_3, β_1, β_2, β_3 represent the angles of the line segments (Figure 5). The condition that $\alpha_1 = \alpha_2 = \alpha_3$ and $\beta_1 = \beta_2 = \beta_3$ should be satisfied when the plane is perfect. The valid seed CLE should also satisfy the condition that the difference between the angles of the point on CLE and that of the cross-point is sufficiently small. A threshold of $\Delta\alpha$ is utilized in this study. Figure 4 shows that the points on the blue plane do not satisfy the condition. $\Delta\alpha$ can be obtained adaptively.

$$\Delta\alpha = \arctan(\frac{d}{l}) \tag{3}$$

In Equation (3), d is the threshold of point to plane distance. l is the minimum line length threshold.

The region growth is the employed to obtain the points of the entire plane after the seed CLEs are extracted.

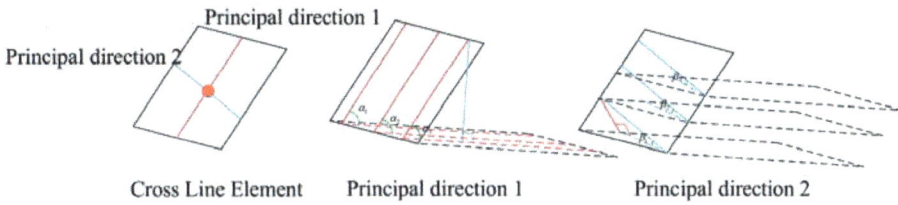

Figure 5. Cross-line element and its characteristic.

2.3. Region Growth

Region growth includes CLE growth and point element growth. Using CLE growth can improve the stability of region growth and accelerate the process.

The disadvantage of the conventional region growth method is the process of obtaining seed points and the reliable similarity measurement of region growth. Researches sometimes use the minimum number of points as the indicator of a valid plane. However, this method may not be stable because of the complex point distribution at tree and noisy areas.

Similar to PCA, the angle limitation is added in the region growth. Subsequently, the angles are more stable than those in the PCA because calculating the angles does not depend on the neighboring relationship. The angles can also be correctly calculated at the edge of the plane, as shown in Figure 1 (*i.e.*, Douglas-Peucker). The angle limitation is that the angles on the horizontal plane of the two principal directions of each candidate point are nearly equal to those of the seed cross-point. The red point in Figure 6 denotes the cross-point. α_0 and β_0 are the angles of CLE in the two principal directions. The angles of the lines crossing the candidate point in the two principal directions should be nearly equal to the cross-point seed when dealing with region growth.

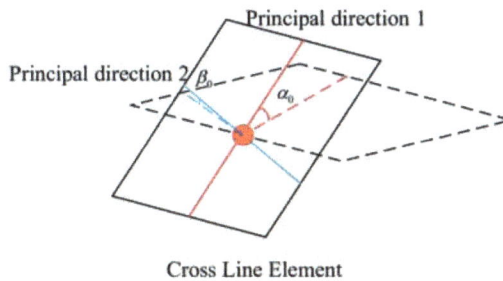

Figure 6. Cross-point of CLE.

Combining CLE growth and point growth can ensure the stability of the region growth. The sequence of adding points in the model fitting procedure influences the results of the region growth. The points on CLE are more stable and have more information than those on the short lines. Therefore, the points on CLE are processed first to ensure the stability of the region growth. The next seed is then processed if no line is added in the CLE growth. The valid seed CLEs are used to calculate the plane function after the stable CLE is obtained.

2.3.1. CLE Growth

After obtaining the seed CLEs, CLE growth is utilized to calculate the principal direction lines which are not the cross-lines of the seed CLE at each seed point and to check whether the candidate CLE is on the plane. This seed CLE is omitted if no CLE is added because the seed CLE is unstable. In Figure 7, the red lines are the seed CLE. The blue and yellow lines are to be grown in step one of

CLE growth. The blue lines on the plane are to be added. The yellow lines are not on the plane. A more stable plane function is obtained thereafter.

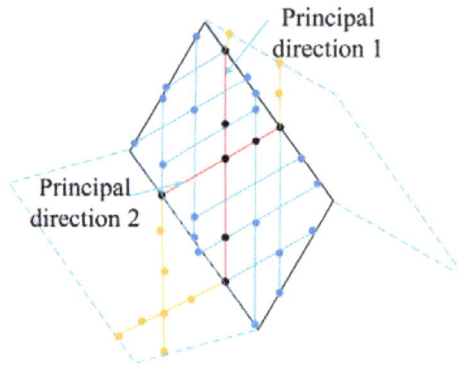

Figure 7. Step one of CLE growth.

All the directions of CLE growth are subsequently employed. The CLE growth principle is similar to that of point growth. The only difference is that the elements are crossing lines. Only the end points of the crossing lines are used in the measurement procedure to analyze whether the crossing lines are on the plane or not.

2.3.2. Point Growth

After CLE growth, some points may be ignored because of noises. The distance of the point to the plane is measured in processing of the point growth. The angles of the lines crossing the candidate points on the two principal directions should be nearly equal to those of the seed cross-point. As shown in Figure 6, α_0 and β_0 of the candidate points should be nearly equal to those of the seed cross-point; otherwise, the length of the line segment is small.

3. Experimental Analysis

3.1. Test Data

The LiDAR point clouds of three different regions are utilized to validate the proposed method. The regions are the Vaihingen area in Germany [48], Wuhan and Guangzhou in China. The description of the datasets is listed in Table 1.

The comparison test consists of roof and area segmentation.

Table 1. Descriptions of test data.

Site	Vaihingen	Wuhan	Guangzhou
Total area size	2,320,000 m^2	127,636,898 m^2	60,115,494 m^2
Point density	4 points/m^2	8 points/m^2	6 points/m^2
Roof type	Mostly gable roof with a large slope	Flat roof and gable roof	Flat and gable roofs with a small slope
Scene type	Urban area with little trees	Urban area with many trees	Urban area with many trees
Feature	The roofs are simple	The roofs are complex	The slope of the roof is small
Used points	3,911,955	2,374,018	15,597,504

3.2. Roof Segmentation

Several of the typical segmentation methods for roof segmentation used in the comparison test are RANSAC [19], 3D Hough transformation [27], PCA + region growth (RG_PCA) [36], iterative PCA + region growth (RG_IPCA) [37], and the global optimization-based algorithm Graphcuts (Global energy) [38]. The algorithms are all implemented with Microsoft Visual C++ under the Microsoft Windows 7 operating system. A personal computer with Intel Core i5, 2.5 GHz CPU, 4GB memory is used for the testing. The ground truth of roof segmentation for quality evaluation is obtained through manual editing.

The seven metrics utilized to evaluate CLEG and the compared algorithms are computation time (time), completeness (comp), correctness (corr) [49], reference cross-lap (RCL), detection cross-lap (DCL) [50,51], boundary precision (BP), and boundary recall (BR) [52].

Completeness is defined as the percentage of reference planes that are correctly segmented. This metric is related to the number of misdetected planes.

$$comp = \frac{TP}{TP + FN} \tag{4}$$

Correctness denotes the percentage of correctly segmented planes in the segmentation results. It indicates the stability of the methods.

$$corr = \frac{TP}{TP + FP} \tag{5}$$

TP in Equations (4) and (5) denotes the number of planes found in both the reference and segmentation results. Only the planes with a minimum overlap of 50% with the reference are true positives. *FN* denotes the number of reference planes not found in the segmentation results (*i.e.*, false negatives). *FP* is the number of detected planes not found in the reference (*i.e.*, false positives).

Reference cross-lap rate is defined as the percentage of reference planes that overlap multiple detected planes. This metric shows the over-segmentation of the methods.

$$RCL = \frac{N_r'}{N_r} \tag{6}$$

N_r in Equation (6), denotes the number of reference planes, and N_r' is the number of reference roof planes that overlap more than one detected plane.

Detection cross-lap rate denotes the percentage of detected planes that overlap multiple reference roof planes. This metric shows the under-segmentation of the methods.

$$DCL = \frac{N_d'}{N_d} \tag{7}$$

N_d in Equation (7) denotes the number of detected planes, and N_d' is the number of detected planes that overlap more than one reference roof plane.

Boundary precision measures the percentage of correct boundary points in the detected boundary points.

$$BP = \left| \frac{B_d \cap B_r}{B_d} \right| \tag{8}$$

Boundary recall measures the percentage of correct boundary points in the reference boundary points.

$$BP = \left| \frac{B_d \cap B_r}{B_r} \right| \tag{9}$$

B_r in Equations (8) and (9) denotes the boundary point set in reference, B_d denotes the boundary point set in the segmentation results, and | | denotes the number of points in a dataset. Over-segmentation may result in a high boundary recall ratio, whereas under-segmentation may lead to high boundary precision. Only when boundary precision and boundary recall are both high can the precision of the method be determined.

The same parameters are utilized in the comparison test to ensure the comparability of the results as shown in Table 2.

Table 2. Parameters used in the comparison test.

Parameter	Value	Methods
Point to plane distance threshold d	0.3 m	RANSAC, Hough, RG_PAC, Global energy, and CLEG
Curvature threshold	0.01	RG_PCA and RG_IPCA
Minimum number of points required for a valid plane	10	RANSAC, Hough, RG_PCA, RG_IPCA, and Global energy
Line segmentation threshold ε	0.25 m	CLEG
Grid size	0.6 m	CLEG
Min line length l	1.8 m	CLEG

Many gable roofs with large slopes are found in Vaihingen. The roof structure is also complex, as shown in Figure 8a. Some noise points also exist (Figure 8b). A complex roof structure with planes that have a small slope difference with its neighbor planes, and also with small structures, is shown in Figure 8c. Many flat and gable roofs are found in Wuhan. The slope of gable roofs is not large. A flat roof is close to the gable roofs, as shown in Figure 9a. A complex symmetric roof structure is shown in Figure 9b. A symmetric trapezoid roof is shown in Figure 9c. Many gable roofs with small slopes are found in Guangzhou. The nearly arc-shaped roofs results in weak edges of the planes, as shown in Figure 10a. Figure 10b,c show several complex structures and roofs close to one another.

Segmentation results of roof points in the Vaihingen area are shown in Figure 8, and the evaluation of precision is listed in Table 3.

Segmentation results of roof points in the Wuhan area are shown in Figure 9, and the evaluation of precision is listed in Table 4.

Table 3. Quality of the segmentation results in the Vaihingen area.

		Time	% Comp	% Corr	% RCL	% DCL	% BP	% BR
(a)	RANSAC	0.016 s	**100**	85.7	0	28.6	72.1	78.7
	3D Hough	59.795 s	75	37.5	75	0	29.8	44.2
	RG_PCA	0.016 s	**100**	72.7	25	0	**100**	5.8
	RG_IPCA	0.015 s	**100**	88.9	12.5	0	94.2	55.1
	Global energy	0.062 s	**100**	88.9	12.5	0	84.9	81.6
	CLEG	**<1 ms**	**100**	**100**	0	0	97.7	**93.7**
(b)	RANSAC	0.046 s	77.8	70	0	80	40.1	88.1
	3D Hough	244.329 s	44.4	7.8	66.7	2.1	17.4	55.9
	RG_PCA	0.046 s	**100**	42.9	44.4	0	41.1	38.4
	RG_IPCA	0.015 s	**100**	62.3	22.2	7.1	48.2	53.6
	Global energy	0.328 s	**100**	81.8	22.2	0	83.4	**84.0**
	CLEG	**<1 ms**	**100**	**100**	0	0	**87.5**	83.4
(c)	RANSAC	0.016 s	**100**	87.5	14.3	12.5	60.4	**87.1**
	3D Hough	84.038 s	71.4	27.8	57.1	5.6	20.5	48.9
	RG_PCA	0.031 s	**100**	**100**	0	0	**84.8**	8
	RG_IPCA	0.015 s	85.7	**100**	0	16.7	50.9	51.4
	Global energy	0.078 s	85.7	**100**	0	16.7	71.5	60.3
	CLEG	**<1 ms**	**100**	**100**	0	0	84.3	86

Table 4. Quality of segmentation results in the Wuhan area.

		Time	% Comp	% Corr	% RCL	% DCL	% BP	% BR
(a)	RANSAC	<1 ms	100	100	0	33.3	19.3	63.4
	3D Hough	136.376 s	33.3	9.1	66.7	41.7	3.7	10.1
	RG_PCA	0.031 s	100	75	33.3	0	40	65.6
	RG_IPCA	0.015 s	100	75	33.3	33.3	26.6	68.3
	Global energy	0.047 s	100	100	0	0	77.5	75.4
	CLEG	<1 ms	100	100	0	0	**87.1**	**84.7**
(b)	RANSAC	0.827 s	15.4	26.7	0	73.3	17.7	60.2
	3D Hough	1434.733 s	46.2	13.4	50	14.1	16.7	54.9
	RG_PCA	0.063 s	100	52	34.6	0	28.3	23.2
	RG_IPCA	0.201 s	100	86.7	3.8	3.3	64.1	61.9
	Global energy	2.325 s	100	88.9	0	7.4	73.3	70.6
	CLEG	**0.015 s**	100	100	0	0	**92.7**	**91.3**
(c)	RANSAC	0.016 s	100	71.4	0	28.5	34.0	66.1
	3D Hough	161.929 s	60	11.1	80	44.4	18.3	61.1
	RG_PCA	0.031 s	100	35.7	60	0	29.2	52.9
	RG_IPCA	0.016 s	100	55.6	60	0	46.5	71.3
	Global energy	0.109 s	100	83.3	20	0	65.6	71.7
	CLEG	<1 ms	100	100	0	0	**88.8**	**87.8**

Segmentation results of roof points in the Guangzhou area are shown in Figure 10, and the evaluation of precision is listed in Table 5.

RANSAC runs fast when the point number is small (Table 4). The time of dataset (a) is less than 1 ms. However, the algorithm runs slow when the point number is large. The voting procedure with all the left points is undertaken afresh when a plane is found. When the roof structure is complex, many errors occur because the spatial relationship of the neighbors is not considered. The results are shown in the black rectangles in Figures 8a–c, 9a–c and 10a–c.

Table 5. Quality of segmentation results in the Guangzhou area.

		Time	% Comp	% Corr	% RCL	% DCL	% BP	% BR
(a)	RANSAC	0.047 s	23.1	37.5	15.4	50	20.0	62.6
	3D Hough	556.970 s	61.5	14.3	76.9	12.5	17.5	54.5
	RG_PCA	0.110 s	84.6	91.7	8.7	0	75.4	21.4
	RG_IPCA	0.031 s	84.6	68.8	8.7	0	69.3	59.0
	Global energy	0.842 s	84.6	78.6	8.7	0	78.0	72.0
	CLEG	**0.016 s**	100	100	0	0	**86.0**	**80.9**
(b)	RANSAC	0.016 s	50	51.4	0	28.6	62.8	77.8
	3D Hough	610.557 s	50	44.4	25	22.2	18.9	9.2
	RG_PCA	0.078 s	100	38.1	37.5	0	66.2	58.8
	RG_IPCA	0.047 s	75	40	12.5	6.7	83.1	57.0
	Global energy	0.374 s	100	100	0	0	86.0	77.8
	CLEG	<1 ms	100	100	0	0	**95.5**	**96.9**
(c)	RANSAC	0.047 s	9.1	14.3	0	100	17.1	57.8
	3D Hough	866.773 s	18.2	3.6	45.5	16.4	15.3	50.5
	RG_PCA	0.078 s	63.6	53.8	18.2	16.4	74.1	29.8
	RG_IPCA	0.046 s	100	84.6	9.1	15.4	40.6	38.9
	Global energy	0.374 s	100	100	0	0	74.5	73.6
	CLEG	**0.015 s**	100	100	0	0	**85.6**	**87.0**

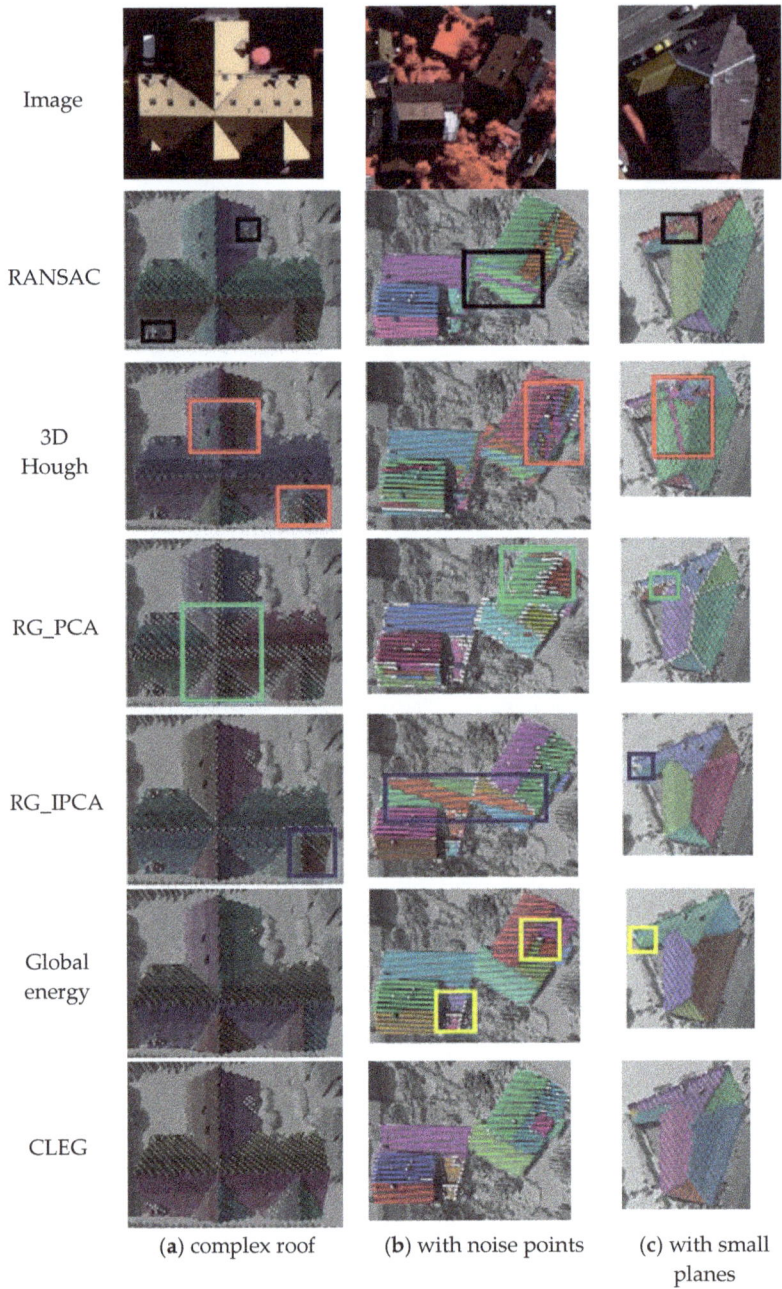

Figure 8. Segmentation of roof points in Vaihingen. (**a**) complex roof; (**b**) with noise points; (**c**) with small planes.

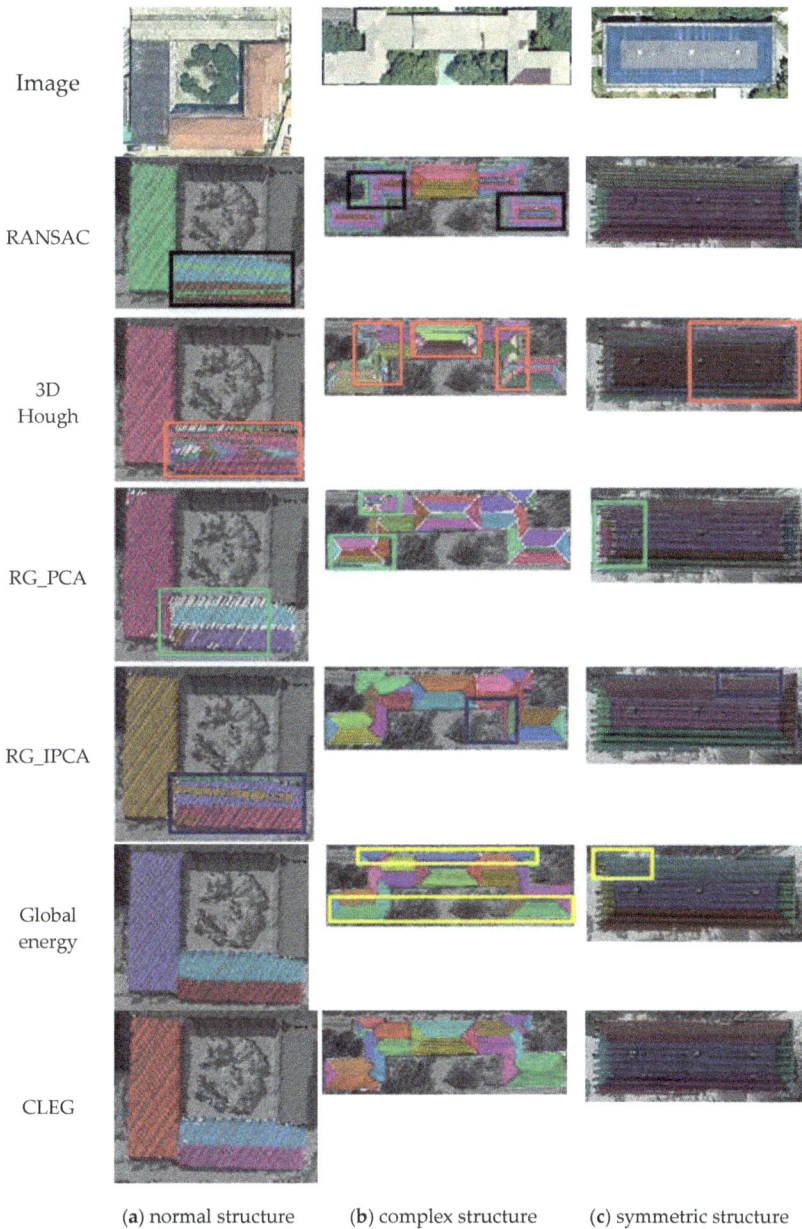

Figure 9. Segmentation of roof points in the Wuhan area. (**a**) normal structure; (**b**) complex structure; (**c**) symmetric structure.

The voting space in 3D Hough transformation is first computed. The votes are then sorted, and the planes are detected in order. The region growth is finally used to obtain an entire plane in the supported points. The results of 3D Hough transformation are sometimes worse than those of RANSAC because one point may support many planes, and the remaining planes may not be the most supported ones. Many false planes are detected, as shown by the red rectangles in Figures 8a–c and 9a–c. 3D Hough transformation has the same disadvantage as RANSAC and causes cross-planes without the use of normal vectors. The terminal condition is difficult to decide, and it uses the ratio of the smallest plane to the largest plane and the ratio of number of remaining points to total points may also lead to missing small planes, as shown by the center red rectangles in Figure 10a–c.

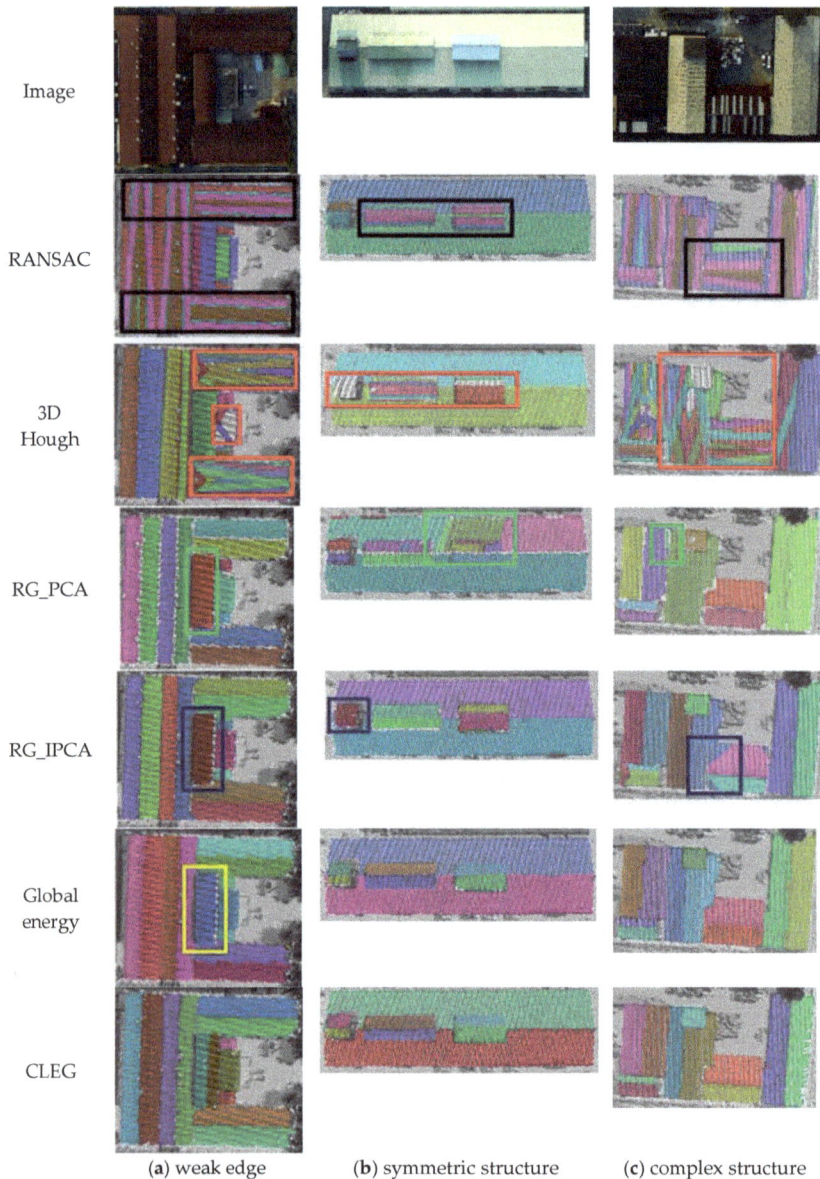

Figure 10. Segmentation of roof points in the Guangzhou area. (**a**) weak edge; (**b**) symmetric structure; (**c**) complex structure.

RG_PCA employs the K-nearest neighbors (KNN) to obtain the neighbor relationship and compute the normal vectors using PCA. The regions are then grown using the normal vectors. PCA may produce unstable results in estimating the normal vector at the edge regions. Therefore, the methods do not perform well in segmenting the points close to the facet boundary, as shown by the green rectangles in Figures 8a–c, 9a and 10c. KNN may produce an unstable neighbor relationship in areas with a largely uneven point density and results in over-segmentation, as shown by the green rectangles in Figure 9b,c. The difference of the normal vectors at the edge areas is small when the slope is small. This causes under-segmentation, as shown by the green rectangles in Figure 10a,b.

RG_IPCA utilized a triangulated irregular network (TIN) to obtain the neighbor relationship, compute the initial normal vectors using PCA, and grow to regions. This method can properly estimate the normal vectors at several boundary regions but may also lead to errors in several areas, as shown by the blue rectangles in Figures 8b,c, 9a,b and 10c. RG_IPCA has the same disadvantage as RG_PCA when the slope is small. The method results is under-segmentation, as shown by the blue rectangles in Figure 10a,b. Over-segmentation also exists in RG_IPCA, as denoted by the blue rectangles in Figures 8a and 9c.

The global energy method utilizes Graphcuts to obtain the minimum energy. This method yields quiet accurate results but depends on a good initial input. Consequently, missed planes will also be missed in the optimization results, as shown by the yellow rectangles in Figures 8c and 10a. The method also causes over-segmentation in noisy areas, as shown by the yellow rectangles in Figures 8b and 9c. The improper neighbor relationship causes under-segmentation, as denoted by the yellow rectangle in Figure 9b. The separated planes are combined because TIN may connect faraway points. The two facets are on the same plane because of symmetry. In other conditions, global energy can perform quite well and obtain complete results with the fewest points left. CLEG can properly handle these complex structures with very few missing points.

The proposed CLEG algorithm also has several disadvantages caused by the strict conditions of seed CLE selection. A seed CLE is not detected when the plane is small. Therefore, the plane may be missed, as shown by the red rectangle in Figure 11.

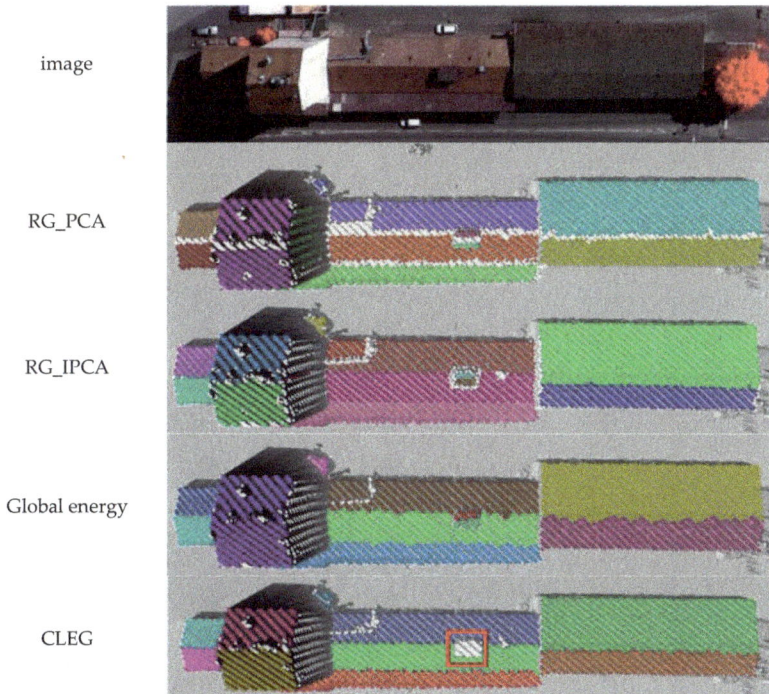

Figure 11. Disadvantage of the proposed method.

3.3. Region Segmentation

The CLEG algorithm can also process the point cloud containing terrains, buildings, trees, *etc.* The proposed method is similar to region growth methods. The comparison methods only include RG_PCA and RG_IPCA. The parameters used are shown in Table 6. The difference is that the minimum number of points required for a valid plane is larger than that in roof segmentation, because if the number is small, there may be many false planes detected in tree areas.

Table 6. Parameters used in the comparison test.

Parameter	Value	Methods
Point to plane distance threshold d	0.3 m	RG_IPCA, CLEG
Curvature threshold	0.01	RG_PCA, RG_IPCA
Minimum number of points required for a valid plane	20	RG_PCA, RG_IPCA
Line segmentation threshold ε	0.25 m	CLEG
Grid size	0.6 m	CLEG
Min line length l	1.8 m	CLEG

Seven datasets are utilized to prove the effectiveness and speed of the proposed method. The description is listed in Table 7.

Building the neighbor relationship possesses the highest computation cost in RG_PCA and RG_IPCA during the comparison test. The methods are different in the two algorithms. KNN is used in RG_PCA, and TIN is used in RG_IPCA. RG_PCA employs PCA to estimate the normal vectors of each point. The results may be unstable at boundary points, which often results in over-segmentation, as denoted by the blue rectangles in Figures 12 and 13. RG_IPCA sometime estimates the false normal vectors and results in some false segmentation, as shown by the yellow rectangles in Figures 12 and 14. Over-segmentation is also found in noisy areas, as shown by the blue rectangle in Figure 13. Although the minimum point of a valid plane is 20, some planes are found in the tree areas, as shown by the green rectangles in Figures 12 and 13. CLEG can handle these cases well with faster speed (Table 7).

Segmentation results of the point cloud Vaighingen using a small dataset.

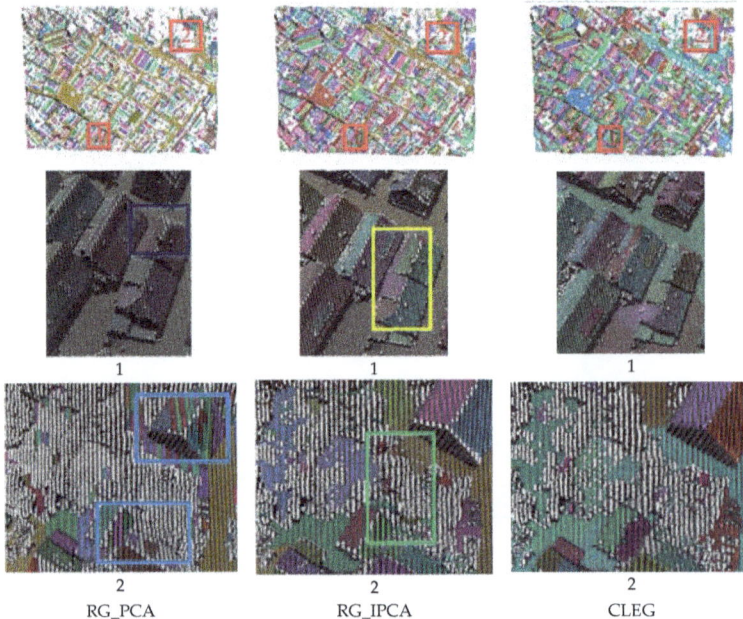

Figure 12. Segmentation results in the Vaighingen area using dataset (a).

Table 7. Computation time in the comparison test.

Dataset	Area	Number of Points	RG_PCA	RG_IPCA	CLEG
(a)	Vaihingen	321,956	70.054 s	1.482 s	0.468 s
(b)	Wuhan	298,666	255.170 s	2.356 s	0.499 s
(c)	Guangzhou	174,830	15.616 s	0.780 s	0.187 s
(d)	Vaihingen	3,582,656	-	17.691 s	6.272 s
(e)	Wuhan	2,058,844	-	10.203 s	2.948 s
(f)	Guangzhou	3,091,547	-	15.116 s	8.580 s
(g)	Guangzhou	12,305,250	-	-	58.126 s

Segmentation results of the point cloud in the Wuhan area using a small dataset.

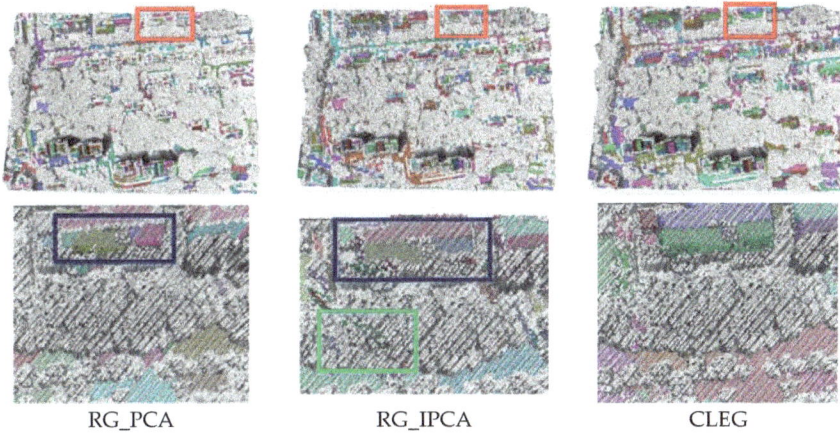

RG_PCA　　　　　　RG_IPCA　　　　　　CLEG

Figure 13. Segmentation results in the Wuhan area using dataset (b).

Segmentation results of the point cloud in the Guangzhou area using a small dataset.

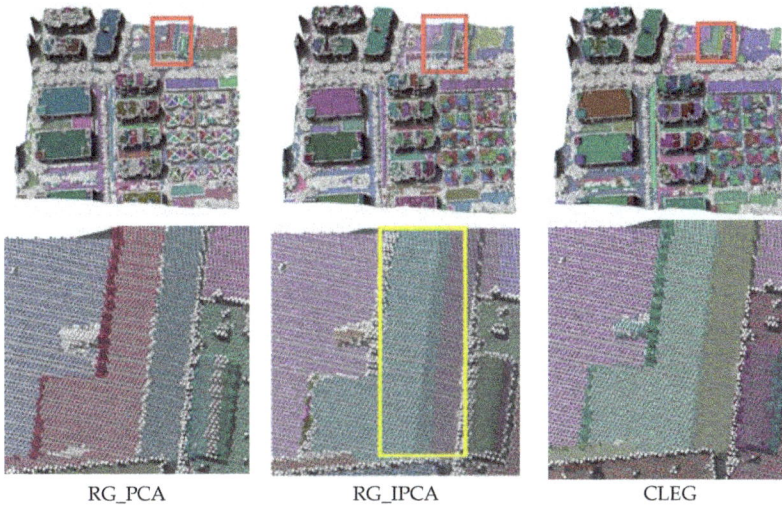

RG_PCA　　　　　　RG_IPCA　　　　　　CLEG

Figure 14. Segmentation results in the Guangzhou area using dataset (c).

KNN is very slow when large datasets are used for segmentation. Therefore, only RG_IPCA is used for comparison. RG_IPCA may result in false segmentation at the roof areas, as shown by the blue rectangles in Figures 15–17. False segmentation is also observed at ground area, as shown by the green rectangle in Figure 16. Under-segmentation is found when the slope is small. This result is denoted by the red rectangles in Figures 15 and 16. A cross-plane is denoted by the black rectangle in Figure 17. Furthermore, many planes are found in the tree areas, as shown by the yellow rectangles in Figures 15–17. CLEG can still handle these areas well with less processing time.

Segmentation results of point cloud in the Vaighingen area using a large dataset.

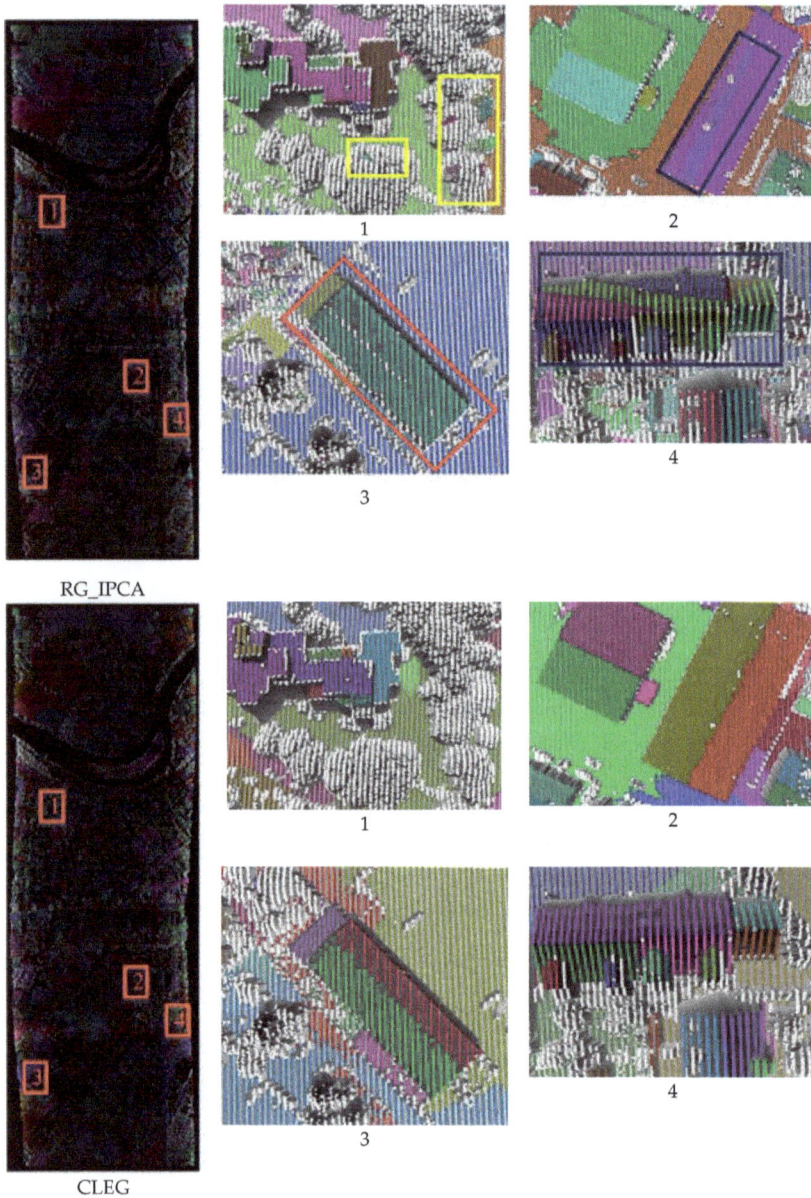

Figure 15. Segmentation results in Vaighingen using dataset (d).

Segmentation results of the point cloud in the Wuhan area using a large dataset.

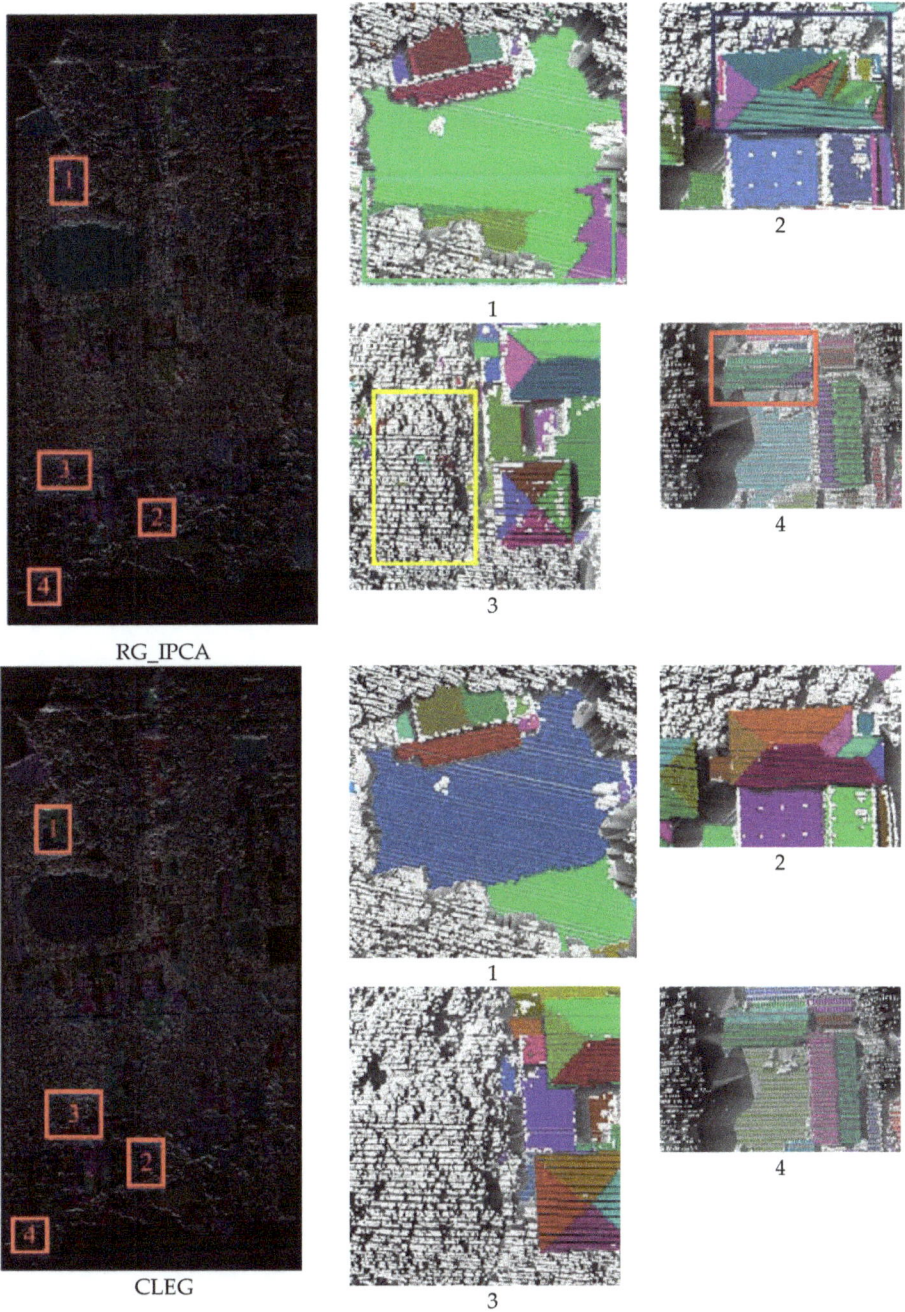

RG_IPCA

CLEG

Figure 16. Segmentation results in the Wuhan area using dataset (e).

Segmentation results of the point cloud in the Guangzhou area using a large dataset.

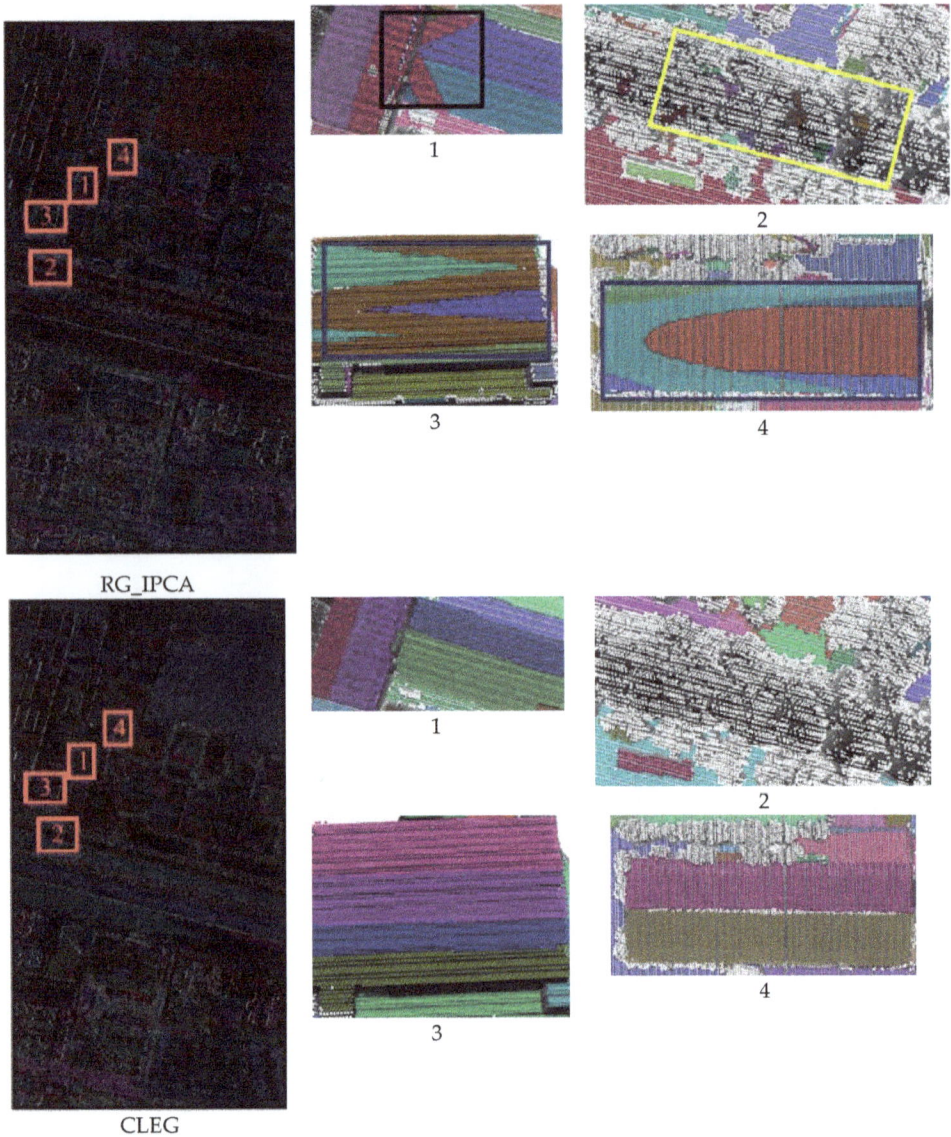

RG_IPCA

CLEG

Figure 17. Segmentation results in the Guangzhou area using dataset (f).

Building TIN in RG_IPCA during the test causes shortage in memory when a large point cloud with 12 million points is used. The CLEG algorithm can handle this large dataset, and completes the segmentation within 1 min (Figure 18). The proposed algorithm uses grid indexing instead of point-based neighbor relationship and CLE growth to overcome the shortage of uneven point cloud density. The process that consumes the most computation time in CLEG is the sorting of the seed points, which can be improved in the future by parallel computing.

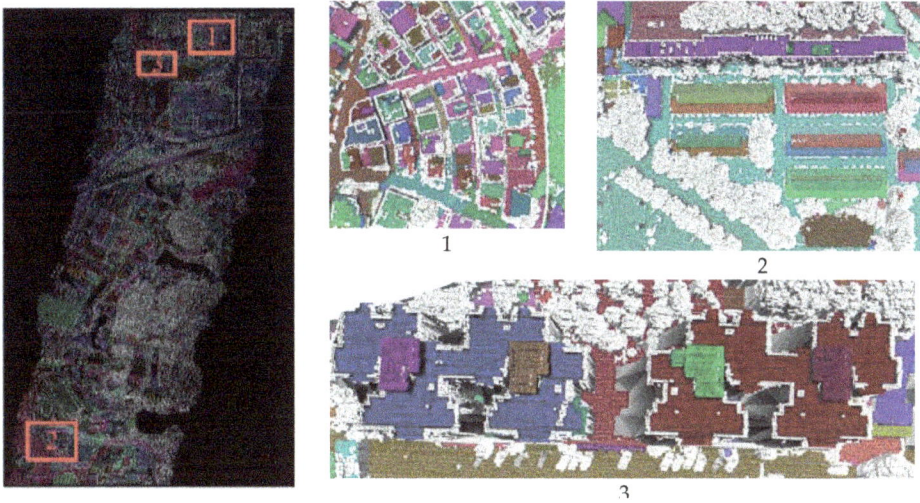

Figure 18. Segmentation results in the Guangzhou area using dataset (g).

3.4. Parameters Setting

The important parameters in CLEG algorithm are grid size and min line length. The grid size can be determined by the average point density.

The threshold of min line length is selected empirically in our experiment. This has an impact on the plane extraction results. The areas with line segments shorter than the threshold are missed. An example is shown in Figure 19.

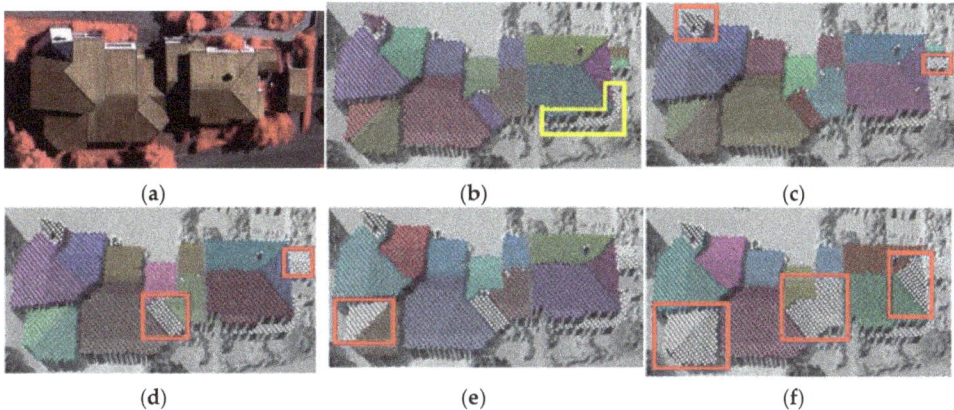

Figure 19. The influence of min line length. (**a**) Corresponding image; (**b**) l = 1.8 m, a narrow plane is missed; (**c**) l = 3.0 m, small planes are missed; (**d**) l = 4.2 m, more small planes are missed; (**e**) l = 6 m, a large plane is missed; (**f**) l = 7.2 m, more large planes are missed.

As shown in Figure 19b, a narrow but long-shaped plane object is missed marked in the yellow box. From Figure 19c–f, with the increase of min line length threshold, more and more planes are omitted as marked in the red boxes. The threshold can be determined by the minimum size of the planes according to the level of detail.

4. Conclusions

Using profiles or scan lines of LiDAR data to segment a surface and classify objects is not new [3–7]. This study focuses on using cross-line elements for plane segmentation. Proper and quality seed selection and region growth based on information derived from CLE are considered for the accurate and stable detection of planes. The pre-segmentation of the point cloud into CLEs eliminates the problem of selecting support points in clustering and model fitting methods, which is the key for the proposed method. With the use of the angle information derived from the CLE, the stages of seed selection and growth become more reliable. Furthermore, the CLEG algorithm is computationally efficient due to simple operations in seed generation and growth. The tests using various datasets show that the proposed algorithm runs much faster than popular methods while producing stable and accurate segmentation results. CLEG has great potential in feature extraction, object classification and 3D modeling of buildings.

However, the CLEG algorithm may still result in missing small facets because of the missing seed CLEs. Furthermore, the parameter of minimum line length has an impact on the plane extraction results; some narrow but long-shaped plane objects are missing. An additional retrieval step may be necessary to find these missing small and narrow planes. Two parallel lines can also determine a plane. In the next study, this could be combined with CLE to detect the missed narrow but long planes. Meanwhile, the CLE-derived features may be utilized in object classification and building detection from point cloud data, which is an important future task in extending the usage of CLEs.

Acknowledgments: This study was partially supported by the National Key Basic Research and Development Program (Project No. 2012CB719904) of China and the research funding by Guangdong province (2013B090400008) of China. The authors thank Guangzhou Jiantong Surveying, Mapping and Geographic Information Technology Ltd. for providing the data used in this research project. The Vaihingen data was provided by the German Society for Photogrammetry, Remote Sensing and Geoinformation (DGPF [Cramer, 2010]: http://www.ifp.uni-stuttgart.de/dgpf/DKEP-Allg.html (In German). The Wuhan data was provided by the National Key Basic Research and Development Program (Project No. 2012CB719904) of China. Thanks to Jixing Yan for providing the source code of RG_IPCA and global energy-based segmentation.

Author Contributions: Teng Wu designed the algorithm in detail, including seed selection and growth, *etc.*, and performed the experimental analysis. He also wrote the paper. Xiangyun Hu originally proposed using cross-lines for segmentation, advised the algorithm design and revised the paper. Lizhi Ye conducted the related initial study on profile-based feature extraction from LiDAR data.

Conflicts of Interest: The authors declare no conflict of interest.

Abbreviations

The following abbreviations are used in this manuscript:

LiDAR	light detection and ranging
DSM	digital surface model
CLE	cross-line element
CLEG	cross-line element growth
RANSAC	random sample census
PCA	principle component analysis
IPCA	iterative PCA
GSD	ground sample distance
RG	region growth
KNN	K-nearest neighbors
TIN	triangulated irregular network
RCL	reference cross-lap
DCL	detection cross-lap
BP	boundary precision
BR	boundary recall

References

1. Jiang, X.; Bunke, H. Edge detection in range images based on scan line approximation. *Comput. Vis. Image Underst.* **1999**, *73*, 183–199. [CrossRef]
2. Sappa, A.D.; Devy, M. Fast Range Image Segmentation by an Edge Detection Strategy. In Proceedings of the Third International Conference on the 3-D Digital Imaging and Modeling, Quebec City, QC, Canada, 28 May–1 June 2001; pp. 292–299.
3. Jiang, X.; Bunke, H. Fast segmentation of range images into planar regions by scan line grouping. *Mach. Vis. Appl.* **1994**, *7*, 115–122. [CrossRef]
4. Wang, J.; Shan, J. Segmentation of lidar point clouds for building extraction. In Proceedings of the American Society for Photogramm Remote Sens Annual Conference, Baltimore, MD, USA, 9–13 March 2009; pp. 9–13.
5. Sithole, G.; Vosselman, G. Automatic structure detection in a point-cloud of an urban landscape. In Proceedings of the 2nd GRSS/ISPRS Joint Workshop on Remote Sensing and Data Fusion over Urban Areas, Berlin, Germany, 22–23 May 2003; pp. 67–71.
6. Sithole, G.; Vosselman, G. Bridge detection in airborne laser scanner data. *ISPRS J. Photogramm. Remote Sens.* **2006**, *61*, 33–46. [CrossRef]
7. Hu, X.; Ye, L. A fast and simple method of building detection from lidar data based on scan line analysis. *ISPRS Ann. Photogramm. Remote Sens. Spat. Inf. Sci.* **2013**, *1*, 7–13. [CrossRef]
8. Wang, M.; Tseng, Y.-H. Automatic segmentation of lidar data into coplanar point clusters using an octree-based split-and-merge algorithm. *Photogramm. Eng. Remote Sens.* **2010**, *76*, 407–420. [CrossRef]
9. Wang, M.; Tseng, Y.H. Incremental segmentation of lidar point clouds with an octree—Structured voxel space. *Photogramm. Rec.* **2011**, *26*, 32–57. [CrossRef]
10. Chehata, N.; David, N.; Bretar, F. Lidar data classification using hierarchical k-means clustering. In Proceedings of the ISPRS Congress, Beijing, China, 3–11 July 2008; pp. 325–330.
11. Morsdorf, F.; Meier, E.; Kötz, B.; Itten, K.I.; Dobbertin, M.; Allgöwer, B. Lidar-based geometric reconstruction of boreal type forest stands at single tree level for forest and wildland fire management. *Remote Sens. Environ.* **2004**, *92*, 353–362. [CrossRef]
12. Sampath, A.; Shan, J. Clustering based planar roof extraction from lidar data. In Proceedings of the American Society for Photogrammetry and Remote Sensing Annual Conference, Reno, NV, USA, 1–5 May, 2006; pp. 1–6.
13. Biosca, J.M.; Lerma, J.L. Unsupervised robust planar segmentation of terrestrial laser scanner point clouds based on fuzzy clustering methods. *ISPRS J. Photogramm. Remote Sens.* **2008**, *63*, 84–98. [CrossRef]
14. Sampath, A.; Shan, J. Segmentation and reconstruction of polyhedral building roofs from aerial lidar point clouds. *IEEE Trans. Geosci. Remote Sens.* **2010**, *48*, 1554–1567. [CrossRef]
15. Melzer, T. Non-parametric segmentation of ALS point clouds using mean shift. *J. Appl. Geod.* **2007**, *1*, 159–170. [CrossRef]
16. Comaniciu, D.; Meer, P. Mean shift: A robust approach toward feature space analysis. *IEEE Trans. Pattern Anal. Mach. Intell.* **2002**, *24*, 603–619. [CrossRef]
17. Ferraz, A.; Bretar, F.; Jacquemoud, S.; Gonçalves, G.; Pereira, L. 3d segmentation of forest structure using a mean-shift based algorithm. In Proceedings of the 2010 IEEE International Conference on Image Processing, Hong Kong, China, 26–29 September 2010; pp. 1413–1416.
18. Yao, W.; Hinz, S.; Stilla, U. Object extraction based on 3D-segmentation of lidar data by combining mean shift with normalized cuts: Two examples from urban areas. In Proceedings of the 2009 Joint Urban Remote Sensing Event, Shanghai, China, 20–22 May 2009; pp. 1–6.
19. Fischler, M.A.; Bolles, R.C. Random sample consensus: A paradigm for model fitting with applications to image analysis and automated cartography. *Commun. ACM* **1981**, *24*, 381–395. [CrossRef]
20. Duda, R.O.; Hart, P.E. Use of the Hough transformation to detect lines and curves in pictures. *Commun. ACM* **1972**, *15*, 11–15. [CrossRef]
21. Medioni, G.; Tang, C.-K.; Lee, M.-S. Tensor Voting: Theory and Applications. Available online: http://159.226.251.229/videoplayer/Medioni_tensor_voting.pdf?ich_u_r_i=32752eaf85f6419f90c3d08468c5e75c&ich_s_t_a_r_t=0&ich_e_n_d=0&ich_k_e_y=1645048929750163052450&ich_t_y_p_e=1&ich_d_i_s_k_i_d=10&ich_u_n_i_t=1 (accessed on 25 February 2016).
22. Brenner, C. Towards fully automatic generation of city models. *Int. Arch. Photogramm. Remote Sens.* **2000**, *33*, 84–92.

23. Bretar, F.; Roux, M. Extraction of 3D planar primitives from raw airborne laser data: A normal driven ransac approach. In Proceedings of the IAPR Conference on Machine Vision Applications, Tsukuba, Japan, 16–18 May 2005.

24. Tarsha-Kurdi, F.; Landes, T.; Grussenmeyer, P. Extended ransac algorithm for automatic detection of building roof planes from lidar data. *Photogramm. J. Finl.* **2008**, *21*, 97–109.

25. Yan, J.; Jiang, W.; Shan, J. Quality analysis on ransac-based roof facets extraction from airborne lidar data. *ISPRS Int. Arch. Photogramm. Remote Sens. Spat. Inf. Sci.* **2012**, *1*, 367–372. [CrossRef]

26. Vosselman, G.; Dijkman, S. 3D building model reconstruction from point clouds and ground plans. *Int. Arch. Photogramm. Remote Sens. Spat. Inf. Sci.* **2001**, *34*, 37–44.

27. Borrmann, D.; Elseberg, J.; Lingemann, K.; Nüchter, A. The 3D Hough transform for plane detection in point clouds: A review and a new accumulator design. *3D Res.* **2011**, *2*, 1–13. [CrossRef]

28. Schuster, H.-F. Segmentation of lidar data using the tensor voting framework. *Int. Arch. Photogramm. Remote Sens. Spat. Inf. Sci.* **2004**, *35*, 1073–1078.

29. Kim, E.; Medioni, G. Urban scene understanding from aerial and ground lidar data. *Mach. Vis. Appl.* **2011**, *22*, 691–703. [CrossRef]

30. Gorte, B. Segmentation of tin-structured surface models. *Int. Arch. Photogramm. Remote Sens. Spat. Inf. Sci.* **2002**, *34*, 465–469.

31. Lee, I.; Schenk, T. Perceptual organization of 3D surface points. *Int. Arch. Photogramm. Remote Sens. Spat. Inf. Sci.* **2002**, *34*, 193–198.

32. Rottensteiner, F. Automatic generation of high-quality building models from lidar data. *IEEE Comput. Graph. Appl.* **2003**, *23*, 42–50. [CrossRef]

33. Pu, S.; Vosselman, G. Automatic extraction of building features from terrestrial laser scanning. *Int. Arch. Photogramm. Remote Sens. Spat. Inf. Sci.* **2006**, *36*, 25–27.

34. Vosselman, G.; Gorte, B.G.; Sithole, G.; Rabbani, T. Recognising structure in laser scanner point clouds. *Int. Arch. Photogramm. Remote Sens. Spat. Inf. Sci.* **2004**, *46*, 33–38.

35. Forlani, G.; Nardinocchi, C.; Scaioni, M.; Zingaretti, P. Complete classification of raw lidar data and 3D reconstruction of buildings. *Pattern Anal. Appl.* **2006**, *8*, 357–374. [CrossRef]

36. Rabbani, T.; van den Heuvel, F.; Vosselmann, G. Segmentation of point clouds using smoothness constraint. *Int. Arch. Photogramm. Remote Sens. Spat. Inf. Sci.* **2006**, *36*, 248–253.

37. Chauve, A.-L.; Labatut, P.; Pons, J.-P. Robust piecewise-planar 3D reconstruction and completion from large-scale unstructured point data. In Proceedings of the 2010 IEEE Conference on Computer Vision and Pattern Recognition (CVPR), San Francisco, CA, USA, 13–18 June 2010; pp. 1261–1268.

38. Yan, J.; Shan, J.; Jiang, W. A global optimization approach to roof segmentation from airborne lidar point clouds. *ISPRS J. Photogramm. Remote Sens.* **2014**, *94*, 183–193. [CrossRef]

39. Kim, T.; Muller, J.-P. Development of a graph-based approach for building detection. *Image Vis. Comput.* **1999**, *17*, 3–14. [CrossRef]

40. Wang, L.; Chu, H. Graph theoretic segmentation of airborne lidar data. *Proc. SPIE* 2008. [CrossRef]

41. Strom, J.; Richardson, A.; Olson, E. Graph-based segmentation for colored 3D laser point clouds. In Proceedings of the 2010 IEEE/RSJ International Conference on Intelligent Robots and Systems (IROS), Taipei, Taiwan, 18–22 October 2010; pp. 2131–2136.

42. Pauling, F.; Bosse, M.; Zlot, R. Automatic segmentation of 3D laser point clouds by ellipsoidal region growing. In Proceedings of the Australasian Conference on Robotics and Automation (ACRA 09), Sydney, New South Wales, Australia, 2–4 December 2009.

43. Golovinskiy, A.; Funkhouser, T. Min-cut based segmentation of point clouds. In Proceedings of the IEEE 12th International Conference on Computer Vision Workshops (ICCV Workshops), Kyoto, Japan, 27 September–4 October 2009; pp. 39–46.

44. Ural, S.; Shan, J. Min-cut based segmentation of airborne lidar point clouds. *ISPRS Int. Arch. Photogramm. Remote Sens. Spat. Inf. Sci.* **2012**, *1*, 167–172. [CrossRef]

45. Kim, K.; Shan, J. Building roof modeling from airborne laser scanning data based on level set approach. *ISPRS J. Photogramm. Remote Sens.* **2011**, *66*, 484–497. [CrossRef]

46. Delong, A.; Osokin, A.; Isack, H.N.; Boykov, Y. Fast approximate energy minimization with label costs. *Int. J. Comput. Vis.* **2012**, *96*, 1–27. [CrossRef]

47. Wu, S.-T.; Marquez, M.R.G. A non-self-intersection douglas-peucker algorithm. In Proceedings of the SIBGRAPI 2003 XVI Brazilian Symposium on Computer Graphics and Image, Brazil, 12–15 October 2003; pp. 60–66.

48. Cramer, M. The dgpf-test on digital airborne camera evaluation–Overview and test design. *Photogramm. Fernerkund. Geoinf.* **2010**, *2010*, 73–82. [CrossRef] [PubMed]

49. Rutzinger, M.; Rottensteiner, F.; Pfeifer, N. A comparison of evaluation techniques for building extraction from airborne laser scanning. *IEEE J. Sel. Top. Appl. Earth Obs. Remote Sens.* **2009**, *2*, 11–20. [CrossRef]

50. Shan, J.; Lee, S.D. Quality of building extraction from ikonos imagery. *J. Surv. Eng.* **2005**, *131*, 27–32. [CrossRef]

51. Awrangjeb, M.; Ravanbakhsh, M.; Fraser, C.S. Automatic detection of residential buildings using lidar data and multispectral imagery. *ISPRS J. Photogramm. Remote Sens.* **2010**, *65*, 457–467. [CrossRef]

52. Estrada, F.J.; Jepson, A.D. Benchmarking image segmentation algorithms. *Int. J. Comput. Vis.* **2009**, *85*, 167–181. [CrossRef]

![remote sensing logo] **remote sensing**

MDPI

Article

Investigation on the Weighted RANSAC Approaches for Building Roof Plane Segmentation from LiDAR Point Clouds

Bo Xu [1], Wanshou Jiang [1,*], Jie Shan [2], Jing Zhang [1] and Lelin Li [3]

[1] State Key Laboratory of Information Engineering in Surveying, Mapping and Remote Sensing,
Wuhan University, Wuhan 430072, China; lmars_xubo@whu.edu.cn (B.X.); jing.zhang@whu.edu.cn (J.Z.)

[2] Lyles School of Civil Engineering, Purdue University, West Lafayette, IN 47907, USA; jshan@purdue.edu

[3] National-Local Joint Engineering Laboratory of Geo-Spatial Information Technology,
Hunan University of Science and Technology, Xiangtan 411201, China; lilelin@hnust.edu.cn

* Correspondence: jws@whu.edu.cn; Tel.: +86-27-6877-8092 (ext. 8321)

Academic Editors: Devrim Akca, Zhong Lu and Prasad Thenkabail
Received: 23 September 2015; Accepted: 15 December 2015; Published: 23 December 2015

Abstract: RANdom SAmple Consensus (RANSAC) is a widely adopted method for LiDAR point cloud segmentation because of its robustness to noise and outliers. However, RANSAC has a tendency to generate false segments consisting of points from several nearly coplanar surfaces. To address this problem, we formulate the weighted RANSAC approach for the purpose of point cloud segmentation. In our proposed solution, the hard threshold voting function which considers both the point-plane distance and the normal vector consistency is transformed into a soft threshold voting function based on two weight functions. To improve weighted RANSAC's ability to distinguish planes, we designed the weight functions according to the difference in the error distribution between the proper and improper plane hypotheses, based on which an outlier suppression ratio was also defined. Using the ratio, a thorough comparison was conducted between these different weight functions to determine the best performing function. The selected weight function was then compared to the existing weighted RANSAC methods, the original RANSAC, and a representative region growing (RG) method. Experiments with two airborne LiDAR datasets of varying densities show that the various weighted methods can improve the segmentation quality differently, but the dedicated designed weight functions can significantly improve the segmentation accuracy and the topology correctness. Moreover, its robustness is much better when compared to the RG method.

Keywords: 3D point clouds; building reconstruction; building roof segmentation; weighted RANSAC

1. Introduction

Numerous studies have been conducted in 3D building reconstruction in the past two decades [1–3]. According to [4,5], reconstruction methods can be divided into two general categories: data-driven and model-driven. For high density point cloud data or complex roof structures, the task often converges on a data-driven process based on segmentation [2]. According to [6,7], there are three data-driven segmentation techniques: edge-based or region growing (RG), feature clustering, and model fitting.

Segmentation methods based on edge or region information [8–12] are relatively simple and efficient but are error-prone in the presence of outliers and incomplete boundaries. When the transitions between two regions are smooth, finding a complete edge or determining a stop criterion for RG becomes difficult [6]. Techniques using feature clustering for segmentation [6,13–18] experience problems in deciding the number of segments; and poor segmentation (over-, under-, no segmentation, or artifacts) can occur when small roof sub-structures exist or tree points close to the building roofs are not completely filtered beforehand. Compared with the above techniques, [7] suggested that model fitting methods can be more efficient and robust in the presence of noise and outliers. RANdom SAmple Consensus (RANSAC) [19] and Hough transform are two well-known algorithms for model fitting. The concept and implementation of the RANSAC method are simple. It simply iterates two steps: generating a hypothesis by random samples and verifying the hypothesis with the remaining data. Given different hypothesis models, RANSAC can detect planes, spheres, cylinders, cones, and tori [20]. Numerous variants have been derived from RANSAC; and a comprehensive review is available in the work of [21]. Those variants (*i.e.*, [22–24]) provide the possibility of improving the methods in both robustness and efficiency. Information like point surface normal [4,7] and connectivity [25] also can be incorporated in RANSAC for better results. Moreover, although the RANSAC method is an iterative process, reference [5] suggests that it is faster than the Hough transform.

LiDAR techniques generate ever increasingly high resolution data. This provides the possibility to recognize subtle roof details and rather complex structures; but in the meantime, it brings challenges to current RANSAC-based segmentation methods. A widely concerning problem is the spurious planes that consist of points from different planes or roof surface [4,6,7,26,27]. A detected plane overlapping multiple reference planes or a plane snatching parts of the points from its neighbor planes are frequent occurrences. Their misidentification and incorrect reconstruction may have a crucial effect on the understanding of the building structure (*i.e.*, topology of the building) [28,29]. To address this issue, many additional processes were designed and used in past studies, such as normal vector consistency validation [4,7], connectivity [26], and standard deviation of the point-plane distances. Those processes need careful fine tuning of their parameters in order to achieve the best performance (e.g., reference [30] suggests that the threshold should be in agreement with the segment scale). This is a difficult task and highly relies on prior knowledge of the data and scene as well as the experience of the operators. Therefore, a more accurate fitting method is needed to suppress the spurious planes.

Although no applications were found in building roof segmentation, the M-estimate SAC (MSAC) and the Maximum Likelihood SAC (MLESAC) in [31] provided a potential solution to the spurious planes problem. In these two methods, the contribution of a point to the hypothesis plane is no longer a constant 0 or 1, but rather a loss function (inversed to weight) according to the point-plane distance. Basically, a large distance is assigned to a large loss, and false hypotheses are suppressed because of the larger total loss. However, we argue that their loss functions were not sufficient to distinguish the spurious plans from the correct hypothesis plane for complex roof segmentation problems. Inclusion of other additional factors into the loss function, such as surface normal, would make the methods more adaptive and robust.

This paper implements the idea of loss function into the popular RANSAC method and proposes a weighted RANSAC framework for roof plane segmentation. In the framework of our new method, the hard threshold voting function which considers both the point-plane distance and the normal vector consistence is transformed into a soft threshold voting function based on two weight functions. New weight functions are introduced based on the error distribution between the proper and improper hypothesis planes. Different forms of weights were tested and compared, yielding a recommended weight form.

The remainder of this paper is organized as follows. Section 2 discusses the related work and the modification of the existing weighted RANSAC into the normalized forms. In Section 3, the design of an ideal weight function is discussed, and several different weight functions are proposed and evaluated. Experimental results are presented and analyzed in Section 4, followed by discussion and concluding remarks in Section 5.

2. Background

2.1. RANSAC-based Segmentation

Although the RANSAC-based segmentation methods have several variations, they consist of three steps [4]: preprocessing, RANSAC, and post-processing. The preprocessing step yields the surface normal for each LiDAR point. The roof points can be separated to a planar set and a nonplanar set (if so, the nonplanar points are excluded from the second step and be retrieved in the final step). The second step is a standard implementation of the RANSAC method [14]. It iteratively and randomly samples points to estimate the hypothesis plane and then tests the plane against the remainder of the dataset. A point is taken as an inlier if the point-plane distance and the angle between the point's normal vector and plane's normal vector (in [6]) are smaller than the given thresholds. After a certain number of iterations, the shape that possessed the largest percentage of inliers relative to the entire data is extracted. The method detects only one plane at a time from the entire point set. Thus, the process has to be implemented iteratively in a subtractive manner, which means that once a plane is detected, the points belonging to the plane are removed and the algorithm continues on the remainder of the dataset until no satisfactory planes are found. To be fast, the constraints of normal vectors [7,32] and local sampling [4,32] are used to avoid the meaningless hypotheses. A fast and rough clustering (or classification) process can be used to decompose the dataset [7,33]. To be robust, validations on normal vector consistency [4,7], connectivity [26], and standard deviation of point-plane distances are also adopted. The main task of post-processing is to refine the segmentation results, retrieve roof points from unsegmented point sets, find missing planes, and remove false spurious planes [27,32].

For classical RANSAC methods, the plane with the maximum inliers is generated when determining the most probable hypothesis plane $\hat{M}$:

$$\hat{M} = \underset{M}{\operatorname{argmax}} \left\{ \sum_{P_i \in U} T(P_i, M) \right\} \tag{1}$$

where U is the set of remaining points, N_U is the number of points in U, and $T(P_i)$ is the inlier indicator:

$$T(P_i, M) = \begin{cases} 1 & d_i < d_t \text{ and } \theta_i < \theta_t \\ 0 & otherwise \end{cases} \tag{2}$$

where d_i is the point-plane distance, θ_i is the angle between point P_i's normal and plane's normal [7], and d_t and θ_t are the corresponding thresholds.

2.2. Spurious Planes

The problem of spurious planes is a widely discussed common problem that has yet to be resolved in RANSAC-based segmentation. Generally, the planes detected by the RANSAC methods may belong to different planes or roof surfaces. As shown in Figure 1a, suppose that the threshold d_t can be well estimated beforehand according to the precision of the point clouds, then a proper segmentation is achieved if the hypothesis planes π_1 and π_2 receive the largest inlier ratio. However, poorly estimated planes may be detected, such as plane π_3 in Figure 1b, whose point count is much larger than that of π_1 or π_2, thus leading to false segmentation. The RANSAC method extracts planes one after the other from LiDAR points so these mistakes may occur at plane transitions. The situation in Figure 1b can further intensify such competitions as the inaccurate hypothesis tends to generate more supports from roof points.

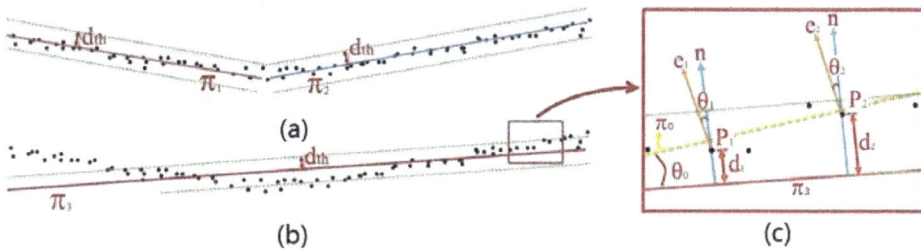

Figure 1. An example of spurious planes. (**a**) The well estimated hypothesis planes (π_1 and π_2); the two green parallel lines are the boundary of the point-to-plane distance threshold; (**b**) A spurious plane (π_3) is generated under the same thresholds; (**c**) A detail view of (**b**), where *n* is the normal vector of the plane π_3, and e_1 and e_2 are the point normal vectors. The d_1, d_2, θ_1, θ_2 are the corresponding observed values of point P_1 and P_2 in Equation (2). θ_0 is the angle between π_3 and the real roof surface (π_0).

2.3. Existing Weighted RANSAC Methods

Instead of using fixed thresholds in the determination of inliers, MSAC and MLESAC [31] use a loss function to count the contribution, which is actually a contribution loss, of the inliers based on the point-to-plane distance. The most probable hypothesis $\hat{M}$ is determined by minimizing the total loss of hypothesis M:

$$\hat{M} = \underset{M}{\text{argmin}} \left\{ \sum_{P_i \in U} Loss\left(d(P_i, M)\right) \right\} \tag{3}$$

The MSAC adopts bounded loss as follows:

$$Loss\left(d\right) = \begin{cases} d^2 & |d| < d_t \\ d_t^2 & otherwise \end{cases} \tag{4}$$

MLESAC utilizes the probability distribution of error by inliers and outliers, models inlier errors as Gaussian distribution and outlier errors as uniform distribution:

$$Loss\left(d\right) = -\log \left(\gamma \frac{1}{\sqrt{2\pi}\sigma_d} \exp(-\frac{d^2}{2\sigma_d^2}) + (1-\gamma)\frac{1}{v} \right) \tag{5}$$

where γ is the prior probability of being an inlier, which is the inlier ratio in Equation (1), σ is the standard deviation of Gaussian noise ($\sigma_d = d_t/1.96$), and v is a constant which reflects the size of available error space.

For the so-called weighted RANSAC, the loss function is transformed as the normalized weight functions for testing points, having values from 0 to 1, so that different weight functions can be easily compared and further applied to more than one factor (*i.e.*, considering both distance and normal directions). As a result, the loss functions of MSAC and MLESAC are normalized as Equation (6).

$$weight\left(d\right) = \frac{loss\left(+\infty\right) - loss\left(d\right)}{loss\left(+\infty\right) - loss\left(0\right)} \tag{6}$$

3. Weighted RANSAC for Point Cloud Segmentation

For the weighted RANSAC methods, the weight value of an inlier reflects its consistency with the hypothesis plane. An ideal weight function is expected to suppress the spurious planes as far as possible without excessively penalizing the proper planes. In this work, the purpose is achieved by comparing the error distribution between the proper and improper hypotheses. In Section 3.1, we discuss the drawback of the existing weighted methods that form the design principle of the ideal weight function. Then, several new weight functions are defined based on the design principle in Section 3.2. In Section 3.3, adding the factor of normal vector errors into the weight functions is considered, and a joint weight function is designed via the multiplication of the two factors (distance and normal vector). Those new weight functions are compared and evaluated in Section 3.4, together with the existing weighted methods.

3.1. Improvements Consideration of the Weighted Function

Figure 2 (Bottom) provides examples of the point-to-plane distance distribution for both the proper hypothesis and the spurious plane. To clarify the discussion that follows, the distance range is divided into three regions, namely A, B, and C. Generally, the inliers of a proper-plane tend to focus on the region with a smaller distance, which follows the normal distribution in theory, while the distribution of the distances to a spurious plane tend to be more dispersed. For traditional RANSAC, the spurious planes are detected instead of the proper plane if there are too many points in region C. An intuitive solution to alleviate the problem is simply using a smaller distance threshold, *i.e.*, changing the threshold d_t to d'_t, which reduces the inliers count of the spurious plane (yellow region). However, a too small threshold will decrease the number of inliers (red region) and eventually result in over-segmentation.

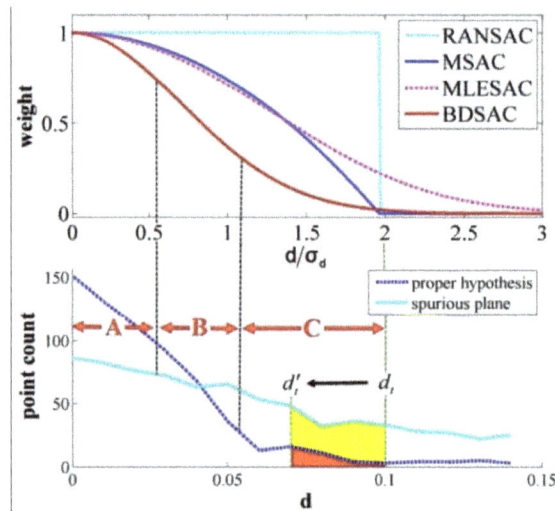

Figure 2. Comparison of point-to-plane distance distribution between proper and improper hypotheses. (**Top**) Plots of the weight functions ($d_t = 1.96\sigma_d$, MLESAC: $\gamma = 0.3$, $v = 3\sigma_d$); (**Bottom**) Examples of distance distribution for the proper hypothesis and the spurious plane. A, B, and C are a rough division of the distance range: A for regions where the proper planes are dominant in the point count, B for regions where the point counts are similar, and C for regions where spurious planes generate more inliers. The red region represents the lost roof points when using a stricter threshold and the yellow region indicates that more points are excluded from the spurious planes than the proper hypothesis. BDSAC is a newly designed weight curve.

Without changing the d_t threshold, MSAC and MLESAC suppress the spurious planes by assigning smaller weights to the inliers with larger distances so that the inliers in area C of Figure 2 will contribute less to the evaluation of the hypothesis plane. The inadequacy of MSAC and MLESAC are mainly caused by its slow decrease of the weight curves. Generally, the weighted methods are expected to suppress the spurious plane as far as possible without excessively penalizing the proper planes. Under such consideration, we expect the curve of the weight function to decrease rapidly in area B and gradually with small weight values in area C (*i.e.*, the curve of BDSAC in Figure 2). However, as shown in Figure 2, there are still a great deal of inliers that have large weight values and gradients in area C for MSAC and MLESAC. MSAC has the largest absolute gradient at the threshold boundary, and the MLESAC has a boundary weight value of over 0.2, which limit their suppressing to spurious planes. To overcome the drawbacks of these two methods, we attempted to modify the weight functions, and the improved versions of weight functions are shown in Section 3.2.

3.2. Modified Weight Functions and New Weight Functions

First, the weight functions of RANSAC, MSAC, and MLESAC were modified. Generally, after a hypothesis plane is accepted, it is expected that all the inliers should be excluded to avoid affecting the detection of other planes; while in the plane detection step, it is wished that as fewer outliers included as possible to decrease the possibility of false plane detection and the absence of minor inliers is acceptable. This reminds us to reduce the thresholds used in the weight functions and to keep the threshold unchanged for inlier exclusion. For such an objective, a reduction ratio μ was applied to the distance threshold d_t in the weight function. For example, the MSAC with a reduction ratio μ is expressed by (denoted by $MSAC_\mu$):

$$weight_\mu(d) = \begin{cases} 1 - \left(\dfrac{d}{\mu \cdot d_t}\right)^2 & |d| < (\mu \cdot d_t) \\ 0 & otherwise \end{cases} \tag{7}$$

Similarly, the reduction ratio μ also was adopted in classical RANSAC and MLESAC to generate the two modified versions, named $RANSAC_u$ and $MLESAC_u$. $RANSAC_u$ uses smaller threshold $\mu \cdot d_t$ for inlier determination; and the σ_d in Equation (5) of $MLESAC_u$ is reduced to $\mu \cdot d_t/1.96$.

As discussed in Section 3.1, two new weight functions stricter in theory can be designed, whose value is close to 1 in region A, close to 0 in region C, and rapidly decreasing in region B. One weight function is a piecewise-linear function, which linearly decreases in Region B (denote by LDSAC).

$$weight(d) = \begin{cases} 1 & |d| \leqslant d_1 \\ \dfrac{d_2 - |d|}{d_2 - d_1} & d_1 < |d| < d_2 \\ 0 & others \end{cases} \tag{8}$$

where d_1 and d_2 are the selected thresholds between 0 and d_t (*i.e.*, $0.2d_t$ and $0.7d_t$ in our test).

Another weight function is a smooth curve decreasing along the "bell" curve (denote by BDSAC):

$$weight(d) = \exp(-\frac{d^2}{\sigma_d{}^2}) \tag{9}$$

The curves of the weight functions and the absolute value of their gradients are illustrated in Figure 3. All the weight functions are inversely proportional to the point-to-plane distance d with a range of from 0 to 1, thus the most probable hypothesis plane $\hat{M}$ is decided similarly with classical RANSAC in Equation (1):

$$\hat{M} = \underset{M}{\operatorname{argmax}} \left\{ \sum_{P_i \in U} weight(d_i) \right\} \tag{10}$$

As the value of the weights is generated by simply mapping the value of d/σ_d into pre-calculated tables, the efficiency of all the methods are similar.

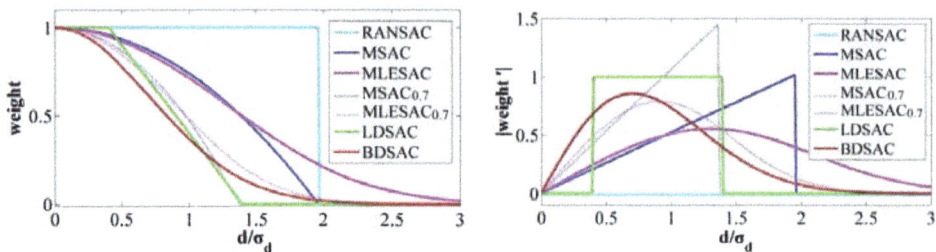

Figure 3. Weight functions of various RANSAC methods: (**Left**) Plots of the weight functions; (**Right**) Plots of the absolute value of gradient.

3.3. Joint Weight Function Regarding Angular Difference

The normal vectors of the inliers are generated by neighborhood analysis [6,10] and often have fine consistency for 2.5D roof surfaces. For poor hypotheses, a systematic deviation of the normal vectors (*i.e.*, θ_0 in Figure 1c) can exist between the hypotheses plane and the real roof surface. As the normal of the points turns out to be in accord with the real roof surface, this deviation will reflect in most roof points. As a result, the angular difference between the points and the hypothesis plane (θ in Equation (2)) has long been used to evaluate the quality of inliers, either as constant thresholds in [7] or as a normal vector consistency validation in [4]. It is very natural for us to consider adding the angular difference into the weight definition.

Suppose the distribution of angular difference θ is independent with the distance and also obeys the normal distribution with a standard deviation of σ_θ. Then, the weight of the angular difference can be defined by using the same form as the distance (simply replace d by θ). For instance, the weight functions of BDSAC for angular difference θ can be defined as:

$$w\left(\theta\right) = \exp(-\frac{\theta^2}{\sigma_\theta^2}) \tag{11}$$

Then, the final weight $weight\left(d_i, \theta_i\right)$, considering both the point-to-plane distances and the angular differences, can be defined as the product of the two weights:

$$weight\left(d_i, \theta_i\right) = w(d_i)w(\theta_i) \tag{12}$$

Similar to Equation (9), the most probable hypothesis plane $\hat{M}$ is determined by maximizing the total weight of all the hypothesis of M

$$\hat{M} = \underset{M}{\mathrm{argmax}} \left\{ \sum_{P_i \in U} weight\left(d_i, \theta_i\right) \right\} \tag{13}$$

To distinguish from methods considering point-to-plane distance only, a subscript of "$_{nv}$" is added to the methods that take the angular difference into account (e.g., $BDSAC_{nv}$ for the improved method of BDSAC).

3.4. Weight Function Evaluation

A proper weight function is expected to suppress the improper hypotheses as much as possible, without excessively penalizing the proper ones. Since the decreasing rates of the total weights for the hypothesis plane under different weight functions are different, an outlier suppression ratio is defined as the evaluation metric here:

$$ratio_{os} = \frac{W_{test}}{W_{ref}} \tag{14}$$

where W_{ref} stands for the total weight of the reference plane (the plane fitted by all the inliers), and W_{test} is the total weight of the test hypothesis plane.

The test hypothesis planes are randomly generated and manually marked as positive or negative, based on whether a correct segmentation can be generated. For a positive hypothesis, we expect that the ratio of a good weight function is stable, which should be close to 1. For negatives hypotheses, we need the ratio to be as small as possible, and a ratio over 1 indicates that a false hypothesis gains larger weights than the proper ones, leading to false segmentation.

To evaluate the weight functions defined in Sections 3.2 and 3.3 10 hypotheses planes are generated from the point cloud of the building in Figure 4, among which three hypotheses are positive and seven hypotheses are negative.

Figure 4. Buildings with both positive and negative hypotheses. The deep blue triangle is a negative hypothesis as it is athwart the two roof planes, and the cyan triangle is a positive hypothesis which can produce a correct segmentation.

The outlier suppression ratios of the 10 hypotheses are shown in Figure 5. As shown in Figure 5a, the ratios of eight methods considering only distances error are compared, and the mean ratio of the 10 hypotheses under different thresholds are illustrated in Figure 5c. As shown in Figure 5b,d, we compare the improvements of the methods after considering both the distance and angular difference in the weight function, corresponding to Figure 5a,c. Several conclusions can be made at this point:

(1) For all the weighted methods, the evaluation of the positive hypotheses (planes 1, 2, and 3) are stable as the ratios in Figure 5a are close to 1.0 and the ratio reductions in Figure 5b are close to 0. Meanwhile, all the weighted methods can significantly decrease the ratios of the negative hypotheses when compared to RANSAC, but their suppressing ability are different.

(2) By comparing the results between the modified weight functions and the original functions (*i.e.*, $MSAC_{0.7}$ and MSAC), it can be concluded that reduction of the inlier threshold can suppress the outliers effectively. The newly designed LDSAC and BDSAC functions have the best performances, which verifies our considerations in Section 3.1.

(3) From Figure 5c, it can be seen that all the methods can be affected by the threshold in some degree, but the newly designed weighted methods are least influenced.

(4)   Figure 5b,d illustrate the improvements after taking the angular differences into the weight functions. All the weighted methods gain positive effects and the effects are not sensitive to the thresholds.

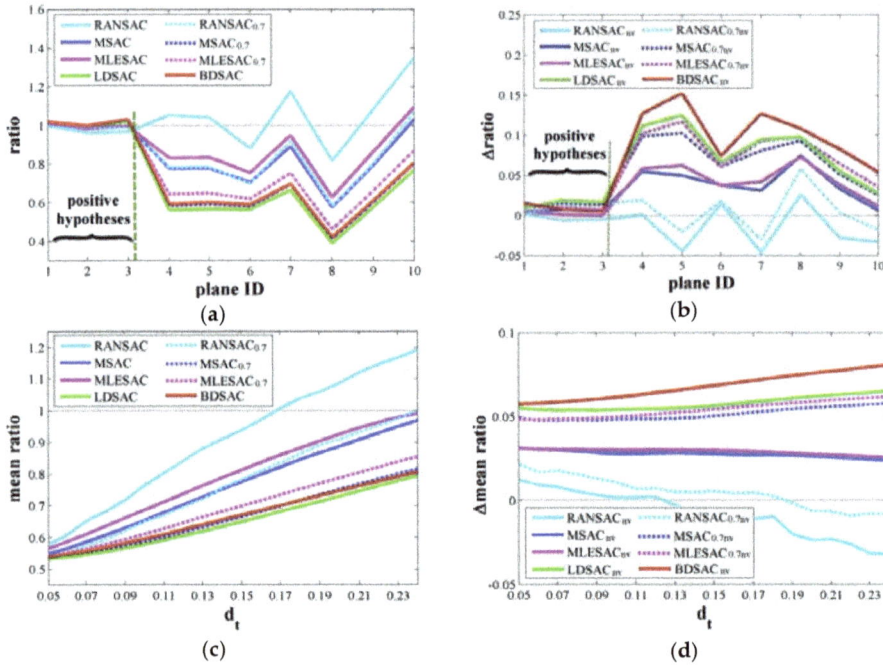

Figure 5. Suppressing ability and threshold sensitivity test. (**a**) Suppressing ratios for planes under different weight forms; (**b**) Ratio reductions after considering the angular difference (*i.e.*, the ratio reduction of BDSACnv is the ratio of BDSAC minus the ratio of BDSACnv); (**c**) Mean ratio of the ten planes under different dt thresholds; (**d**) Mean ratios reduction after considering the angular difference (the reduction approach is similar to (**b**)).

As the performances of the segmentation methods are greatly influenced by the complexity of the input data and the threshold parameters, we simulate the data in Figure 6 to test the robustness of the algorithm on a variety of conditions. The data consists of two adjacent horizontal planes, both 10 m × 5 m with an average point distance is 0.5 m. The height difference between the planes (Δd) and the added Gaussian noise (with a standard deviation of σ) are both changeable. The thresholds dt for the methods are tested from 0.02 m to 0.2 m, every 0.01 m a trail.

The difficulty of segmentation will obviously increase when Δd decreased or σ increased, which will influence the selection of the d_t thresholds. For data with a larger σ, the d_t needs to be larger in order to include all the plane inliers; otherwise, over-segmentation may occur. As a result, nearly all the methods fail when d_t is smaller than 2σ in Figure 6c (the value need to be even larger in real applications). The value of Δd reflects the separability of the two planes and stricter thresholds are needed for a successful separation. The setting of thresholds needs to consider both factors and find a proper value between the two limitations, finally forming the acceptable areas for different weighted methods in Figure 6. For classical RANSAC, the results are rather disappointing and a proper threshold is difficult to generate. However, for the weighted methods, as the spurious planes are suppressed, much looser thresholds are allowed which result in larger areas in Figure 6. It also can be seen that both adding new weight forms and considering the angular difference in the weights produce positive effects on the acceptable areas. This decreases the difficulty of threshold selection and allows the possibility of processing more complex data. For instance, when Δd equals 0.15 m and 0.2 m in Figure 6b or when σ equals 0.03 m and 0.04 m in Figure 6c, the classical RANSAC methods will always fail while our new weighted methods can produce a correct segmentation. Intuitively, a spurious plane that passes through the middle of the two planes will include all the points if d_t is larger than $\Delta d/2$ for classical RANSAC and cannot distinguish the two planes well when d_t is larger than $\Delta d/3$ in our experiments. In comparison, proper results are produced by $BDSAC_{nv}$ even when d_t is larger than $2\Delta d/3$.

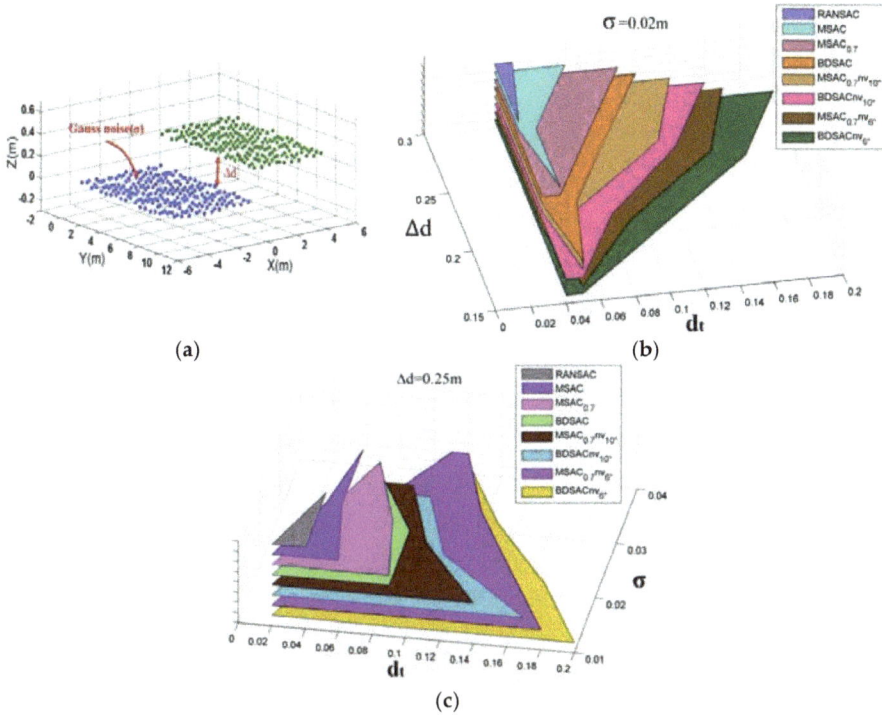

Figure 6. Data sensitivity test. (**a**) Simulated data, with changeable Δd and σ; (**b**) Segmentation results under different Δd; (**c**) Segmentation results under different σ. The colored regions depict the range of d_t that can produce a correct segmentation.

4. Experiments and Evaluation

After comparing the effects of different weight functions in suppressing spurious planes and their sensitivity to thresholds and input data, this section presents the stability and robustness segmentation results and the optimal weight recommendations. The various assessment metrics are introduced, and the experiments on various datasets to test the overall performance of the methods are presented.

4.1. Datasets and Fundamental Algorithm

The experiments utilized two datasets. The first dataset was collected in the city of Vaihingen, ISPRS dataset [34] and the other set, which has a higher point density, was collected on the Wuhan University campus, China. In the quantitative tests, the reference data were created manually based on the initial segmentation results and aerial images. Since our segmentation algorithms initiates from the classified building roof points, the points on the ground, walls, and vegetation were filtered beforehand and excluded from the quantitative tests. The results of a RG-based method [11] are also used for comparison purposes.

To evaluate the effects of the weighted methods, a fundamental RANSAC-based segmentation algorithm is needed. Since this paper focuses on the effects of the different weight functions, only a brief introduction to the algorithm implementation is provided here. The main framework of the algorithm follows the work of [4], but we also refer to the work of [7,25] (described in Section 2.1). In the pre-processing stage, the points normal are estimated through the tensor voting algorithm [10], which also divides the points into planar and nonplanar sets. In the second step (standard RANSAC stage), the density-based connectivity clustering is implemented [22] to ensure the spatial connectivity of the detected planes. Some speed-up techniques also are also utilized: a fast and rough connectivity clustering to decompose the integral data [35] (connectivity of the octree cells) and the ND-RANSAC [32] and Local RANSAC [4] to avoid meaningless hypotheses. The post-processing mainly included the following aspects: (1) completion of the roof plane by searching the points from the unsegmented points; (2) clustering of the remaining points set and an extra searching process to detect the lost segments; and (3) to avoid over-segmentation or detecting a plane twice, a region merging process [32] was adopted among the neighbor planes, which required the total weights of the merged plane to be larger than that of either single plane. Some basic specifications for the datasets are provided in Table 1, and the main parameters are shown in Table 2.

Table 1. Properties of the two datasets.

Site	Vaihingen	Wuhan
Acquisition Date	22 August 2008	22 July 2014
Acquisition System	Leica ALS 50	Trimble Harrier 68i
Fly Height	500 m	1000 m (cross flight)
Point Density	~4/m^2	>15/m^2

Table 2. Parameters used in the experiments.

	MinPt	MinLen	Angle	d_t	θ_t	Ncc	Dcc	P_0	NbPt1	NbPt2
Vaihingen	5	1 m	15°	0.15 m	10°	5	1.5 m	0.99	5	20
Wuhan	20	1 m	15°	0.20 m	10°	5	0.75 m	0.99	10	30

MinPt: the minimum number of points for a plane; MinLen: the minimum length of the detected edge. Angle: the angle threshold between the three sample points' normal and the plane's normal used in hypothesis generation (ND-RANSAC, see [31]). Ncc (least number of points) and Dcc (searching distance) are the parameters used in the density-based connectivity clustering [24]. P_0 is the confidence probability to select the positive hypotheses at least once. NbPt1 and NbPt1 are the two parameters (nearest n points) used in the tensor voting-based method [10] (two rounds of voting).

4.2. Evaluation Metrics

The evaluation metrics consisted of two parts: the object-level evaluation metrics provided in [36] and the quality of the roof ridges detected after segmentation. *Completeness (Comp)*, *Correctness (Corr)* and *Quality* in [36] are used to assess the segmentation results:

$$comp = \frac{||TP||}{||TP|| + ||FN||}$$
$$corr = \frac{||TP||}{||TP|| + ||FP||} \tag{15}$$
$$Quality = \frac{||TP||}{||TP|| + ||FN|| + ||FP||}$$

where *TP* (True Positive) is the number of objects found both in the reference and segmentation, *FN* (False Negative) is the number of reference objects not found in segmentation, and *FP* (False Positive) is the number of detected objects not found in the reference. Different from the metrics defined in [37], which are widely adopted for the ISPRS benchmark dataset, the metrics in [36] found the correspondences between the reference and the segmented data by using the "maximum overlap" instead of the "overall coverage". As they only establish one-to-one correspondences, the TP values in the reference and segmented data are always the same. This can be more convenient for distinguishing the segmentation errors when the relationships of one-to-many, many-to-one, or many-to-many occurred. For example, if one segmented plane corresponds to two reference planes (one-to-many), the two reference planes will all be taken as TPs for the metrics in [37] (fail to detect under-segmentation), while the smaller reference plane will be detected as FN in [36] instead.

Even a small number of incorrectly segmented points sometimes can have a very large influence on the identification of building structures (*i.e.*, false division of roof boundary points can affect the roof topology). Such errors may not be easily detected by the segmentation-based metrics as they only offer a quick assessment at plane level, (*i.e.*, a minimum overlap of 50% with the reference is required to be a TP). Consequently, a result-driven metrics is designed based on whether the segmentation results influences the extraction of roof ridges. The intersection line is calculated using the method and parameters provided in [28]. Considering that the intersections of roofs ridges in corners (*i.e.*, using the close-circle analysis in [38]) may cover up some mistakes in segmentation, only the original ridges are compared in the experiments.

Two ridge-based metrics are utilized in our experiments. One metric is based on the roof topology graph (RTG) which mainly considers the existing ridges between planes. In the above metrics, the one-to-one correspondences among the reference planes and roof planes have been established. A detected ridge is related to two extracted planes and is taken as a TP only when two correspondences between planes can be found and a reference ridge exists between the two planes. The second metric is much stricter and accepts a TP only when the corresponding ridges are similar enough. To achieve such a goal, the similarity between the reference ridges and the test ridges are defined (Figure 7), which consists of three aspects: distance consistence (*dc*), orientation consistence (*oc*), and projection consistence (*pc*):

$$dc = \exp\left\{-\left(\frac{|CC_1| + |DD_1|}{2 \cdot dis_0}\right)^2\right\}$$
$$oc = \exp\left\{-(\frac{\alpha}{\alpha_0})^2\right\} \tag{16}$$
$$pc = \frac{|AB| \cap |C_1D_1|}{|AB| \cup |C_1D_1|} = \frac{|C_1B|}{|AD_1|}$$

where α_0 and dis_0 are two previously established values (*i.e.*, 5° and 0.2 m). As the *oc*, *pc* and *dc* are values between 0 and 1, the large the better, the integral consistence is set as the product of the three values:

$$ic = oc \cdot dc \cdot pc \tag{17}$$

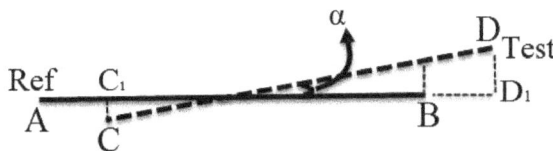

Figure 7. Definition of ridge similarity. Line **AB**: the reference ridge (Ref); line **CD**: the detected ridge (Test), where C1 and D1 are the corresponding projection points of C and D; and α is the intersect angle.

4.3. Experiments

In this section, the improvements from using the weighted approach experimentally are verified. First, the methods for typical scenes that are error-prone for classical RANSAC are presented. Then, the results for the Vaihingen and Wuhan University datasets are evaluated by the metrics given in Section 4.2.

4.3.1. Local Data

For the RANSAC-based methods, spurious planes that consist of points from several roof surfaces are easily generated when adjacent planes have very similar heights or normal orientations. As shown in Figure 8, we select eight typical buildings (a)–(h) to examine the new weighted methods, with the error-prone regions numbered from 1 to 12. In regions 1–2, 5, and 11, a detected plane is possibly be overlapping multiple reference planes; for regions 3–4 and 9–10, poor segmentation may occur when inaccurate hypothesis planes snatch points from neighbor planes; for regions 6–8, two planes are shown as merged into one; and the roof in region 12 is not complanate, thus the segmented results are likely fragmentized. Due to the limited space, we present only the segmentation results for RANSAC, BDSAC, and RG. The MSAC and MLESAC results are very similar to the conventional RANSAC results and can hardlyable to distinguish the poorly estimated planes. Further discussion will be provided in Figure 9.

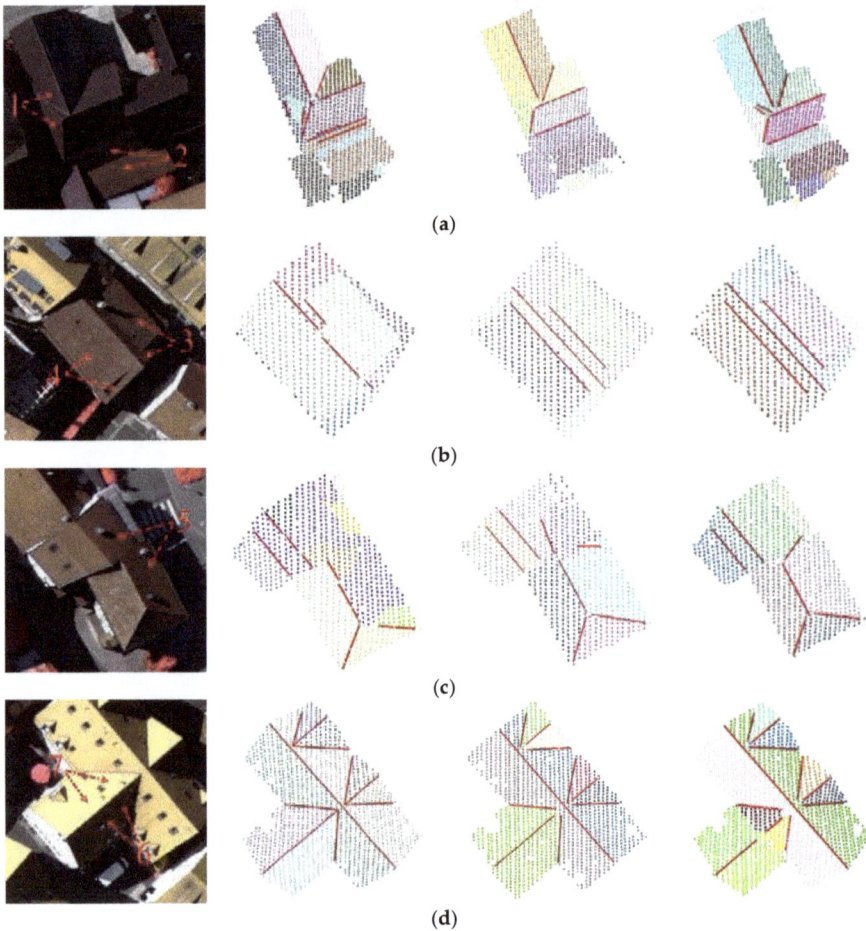

(a)

(b)

(c)

(d)

Figure 8. *Cont.*

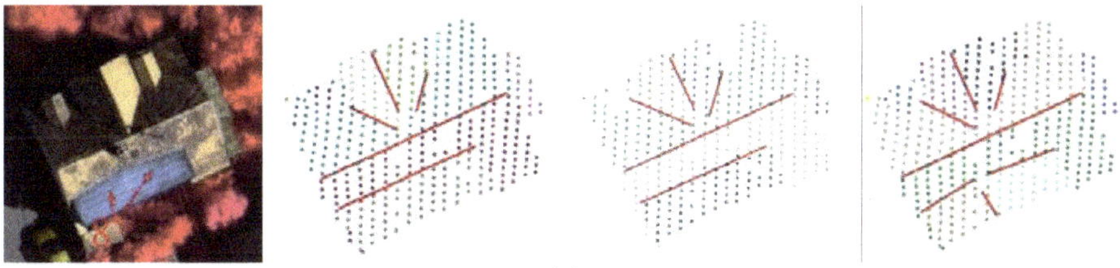

(e)

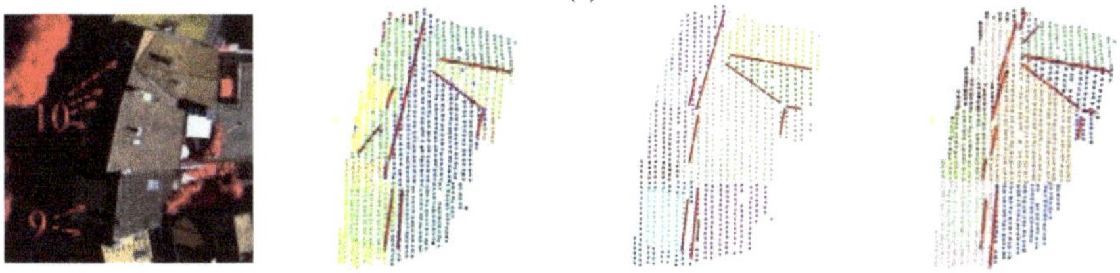

(f)

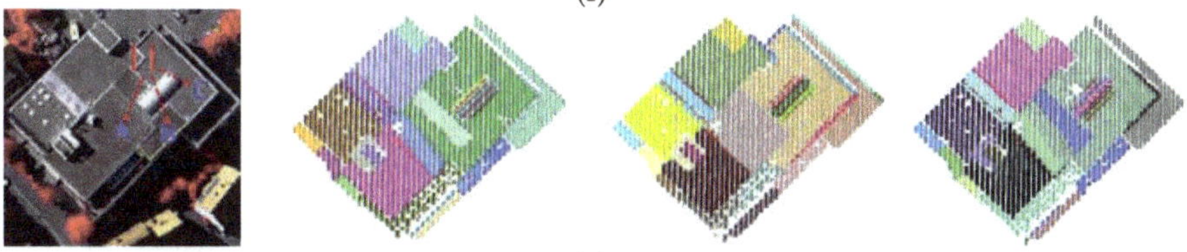

(g)

(h)

Figure 8. Results of segmentation and ridge detection for error-prone buildings. (**a–h**) are eight selected buildings containing error-prone regions. From left to right: reference images, results by classical RANSAC, results by RG, and results by BDSAC$_{nv}$.

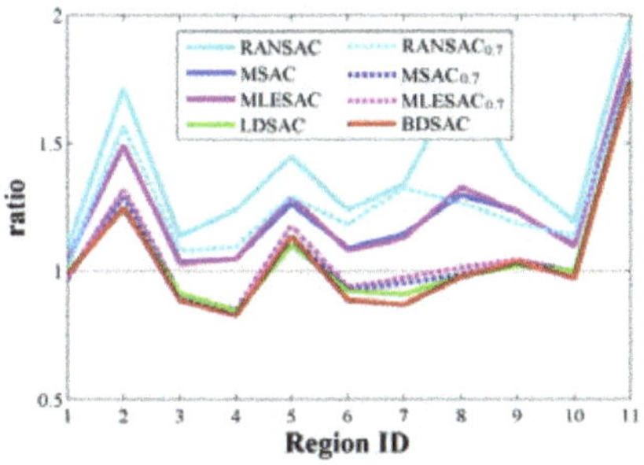

(a)

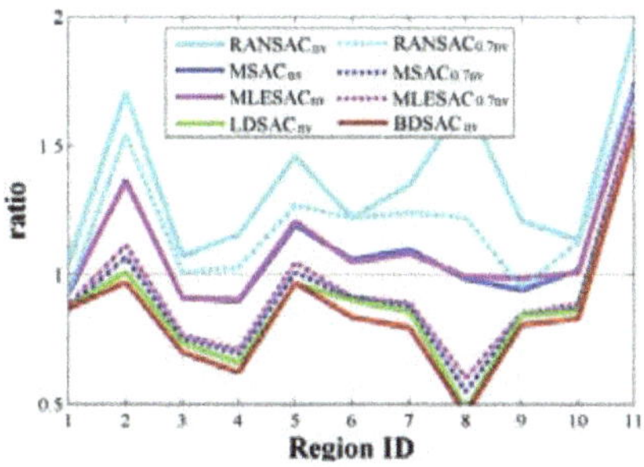

(b)

Figure 9. Suppressing ratios comparison. The spurious planes detected by RANSAC in Figure 8 (regions 1–11). (**a**) Suppressing ratio for methods that only consider point-plane in weight functions; (**b**) Ratios for methods considering both distance and angular difference.

As shown in Figure 8, our new weighted method significantly improved the segmentation and ridge detection results. In regions 1–9, most of the segmentation errors for the RANSAC method, which are also common for the RG method (regions 5–9), are properly solved. In regions 10–12, all the methods fail to create ideal results. The errors in region 10 are mainly caused by sparse data; and in region 11, the normal difference between planes A and B is about 3°and the height difference between B and C is only about 0.15 m, which are too small to distinguish under the current thresholds. The RG method successfully distinguished roofs A and B while it fail to separate B and C. For region 12, all the methods fail because the origin data is not complanate. As a result, our method, compared to the RG method, is slightly better in region 10 but worse in region 11. The quantitative results in Table 3 support our conclusion. Comparing the results of our method to classical RANSAC, the overall segmentation quality increases from 61.3% to 77.2%, and the two ridge-based metrics also increase from 51.8% to 81.7% and 41.6% to 69.3%. Meanwhile, our results are also better than the RG method by the metrics. It can be seen that for regions like 1–4 and 9–0, the incorrectly classified points may not be significant in point count but had strong influences on the identification of roof topology. Such errors are not distinguished by the segmentation-based metrics (*i.e.*, the three planes in region 9 are considered as TPs). Our ridge based metrics show more reasonable evaluation under such situations as the errors will damage the distinguishing of roof ridges. The metrics based on ridge similarity are stricter than those based simply on RTG and exclude some ambiguous or incomplete ridges, such as the ridges in building (b) in Table 3, and thus are more reasonable in some situations.

Table 3. Quality of segmentation results for data in Figure 8.

ID	nPls	nRidges	Method	Segmentation			Ridges (RTG)			Ridges (ic > 0.3)		
				%Cm	%Cr	%Qua	%Cm	%Cr	%Qua	%Cm	%Cr	%Qua
a	10	7	RANSAC	80	100	80	71.4	55.5	45.5	57.1	44.4	33.3
			RG	100	100	100	71.4	100	71.4	71.4	100	71.4
			BDSAC$_{nv}$	100	100	100	100	100	100	71.4	100	71.4
b	5	3	RANSAC	80	57.1	50	66.7	50.0	40.0	0	0	0
			RG	100	100	100	100	100	100	100	100	100
			BDSAC$_{nv}$	100	100	100	100	100	100	100	100	100
c	7	5	RANSAC	85.7	60	51.5	60.0	37.5	30.0	40.0	25.0	18.2
			RG	85.7	75	66.7	100	71.4	71.4	100	71.4	71.4
			BDSAC$_{nv}$	100	100	100	100	100	83.3	100	83.3	83.3
d	10	11	RANSAC	80.0	100	80.0	90.9	100	90.9	90.9	100	90.9
			RG	80.0	100	80.0	90.9	100	90.9	90.9	100	90.9
			BDSAC$_{nv}$	100	100	100	100	100	100	100	100	100
e	9	7	RANSAC	88.9	100	88.9	71.4	100	71.4	57.1	80	50
			RG	60	100	60	71.4	100	71.4	57.1	80	50
			BDSAC$_{nv}$	100	100	100	100	100	100	100	100	100
f	12	12	RANSAC	66.7	66.7	50	33.3	50.0	25.0	33.3	50	25.0
			RG	83.3	90.9	76.9	41.7	55.5	31.2	41.7	55.5	31.2
			BDSAC$_{nv}$	91.7	91.7	84.6	91.7	91.7	84.6	66.7	66.7	50.0
g	23	5	RANSAC	69.6	64.0	50.0	100	62.5	62.5	100	62.5	62.5
			RG	78.3	72.0	60.0	100	71.4	71.4	100	71.4	71.4
			BDSAC$_{nv}$	69.6	66.7	51.6	100	50.0	50.0	100	50.0	50.0
h	11	10	RANSAC	90.9	71.4	66.7	90.0	64.3	60.0	80.0	57.1	50.0
			RG	81.8	69.2	60.0	70.0	46.7	38.9	60.0	40.0	31.6
			BDSAC$_{nv}$	90.9	66.7	62.5	90.0	69.2	64.3	80.0	61.5	53.3
sum	87	60	RANSAC	78.2	73.9	61.3	71.7	65.2	51.8	61.7	56.1	41.6
			RG	82.8	83.7	71.3	75.0	73.8	59.2	71.7	70.5	55.1
			BDSAC$_{nv}$	89.7	84.8	77.2	96.7	84.1	81.7	86.7	77.6	69.3

Figure 9 depicts the performance results of the different weighted methods based on the data in Figure 8, via the outlier suppression ratio (Equation (14)). In each error-prone region, we utilize the largest spurious plane by RANSAC as the test hypothesis plane, whose total weight is W_{test}, and the total weight of the largest reference plane in the corresponding region is W_{ref}. Regions 1–11 in Figure 8 are evaluated. Region 12 is omitted because the roof surface is nonplanar.

Although the ratios are smaller for MSAC and MLESAC than for classical RANSAC, all of them are over 1; thus, the two methods will still accept all the spurious planes that result in false segmentation. As a result, simply using MSAC and MLESAC cannot improve the segmentation results. For our new method, both the new weight forms and the weights regarding angular difference have distinct positive effects on the final results. Considering only one factor may fail in some situations, such as regions 4 and 7 for BDSAC and $MSAC_{nv}$. Meanwhile, the extent of the improvements by angular difference in the weights may be different for the planes. For planes with distinct biases in both the distance and normal vectors, such as regions 2 and 8, the suppressing of the total weights can be larger than in other regions. Again, $BDSAC_{nv}$ provides the best results. In addition, although our methods fail in region 11, the ratio of the $BDSAC_{nv}$ is still smaller than the other weighted methods.

4.3.2. Vaihingen (Germany)

Figure 10 illustrates the segmentation results for the Vaihingen data; specifically, Figure 10a–c are the benchmark data of the "ISPRS Test Project on Urban Classification and 3D Building Reconstruction", in which A and B have been tested in Figure 8e,f, respectively. Other error-prone regions for the classical methods also are indicated in the Figures. For region E, the situation is similar to Figure 8b,f, where spurious planes overlapping multiple roof planes can be produced. In region G, several planes intersect at the same roof corner, which requires more accurate segmentation methods and neighbor competition in the post-processing to better divide the roof boundary points. Our weighted methods show advantages in those regions as well, as the planes with smaller distance errors are more likely to be accepted. Such differences can be detected by the ridge-based metrics. When the transitions between the neighbor regions are smooth, as shown in E, the RG-based methods may fail. Some of the errors are caused by the processing before roof segmentation; for example, the points in region C are classified as vegetation and parts of the points in region D are lost in the original data. All the methods fail in those regions and the related roof ridges are also lost. For E, F, and G, the segmentation results of the different methods are illustrated in Appendix I for comparison.

The quantitative results of the Vaihingen data are shown in Figure 11. It can be seen that our $BDSAC_{nv}$ method generates significant improvements compared to the traditional methods RANSAC and MSAC. Higher scores are achieved by our methods when using either the segmentation-based metrics or the two ridge-based metrics. The improvements of MSAC and MLESAC to RANSAC are not evident in the test data, and many spurious planes are still detected. It should be noted that some error-prone regions are also difficult for the RG method because regions with small angular or height differences often have very smooth transitions (e.g., B and E). Besides, the RG methods seem to be unstable in a few regions, such as the over-segmentation that unexpectedly occurs in F (see Appendix I).

Figure 10. *Cont.*

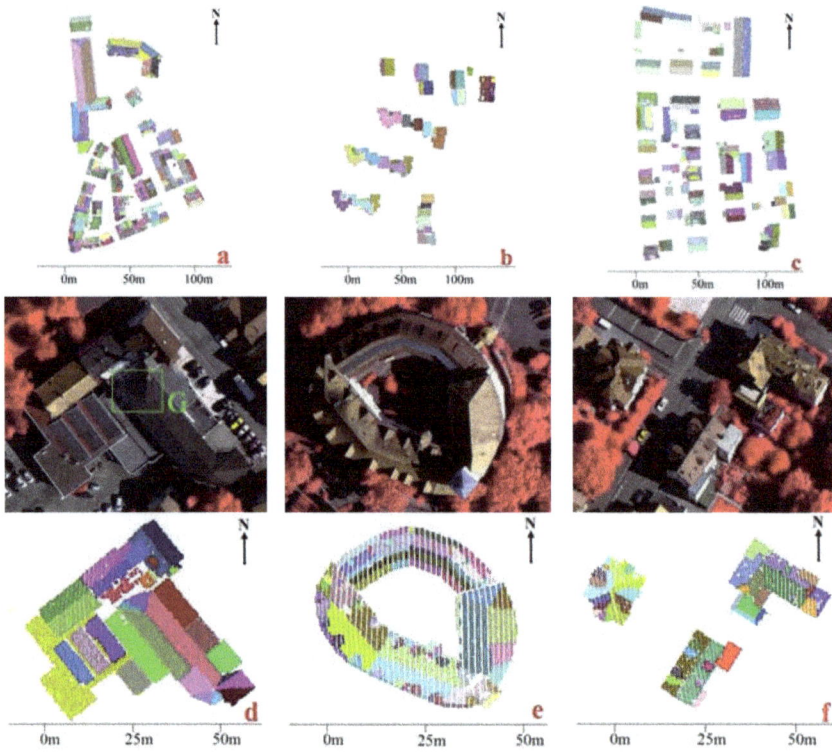

Figure 10. Segmentation of Vaihingen data. (**a–f**) are six selected areas from the data. (top: image, bottom: results of BDSAC$_{nv}$).

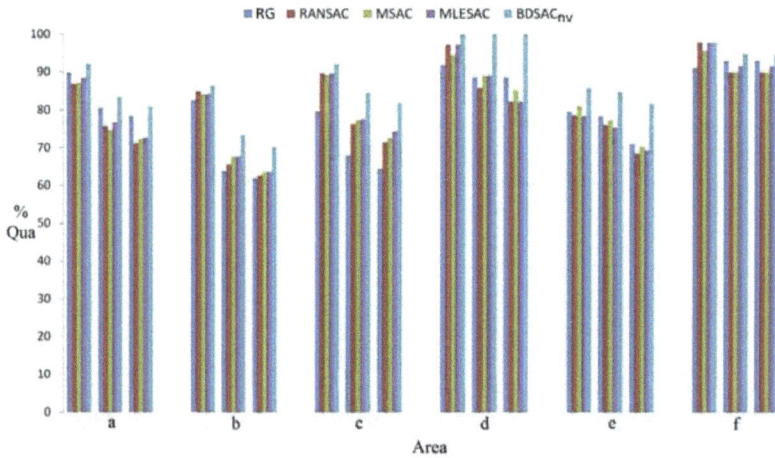

Figure 11. Quantitative results of the Vaihingen data. (**a–f**) are the six areas selected in Figure 10. Three metrics are used, from left to right: quality of segmentation and quality of two ridge based metrics.

4.3.3. Wuhan University (China)

The segmentation results of the Wuhan University data are illustrated in Figure 12. Some error-prone regions are designated. For L and M, spurious planes that overlap multiple roof planes can be produced. For J and O, the small roof planes or short roof ridges may be lost because of small point counts and the competition of roof points from large neighbor planes. In areas (g)–(l), there are many roof details (e.g., Figure 10e), including small windows, eaves, and even guard bars made of glazed tiles, which greatly increase the segmentation difficulties and ultimately results in small plane pieces and short false ridges. In K, a horizontal plane is produced passing through the four planes because the normal errors are not considered in the weight functions. The weighted methods demonstrate great robustness under those situations and therefore significantly improve the segmentation results. Our methods also encounter problems which are unable to resolve. For example, since our weighted methods are not yet adaptable to a curved surface, they divide H and N into several broken pieces. In addition, the RG-based methods fail in I because the points from the upper structure divide the bottom plane into many pieces. The segmentation results for I, K, L and M using different methods are also shown in Appendix I.

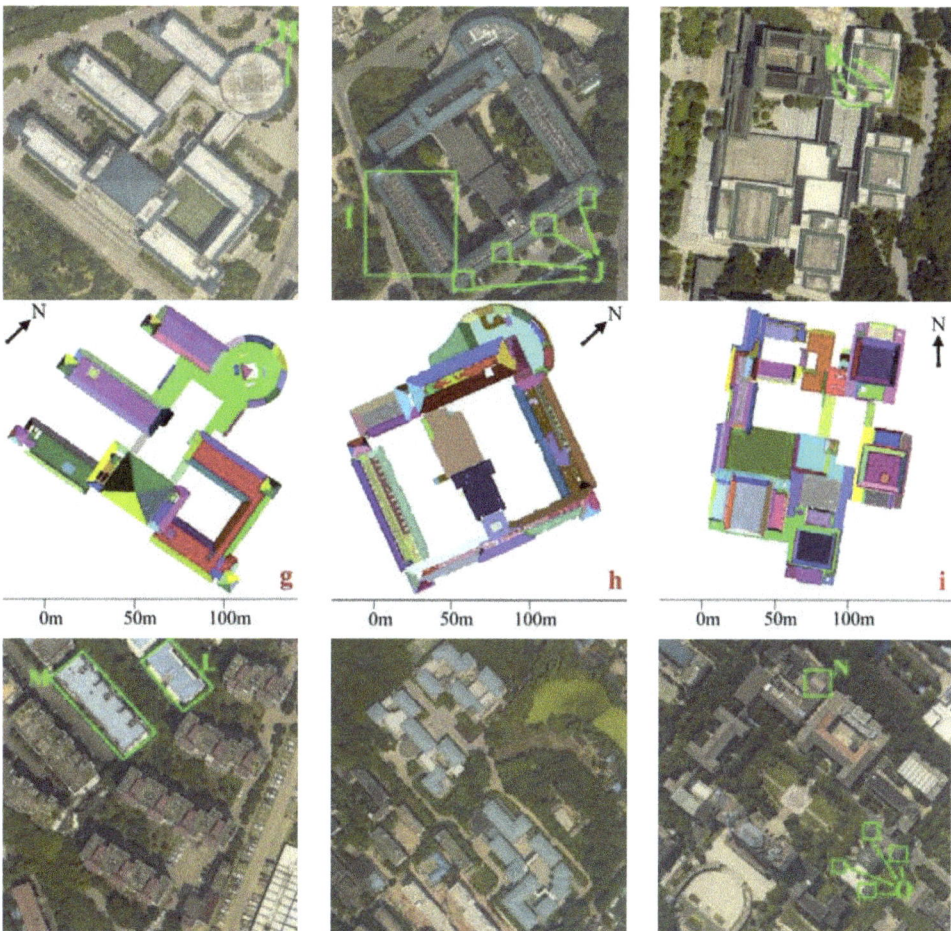

Figure 12. *Cont.*

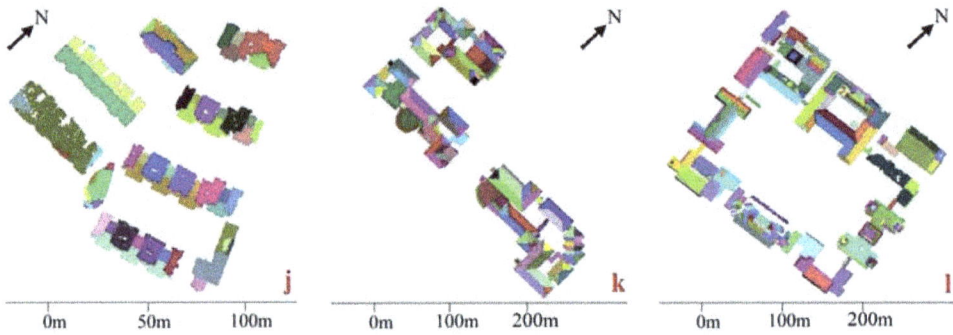

Figure 12. Segmentation results of Wuhan University data. (**g–l**) are six selected areas from the data. (top: image, bottom: results of BDSAC$_{nv}$).

The quantitative results of the Wuhan University data are illustrated in Figure 13. Similar to Figure 11, our method makes significant improvements compared to the other weighted methods. For areas (g) and (h), the RG method's performance is unsatisfactory because many broken fragments exist. For I in area (h) especially, the bottom planes become numerous broken fragments. The RANSAC-based method can be more robust in those situations. Since the RG method considers the roof slope, it can distinguish very small angular differences, which makes it better than the original RANSAC method results in L and M; and over-segmentation also occurred using RG in L and M. A detailed comparison is available in Appendix I.

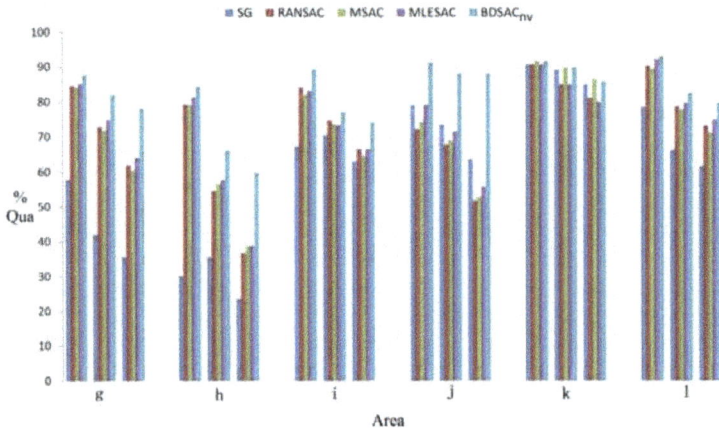

Figure 13. Quantitative results of the Wuhan University data. (**g–l**) are the six areas selected in Figure 12. Three metrics are used, from left to right: quality of segmentation and quality of two ridge based metrics.

The overall quantitative results are shown in Figure 14, which includes the results of Figures 8, 10 and 12. It can be seen that, while the improvements were not very obvious, the results of MSAC and MLESAC are slightly better than that of RANSAC. Our BDSAC$_{nv}$ generate significant improvements compared to both classical RANSAC and the existing weighted methods. Compared to RANSAC, BDSAC$_{nv}$ improves the overall segmentation quality from 85.7% to 90.1%, as well as the two ridge-based metrics from 75.9% to 83.6% and 68.9% to 80.2%. The quality of the RG method is lower than the RANSAC-based methods, mainly due to their instability in areas (c), (g) and (h).

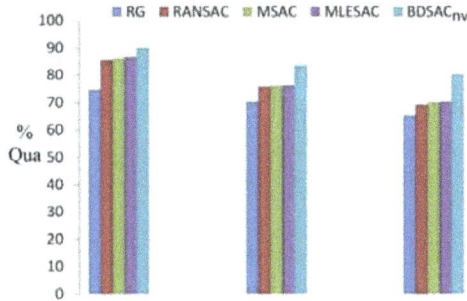

Figure 14. Integral quantitative results. Three metrics are used, from left to right: quality of segmentation and quality of two ridge based metrics.

5. Conclusions

A new weighted RANSAC algorithm for roof point cloud segmentation is introduced in this paper, in which the hard threshold voting function considering both the point-plane distance and the normal vector consistence is transformed into a soft threshold voting function based on two weight functions. Our method utilizes a new strategy to design the ideal weight functions based on the error distribution between the proper and improper hypotheses. Several different weight functions are defined using this strategy, and an outlier suppression ratio is put forward to compare the performance of different weight functions. Preliminary experiments comparing the suppression ratios of different weight functions demonstrated that the $BDSAC_{nv}$ method is able to effectively suppress the outliers from spurious planes. As a result, we chose $BDSAC_{nv}$ for the further experiments and compare its performance with other existing segmentation methods, including original RANSAC, MSAC, MLESAC, and a representative RG method. A set of local data with error-prone regions and two large area datasets of varying densities are used to evaluate the performance of the different methods. The quantitative results of both the segmentation-based metrics and the ridge-based metrics indicated that the different weighted methods improve the segmentation quality differently, but $BDSAC_{nv}$ significantly improve the segmentation accuracy and topology correctness. When compared with RANSAC, $BDSAC_{nv}$ improved the overall segmentation quality from 85.7% to 90.1%; and the two ridge-based metrics also improved from 75.9% to 83.6% and 68.9% to 80.2%. Moreover, the robustness of $BDSAC_{nv}$ is better compared to the RG method. As a result, we believe there is potential for the wide adoption of $BDSAC_{nv}$ as an upgrade to or replacement of classical RANSAC in roof plane segmentation.

However, our method has several limitations. First, although the weighted RANSAC approach is robust to parameters, a small amount of post-processing is still needed to avoid false segmentation or artifacts (see Section 4.1). Second, the weight definition of our method requires a robust estimate of point surface normal, which can be problematic for small buildings or when the point density is low with regard to the roof dimensions. Third, the issue of spurious planes is efficiently suppressed by our method but not completely solved; therefore, spurious planes still may occur in extreme conditions (*i.e.*, Figure 8g).

There are also some possible improvement directions for future work. The number of iterations for RANSAC increases rapidly when the inlier ratio decreases, thus a combination of cluster and fitting to decompose the input data step by step could greatly improve the algorithm's efficiency and robustness. Meanwhile, RANSAC is a one-at-a-time process so adopting the competition approach among neighbor planes could improve the accuracy of segmentation. Finally, only the segmentation of roof planes was considered in this paper, but applying the weighted methods to other roof shapes is possible as the methods mainly are concerned with the procedure of hypothesis verification and do not change the generation of the hypothesis.

Acknowledgments: Acknowledgments: This work was partially supported by the National Basic Research Program of China under Grant 2012CB719904 as well as the science and technology plan of the Sichuan Bureau of Surveying, Mapping and Geoinformation under Grant of J2014ZC02. The Vaihingen dataset was provided by the German Society for Photogrammetry, Remote Sensing, and Geoinformation (DGPF) [Cramer, 2010]: http://www.ifp.uni-stuttgart.de/dgpf/DKEP-Allg.html. The segmentation results by the RG-based method were provided by Biao Xiong.

Author Contributions: Author Contributions: Bo Xu, Wanshou Jiang and Jie Shan contributed to the study design and manuscript writing. Bo Xu and Wanshou Jiang conceived and designed the experiments; Bo Xu performed the experiments; Jing Zhang and Lelin Li contributed to the initial data, the analysis tools and partial codes of the algorithm.

Conflicts of Interest: The authors declare no conflict of interest.

Appendix I

Appendix I Results compare for some regions marked in Figures 10 and 12.

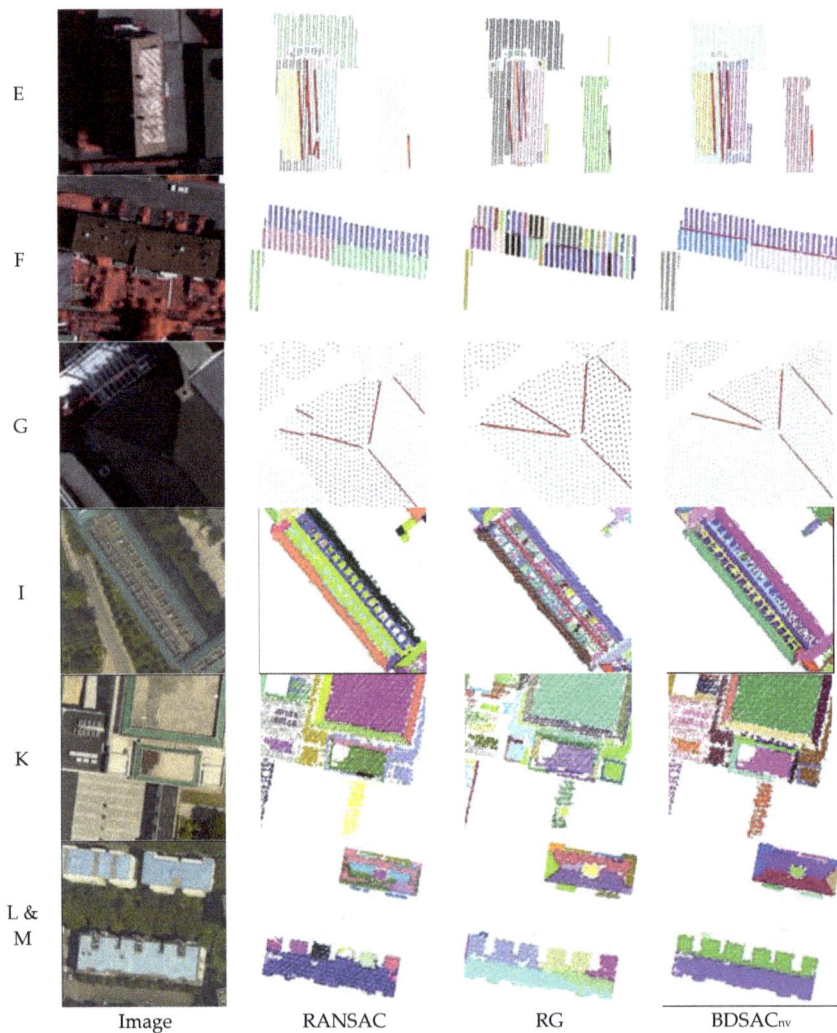

| Image | RANSAC | RG | BDSAC_nv |

References

1. Baltsavias, E.P. Object extraction and revision by image analysis using existing geodata and knowledge: Current status and steps towards operational systems. *ISPRS J. Photogramm. Remote Sens.* **2004**, *58*, 129–151. [CrossRef]
2. Haala, N.; Kada, M. An update on automatic 3D building reconstruction. *ISPRS J. Photogramm. Remote Sens.* **2010**, *65*, 570–580. [CrossRef]
3. Rottensteiner, F.; Sohn, G.; Gerke, M.; Wegner, J.D.; Breitkopf, U.; Jung, J. Results of the ISPRS benchmark on urban object detection and 3D building reconstruction. *ISPRS J. Photogramm. Remote Sens.* **2014**, *93*, 256–271. [CrossRef]
4. Chen, D.; Zhang, L.Q.; Li, J.; Liu, R. Urban building roof segmentation from airborne Lidar point clouds. *Int. J. Remote Sens.* **2012**, *33*, 6497–6515. [CrossRef]
5. Tarsha-Kurdi, F.; Landes, T.; Grussenmeyer, P.; Koehl, M. Model-driven and data-driven approaches using Lidar data analysis and comparison. In Proceedings of the ISPRS, Workshop, Photogrammetric Image Analysis (PIA07), Munich, Germany, 19–21 September 2007.
6. Sampath, A.; Shan, J. Segmentation and reconstruction of polyhedral building roofs from aerial Lidar point clouds. *IEEE Trans. Geosci. Remote Sens.* **2010**, *48*, 1554–1567. [CrossRef]
7. Awwad, T.M.; Zhu, Q.; Du, Z.; Zhang, Y. An improved segmentation approach for planar surfaces from unconstructed 3D point clouds. *Photogramm. Rec.* **2010**, *25*, 5–23. [CrossRef]
8. Fan, T.J.; Medioni, G.; Nevatia, R. Segmented descriptions of 3-D surfaces. *IEEE Trans. Rob. Autom.* **1987**, *3*, 527–538.
9. Alharthy, A.; Bethel, J. Detailed building reconstruction from airborne laser data using a moving surface method. *Int. Arch. Photogramm. Remote Sens.* **2004**, *35*, 213–218.
10. You, R.J.; Lin, B.C. Building feature extraction from airborne Lidar data based on tensor voting algorithm. *Photogramm. Eng. Remote Sens.* **2011**, *77*, 1221–1231. [CrossRef]
11. Vosselman, G. Automated planimetric quality control in high accuracy airborne laser scanning surveys. *ISPRS J. Photogramm. Remote Sens.* **2012**, *74*, 90–100. [CrossRef]
12. Lagüela, S.; Díaz-Vilariño, L.; Armesto, J.; Arias, P. Non-destructive approach for the generation and thermal characterization of an as-built BIM. *Constr. Build. Mater.* **2014**, *51*, 55–61. [CrossRef]
13. Hoffman, R.; Jain, A.K. Segmentation and classification of range images. *IEEE Trans. Pattern Anal. Mach. Intell.* **1987**, *9*, 608–620. [CrossRef] [PubMed]
14. Filin, S. Surface clustering from airborne laser scanning data. *Int. Arch. Photogramm. Remote Sens.* **2002**, *34*, 119–124.
15. Filin, S.; Pfeifer, N. Segmentation of airborne laser scanning data using a slope adaptive neighborhood. *ISPRS J. Photogramm. Remote Sens.* **2006**, *60*, 71–80. [CrossRef]
16. Biosca, J.M.; Lerma, J.L. Unsupervised robust planar segmentation of terrestrial laser scanner point clouds based on fuzzy clustering methods. *ISPRS J. Photogramm. Remote Sens.* **2008**, *63*, 84–98. [CrossRef]
17. Dorninger, P.; Pfeifer, N. A comprehensive automated 3D approach for building extraction, reconstruction, and regularization from airborne laser scanning point clouds. *Sensors* **2008**, *8*, 7323–7343. [CrossRef]
18. Awrangjeb, M.; Fraser, C.S. Automatic segmentation of raw Lidar data for extraction of building roofs. *Remote Sens.* **2014**, *6*, 3716–3751. [CrossRef]
19. Fischler, M.A.; Bolles, R.C. Random sample consensus—A paradigm for model-fitting with applications to image-analysis and automated cartography. *Commun. ACM* **1981**, *24*, 381–395. [CrossRef]
20. Schnabel, R.; Wahl, R.; Klein, R. Efficient RANSAC for point-cloud shape detection. *Comput. Graph. Forum* **2007**, *26*, 214–226. [CrossRef]
21. Choi, S.; Kim, T.; Yu, W. Performance evaluation of RANSAC family. In Proceedings of the British Machine Vision Conference, London, UK, 7–10 September 2009.
22. Frahm, J.-M.; Pollefeys, M. RANSAC for (Quasi-) degenerate data (QDEGSAC). In Proceedings of the IEEE Computer Society Conference on Computer Vision and Pattern Recognition, New York, NY, USA, 17–22 June 2006.
23. Chum, O.R.; Matas, J.R.I. Matching with PROSAC-progressive sampling consensus. In Proceedings of the IEEE Computer Society Conference on Computer Vision and Pattern Recognition, Miami, FL, USA, 20–25 June 2005.

24. Chum, O.; Matas, J. Optimal randomized RANSAC. *IEEE Trans. Pattern Anal. Mach. Intell.* **2008**, *30*, 1472–1482. [CrossRef] [PubMed]
25. Berkhin, P. *A Survey of Clustering Data Mining Techniques*; Springer Berlin Heidelberg: New York, NY, USA, 2006.
26. Gallo, O.; Manduchi, R.; Rafii, A. CC-RANSAC: Fitting planes in the presence of multiple surfaces in range data. *Pattern Recognit. Lett.* **2011**, *32*, 403–410. [CrossRef]
27. Yan, J.X.; Shan, J.; Jiang, W.S. A global optimization approach to roof segmentation from airborne Lidar point clouds. *ISPRS J. Photogramm. Remote Sens.* **2014**, *94*, 183–193. [CrossRef]
28. Xiong, B.; Elberink, S.O.; Vosselman, G. A graph edit dictionary for correcting errors in roof topology graphs reconstructed from point clouds. *ISPRS J. Photogramm. Remote Sens.* **2014**, *93*, 227–242. [CrossRef]
29. Elberink, S.O.; Vosselman, G. Building reconstruction by target based graph matching on incomplete laser data: Analysis and limitations. *Sensors* **2009**, *9*, 6101–6118. [CrossRef] [PubMed]
30. Hesami, R.; BabHadiashar, A.; HosseinNezhad, R. Range segmentation of large building exteriors: A hierarchical robust approach. *Comput. Vis. Image Underst.* **2010**, *114*, 475–490. [CrossRef]
31. Torr, P.H.S.; Zisserman, A. Mlesac: A new robust estimator with application to estimating image geometry. *Comput. Vis. Image Underst.* **2000**, *78*, 138–156. [CrossRef]
32. Bretar, F.; Roux, M. Hybrid image segmentation using Lidar 3D planar primitives. In Proceedings of the ISPRS Workshop Laser Scanning, Enschede, The Netherlands, 12–14 September 2005.
33. López-Fernández, L.; Lagüela, S.; Picón, I.; González-Aguilera, D. Large-scale automatic analysis and classification of roof surfaces for the installation of solar panels using a multi-sensor aerial platform. *Remote Sens.* **2015**, *7*, 11226–11248. [CrossRef]
34. Wang, C.; Sha, Y. A designed beta-hairpin forming peptide undergoes a consecutive stepwise process for self-assembly into nanofibrils. *Protein Pept. Lett.* **2010**, *17*, 410–415. [CrossRef] [PubMed]
35. Girardeau-Montaut, D. Detection de Changement sur des Données Géométriques 3D. Ph.D. Thesis, Télécom ParisTech, Paris, France, 2006.
36. Awrangjeb, M.; Fraser, C.S. An automatic and threshold-free performance evaluation system for building extraction techniques from airborne Lidar data. *IEEE J. Sel. Top. Appl. Earth Observ. Remote Sens.* **2014**, *7*, 4184–4198. [CrossRef]
37. Rutzinger, M.; Rottensteiner, F.; Pfeifer, N. A comparison of evaluation techniques for building extraction from airborne laser scanning. *IEEE J. Sel. Top. Appl. Earth Observ. Remote Sens.* **2009**, *2*, 11–20. [CrossRef]
38. Perera, G.S.N.; Maas, H.G. Cycle graph analysis for 3D roof structure modeling: Concepts and performance. *ISPRS J. Photogramm. Remote Sens.* **2014**, *93*, 213–226. [CrossRef]

![remote sensing logo] *remote sensing*

[MDPI]

Article

Three-Dimensional Reconstruction of Building Roofs from Airborne LiDAR Data Based on a Layer Connection and Smoothness Strategy

Yongjun Wang [1,2,3], Hao Xu [1,4,5,*], Liang Cheng [1,4,5,6,*], Manchun Li [1,4,6], Yajun Wang [4], Nan Xia [4], Yanming Chen [4,6] and Yong Tang [1,2,3]

[1] Jiangsu Center for Collaborative Innovation in Geographical Information Resource Development and Application, Nanjing 210023, China; wangyongjun@njnu.edu.cn (Y.W.); limanchun@nju.edu.cn (M.L.); 14751775121@163.com (Y.T.)
[2] Key Laboratory of Virtual Geographic Environment, Ministry of Education, Nanjing Normal University, Nanjing 210093, China
[3] State Key Laboratory Cultivation Base of Geographical Environment Evolution, Nanjing 210093, China
[4] Jiangsu Provincial Key Laboratory of Geographic Information Science and Technology, Nanjing University, Nanjing 210093, China; wangyajun_csu@163.com (Y.W.); mg1427065@smail.nju.edu.cn (N.X.); chenyanming@nju.edu.cn (Y.C.)
[5] Collaborative Innovation Center of Novel Software Technology and Industrialization, Nanjing University, Nanjing 210093, China
[6] Collaborative Innovation Center for the South Sea Studies, Nanjing University, Nanjing 210093, China
* Correspondence: xuhao_nju@smail.nju.edu.cn (H.X.); lcheng@nju.edu.cn (L.C.);
   Tel.: +86-25-8359-7359 (H.X.); +86-25-8359-5336 (L.C.); Fax: +86-25-8359-7359 (H.X.)

Academic Editors: Juha Hyyppä, Jie Shan, Guoqing Zhou and Prasad S. Thenkabail
Received: 17 January 2016; Accepted: 12 April 2016; Published: 16 May 2016

Abstract: A new approach for three-dimensional (3-D) reconstruction of building roofs from airborne light detection and ranging (LiDAR) data is proposed, and it includes four steps. Building roof points are first extracted from LiDAR data by using the reversed iterative mathematic morphological (RIMM) algorithm and the density-based method. The corresponding relations between points and rooftop patches are then established through a smoothness strategy involving "seed point selection, patch growth, and patch smoothing." Layer-connection points are then generated to represent a layer in the horizontal direction and to connect different layers in the vertical direction. Finally, by connecting neighboring layer-connection points, building models are constructed with the second level of detailed data. The key contributions of this approach are the use of layer-connection points and the smoothness strategy for building model reconstruction. Experimental results are analyzed from several aspects, namely, the correctness and completeness, deviation analysis of the reconstructed building roofs, and the influence of elevation to 3-D roof reconstruction. In the two experimental regions used in this paper, the completeness and correctness of the reconstructed rooftop patches were about 90% and 95%, respectively. For the deviation accuracy, the average deviation distance and standard deviation in the best case were 0.05 m and 0.18 m, respectively; and those in the worst case were 0.12 m and 0.25 m. The experimental results demonstrated promising correctness, completeness, and deviation accuracy with satisfactory 3-D building roof models.

Keywords: airborne LiDAR; building roof; three-dimensional (3-D) reconstruction; layer-connection points; smoothness strategy

---

1. Introduction

The three-dimensional (3-D) reconstruction of building models is an important means of obtaining 3-D structural information of urban scenes. Such reconstructions are applicable in fields such as urban planning, change detection research, and solar mapping [1–3]. The 3-D modeling solutions enable users to rapidly construct 3-D maps of surrounding areas that are suitable for professional visualization systems. As the most important and challenging task in digital city construction, the 3-D reconstruction of building models has received considerable attention over the past few decades [4,5]. Traditionally, photogrammetry is the primary approach used for deriving geo-spatial information, and it is implemented through the use of single or multiple optical images; often, aerial stereo images have been used for 3-D building model reconstruction. Some detailed reviews of the techniques for building reconstruction from aerial imagery have been published in the literature [6]. However, considerable manual assistance is required, which results in a low degree of automation.

Airborne light detection and ranging (LiDAR) technology has developed rapidly, and has been found useful for many applications in various fields [7–10]. In particular, airborne LiDAR technology presents a new avenue for the 3-D reconstruction of building models [11], and relevant methods have been reviewed in the literature [12]. Tomljenovic *et al.* [13] provided an overview of building extraction approaches applied to airborne LiDAR data from several aspects, such as dataset area, accuracy measures, reference data for accuracy assessment, and the use of auxiliary data. Presently, detailed 3-D information about the ground surface can be obtained by the airborne LiDAR equipment with a high degree of automation; however, the massive number of irregularly distributed points brings new challenges for the reconstruction work. Therefore, full utilization of the advantages of LiDAR points for high-quality building model reconstruction remains an important research topic.

Most previous approaches related to building model reconstruction with airborne LiDAR data can be divided into the following two categories: model-driven (parametric or top-down strategy) and data-driven (non-parametric or bottom-up strategy) methods. The benefits and drawbacks of the model- and data-driven methods have been discussed in a previous study [14]. In the model-driven methods, a predefined catalog of roof shapes is prescribed (e.g., flat, hip, gambrel). One of the advantages of model-driven methods is that the final roof shape is always topologically correct according to the predefined shapes. However, failure is possible when reconstructing complex building characteristics and building models that are excluded in the predefined shapes [12]. In addition, the level of detail in the reconstructed buildings is compromised as the input models usually consist of rectangular footprints and the current level of automation is comparatively low. In contrast to model-driven methods, data-driven approaches are more flexible and do not require prior knowledge, in which a building roof is reassembled from roof parts found by a segmentation algorithm. A challenging feature of these methods is to identify the relationship between the neighboring rooftop patches; for example, coplanar patches, intersection lines, or step edges between neighboring planes. The main advantage of these methods is that polyhedral buildings of arbitrary shape may be reconstructed [15]. The main drawback of data-driven methods is their susceptibility to the point density of the point clouds.

In the past studies, much of the work on the reconstruction of building models using airborne LiDAR data focuses on the extraction of rooftop contours [11,16]. Cheng *et al.* [17] combined airborne LiDAR data and optical remotely sensed images for the reconstruction of 3-D building models. They developed an integration mechanism that incorporates the segmented roof points and two-dimensional (2-D) lines extracted from optical multi-view aerial images to enable 3-D step line determination, from which 3-D roof models could be reconstructed. Similarly, Susaki [18] achieved 3-D building model reconstruction through a combination of airborne LiDAR data and high-spatial resolution aerial images. Verma *et al.* [19] introduced a new method for the detection and reconstruction of complex building models in which no a priori hypotheses are required; with this method, the topology of complex roof shapes is determined by using the roof-topology graph. Sohn *et al.* [20] used a binary space partitioning tree to reconstruct the global geometric topology of polyhedral buildings from adjacent linear features by using airborne LiDAR data. Zhang *et al.* [21] derived the building footprints through the combination of a region-growing algorithm and a boundary extraction method before building model reconstruction. Kada and McKinley [22] proposed an approach for decomposition of footprint with an additional generalization of the footprint. The building models were reconstructed by assembling building blocks from a library of parameterized standard shapes. Further, Vallet *et al.* [23] introduced an approach where the footprint decomposition is triggered by a digital surface model derived from the laser points. However, the reconstruction of 3-D building models with rooftop contours extracted from airborne LiDAR data is usually difficult. This is because the topological relationship between rooftop contours of different roof layers is difficult to confirm and, additionally, the extraction of rooftop contours is strongly affected by noise.

The extraction of rooftop patches, a prerequisite of plane-based methods, is another way to obtain building models from airborne LiDAR data. Common methods of extracting rooftop patches include the 3-D Hough transformation [24,25], the region growing technique [26,27], and application of the random sample consensus (RANSAC) algorithm [28,29]. Fan *et al.* [30] proposed the hierarchical decomposition of ridge lines for rooftop patch extraction. Awrangjeb and Fraser [31] classified original airborne LiDAR points into ground points and non-ground points. Coplanarity and local characteristics of each point were then used to segment the building rooftops from the non-ground points. Chen *et al.* [32] conducted a sequential process of morphological filtering, region growing, and adaptive RANSAC algorithm calculations to segment the rooftop points, whereas Kim and Shan [33] considered the optimization of an energy function and introduced a global segmentation strategy for rooftop patches that guaranteed the topological consistency of the extracted patches. Sampath and Shan [34] applied a fuzzy *k*-means algorithm to cluster the rooftop points to each patch and distinguished parallel and coplanar patches based on distance and connectivity. The 3-D building models were then obtained by using an adjacency matrix. Although the rooftop patches can be segmented well by using the above mentioned methods, the patches are not fit to construct 3-D building models directly. This is because the airborne LiDAR points representing these rooftop patches are usually irregularly distributed. Therefore, some researchers considered combining the model- and data-driven methods to reconstruct building roofs from airborne LiDAR data. This hybrid approach, also known as the global strategy, exhibits both model- and data-driven characteristics. For example, Satari *et al.* [35] applied the data-driven method to reconstruct cardinal planes and the model-driven method to reconstruct dormers. Lafarge *et al.* [36] presented a structural approach for building reconstruction from a single DSM, which treats buildings as an assemblage of simple urban structures extracted from a library of 3D parametric blocks. The Gibbs model and the Bayesian decision approach were used to control the block assemblage and to find the optimal configuration of 3-D blocks. To reflect the orientation and placement similarities between planar elements in building structures, Zhou and Neumann [37] emphasized global regularity during the construction of planar rooftops. This approach improved the reliability of the final results and decreased the complexity of the building models. Similarly, Zhang *et al.* [38] proposed a novel method that represents building roofs by geometric primitives and constructs a cost function for the final 3-D model reconstruction. In addition, Chen *et al.* [39] used a multiscale grid method for the detection and reconstruction of building roofs from airborne LiDAR data. Although it is beneficial to the plane-based methods that LiDAR data provide a high density of 3-D points, the discrete and irregular distribution of these points may lead to low geometrical accuracy for building models. Especially, it is difficult to determine accurate boundaries and the connection relationships among roof faces with a height jump [34,40].

Here, we present a new point-based approach for 3-D reconstruction of building roofs from airborne LiDAR data. The overall idea is as follows.

1. *Smoothness-oriented rooftop patch extraction.* For airborne LiDAR points of buildings, rooftop patches are segmented by using a region growing approach. To reduce noise interference and eliminate the effect of irregularly- distributed points, the rooftop points are smoothed before the building roofs are reconstructed.

2. *Determination of layer-connection points and calculations for building roof reconstruction.* Layer-connection points are generated from a 2-D grid to guarantee consistency between the boundary footprints of different roof layers. Building roofs are then reconstructed by connecting neighboring

layer-connection points. The generation of layer-connection points helps to establish the relationships among different rooftop patches effectively and efficiently.

2. Methodology

The proposed approach for 3-D building roof reconstruction from airborne LiDAR data consists of four steps. These four steps are as follows.

1. *Preprocessing.* Building rooftop points are extracted from airborne LiDAR data by using the reversed iterative mathematic morphological (RIMM) algorithm and the density-based method.

2. *Smoothness-oriented rooftop patch extraction.* A strategy of "seed point selection, patch growth, and patch smoothing" is introduced during the rooftop patch extraction to smooth the building rooftop points.

3. *Generation of layer-connection points.* Layer-connection points are generated from a 2-D grid, thus guaranteeing consistency between the boundary footprints of different roof layers.

4. Building *model reconstruction.* By connecting neighboring layer-connection points together, the building roofs are reconstructed.

2.1. Extraction of Building Rooftop Points

As a precondition for 3-D building roof reconstruction, building roof points need to be detected and extracted from airborne LiDAR data. We applied Cheng's RIMM algorithm to extract the building points [41]. The RIMM method first employs a morphological opening operation, and this opening operation is iterated by gradually decreasing the window size at a fixed step length (3 m). The elevation difference between two adjacent iterations is then compared, and parts with elevation differences exceeding the minimum building height (3 m in this study) are regarded as building point clouds. However, in the detected building point clouds, there may be some dense tree points, and the tree points are removed by using a threshold roughness value (we set this to 0.8 m), *i.e.*, the standard deviation of height values of LiDAR points. The algorithm has been described in detail in the literature [41].

Airborne laser scanners not only acquire the laser measurements from building roofs, but also obtain partial reflected pulses from building walls. Therefore, wall points still exist in the building point clouds obtained by the RIMM algorithm. In order to retain only the rooftop points, we used Awrangjeb's method to remove wall points [42]. The experimental results demonstrated that the wall points were eliminated by this method effectively.

2.2. Smoothness-Oriented Rooftop Patch Segmentation

Once we have extracted the building roof LiDAR points from the raw data, the different rooftop patches must be determined, *i.e.*, rooftop patches must be segmented. The segmentation process follows a strategy of seed point selection, patch growth, and patch smoothing.

2.2.1. Rooftop Patch Segmentation

The input for the rooftop patch segmentation algorithm was implemented based on the classified individual buildings, and for this, the region growing segmentation algorithm proposed by Sun and Salvaggio [26] was used in our study; this algorithm uses the point normals and their curvatures. First, the normal and curvature values of each LiDAR point are calculated and the point with the smallest curvature value is selected as the seed point. Within a small neighborhood of this seed point, the direction of the normal vector of any other point with the normal direction of this seed point are compared. If the directional difference is larger than a predetermined threshold, the point being examined does not belong to the group initiated by the seed point, and otherwise, it does. In those points that have been grouped together by the seed point, points with curvature values lower than a predetermined threshold are chosen as future seed points. The procedure is iteratively executed until all LiDAR points have been visited.

2.2.2. Smoothness-Oriented Rooftop Patch Optimization

Rooftop patch optimization involves (1) smoothing of the rooftop patch points and (2) eliminating the interference of omissive LiDAR points.

The red box in Figure 1a contains a number of protuberant points (*i.e.*, a small number of points above a large rooftop patch); however, these protuberant points were not seen in the rooftop patch segmentation results (Figure 1b). If these protuberant points are directly discarded before building roof reconstruction, some holes will appear on the building roofs. As these points are usually distributed above the real rooftop patches, we set a distance threshold (2 m) to recognize them. After all the protuberant points have been found, they are projected to the corresponding segmented rooftop patch by using the plane equation calculated by the RANSAC algorithm [29].

Furthermore, even if such points belong to the same rooftop patch, they may not be precisely distributed on the same plane. Hence, a smoothing operation must be conducted prior to the building roofs being reconstructed. Here, for each segmented rooftop patch, the RANSAC algorithm was applied to fit a virtual plane from the candidate points, and then the points were forced to move on to this estimated plane in order to assign a perfect flatness property to each surface. The smoothing procedures described above were conducted iteratively for all rooftop patches. Figure 1c illustrates the point clouds after smoothing.

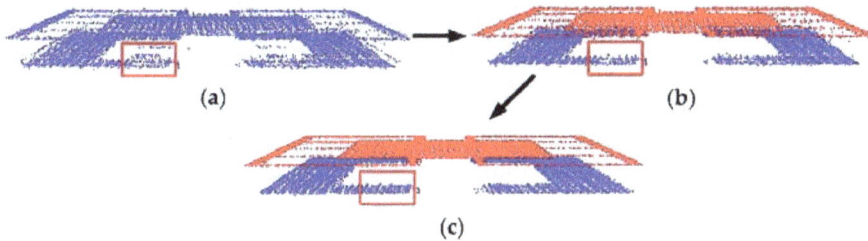

Figure 1. Smoothing of building rooftop points: (**a**) original points; (**b**) segmented points; and (**c**) smoothed points.

2.3. Generation of Layer-Connection Points

According to the derived rooftop patches, contour-based approaches are usually used for the reconstruction of 3-D building roofs. However, some major flaws impede the use of 3-D contours in building model reconstruction. These limitations are as follows: (1) the fitting of 3-D contours is strongly affected by noise; (2) the extraction of 3-D contours demands a large number of points, so the 3-D contours may be broken when not enough points are provided; and (3) the topological relationships of 3-D contours among different roof layers are difficult to confirm.

Therefore, a point-based method for 3-D building roof reconstruction is proposed here. The core objective of this method is to generate points to represent and connect roof layers (see Figure 2), which are named as layer-connection points, for a building rooftop. Layer-connection points have two purposes, namely, to represent a roof layer in the horizontal direction and to connect different roof layers in the vertical direction. In the horizontal direction, as shown in Figure 2a,b, yellow points represent the first layer (ground), blue points represent the second layer, and red points represent the third layer. These points with different colors are defined as layer-points. In the vertical direction, as shown in Figure 2c, a yellow point and red point form a line to connect the corresponding local region of the first layer and the third layer. These two layer-points, which have the same x–y coordinates but different z values, as a whole, are called a layer-connection point. Similarly, in Figure 2d, a yellow point and blue point form a line to connect the first layer and the second layer. Moreover, in Figure 2e, a yellow point, blue point, and red point form a line to connect the first layer, the second layer, and the third layer. These three layer-points, as a whole, are also called a layer-connection point.

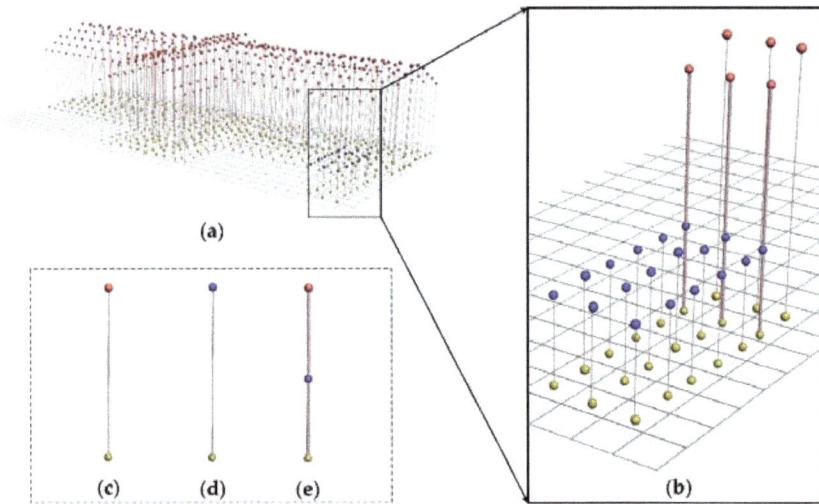

Figure 2. An example of layer-connection points: (**a**) yellow points represent the first layer (ground), blue points represent the second layer, and red points represent the third layer; (**b**) an enlarged view; (**c**); (**d**) lines to connect points from two layers; and (**e**) line to connect points from three layers.

Before calculation of layer-connection points, we need to merge the rooftop patches into several roof layers. First, we give a clear definition of the *roof layer*. A roof layer is defined as being made up of several rooftop patches, which are adjacent and intersected with one another. For example, in Figure 2a, the red points represent the third layer, but the layer has four different rooftop patches, and, therefore, the four rooftop patches should be merged into a roof layer. Airborne LiDAR points corresponding to different rooftop patches are used as source data for merging. Here, we first introduce the principles for merging rooftop patches. The principles include whether there is an intersection line and an adjacent relationship between two rooftop patches. For example, in Figure 3a, there are two rooftop patches, *i.e.*, *S1* (red points in Figure 3a) and *S2* (yellow points in Figure 3a). When *S1* and *S2* have an intersection line (blue line in Figure 3a), and the two rooftop patches have an adjacent relationship, they can be merged into the same roof layer. After the completion of judgments for all rooftop patches in accordance with the above principles, we can obtain the final merging results, as illustrated in Figure 3b. After the merging of rooftop patches, the layer-connection points can be calculated.

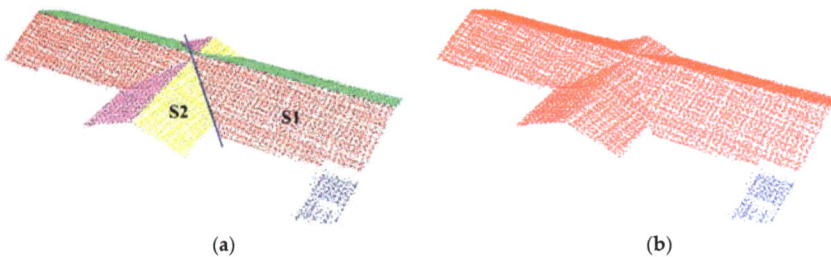

Figure 3. Example of merging rooftop patches into a layer. (**a**) Points with different colors represent different rooftop patches; the blue line represents the intersection line between S1 and S2; and (**b**) red points and blue points represent a roof layer.

2.3.1. Construction of the 2-D Grid System

Before calculating the layer-connection points, we need to create a 2-D grid system. Thus, the scale and grid cell size of the 2-D grid system need to be determined. The scale of the grid system is set according to the maximum and minimum values of the point clouds in the X and Y directions. To guarantee that a reasonable number of points lie inside each grid cell, the grid cell size is set to 2–3 times the average point spacing (we set it to 1.0 m in this study). The grid cells record the serial numbers of each LiDAR point within them, and a cell with no points is said to be empty. For each LiDAR point, we record the row and column number of its corresponding cell to construct a two-way index. The index of a point with coordinates (x, y, z) can be obtained from the following formula:

$$i = \text{int}((y - y_{\min})/Gridsize)$$
$$j = \text{int}((x - x_{\min})/Gridsize) \tag{1}$$

where i and j represent the number of row and column of a grid cell, respectively, $(x_{\min}, y_{\min})$ represents the minimum coordinates of the building points, and *Gridsize* represents the size of a grid cell.

2.3.2. Calculation of Layer-Connection Points

During the calculation of layer-connection points, a center cell and its neighboring four cells are taken into consideration. Aforementioned layer results are used to determine whether the LiDAR points in the neighboring four cells are on the same roof layer with that in the center cell. There are five potential situations.

In the first situation, as illustrated in Figure 4a, where all LiDAR points belong to the same roof layer, the center cell (red box in Figure 4a) does not contain any connections between different roof layers. In this situation, the cell will only generate a layer-point for the layer connection. The coordinates of this cell's center (blue rectangle in Figure 4a) are set as the x–y coordinates of this layer-point, and the average height value of all LiDAR points inside the center cell is set as its height.

In the second situation, as illustrated in Figure 4b, where the LiDAR points inside the center cell and four neighboring cells do not belong to the same roof layer, the violet cell may contain connections between different roof layers. A line splitting different roof layers can be calculated on which a point to connect different layers can be located. If the cell (violet box in Figure 4b) with LiDAR points from different roof layers lies to the left or right of the center cell (red box in Figure 4b), a black dotted horizontal line is generated to intersect the splitting line (black solid line in Figure 4b). The point of intersection (blue rectangle in Figure 4b) is taken as the x–y coordinates of the layer-connection point. Simultaneously, the cell containing the planimetric coordinates of the intersection point is confirmed and the LiDAR points in it are selected. Based on these points, height values of each layer-point are determined by the average height of the LiDAR points of the corresponding roof layer.

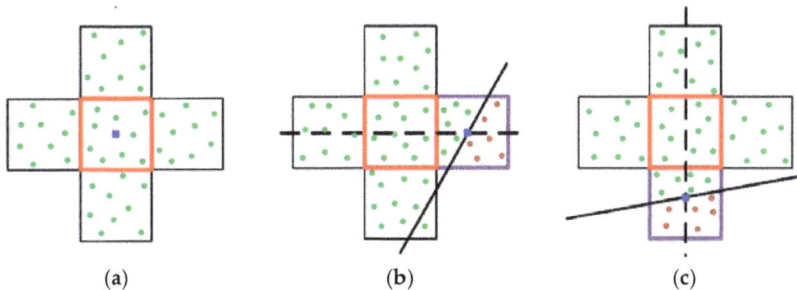

| (a) | (b) | (c) |

Figure 4. *Cont.*

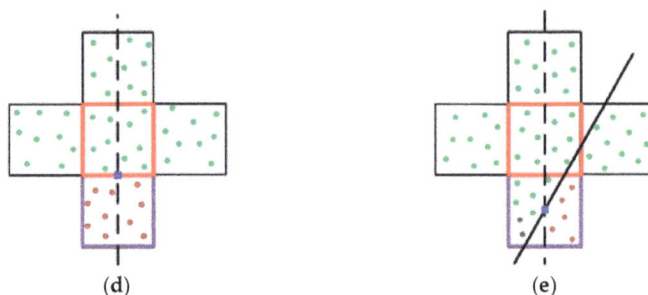

Figure 4. Calculation of layer-connection points, points with different colors representing different roof layers, and the blue rectangle representing the *x*–*y* coordinates of the derived layer-connection point: (**a**) the points inside the five cells belonging to the same roof layer; and (**b**), (**c**), (**d**), (**e**) the points inside the five cells belonging to different roof layers.

In the third situation, as illustrated in Figure 4c, if the cell (violet box in Figure 4c) with LiDAR points from different roof layers lies above or below the center cell (red box in Figure 4c), a black dotted vertical line is generated to intersect with the splitting line (black solid line in Figure 4c). The point of intersection (blue rectangle in Figure 4c) gives the *x*–*y* coordinates of the layer-connection point. Similarly, the cell containing the planimetric coordinates of the intersection point is confirmed and height values of each layer-point is determined by the average height of the LiDAR points of the corresponding roof layer.

In the fourth situation, as illustrated in Figure 4d, if the *x*–*y* coordinates of the layer-connection point are exactly on the boundary of the center cell (red box in Figure 4d), a black dotted vertical line intersecting the boundary gives the *x*–*y* coordinates of the layer-connection point (blue rectangle in Figure 4d). Again, height values of each layer-point are set according to the average height value of the LiDAR points (red and violet cells) of the corresponding roof layer.

In the fifth situation, as illustrated in Figure 4e, if the LiDAR points inside a cell (violet box in Figure 4e) come from more than two roof layers, the number of points from each roof layer is counted, and only two major groups of points are used to determine the *x*–*y* coordinates. The subsequent calculations are the same as those used in situation two, three, or four.

2.3.3. Optimization of Layer-Connection Points

There are a number of layer-connection points distributed along building contours. If these layer-connection points are directly connected, a series of zigzag contours will be produced. Thus, the layer-connection points along building contours need to be smoothed. We applied Zhou's method [43], in which the principal orientations are derived from boundary points, and the points are iteratively fitted to a line running along the principal orientations.

2.4. Building Model Reconstruction

Figure 5a shows a sample of the layer-connection points generated by the proposed approach. As shown in Figure 5a, a building model can be obtained by connecting neighboring layer-connection points.

For building roof reconstruction, the three neighboring cells should be searched, which correspond to three layer-connection points. If the layer-points of the layer-connection points inside the three neighboring cells are located on the same roof layer, these layer-points are connected to construct a triangle mesh (red box in Figure 5b). By traversing all of the cells, the construction of rooftops can be completed.

This paper only focuses on building roof reconstruction. Therefore, building walls are replaced with vertical planes. A similar operation is performed to construct the building walls. Briefly, layer-connection points including multiple layer-points inside neighboring or diagonally adjacent cells indicate the existence of a building wall. By connecting the layer-points belonging to different roof layers (red box in Figure 5c), this building wall can be obtained. In addition, the layer-connection points located on the boundary of a building must be connected to enable the construction of whole building walls (blue box in Figure 5c).

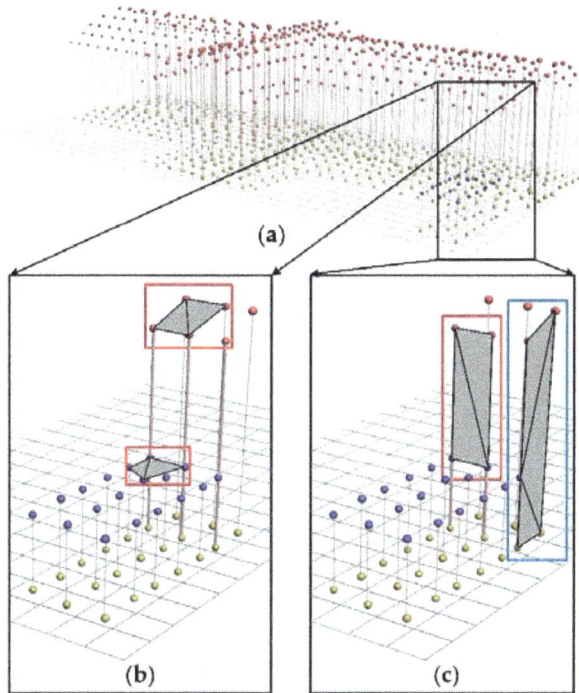

Figure 5. An example of building model reconstruction: (**a**) layer-connection points; (**b**) rooftop construction; and (**c**) wall construction.

2.5. Sensitivity Analysis of the Key Parameters

Given that a few parameters were used in this study, a summary of the setting procedures used for the associated key thresholds is necessary. The setting basis of these thresholds involved two types of information; namely, the data source and empirical results. The term "data source" means that a threshold is set according to the real data. If this method is applied to some other 3-D building model reconstruction, the "data source" thresholds can be determined easily, which increases the applicability of the proposed method. The term "empirical" means that the thresholds are set empirically, and in most cases, they can be set directly as we have proposed here.

During the process of extracting building rooftop points, the RIMM algorithm is employed. The parameters used during this step are shown in Table 1. The initial window size I_w and the height difference T_h were set according to the data source, and these values usually refer to the length of the largest building (106 m) and the height of the lowest building (3 m) in the experimental area, respectively. The fixed step length L_c, and roughness value R_v were, respectively, set to 3 m and 0.8 m, empirically. Two sets of different values (in meters) were tested while setting the values for the two parameters L_c and R_v. For L_c the test values were 1, 3, 5, 7 m; for R_v the test values were 0.4, 0.6, 0.8, 1.0, and 1.2 m. According to the completeness and correctness of the segmented rooftop patches, we found that the smaller L_c and the larger R_v could lead to that the extracted building point clouds contained some tree LiDAR points. Conversely, there could be some missing building points if L_c was too large and R_v was too small. The optimal extraction results were observed at $L_c = 3$ m and $R_v = 0.8$ m.

In the process of smoothness-oriented rooftop patch segmentation, an extraction method based on region growth and rooftop patch optimization is used. The key parameters are shown in Table 1. As we can see, all parameters used in this step can be set empirically. The search radius R_s and the number of inner points N are related to the input LiDAR data. To guarantee that there were more than ten points to calculate the normal of each LiDAR point, R_s is suitable for 2–3 times average of point spacing. Elaborate consideration was given to the value of N as follows. We assumed that the area of a minimum rooftop patch that could be detected was 4 m², *i.e.*, 2 m × 2 m. According to the point density of LiDAR data, a threshold for N can then be calculated easily. The distance threshold T_d were set to 0.2, 0.3, 0.4, 0.5, and 0.6 m to find the optimal value. In the process of patch optimization, the phenomenon of over-smoothing will be occurred, if T_d is set too large. At $T_d = 0.5$ m the effect of smoothing is moderate. The probability is a minimum probability of finding at least one good set of observations in all iterative procedures. It usually lies between 0.90 and 0.99. In our experiments, the probability was set to 0.98. During the generation of layer-connection points, a grid-based method is introduced, and the cell size is set empirically.

Table 1. Key parameters in the proposed approach.

Procedure		Threshold	Scale	Setting Basis
Extraction of building rooftop points		Initial window I_w	The length of the largest building	Data source
		Fixed step length L_c	3 m	Empirical
		Height difference T_h	The minimum building height	Data source
		Roughness value R_v	0.8 m	Empirical
Smoothness-oriented rooftop patch segmentation	Patch segmentation	Search radius R_s	2–3 times average of point spacing	Empirical
	Patch optimization	Distance threshold T_d	0.5 m	Empirical
		Number of inner points N	2 × 2 × point density	Empirical
		Probability P	0.98	Empirical
Generation of layer-connection points	Construction of the 2-D grid system	Cell size C	2–3 times average of point spacing	Empirical

3. Experiments and Analysis

3.1. Experimental Data

The airborne LiDAR data used in this paper were collected over Nanjing City, China, by using an Optech ALTM Gemini laser scanning system from a flying altitude of about 1000 m on 26 November 2011. The average point density was about 10 points per m², its average point spacing is about 0.25 m, and data had a vertical accuracy of 0.15 m and a horizontal accuracy of 0.20 m. We used the campus of Nanjing University, China, as the experimental Region 1 (Figure 6); this region covered an area of about 900 m × 500 m and contained 4.2 million LiDAR points. Figure 6a,b shows the aerial orthophotos with 0.3 m resolution and the LiDAR data from a side view, respectively. Figure 6c shows no-data areas where very sparse LiDAR points (one point in 30 m²) were collected as a result of the particular color and special structures of the corresponding building tops. The buildings in these no-data areas were not involved in the 3-D reconstruction process. The experimental Region 2 (Figure 7) was a residential area in the Jianye district, Nanjing City, China; this region covered an area of about 900 m × 600 m and contained 4.5 million LiDAR points. Figure 7a,b shows the aerial orthophotos and the LiDAR data, respectively. There were many buildings with various sizes and spatial distributions in Region 2.

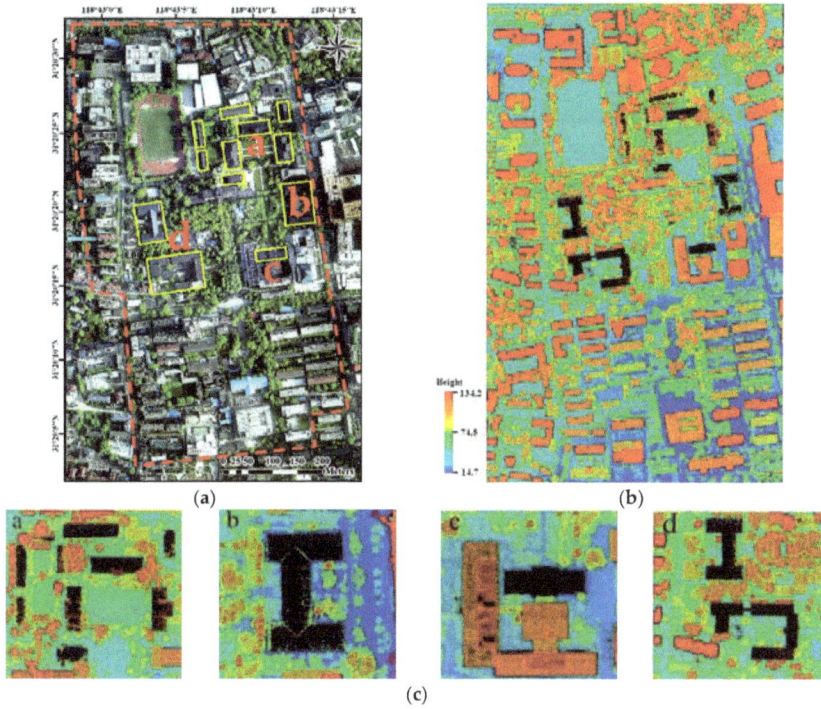

Figure 6. Experimental Region 1: (**a**) aerial orthophotos with 0.3 m resolution (no-data areas are shown by yellow boxes); (**b**) airborne LiDAR data; and(**c**) no-data areas (black), corresponding the yellow boxes in (**a**) with letters.

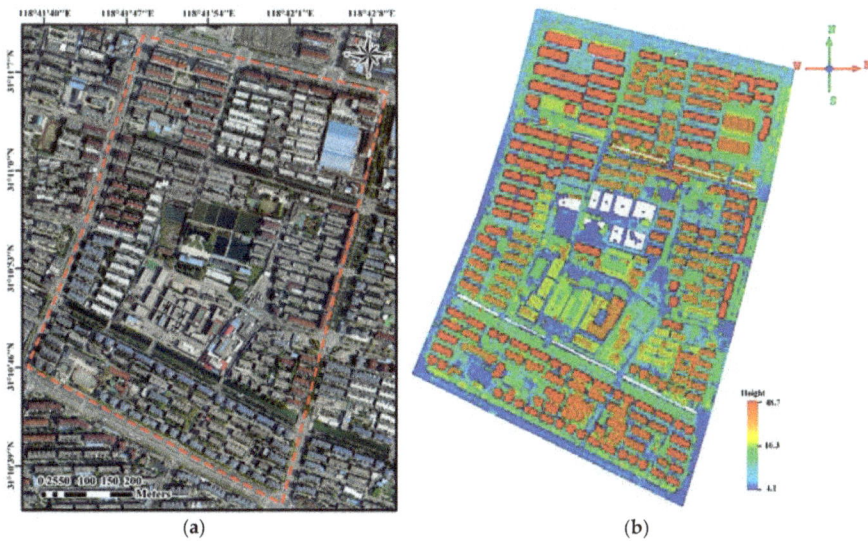

Figure 7. Experimental Region 2: (**a**) aerial orthophotos with 0.3 m resolution; and (**b**) airborne LiDAR data.

3.2. Experimental Results

Figures 8a and 9a demonstrate the reconstructed 3-D building roof models in Regions 1 and 2, respectively. Details of the roof models in Region 1 can be seen in Figure 8b–d. Several building models were selected to illustrate the details of the roofs in Region 2 (see Figure 9b–d). In Figures 8 and 9 the building roofs with different structures, different directions, and different levels of complexity were well-built and the results were satisfactory.

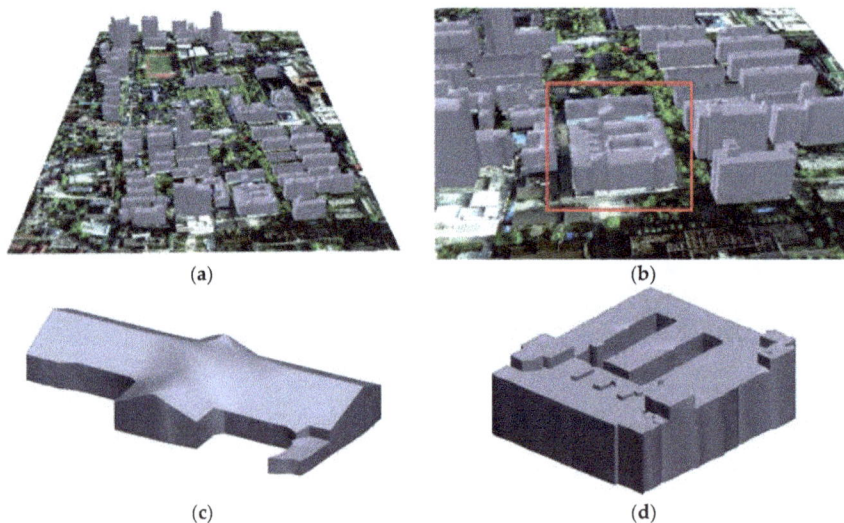

(a)   (b)

(c)   (d)

Figure 8. Reconstruction results in Region 1: (**a**) an overview; (**b**) a side view of the local reconstructed roof models; and (**c**) and (**d**), building roof models for the red box in (**b**).

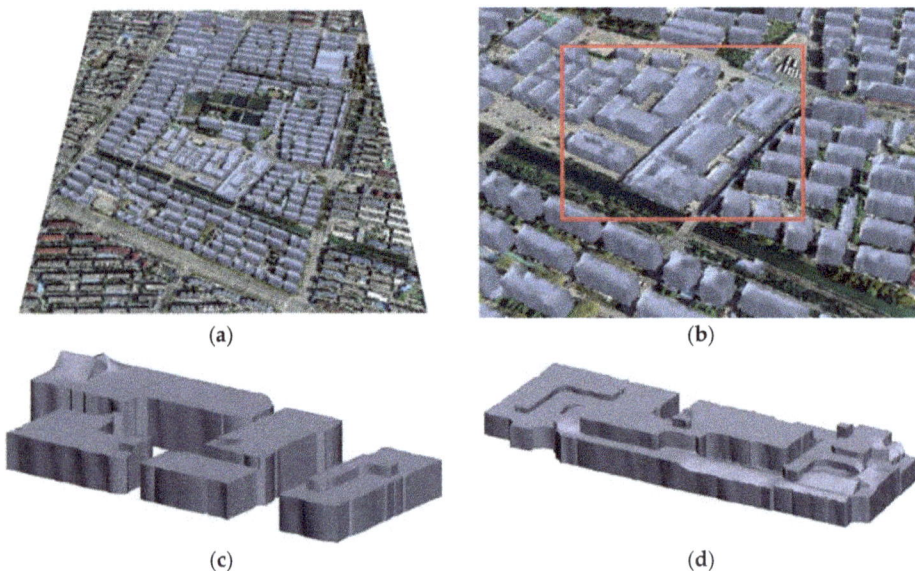

(a)   (b)

(c)   (d)

Figure 9. Reconstruction results in Region 2: (**a**) an overview; (**b**) a side view of the local reconstructed roof models; and (**c**) and (**d**), building roof models for the red box in (**b**).

3.3. Experimental Analysis

Evaluation of the experimental results was conducted according to (1) correctness and completeness of the reconstruction results and (2) deviation distances between model points and their nearest points in the laser data. These assessment criteria have been widely used in previous studies to analyze the reconstruction performance of building models [39,44–46].

3.3.1. Correctness and Completeness

This section employs the correctness and completeness information to quantitatively evaluate the quality of the 3-D reconstructions. The LiDAR point clouds, which contain building points, and the aerial orthophotos were used as reference data. Here, the unit of evaluation were the rooftop patches of the buildings. The rooftop patches in the reference data were extracted manually according to the airborne LiDAR building point clouds and aerial orthophotos. When the area of a rooftop patch was more than 2 m, the rooftop patch was determined and extracted. The correctness and completeness of the experimental results were evaluated from two aspects; namely, the number of rooftop patches (called the number evaluation) and the area of rooftop patches (called the area evaluation). For the aspect of number evaluation, the reconstructed results and the reference data were then put together as follows: (1) overlaid rooftop patches, in which the ratio of the overlapping area was more than 80%, were taken as correct reconstructions; (2) rooftop patches only existing in the reference data were taken as missing reconstructions; and (3) rooftop patches only existing in the reconstructed results, or rooftop patches whose ratios of the overlapping areas were less than 20%, were taken as wrong reconstructions. For the aspect of area evaluation, the accuracy is computed by the accumulated statistics on the correct, missing, and wrong areas of the reconstructed rooftop patches.

$$Completeness = \frac{TP}{TP + FN} \tag{2}$$

$$Correctness = \frac{TP}{TP + FP} \tag{3}$$

where TP (true positives) represents the number or area of correct reconstructions for rooftop patches, FN (false negatives) represents the number or area of missing reconstructions for rooftop patches, and FP (false positives) represents the number or area of wrong reconstructions for rooftop patches.

Table 2 lists the detailed evaluation values. The evaluation results demonstrate that the proposed method has high correctness and completeness. In regards to the number evaluation, the completeness and correctness of Region 1 were 89.39% and 97.12%, respectively, and the completeness and correctness of Region 2 were 90.37% and 95.42%, respectively. In regards to the area evaluation, the completeness and correctness of Region 1 were 90.64% and 97.19%, respectively, and the completeness and correctness of Region 2 were 91.85% and 93.26%, respectively. As can be seen from the results of the experimental evaluation, the correctness was higher than the completeness. This was mainly due to the use of a point-based method in the proposed approach whereby it is difficult to produce erroneous commissions. The missing building rooftop patches were mainly caused by the fact that some small building rooftop patches were not extracted during the process of segmentation because the number of LiDAR points representing them was less than the pre-set threshold.

Table 2. Correctness and completeness of the reconstructed buildings' rooftop patches.

Reconstructed Results	Correct Quantity		Missing Quantity		False Quantity		Completeness (%)		Correctness (%)	
	Number	Area (m²)	Number	Area (m²)	Number	Area (m²)	Number	Area	Number	Area
Region 1	236	84,183.25	28	8695.64	7	2432.90	89.39	90.64	97.12	97.19
Region 2	1145	145,659.82	122	12,924.63	55	10,526.89	90.37	91.85	95.42	93.26

3.3.2. Deviation Analysis of the Reconstructed Building Roofs

The deviation analysis of the reconstructed building roofs was performed by calculating the deviation distances between the reconstructed building roofs and the corresponding airborne LiDAR data. The airborne LiDAR points of buildings were used as reference data for the quantitative evaluation of the reconstructed building roofs' accuracy. In the reconstructed building model M, we selected a point R_i from reference data R and searched for the most neighboring triangular polygon M_i by using a method that has been described previously in the literature [47]. Then, the deviation distance between each LiDAR point and its corresponding patch was calculated. The statistical results of the deviation distances for Regions 1 and 2 were computed based on the validation point set (as shown in Figure 10a,b).

Table 3 lists the statistical results of the deviation analysis for Regions 1 and 2. The average deviation distance and standard deviation (abbreviated as Std. Dev. in Table 3) of all reconstructed building roofs in Region 1 were 0.05 m and 0.18 m, respectively. The average deviation distance and standard deviation of all roofs in Region 2 were 0.12 m and 0.25 m, respectively. Compared with Region 1, Region 2 had a larger average deviation distance and a higher standard deviation, which was due to the fact that the roof structures of residential buildings were irregular and contained some small objects. Overall, the evaluation results demonstrated that the reconstructed building roofs were well matched with the reference data. Certainly, the deviation distances were mainly concentrated at the average value. For example, in Region 1, about 96.61% points of the deviation distances are distributed in the range from 0 to 0.3 m, and in Region 2, about 93.28% of the points are distributed in the range from 0 to 0.3 m. The points with large deviation distances were mainly distributed in the contour regions of buildings or in air conditioning and chimney areas.

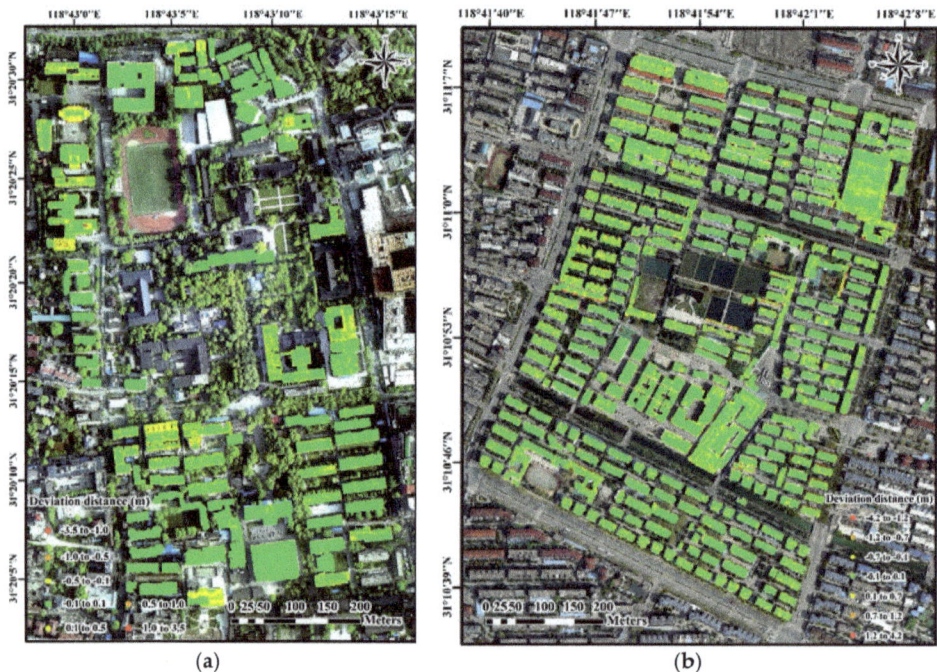

(a)   (b)

Figure 10. Deviation distances between the reconstructed building roof models and the LiDAR-derived validation data, as represented by points with different colors: (**a**), (**b**) Region 1 and Region 2, respectively.

Table 3. Deviation distances of the reconstructed building roofs.

Reconstructed Results	Number of Points	Maximum	Average	Std. Dev.	Skewness	Kurtosis	Percentage of Less Than 0.3 m (%)
Region 1	743,502	3.53	0.05	0.18	4.78	31.47	96.61
Region 2	1,672,006	4.15	0.12	0.25	4.28	34.92	93.28

3.3.3. Influence of Elevation to 3-D Roof Reconstruction

This study also analyzed the influence of elevation on the reconstructed building roofs. First, the airborne LiDAR points of buildings were divided into different groups according to the elevation value (the elevation interval was set to 1 m). Afterward, we calculated the deviation distance under each elevation range. The statistical results of Regions 1 and 2 are shown in Figure 11a,b, where the solid squares in the figures represent the average values of the deviation distances under each elevation range and the error bars represent the positive and negative deviations of each average value. From Figure 11a,b, we can draw some conclusions. First, most of the average values of the deviation distances were located in the range from −0.05 m to 0.05 m. Second, in regards to different ranges of elevations, when the elevation was more than the height of a specific value (80 m in Region 1; 37 m in Region 2), fluctuations of the average values and standard deviations were larger than those where the elevation was less than a specific value (70 m in Region 1; 33 m in Region 2). The reason for this phenomenon is that for modern high-rise buildings, there are many central air conditioning units on the building roofs, but there are not reflected in the reconstructed building roofs. Therefore, the average values and standard deviations are more sensitive to the presence of modern high-rise buildings.

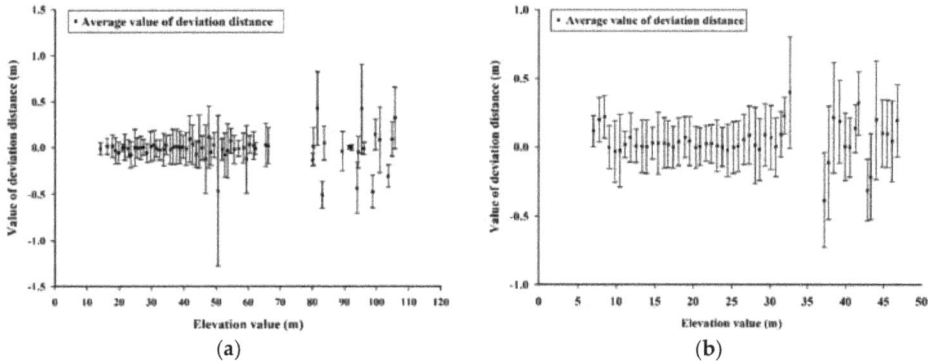

Figure 11. Evaluation of the building roofs' deviations under different elevations, where the solid squares in the figures represent the average values of the deviation distances under each elevation range, and the error bars represent the positive and negative deviations of each average value: (**a**), (**b**) Region 1 and Region 2, respectively.

3.4. Experimental Discussion

To further verify the effectiveness and reliability of the proposed method, several analysis processes were conducted from two aspects of evaluation from the ISPRS test project and comparison with other methods.

3.4.1. Evaluation from the ISPRS Test Project

The ISPRS benchmark test on urban object detection and reconstruction offers a unique possibility to compare state-of-the-art methods [48,49]. Therefore, we used the test data provided by the ISPRS benchmark test to evaluate the proposed method. The airborne LiDAR data for Vaihingen Areas 1, 2, and 3 were selected. The focus of the evaluation is on the quality of the roof plane segmentation and on the geometrical accuracy of the roof polygons. The properties of datasets and the evaluation metrics have been described in literature [48,50]. For the purpose of comparison with the state-of-the-art methods, six different approaches from the ISPRS test project were selected [16,20,51–53], and the overview of these methods is showed in Table 4, in which the proposed method is abbreviated as NUC.

The evaluation of the building reconstruction results for Vaihingen Areas 1–3 are summarized in Table 5. The quality metrics are explained in literature [48]. For Area 1, the completeness and the correctness of the extracted roof planes by using the proposed method were 73.6% and 99.2%, respectively; the completeness and the correctness of the roof planes covering an area of at least 10 m^2 were 75.5% and 99.0%, respectively. For Area 2, the completeness and the correctness of the extracted roof planes by using the proposed method were 71.0% and 100.0%, respectively; the completeness and the correctness of the roof planes covering an area of at least 10 m^2 were 89.6% and 100.0%, respectively. For Area 3, the completeness and the correctness of the extracted roof planes were 74.9% and 100.0%, respectively; the completeness and the correctness of the roof planes covering an area of at least 10 m^2 were 85.5% and 100.0%, respectively. Although there was a clear difference between the quality metrics for all planes and for roof planes larger than 10 m^2, the reconstruction quality of the proposed method could reach about or above the average accuracy of the six state-of-the-art methods. In addition, the geometrical errors (*RMS*, *RMSZ*) caused by the proposed approach were also in the range of the six previous methods. Therefore, the proposed method produced a good reconstructions for Areas 1–3.

Table 4. Overview of the reconstruction methods. *ID*: Identifier of the method used in this paper. *Researcher/Affiliation*: name and affiliation of the person submitting the results. *Reference*: a reference where the method is described.

ID	Researcher	Affiliation	Reference
CKU	J.-Y. Rau	N. Cheng-Kung U., Taiwan	(Rau and Lin, 2011)
ITCE1	S. Oude Elberink	ITC, The Netherlands	(Oude Elberink and Vosselman, 2009)
ITCE2	S. Oude Elberink	ITC, The Netherlands	(Oude Elberink and Vosselman, 2009)
ITCX	B. Xiong	ITC, The Netherlands	(Xiong *et al.*, 2014)
VSK	P. Dorninger	TU Vienna, Austria	(Dorninger and Pfeifer, 2008)
YOR	G. Sohn	York University, Canada	(Sohn *et al.*, 2008)
NUC	Y.J. Wang	Nanjing University, China	This paper

Table 5. Evaluation of building reconstruction results in Areas 1, 2, and 3. The best values per column are printed in bold font.

ID	Cm_{ob}/Cr_{ob} [%]	Cm_{10}/Cr_{10} [%]	$N_{1:M}/N_{N:1}/N_{N:M}$	RMS [m]	RMSZ [m]
\multicolumn{6}{c}{*Area* **1** (288 roof planes)}
CKU	86.7/**98.9**	86.7/**99.3**	10/36/3	**0.66**	0.70
ITCE1	60.8/94.6	58.5/94.0	16/**26**/17	0.91	0.55
ITCE2	65.3/97.3	63.3/97.3	0/38/3	0.94	0.55
ITCX	76.0/94.5	72.9/95.1	2/40/**2**	0.84	0.53
VSK	72.2/96.7	77.7/96.5	7/42/6	0.79	0.65
YOR	**88.2**/98.5	**89.9**/98.2	5/36/14	0.75	0.58
NUC	73.6/**99.2**	75.5/99.0	2/42/3	0.92	**0.45**
\multicolumn{6}{c}{*Area* **2** (69 roof planes)}
CKU	78.3/93.1	90.0/93.7	8/4/0	0.85	1.02
ITCE1	**79.7**/73.7	**94.0**/73.7	0/7/0	1.11	3.33
ITCE2	**79.7**/95.0	**94.0**/100.0	0/7/0	1.16	3.31
ITCX	62.3/92.9	74.0/92.7	2/4/0	0.79	**0.44**
VSK	73.9/**100.0**	88.0/**100.0**	3/5/1	1.03	0.88
YOR	73.9/**100.0**	90.0/**100.0**	5/3/0	**0.77**	1.04
NUC	71.0/**100.0**	89.6/**100.0**	3/7/1	0.83	0.62

Table 5. *Cont.*

ID	Cm$_{ob}$/Cr$_{ob}$ [%]	Cm$_{10}$/Cr$_{10}$ [%]	N$_{1:M}$/N$_{N:1}$/N$_{N:M}$	RMS [m]	RMSZ [m]
	Area 3 (235 roof planes)				
CKU	81.3/98.4	82.2/98.3	4/48/2	**0.76**	0.65
ITCE1	67.7/**100.0**	62.8/**100.0**	0/47/2	0.96	**0.29**
ITCE2	64.3/**100.0**	55.9/**100.0**	0/46/0	1.04	0.42
ITCX	70.2/**100.0**	62.8/**100.0**	1/48/0	0.87	0.30
VSK	76.6/99.1	74.5/99.1	3/50/0	0.84	0.38
YOR	**84.7**/**100.0**	**89.0**/**100.0**	2/51/1	0.77	0.35
NUC	74.9/**100.0**	85.5/**100.0**	0/49/0	0.91	0.36

Compared to Area 2, Areas 1 and 3 was more difficult to reconstruct roofs because various buildings with gabled roofs and small superstructures were present in the scene [50]. There was more under-segmentation in Areas 1 and 3, which might be explained by a large number of small attachments to the houses that were erroneously merged with neighboring roof planes.

3.4.2. Comparison with other Methods

We conducted a comparative experiment on six buildings, *i.e.*, Buildings 1–6, which were covered with all the rooftop types of the experimental regions. One innovation of the proposed method is that it smooths segmented rooftop patches before reconstructing the building roofs. Therefore, this experiment compared the proposed approach (Approach A) with the ordinary building reconstruction approach (Approach B). In Approach B, the segmented rooftop patches along with the region growing algorithm were directly used for the reconstruction task without smoothing the irregularly distributed rooftop points and eliminating the interference of omissive LiDAR points. Furthermore, a grid-based approach (Approach C) was selected to compare with the proposed method. Details regarding Approach C can be found in the literature [39]. Figures 12 and 13 show the six building roof models reconstructed by use of Approaches A, B, and C.

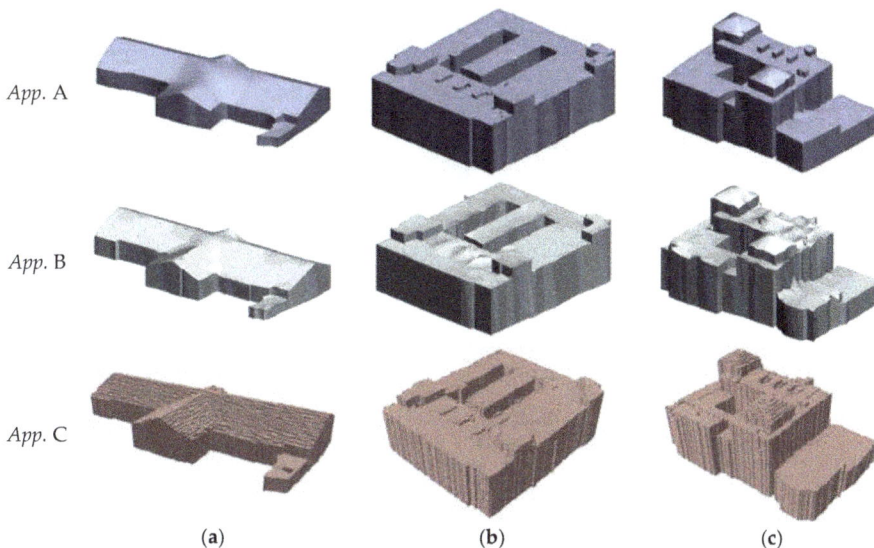

App. A

App. B

App. C

(a)   (b)   (c)

Figure 12. Comparison of Approaches (abbreviated as *App.*) A, B, and C: (**a**), (**b**), and (**c**) the reconstructed roof models of Buildings 1, 2, and 3, respectively.

From the visual perspective, the reconstructed models of Building 1 that were obtained by using Approaches A and B had similar reconstruction accuracies (see Figure 12a). The other building roof models reconstructed by using the proposed approach were obviously superior to those reconstructed by using Approach B, thus demonstrating the importance of smoothing before model reconstruction. The superior behavior of Approach A is understandable because, as mentioned previously, even if points belong to the same rooftop patch, small projections usually exist. If layer-connection points are generated directly from the rough points, the reconstructed results will be fractured. In particular, when it comes to interference information, the building roof models will be incoherent and unable to accurately reflect the roofs of the buildings. Here, we must notice that the reconstruction of sloped roofs with Approach C was unsatisfactory. This was mainly due to the fact that stair-step shapes were created during the segmentation of depth imagery. For flat roofs, Approach C can guarantee satisfactory reconstruction results, except irregular boundaries still exist in the roof models. The difficulty of determining the optimal segmentation scale limits the application of the grid-based method for 3-D reconstruction of building roofs with complex top structures in large urban areas. However, some small mistakes were encountered in the model of Building 4 reconstructed by the proposed approach (see Figure 13a); in particular, there was a lack of building fences. The reason for these errors lies in the point spacing of the acquired LiDAR data. The width of the fence pickets was only a few tens of centimeters, and the average point spacing was about 0.4 m. Thus, only one or two points could be acquired along the fences, which made it difficult to accurately reconstruct them.

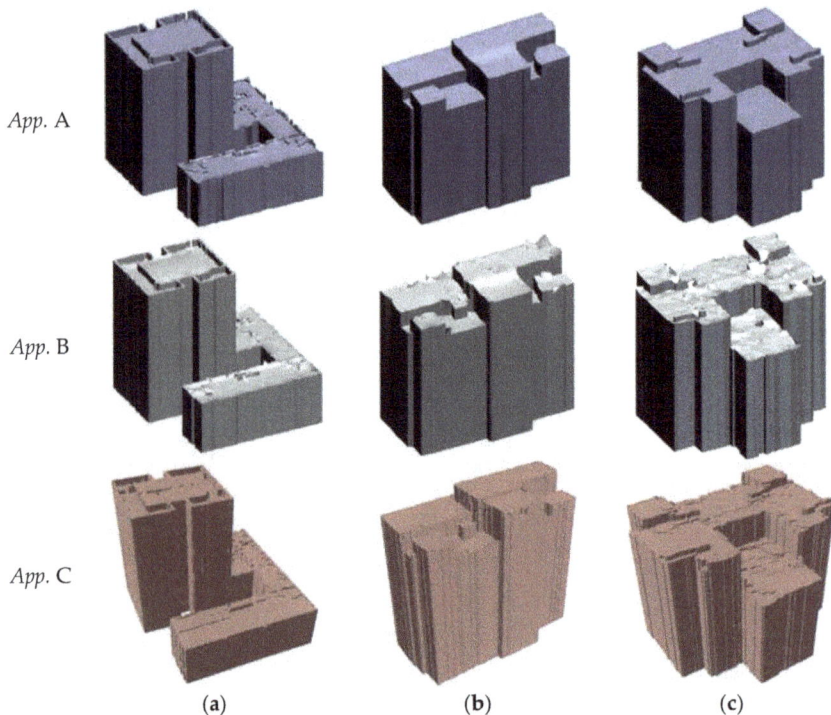

Figure 13. Comparison of Approaches (abbreviated as *App.*) A, B, and C: (**a**), (**b**), and (**c**) The reconstructed roof models of Buildings 4, 5, and 6, respectively.

To quantitatively evaluate the modeling accuracy of the different approaches, Table 6 lists the comparison results of the deviation distances for Buildings 1–6. The statistical results demonstrate that the performance of building roof models reconstructed by Approach A was superior to those reconstructed by Approaches B and C. Approach A was associated with smaller average deviation distances (less than 0.14 m) and lower standard deviations than those derived from the other approaches. Specifically, the average deviation distances and standard deviations of the six building roof models from Approaches B and C were larger than those from Approach A. This situation illustrates that the roof models reconstructed by the latter two methods are unstable. For Buildings 1 and 2, Approaches A and B had similar average deviation distances and standard deviations, thus indicating that the distribution of original airborne LiDAR data was smooth. When there were sloped roof structures on building roofs, the average deviation distances and standard deviations from Approach C were higher than those from Approaches A and B. For example, the average deviation distance and standard deviation of Building 1 reconstructed by Approach C were 0.28 m and 0.56 m, respectively. However, the average deviation distance and standard deviation of Building 1 reconstructed by Approach A were only 0.02 m and 0.11 m.

From Figures 12 and 13 we found that the building roof models reconstructed by using Approach B were much rougher than those reconstructed by using Approach A. To quantitatively analyze the experimental results of the two approaches, we calculated the roughness for each building. However, because the roughness of a roof model cannot be calculated directly, we had to resample the roof model to a high density point cloud (200 points/m^2 in this paper) as replacement data. First, the roughness, *i.e.*, standard deviation, needs to be calculated for each point p. In the process of determining roughness, the neighboring points (20 points in this paper) of point p are defined. Then, the roughness of point p can be determined based on Euclidean distances from all neighboring points to the fitted plane. Finally, the roughness of each building can be derived according to the calculated roughness of each point.

Table 6. Comparison of deviation distances for Buildings 1–6 reconstructed by Approaches A, B, and C.

Reconstructed Results	Number of Points	Approach A		Approach B		Approach C	
		Average	Std. Dev.	Average	Std. Dev.	Average	Std. Dev.
Building 1	15,163	0.02	0.11	0.03	0.13	0.28	0.56
Building 2	29,480	0.06	0.18	0.09	0.19	0.10	0.27
Building 3	23,838	0.08	0.25	0.19	0.25	0.21	0.35
Building 4	28,151	0.14	0.22	0.17	0.23	0.15	0.29
Building 5	17,612	0.05	0.16	0.12	0.31	0.06	0.17
Building 6	18,705	0.07	0.13	0.13	0.14	0.13	0.22

Table 7 presents the roughness comparisons for Buildings 1–6. Generally speaking, the roughness of building roof models reconstructed by using Approach A was about 0.005 m. Ideally, the roughness values of all building roof models should be zero, as the rooftop patches are strict planes. However, during the process of resampling roof models and calculating roughness, the loss of sampled data accuracy may not lead to a roughness value of zero. As a comparison, the roughness values of building roof models reconstructed by using Approach B were in the range of 0.10 m to 0.25 m. From Table 7, we can see that the roughness values of Building 1 reconstructed by using the two approaches were almost the same. This was because the roof of Building 1 contained no interference information, and the original LiDAR point cloud was relatively smooth. Figure 14 shows the specific distribution of roughness for Buildings 4 and 6, which supports the experimental analysis in Table 7. Therefore, we can conclude that the building roof models reconstructed by using the proposed approach, which were associated with smaller fluctuations and better smoothness, were superior to those reconstructed by using Approach B. Thus, the importance of smoothing before model reconstruction is further verified here.

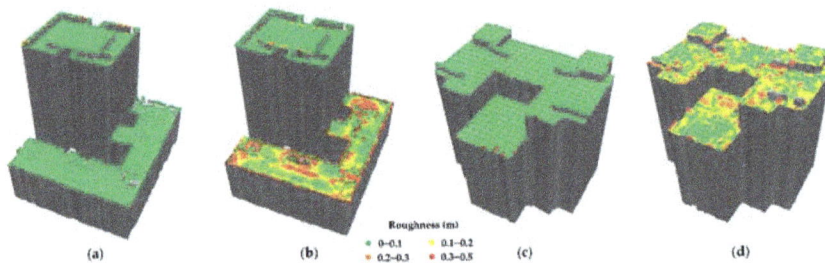

Figure 14. Roughness comparison between Approaches A and B: (a), (c) roughness of roof models reconstructed using Approach A for Buildings 4 and 6; (b), (d) roughness of roof models reconstructed using Approach B for Buildings 4 and 6, respectively. Data are represented by points with different colors.

Table 7. Roughness comparison of building roof models for Approaches A and B.

Reconstructed Buildings	Number of Rooftop Patches	Approach A	Approach B
Building 1	5	0.006	0.009
Building 2	13	0.004	0.128
Building 3	14	0.004	0.237
Building 4	4	0.005	0.221
Building 5	5	0.005	0.189
Building 6	3	0.002	0.171

4. Conclusions

This paper has presented a new approach involving a layer connection and smoothness strategy for the reconstruction of building roof models from airborne LiDAR data. The proposed approach consists of building rooftop point extraction, smoothness-oriented rooftop patch extraction, layer-connection point generation, and building model reconstruction. The main contributions of the proposed approach are as follows. (1) During the rooftop patch extraction, a "seed point selection, patch growth, and patch smoothing" strategy is used to smooth building points, eliminate interference information, and ensure the integrity of the point cloud data; and (2) layer-connection points are proposed to guarantee consistency between the boundary footprints of different roof layers. By connecting neighboring layer-connection points, the building roofs are reconstructed. Through the calculation of layer-connection points, different roof layers are connected in a simple and fast way. In the two experimental regions used in this paper, the completeness and correctness of the reconstructed rooftop patches were about 90% and 95%, respectively. For the deviation accuracy, the average deviation distance and standard deviation in the best case were 0.05 m and 0.18 m, respectively, and those in the worst case were 0.12 m and 0.25 m. Our experiments prove that this method has good applicability for model reconstruction of buildings in urban environments.

However, too many types of geometric shapes may exist on building roofs, such as artistic sculptures, curved surfaces, and so forth. Therefore, there could be some phenomena of over-smoothing for not-flat roofs by using the proposed method. To reconstruct building roofs with very complex structures, further investigations will be necessary. In addition, small mistakes can persist in certain tiny structures such as fences, air conditioning vents, and chimneys. The proposed method cannot deal with roof overhangs, which are also difficult to reconstruct by the most previous methods. In our future work, we will consider to reconstruct these roof parts with the aid of the auxiliary data (e.g., terrestrial LiDAR data). This paper has concentrated on the reconstruction of building roofs from airborne LiDAR data, and little work was done for the subtle reconstruction of building façades.

To achieve the reconstruction of complete building models, further work is needed to allow for the subtle reconstruction of façade models from terrestrial LiDAR data.

Acknowledgments: The authors thank the anonymous reviewers and members of the editorial team for their comments and contributions. We also thank M. Gerke (University of Twente) and U. Breitkopf (University Hannover) for the accuracy analysis of the building reconstruction results of the ISPRS benchmark test project. This work was supported by the National Natural Science Foundation of China (NSFC) Project (Grant Nos. 41071244 and 41371017).

Author Contributions: Hao Xu proposed and developed the research design, collected airborne LiDAR data. Yongjun Wang performed the data analysis, results interpretation. Yajun Wang assisted with developing the research design. Nan Xia and Yong Tang assisted with the experimental result analysis. Liang Cheng and Yanming Chen assisted with refining the research design. Manchun Li checked the manuscript. All authors contributed to manuscript writing.

Conflicts of Interest: The authors declare no conflict of interest.

References

1. Groger, G.; Plumer, L. CityGML—Interoperable semantic 3D city models. *ISPRS J. Photogramm. Remote Sens.* **2012**, *71*, 12–33. [CrossRef]
2. Qin, R.J. Change detection on LOD 2 building models with very high resolution spaceborne stereo imagery. *ISPRS J. Photogramm. Remote Sens.* **2014**, *96*, 179–192. [CrossRef]
3. Santos, T.; Gomes, N.; Freire, S.; Brito, M.; Santos, L.; Tenedório, J. Applications of solar mapping in the urban environment. *Appl. Geogr.* **2014**, *51*, 48–57. [CrossRef]
4. Volk, R.; Stengel, J.; Schultmann, F. Building Information Modeling (BIM) for existing buildings—Literature review and future needs. *Autom. Constr.* **2014**, *38*, 109–127. [CrossRef]
5. Yue, H.; Chen, W.; Wu, X.; Liu, J. Fast 3D modeling in complex environments using a single Kinect sensor. *Opt. Laser Eng.* **2014**, *53*, 104–111. [CrossRef]
6. Remondino, F.; El-Hakim, S. Image-based 3D modelling: A review. *Photogramm. Rec.* **2006**, *21*, 269–291. [CrossRef]
7. Cheng, L.; Wu, Y.; Wang, Y.; Zhong, L.S.; Chen, Y.M.; Li, M.C. Three-dimensional reconstruction of large multilayer interchange bridge using airborne LiDAR data. *IEEE J. Sel. Top. Appl. Earth Observ. Remote Sens.* **2015**, *8*, 691–708. [CrossRef]
8. Lu, Z.Y.; Im, J.; Rhee, J.; Hodgson, M. Building type classification using spatial and landscape attributes derived from LiDAR remote sensing data. *Landsc. Urban Plan.* **2014**, *130*, 134–148. [CrossRef]
9. Xu, H.; Cheng, L.; Li, M.C.; Chen, Y.M.; Zhong, L.S. Using octrees to detect changes to buildings and trees in the urban environment from airborne LiDAR data. *Remote Sens.* **2015**, *7*, 9682–9704. [CrossRef]
10. Shayeganrad, G. On the remote monitoring of gaseous uranium hexafluoride in the lower atmosphere using LiDAR. *Opt. Laser Eng.* **2013**, *51*, 1192–1198. [CrossRef]
11. Kabolizade, M.; Ebadi, H.; Mohammadzadeh, A. Design and implementation of an algorithm for automatic 3D reconstruction of building models using genetic algorithm. *Int. J. Appl. Earth Observ.* **2012**, *19*, 104–114. [CrossRef]
12. Haala, N.; Kada, M. An update on automatic 3D building reconstruction. *ISPRS J. Photogramm. Remote Sens.* **2010**, *65*, 570–580. [CrossRef]
13. Tomljenovic, I.; Höfle, B.; Tiede, D.; Blaschke, T. Building extraction from airborne laser scanning data: An analysis of the state of the art. *Remote Sens.* **2015**, *7*, 3826–3862. [CrossRef]
14. Tarsha-Kurdi, F.; Landes, T.; Grussenmeyer, P.; Koehl, M. Model-driven and data-driven approaches using LiDAR data: Analysis and comparison. *Int. Arch. Photogramm. Remote Sens. Spat. Inf. Syst.* **2007**, *36*, 87–92.
15. Awrangjeb, M.; Zhang, C.; Fraser, C.S. Automatic extraction of building roofs using LiDAR data and multispectral imagery. *ISPRS J. Photogramm. Remote Sens.* **2013**, *83*, 1–18. [CrossRef]
16. Dorninger, P.; Pfeifer, N. A comprehensive automated 3D approach for building extraction, reconstruction, and regularization from airborne laser scanning point clouds. *Sensors* **2008**, *8*, 7323–7343. [CrossRef]
17. Cheng, L.; Tong, L.H.; Chen, Y.M.; Zhang, W.; Shan, J.; Liu, Y.X.; Li, M.C. Integration of LiDAR data and optical multi-view images for 3D reconstruction of building roofs. *Opt. Laser Eng.* **2013**, *51*, 493–502. [CrossRef]

18. Susaki, J. Knowledge-based modeling of buildings in dense urban areas by combining airborne LiDAR data and aerial images. *Remote Sens.* **2013**, *5*, 5944–5968. [CrossRef]

19. Verma, V.; Kumar, R.; Hsu, S. 3D building detection and modeling from aerial LiDAR data. In Proceedings of the IEEE Computer Society Conference on Computer Vision and Pattern Recognition, New York, NY, USA, 17–22 June 2006; pp. 2213–2220.

20. Sohn, G.; Huang, X.F.; Tao, V. Using a binary space partitioning tree for reconstructing polyhedral building models from airborne LiDAR data. *Photogramm. Eng. Remote Sens.* **2008**, *74*, 1425–1438. [CrossRef]

21. Zhang, K.; Yan, J.H.; Chen, S.C. Automatic construction of building footprint's from airborne LiDAR data. *IEEE Trans. Geosci. Remote Sens.* **2006**, *44*, 2523–2533. [CrossRef]

22. Kada, M.; McKinley, L. 3D building reconstruction from LiDAR based on a cell decomposition approach. *Arch. Photogramm. Remote Sens. Spat. Inf. Sci.* **2009**, *38*, 47–52.

23. Vallet, B.; Pierrot-Deseilligny, M.; Boldo, D.; Bredif, M. Building footprint database improvement for 3D reconstruction: A split and merge approach and its evaluation. *ISPRS J. Photogramm. Remote Sens.* **2011**, *66*, 732–742. [CrossRef]

24. Vosselman, G.; Dijkman, S. 3D building model reconstruction from point clouds and ground plans. *Int. Arch. Photogramm. Remote Sens. Spat. Inf. Sci.* **2001**, *34*, 37–44.

25. Sumer, E.; Turker, M. Automated extraction of photorealistic facade textures from single ground-level building images. *Int. J. Pattern Recogn.* **2014**, *28*, 1455007. [CrossRef]

26. Sun, S.H.; Salvaggio, C. Aerial 3D building detection and modeling from airborne LiDAR point clouds. *IEEE J. Sel Top. Appl. Earth Observ. Remote Sens.* **2013**, *6*, 1440–1449. [CrossRef]

27. Vo, A.V.; Truong-Hong, L.; Laefer, D.F.; Bertolotto, M. Octree-based region growing for point cloud segmentation. *ISPRS J. Photogramm. Remote Sens.* **2015**, *104*, 88–100. [CrossRef]

28. Henn, A.; Groger, G.; Stroh, V.; Plumer, L. Model driven reconstruction of roofs from sparse LiDAR point clouds. *ISPRS J. Photogramm. Remote Sens.* **2013**, *76*, 17–29. [CrossRef]

29. Tarsha-Kurdi, F.; Landes, T.; Grussenmeyer, P. Extended RANSAC algorithm for automatic detection of building roof planes from LiDAR data. *Photogramm. J. Finl.* **2008**, *21*, 97–109.

30. Fan, H.C.; Yao, W.; Fu, Q. Segmentation of sloped roofs from airborne LiDAR point clouds using ridge-based hierarchical decomposition. *Remote Sens.* **2014**, *6*, 3284–3301. [CrossRef]

31. Awrangjeb, M.; Fraser, C.S. Automatic segmentation of raw LiDAR data for extraction of building roofs. *Remote Sens.* **2014**, *6*, 3716–3751. [CrossRef]

32. Chen, D.; Zhang, L.Q.; Li, J.; Liu, R. Urban building roof segmentation from airborne LiDAR point clouds. *Int. J. Remote Sens.* **2012**, *33*, 6497–6515. [CrossRef]

33. Kim, K.; Shan, J. Building roof modeling from airborne laser scanning data based on level set approach. *ISPRS J. Photogramm. Remote Sens.* **2011**, *66*, 484–497. [CrossRef]

34. Sampath, A.; Shan, J. Segmentation and reconstruction of polyhedral building roofs from aerial LiDAR point clouds. *IEEE Trans. Geosci. Remote Sens.* **2010**, *48*, 1554–1567. [CrossRef]

35. Satari, M.; Samadzadegan, F.; Azizi, A.; Maas, H.G. A multi-resolution hybrid approach for building model reconstruction from LiDAR data. *Photogramm. Rec.* **2012**, *27*, 330–359. [CrossRef]

36. Lafarge, F.; Descombes, X.; Zerubia, J.; Pierrot-Deseilligny, M. Structural approach for building reconstruction from a single DSM. *IEEE Trans. Pattern. Anal.* **2010**, *32*, 135–147. [CrossRef] [PubMed]

37. Zhou, Q.Y.; Neumann, U. 2.5D building modeling by discovering global regularities. In Proceedings of the IEEE Conference on Computer Vision and Pattern Recognition, Providence, RI, USA, 16–21 June 2012; pp. 326–333.

38. Zhang, W.; Wang, H.; Chen, Y.; Yan, K.; Chen, M. 3D building roof modeling by optimizing primitive's parameters using constraints from LiDAR data and aerial imagery. *Remote Sens.* **2014**, *6*, 8107–8133. [CrossRef]

39. Xiong, B.; Jancosek, M.; Oude Elberink, S.; Vosselman, G. Flexible building primitives for 3D building modeling. *ISPRS J. Photogramm. Remote Sens.* **2015**, *101*, 275–290. [CrossRef]

40. Chen, Y.M.; Cheng, L.; Li, M.C.; Wang, J.C.; Tong, L.H.; Yang, K. Multiscale grid method for detection and reconstruction of building roofs from airborne LiDAR data. *IEEE J. Sel. Top. Appl. Earth Observ. Remote Sens.* **2014**, *7*, 4081–4094. [CrossRef]

41. Cheng, L.; Zhao, W.; Han, P.; Zhang, W.; Shan, J.; Liu, Y.X.; Li, M.C. Building region derivation from LiDAR data using a reversed iterative mathematic morphological algorithm. *Opt. Commun.* **2013**, *286*, 244–250. [CrossRef]

42. Awrangjeb, M.; Lu, G.; Fraser, C. Automatic building extraction from LiDAR data covering complex urban scenes. *ISPRS Ann. Photogramm. Remote Sens. Spat. Inf. Sci.* **2014**, *1*, 25–32. [CrossRef]

43. Zhou, Q.Y.; Neumann, U. Fast and extensible building modeling from airborne LiDAR data. In Proceedings of the 16th ACM SIGSPATIAL International Conference on Advances in Geographic Information Systems, Irvine, CA, USA, 5–7 November 2008; pp. 1–8.

44. Zhu, L.; Lehtomäki, M.; Hyyppä, J.; Puttonen, E.; Krooks, A.; Hyyppä, H. Automated 3D scene reconstruction from open geospatial data sources: Airborne laser scanning and a 2D topographic database. *Remote Sens.* **2015**, *7*, 6710–6740. [CrossRef]

45. Awrangjeb, M.; Fraser, C.S. An automatic and threshold-free performance evaluation system for building extraction techniques from airborne LiDAR data. *IEEE J. Sel. Top. Appl. Earth Observ. Remote Sens.* **2014**, *7*, 4184–4198. [CrossRef]

46. Elberink, S.O.; Vosselman, G. Quality analysis on 3D building models reconstructed from airborne laser scanning data. *ISPRS J. Photogramm. Remote Sens.* **2011**, *66*, 157–165. [CrossRef]

47. Besl, P.J.; Mckay, N.D. A method for registration of 3D shapes. *IEEE Trans. Pattern Anal. Mach. Intell.* **1992**, *14*, 239–256. [CrossRef]

48. Rottensteiner, F.; Sohn, G.; Jung, J.; Gerke, M.; Baillard, C.; Benitez, S.; Breitkopf, U. ISPRS benchmark on urban object classification and 3D building reconstruction. *ISPRS Ann. Photogramm. Remote Sens. Spat. Inf. Sci.* **2012**, *1*, 293–298. [CrossRef]

49. Web Site of the ISPRS Test Project on Urban Classification and 3D Building Reconstruction. Available online: http://www2.isprs.org/commissions/comm3/wg4.html (accessed on 28 February 2016).

50. Rottensteiner, F.; Sohn, G.; Gerke, M.; Wegner, J.D.; Breitkopf, U.; Jung, J. Results of the ISPRS benchmark on urban object detection and 3D building reconstruction. *ISPRS J. Photogramm. Remote Sens.* **2014**, *93*, 256–271. [CrossRef]

51. Rau, J.-Y.; Lin, B.-C. Automatic roof model reconstruction from ALS data and 2D ground planes based on side projection and the TMR algorithm. *ISPRS J. Photogramm. Remote Sens.* **2011**, *66*, s13–s27. [CrossRef]

52. Oude Elberink, S.; Vosselman, G. Building reconstruction by target based graph matching on incomplete laser data: Analysis and limitations. *Sensors* **2009**, *9*, 6101–6118. [CrossRef] [PubMed]

53. Xiong, B.; Oude Elberink, S.; Vosselman, G. A graph edit dictionary for correcting errors in roof topology graphs reconstructed from point clouds. *ISPRS J. Photogramm. Remote Sens.* **2014**, *93*, 227–242. [CrossRef]

remote sensing

MDPI

Article

Automated Reconstruction of Building LoDs from Airborne LiDAR Point Clouds Using an Improved Morphological Scale Space

Bisheng Yang [1],*, Ronggang Huang [1],*, Jianping Li [1], Mao Tian [1], Wenxia Dai [1] and Ruofei Zhong [2]

[1] State Key Laboratory of Information Engineering in Surveying, Mapping and Remote Sensing, Wuhan University, Wuhan 430079, China; lijianping@whu.edu.cn (J.L.); mtian@whu.edu.cn (M.T.); daiwenxia@whu.edu.cn (W.D.)
[2] Beijing Advanced Innovation Center for Imaging Technology, Capital Normal University, Beijing 100048, China; zrfsss@163.com
* Correspondence: bshyang@whu.edu.cn (B.Y.); gang3217@whu.edu.cn (R.H.); Tel.: +86-27-6877-9699 (B.Y.); +86-130-0632-6261 (R.H.)

Academic Editors: Jie Shan, Juha Hyyppä, Xiaofeng Li and Prasad S. Thenkabail
Received: 17 October 2016; Accepted: 22 December 2016; Published: 27 December 2016

Abstract: Reconstructing building models at different levels of detail (LoDs) from airborne laser scanning point clouds is urgently needed for wide application as this method can balance between the user's requirements and economic costs. The previous methods reconstruct building LoDs from the finest 3D building models rather than from point clouds, resulting in heavy costs and inflexible adaptivity. The scale space is a sound theory for multi-scale representation of an object from a coarser level to a finer level. Therefore, this paper proposes a novel method to reconstruct buildings at different LoDs from airborne Light Detection and Ranging (LiDAR) point clouds based on an improved morphological scale space. The proposed method first extracts building candidate regions following the separation of ground and non-ground points. For each building candidate region, the proposed method generates a scale space by iteratively using the improved morphological reconstruction with the increase of scale, and constructs the corresponding topological relationship graphs (TRGs) across scales. Secondly, the proposed method robustly extracts building points by using features based on the TRG. Finally, the proposed method reconstructs each building at different LoDs according to the TRG. The experiments demonstrate that the proposed method robustly extracts the buildings with details (e.g., door eaves and roof furniture) and illustrate good performance in distinguishing buildings from vegetation or other objects, while automatically reconstructing building LoDs from the finest building points.

Keywords: airborne LiDAR point clouds; building point extraction; building LoDs; the morphological scale space; point cloud segmentation

1. Introduction

Three-dimensional (3D) building models play an important role in urban planning and management, telecommunications, tourism, disaster relief and evaluation, environmental simulation, vehicle navigation, and so on [1]. Automatically reconstructing building models at different levels of detail (LoDs) is important for various applications. For example, the finest model would be taken as the basis for assessing solar potential of rooftops [2], and a coarser model could satisfy personal navigation in a mobile device [3].

The LoDs of buildings are the multiple representations of 3D building models. In the past decade, many researchers have concentrated on the generation of LoDs from the finest 3D building models [4–8]. Generally, most methods derive coarse LoD models by employing the operators of simplification and aggregation on a fine-scale 3D building model [5,7] or on the 2D ground plans [4,6,9]. However, there are many definitions for LoDs, and the standard is still not unified [4,10]. After the CityGML (OGC City Geography Markup Language) standard was published [1], many studies focused on deriving coarse models from a fine-scale 3D model according to the framework of CityGML. Mao et al. generated CityGML models by simplification and aggregation, and then transformed the generated CityGML models to a CityTree for realizing dynamic zoom functionality in real time [6]. Fan and Meng proposed a three-step approach to simplify and aggregate 2D ground plans and generalize roof structures [4]. Verdie et al. generated building LoDs from the finest LoD to the coarsest LoD based on surface meshes [11]. In a word, the above-reported methods generate 3D building models at different LoDs from a fine-scale building model. However, reconstructing a fine-scale building model is quite expensive and may not be relevant for many applications. Moreover, the number of levels for discrete LoDs is fixed and thus limited in the framework of CityGML, and the large difference between two adjacent building LoDs could cause a big jump from one level to another level in the visualization [3,10]. Hence, automatically reconstructing a 3D building model at desired levels from 3D information of buildings rather than from a fine-scale 3D building model is an economical and flexible way to meet the user's requirements.

Airborne Light Detection and Ranging (LiDAR) has become a mature technology for capturing 3D information of buildings [12], which could be taken as the basis for generating building LoDs. At present, robustly extracting building points from various and complex urban scenes is still a challenging issue [13,14]. In the last decade, numerous methods have been reported for extracting building information from airborne laser scanning points, including DSM (Digital Surface Model)-based methods [15], point cloud-based methods [16] and methods based on imagery-fusing point clouds [17]. With the improvement of point density and the penetrating capacity of commercial LiDAR systems (e.g., Full Waveform LiDAR systems), the point cloud-based methods could be more suitable for complicated urban scenes. In general, segmentation-based methods and supervised learning-based methods are two main solutions for building extraction based on point clouds. Supervised learning-based methods [18–23] first select some building and non-building data as samples for training classifiers, and then extract building points. However, it is time consuming in selecting samples, and the result is highly dependent on samples [14]. Segmentation-based methods begin by splitting point clouds into disjointed segments, and then extract building segments with some prior knowledge or assumptions [16,24–27]. Generally, segmentation-based methods are widely utilized in various engineering applications. These methods take each segment as an individual unit, although many features derived from a single local segment cannot describe the differences between buildings and other objects properly, causing classification errors. Fortunately, it can perform better when the method combines features derived from the entire object with features derived from the local neighbors, just like the human visual system distinguishes different objects from the whole to the local [28]. The key step is to link the relationship between the segments of a building and the entire building, and it is of great importance to generate building LoDs from extracted segments.

Scale-space theory lays a sound foundation for representing one object from a finer level to a coarser level [29]. It gradually ignores the details and merges parts of an object into a group with the increasing of the scale and could directly generate an arbitrary level from the finest point clouds when the corresponding scale is given. Moreover, it maintains the spatial relations between adjacent scales, and provides a good way to imitate the human visual system (HVS) for perceiving objects ranging from whole to local details [28]. Generally, scale spaces can be constructed by wavelet transform [30], Gaussian smoothing [31], and mathematical morphology [32]. The scale space constructed by mathematical morphology is non-linear, and it is good for maintaining the shape of an object. It has been widely used in various fields, such as signal processing and image processing [29]. Vu et al. generated a DSM from airborne laser scanning point clouds and constructed the scale space with area morphology for building extraction by fusing spectral imageries, and providing the simple models with multi-scale representation [33]. Nevertheless, loss of information (e.g., the multiple returns) in the generation of DSMs affect the extraction of buildings, and the method ignores the local details (e.g., dormers and other roof elements) of the building model in the multi-scale representation. Fortunately, the scale space constructed by the morphological reconstruction (e.g., opening and closing by reconstruction) generates the LoDs of an object by controlling the scale (e.g., the size of a structuring element in morphology). It could better describe the local changes of objects across different levels by the smoothing operators of opening and closing [32]. Hence, we propose a novel method to extract building points and generate 3D building LoDs from airborne LiDAR point clouds by applying the morphological scale space, where each level is directly generated from point clouds by the morphological reconstruction. The main contributions of the proposed method are as follows.

- Directly construct the scale space from airborne laser scanning point clouds by applying the morphological reconstruction with planar segment constraints for feature preservation, and a TRG (topological relationship graph) is created for representing the spatial relations between segments across levels;
- Generate 3D building LoDs from the extracted building points based on the TRG, and the building LoD with a specified level could be automatically reconstructed from the finest building points.

The remainder of this paper is organized as follows. An improved morphological scale-space for point clouds is elaborated in Section 2. Section 3 describes the generation of building LoDs from airborne laser scanning point clouds based on the improved morphological scale space. In Section 4, the experimental studies that were undertaken to evaluate the proposed method are outlined. Finally, conclusions are drawn at end of this paper.

2. An Improved Morphological Scale Space for Point Clouds

The improved morphological scale space is iteratively constructed by a morphological reconstruction with planar segment constraints with the increasing of scale. Moreover, the topological relationship graph (TRG) describing the spatial relations between different levels of one object is generated for extracting building points and reconstructing 3D building LoDs.

2.1. A Morphological Reconstruction for Each Level with Planar Segment Constraints

The improved morphological reconstruction on the point clouds includes two steps, the opening by reconstruction and the closing by reconstruction. Although the exterior shape of an object could be maintained, part of an inclined roof may be flattened during the morphological reconstruction. It leads to a failure in linking the topology between different levels. To overcome the drawback, the result of a plane segmentation is adopted as constraints. The improved morphological reconstruction is described as follows.

Let $P = \{p_0, p_1, \ldots, p_n\}$ be the point clouds. P is segmented by the plane segmentation method of [34] and small segments are removed by the threshold t_N, which is defined as the number of points in one segment. The remained segments are denoted as $PS = \{PS_0, PS_1, PS_2, \ldots\}$, and all points in the removed segments are pushed into one set of individual points. Moreover, the slope of each segment is calculated, and each segment is robustly labeled as horizontal or inclined by Equation (1) to avoid the disturbance of noises.

$$L_{ps_i} = \begin{cases} 1 & if\ S_{ps_i} \geq t_S \\ 0 & if\ S_{ps_i} < t_S \end{cases} \tag{1}$$

where S_{ps_i} is the slope of the segment ps_i; t_S is the slope threshold; L_{ps_i} is assigned to 1 or 0 for marking one segment to be horizontal or inclined.

Then, the opening reconstruction operator is defined as follows: Set an arbitrary value s as the current scale, which is taken as the radius of a window B_s, and perform an opening operator on point clouds P according to Equation (2) to flatten the sharp details, which are smaller than two times s, and the result is denoted as P_{OPEN}. P_{OPEN} is taken as the marker point clouds, and P is the mask point clouds. A geodesic dilation with a window B_I is adopted iteratively according to Equations (3) and (4) until the result is stable [32]. The result of the opening by reconstruction is denoted as $P_{OPEN\_REC} = \delta_P^{(n)}(P_{OPEN})$.

$$P_{OPEN} = (P \ominus B_s) \oplus B_s \tag{2}$$

$$\delta_P^{(1)}(P_{OPEN}) = (P_{OPEN} \oplus B_I) \wedge P \tag{3}$$

$$\delta_P^{(n)}(P_{OPEN}) = \delta_P^{(1)} \circ \delta_P^{(1)} \circ \ldots \circ \delta_P^{(1)}(P_{OPEN}) \tag{4}$$

where $\oplus$ is the operator of the dilation; $\ominus$ is the operator of the erosion; δ is the operator of the geodesic dilation; $\wedge$ stands for the point-wise minimum; n is the iteration number.

The closing reconstruction operator is defined as follows: Perform a closing operator with the disc window (B_s) on P_{OPEN_REC} according to Equation (5) to remove lower details, which are smaller than two times s, and the result is denoted as P_{CLOSE}. P_{CLOSE} is taken as the marker point clouds, and P_{OPEN_REC} is the mask point clouds. A geodesic erosion is adopted iteratively according to Equations (6) and (7) until the result is stable [32]. The result of the closing by reconstruction $P_{CLOSE\_REC} = \varepsilon_{P_{OPEN\_REC}}^{(n)}(P_{CLOSE})$ is regarded as the reconstruction result at the level of s.

$$P_{CLOSE} = (P_{OPEN\_REC} \oplus B_s) \ominus B_s \tag{5}$$

$$\varepsilon_{P_{OPEN\_REC}}^{(1)}(P_{CLOSE}) = (P_{CLOSE} \ominus B_I) \vee P_{OPEN\_REC} \tag{6}$$

$$\varepsilon_{P_{OPEN\_REC}}^{(n)}(P_{CLOSE}) = \varepsilon_{P_{OPEN\_REC}}^{(1)} \circ \varepsilon_{P_{OPEN\_REC}}^{(1)} \circ \ldots \circ \varepsilon_{P_{OPEN\_REC}}^{(1)}(P_{CLOSE}) \tag{7}$$

where ε is the operator of geodesic erosion, $\vee$ stands for the point-wise maximum, and n is the number of iterations.

For example, one building is illustrated in Figure 1a, and Figure 1b shows a cross-section of that building. Figure 1c shows the result of the morphological reconstruction at the scale of 2 m. It shows that T_6^0 is flattened onto the larger segment, but T_3^0 and T_4^0 are erroneously processed as three segments (T_3^1, T_4^1 and T_5^1). The phenomenon is the canonical cut-off problem in the morphological operator [35] and will result in a failure of relinking the relationships between these segments from adjacent levels. In order to address the problem, a segment is restricted to two states after the morphological reconstruction: one horizontal segment or itself. The result of plane segmentation (PS) is adopted to correct the result of morphological reconstruction. First, we design an indicator to check whether a segment becomes horizontal or not after the morphological reconstruction. If the elevation difference h_{ps_i} as described by Equation (8) is less than a threshold t_{SH}, the segment is marked as horizontal in Equation (9). Otherwise, the elevations of the point in the segment are recovered by the corresponding value after the morphological reconstruction. Figure 1d is the result after the modification.

$$h_{ps_i} = (h_{MAX} - h_{MIN}) \times L_{ps_i} \tag{8}$$

$$L_{ps_i}^* = \begin{cases} 1 & if \; h_{ps_i} \geq t_{SH} \\ 0 & if \; h_{ps_i} < t_{SH} \end{cases} \tag{9}$$

where h_{MAX} and h_{MIN} are the maximum and minimum elevation in the segment ps_i after morphological reconstruction; h_{ps_i} is the indicator; t_{SH} is the threshold; and $L_{ps_i}^*$ is the judged result by the indicator.

Additionally, although some segments are smaller than twice the scale, they may fail to be removed. For example, there are two small segments T_1^1 and T_5^1 in Figure 1d, which is the morphological reconstruction result at the scale of 2 m, and they fail to flatten into the segments T_0^1 and T_4^1. Therefore, the method automatically edits these false segments through two steps. The first step is that the method detects these false segments according to their size and relationship with the neighboring segments. For the first case T_1^1, the method first groups all segments into different clusters according to adjacent segments with a minor elevation difference in the vertical direction, and each cluster is taken as an individual structure. Then, segments of any cluster which is smaller than two times the scale are detected. For the second case T_5^1, the method checks each small segment with a width less than twice the scale. If a small segment is included in another segment which is larger than two times the scale, the small segment will be detected. After detection of false segments, the method searches a neighboring segment with the width larger than twice the scale to modify each false segment. Figure 1e is the final result of the improved morphological reconstruction at the scale of 2 m.

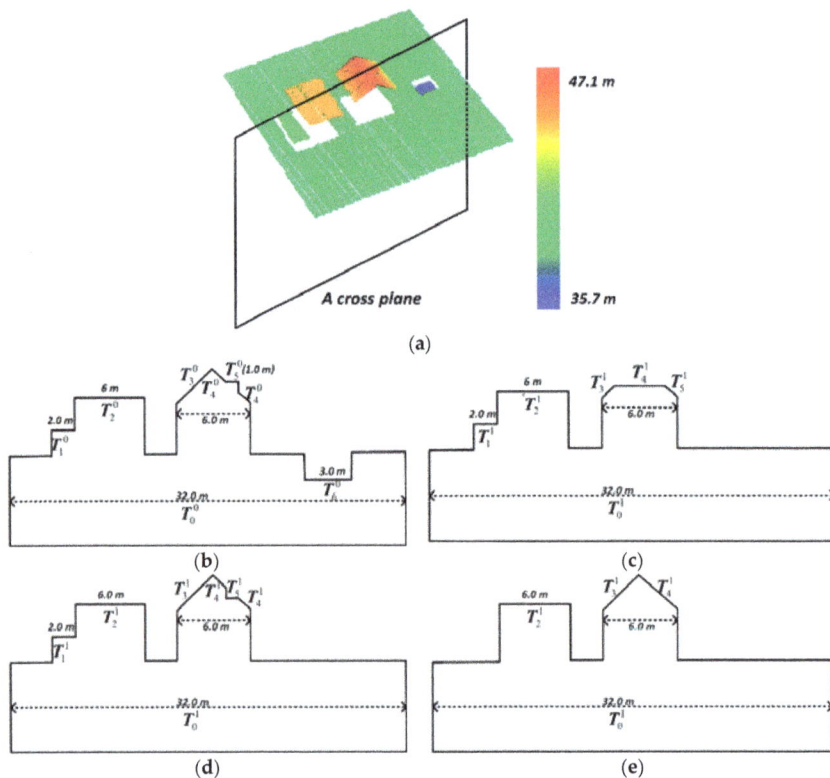

Figure 1. Improved morphological reconstruction for a building. (**a**) Raw point clouds of a building; (**b**) A cross-section of the raw point clouds, where the cross plane is illustrated in (**a**), and the width of each segment is annotated; (**c**) Result of morphological reconstruction at the scale of 2 m, where parts of the inclined roofs T_3^0 and T_4^0 are flattened; (**d**) Result of recovering the inclined segments T_3^0 and T_4^0; (**e**) Result of modifying false segments, where T_1^1 and T_5^1 are flattened onto the larger segment.

2.2. Generating the Morphological Scale Space and Constructing the Topological Relationship Graph (TRG)

To generate the scale space for an object, the improved morphological reconstruction is iteratively executed with the increasing of the scale. Hence, for one object, a scale space is constructed by employing a series of scale values ($S = \{s_0, s_1, s_2, \ldots, s_n\}$) until all points of the object are located on a horizontal plane, and each scale value indicates one level. It is clear that the points of one object have been portioned as different segments at each level. Sequentially, topological relationship graphs (TRGs) across levels can thus be created by linking the spatial relations between segments of adjacent levels, and each segment of one level is taken as a node. The rule of linking is that if most points from one segment in a fine level can be found in another segment of the next coarse level by the point index, the spatial relation between them is recorded. Figure 2 is an example of generating scale space and constructing topological relationship graphs within one building, and the scales are defined as $S = \{2, 4, 8, 16, \ldots\}$, where the former scale is half of the latter scale. Figure 2a is the raw point clouds, and Figure 2d is a corresponding cross-section. There are seven segments, and the size of each segment is annotated. Figure 2b,e are the results of the improved morphological reconstruction at the scale of 2 m. T_1^0, T_5^0 and T_6^0 are flattened, and T_3^0 and T_4^0 are preserved. Figure 2c,f are the final results of the improved morphological reconstruction, and the maximum scale is 4 m. Figure 2g is the generated TRG.

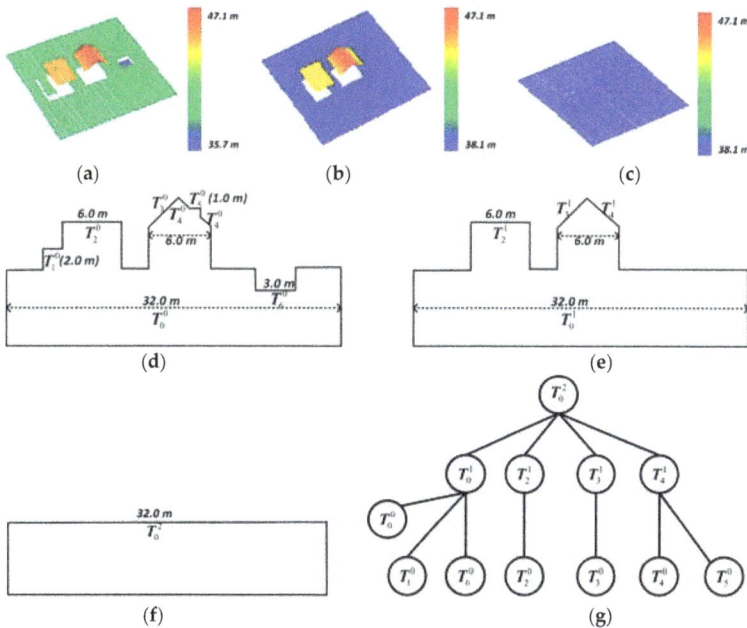

Figure 2. Generation of scale-space and the topological relationship graph within a building. (**a–c**) Results of morphological reconstruction at the first scale ($s = 0$ m), the second scale ($s = 2$ m) and the third scale ($s = 4$ m); (**d–f**) Results of scale space are displayed by cross-sections, and the location of the cross plane is illustrated in Figure 1a; (**g**) Topological relationship graph, which is generated by relinking the relationship between two segments from adjacent levels.

Moreover, the proposed method labels the topological relationship between each two adjacent segments for each level in the generated TRG, where the level is in order from the coarsest to the finest. For the labeling, four types of situations are designed, namely, *INTERSECTION*, *STEP*, *INTERSECTION and INCLUSION*, *STEP and INCLUSION*, as shown in Figure 3. The steps of labeling are described as follows:

Step 1: all segments are grouped into different clusters according to their father node. For example, segments of the third level in Figure 2g would be grouped into four clusters, which are the set of $\{\{T_0^0, T_1^0, T_6^0\}, \{T_2^0\}, \{T_3^0\}, \{T_4^0, T_5^0\}\}$.

Step 2: arbitrary two segments in one cluster are judged whether they are neighboring in the horizontal direction. Two neighboring segments are denoted as a segment pair. For example, the cluster $\{T_0^0, T_1^0, T_6^0\}$ would result in one set of two pairs $\{\{T_0^0, T_1^0\}, \{T_0^0, T_6^0\}\}$.

Step 3: traverses the segment pairs in each cluster one by one, derives an intersection line from the segment pair, and labels the relationship of the pair as either *INTERSECTION* or *STEP*. More specifically, if the distance between the points in the segment pair and the intersection line is less than one threshold (e.g., two times the point spacing), the relationship is labeled as *INTERSECTION*. Otherwise, it is labeled as *STEP*. On the other hand, if points of one segment are fully located in the exterior boundary of another segment in a segment pair, the relationship is labeled as *INCLUSION* as well. Additionally, the relationship between the segments from different clusters could be derived from their father nodes. For example, the labeled result of Figure 2g is illustrated in Figure 4.

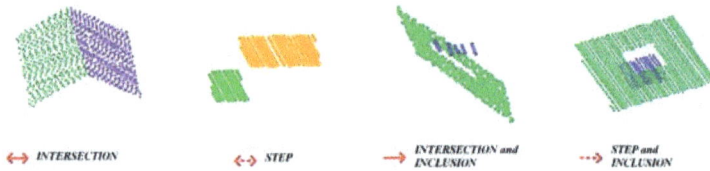

Figure 3. Four types of the relationship between two adjacent segments, which are dotted in different colors.

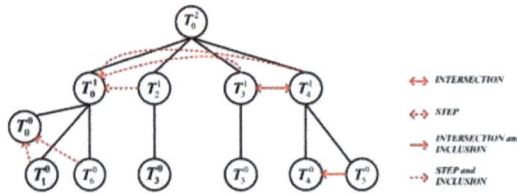

Figure 4. Labeling the relationship between the pair of two adjacent segments from each level of the generated topological relationship graph (TRG), and two adjacent segments should be from the same father node. When two adjacent segments are from different father nodes, the relationship is not labeled, and it could be derived from their father nodes.

3. Generating Building Levels of Detail (LoDs) Based on the Improved Morphological Scale Space

Figure 5 illustrates the flowchart of the proposed method. Four key steps are integrated to generate 3D building LoDs from airborne laser scanning point clouds, namely, detection of building candidate regions, generation of the improved morphological scale space, detection of building points, and generation of building LoDs.

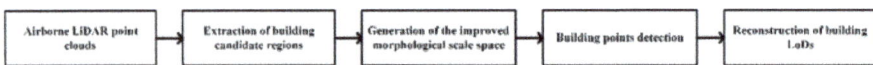

Figure 5. Flowchart of generating building LoDs from Airborne LiDAR point clouds.

3.1. Building Candidate Region Extraction and Generation of the Morphological Scale Space

Buildings in an urban scene have different structures with highly variable sizes and stories. In general, the maximum scale is determined by their sizes and structures in the scale space. That is to say, different buildings may be assigned different values for the maximum scales. Hence, the candidate region of each building is first detected from the point clouds for adaptively tuning the maximum value, and then the morphological scale space is generated for each candidate region respectively.

Step 1: The filtering method of [36] is utilized to separate ground points from non-ground points. The filtering method classifies the points into a set of segments and one set of individual points by point cloud segmentation, which are filtered by segment-based filtering and multi-scale morphological filtering, respectively. Therefore, the non-ground points include two sets, non-ground segments and non-ground individual points. Figure 6b is the filtering result of Figure 6a, and Figure 6c is the non-ground segments.

Step 2: Extract the candidate region of each building. First, the non-ground segments are clustered via a region-growing method with the constraint of two-dimensional Euclidean distance, and the distance threshold is specified as two times that of the point spacing. Then, with the assumption that one building has a certain area, width and large elevation differences with its neighboring terrain areas, the clusters are classified by Equation (10) to obtain the candidate building clusters. Finally, each candidate building cluster is buffered with a distance (e.g., 3 m) to obtain a buffer area, which is regarded as the candidate region of each building object. The buffer operator aims to ensure the completeness of an object. For example, Figure 6d is the result of extracting candidate regions.

$$cBuilds = \left\{ SC_i \in SC \left| \begin{array}{c} Rule1: \ Num(Bound(SC_i) > t_H) > 0.25 \times Num(Bound(SC_i)) \ \&\& \\ Rule2: \ Width(SC_i) > t_W \ \&\& \\ Rule3: \ Area(SC_i) > t_A \end{array} \right. \right\} \quad (10)$$

where $cBuilds$ denotes the candidate building clusters; SC is the set of clusters, and SC_i is the ith cluster; $Bound()$ is used to extract boundary points of each cluster; $Num()$ is a counter of points satisfying the condition of the elevation difference; $Width()$ and $Area()$ calculate the width and area of a cluster; t_W, t_A, t_H are three thresholds of the width, area and elevation difference respectively. t_W and t_A should be tuned according to the scene (e.g., a modern megacity or a village), where t_A could be specified as 2.0–100.0 m^2, and t_W could be specified as 2.0–10.0 m. t_H could be specified as a value in consideration of a building no lower than 1.5 m.

Step 3: Generate the morphological scale space and the corresponding TRGs. Once the candidate region of one building object is determined, the morphological scale space is generated according to Section 2, and the corresponding TRGs are recorded as well. Generally, the root of a TRG represents the entire object region, and the leaf nodes of a TRG are the segments of an object region in the minimum scale. The relationships between segments from the same level are also labeled in the generated TRG. For the generation of the morphological scale space, in consideration of time efficiency, a set of scales $S = \{2, 4, 8, 16, \ldots\}$ is specified for iteratively generating each level of scale space, whereby the former scale value is half of the later. An example is illustrated in Figure 7.

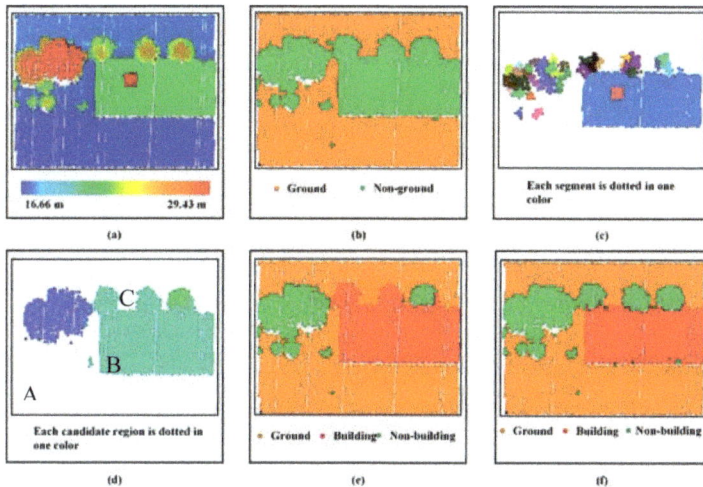

Figure 6. An example for building point detection. (**a**) Raw point cloud. There are a building and several trees, and three trees are near to the building; (**b**) Filtering result. Ground and non-ground points are separated; (**c**) Some non-ground segments; (**d**) Generated building candidate regions by grouping non-ground segments; (**e**) Result of TRG classification. Only one candidate region is labeled as a building; (**f**) Non-building points near the building are removed, and the remained points are classified as building points.

(a) A side local view for the raw point cloud

(b) Segmentation result of *s* = 0 m

(c) Segmentation result of *s* = 2 m

(d) Segmentation result of *s* = 4 m

(e) Generated TRG

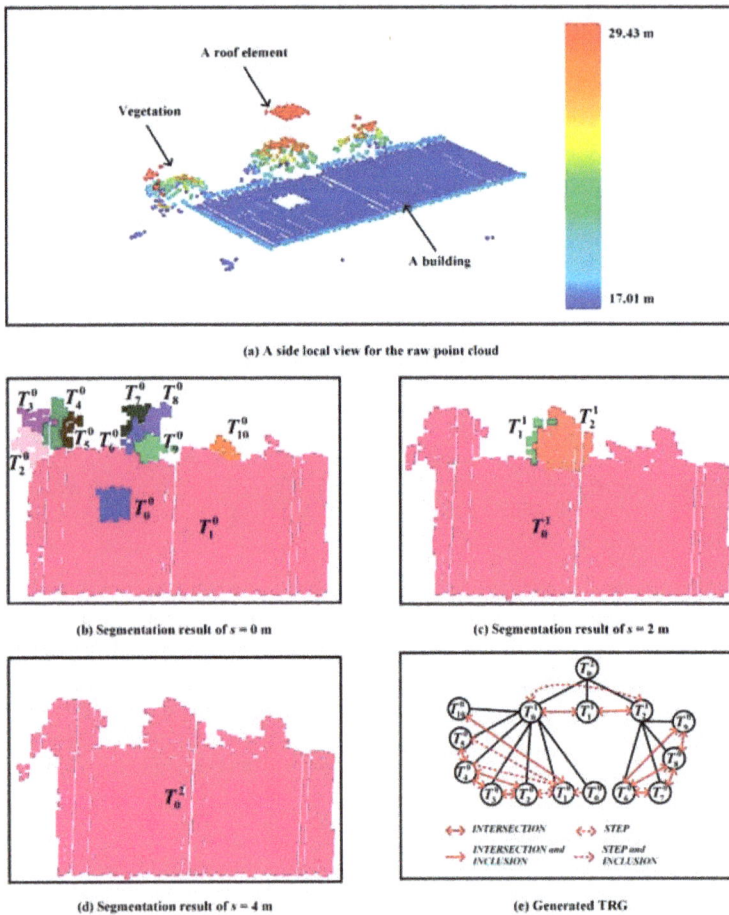

Figure 7. Generating the TRG for the building candidate region B of Figure 6d. (**a**) Raw point cloud; (**b–d**) Segmentation results of three scales. Each segment is dotted in one color, and each segment is annotated with a unique identification; (**e**) Generating TRG according to the method in Section 2.

3.2. Building Point Detection

A method based on the generated TRGs is employed to extract building points from each building candidate region. The method distinguishes buildings from other points in consideration of the entire object and its changes across scales. The method includes two steps. The first step is to label the building TRGs, and the second step is to remove non-building points from the building TRG.

3.2.1. Classification of TRGs

The method first classifies all TRGs into building TRGs and non-building TRGs by five features, as listed in Table 1. The five features are mainly related to geometrical sizes, surface characteristics, the penetrating capacities within different objects, and the changing characteristics of objects across scales. The classification rules are defined in Equation (11). For example, Figure 6e is the result of TRG classification.

$$buildTRGs = \left\{ pTRG_i \in ATRGs \left| \begin{array}{l} Rule1: \ A > t_A \ \&\& \\ Rule2: \ W > t_W \ \&\& \\ Rule3: \ AR_{MIN-MAX} > t_{ARMM} \ \&\& \\ Rule4: \ AR_{G-O} < t_{ARGO} \ \&\& \\ Rule5: \ PNR_{MIN-MAX} > t_{PNRMM} \end{array} \right. \right\} \quad (11)$$

where *buildRegions* is the set of building TRGs, *ATRGs* is the set of all TRGs, $pTRG_i$ is *i*th TRG, t_A, t_W, t_{ARMM}, t_{ARGO} and t_{PNRMM} are thresholds of five features, respectively. The threshold t_{ARMM} should be determined by several factors, such as the flatness of an object. Generally, it should not be lower than 0.5. For the threshold t_{ARGO}, in theory, it should be near zero. However, because of the structure and material of a building, there may be a lot of ground points below roofs. Therefore, the value of t_{ARGO} ranges from 0.2 to 0.6. The threshold of t_{PNRMM} is mainly relevant to the penetrating capacity and the surface characteristics, and it could be larger than 0.5.

Table 1. Five features based on the TRG.

Features	Descriptions	Characteristics
The area of the TRG (A)	The area of the TRG	The areas of buildings and large trees are large, and the areas of small objects (e.g., vehicles, low vegetation and street furniture) are small
The width of the TRG (W)	The width of the TRG	The widths of buildings and large trees are large, and the widths of small objects (e.g., vehicles, low vegetation and street furniture) are small
The area ratio of the segments ($AR_{MIN-MAX}$)	The value is the ratio between the minimum and the maximum area of segments across scales. It reflects the result of segmentation for objects in different scales	The value of a building is large, and that of a tree may be small
The area ratio of ground points (AR_{G-O})	The ratio in areas between the entire object and the ground points in the corresponding region. It reflects the penetrating capacities in different objects	The value of a building generally approximates zero, and it may be higher in the area of vegetation
The ratio of segmented points ($PNR_{MIN-MAX}$)	The ratio in the number of segmented points between the minimum scale and the maximum scale. It reflects the changing of surface characteristics across scales and the penetrating capacities in different objects	The value of a building is large, and it is small for vegetation

3.2.2. Extraction of the Final Building Points from Each Building TRG

Although TRGs have been classified, there may be some other objects (e.g., vegetation, vehicles) in the building TRGs, and these objects should be removed. Generally, these objects consist of small segments or individual points in the minimum scale, and they are near the border of the building region. Therefore, the process is described as follows.

Step 1: The method detects the small segments by an area threshold t_{SA} in the minimum scale, and non-ground points removed in the process of segmentation are also detected. The detected points are labeled as unclassified points. Generally, the threshold t_{SA} is specified as 3.0–5.0 m^2.

Step 2: The unclassified points are grouped into different clusters by a region-growing method with the constraint of two-dimensional Euclidean distance.

Step 3: For each cluster, the distance between its boundary and the border of the building region is calculated. And then, the cluster will be determined whether it locates inside the building region or near the border of the building region by one distance threshold, which is also specified as two times that of the point spacing. If a cluster locates inside the building region, it would be classified as building. Otherwise, five features are calculated after reconstructing a new TRG for each unclassified cluster, and each unclassified cluster is labeled as building or non-building by Equation (11).

Figure 6f is the result of extracting the final building points of Figure 6a. Two trees near the building are removed. Based on the detection result, the nodes of non-building segments would be removed from the finest level to the coarsest level, and the relationships of these segments are also removed at the same time. The process is illustrated in Figure 8. Moreover, if there are non-building child nodes, the non-building points should also be removed the father node. For example, points of T_{10}^0 should be removed from the segments T_0^1 and T_0^2.

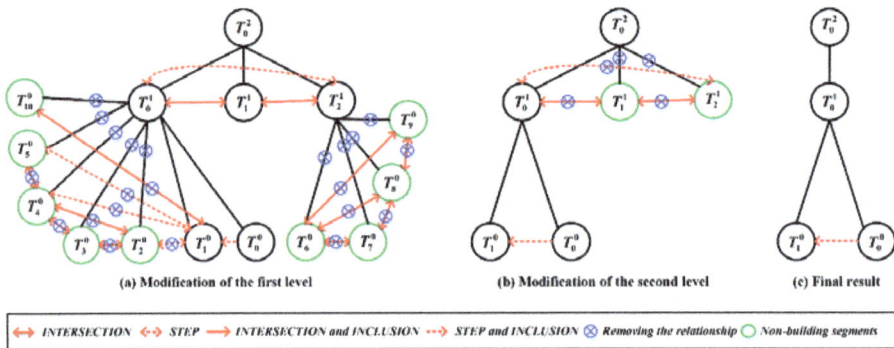

(a) Modification of the first level   (b) Modification of the second level   (c) Final result

↔ *INTERSECTION* ↔ *STEP* → *INTERSECTION and INCLUSION* ⟶ *STEP and INCLUSION* ⊗ *Removing the relationship* ○ *Non-building segments*

Figure 8. Modifying one TRG (i.e., Figure 7e) from the finest level to the coarsest level after building point detection. If all points in the segment are classified as non-building, the segment node and its relationships are removed from the TRG.

3.3. Building LoD Generation

After automatically extracting building points and modifying the corresponding TRG, the method reconstructs each building in each scale by the cycle graph analysis method of [37] to obtain the corresponding building LoDs. The main steps of reconstructing a building model are as follows. First, a graph about the topological relationship between each two adjacent segments from the same level is constructed, and it is derived from the labeled TRG. Simultaneously, the feature lines are also derived from each two adjacent segments. Then, roof corners are obtained by a strategy for detecting closed cycles in the graph. The corner points are used to fix the ending points of the corresponding feature lines. Finally, one building model is reconstructed by the combination of feature lines. Two cases are illustrated in Tables 2 and 3.

Table 2. First example of generating the LoDs for the building of Figure 6.

Scale Values	Multi-Scale Roof Data	Plane Segmentation Results	Building LoDs
0 m			
2 m			

Table 3. Second example of generating the LoDs for a building with gable roofs and dormers.

Scale Values	Multi-Scale Roof Data	Plane Segmentation Results	The Final TRG	Building LoDs
0 m				
2 m				
4 m				

4. Experimental Results and Analysis

The experiment was conducted on the Toronto dataset (as shown in Figure 9) provided by International Society for Photogrammetry and Remote Sensing (ISPRS) to validate the performance of the proposed method. The area of this dataset is about 403 m × 532 m, and the elevation ranges from 40 to 190 m. There are 58 buildings larger than 2.5 m², and the corresponding building area is 88,249.8 m². The dataset is located in a commercial zone with representative scene characteristics of a modern megacity. Moreover, the area is covered by the high-rise and multi-story buildings with complex rooftop structures, which are very suitable for verifying the proposed method.

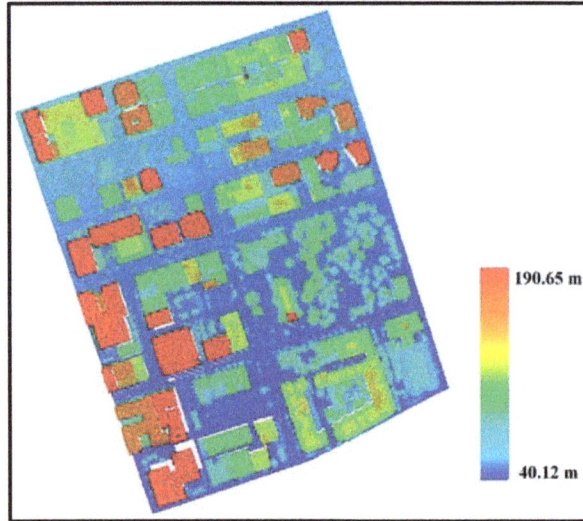

Figure 9. Raw point clouds of the Toronto dataset provided by International Society for Photogrammetry and Remote Sensing (ISPRS).

The procedure of the proposed method was executed to extract building points and generate the LoDs for each building. The parameters involved in the proposed method are listed in Table 4. The point clouds were first filtered into non-ground points and ground points. The result is shown in Figure 10a, and the non-ground segments are showed in Figure 10b. Then, building candidate regions were extracted, as illustrated in Figure 10c. For each building candidate region, the improved morphological scale space and the labeled TRG were generated, TRGs were classified, and building points were detected, as shown in Figure 10d–f. Then, the building LoDs were reconstructed. Figures 11–13 are the processes for extracting building points and reconstructing building LoDs from the building candidate region PB of Figure 10c. The result of reconstructing the building LoDs for the entire scene is shown in Figure 14. It can be seen that roof structures change from complicated to simple with the increasing scale until each roof becomes a plane. Therefore, the reconstructed building models at different LoDs can serve various urban monitoring and analysis applications. Moreover, the number of levels for each building self-adapts to its size and roof structures, ranging from three levels to six levels. More importantly, the proposed method can reconstruct the roof model with any one scale from the finest building points by the morphological reconstruction. The coarser roof model does not need to be generated from the finest roof model. For example, the proposed method could directly generate the level $s = 4$ m from the raw building points. This is very helpful to save the cost and satisfy the user's requirement.

Table 4. Parameter settings.

Parameters	Values	Description	Steps
t_A/m^2	50	The area threshold	
t_W/m	5	The width threshold	Building candidate region extraction
t_H/m	1.5	The threshold of describing the elevation difference between the boundary points of a building and the DEM	
t_N	10	This parameter is used to remove very small segments in plane segmentation	
$t_S/^\circ$	10	A threshold for the slope parameter	The generation of the scale space
t_{SH}/m	0.2	It is a threshold of the elevation difference for determining a segment is inclined or horizontal after morphological reconstruction	
t_{ARMM}	0.5	The area ratio of the segments across levels of a TRG	
t_{ARGO}	0.5	The area ratio of ground points within a TRG	Building point detection
t_{PNRMM}	0.5	The ratio of segmented points across levels of a TRG	
t_{SA}/m^2	5	An area threshold for detecting small segments near buildings	

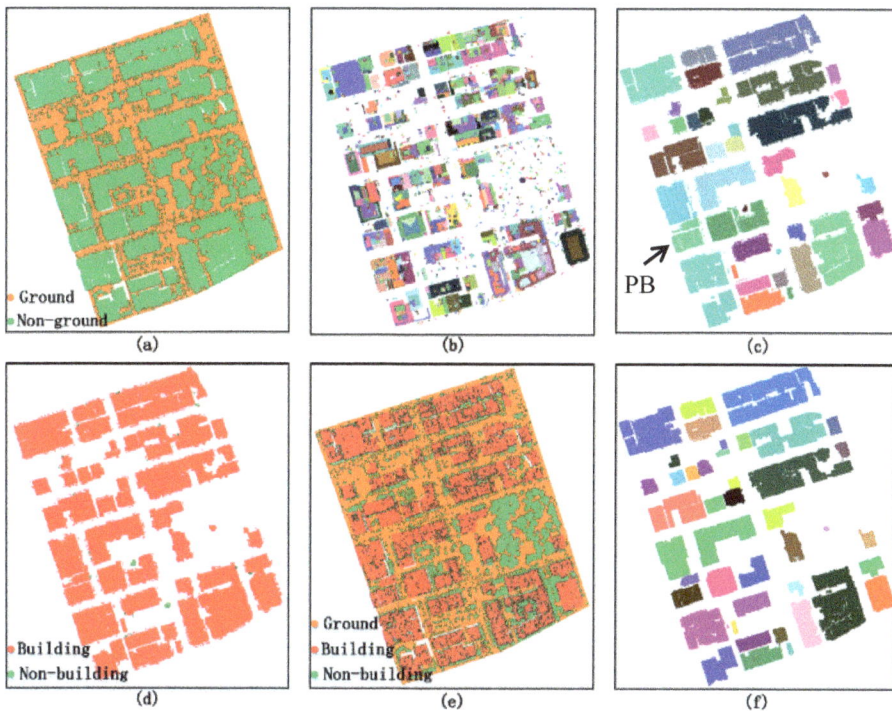

Figure 10. Detecting buildings from the Toronto dataset. (**a**) Filtering result; (**b**) Non-ground segments, and each segment is dotted in one color; (**c**) Result of generating building candidate regions, where each region is dotted in one color; (**d**) Result of TRG classification; (**e**) Result of extracting buildings; (**f**) Result of the extracted buildings, and different buildings are dotted in different colors.

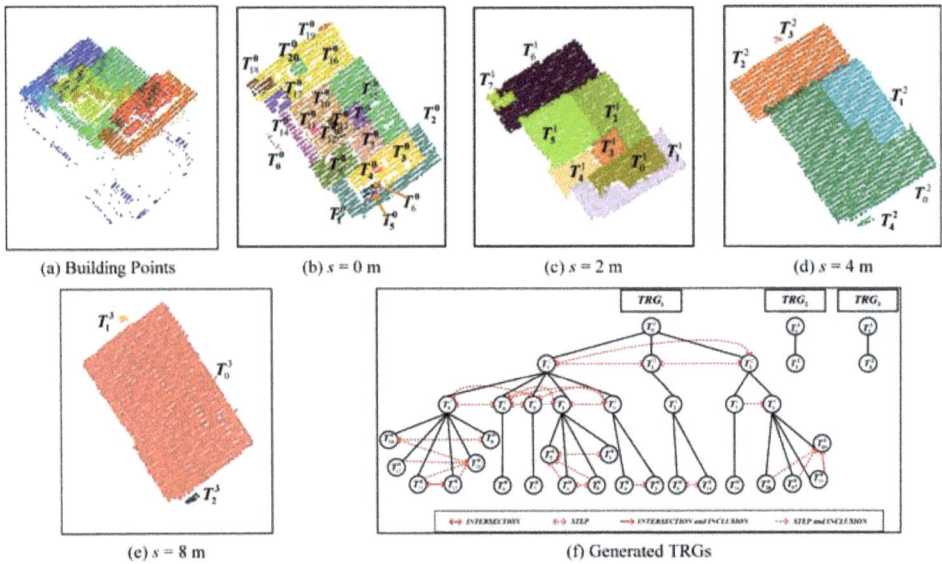

(a) Building Points   (b) $s = 0$ m   (c) $s = 2$ m   (d) $s = 4$ m

(e) $s = 8$ m   (f) Generated TRGs

Figure 11. Generating the scale space and the corresponding TRG for a building candidate region PB in Figure 10c. (**a**) Point clouds of the building candidate region; (**b–e**) Segmentation results at four scales, where different segments are dotted in different colors. Additionally, each segment is annotated with a unique identification; (**f**) Generated TRGs.

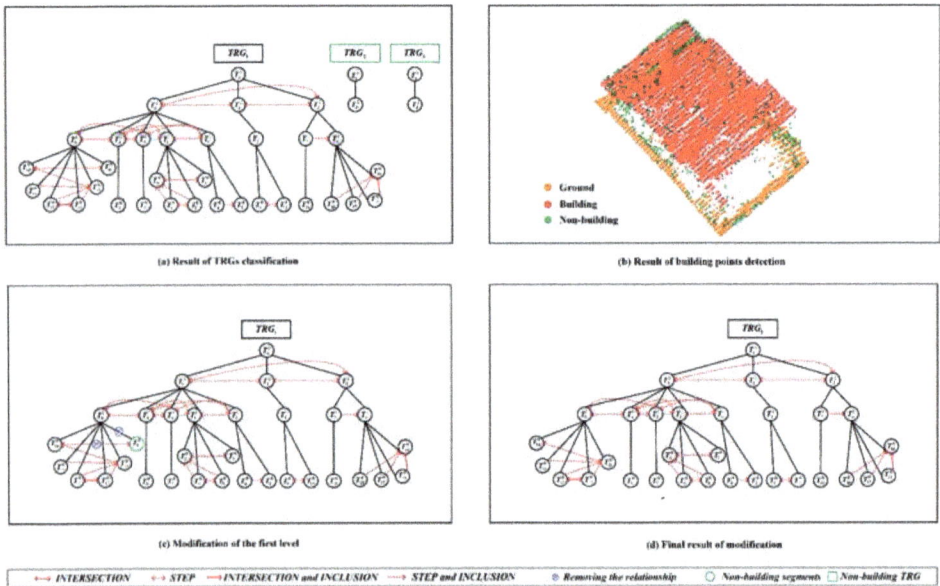

(a) Result of TRGs classification   (b) Result of building points detection

(c) Modification of the first level   (d) Final result of modification

Figure 12. Extracting building points and modifying the TRG from the building candidate region PB in Figure 10c. (**a**) TRG classification. Two TRGs are classified as non-building, and one TRG is labeled as a building; (**b**) Final result of building point detection; (**c,d**) Process of modifying the TRG according to the result of building point detection, and only one segment node is removed.

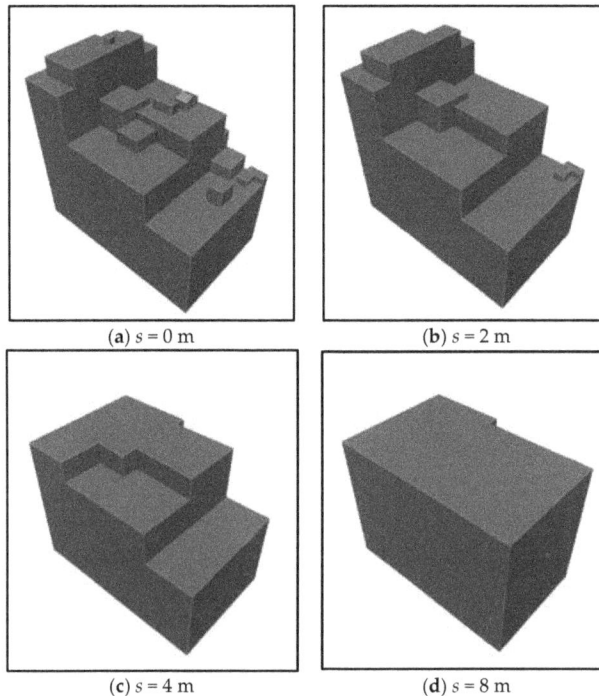

Figure 13. Reconstructing the building LoDs of the building candidate region PB in Figure 10c, where there are four levels. The roof structures are changed from complicated to simple with the increasing scale. (**a**) The building model within the scale of 0 m. (**b**) The building model within the scale of 2 m. (**c**) The building model within the scale of 4 m. (**d**) The building model within the scale of 8 m.

The high quality of building point detection results was the prerequisite for reconstructing building LoDs in a scene. Hence, the result of building point detection was submitted to the organization ISPRS for evaluation [38]. The evaluation result is shown in Figure 15, and the details can be found on the website [39]. Several indicators are adopted for quantitative evaluation, including Completeness (CP), Correctness (CR) and Quality (Q) at the pixel or object level, and the total Root Mean Square (RMS) of reference boundaries. The result is listed in Table 5. It can be seen that the Correctness values are 95.5% at the pixel level, and 96.6% at the object level. The high values show that the proposed method can robustly distinguish buildings from vegetation or other objects. It may benefit from the combination of features derived from the local and the whole of an object. In this result, there are only two false positives at the object level. The false positives are large objects with smooth surfaces, which are very easily classified as buildings. The Completeness values are 94.7% at the pixel level, and 98.3% at the object level. The values indicate the method could robustly extract buildings, as shown in the yellow areas of Figure 15. Additionally, the proposed method could also preserve annex structures and rooftop furniture well by taking large parts of the building and small structures as a whole in the process of detecting buildings, and robustly removing noise and vegetation points on the roofs, as illustrated in Figure 16. In order to further analyze the performance of the proposed method, the comparison between the proposed method and the other methods [13] is listed in Table 5, showing that the proposed method has the best qualities in detecting buildings at the pixel level and the total RMS, and only the method of FIE [40] obtained a better performance than the proposed method at the object level. Therefore, the result of building point detection could provide a good foundation for the reconstruction of building LoDs. However, some small segments near the boundary of a building may be erroneously removed, as shown in Figure 15 (dotted in blue), thereby resulting in incorrect building LoDs reconstruction. Figure 17 shows an example of reconstructing building LoDs for a building with complex rooftop structure. Because the points of dormers are few, they failed to be detected, as illustrated in Figure 17a. The model of $s = 0$ m also missed several dormers, as shown in Figure 17b.

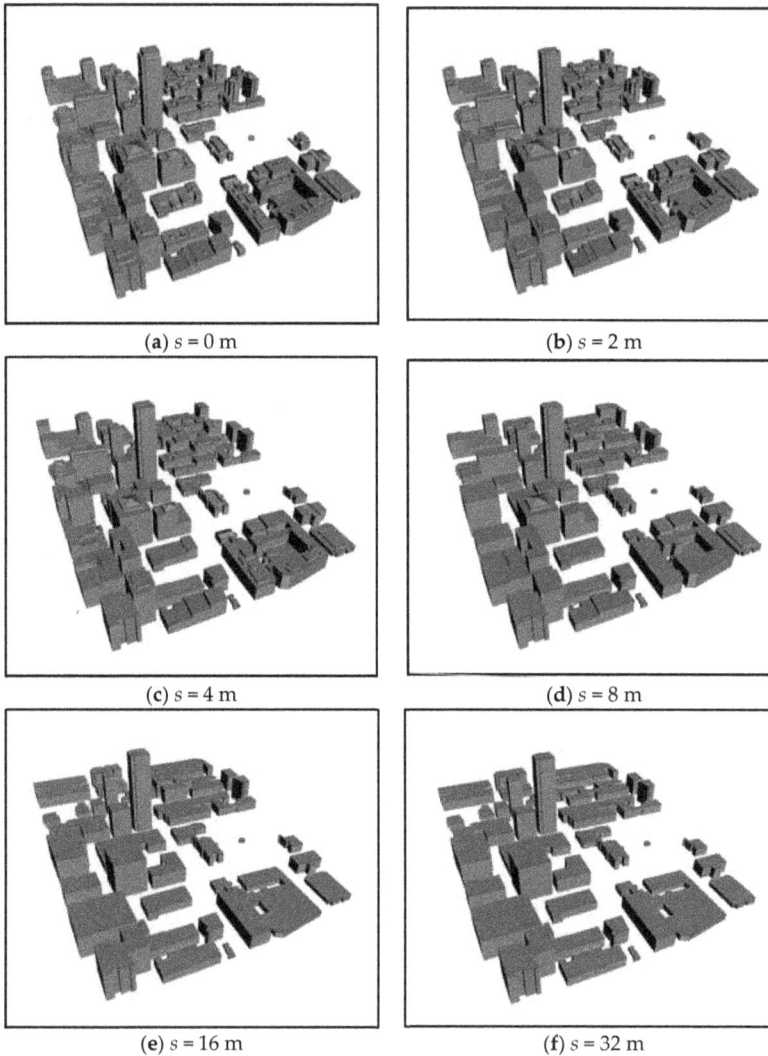

(a) $s = 0$ m

(b) $s = 2$ m

(c) $s = 4$ m

(d) $s = 8$ m

(e) $s = 16$ m

(f) $s = 32$ m

Figure 14. Results of reconstructing the building LoDs in the entire scene. The roof structures are changed from complicated to simple with the increasing of the scale. Because different buildings have different levels, the model of the maximum scale is utilized at a larger scale. (a) The building models within the scale of 0 m. (b) The building models within the scale of 2 m. (c) The building models within the scale of 4 m. (d) The building models within the scale of 8 m. (e) The building models within the scale of 16 m. (f) The building models within the scale of 32 m.

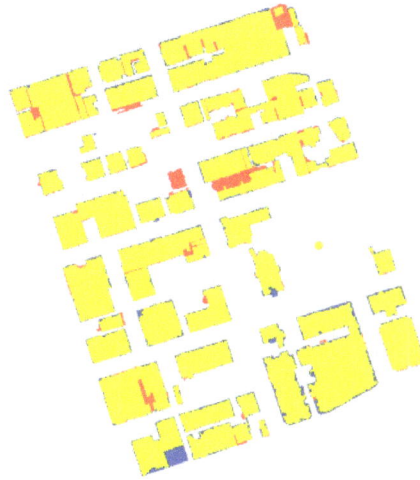

Figure 15. Elevation result provided by ISPRS. Yellow pixels are true positives, red pixels are false positives, and blue pixels are false negatives.

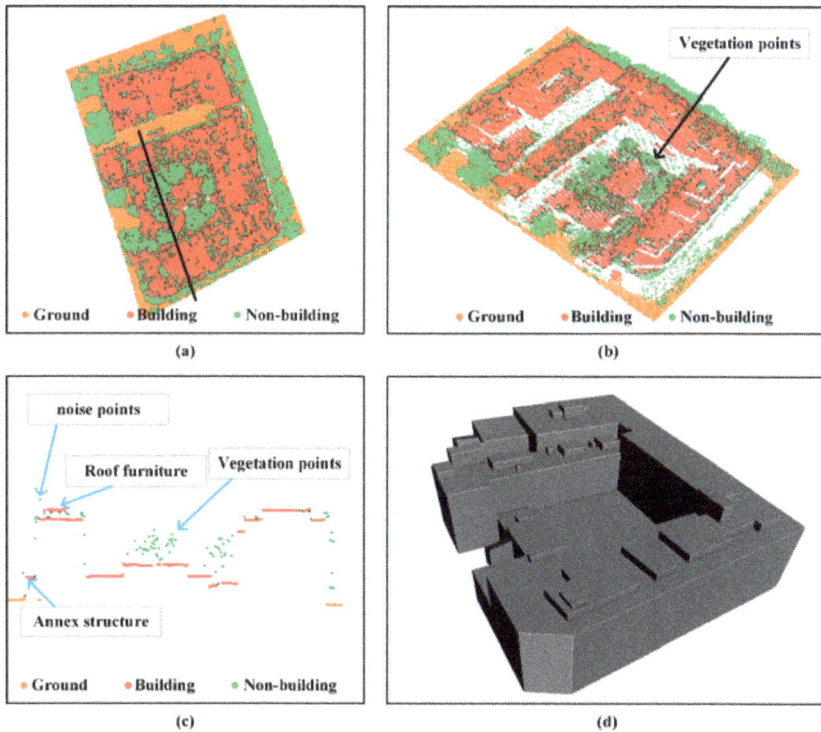

Figure 16. A result of detecting a building. (**a**) top-view of the building detection result; (**b**) side-view of the building detection result; (**c**) cross-section of the black line in (**a**) for detailed description of the building detection result, where roof furniture and annex structures are preserved, and vegetation points and noise points are removed; (**d**) corresponding building model at the scale of 0 m.

Table 5. Evaluation result by ISPRS; the best results are highlighted.

Methods	Per_Area/%			Per_Object/%			RMS/m
	CP	CR	Q	CP	CR	Q	
The proposed method	94.7	95.5	**90.6**	98.3	96.6	95.0	**0.8**
WHUY2 [34]	95.1	89.3	85.4	96.6	94.6	91.6	1.2
TUM [41]	85.1	80.0	70.1	86.2	92.3	80.4	1.6
FIE [40]	96.6	90.6	87.8	98.3	98.2	**96.6**	1.2
ITCM [42]	80.5	82.1	68.5	96.6	22.9	22.7	1.5
MAR2 [15]	93.7	94.9	89.2	98.3	94.9	93.4	2.8
MON2 [43]	95.1	91.1	87.0	100	83.6	83.6	1.1
Z_GIS [44]	93.0	94.5	88.2	96.6	96.5	93.3	1.0
MIN	80.5	80	68.5	86.2	22.9	22.7	0.8
MAX	96.6	95.5	90.6	100	98.2	96.6	2.8

Finally, we selected two cases to describe the results of reconstructing building LoDs in the local view, and the results were compared with the building LoDs in the framework of CityGML. The first case is a connected building in Figure 18, where there is an ensemble of three parts with annex structures and roof furniture. Figure 18a is the result of the proposed method, and Figure 18b is the result based on CityGML. It shows that the proposed method could find the corresponding model matching to each level of building LoDs based on CityGML. For example, the model of $s = 0$ m is the same with LoD2, the model of $s = 2$ m approximates LoD1, and the model of $s = 16$ m is similar to the LoD0, where only the elevation of the model of $s = 16$ m is assigned the minimum elevation of the building points. Moreover, the models based on CityGML have only three levels, but the models of the proposed method have five levels with a more gradual change for reducing the difference between two adjacent levels. In the visualization of multi-scale representations for a building, the result of the proposed method could have a smaller jump between two adjacent levels. The second case is a building with multiple stories, where there are various types of roof structures (e.g., flat roofs and gable roofs), as shown in Figure 19. Figure 19a is the result of the proposed method, and Figure 19b is the result based on the CityGML. It also shows the models of the proposed method have a smaller change between two adjacent levels. In addition, the inclined roofs are preserved in the model of $s = 2$ m from the proposed method, while each roof is flat in the LoD1 of the CityGML.

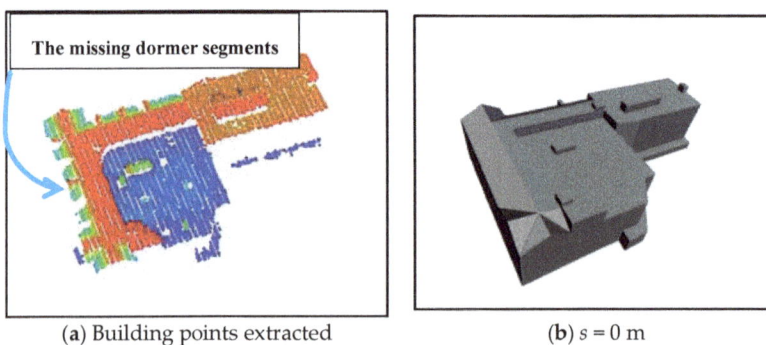

(a) Building points extracted   (b) $s = 0$ m

The missing dormer segments

Figure 17. *Cont.*

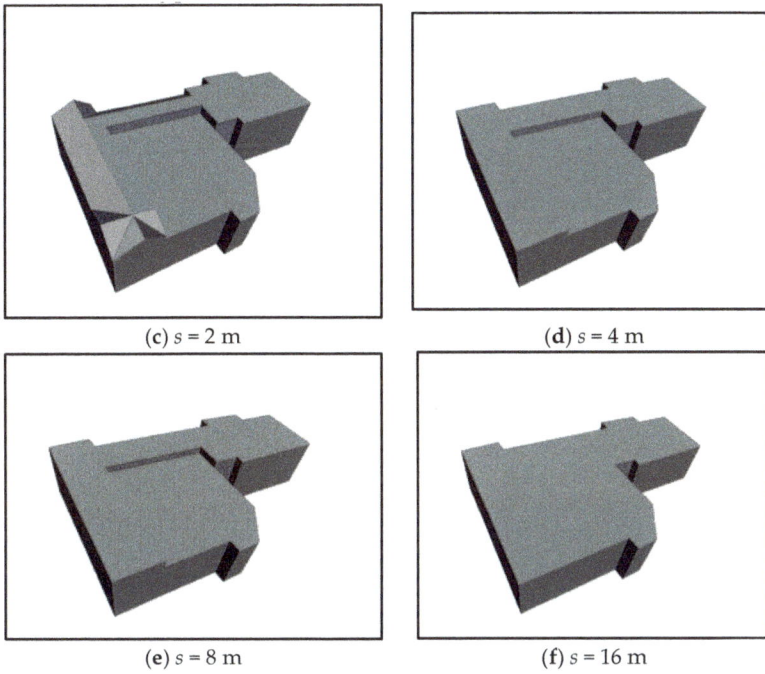

(c) *s* = 2 m   (d) *s* = 4 m

(e) *s* = 8 m   (f) *s* = 16 m

Figure 17. An example for describing some problems in the result of building LoDs. Because some dormers are missed in the building point detection, the models of some levels may be incomplete. (**a**) Extracted building points. (**b**) The building model within the scale of 0 m. (**c**) The building model within the scale of 2 m. (**d**) The building model within the scale of 4 m. (**e**) The building model within the scale of 8 m. (**f**) The building model within the scale of 16 m.

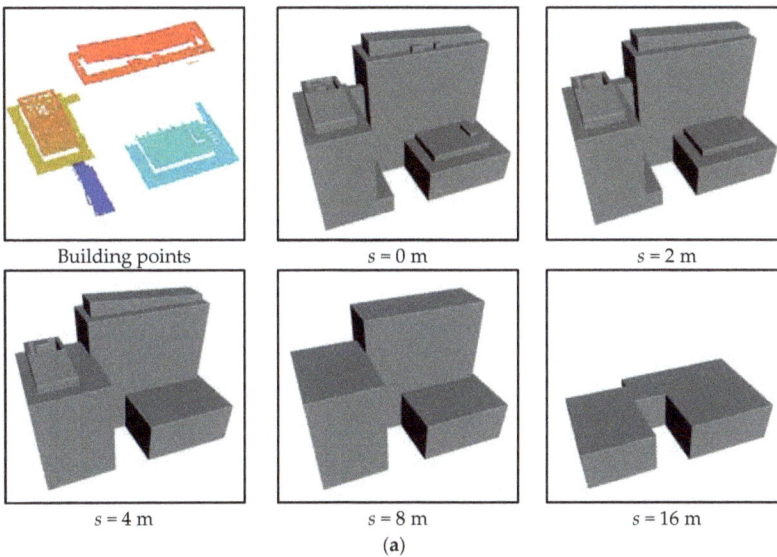

Building points   *s* = 0 m   *s* = 2 m

s = 4 m   *s* = 8 m   *s* = 16 m

(**a**)

Figure 18. *Cont.*

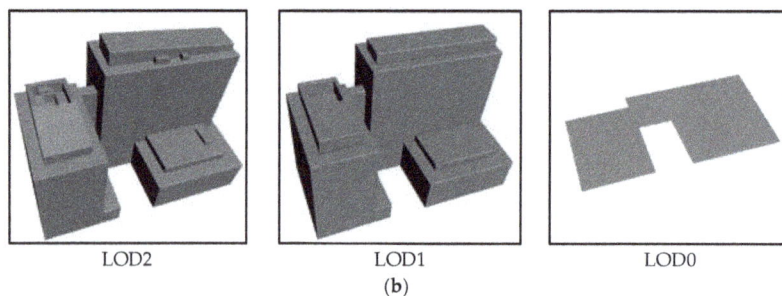

LOD2   LOD1   LOD0

(b)

Figure 18. Comparison of LoDs from CityGML and the proposed method for a connected building. (**a**) Building LoDs from the proposed method; (**b**) Building LoDs from CityGML.

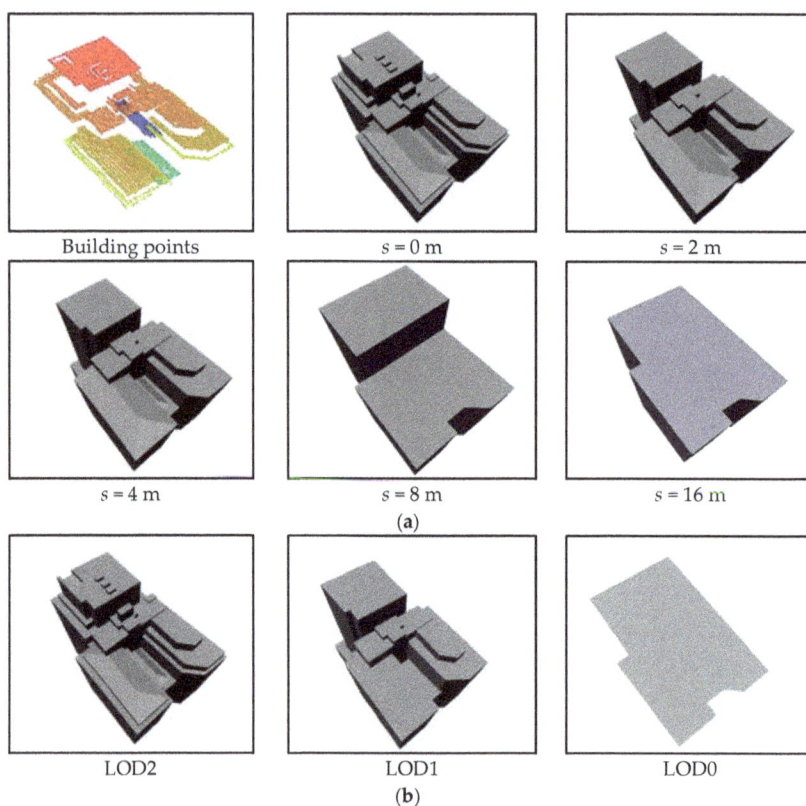

Building points $s = 0$ m $s = 2$ m

$s = 4$ m $s = 8$ m $s = 16$ m

(a)

LOD2   LOD1   LOD0

(b)

Figure 19. Comparison of LoDs from CityGML and the proposed method for a building with multiple stories. (**a**) Building LoDs from the proposed method; (**b**) Building LoDs from CityGML.

5. Conclusions

In this study, we propose a method to reconstruct building levels of detail (LoDs) by using an improved morphological scale-space. After separating ground and non-ground points, the candidate region of each building is detected. The scale-space of each building candidate region is obtained by iteratively using the improved morphological reconstruction. Topological relationship graphs (TRGs) are generated by relinking the relationships of segments between two adjacent scales. Then, building points are detected by features based on TRG, and the TRG will be modified after detection. Finally, the proposed method reconstructs the roof model for each building at each scale. To verify the validities and the robustness of the proposed method, the Toronto dataset from International Society for Photogrammetry and Remote Sensing (ISPRS) was selected to extract building points and reconstruct building LoDs. The results of building point detection were submitted to ISPRS for evaluation, and the building LoDs were compared with the building LoDs based on the CityGML. The results demonstrate that the proposed method has a good performance in robustly extracting the buildings with details (e.g., roof furniture) and distinguishing buildings from vegetation or other objects. More importantly, the proposed method can directly reconstruct building LoDs from airborne Light Detection and Ranging (LiDAR) point clouds with the adaptive number of levels while maintaining the spatial relations between adjacent levels. However, some small parts of buildings may be missed, which affects the quality of building LoDs. In the future, we will incorporate spatial reasoning to improve the performance of extracting building details.

Acknowledgments: This study was jointly supported by the National Key Technology R&D Program (No. 2014BAL05B07), NSFC project (No. 41531177, 41371431), National Key Research and Development Program of China (No. 2016YFF0103501), and Public science and technology research funds projects of ocean (No. 2013418025).

Author Contributions: Bisheng Yang and Ronggang Huang designed the algorithm, and they wrote the paper. Jianping Li implemented the regularization of the outline for each building. Mao Tian performed the study of scale-space. Wenxia Dai and Ruofei Zhong implemented the segmentation of roof facets.

Conflicts of Interest: The authors declare no conflict of interest.

References

1. Gröger, G.; Kolbe, T.H.; Nagel, C.; Häfele, K.H. *OGC City Geography Markup Language (CityGML) Encoding Standard*; Open Geospatial Consortium: Wayland, MA, USA, 2012.
2. Jochem, A.; Hofle, B.; Rutzinger, M.; Pfeifer, N. Automatic roof plane detection and analysis in airborne lidar point clouds for solar potential assessment. *Sensors* **2009**, *9*, 5241–5262. [CrossRef] [PubMed]
3. Biljecki, F.; Ledoux, H.; Stoter, J.; Zhao, J. Formalisation of the level of detail in 3D city modelling. *Comput. Environ. Urban Syst.* **2014**, *48*, 1–15. [CrossRef]
4. Fan, H.; Meng, L. A three-step approach of simplifying 3D buildings modeled by CityGML. *Int. J. Geogr. Inf. Sci.* **2012**, *26*, 1091–1107. [CrossRef]
5. Forberg, A. Generalization of 3D building data based on a scale-space approach. *ISPRS J. Photogramm. Remote Sens.* **2007**, *62*, 104–111. [CrossRef]
6. Mao, B.; Ban, Y.; Harrie, L. A multiple representation data structure for dynamic visualisation of generalised 3D city models. *ISPRS J. Photogramm. Remote Sens.* **2011**, *66*, 198–208. [CrossRef]
7. Thiemann, F.; Sester, M. Segmentation of buildings for 3D-generalisation. In Proceedings of the ICA Workshop on Generalisation and Multiple Representation, Leicester, UK, 20–21 August 2004.
8. Kada, M. 3D building generalization based on half-space modeling. *Int. Arch. Photogramm. Remote Sens. Spat. Inf. Sci.* **2006**, *36*, 58–64.
9. Sester, M. Generalization based on least squares adjustment. *Int. Arch. Photogramm. Remote Sens. Spat. Inf. Sci.* **2000**, *33*, 931–938.
10. Biljecki, F.; Ledoux, H.; Stoter, J. An improved LOD specification for 3D building models. *Comput. Environ. Urban Syst.* **2016**, *59*, 25–37. [CrossRef]
11. Verdie, Y.; Lafarge, F.; Alliez, P. LOD Generation for urban scenes. *ACM Trans. Graph.* **2015**, *34*, 1–14. [CrossRef]
12. Shan, J.; Toth, C.K. *Topographic Laser Ranging and Scanning, Principles and Processing*; CRC Press: London, UK, 2008; Volume 15, pp. 423–446.
13. Rottensteiner, F.; Sohn, G.; Gerke, M.; Wegner, J.D.; Breitkopf, U.; Jung, J. Results of the ISPRS benchmark on urban object detection and 3D building reconstruction. *ISPRS J. Photogramm. Remote Sens.* **2014**, *93*, 256–271. [CrossRef]
14. Tomljenovic, I.; Höfle, B.; Tiede, D.; Blaschke, T. Building extraction from airborne laser scanning data, an analysis of the state of the art. *Remote Sens.* **2015**, *7*, 3826–3862. [CrossRef]

15. Mongus, D.; Lukač, N.; Žalik, B. Ground and building extraction from LiDAR data based on differential morphological profiles and locally fitted surfaces. *ISPRS J. Photogramm. Remote Sens.* **2014**, *93*, 145–156. [CrossRef]

16. Jochem, A.; Höfle, B.; Wichmann, V.; Rutzinger, M.; Zipf, A. Area-wide roof plane segmentation in airborne LiDAR point clouds. *Comput. Environ. Urban Syst.* **2012**, *36*, 54–64. [CrossRef]

17. Zhao, Z.; Duan, Y.; Zhang, Y.; Cao, R. Extracting buildings from and regularizing boundaries in airborne lidar data using connected operators. *Int. J. Remote Sens.* **2016**, *37*, 889–912. [CrossRef]

18. Xu, S.; Vosselman, G.; Oude Elberink, S. Multiple-entity based classification of airborne laser scanning data in urban areas. *ISPRS J. Photogramm. Remote Sens.* **2014**, *88*, 1–15. [CrossRef]

19. Zhang, J.; Lin, X.; Ning, X. SVM-based classification of segmented airborne LiDAR point clouds in urban areas. *Remote Sens.* **2013**, *5*, 3749–3775. [CrossRef]

20. Chehata, N.; Guo, L.; Mallet, C. Airborne lidar feature selection for urban classification using random forests. *Int. Arch. Photogramm. Remote Sens. Spat. Inf. Sci.* **2009**, *39*, 207–212.

21. Niemeyer, J.; Rottensteiner, F.; Soergel, U. Contextual classification of lidar data and building object detection in urban areas. *ISPRS J. Photogramm. Remote Sens.* **2014**, *87*, 152–165. [CrossRef]

22. Guo, B.; Huang, X.; Zhang, F.; Sohn, G. Classification of airborne laser scanning data using JointBoost. *ISPRS J. Photogramm. Remote Sens.* **2015**, *100*, 71–83. [CrossRef]

23. Gu, Y.; Wang, Q.; Xie, B. Multiple kernel sparse representation for airborne LiDAR data classification. *IEEE Trans. Geosci. Remote Sens.* **2016**, 1–21. [CrossRef]

24. Awrangjeb, M.; Fraser, C. Automatic segmentation of raw LIDAR data for extraction of building roofs. *Remote Sens.* **2014**, *6*, 3716–3751. [CrossRef]

25. Richter, R.; Behrens, M.; Döllner, J. Object class segmentation of massive 3D point clouds of urban areas using point cloud topology. *Int. J. Remote Sens.* **2013**, *34*, 8408–8424. [CrossRef]

26. Sánchez-Lopera, J.; Lerma, J.L. Classification of lidar bare-earth points, buildings, vegetation, and small objects based on region growing and angular classifier. *Int. J. Remote Sens.* **2014**, *35*, 6955–6972. [CrossRef]

27. Yan, J.; Zhang, K.; Zhang, C.; Chen, S.-C.; Narasimhan, G. Automatic construction of 3-D building model from Airborne LIDAR data through 2-D snake algorithm. *IEEE Trans. Geosci. Remote Sens.* **2015**, *53*, 3–14.

28. Ullman, S.; Vidal-Naquet, M.; Sali, E. Visual features of intermediate complexity and their use in classification. *Nat. Neurosci.* **2002**, *5*, 682–687. [CrossRef] [PubMed]

29. Goutsias, J.; Vincent, L.; Bloomberg, D.S. *Mathematical Morphology and Its Applications to Image and Signal Processing*; Computational Imaging and Vision; Kluwer: Dordrecht, The Netherlands, 2000.

30. Jung, C.R.; Scharcanski, J. Adaptive image denoising and edge enhancement in scale-space using the wavelet transform. *Pattern Recognit. Lett.* **2003**, *24*, 965–971. [CrossRef]

31. Lopez-Molina, C.; De Baets, B.; Bustince, H.; Sanz, J.; Barrenechea, E. Multiscale edge detection based on Gaussian smoothing and edge tracking. *Knowl. Based Syst.* **2013**, *44*, 101–111. [CrossRef]

32. Vincent, L. Morphological grayscale reconstruction in image analysis, applications and efficient algorithms. *IEEE Trans. Image Process.* **1993**, *2*, 176–201. [CrossRef] [PubMed]

33. Vu, T.T.; Yamazaki, F.; Matsuoka, M. Multi-scale solution for building extraction from LiDAR and image data. *Int. J. Appl. Earth Obs. Geoinf.* **2009**, *11*, 281–289. [CrossRef]

34. Yang, B.; Xu, W.; Dong, Z. Automated extraction of building outlines from airborne laser scanning point clouds. *IEEE Geosci. Remote Sens.* **2013**, *10*, 1399–1403. [CrossRef]

35. Cui, Z.; Zhang, K.; Zhang, C.; Chen, S.C. A multi-pass generation of DEM for urban planning. In Proceedings of the International Conference on Cloud Computing and Big Data (CloudCom-Asia), Fuzhou, China, 16–19 December 2013.

36. Yang, B.; Huang, R.; Dong, Z.; Zang, Y.; Li, J. Two-step adaptive extraction method for ground points and breaklines from lidar point clouds. *ISPRS J. Photogramm. Remote Sens.* **2016**, *119*, 373–389. [CrossRef]

37. Perera, G.S.; Maas, H.G. Cycle graph analysis for 3D roof structure modelling, Concepts and performance. *ISPRS J. Photogramm. Remote Sens.* **2014**, *93*, 213–226. [CrossRef]

38. Rutzinger, M.; Rottensteiner, F.; Pfeifer, N. A comparison of evaluation techniques for building extraction from airborne laser scanning. *IEEE J. Sel. Top. Appl. Earth Obs. Remote Sens.* **2009**, *2*, 11–20. [CrossRef]

39. ISPRS Benchmark Test Results—WHU_YD. Available online: http://www2.isprs.org/commissions/comm3/wg4/results.html (accessed on 10 October 2016).

40. Bulatov, D.; Rottensteiner, F.; Schulz, K. Context-based urban terrain reconstruction from images and videos. In Proceedings of the XXII ISPRS Congress of the International Society for Photogrammetry and Remote Sensing ISPRS Annals, Melbourne, Australia, 25 August–1 September 2012.

41. Wei, Y.; Yao, W.; Wu, J.; Schmitt, M.; Stilla, U. Adaboost-based feature relevance assessment in fusing lidar and image data for classification of trees and vehicles in urban scenes. *Int. Arch. Photogramm. Remote Sens. Spat. Inf. Sci.* **2012**, *1*, 323–328. [CrossRef]

42. Gerke, M.; Xiao, J. Fusion of airborne laserscanning point clouds and images for supervised and unsupervised scene classification. *ISPRS J. Photogramm. Remote Sens.* **2014**, *87*, 78–92. [CrossRef]

43. Awrangjeb, M.; Lu, G.; Fraser, C. Automatic building extraction from LiDAR data covering complex urban scenes. *Int. Arch. Photogramm. Remote Sens. Spat. Inf. Sci.* **2014**, *40*, 25–32. [CrossRef]

44. Tomljenovic, I.; Blaschke, T.; Höfle, B.; Tiede, D. Potential and idiosyncrasy of object-based image analysis for airborne Lidar-based building detection. *South-East. Eur. J. Earth Obs. Geomat.* **2014**, *3*, 517–520.